Environmental Science and Engineering

Series Editors

The ultimate goal of this series is to contribute to the protection of our environment, which calls for both profound research and the ongoing development of solutions and measurements by experts in the field. Accordingly, the series promotes not only a deeper understanding of environmental processes and the evaluation of management strategies, but also design and technology aimed at improving environmental quality. Books focusing on the former are published in the subseries Environmental Science, those focusing on the latter in the subseries Environmental Engineering.

Yongping Li · Chaotian Xie · Guohe Huang ·
Peng Zhang
Editors

16th International Conference on Environmental Science and Development

Springer

Editors
Yongping Li
School of Environment
Beijing Normal University
Beijing, China

Chaotian Xie
Fisheries College
Jimei University
Xiamen, Fujian, China

Guohe Huang
International Society for Environmental
Information Sciences
Regina, SK, Canada

Peng Zhang
Institute for Energy, Environment
and Sustainable Communities
Regina, SK, Canada

ISSN 1863-5520 ISSN 1863-5539 (electronic)
Environmental Science and Engineering
ISBN 978-3-032-15079-0 ISBN 978-3-032-15080-6 (eBook)
https://doi.org/10.1007/978-3-032-15080-6

This Springer imprint is published by the registered company Springer Nature Switzerland AG
The registered company address is: Gewerbestrasse 11, 6330 Cham, Switzerland

Committees

Honorary Chair

Gordon Huang, University of Regina, Canada

Conference Chairs

Chaotian Xie, Jimei University, China
Yongping Li, Beijing Normal University, China

Organizing Committees

Guorong Cai, Jimei University, China (Co-chair)
Minhui Lai, Jimei University, China (Co-chair)
Qiuhua Tao, Jimei University, China (Co-chair)
Yunfeng Huang, Jimei University, China
Yanfeng Li, Beijing Normal University, China
Chenchen Lin, Jimei University, China
Feng Wang, Beijing Normal University, China
Qihui Wu, Jimei University, China
Shenghua Zhang, Jimei University, China
Xiong Zhou, Beijing Normal University, China

International Programme Committee

Preface

2025 16th International Conference on Environmental Science and Development (ICESD 2025) was held successfully during June 6–8, 2025 in Xiamen, China. It aims to provide a unique platform for researchers, scientists, engineers and professionals from all over the world to present their latest research results and new ideas in terms of environmental science and development. We deeply appreciate their active participation in ICESD 2025.

ICESD 2025 was organized by the International Society for Environmental Information Sciences (ISEIS), co-organized by Xiamen Society for Environmental Sciences and United Nations Development Programme (UNDP), and hosted by Jimei University, China. This conference was supported by the National Key R&D Program of China (2023YFC3206503) and the Innovative Research Group of the National Natural Science Foundation of China (52221003).

The conference received strong support from experts, practitioners and researchers from various countries. During the conference, our conference honorary chair, president Guohe Huang of ISEIS hosted the opening ceremony for the conference. Our conference chair, Prof. Chaotian Xie from Jimei University, China, and Dr. Weidong Zhang from UNDP China were invited to give Opening Remarks for the conference, respectively. Prof. Huining Xiao from the University of New Brunswisk, Canada, Prof. Guangwei Huang from Sophia University, Japan and Prof. Shin-Cheng Yeh from National Taiwan Normal University further provided wonderful keynote presentations.

ICESD 2025 received submissions contributed by researchers from many countries, such as China, Thailand, United States, Australia, Egypt, Malaysia and other countries. After two rounds of rigorous peer reviews, a total of 47 papers were selected in this proceeding, within the scope of environmental science and development such as (1) Water Treatment and Quality Control: Technologies, Mechanisms, and Applications, (2) Watershed Hydrological Modeling and Water-Ecology-Energy-Food Nexus Management, (3) Low-Carbon Technology and Carbon Management Innovation, and others. We sincerely thank all the authors, the conference committee members, and the reviewers for their participation and support.

Thanks for your contributions to the ICESD again. We will continue to organize this conference in the future to provide a much effective platform for further exchange of new knowledge and perhaps potential collaboration in research areas of environmental science and development.

Beijing, China
June 2025

Prof. Yongping Li
Conference Co-chair

Contents

Part I
Water Treatment and Quality Control: Technologies, Mechanisms, and Applications

Chapter 1
Study on Hydrogen Production Performance of Iron-Modified Hydrothermal Carbon Enhanced Sludge Anaerobic Fermentation

Ting He, Tian Chai, Jing Zhang, Zuocheng Yin, and Longjun Hua

Abstract The resource utilization of surplus sludge can effectively alleviate the current global energy crisis. In order to improve the efficiency of hydrogen production, the effects of Fe-WHC in different concentrations on the final fermentation product and cumulative hydrogen production were investigated by using iron-modified wheat flour hydrothermal carbon Fe-WHC as an exogenous additive. The mechanism of hydrogen production in sludge fermentation was explored through kinetic analysis of the cumulative hydrogen production and microbial community analysis. The results showed that the optimal hydrogen production of 0.3 mg/g-VSS-Fe-WHC was 1.813 mL/g-VSS, which was 34.08% higher than that of the blank control group. The output value of energy is 1.78 kJ/g-VSS. The maximum hydrogen production was 1.88 mL/g-VSS in the experimental group, and the maximum hydrogen production rate was 0.38 mL/h. After fermentation, the maximum accumulation of short-chain fatty acids in the experimental group was 1093.134 mg/L, which was increased by 51.10% compared with the blank control group. *Bacillota* was the most dominant strain in this study. Compared with the blank control group, *Bacillota* increased from 61.32% to 78.06%, an increase of 16.74%, with a growth rate of 27.30%. The results of this study can provide an important reference for further promoting sludge recycling.

Keywords Anaerobic fermentation · Hydrogen production · Biochar

T. He · T. Chai (✉) · J. Zhang · Z. Yin · L. Hua
Xiamen University of Technology, Xiamen, China
e-mail: chait@xmut.edu.cn

Y. Li et al. (eds.), *16th International Conference on Environmental Science and Development*, Environmental Science and Engineering,
https://doi.org/10.1007/978-3-032-15080-6_1

1.1 Introduction

In order to reduce the dependence on fossil energy and improve the in-depth research and utilization of new clean energy, energy transformation has become one of the important measures. Hydrogen is a clean energy source and can be used in a wide range. Current hydrogen production methods include hydroelectrolysis, fossil fuel decomposition and biohydrogen synthesis (Ramprakash et al. 2022). Among many common hydrogen production methods, biological anaerobic fermentation can realize the resource utilization of urban organic waste more effectively. It is noteworthy that dark fermentation has made great progress in biological hydrogen production (Hallenbeck and Ghosh 2009). Hydrogen is produced in the initial stages of anaerobic fermentation (Huang et al. 2024), but is usually not very efficient at producing hydrogen. In order to improve the hydrogen production rate, relatively early studies focused on the pretreatment of residual sludge in different ways. In recent years, many scholars have improved the stability of anaerobic fermentation and the output of biogas by adding trace elements or carbon sources to the anaerobic fermentation system, but the efficiency of a single carbon source to increase hydrogen production is low. In this paper, a new type of Fe modified carbon source, iron modified wheat flour, was successfully synthesized by hydrothermal method and applied to anaerobic fermentation to improve its hydrogen production efficiency. The performance and mechanism of carbon material promoting hydrogen production by anaerobic fermentation of residual sludge were analyzed in detail. For the reference of relevant researchers, it can also provide research help in green energy, greenhouse gas emissions, sustainable development and other aspects, so it has certain practical significance.

1.2 Materials and Methods

1.2.1 Fermentation Experiment

The hydrothermal carbon of iron modified wheat flour was prepared by hydrothermal method. The fermentation test was conducted in sealed fermentation glass reaction bottles. At the beginning of the experiment, 0.1 L of acclimated concentrated sludge and 0.3 L of pure water were injected into each test bottle respectively. Add 0.1 g of peptone to each reaction bottle before reaction. The dosage concentration can be calculated by formula (1.1–1.4). The weight of the filter paper after drying is m_1, and the weight of the filter paper and the sludge to be measured is m_2. The weight of the crucible burned at 600 °C for 1 h is recorded as m_3. The filter paper and the sludge as a whole were moved to the crucible and burned at 600 °C for 2 h. The weight of the crucible and sludge was recorded as m_4. AC is ash. The VSS of the original sludge sample was 9684.9 mg/L, and L1 was 0.1 L of experimental sludge. Input 0, 0.1, 0.2, 0.3, 0.4 mg/g-VSS carbon materials respectively. Before anaerobic fermentation, the dissolved oxygen in the water is purged with N_2(99.99%) $\leq$ 0.2 mg/L to achieve anaerobic conditions, then sealed and stirred in an air bath oscillator (170 rpm and 35 ± 2 °C). The batch was tested for 48 h, during which time gas measurements were performed every 3 h and short-chain fatty acids every 9 h. Through the detection of gas index, the best hydrogen producing material and its concentration were explored.

$$TSS(g/L) = \frac{m_2 - m_1}{V} \times 1000 \tag{1.1}$$

$$AC = \frac{m_4 - m_3}{V} \times 1000 \tag{1.2}$$

$$VSS(g/L) = TSS - AC \tag{1.3}$$

$$\rho = \frac{VSS}{\mathrm{L}_1} \tag{1.4}$$

1.2.2 *Hydrogen Detection and Analysis*

The gas produced by fermentation was determined by gas chromatography. The energy output value of hydrogen in the experiment is calculated from formula (1.5), where the density ρ of hydrogen is 0.09 kg/m^3, and the calorific value q is 142.35 kJ/kg (Haoming et al. 2023), where Q (kJ/g-VSS) is the energy output value of hydrogen. P (mL/g-VSS) is the hydrogen production. The cumulative hydrogen production in the experiment was fitted using Gompertz model (Chang et al. 2011) modified by formula (1.6). Where: H (mL) is the cumulative hydrogen production at the reaction time t(h); P_{max}(mL/g-VSS) is the maximum hydrogen production. R_{max}(mL/h) is the maximum hydrogen production rate. T(h) is the delay time.

$$Q = m \times q = \rho \times P \times q \tag{1.5}$$

$$H = P_{\max} \times \exp\left\{-\exp\left[\frac{R_{\max} \times e}{P_{\max}}(\lambda - t) + 1\right]\right\} \tag{1.6}$$

1.2.3 *Microbial Community Analysis Methods*

Both DNA concentration and DNA purity were measured by NanoDrop2000. PCR amplification was performed using a Pro Taq, 20 μl reaction system, and the PCR instrument was ABI GeneAmp® 9700. Illumina's Miseq PE300 platform was used for sequencing.

1.3 Results and Discussion

1.3.1 Effect of Carbon Materials on Hydrogen Production

The cumulative hydrogen production of Fe-WHC on the sludge anaerobic fermentation system for 48 h is shown in Fig. 1.1, The hydrogen production of the anaerobic fermentation system in the experimental group has reached the peak value at 21 h of fermentation, and the hydrogen production is gradually decreasing. The cumulative total hydrogen production of the original sludge group was 1.353 mL, and the cumulative total hydrogen production of the four experimental groups supplemented with Fe-WHC was 1.609 mL, 1.718 mL, 1.813 mL and 1.752 mL, respectively. With the increase of Fe-WHC concentration, the cumulative hydrogen production of anaerobic fermentation system continued to increase, especially when the Fe-WHC concentration was 0.3 mg/g-VSS, the cumulative hydrogen production for 48 h was 34.08% higher than that of the blank control group. When the Fe-WHC concentration increased to 0.4 mg/g-VSS, the cumulative hydrogen production decreased by 4.57% compared with that at 0.3 mg/g-VSS. It can be seen that an appropriate concentration of Fe-WHC can promote hydrogen production by hydrogen-producing bacteria, but when the concentration of Fe-WHC is too high, excessive iron metal ions may come into contact with microorganisms, resulting in the destruction of microbial structure and ultimately the death of microorganisms (Yin and Wang 2019). In addition, anaerobic bacteria can produce oxidative stress under the action of excessive metal iron ions, resulting in cell damage (Wang and Chen 2016). When the concentration is too low, Fe-WHC is not enough to promote the activity of hydrogen-producing bacteria, thus promoting hydrogen production.

As shown in Fig. 1.2, when the dosage was 0.3 mg/g-VSS, the hydrogen production in the experimental group reached a maximum of 23.233 kJ/g-VSS, which was 34.08% higher than that in the blank control group. The fitting coefficient (R^2) of all anaerobic fermentation systems ranged from 0.995 to 0.999, which indicated that the model was reasonable to simulate the process of cumulative hydrogen production, and the experimental data were reliable. When the dosage was 0.3 mg/g-VSS, (P_{max}) 1.88 mL/g-VSS was the highest in the experimental group, and the maximum hydrogen production rate (R_{max}) was 0.38 mL/h, The lag time λ (h) of 8.7 h was 0.5 h longer than that of 8.2 h in the control group.

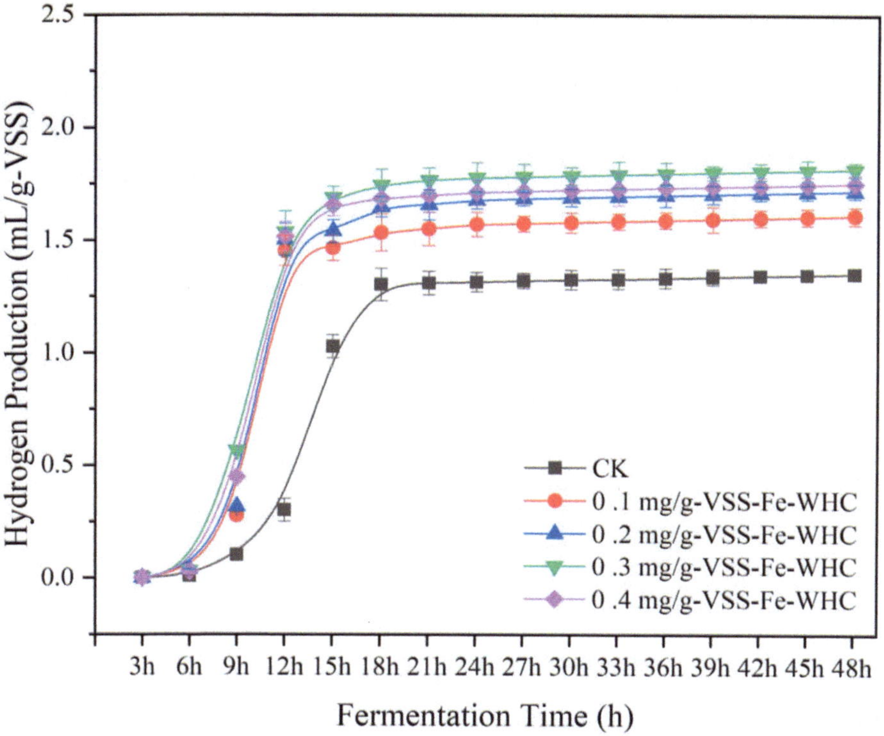

Fig. 1.1 Hydrogen accumulation in 48 h anaerobic fermentation system

1.3.2 Effects of Carbon Materials on Dissolved Components

The soluble components of anaerobic fermentation systems can reflect the process of fermentation hydrolysis to a certain extent, and the yield of short-chain fatty acids in fermentation systems with different dosage concentrations is different. The fluctuation of ecological factors in the acid-producing stage of dark fermentation will have a significant impact on the microbial community, and then change the physiological metabolism of the microorganisms, resulting in changes in the composition of metabolites (Chen et al. 2024; Jin et al. 2024; Nabaterega et al. 2022). Figure 1.3 shows the concentrations of acetic acid, caproic acid and isovaleric acid in Fe-WHC anaerobic fermentation system with different concentrations for 48 h. As can be seen from the Fig. 1.3, with the increase of Fe-WHC dosage, there was no significant difference in the changes of isovaleric acid and caproic acid in the fermentation system, indicating that it had no significant effect on isovaleric acid and caproic acid. The yield of acetic acid increased with the increase of Fe-WHC dosage. The reason may be that the pre-treated sludge released more soluble organic matter in the early stage of fermentation, and acidifying bacteria fully converted organic matter into

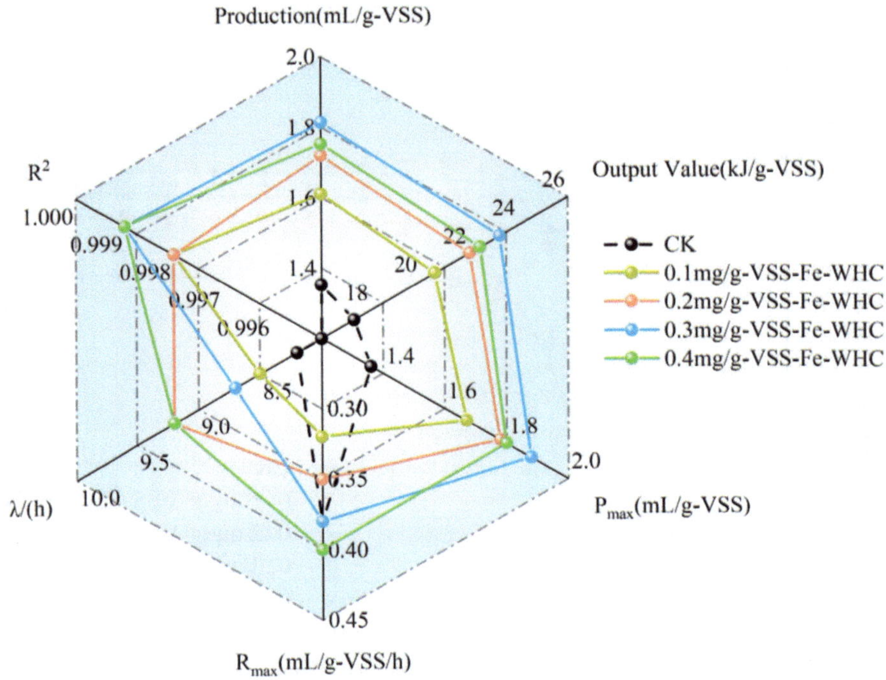

Fig. 1.2 Gompertz model fitting results

short-chain fatty acids under the stimulation of hydrothermal carbon and its metal modified hydrothermal carbon.

The acetic acid production stage is mainly by the hydrogen-producing acetic acid producing bacteria using propionic acid, butyric acid, valeric acid and alcohol as raw materials, which is oxidized and decomposed into acetic acid, hydrogen and so on. As shown in Fig. 1.4, when the dosage of Fe-WHC was 0.3 mg/g-VSS, the acetic acid content reached a peak of 1643.134 mg/L at 18 h in the initial fermentation stage. However, with the progress of fermentation, some acidifying bacteria can not withstand the impact of water and hot carbon, resulting in death.

This is demonstrated by the leaching concentrations of zinc ions at different dosing concentrations in Fig. 1.5. With the development of anaerobic fermentation, the acid produced by hydrolysis is utilized by hydrogen-producing bacteria, resulting in a significant decrease in the yield of acetic acid. It can be seen that hydrothermal carbon has different effects on short-chain fatty acids produced in anaerobic fermentation, but it can promote the production of acetic acid, and the performance of hydrothermal carbon modified by iron is better.

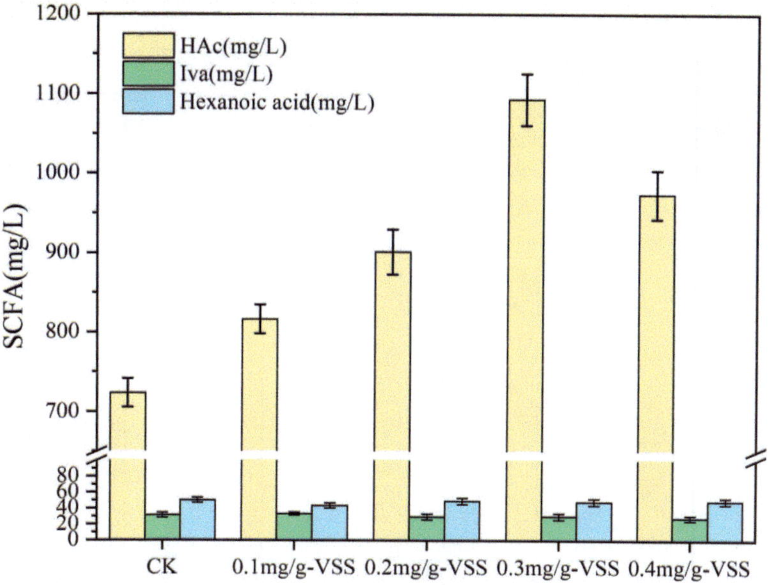

Fig. 1.3 SCFA content at different dosages 48 h after fermentation

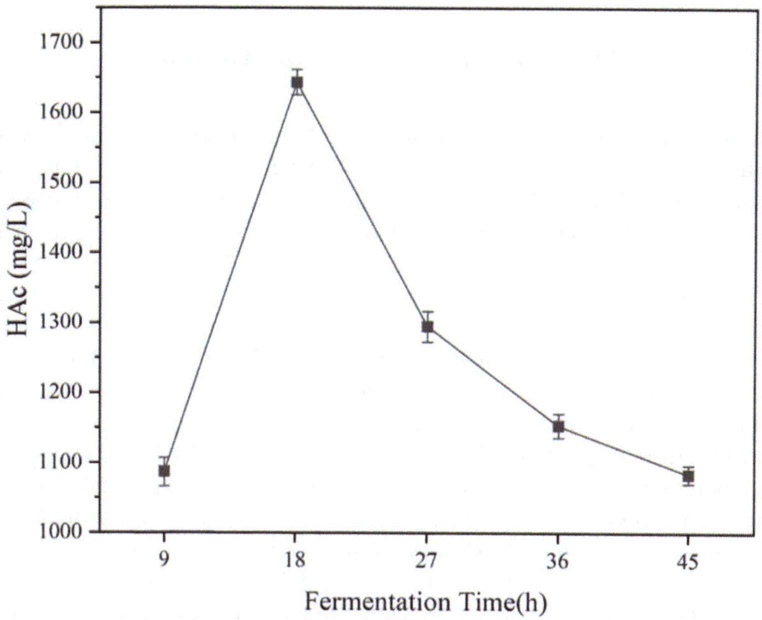

Fig. 1.4 Acetic acid content change during fermentation at 0.3 mg/g-VSS-Fe-WHC

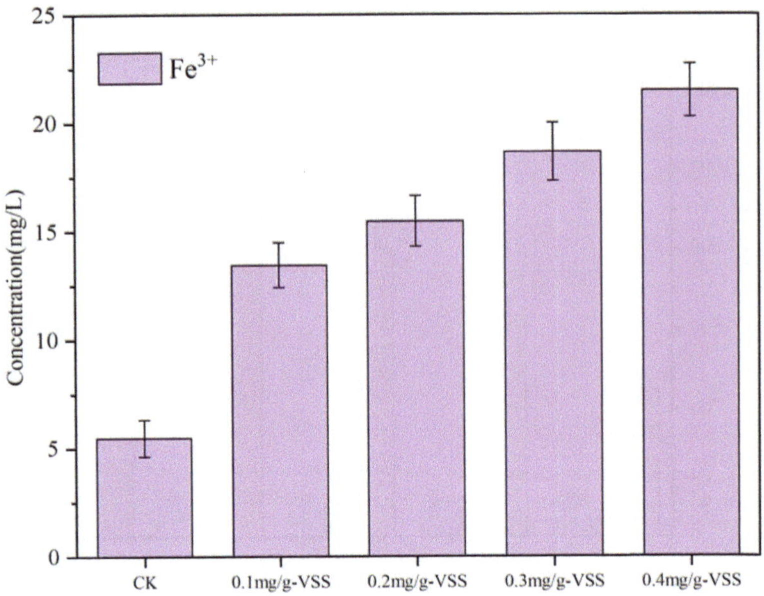

Fig. 1.5 Fe^{3+} ion leaching concentration after Fe-WHC fermentation at different concentrations

1.3.3 Microbial Community Composition Analysis

The fermentation system of the blank control group and the sample with the largest hydrogen production was used for gate level bacterial community analysis (0.3 mg/g-VSS-Fe-WHC). *Bacillota*, *Proteobacteria*, *Chloroflexi*, *Actinobacteriota* and *Bacteroidota* are relatively common bacterial groups in the anaerobic fermentation process of sludge (Cordova et al. 2018; Luo et al. 2018). As shown in Fig. 1.6, it is the dominant gate in the experimental group sludge anaerobic fermentation hydrogen production system.

As shown in Fig. 1.6, *Bacillota* is the most dominant strain in this study, and *Proteobacteria*, *Chlorocurvula*, *Actinobacteria* and *Bacteroidetes* are its favorable competitors, which can promote the decomposition of non-degradable organic matter in materials. In 0.3mg/g-VSS-Fe-WHC, compared with the blank control group, *Bacillus* increased from 61.32% to 78.06%, an increase of 16.74% and a growth rate of 27.30%. *Actinobacteriota* increased from 5.44% to 7.01%, increasing by 1.57%, with a growth rate of 28.89%. *Bacteroidota* decreased from 14.50% to 2.68%, a decrease of 11.83%, with a reduction rate of 81.53%. *Proteobacteria* decreased from 8.35% to 6.07%, a decrease of 2.28%, with a reduction rate of 27.35%. *Proteobacteria* are mostly facultative or obligate anaerobic bacteria, *Bacteroidetes* are obligate anaerobic bacteria, both are heterotrophic bacteria.

In summary, it is proved that hydrothermal carbon can significantly promote the activity of *Bacillus* and actinomycetes, greatly enhance their advantages in biological

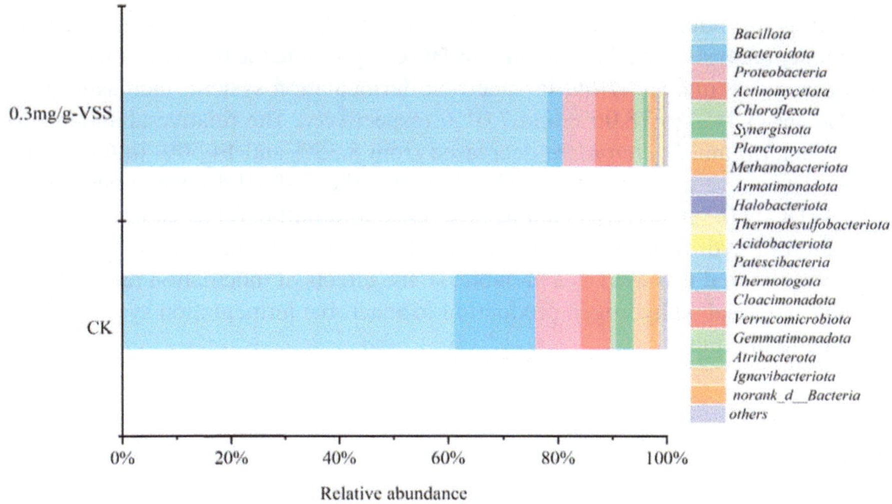

Fig. 1.6 Phylum level bacterial community analysis diagram

anaerobic fermentation system, and promote the enrichment of a series of dominant strains. At the same time, hydrothermal carbon and its iron-modified hydrothermal carbon can inhibit the growth and activity of heterotrophic bacteria.

1.4 Conclusion

In this chapter, the effect of adding Fe-WHC to anaerobic fermentation system on hydrogen production capacity of the whole system was studied. The whole fermentation process and results were studied by combining the cumulative hydrogen production, volatile organic acid, microbial community and other important indexes. The results showed that Fe-WHC could promote the hydrogen production of anaerobic sludge fermentation system, and the optimal dosage was 0.3 mg/g-VSS, and the cumulative hydrogen production could reach 1.813 mL/g-VSS. At this time, the maximum hydrogen energy production was 23.233 kJ/g-VSS, which was 34.08% higher than that of the control group. The variation trend of hydrogen output value is consistent with that of hydrogen production. However, when the Fe-WHC concentration is too high, the hydrogen producing bacteria will be inhibited and the hydrogen production will be reduced. The cumulative hydrogen production of the anaerobic fermentation system in this study was analyzed dynamically, and the hydrogen production effect was the best when the Fe-WHC concentration was 0.3 mg/g-VSS, P_{max} was 1.88 mL/g-VSS, but the hysteresis time was delayed by 0.5 h, which proved that the fermentation cycle and time could be properly extended under the premise of ensuring sufficient nutrient substrate. Fe-WHC provides a good living environment for hydrogen-producing bacteria, From the perspective of phylum level bacterial community analysis, the addition of Fe-WHC can promote the relative abundance of *Bacillota* and *Actinobacteriota* in anaerobic fermentation system, increasing from 61.32% and 5.44% to 78.06% and 7.01% respectively. The relative abundance of *Proteobacteria* and *Bacteroidota* decreased from 8.35% and 14.50% to 6.07% and 2.68%. The above experimental results show that the optimal dosage concentration of Fe-WHC is 0.3 mg/g-VSS, and at this concentration Fe-WHC can promote the generation of hydrogen to the greatest extent. In this study, only the measurement of carbon material was used as a variable, so the effects of inoculation ratio, pH and REDOX potential on hydrogen production in anaerobic fermentation system can be explored in the future.

References

Chang S, Li J-Z, Liu F (2011) Evaluation of different pretreatment methods for preparing hydrogen-producing seed inocula from waste activated sludge[J]. Renew Energy 36(5):1517–1522

Chen R, Zhou X, Huang L et al (2024) Effects of yeast inoculation methods on caproic acid production and microbial community during anaerobic fermentation of Chinese cabbage waste[J]. J Environ Manage 356:120632

Cordova O, Chamy R, Guerrero L et al (2018) Assessing the effect of pretreatments on the structure and functionality of microbial communities for the bioconversion of microalgae to biogas[J]. Front Microbiol 9:1388

Hallenbeck PC, Ghosh D (2009) Advances in fermentative biohydrogen production: the way forward? [J]. Trends Biotechnol 27(5):287–297

Haoming S, Hongyu R, Bingfeng L et al (2023) Research progress of adding different substrates to promote hydrogen production by sludge co-fermentation [J]. Environ Sci China 43(S1):70–78

Huang JL, Mu L, Wu YT et al (2024) Screening and hydrogen production performance of four mixed photosynthetic anaerobic hydrogen-producing bacteria [J]. Environ Eng 42(09):301–310. https://doi.org/10.13205/j.hjgc.202409030

Jin B, Jia Y, Cheng K et al (2024) Facilitating effects of the synergy with zero-valent iron and peroxysulfate on the sludge anaerobic fermentation system: combined biological enzyme, microbial community and fermentation mechanism assessment [J]. Chemosphere 355:141805

Luo J, Zhang Q, Wu L et al (2018) Improving anaerobic fermentation of waste activated sludge using iron activated persulfate treatment[J]. Bioresour Technol 268:68–76

Nabaterega R, Kieft B, Hallam SJ et al (2022) Fractional factorial experimental design for optimizing volatile fatty acids from anaerobic fermentation of municipal sludge: microbial community and activity investigation [J]. Renew Energy 199:733–744

Ramprakash B, Lindblad P, Eaton-Rye JJ et al (2022) Current strategies and future perspectives in biological hydrogen production: a review [J]. Renew Sustain Energy Rev 168:112773

Wang DB, Chen YG (2016) Critical review of the influences of nanoparticles on biological wastewater treatment and sludge digestion[J]. Crit Rev Biotechnol 36(5):816–828

Yin Y, Wang J (2019) Enhanced biohydrogen production from macroalgae by zero-valent iron nanoparticles: Insights into microbial and metabolites distribution [J]. Biores Technol 282:110–117

Chapter 2
Degradation of Tetracycline in Hydrodynamic Cavitation/Persulfate/Cu (II) System

Zheng Li, Xiufeng Zhu, Sivakumar Manickam, and Xun Sun

Abstract In this study, hydrodynamic cavitation (HC) combined with Cu (II) was used to activate peroxydisulfate (PDS) for degradation of tetracycline (TC). The results show that the degradation rate of TC in 10 L model wastewater reached 100% in 60 min, at the optimal conditions of an inlet pressure of the Venturi of 0.42 MPa, a $Na_2S_2O_8$ concentration of 0.40 mM, and a concentration ratio of Cu (II) to $Na_2S_2O_8$ of 2:1. Moreover, the synergistic index of the HC/Cu (II)/PDS system was found to be as high as 10.21. Therefore, the HC/Cu (II)/PDS system provides a new idea for the degradation of antibiotic wastewater.

Keywords Tetracycline · Degradation · Hydrodynamic cavitation · Peroxydisulfate · Divalent copper

2.1 Introduction

In recent years, with the massive use of antibiotics in biomedicine, agricultural farming, and other fields, the problem of antibiotic contamination in water environment and bacterial resistance to antibiotics is becoming a worldwide concern (Xu et al. 2024; Fan 2023). At the same time, long-term retention of antibiotics in external environments leads to the emergence of antibiotic-resistant genes (ARGs). They can spread widely in the environment, contaminate the water source, and damage the ecosystem (Zhou et al. 2023). The above issues have raised deep concerns about the environmental safety and human health.

Z. Li · X. Zhu · X. Sun (✉)
School of Mechanical Engineering, Shandong University, Jinan, China
e-mail: xunsun@sdu.edu.cn

S. Manickam
Faculty of Engineering, Universiti Teknologi Brunei, Bandar Seri Begawan, Brunei Darussalam, Brunei

Y. Li et al. (eds.), *16th International Conference on Environmental Science and Development*, Environmental Science and Engineering,
https://doi.org/10.1007/978-3-032-15080-6_2

The advanced oxidation process (AOP), based on hydroxyl radicals (·OH), has an excellent performance in treating refractory organic pollutants in wastewater (Mukherjee et al. 2023). Recently, the advanced oxidation process based on sulfate radicals (SR-AOP) has been rapidly developed, due to high oxidation potential, stability, wide range of degradation selectivity, and low cost (Roy and Moholkar 2024). SR-AOP often requires external energy or catalysts to be activated, producing sulfate radicals ($SO_4^{2-}\cdot$) to achieve degradation effectiveness (Fedorov et al. 2020).

Hydrodynamic cavitation (HC), a widely used green process for wastewater treatment (Patil et al. 2022), is capable of producing high-temperature and high-pressure environment that can be used to activate persulfate (Brujan 2017; Wang and Wang 2022). In addition, the transition metals have been also found to be effective for activating persulfate (Zhou et al. 2016). Many researchers have used HC in combination with transition metals to activate persulfate for the degradation of pollutants (Roy and Moholkar 2020; Weng et al. 2022). This process is still in the preliminary research stage.

In this study, Cu (II) was chosen to combine with HC for the activation of peroxydisulfate (PDS); and tetracycline (TC), a representative antibiotic in the aqueous environment, was selected as a model pollutant to be degraded by the system. The effectiveness of the HC/Cu (II)/PDS system was evaluated under the optimal operating condition.

2.2 Experimental Method

2.2.1 Materials

Tetracycline ($C_{22}H_{24}N_2O_8$), sodium peroxydisulfate ($Na_2S_2O_8$), copper sulfate pentahydrate ($CuSO_4\cdot5H_2O$), acetonitrile (C_2H_3N), oxalic acid ($H_2C_2O_4$), methanol (CH_4O) were purchased from Shandong Chengchuan Life Science and Technology Co. Ltd. The purity of C_2H_3N was LC grade, and the purity of the rest reagents was AR grade, which were used as received without further purification. Ultrapure water was used for all experiments.

2.2.2 Experimental System

The HC experimental system and the Venturi reactor used in this study are shown in Fig. 2.1. A stainless steel tank with a volume of 15 L was used to retain the model wastewater. The wastewater was supplied to the Venturi by a pump (3 kW). Two pressure sensors were used to measure the upstream and downstream pressure of the Venturi.

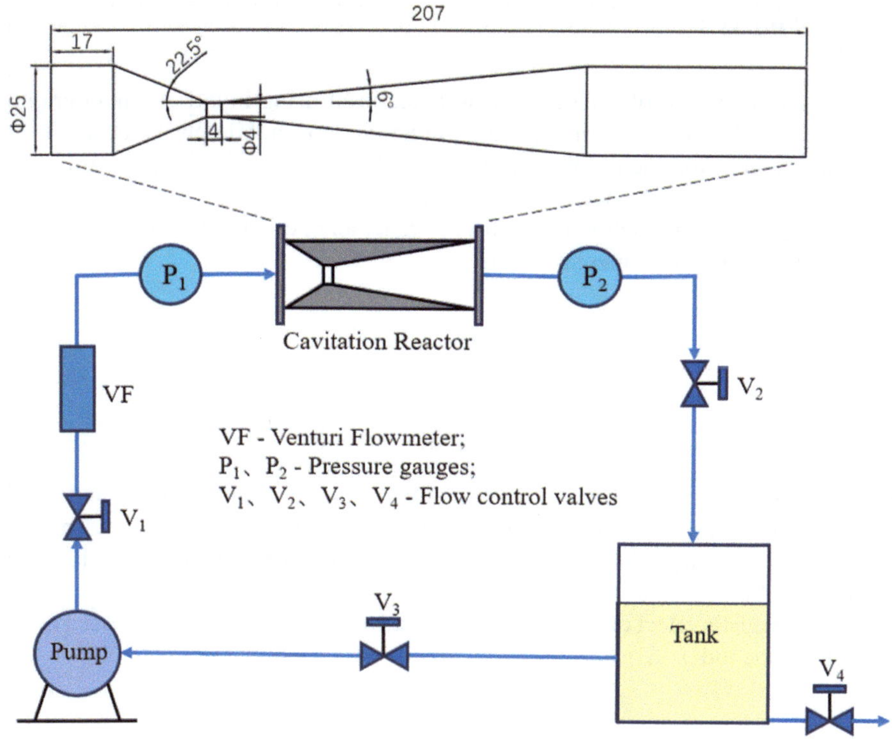

Fig. 2.1 Schematic diagram of the HC system and the Venturi reactor (in millimeter)

The Venturi tube was made of PMMA material which has a refractive index close to water. High-speed images of cavitating flow inside the Venturi were taken by a high-speed camera (Phantom VEO 1310, Phantom High Speed Co.) with an atx-i 100 mm F2.8 FF lens from Tokina.

2.2.3 *Experimental Procedure*

For each experiment, 10 L of TC solution with a concentration of 30 mg/L was prepared, and then PDS and $CuSO_4 \cdot 5H_2O$ were added, followed by switching on the HC device to start the experiment. The total treatment duration was set to 60 min, and the samples were taken at fixed time intervals. The reaction was terminated by adding excess methanol to quench the radicals in the samples. Water cooling was achieved using circulating water to ensure that the temperature was maintained at 25 ± 5 °C. Each experiment was repeated two times. Calculate the standard deviation of the data at the same moment to obtain the error bars.

2.2.4 Analysis

The TC concentration of each sample was analyzed using liquid chromatography (Shimazu LC-2050C), with the operating conditions of mobile phase: acetonitrile, methanol, and oxalic acid solution (v/v/v = 22: 11: 67); flow rate: 1 mL/min; temperature: 25 °C; and wavelength: 355 nm.

Use the first-order reaction rate constant, K_{obs}, and per-pass degradation models to characterize the experimental results:

$$\ln\frac{C}{C_0} = -K_{obs}t = -\phi_0 n_p, \tag{2.1}$$

$$n_p = \frac{Q}{V}t, \tag{2.2}$$

where C and C_0 are the TC concentration after sampling at any moment and at the initial moment, respectively; K_{obs} (min^{-1}) is the first-order reaction rate constant; ϕ_0 is the per-pass degradation factor; n_p is the number of passes; Q is the flow rate through the Venturi; and V is the working volume.

The synergistic effect of the combined degradation systems can be assessed using the synergistic index (SI):

$$SI = \frac{K_{HC/Cu(II)/PDS}}{K_{HC} + K_{Cu(II)/PDS}} \tag{2.3}$$

where $K_{HC/Cu\ (II)/PDS}$ is the first-order reaction rate constant for the HC/Cu (II)/PDS system, K_{HC} is the first-order reaction rate constant of HC alone, and $K_{Cu\ (II)/PDS}$ is the first-order reaction rate constant for the Cu (II)/PDS system.

2.3 Results and Discussion

2.3.1 Cavitation Visualisation

Figure 2.2 shows the instantaneous image of cavitating flow in the Venturi reactor at different inlet pressures (the fluid flows from left to right). It can be observed that the maximum length of the cavity increased with increasing inlet pressure. According to complete development process of cavitation, it is found that the higher the inlet pressure, the shorter the time between the appearance and collapse of the cavity. Because the choked cavitation was not observed at the maximum inlet pressure of 0.42 MPa in the Venturi, this pressure was considered to be optimal and was used in the degradation experiment.

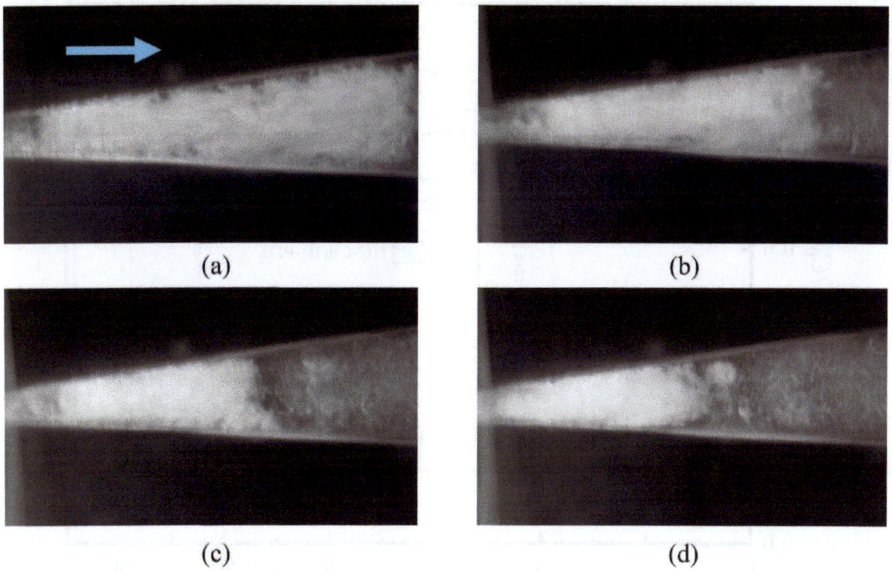

Fig. 2.2 Cavitating flow in the Venturi at different inlet pressures: **a** 0.42 MPa, **b** 0.37 MPa, **c** 0.32 MPa, and **d** 0.27 MPa

2.3.2 Degradation of Tetracycline via Different Systems

The effect of three systems, i.e., sole use of HC and PDS and HC/Cu (II)/PDS, on the TC degradation was investigated. As can be seen in Fig. 2.3, the TC degradation rates of the above systems after a 60-min treatment were 14.37%, 4.05%, and 100%, respectively. The degradation effects of the above systems are consistent with the reaction rate constant and per-pass degradation factor presented in Table 2.1. It can be concluded that the synergistic effect of HC, Cu (II), and PDS is considerably high, with a SI of 10.21. This indicates that the energy generated by the transition metals and HC can help to activate the PDS, leading to large production of $SO_4^{2-}\cdot$. At the same time, HC promotes the production of ·OH, and its mass-transfer effect enhances the contact between various substances in the solution, promotes the reaction between TC and various radicals.

2.3.3 Effect of Inlet Pressures

The inlet pressure of the HC reactor directly affects the formation and collapse of cavitation bubbles (Dehghani et al. 2024). To further clarify it, the effect of inlet pressure (0.27, 0.32, 0.37, and 0.42 MPa) on TC degradation was investigated, which yielded the same results as the visualization images. From Fig. 2.4, the degradation

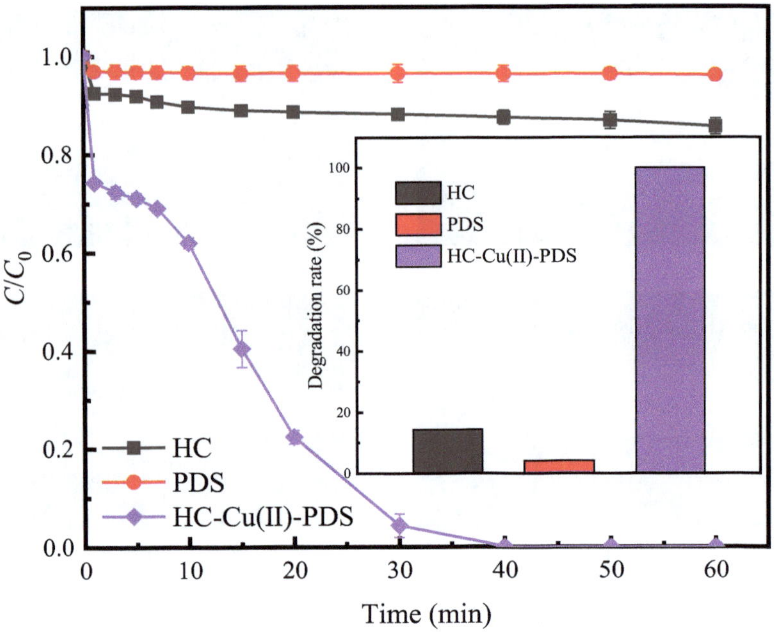

Fig. 2.3 Different systems for degrading TC. Reaction conditions: $C_{TC} = 30$ mg/L, $p = 0.42$ MPa, $C_{PDS} = 0.40$ mM, $C_{Cu\ (II)} = 0.80$ mM, $T = 25$ °C

Table 2.1 Reaction rate constant and per-pass degradation factors of different systems for degradation of TC

Factor	Degradation system		
	HC	*PDS*	HC/Cu (II)/*PDS*
K_{obs} ($\times 10^3$)	3.41	0.97	86.76
ϕ_0 ($\times 10^6$)	9.05	2.58	230.08

speed is directly proportional to the inlet pressure, indicating that the increase in inlet pressure enhanced the energy generated during the bubble collapse and the cavitation intensity.

2.3.4 *Effect of PDS Concentration*

The $SO_4^{2-}\cdot$, which plays a major role in TC degradation, is derived from PDS, and due to the self-scavenging reaction of $SO_4^{2-}\cdot$, the amount of PDS added must be moderate (Choi et al. 2018). Therefore, it is necessary to change the concentration of PDS to determine the optimum dose for the reaction. Figure 2.5 represents the degradation curves when $C_{Cu(II)}$: C_{PDS} was 1: 2 and the PDS concentration was 0.3, 0.4, and 0.5 mM. The removal of TC increased from 95.47% to 100% with the increase of

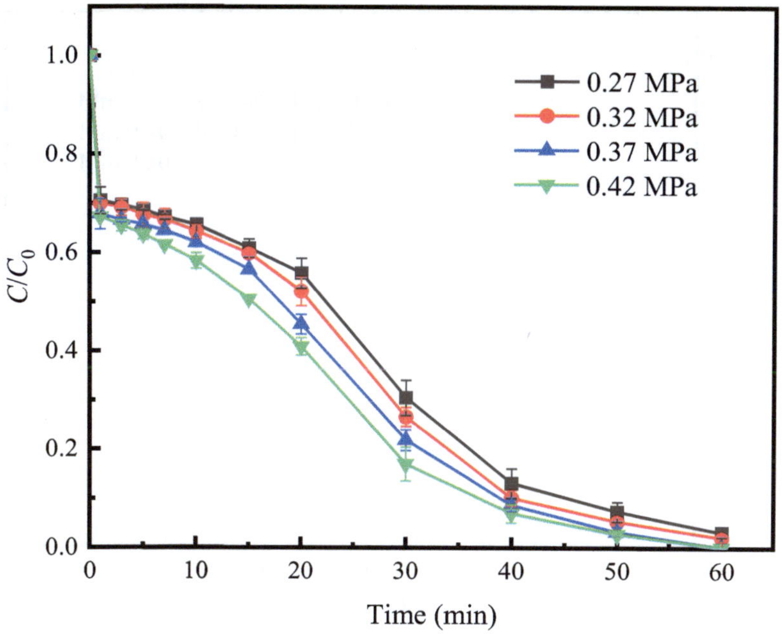

Fig. 2.4 Effect of different inlet pressures on TC degradation. Reaction conditions: $C_{TC} = 30$ mg/L, $C_{PDS} = 0.40$ mM, $C_{Cu(II)} = 0.20$ mM, $T = 25$ °C

PDS concentration. K_{obs} of 0.0479, 0.0659, and 0.0677 min^{-1}, respectively. The first-order rate reaction constants were basically the same when the PDS concentrations were 0.4 and 0.5 mM. So the concentration of 0.4 mM was selected as the optimal concentration for the subsequent experiments from the point of view of economy.

2.3.5 *Effect of Concentration Ratio of Cu (II) and PDS*

Copper ions undergo complex redox reactions with substances such as PDS in aqueous solution (Chen et al. 2019). The dose ratio between each reactant has an inescapable effect on the degradation of TC. Therefore, the degradation effect of TC was investigated when the Cu (II) and PDS had different concentration ratios. From Fig. 2.6, it can be seen that as the concentration ratio of Cu (II) and PDS was increased from 1: 5 to 2: 1, the removal of TC was increased from 89.70% to 100%. It indicates that Cu (II) plays an important role in activating PDS.

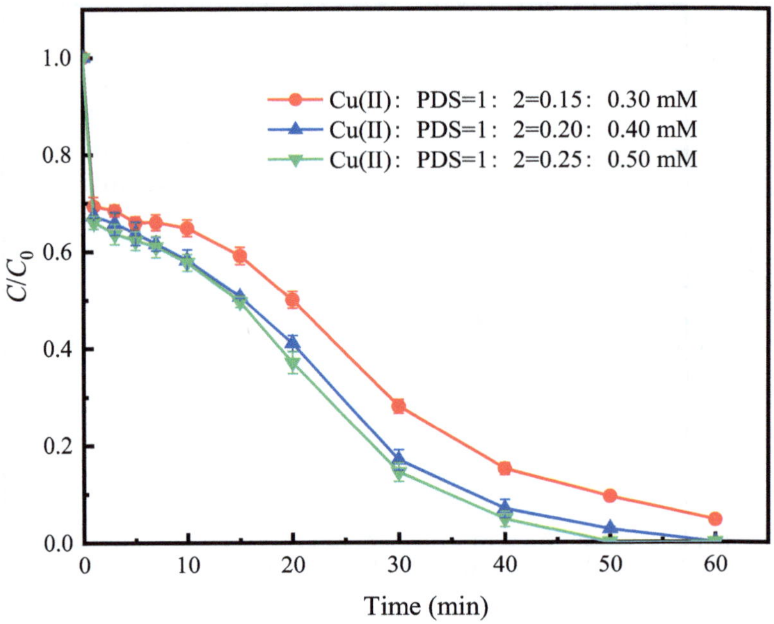

Fig. 2.5 Effect of PDS concentration on TC degradation. Reaction conditions: $C_{TC} = 30$ mg/L, $p = 0.42$ MPa, $C_{Cu(II)}$: $C_{PDS} = 1: 2$, $T = 25$ °C

2.3.6 Discussion of Experimental Results

According to the above experimental results, the system of HC/Cu (II)/PDS has a high synergistic effect on TC degradation, which is due to the "hot spot" effect of the cavitation bubbles and the energy released by the transition metals destroys the "–O–O–" bonds of the PDS, and promotes the production of $SO_4^{2-}\cdot$, resulting in the rapid degradation of TC. The optimal value of the inlet pressure is related to the configuration of the experimental equipment and the type of pollutants. In this experiment, as the inlet pressure increased, there was no decrease in the TC degradation rate, which means that there was no chocked cavitation in the experiment, therefore, as the inlet pressure increased, the energy produced by the rupture of the cavitation bubbles increased, which led to the continuous increase in the TC degradation rate. Theoretically, the higher the concentration of PDS, the higher the SO_4^{2-} production rate, however, too much PDS will lead to the reduction of the content of $SO_4^{2-}\cdot$ and $\cdot OH$, affect the degradation of TC, and after the experiments to determine the optimal concentration of PDS, enough Cu (II) was needed to activate PDS fully, so it was experimentally determined that the concentration ratio of Cu (II) to PDS reaches 2: 1, which can completely degrade TC.

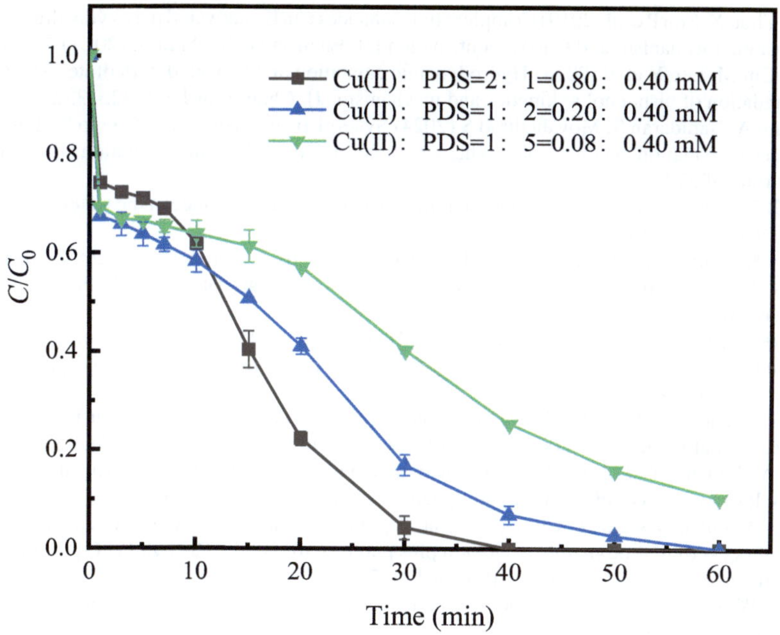

Fig. 2.6 Effect of concentration ratio of Cu (II) and PDS on TC degradation. Reaction conditions: $C_{TC} = 30$ mg/L, $p = 0.42$ MPa, $C_{PDS} = 0.40$ mM, $T = 25$ °C

2.4 Conclusion

In this study, the effectiveness of three systems, HC, PDS, and HC/Cu (II)/PDS, for degrading TC were investigated. The effects of different operating conditions on the TC degradation were studied. It was found that the HC/Cu (II)/PDS system had a considerably high synergistic effect, and HC could promote the circulation of fluid and increase the activation rate of PDS, which effectively improved the TC degradation. Meanwhile, the degradation effect was positively correlated with the inlet pressure, the concentration of PDS, and the concentration ratio of Cu (II) to PDS.

Acknowledgements This work was supported by the National Natural Science Foundation of China (Grant Nos. 52276032 and 52111540266) and "Youth Innovation Team Program" of Shandong Higher Education Institution (Grant No. 2023KJ006).

References

Brujan E-A (2017) Jets from pulsed-ultrasound-induced cavitation bubbles near a rigid boundary [J]. J Phys D-Appl Phys 50(21)

Chen J, Zhou X, Sun P et al (2019) Complexation enhances Cu(II)-activated peroxydisulfate: a novel activation mechanism and Cu(III) contribution [J]. Environ Sci Technol 53(20):11774–11782
Choi J, Cui M, Lee Y et al (2018) Hydrodynamic cavitation and activated persulfate oxidation for degradation of bisphenol a: kinetics and mechanism [J]. Chem Eng J 338:323–332
Dehghani A, Baradaran S, Movahedirad S (2024) Synergistic degradation of Congo Red by hybrid advanced oxidation via ultraviolet light, persulfate, and hydrodynamic cavitation [J]. Ecotoxicol Environ Saf 272
Fan S (2023) On the generation, impact and removal of antibiotic resistance in the water environment [J]. BIO Web Conf 02008, 7
Fedorov K, Plata-Gryl M, Khan JA, et al (2020) Ultrasound-assisted heterogeneous activation of persulfate and peroxymonosulfate by asphaltenes for the degradation of BTEX in water [J]. J Hazard Mater 397
Mukherjee J, Lodh BK, Sharma R et al (2023) Advanced oxidation process for the treatment of industrial wastewater: a review on strategies, mechanisms, bottlenecks and prospects [J]. Chemosphere 345:140473
Patil PB, Thanekar P, Bhandari VM (2022) Intensified hydrodynamic cavitation using vortex flow based cavitating device for degradation of ciprofloxacin [J]. Chem Eng Res des 187:623–632
Roy K, Moholkar VS (2020) Sulfadiazine degradation using hybrid AOP of heterogeneous Fenton/persulfate system coupled with hydrodynamic cavitation [J]. Chem Eng J 386
Roy K, Moholkar VS (2024) Carbamazepine degradation using ternary hybrid advanced oxidation process of hydrodynamic cavitation plus photocatalysis (UV/ZnO/$ZnFe_2O_4$) + persulfate: kinetic investigations [J]. J Water Process Eng 58
Wang B, Wang Y (2022) A comprehensive review on persulfate activation treatment of wastewater [J]. Sci Total Environ 831
Weng M, Cai M, Xie Z et al (2022) Hydrodynamic cavitation-enhanced heterogeneous activation of persulfate for tetracycline degradation: synergistic effects, degradation mechanism and pathways [J]. Chem Eng J 431
Xu Q, Liu S, Lou S et al (2024) Typical antibiotic resistance genes and their association with driving factors in the coastal areas of Yangtze River Estuary [J]. Environ Sci Pollut Res Int 31(21):30440–30453
Zhou P, Zhang J, Liu J et al (2016) Degradation of organic contaminants by activated persulfate using zero valent copper in acidic aqueous conditions [J]. RSC Adv 6(101):99532–99539
Zhou Y, Leong SY, Li Q (2023) Modified biochar for removal of antibiotics and antibiotic resistance genes in the aqueous environment: a review [J]. J Water Process Eng 55

Chapter 3
Study of pH Dynamic Equilibrium and Dissolved Oxygen Synergistically Regulating the Performance of SNAD Process for Nitrogen Removal

Xu Chen, Lihui Zhan, Zhouzhenyang Xia, Jiajun Luo, Ze Bao, Yicheng Wu, and Haiyan Fu

Abstract The study utilized a sequencing batch reactor (SBR) to initiate the Simultaneous nitritation, anammox and denitrification (SNAD) process, using low-concentration municipal wastewater as the treatment object, to investigate the synergistic regulation mechanism between pH and dissolved oxygen (DO). A DO gradient (0–10 days: 1.5–2.0 mg/L; after 10 days: 0.5–1.0 mg/L) and pH dynamic equilibrium (7.0–7.5) were set to analyze the effects on ammonia nitrogen removal efficiency. The results showed that the total nitrogen (TN) removal efficiency fluctuated from 20 to 40%, while the ammonia nitrogen removal efficiency was poor. The reason was that the high DO in the start-up period caused excessive oxidation of ammonia nitrogen and inhibited anaerobic ammonia oxidizing bacteria, and the microbial community was slow to adapt and synergistic effects were not effectively established in the stabilization period with low DO, and the fluctuation of pH control was still difficult to be controlled by the addition of $NaHCO_3$, affecting the activity of the key enzymes and leading to the low efficiency of nitrogen removal. This study revealed the problems in the synergistic control of DO and pH, and provided a direction for the subsequent optimization of the SNAD process.

Keywords pH · Dissolved oxygen · SNAD process · Denitrification performance · Synergistic regulation

---

X. Chen · L. Zhan · Z. Xia · Z. Bao · Y. Wu · H. Fu (✉)
Key Laboratory of Environmental Biotechnology of Fujian University, Xiamen University of Technology, Xiamen, PR China
e-mail: fuhy@xmut.edu.cn

J. Luo
Wanjia Qingshui (Xiamen) Environmental Technology Co., LTD, Xiamen, China

Y. Li et al. (eds.), *16th International Conference on Environmental Science and Development*, Environmental Science and Engineering,
https://doi.org/10.1007/978-3-032-15080-6_3

3.1 Introduction

The synchronous nitrification, denitrification and anammox process (SNAD) achieves low-carbon and high-efficiency treatment of high-ammonia wastewater by coupling nitritation, anammox reaction and denitrification. Its core advantages include saving 60% aeration energy consumption and complete elimination of carbon source dependence (Ding et al. 2018). However, process stability is strictly constrained by dissolved oxygen (DO) and pH levels: excessive DO promotes over-oxidation of nitrite (NO_2^--N) to nitrate (NO_3^--N), inhibiting anammox activity, while pH fluctuations impair microbial enzymatic functions, example the optimal pH for anammox bacteria is 6.5–8.0, thereby disrupting nitrogen conversion efficiency (Zhou and Wang 2024). Currently, the interactive effects of DO and pH co-regulation on microbial community structure and nitrogen removal pathways remain insufficiently understood, particularly regarding their combined influence on denitrification efficiency through substrate supply NO_2^--N availability and metabolic conditions (acid–base equilibrium).

This study utilized a lab-scale sequencing batch reactor (SBR) as the experimental subject. Referencing established SNAD process start-up strategies from similar studies, and considering the differential DO responses between nitritation bacteria (Nitrosomonas) and nitrifying bacteria (Nitrobacter) (nitritation bacteria thrive at DO = 1.0–2.0 mg/L while nitrifying bacteria activity is inhibited at DO < 0.8 mg/L (Xie et al. 2016), we implemented a DO gradient protocol: during the start-up phase (0–10 d) DO was maintained at 1.5–2.0 mg/L to enrich nitritation bacteria, then reduced to 0.5–1.0 mg/L in the stable phase (after 10 d) to suppress nitrifying bacteria growth and create a microaerobic environment for anammox. By establishing this DO gradient coupled with a dynamic pH buffering system (simulating anammox acid production and alkalinity compensation), we monitored temporal variations in removal efficiencies of TN, NO_4^+-N, NO_2^--N and NO_3^--N. The study aims to: (1) Determine how low DO (0.5–1.0 mg/L) selectively suppresses nitrifying bacteria while enriching nitrite-oxidizing bacteria, thereby providing optimal substrate for anammox; (2) Investigate how alkalinity dosing counteracts acid production during anammox (pH decline from 7.5 to 7.0) to maintain microbial metabolic activity; (3) Elucidate the synergistic effects of DO-pH interactions on NO_3^--N accumulation and denitrification performance. The findings will provide critical control parameters for SNAD process optimization, facilitating its broader application in high-ammonia wastewater treatment.

3.2 Materials and Methods

3.2.1 Reactor Set-Up

The study used a sequencing batch reactor (SBR, effective volume 5 L) made of plexiglass, equipped with an aeration system (DO controlled by rotameter), automatic pH regulation device ($NaHCO_3$ solution dosed by peristaltic pump), and water bath heating system (temperature controlled at 28 ± 1 °C).

3.2.2 Wastewater and Inoculation

The inoculated sludge was collected from Xinglin Wastewater Treatment Plant, with an initial MLSS of 3000 ± 200 mg/L and MLVSS/MLSS ratio of 0.7 ± 0.05. Synthetic wastewater was used throughout the experiment, consisting of conventional nutrients and trace elements. The conventional nutrients included carbon sources (sucrose, sodium acetate, sodium bicarbonate, and potassium bicarbonate) and nitrogen sources (ammonium chloride), corresponding to pollutants such as COD and nitrogen species (NH_4^+-N and TN). NH_4Cl served as the nitrogen source with an NH_4^+-N concentration of 50 mg/L, while $NaHCO_3$ was used to maintain alkalinity at a concentration of 500 mg/L. Additionally, to provide favorable growth conditions for microorganisms, a defined quantity of trace element solution was added to the influent at a dosage of 1 mL/L. The trace element composition included: 1500 mg/L $FeCl_3 \cdot 6H_2O$, 120 mg/L $MnCl_2 \cdot 4H_2O$, 26 mg/L $CuSO_4 \cdot 5H_2O$, 105 mg/L $ZnSO_4 \cdot 7H_2O$, 130 mg/L $CoCl_2 \cdot 6H_2O$, 60 mg/L $Na_2MoO_4 \cdot 2H_2O$, 200 mg/L H_3BO_4, and 5000 mg/L EDTA (Singh et al. 2022).

3.2.3 Experimental Methods

Influent: Controlled by a peristaltic pump at a flow rate of 500 mL/h, filling the reactor with synthetic wastewater in 1 h. Aeration: The start-up phase DO gradient was determined based on pre-experimental results: by comparing nitrite accumulation rates (measured by NO_2^--N/NO_3^--N ratio) under DO = 1.0, 1.5, and 2.0 mg/L conditions, it was found that DO = 1.5–2.0 mg/L yielded the highest nitrite accumulation rate (> 60%), meeting partial nitrification requirements. The stable phase DO was set at 0.5–1.0 mg/L, referencing the microaerobic preference of anammox bacteria (Candidatus Brocadia) (DO < 1.0 mg/L) combined with the critical value for nitrifying bacteria inhibition (DO < 0.8 mg/L). pH adjustment: Monitored by an online pH sensor (±0.05 accuracy). Excessive nitrification consumed alkalinity, causing pH to drop. When pH < 7.0, the peristaltic pump automatically dosed 5% $NaHCO_3$ solution (5–10 mL each time) to maintain alkalinity at 200–300 mg/L (as $CaCO_3$) and

Table 3.1 Water quality indices and detection methods

Water quality parameter	Detection method	Detection instrument
NH_4^+-N	Nessler's reagent spectrophotometry	UV-1000 UV–Vis spectrophotometer
NO_2^--N	N-(1-naphthyl)-ethylenediamine spectrophotometry	UV-1000 UV–Vis spectrophotometer
NO_3^--N	UV spectrophotometry	Hach DR3900 spectrophotometer
TN	Potassium persulfate oxidation-UV spectrophotometry	UV-1000 UV–Vis spectrophotometer
DO	Online meter detection	Water quality analyzer (HACH HQ40d)
Temperature	Online meter detection	Mercury thermometer

stabilize pH at 7.0–7.5. Mixing: During aeration, a magnetic stirrer mixed the reactor liquid at 150 ± 10 rpm to ensure substrate-microorganism contact and enhance mass transfer. Settling: 1 h settling phase with no aeration/stirring, allowing gravity-based sludge-water separation. Decanting: 0.5 h drainage phase; treated supernatant was discharged via an upper outlet at 50% exchange ratio. Idle phase: 0.5 h standby with no operations, preparing for the next cycle. Sludge retention time (SRT): Controlled through periodic sludge wasting, maintaining a short SRT of 9 ± 2 d throughout the experiment to ensure stability and activity of microbial communities in the reactor. Simultaneously, online pH and DO sensors were used to monitor real-time changes in pH and DO within the reaction system, automatically triggering the pH adjustment device and aeration system start/stop based on set thresholds (Jiang et al. 2020).

3.2.4 Test Items and Measurement Methods

The water quality indicators and the corresponding test methods are shown in Table 3.1.

3.3 Results and Analysis

3.3.1 Analysis of DO and pH Co-Regulation

Dissolved Oxygen (DO), During the initial reactor operation (0–10 days), DO was maintained at 1.5–2.0 mg/L to establish the microaerobic environment required for Nitrosomonas enrichment. After day 10, DO was progressively reduced to 0.5–1.0 mg/L and maintained at this level to suppress Nitrobacter growth while creating

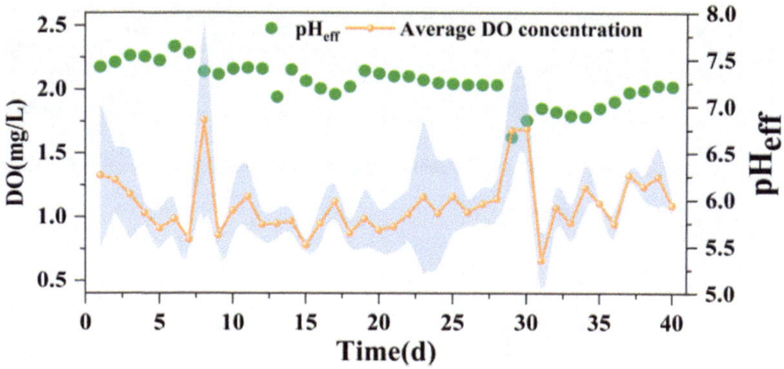

Fig. 3.1 Effluent pH and DO ranges in the reactor

the anoxic microenvironment necessary for Candidatus Brocadia proliferation. The temporal DO variation in the reactor is shown in Fig. 3.1. Under high DO conditions (1.5–2.0 mg/L), active nitrifying bacteria promote excessive ammonia oxidation to nitrate, resulting in a conventional nitrification–denitrification pathway. This process increases both oxygen demand and external carbon requirement for denitrification. In contrast, at low DO levels (0.5–1.0 mg/L), nitrifying bacteria are suppressed while ammonia-oxidizing bacteria dominate, facilitating partial nitritation to nitrite. This creates ideal substrate conditions for anammox, enabling the energy-efficient "nitritation-anammox-denitrification" pathway that significantly reduces both aeration energy and carbon dependency (James 2022).

pH control strategy, The initial influent pH was set at 7.5. Excessive nitrification in the reaction system consumed alkalinity, causing pH to drop to 7.0 ± 0.2. Intermittent $NaHCO_3$ dosing maintained alkalinity at 200–300 mg/L as $CaCO_3$ to prevent further pH decline, ultimately stabilizing pH fluctuations between 7.0 and 7.5. The pH variations in the reactor are shown in Fig. 3.1. pH significantly influences microbial metabolic activity. Anammox bacteria thrive within a pH range of 6.5–8.0. When pH falls below 7.0, the proton gradient across their cell membranes becomes disrupted, reducing hydrazine synthase (HZS) activity and impairing the coupled reaction of ammonium and nitrite. In this study, pH was maintained at 7.0–7.5 through $NaHCO_3$ dosing, effectively preventing acidic inhibition and sustaining anammox metabolic activity. Additionally, low pH intensifies competition between nitrate (NO_3^-) and nitrite (NO_2^-) for enzyme binding sites in anammox bacteria. Maintaining neutral pH mitigates this competitive inhibition, thereby enhancing nitrogen removal efficiency.

As shown in Fig. 3.2, the TN removal efficiency declined during the later operational phase. During the start-up period with higher DO (1.5–2.0 mg/L), partial over-oxidation of ammonia occurred, as evidenced by the notable increase in NO_3^- concentration by day 10. After DO reduction to 0.5–1.0 mg/L (post day 10), while nitrifying bacteria were effectively suppressed, the microbial community restructuring adversely affected the substrate transfer efficiency between nitrite-oxidizing bacteria and anammox bacteria. This is reflected in the TN removal efficiency

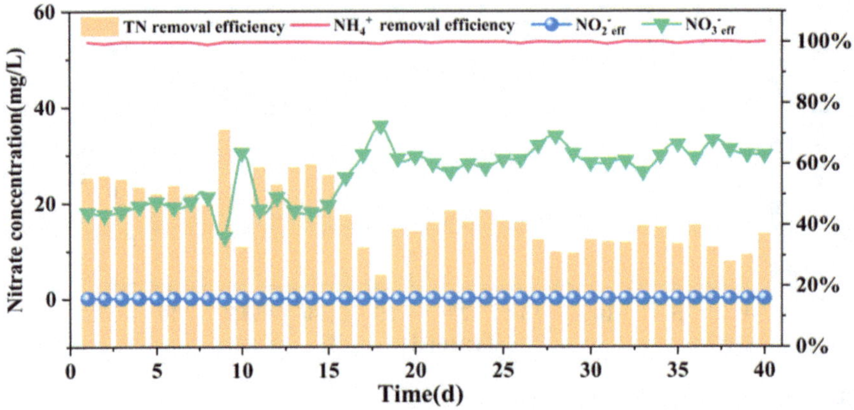

Fig. 3.2 Variations in nitrogen removal within the reactor

decrease from 40 to 30% during days 25–40. The pH was maintained at 7.0–7.5. It is speculated that excessive nitrification during the reaction process consumed alkalinity, causing pH instability and affecting microbial enzyme activity. When pH deviated from the optimal range, the hydrazine oxidase activity of anammox bacteria decreased, reducing the coupling reaction rate between ammonium and nitrite (Yao et al. 2020). For instance, during periods of significant pH fluctuation, the ammonium removal efficiency exhibited minor variations (Fig. 3.5). Furthermore, pH variations affected microbial substrate affinity and uptake capacity, collectively contributing to the decline in total nitrogen (TN) removal efficiency. In contrast, ammonium removal remained relatively stable, with effluent NO_2^- concentrations consistently maintained at low levels (approaching 0 mg/L), while NO_3^- concentrations fluctuated between 10 and 40 mg/L in the effluent. Under low DO (0.5–1.0 mg/L) and suitable pH (7.0–7.5) conditions, it is speculated that a stable symbiotic relationship formed between nitrosomonas and anammox bacteria in the reactor. The nitrosomonas secreted extracellular polymeric substances (EPS) that encapsulated the anammox bacteria, forming microfloc structures. These microflocs created anoxic microzones inside, providing suitable living conditions for the anammox bacteria (Luo et al. 2024). Simultaneously, this structure facilitated substrate transfer between microbial communities, promoted the anammox reaction process, and improved nitrogen removal performance.

3.3.2 Analysis of DO and pH Co-Regulation

During the operational period, TN removal efficiency showed significant variations (Fig. 3.3). In the start-up phase (0–10 days), TN removal remained low (20–30%) due to microbial community adaptation, low anammox activity, and excessive ammonia

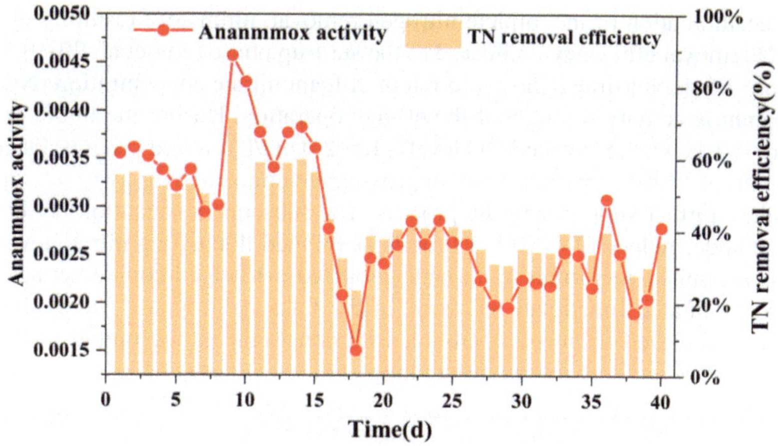

Fig. 3.3 Anammox bacteria activity and TN removal efficiency variations

oxidation to nitrate under high DO conditions, with denitrification being the main removal pathway albeit with inefficient carbon source utilization. From days 10–25, TN removal efficiency rapidly increased to around 40% as the microbial community adapted and the synergistic nitrogen removal pathways became established (Guo et al. 2016). This improvement was attributed to the enhanced synergistic relationship between nitrifying bacteria and anammox bacteria under low DO conditions, where nitrifiers provided sufficient nitrite substrate for the anammox process. After day 25, TN removal efficiency stabilized at approximately 35%, indicating that while the microbial community had reached equilibrium, the overall biological activity remained suboptimal. During this phase, the anammox reaction proceeded at reduced efficiency, with anammox bacteria maintaining only modest metabolic activity.

As shown in Fig. 3.4, the final TN removal efficiency was lower than during the start-up phase, while the effluent ammonium concentration ($NH_4^+{}_{eff}$) remained stable despite variations in influent ammonium ($NH_4^+{}_{inf}$). Analysis of the DO and pH operating ranges revealed that during the start-up phase (DO: 1.5–2.0 mg/L), although partial over-oxidation of ammonium to nitrate occurred, the microbial community underwent an adaptation process to the influent characteristics. This enabled partial utilization of substrates for nitrogen removal, resulting in relatively better TN removal performance during this phase. Subsequent reduction of DO to 0.5–1.0 mg/L aimed to suppress nitrifying bacteria while promoting anammox activity. However, the system performance was compromised by several factors: (1) fluctuating influent ammonium concentrations ($NH_4^+{}_{inf}$ variations) challenged microbial adaptation to new substrate levels; (2) pH oscillations within 7.0–7.5 created metabolic instability in anammox bacteria and other functional microbes; and (3) the synergistic relationship between denitrifiers and anammox bacteria weakened under low DO conditions due to substrate and environmental variations. These factors collectively led to either nitrate accumulation or incomplete nitrogen removal, ultimately resulting in lower final TN removal efficiency compared to the start-up phase (Xie et al. 2019).

Figure 3.5 demonstrates the evolution of effluent nitrate concentration ($NO_3{}^-{}_{eff}$) and anammox activity throughout the 40-day operation. During the initial start-up phase (0–10 days), the elevated DO levels (1.5–2.0 mg/L) caused progressive nitrate accumulation while only marginally improving anammox activity, with the gradual pH decline further suppressing the process. The subsequent transition period (10–25 days) under reduced DO (0.5–1.0 mg/L) exhibited fluctuating nitrate levels and variable anammox performance, though stabilized pH helped counteract acidification and facilitated functional recovery. In the final stable phase (after 25 days), the system achieved balanced operation with consistent nitrate output and steady

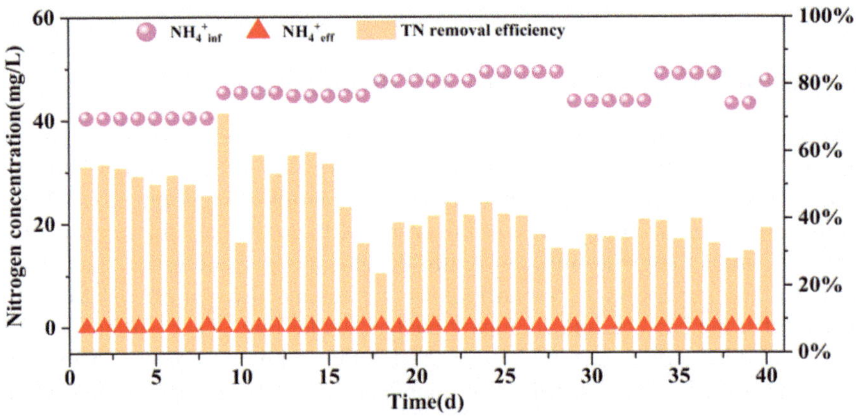

Fig. 3.4 Influent and effluent $NH_4{}^+$-N concentrations and TN removal efficiency

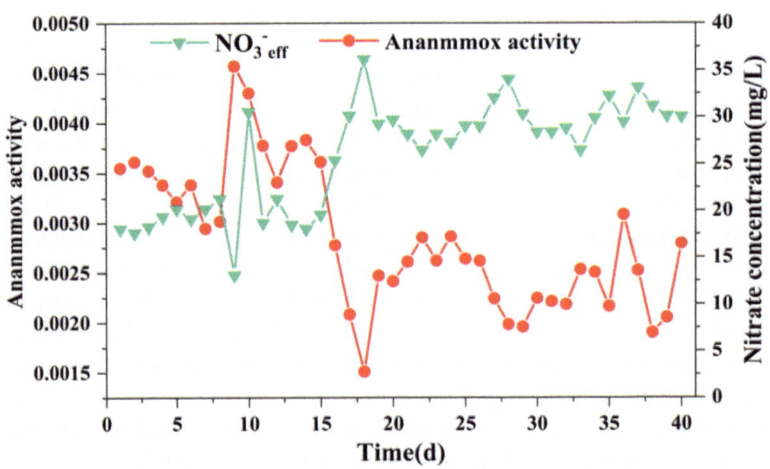

Fig. 3.5 Variations of effluent nitrate concentration and anammox bacteria activity

anammox activity under maintained low DO and optimal pH conditions (7.0–7.5), clearly demonstrating the crucial synergistic relationship between dissolved oxygen control, pH stability, and microbial functionality in determining the overall nitrogen removal efficiency of the SNAD process. The data provides empirical validation of the operational principle that stable performance requires integrated parameter management rather than isolated variable optimization.

3.4 Conclusions

The high DO (1.5–2.0 mg/L) during the start-up phase caused excessive ammonia oxidation, while subsequent low DO (0.5–1.0 mg/L) affected the synergy between nitrosomonas and anammox bacteria, leading to decreased TN removal efficiency. Although $NaHCO_3$ addition maintained pH at 7.0–7.5, excessive nitrification in the reactor reduced alkalinity, resulting in pH instability that affected microbial enzyme activity and decreased denitrification rates. Ammonia removal efficiency fluctuated under the combined influence of DO and pH, with low effluent nitrite concentrations but significant fluctuations in nitrate concentrations, indicating an imbalance between nitrification and denitrification.

Acknowledgements This research was financially supported by Fujian Provincial Industry-university Research Collaborative Innovation (2021Y4005), and Key Technology Collaborative Innovation Platform Project for Improving Urban and Rural Ecological Environment Quality in Fu-Xia-Quan National Independent Innovation Demonstration Zone (Grant No. 2022-P-024). In addition, we acknowledge support of Fujian Engineering and Research Center of Rural Sewage Treatment and Water Safety, and the Key Laboratory of Environmental Biotechnology (XMUT), Fujian Province University.

References

Ding S, Bao P, Wang B, Zhang Q, Peng Y (2018) Long-term stable simultaneous partial nitrification, anammox and denitrification (SNAD) process treating real domestic sewage using suspended activated sludge. Chem Eng J 339:180–188

Guo Y, Peng Y, Wang B, Li B, Zhao M (2016) Achieving simultaneous nitrogen removal of low C/N wastewater and external sludge reutilization in a sequencing batch reactor. Chem Eng J 306:925–932

James SN, Vijayanandan A (2022) Anoxic-aerobic-anoxic sequencing batch reactor for enhanced nitrogen removal. Bioresour Technol 363:127892

Jiang H, Peng Y, Li X, Zhang F, Wang Z, Ren S (2020) Advanced nitrogen removal from mature landfill leachate via partial nitrification-anammox biofilm reactor (PNABR) driven by high dissolved oxygen (DO): protection mechanism of aerobic biofilm. Bioresour Technol 306:123119

Luo J, Wu Y, Fu H, Fu M, Liu M, Guo H, Jin L, Wang S (2024) Shift in microorganism and functional gene abundance during completely autotrophic nitrogen removal over nitrite (CANON) process. J Environ Manage 359:121009

Singh V, Ormeci B, Mishra S, Hussain A (2022) Simultaneous partial nitrification, anammox and denitrification (SNAD)—a review of critical operating parameters and reactor configurations. Chem Eng J 433(3):133677

Xie Y, Wan A, Wang X, Dong H, Wu Y (2019) An iron-carbon-activated carbon and zeolite composite filter, anaerobic-aerobic integrated denitrification device for nitrogen removal in low C/N ratio sewage. Water Sci Technol 80(2):223–231

Yao R, Yuan Q, Wang K (2020) Enrichment of denitrifying bacterial community using nitrite as an electron acceptor for nitrogen removal from wastewater. Water 12(1):48

Yin Z, dos Santos CED, Vilaplana JG, Sobotka D, Czerwionka K, Damianovic MHRZ, Xie L, Morales FJF, Makinia J (2016) Importance of the combined effects of dissolved oxygen and pH on optimization of nitrogen removal in anammox-enriched granular sludge. Process Biochem 51(9):1274–1282

Zhou Y, Wang C, Xu X, Liu L, Zhang G, Yang F (2024) Advance nitrogen removal from anaerobic sludge digestion liquor using partial nitrification and denitrification coupled with simultaneous partial nitrification, anammox, and denitrification process. Bioresource Technol 393:130117

Chapter 4
Hydrochemical Characteristics and Controlling Factors of Water Quality in Typical Rivers of the Jiziwan Region, Yellow River Basin

Yiting Wei, Guohe Huang, Yongping Li, Yanfeng Li, Feng Wang, Changping Chen, Zhirui Yang, and Zhipeng Xu

Abstract The chemical component of natural water reflects the process of water transformation, which is crucial for water resource sustainable development in arid and semi-arid regions. However, research on the hydrochemical characteristics of water environment in arid and semi-arid regions remains limited. This study investigated the hydrochemical characteristics of the Kuye River and Wuding River and analyzed the controlling factors. Groundwater exhibits superior quality compared to surface water and both of them are weakly alkaline. The main ion compositions of groundwater are characterized by $Ca2^{+} > Na^{+} > Mg2^{+} > K^{+}$ in cations and $SO_{4}2^{-} > Cl^{-} > NO_{3}^{-}$ in anions, while that of surface water are $Na^{+} > Ca2^{+} > Mg2^{+} > K^{+}$ in cations and $SO_{4}2^{-} > Cl^{-} > NO_{3}^{-}$ in anions. Water quality of the Kuye River is inferior than that of the Wuding River and ion compositions in the two basins exhibit distinct spatial distributions. The ion sources of surface water and groundwater are different. In surface water, the sources of Na^{+}, K^{+}, and Cl^{-} likely are evaporation and irrigation, while that of $Ca2^{+}$, $SO_{4}2^{-}$ and NO_{3}^{-} may be carbonate rock dissolution, sulfate mineral and fertilizer application. In groundwater, $Ca2^{+}$ and $Mg2^{+}$ primarily come from carbonate mineral, while Na^{+}, K^{+}, Cl^{-} and $SO_{4}2^{-}$ are mainly from mine drainage, evaporative concentration, and the dissolution of sodium and sulfate minerals. This study is meaningful to understand the hydrogeochemical process, providing the scientific basis and reference for water resource sustainable development in arid/semi-arid regions.

Y. Wei · G. Huang (✉) · Y. Li · Y. Li · F. Wang · Z. Yang · Z. Xu
State Key Laboratory of Regional Environment and Sustainability, China-Canada Center for Energy, Environment and Ecology Research, UR-BNU, School of Environment, Beijing Normal University, Beijing, China
e-mail: huang@iseis.org

G. Huang
Environmental Systems Engineering Program, University of Regina, Regina, Canada

C. Chen
Quanzhou Normal University, Quanzhou, China

Y. Li et al. (eds.), *16th International Conference on Environmental Science and Development*, Environmental Science and Engineering,
https://doi.org/10.1007/978-3-032-15080-6_4

Keywords Hydrochemical characteristic · Water quality · Spatial distribution · Ion sources

4.1 Introduction

Water scarcity and its sustainable utilization pose significant challenges in arid and semi-arid regions (Yang et al. 2018). Comprehensive understanding of the water cycle is critical for achieving sustainable water resource sustainable development (Kotowski et al. 2023). The chemical component of natural water is shaped by prolonged interactions with the surrounding environment, effectively recording the history of water movement and transformation (Nie et al. 2005). Hydrochemical analysis serves as a powerful tool for investigating exchange process of groundwater and surface water (Gao et al. 2025). Identifying the hydrochemical characteristics and transformation processes within the water cycle is crucial for maintaining the stability and health of ecosystems in arid and semi-arid regions (Tang et al. 2023).

Jiziwan region is a typical arid and semi-arid area, located in the middle reaches of the Yellow River. Restricted by natural conditions, the region is faced with problems including lack of water resources, soil erosion and fragile ecological environment (Zhao et al. 2024). The Kuye River and Wuding River are representative basins in Jiziwan region, characterized by complex hydrogeological conditions and significant anthropogenic influences (Lei et al. 2013; Zhang and Xi 2019). Interactions between groundwater and surface water in these basins are shaped by natural hydrochemical processes and anthropogenic activities. In recent years, intensified anthropogenic activities such as coal mining, agricultural irrigation, and industrial expansion have led to water quality degradation and potential contamination (Wang et al. 2008; Li et al. 2022). Research on the hydrochemical characteristics of groundwater and surface water in arid and semi-arid basins remains limited. Investigating the hydrochemical characteristics and controlling factors of typical basins is essential for effective water resource management and ecological conservation (Chen et al. 2020).

This study investigated the hydrochemical characteristics of the Kuye River and Wuding River basins and analyzed the potential controlling factors based on field-collected samples and measured hydrochemical parameters. The objectives of this study are to: (1) characterize the hydrochemical properties of surface water and groundwater in the two basins; (2) identify the spatial distribution patterns of hydrochemistry; (3) analyze ion sources and hydrochemical formation mechanisms. This study contributes to a meaningful understanding of the hydrochemical processes in the typical basins of Jiziwan region, offering a scientific reference for environmental management strategies and water resource sustainable development.

4.2 Materials and Methods

4.2.1 Study Area

The Jiziwan region is located in the east of Northwest China, middle reaches of the Yellow River. The Kuye River and Wuding River are key tributaries draining the semi-arid and arid area in this region. The Kuye River basin, with a drainage area of approximately 8706 km^2, is characterized by a loess-covered landscape with extensive coal mining activities, which significantly impact regional hydrology and water quality (Su et al. 2007; Zhang et al. 2024). The Wuding River basin, covering an area of around 30261 km^2 is known for its severe soil erosion (Zhang and Xi 2019), and faces significant water quality challenges (Li et al. 2022).

The two rivers originate in the Loess Plateau and experience a continental monsoon climate, with average annual precipitation ranging from 300 to 500 mm (Li 2020; Bai et al. 2015). Due to the semi-arid conditions, evaporation rates often exceed annual precipitation, leading to substantial water loss and salinization effects in certain areas. Groundwater resources in these basins play a critical role in sustaining local ecosystems and agricultural activities, yet they are increasingly stressed by coal mining, industrial expansion, and irrigation demands (Xie and Bai 2023; Xu et al. 2025). The exchange process of groundwater and surface water is complex, influenced by lithological variations, anthropogenic activities, and climate variability.

4.2.2 Sample Collection

Water samples were collected from the Kuye River and Wuding River basins of the Jiziwan region in July 2024, with sampling locations shown in Fig. 4.1. Surface water samples were obtained using a water sampler from rivers, lakes, reservoirs, and wetlands, while groundwater samples were collected from domestic wells used for irrigation, drinking, and livestock. A total of 40 water samples were collected, including 30 surface water and 10 groundwater samples. Each site was sampled with 200 mL of water, sealed, and stored at − 4 °C in the laboratory prior to analysis.

4.2.3 Sample Analysis

The physicochemical parameters of water samples, including total dissolved solids (TDS), electrical conductivity (EC), pH, and major ions (K^+, Na^+, $Ca2^+$, $Mg2^+$, NO_3^-, Cl^-, SO_42^-) were analyzed. pH, EC, and TDS were measured in situ. Ion concentrations were tested using an ion chromatograph (ICS-2100, USA). To minimize the impact of high ion concentrations on measurement accuracy, samples were diluted sixfold before analysis, based on in situ EC values. A 20 mL aliquot of the

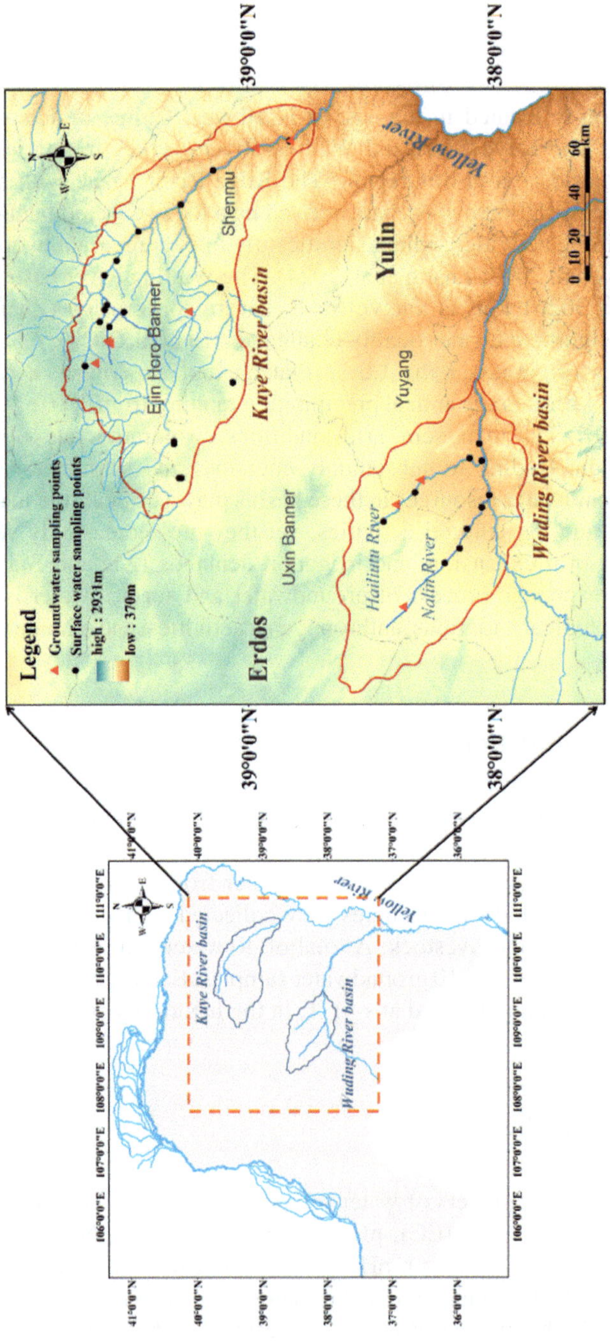

Fig. 4.1 Location of study area and sampling points

diluted sample was used for each measurement. To prevent cross-contamination, the injection system was rinsed with ultrapure water before each analysis. Ion concentrations were obtained through peak processing of chromatograms.

4.3 Result and Discussion

4.3.1 Overall Hydrochemical Characteristics of the Study Area

Basic physicochemical characteristics. In the whole study area, groundwater exhibits superior quality compared to surface water (Fig. 4.2), as indicated by significantly lower pH, electrical conductivity (EC), and total dissolved solids (TDS). This disparity was attributed to prolonged percolation and mineralization within soil and aquifer, which enhance natural purification of the groundwater (Hashimi et al. 2021). Besides, the relatively enclosed hydrogeological conditions of groundwater limit direct input of external contaminants. Additionally, groundwater recharge mainly originates from precipitation and surface water infiltration, undergoing adsorption, ion exchange, and microbial degradation during percolation, thereby reducing dissolved pollutant concentrations (Cheng et al. 2021; Barba et al. 2019).

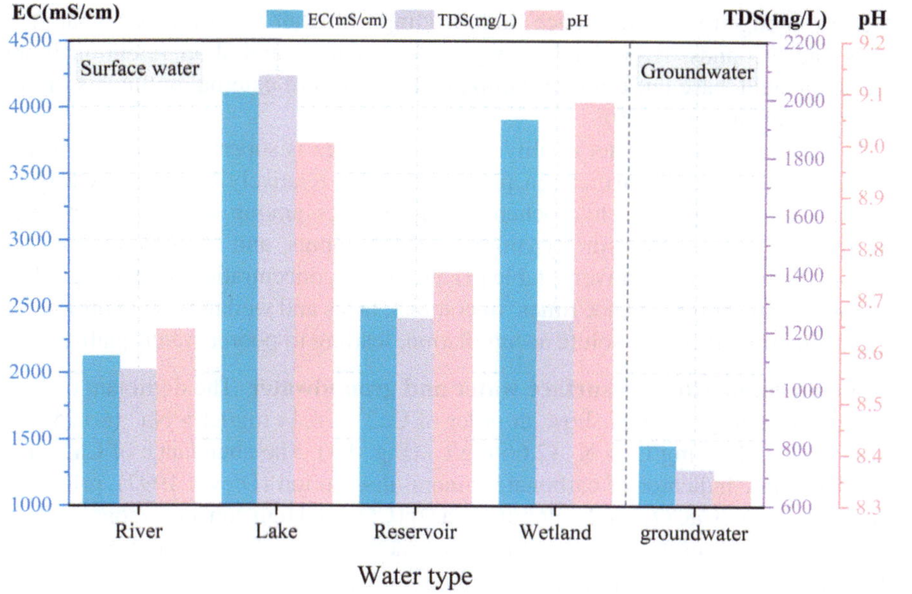

Fig. 4.2 Basic physicochemical indicators of different water types

Surface water exhibits distinct hydrochemical characteristics between different water types. Both of EC and TDS follows the order of lakes > wetlands > reservoirs > rivers. Lower-flow systems, such as wetlands and lakes, show higher EC and TDS. Lakes exhibit the highest EC and TDS due to their relative enclosure and prolonged hydraulic retention. In arid and semi-arid regions, intense evaporation further concentrates dissolved inorganic ions. Additionally, lakes serve as the confluence of inland rivers, often carrying solutes from agricultural, industrial, and domestic sources, further elevating TDS levels (Ren et al. 2023). In contrast, wetlands are highly covered by vegetation, thus biogeochemical processes such as microbial degradation, sediment adsorption, and plant uptake can mitigate dissolved solute concentrations (Mustafa et al. 2024; Yu et al. 2024). Reservoirs, subjecting to anthropogenic regulation, exhibit intermediate water exchange rates and thus moderate EC and TDS levels. Rivers have the lowest EC and TDS due to short hydraulic retention times, which enhance dilution and limit dissolved substances accumulation (Turunen et al. 2020; Tian et al. 2023).

Surface water and groundwater in the study area are generally weakly alkaline (pH > 8.3), reflecting the influence of the geological setting and carbonate mineral weathering. The widespread presence of loess and carbonate rocks (e.g., limestone and dolomite) facilitates water–rock interactions, leading to the release of bicarbonate (HCO_3^-). This process increases pH by neutralizing acidity and enhancing the buffering capacity of the water. Besides, Phytoplankton photosynthesis further contributes to pH elevation by consuming dissolved CO_2, which reduces carbonic acid formation. This effect is particularly pronounced in lentic systems such as wetlands and lakes (Guildford et al. 2007), where biological activity is more sustained due to prolonged water residence time, leading to higher pH levels. Compared to lakes and wetlands, rivers have strong hydrodynamics and short retention times accelerate acid–base buffering, minimizing the impact of external pollutants on pH levels (Turunen et al. 2020).

In summary, groundwater quality in the study area is superior to surface water, mainly due to natural purification processes and a relatively enclosed hydrogeological environment. The hydrochemical differences among surface water types are primarily driven by hydrodynamics, external inputs, and evaporation concentration. Rivers exhibit relatively stable pH and solute concentrations due to high flow dynamics and short residence times. In contrast, lakes and wetlands, with prolonged water retention, facilitate solute accumulation, leading to poorer water quality.

Main ion composition in surface water and groundwater. The dominant cations of groundwater in the area show an order of $Ca2^+$ (74.44 mg/L) > Na^+ (60.25 mg/L) > $Mg2^+$ (21.72 mg/L) > K^+ (2.66 mg/L) (Fig. 4.3). The abundance of $Ca2^+$ and $Mg2^+$ is a key indicator of carbonate mineral dissolution (Drever 1997), primarily from limestone ($CaCO_3$) and dolomite ($CaMg(CO_3)_2$). High concentrations of $Ca2^+$ and $Mg2^+$ reflect prolonged water–rock interactions of carbonates in groundwater. Among anions in groundwater, the dominant order is SO_42^- (54.95 mg/L) > Cl^- (41.51 mg/L) > NO_3^- (14.67 mg/L). The accumulation of SO_42^- is primarily driven by sulfate mineral dissolution (Spoelstra et al. 2021), while the concentration of

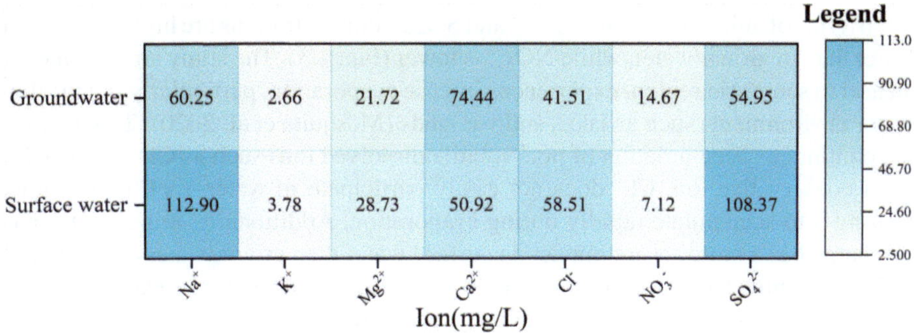

Fig. 4.3 Concentration of main ions in surface water and groundwater

Cl^- is influenced by evaporite dissolution like halite NaCl and anthropogenic pollution (Gutchess et al. 2016). Its elevated levels suggest groundwater is affected by evaporation-induced concentration and external pollutant inputs.

In surface water, the concentrations of cations are characterized by the order of Na^+ (112.90 mg/L) > $Ca2^+$ (50.92 mg/L) > $Mg2^+$ (28.73 mg/L) > K^+ (3.78 mg/L) (Fig. 4.3). The enrichment of Na^+ in surface water is likely driven by a combination of evaporation and leaching of saline-alkali soils, which introduce additional saline inputs into the system. The concentrations of anions follow the order of SO_42^- (108.37 mg/L) > Cl^- (58.51 mg/L) > NO_3^- (7.12 mg/L). This distribution pattern is influenced by mineral weathering, anthropogenic discharge, and evaporation, which collectively regulate the ionic composition of surface water. The elevated SO_42^- levels in surface water likely result from the dissolution of upstream sulfate-bearing minerals and industrial pollution, while the enrichment of Cl^- may be attributed to mining wastewater discharge and evaporation-induced concentration (Rinder et al. 2020).

Comparative analysis reveals significant differences in the ionic composition of groundwater and surface water. The concentrations of $Mg2^+$ and Na^+ in surface water are higher than in groundwater, whereas $Ca2^+$ is lower (Fig. 4.3). This suggests that cation exchange and mineral dissolution may occur during surface water recharge to groundwater. The stronger hydrodynamic conditions in surface water promote mineral dissolution and ion exchange processes (Turunen et al. 2020). As a more mobile ion, Na^+ in groundwater is likely influenced by cation exchange (Li et al. 2023), where Na^+ is partially adsorbed by aquifer media, leading to its lower concentration. In contrast, the higher $Ca2^+$ concentration in groundwater is primarily controlled by rock weathering and mineral dissolution. Prolonged contact with carbonate rocks (e.g., limestone, dolomite) and silicate minerals facilitates $Ca2^+$ release into groundwater (Drever 1997). Additionally, as surface water infiltrates into groundwater, water–rock interactions may further enrich $Ca2^+$. In comparison, surface water has shorter residence time and stronger mobility, limiting rock-water interaction and resulting in lower $Ca2^+$ concentrations (Cheng et al. 2021).

In terms of anion composition, Cl^- and SO_42^- concentrations are higher in surface water than in groundwater, while NO_3^- is lower (Fig. 4.3). The study area, located in an arid to semi-arid region, experiences intense evaporation, particularly in low-flow water environments such as lakes and wetlands (Mesquita et al. 2020). This leads to the continuous accumulation of non-volatile dissolved ions such as Cl^- and SO_42^-. As a conservative ion, Cl^- does not easily participate in water–rock interactions, leading it to accumulate rapidly during evaporation. Additionally, mine wastewater discharge has a more severe impact on surface water than groundwater (Rinder et al. 2020). In contrast, the closed nature of the groundwater system limits external inputs, leading to relatively lower Cl^- and SO_42^- concentrations. The higher NO_3^- concentration in groundwater may result from nitrification-driven nitrate leaching from soils into the aquifer (Dahan et al. 2014). In surface water, NO_3^- levels remain lower due to plant uptake, photosynthetic activity, and microbial denitrification (Kreiling et al. 2011).

The results indicate distinct differences in ion concentrations and controlling mechanisms between groundwater and surface water. Surface water chemistry is strongly influenced by precipitation, evaporation-driven concentration, and external inputs, while groundwater composition is primarily governed by rock weathering, mineral dissolution, and long-term water–rock interactions. Additionally, anthropogenic activities, such as agricultural fertilization, mine water discharge, and wastewater infiltration, could significantly impact water chemistry and should not be overlooked.

4.3.2 Spatial Hydrochemical Characteristics in the Wuding River and Kuye River Basins

Basic physicochemical characteristics of two basins. The surface water quality in the Kuye River basin is generally inferior than that in the Wuding River basin, as indicated by significantly higher EC, pH, and TDS in the Kuye River Basin (Fig. 4.4). This contrast is likely attributed to differences in geological settings, hydrological conditions, and anthropogenic influences. Kuye River located in the arid Ordos Plateau, experiences intense evaporation, which enhances mineral dissolution and increases water salinity (Su et al. 2007). Its relatively enclosed topography and limited water flow further amplify evaporation-driven concentration, elevating EC and TDS levels. Additionally, the basin is heavily influenced by energy development and industrial activities, receiving substantial inputs of dissolved salts and ions from mining operations and industrial wastewater discharge (Zhang et al. 2024). In contrast, the Wuding River basin, situated in a more humid region, benefits from greater precipitation recharge, strong hydrodynamics, diverse recharge sources, and active runoff exchange, facilitating dilution and lowering EC and TDS (Li et al. 2022). The higher pH in the Kuye River Basin may result from intensified evaporation and inputs of alkaline substances, such as carbonates and bicarbonates.

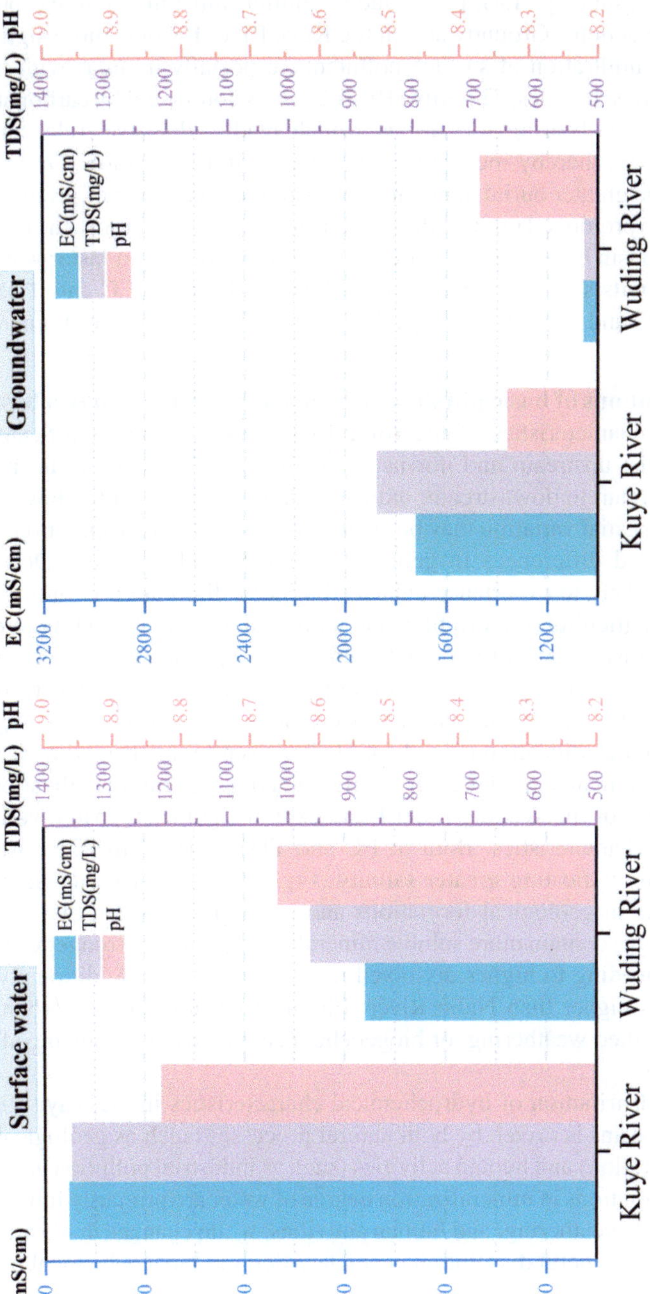

Fig. 4.4 Overall basic physicochemical indicators of two basins

Groundwater in the Kuye River Basin exhibits higher EC and TDS than in the Wuding River Basin (Fig. 4.4), likely due to aquifer mineralization and prolonged water–rock interactions. Groundwater in the Kuye River Basin is more significantly affected by the infiltration of surface pollutants, especially in areas with intensive industrial activities. Besides, The Kuye River Basin is dominated by carbonate rocks, mudstone, and coal-bearing formations, which readily dissolve, releasing $Ca2^{+}$, $Mg2^{+}$, and SO_42^{-}, thereby increasing groundwater mineralization (Drever 1997). Additionally, the greater burial depth and lower renewal capacity of groundwater in this basin lead to prolonged mineral dissolution and ion accumulation. In contrast, the Wuding River Basin benefits from better recharge conditions and faster water circulation, which limits solute accumulation and results in lower EC and TDS levels. The elevated EC and TDS in the Kuye River Basin indicate greater human impact on water quality.

Spatial distributions of basic physicochemical characteristics in two basins. The hydrochemical characteristics of the Kuye River Basin exhibit significant spatial differences in the upstream and downstream. EC and TDS exhibit an increasing trend from upstream to downstream, indicating higher salinity in the lower reaches (Fig. 4.5). This spatial variation may be attributed to intensified evaporation, reduced flow velocity, and differences in geological settings and land use. Downstream areas are more likely to experience enhanced mineral dissolution or anthropogenic inputs, such as agricultural return flow and industrial discharge, leading to elevated salinity. In contrast, upstream regions benefit from expanded catchment areas and external recharge such as precipitation and low-salinity groundwater, resulting in lower salinity. Additionally, pH shows a decreasing trend downstream, suggesting more pronounced acidification (Fig. 4.5), likely driven by intensified rock and soil weathering and human activities such as mining and agricultural fertilization.

The tributaries of the Wuding River basin exhibit significant spatial variation in hydrochemical characteristics. Both of EC and TDS in Hailiutu River are higher than Nalin River, indicating greater salinity (Fig. 4.5). This disparity likely stems from differences in geological formations and soil types between two tributaries. Hailiutu River may contain more soluble minerals or experience stronger evaporative concentration, leading to higher dissolved ion concentrations. Additionally, pH of Hailiutu river is higher than Nalin River (Fig. 4.5), resulting from carbonate-rich lithology, intensified weathering, or biogeochemical processes enhancing alkalinity (Lyu et al. 2018).

The spatial distribution of hydrochemical characteristics in the Kuye River and Wuding River basins is driven by both natural processes (such as geological conditions and evaporation) and human activities (such as industrial pollution and agricultural runoff). Variations in mineralization degree of water are primarily influenced by evaporation, rock weathering, and human activities, while changes in pH are closely related to distinct mineral dissolution characteristics and biogeochemical processes within the basins.

Main ion composition in two bains. The compositions of Na^{+}, $Mg2^{+}$, and $Ca2^{+}$ in surface water and groundwater is consistent across the Kuye River and Wuding River

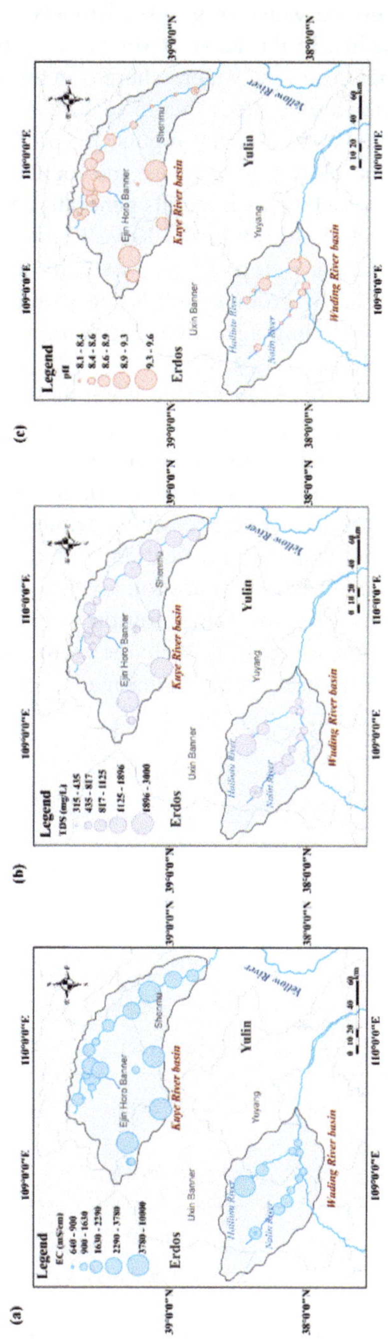

Fig. 4.5 Spatial distributions of EC **a**, TDS **b** and pH **c** in the Kuye River basin and Wuding River basin

basins, with lower Na^+ and $Mg2^+$ concentrations and higher $Ca2^+$ concentrations in groundwater compared to surface water (Fig. 4.6). However, K^+ exhibits distinct patterns between the two basins. In the Kuye River basin, concentration of K^+ in groundwater is lower than that in surface water, whereas in the Wuding River basin, concentration of K^+ in groundwater exceeds that in surface water, indicating differences in water–rock interactions and external inputs. K^+ primarily originates from the weathering of feldspar minerals (e.g., orthoclase, microcline), but its low mobility in water and tendency for fixation by clay minerals generally result in lower groundwater concentrations (White et al. 2001). In the Kuye River basin, limited contact between groundwater and feldspar-rich rocks may explain the lower K^+ levels. In contrast, in the Wuding River basin, groundwater K^+ enrichment may be influenced by anthropogenic activities, including agricultural irrigation and fertilization, and enhanced water–rock interactions.

The compositions of Cl^- and SO_4^{2-} in surface water and groundwater is consistent across the Kuye River and Wuding River basins, with lower concentrations in groundwater than in surface water (Fig. 4.6). However, NO_3^- exhibits distinct patterns between the two basins. In the Kuye River basin, the concentration of NO_3^- in groundwater exceed that in surface water, possibly attributed to agricultural fertilization, wastewater infiltration, and soil nitrification. Extensive nitrogen fertilizer application may facilitate nitrate leaching into groundwater (Dahan et al. 2014). In contrast, in the Wuding River basin, the concentration of NO_3^- in groundwater is lower than that in surface water, suggesting recharge from atmospheric precipitation or relatively clean groundwater sources. Additionally, microbial denitrification within the aquifer may further reduce groundwater NO_3^- levels (Kreiling et al. 2011).

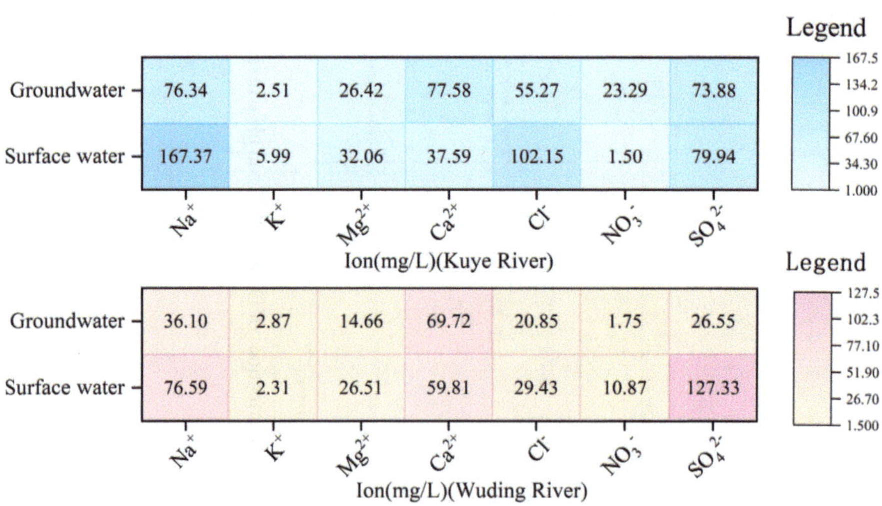

Fig. 4.6 Concentration of main ions in surface water and groundwater in the Kuye River basin and Wuding River basin

Spatial distributions of main ions in two basins. The Kuye River basin exhibits significant spatial variation in $Ca2^{+}$, with higher levels downstream (Fig. 4.7). This trend is likely influenced by the presence of carbonate or evaporite-rich formations in the lower reaches. Water–rock interactions, particularly the dissolution of carbonate rocks such as dolomite and calcite may release $Ca2^{+}$, contributing to its elevated concentrations downstream (Drever 1997). In contrast, the spatial distribution of Na^{+}, $Mg2^{+}$, and K^{+} remains relatively stable. Regarding anions, SO_42^{-} and Cl^{-} concentrations are markedly higher downstream (Fig. 4.7). This pattern may result from combined effects of water rock interaction, evaporation, and human activities. There may be more formations rich in sulfate minerals (such as gypsum and hard gypsum) and halide minerals (such as rock salt) in downstream areas, which release SO_42^{-} and Cl^{-} through water rock interactions (Spoelstra et al. 2021; Gutchess et al. 2016). Furthermore, the enhancement of downstream evaporation will lead to the accumulation of these soluble ions. In addition, human activities such as coal mining, agricultural fertilization, and industrial emissions may input additional SO_42^{-} and Cl^{-} into water environments, exacerbating the mineralization level of downstream water (Rinder et al. 2020).

In the Wuding River basin, the concentrations of Na^{+} and $Ca2^{+}$ are generally higher in the Nalin River than in the Hailiutu River (Fig. 4.7), likely due to differences in geological settings and recharge sources. The Nalin River may flow through sodium-rich deposits, such as feldspar-bearing rocks and evaporites, whereas the Hailiutu River receives more low-mineralized surface runoff, resulting in relatively lower Na^{+} and $Ca2^{+}$ levels. As for anions, the concentrations of Cl^{-} in Nalin River is higher than that in Hailiutu River (Fig. 4.7). This difference may stem from two aspects. On the one hand, increased evaporation in the Nalin River, leading to Cl^{-} enrichment. On the other hand, there may be saline alkali land or evaporite rock layers in the Nalin River basin, contributing additional Cl^{-}. Furthermore, emissions from agricultural irrigation or industrial activities may further increase the concentration of Cl^{-} in the Nalin River water (Xie et al. 2017).

The hydrochemical characteristics of the Kuye and Wuding River basins are influenced by water–rock interactions, evaporative concentration, recharge sources, and human activities. In the Kuye River basin, $Ca2^{+}$, SO_42^{-}, and Cl^{-} are significantly enriched downstream, likely driven by mineral dissolution, evaporation, and anthropogenic inputs. In the Wuding River basin, higher Na^{+}, $Ca2^{+}$, and Cl^{-} concentrations in the Nalin River compared to the Hailiutu River may reflect differences in geological settings, groundwater recharge, and hydrological processes. The spatial variability of these hydrochemical patterns not only highlights natural geochemical processes but also underscores the impact of human activities.

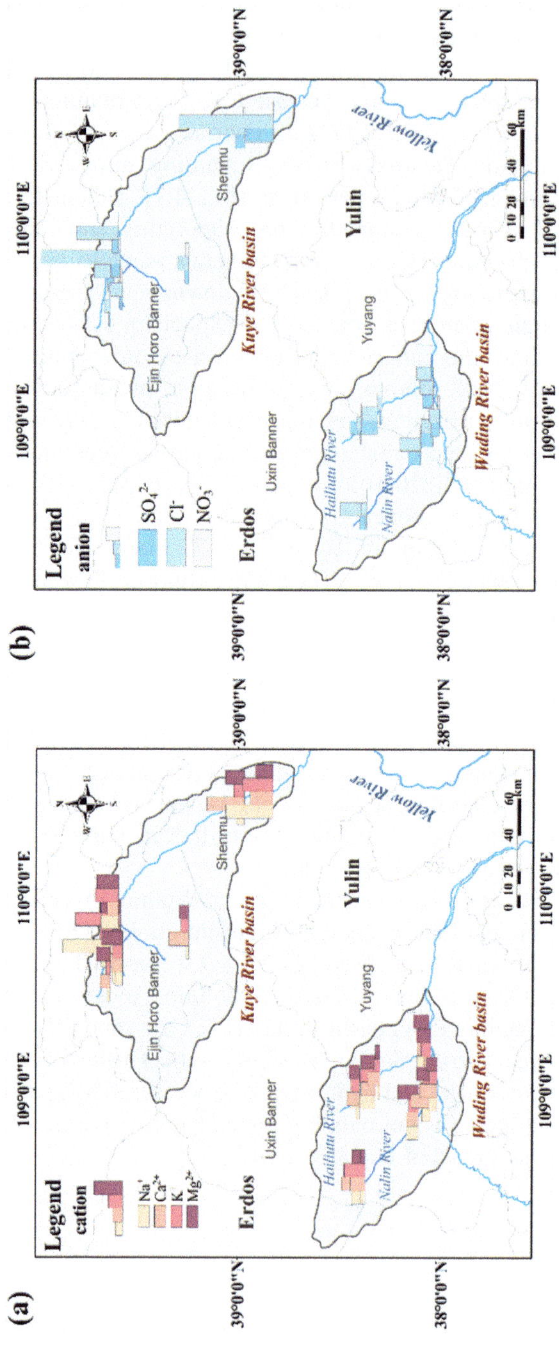

Fig. 4.7 Spatial distributions of cations **a** and anions **b** in the Kuye River basin and Wuding River basin

4.3.3 Correlation of Hydrochemical Indicators and Analysis of Ion Sources

The correlation among hydrochemical indicators is a complex phenomenon, often indicating a shared source or similar external influences (Kong et al. 2020). Pearson correlation analysis of hydrochemical components can reveal relationships between different indicators, providing insights into their origins, similarities, and variations (Kong et al. 2020). Correlations between hydrochemical indicators are illustrated in Fig. 4.8.

Pearson correlation analysis reveals distinct differences in ion relationships between surface water and groundwater. In surface water, TDS shows a strong positive correlation with Na^+, K^+, and Cl^- ($p < 0.001$), indicating their dominant role in dissolved salt contributions. pH is positively correlated with Na^+ and K^+ but negatively correlated with Ca^{2+} ($p < 0.05$), suggesting carbonate equilibrium as a key regulator of water alkalinity (Drever 1997). Under high pH conditions, the concentration of $Ca2^+$ may decrease due to the precipitation of calcium carbonate, while Na^+ and K^+ likely increase due to evaporation, thus exhibiting the opposite correlation (Kolpakova and Gaskova 2018). The highly significant positive correlation ($p < 0.001$) between Na^+, K^+, and Cl^- indicates that they come from the same source. This further suggests chloride enrichment through evaporation, and irrigation return flow (Rinder et al. 2020). Additionally, the positive correlation between Ca 2^+ and NO_3^- ($p < 0.05$) may be related to the combined effects of fertilizer application, soil leaching, or carbonate rock dissolution (Dahan et al. 2014; Kreiling et al. 2011), indicating that nitrate pollution may be driven by both natural processes and human activities.

In groundwater, TDS exhibits a strong positive correlation with $Ca2^+$ ($p < 0.001$), indicating that groundwater mineralization is primarily controlled by this ion. The significant negative correlation between pH and $Ca2^+$ ($p < 0.05$) suggests that carbonate dissolution–precipitation equilibrium is pH-dependent, where higher pH promotes $CaCO_3$ precipitation, reducing $Ca2^+$ concentrations (Drever 1997). Additionally, Na^+ and K^+ show strong positive correlations with Cl^- and SO_4^{2-}, indicating that the dissolution of sodium and sulfate minerals is a key source of these ions. This correlation may also reflect cation exchange processes (Huang et al. 2023). Furthermore, Cl^- and SO_42^- accumulation may be influenced by mine drainage and evaporative concentration (Spoelstra et al. 2021; Rinder et al. 2020).

Overall, variation of TDS in surface water is primarily controlled by evaporation and external inputs, whereas that in groundwater is mainly driven by rock weathering and water–rock interactions. The strong correlation between Na^+, K^+, and Cl^- in surface water highlights the influence of evaporation and irrigation return flow on hydrochemistry, while the strong association of Na^+ and K^+ with Cl^- and SO_42^- in groundwater suggests significant contributions from evaporite dissolution and mining drainage. The positive correlation between $Ca2^+$ and NO_3^- in surface water indicates agricultural inputs as a key source of nitrate, whereas the negative correlation between

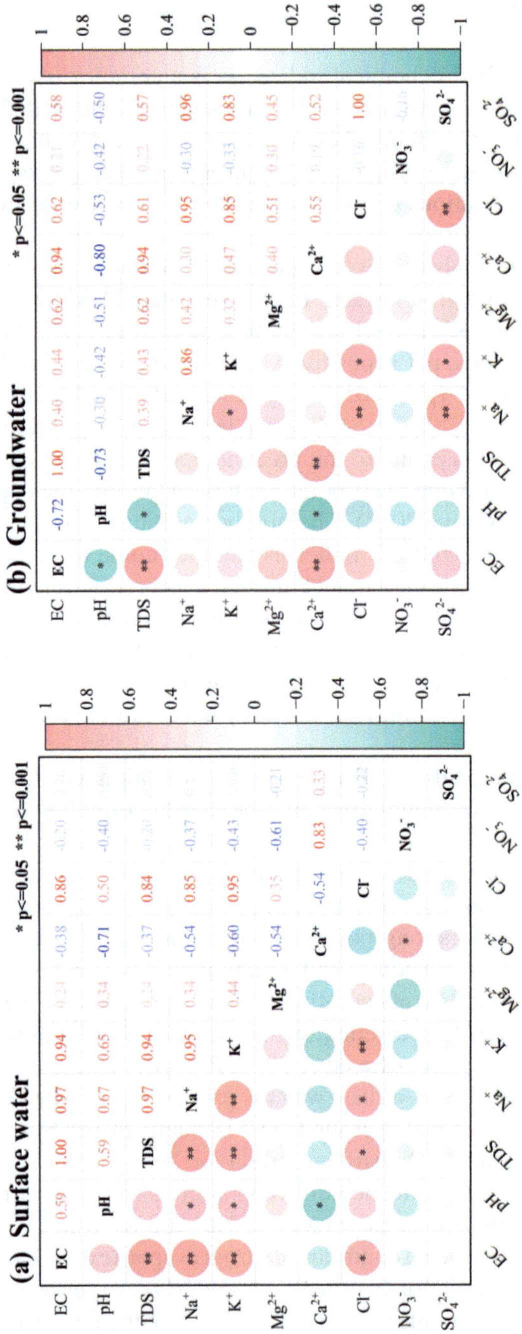

Fig. 4.8 Correlation between hydrochemical indicators of surface water and groundwater

pH and $Ca2^+$ in groundwater underscores the critical role of carbonate equilibrium in groundwater systems.

4.4 Conclusion

This study investigated the hydrochemical characteristics of the Kuye River and Wuding River basins and analyzed the controlling factors. Groundwater quality is superior than surface water due to natural purification processes and enclosed hydrogeological environment. Lower-flow systems such as wetlands and lakes show higher EC and TDS than rivers. Besides, both of surface water and groundwater are weakly alkaline. The main ion compositions of groundwater are characterized by $Ca2^+ > Na^+ > Mg2^+ > K^+$ in cations and $SO_42^- > Cl^- > NO_3^-$ in anions, while that of surface water are $Na^+ > Ca2^+ > Mg2^+ > K^+$ in cations and $SO_42^- > Cl^- > NO_3^-$ in anions. The concentrations of Na^+, $Mg2^+$ and Cl^-, SO_42^- in surface water of the study area are higher than in groundwater, whereas $Ca2^+$ and NO_3^- is lower.

EC and TDS in the Kuye River basin are higher than those in the Wuding River basin. EC and TDS in the Kuye River basin exhibit an increasing trend from upstream to downstream, while pH shows a decreasing trend. In Wuding River basin, Hailiutu River shows higher EC, TDS, and pH compared to Nalin River. Moreover, K^+ and NO_3^- exhibits distinct patterns between the two basins. $Ca2^+$, SO_42^-, and Cl^- are enriched downstream in the Kuye River basin. In the Wuding River basin, the concentrations of Na^+, $Ca2^+$, and Cl^- in the Nalin River are higher than Hailiutu River.

Carbonate equilibrium is a key regulator of surface water and groundwater alkalinity. The positive correlation between Na^+, K^+, and Cl^- in surface water highlights the influence of evaporation and irrigation, while the strong association of Na^+ and K^+ with Cl^- and SO_42^- in groundwater suggests significant contributions from evaporite dissolution and mining drainage. The positive correlation between $Ca2^+$ and NO_3^- in surface water indicates agricultural inputs as a key source of nitrate, whereas the negative correlation between pH and $Ca2^+$ in groundwater underscores the critical role of carbonate equilibrium in groundwater systems. These results are insightful to understand the hydrogeochemical process, providing scientific basis for water resource management in arid/semi-arid areas of the Loess Plateau and reference for regional sustainable development.

4.5 Shortcomings and Future Work

This study is just based on a single sampling event with a limited number of surface water and groundwater samples. Future research will enrich sampling points and increase frequency for long-term monitoring to further analyze hydrochemical dynamics. Besides, more hydrochemical models and analytical methods can be employed to provide deeper insights into the spatiotemporal variations and driving mechanisms of hydrochemical in arid/semi-arid areas.

Acknowledgements This research was supported by the National Key R&D Program of China (2023YFC3206503) and the Innovative Research Group of the National Natural Science Foundation of China (52221003).

References

Bai L, Li H, He H (2015) Analysis on detection and attribution of runoff change in Kuye River Basin. J Hydroelectric Eng 34(2):15–22

Barba C, Folch A, Gaju N, Sanchez-Vila X, Sanchez-Vila X, Carrasquilla M, Grau-Martínez A, Martínez-Alonso M (2019) Microbial community changes induced by managed aquifer recharge activities: linking hydrogeological and biological processes. Hydrol Earth Syst Sci 23(1):139–154

Chen J, Qian H, Gao Y, Wang H, Zhang M (2020) Insights into hydrological and hydrochemical processes in response to water replenishment for lakes in arid regions. J Hydrol 581:124386

Cheng K, Xu X, Cui L, Li Y, Zheng J, Wu W, Sun J, Pan G (2021) The role of soils in regulation of freshwater and coastal water quality. Philos Trans Royal Soc B: Biol Sci 376:1834

Dahan O, Babad A, Lazarovitch N, Russak EE, Kurtzman D (2014) Nitrate leaching from intensive organic farms to groundwater. Hydrol Earth Syst Sci 18(1):333–341

Drever JI (1997) The geochemistry of natural waters: surface and groundwater environments. J Environ Qual 27(1):245–246

Gao Z, Huang T, Chen J, Tian H, Tan M, Niu Y, Lou K (2025) Comprehensive hydrochemical analysis, controlling mechanisms, and water quality assessment of surface and groundwater in a typical intensive agricultural area, Northern China. Water 17(2):276

Guildford SJ, Bootsma HA, Taylor WD, Hecky RE (2007) High variability of phytoplankton photosynthesis in response to environmental forcing in Oligotrophic Lake Malawi/Nyasa. J Great Lakes Res 33(1):170–185

Gutchess K, Jin L, Lautz L, Shaw SB, Zhou X, Lu Z (2016) Chloride sources in urban and rural headwater catchments, central New York. Sci Total Environ 565(15):462–472

Hashimi OAA, Hashim K, Loffill E, Ansari N (2021) A comprehensive review for groundwater contamination and remediation: occurrence, migration and adsorption modelling. Molecules 26(19):5913

Huang J, Yan B, Liu P (2023) Analysis of hydrochemical characteristics and main controlling factors in the Bangong Lake Basin. Geol Resour 32(04):453–461

Kolpakova MN, Gaskova OL (2018) Major ions behaviour during evaporation of different saline type water of western Mongolian lakes (geochemical modelling). Hydrol Res 49(1–2):163–176

Kong L, Wang Z, Wang B (2020) Hydrochemical characteristics and cause analysis of centralized deep groundwater drinking water source in Fuyang City. China Rural Water Conservancy Hydropower 3:78–82

Kotowski T, Najman J, Nowobilska, Luberda A, Bergel T, Kaczor G (2023) Analysis of the interaction between surface water and groundwater using gaseous tracers in a dynamic test at a riverbank filtration intake. Hydrol Process 37(4):14862

Kreiling MR, Richardson BW, Cavanaugh CJ, Bartsch LA (2011) Summer nitrate uptake and denitrification in an upper Mississippi River backwater lake: the role of rooted aquatic vegetation. Biogeochemistry 104(1–3):309–324

Lei Y, Zhang X, Zhang J, Liu E (2013) Change trends and driving factors of base flow in Kuye River catchment. Acta Ecol Sin 33(5):1559–1567

Li X (2020) Innovative research on ecological co-culture technology of rice and crab in Wuding River Basin. Modern Agric Res 26(3):114–115
Li S, Han X, Wang W, Li Z (2022) Hydrochemical characteristics and controlling factors of surface water and groundwater in Wuding River Basin. Environ Sci 43(1):220–229
Li J, He P, Chen J, Hamad AAA, Dai X, Jin Q, Ding S (2023) Tomato performance and changes in soil chemistry in response to salinity and Na/Ca ratio of irrigation water. Agric Water Manag 285:108363
Lyu X, Tao Z, Gao Q, Peng H, Zhou M (2018) Chemical weathering and riverine carbonate system driven by human activities in a subtropical Karst Basin, South China. Water 10(11):1524
Mesquita JBDF, Neto IEL, Raabe A, Raabe A, Araújo JC (2020) The influence of hydroclimatic conditions and water quality on evaporation rates of a tropical lake. J Hydrol 590:125456
Mustafa G, Hussain S, Liu Y, Ali I, Liu J, Bano H (2024) Microbiology of wetlands and the carbon cycle in coastal wetland mediated by microorganisms. Sci Total Environ 954:175734
Nie Z, Chen Z, Cheng X, Hao M, Zhang G (2005) The chemical information of the interaction of unconfined groundwater and surface water along the Heihe River, Northwestern China. J Jilin Univ (Earth Sci Ed) 35(1):48–53
Ren X, Zhang Z, Yu R, Li Y, Li Y, Zhao Y (2023) Hydrochemical variations and driving mechanisms in a large linked river-irrigation-lake system. Environ Res 225:115596
Rinder T, Dietzel M, Stammeier AJ, Leis A, Bedoya-González D, Hilberg S (2020) Geochemistry of coal mine drainage, groundwater, and brines from the Ibbenbüren mine, Germany: a coupled elemental-isotopic approach. Appl Geochem 121:104693
Spoelstra J, Leal KA, Senger ND, Schiff SL, Post R (2021) Isotopic characterization of sulfate in a shallow aquifer impacted by agricultural fertilizer. Groundwater 59(5):658–670
Su N, Ma W, Cai W, Cao S, Bai Y, Luo R (2007) Analysis on hydrological characteristic of Kuye River. J Water Resour Water Eng 18(2):65–68
Tang L, Yao R, Zhang Y, Ding W, Wang J, Kang J, Liu G, Zhang W, Li X (2023) Hydrochemical analysis and groundwater suitability for drinking and irrigation in an arid agricultural area of the Northwest China. J Contam Hydrol 259:104256
Tian H, Ren H, Li X, Zhang X, Xu X, Wang S (2023) Specificality, quality variation, assessment and treatment of estuarine water in the Pearl River Delta, South China. J Environ Inf 42(1):13–24
Turunen K, Rsnen T, Hmlinen E, Hämäläinen M, Pajula P, Nieminen SP (2020) Analysing contaminant mixing and dilution in river waters influenced by mine water discharges. Water Air Soil Pollut 231(6):1–15
Wang X, Cai H, Zhang X, Wang J, Liu Y, Zhai J (2008) Analysis on the seasonal drying-up causes of Kuye River. Resour Sci 30(3):475–480
White FA, Bullen DT, Schulz SM, Blum AE, Huntington TG, Peters NE (2001) Differential rates of feldspar weathering in granitic regoliths. Geochim Cosmochim Acta 65(6):847–869
Xie X, Bai X (2023) Ecological security assessment and driving factor analysis of Kuye River Basin based on structural equation model. J Ecol Rural Environ 39:864–873
Xie F, Cai H, Zhao C, Wang Z, Wang J, Xu J (2017) Chemical characteristics of groundwater in Jinghuiqu irrigation area and its response to irrigation with different water sources. J Irrig Drainage 36(04):77–82
Xu G, Zhi Z, Li Z, Zhuang J, Guo M, Fang K, Yi J, Ren Z, Gao H, Jia L (2025) Attribution of sediment changes and influencing factors of microbial communities in sediment in the Wuding River Basin of the Yellow River. CATENA 249:108626
Yang N, Wang G, Shi Z, Zhao D, Jiang W, Guo L, Liao F, Zhou P (2018) Application of multiple approaches to investigate the hydrochemistry evolution of roundwater in an arid region: Nomhon, Northwestern China. Water 10(11):1667
Yu ZB, Yin S, Bai JH, Ali I, Liu J, Bano H (2024) Suaeda salsa in relation to hydrological connectivity: from the view of plant trait networks. J Environ Inf 43(1):41–49
Zhang Y, Xi Z (2019) Assessment of water quality of Wuding River basin in Yulin city. Agric Eng 9(2):31–33

Zhang J, Wang J, Zhao N, Shi J, Wang Y (2024) Analysis of changes in runoff and sediment load and their attribution in the Kuye River Basin of the Middle Yellow River based on the slope change ratio of cumulative quantity method. Water 16(7):944

Zhao J, Li G, Zhu Z, Hao Y, Hao H, Yao J, Bao T, Liu Q, Yeh TCJ (2024) Analysis of the spatiotemporal variation of groundwater storage in Ordos Basin based on GRACE gravity satellite data. J Hydrol 632:130931

Chapter 5
Degradation of Tetracycline (TC) by Combining Hydrodynamic Cavitation (HC), Ultraviolet (UV), and Persulfate

Zhenxing Wang, Shuai Liu, Sivakumar Manickam, and Xun Sun

Abstract In this study, hydrodynamic cavitation (HC) and ultraviolet (UV) techniques were employed to activate peroxydisulfate ($Na_2S_2O_8$) to degrade tetracycline (TC) model wastewater with an initial concentration of 30 mg/L and a volume of 15 L. The results demonstrate that over 95% of TC was degraded within 60 min at an inlet pressure of the Venturi of 0.35 MPa, a $Na_2S_2O_8$ concentration of 1 mM, and a pH of 3. The synergistic index of the HC/UV/$Na_2S_2O_8$ system was found to be as high as 7.5. Therefore, this work provides a highly feasible method for high-efficiency treatment of antibiotic wastewater.

Keywords Tetracycline · Hydrodynamic cavitation · Ultraviolet · Peroxydisulfate

5.1 Introduction

Tetracycline (TC) is a widely used class of pharmaceuticals for treatment of both human and animal diseases (Jeong et al. 2010). Due to the misuse and improper treatment of TC, its residues in aquatic environments have potential to induce resistance in microorganisms, which can subsequently affect ecosystem stability (Braeutigam et al. 2012). Conventional water treatment methods are ineffective in completely removing TC, and they are associated with low treatment efficiency, high cost, and low safety (Batt et al. 2007). Therefore, there is a clear need for the development of new and efficient water treatment technologies.

In recent years, advanced oxidation process (AOP) has been demonstrated to be an effective method for degradation of organic pollutants. Of particular interest is the

Z. Wang · S. Liu · X. Sun (✉)
School of Mechanical Engineering, Shandong University, Jinan, China
e-mail: xunsun@sdu.edu.cn

S. Manickam
Faculty of Engineering, Universiti Teknologi Brunei, Bandar Seri Begawan, Brunei

Y. Li et al. (eds.), *16th International Conference on Environmental Science and Development*, Environmental Science and Engineering,
https://doi.org/10.1007/978-3-032-15080-6_5

sulfate radical-advanced oxidation process (SR-AOP). Many studies have demonstrated that SR-AOP is more efficient than hydroxyl radicals or ozone in degradation of recalcitrant organics (Gao et al. 2012; Huang et al. 2005). Nevertheless, the peroxydisulfate (PDS) is a neutral oxidizer, requiring to be activated to produce highly-active sulfate radicals (Tsitonaki et al. 2010). Furthermore, existing technologies such as thermal activation, alkali activation, transition metal activation, and electrochemical activation not only consume large amounts of energy but also use expensive precious metal equipment, hindering industrial application of this technology.

To address this issue, a few researchers have successfully used hydrodynamic cavitation (HC) or ultraviolet (UV) to activate PDS, thereby generating highly reactive free radicals to degrade organic pollutants (Bethi et al. 2016; Nihemaiti et al. 2018). HC and UV activation technologies are distinguished by their simplicity, ease of maintenance, and broad applicability, with potential for large-scale applications (Bethi et al. 2016). In this study, we further explored the degradation effect of the combined activation of PDS by HC and UV systems on the target pollutant TC, determined the optimal parameters, and investigated the synergistic effect between them for further application in engineering.

5.2 Experimental Method

5.2.1 Materials

Tetracycline ($C_{22}H_{24}N_2O_8$), sodium peroxydisulfate ($Na_2S_2O_8$), acetonitrile (C_2H_3N), oxalic acid ($H_2C_2O_4$), and methanol (CH_4O) were purchased from Shanghai McLean Biochemistry and Technology Co. All the used chemical reagents were AR grade purity. Ultrapure water was used for each experiment.

5.2.2 Experimental Setup

The degradation experiment was conducted on the HC experimental system depicted in Fig. 5.1a. The HC reactor is a transparent Venturi constructed from PMMA material with a refractive index that is relatively close to that of water, as illustrated in Fig. 5.1b. A stainless-steel water tank, with a capacity of 30 L, equipped with an outer cooling jacket was used. The TC solution temperature was maintained at 25° ± 2 °C.

To achieve a synergistic effect between HC and UV processes, four UV lamps with a power of 8 W were fixed in the water tank and completely submerged in the TC model wastewater. The water circulation of the system was achieved by means of a centrifugal pump. The flow rate was controlled by two ball valves, with an adjustable range from 0 to 1.2 m^3/h, and was measured by a flow meter. Two

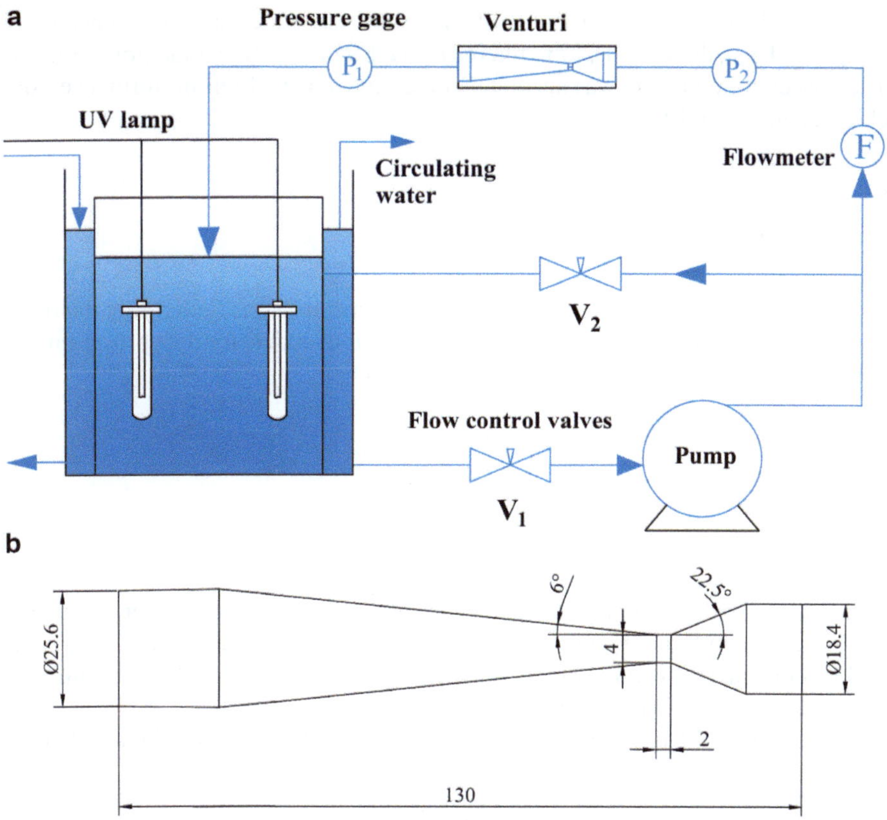

Fig. 5.1 **a** Schematic diagram of HC/UV treatment system. **b** Venturi structure (in millimeter)

hydrostatic pressure sensors were placed at 20 cm upstream and downstream side of the Venturi, respectively.

5.2.3 Experimental Methodology

All experiments were conducted using 15 L of aqueous TC solution, with an initial concentration of 30 mg/L, and continued for 60 min. To ensure the accuracy of the results, each experiment was repeated twice. Before the commencement of the experiment, a 5-mL sample was collected from the original solution. During the experiment, the pH of the TC solution was adjusted as required by the experimental conditions, using either sulfuric acid or sodium hydroxide.

Four UV lamps were immersed in the TC solution. The pump and UV lamps were started simultaneously, and a certain amount of $Na_2S_2O_8$ was added to the solution. Subsequently, the solution samples were collected at predefined time intervals. To capture hydroxyl radicals, 100 μL of methanol was added as a quencher before sampling. Then, the samples were filtered through a filter before undergoing analysis. The concentration of TC was accurately determined using high-performance liquid chromatography (HPLC).

5.2.4 Analysis

Kinetic studies facilitate the investigation of the mechanisms underlying the degradation of pollutants, employing per-pass degradation models for data processing and conducting kinetic analyses (Ranade et al. 2021):

$$\mathrm{In}\left(\frac{C_t}{C_0}\right) = -\phi n_p \tag{5.1}$$

$$n_p = \frac{Q}{V} t \tag{5.2}$$

where ϕ denotes the degradation factor per pass, n_p denotes the number of passes of the batch system through the cavitation unit, Q is the flow rate through the Venturi, and V is the internal volume of the Venturi, and C_0 and C_t are the TC concentrations at initial and arbitrary moments, respectively.

We also used the first-order reaction rate constant, K_{obs}, to describe the chemical reaction rate:

$$-\mathrm{In}\frac{C}{C_0} = K_{obst} \tag{5.3}$$

where t is the given time, C is the concentration of pollutants at the given time, and C_0 is the initial concentration.

The synergistic index (SI) was calculated to assess the synergistic effect of different processes:

$$\text{Synergistic index}(\xi) = \frac{K_{obs(1,2)}}{K_{obs(1)} + K_{obs(2)}} \tag{5.4}$$

where subscripts 1 and 2 represent different methods, and subscript (1, 2) represents the simultaneous application of two methods.

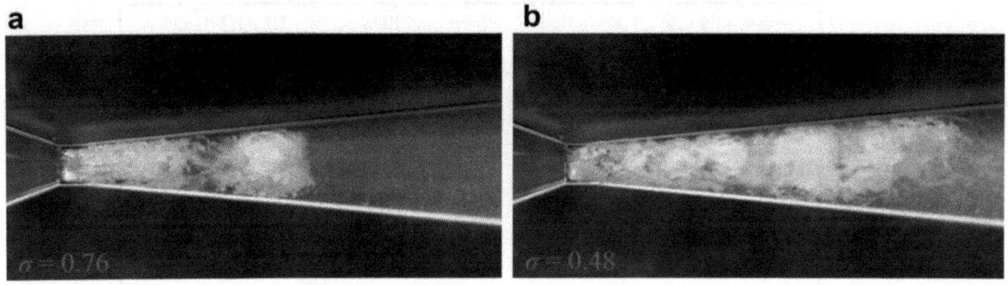

Fig. 5.2 Snapshots of local cavitation in Venturi at different cavitation numbers: **a** image at cavitation number of 0.76, **b** image at cavitation number of 0.48

5.3 Result and Discussion

5.3.1 Visualization of HC

Utilizing the high-speed camera (Phantom VEO 1310), the cavitation images in the Venturi were recorded at 40,000 frames per second. When cavitation occurs within the Venturi tube, the distinct refractive indices of water and the cavitation zone can be clearly identified. Specifically, the water and Venturi tube appear dark, whereas the cavitation zone appears bright, facilitating precise determination of the cavity's morphology and location.

Figure 5.2 illustrates the cavitation phenomena observed at the narrow throat of the Venturi, captured using the high-speed camera. The water flows from left to right, and as it traverses the narrowest point of the Venturi, the cavitation bubbles were formed. Figure 5.2a and b show the cavitation generated at a cavitation number of 0.76 and 0.48, respectively, both at the same stage of full development of cloud cavitation. It can be seen that cavitation became more intense as the cavitation number decreased.

5.3.2 Degradation of Tetracycline via Different Systems

The effects of sole use of HC, UV, and PDS and the coupling systems on TC degradation were investigated, as presented in Fig. 5.3. The HC, UV, and PDS resulted in TC degradation rates of only 19.53, 13.47, and 2.13% after 60 min, respectively. It can be observed that the sole use of the above three methods were less effective in the degradation of TC and were unable to effectively remove TC from the wastewater.

As illustrated in Fig. 5.3 and Table 5.1, when HC, UV, and PDS were employed in conjunction, the TC degradation rate reached 88.51% within 60 min, and the SI was 7.5, indicating markable enhancement. This can be attributed to the fact that the energy generated by UV irradiation helped to activate PDS, which produced more $SO_4^{-\cdot}$ (Zhang et al. 2016). Moreover, the extremely high temperature and high-pressure environment created by HC (Badve et al. 2013) enhanced the turbulence

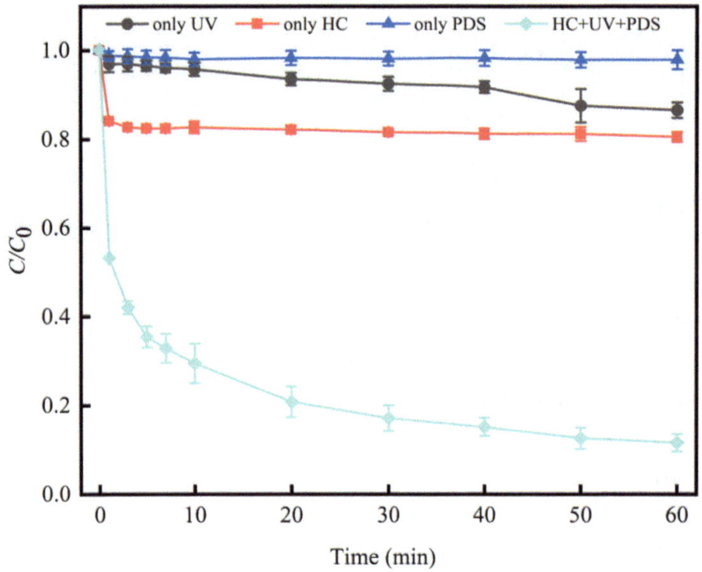

Fig. 5.3 Degradation effect of TC via different systems ($p_{in} = 0.35$ MPa, $c(Na_2S_2O_8) = 1$ mM, pH = unadjusted, $T = 25°C$)

Table 5.1 Degradation factor, first-order rate constant, and synergistic index for TC degradation

System	Factor		
	$\phi \times 10^{-6}$	$K_{obs} \times 10^{-3}$ (min^{-1})	SI
HC	1.5	1.3	–
UV	–	2	–
PDS	–	0.2	–
HC/UV/PDS	32	26.3	7.5

effect of the system, which further improved the mass transfer effect of the reaction system and facilitated the generation of ·OH as well as $SO_4^{-\cdot}$, which significantly enhanced the TC degradation (Fedorov et al. 2022).

5.3.3 Effect of Persulfate Concentration

By applying different initial concentrations of $Na_2S_2O_8$ in the HC/UV/$Na_2S_2O_8$ system, this study thoroughly investigated its impact on the degradation efficiency of TC. As shown in Fig. 5.4, with the gradual increase of $Na_2S_2O_8$ concentration from 0.5 to 2.0 mM, the degradation rate exhibited an increasing trend, and the corresponding removal rate also significantly increased from 63.14 to 94.84%.

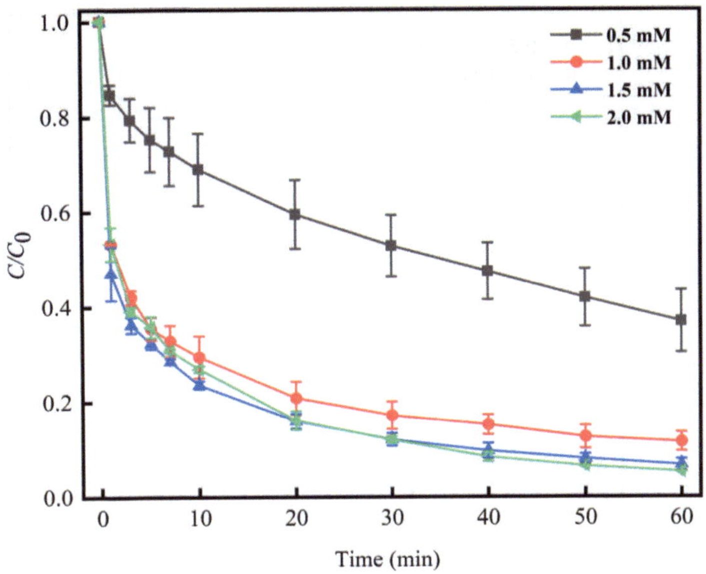

Fig. 5.4 Effect of $Na_2S_2O_8$ concentration on the degradation of TC in the HC/UV/$Na_2S_2O_8$ system (p_{in} = 0.35 MPa, pH = unadjusted, T = 25°C)

Notably, when the concentration was increased from 0.5 to 1 mM/L, the degradation rate showed a significant improvement. However, as the concentration was further increased to 1.5 and 2 mM/L, although the degradation rate remained elevated, the increase was already significantly constrained. This reveals an important phenomenon: within a certain concentration range, increasing the concentration of $Na_2S_2O_8$ can significantly improve the degradation rate of TC; however, once the concentration exceeds a certain threshold, the enhancement effect on TC degradation rate became relatively limited when further increasing the concentration of $Na_2S_2O_8$.

Furthermore, as illustrated in Fig. 5.5, the TC degradation speed demonstrates an increase by the rise in $Na_2S_2O_8$ concentration. At concentrations of $Na_2S_2O_8$ of 1.5 and 2.0 mM, the corresponding reaction rate constants are more similar, and their growth speed gradually slowed down. This demonstrates that the increase in $Na_2S_2O_8$ concentration also has a more limited effect on the degradation speed of TC.

5.3.4 *Effect of pH*

pH is a key factor influencing the activation of $Na_2S_2O_8$ in the HC/UV/$Na_2S_2O_8$ system (Sha et al. 2024), while it has an impact on the morphology of TCs, the formation, and stability of free radicals, and reaction intermediates (Wang et al. 2022). As shown in Fig. 5.6, the degradation rate of TC was increased after adjusting

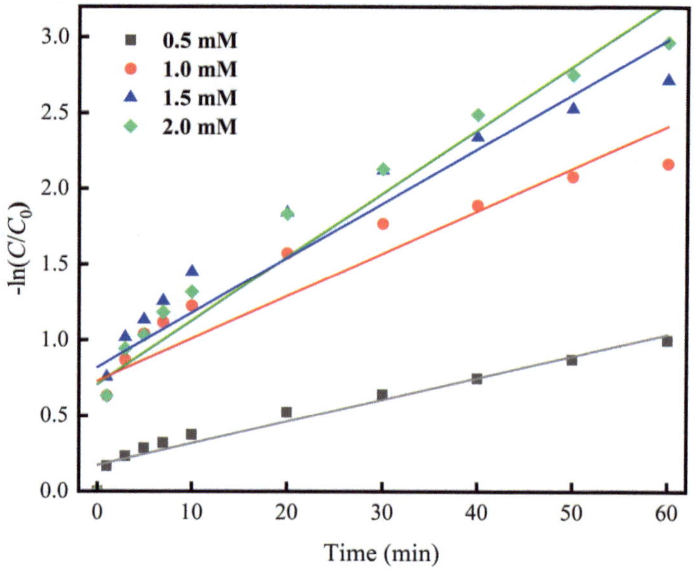

Fig. 5.5 Degradation rate curves for the effect of $Na_2S_2O_8$ concentration on TC degradation (p_{in} = 0.35 MPa, pH = unadjusted, T = 25°C)

the solution pH. In an acidic environment, when the pH values were 3 and 5, the TC degradation rates were 96.96 and 94.44%, respectively. In an alkaline environment, when the pH value was 11, the TC degradation rate was 96.10%. Thus, appropriate pH can effectively promote the degradation process of TC and enhance degradation rate.

5.3.5 Discussion of Experimental Results

From the above experimental results, it can be seen that there is a significant synergistic effect between HC, UV, and $Na_2S_2O_8$, and the synergistic index is as high as 7.5 when these three processes are coupled, which is much more than the effect of any two of them coupled. In the HC/UV/$Na_2S_2O_8$ system, with the increase of $Na_2S_2O_8$ concentration from 0.5 to 2 mM, the overall degradation effect was significantly enhanced, and the degradation rate increased to 94.84% at 2 mM. However, when the concentration of $Na_2S_2O_8$ reaches a certain level, the $SO_4^{-\cdot}$ in the system may reduce its contribution to the TC degradation due to the self-scavenging effect, and the increase in the TC degradation rate becomes less obvious. pH adjustment has a favorable impact on the TC degradation due to the fact that the activation of $Na_2S_2O_8$ can be activated under both acidic and alkaline conditions and the activation effect is slightly better than that of alkaline under acidic conditions. Under acidic conditions, the concentration of hydrogen ions in the solution is higher, which is more likely to

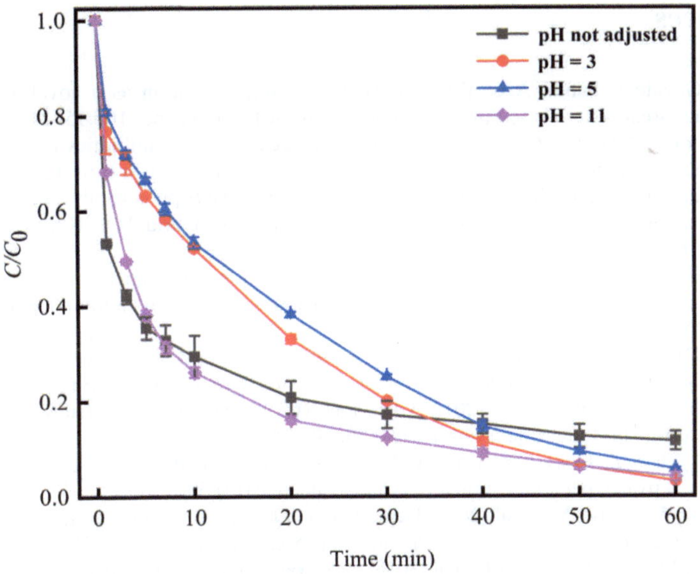

Fig. 5.6 Effect of pH on the HC/UV/$Na_2S_2O_8$ system (p_{in} = 0.35 MPa, $c(Na_2S_2O_8)$ = 1 mM, T = 25°C)

generate $SO_4^{-\cdot}$; under alkaline conditions, the concentration of hydroxide ions in the solution is higher, which is more likely to generate OH·. and $SO_4^{-\cdot}$ is more reactive to TC than OH·.

5.4 Conclusion

In this study, we evaluated the synergistic effects and the impact of various operational parameters of the HC/UV/$Na_2S_2O_8$ system in the degradation of TC. Significant synergistic effects were observed with a SI reaching 7.51. The optimal condition for TC degradation was identified as p_{in} = 0.35 MPa, $c(Na_2S_2O_8)$ = 1 mM, and pH = 3. Under these conditions, a maximum degradation rate of 96.96% for TC was achieved within 60 min. In conclusion, the novel HC/UV/$Na_2S_2O_8$ process for TC can be considered an efficient water treatment method for practical applications.

Acknowledgements This work was supported by the National Natural Science Foundation of China (Grant Nos. 52276032 and 52111540266) and "Youth Innovation Team Program" of Shandong Higher Education Institution (Grant No. 2023KJ006).

References

Badve M, Gogate P, Pandit A et al (2013) Hydrodynamic cavitation as a novel approach for wastewater treatment in wood finishing industry. Separ Purif Technol 106:15–21

Batt AL, Kim S, Aga DS (2007) Comparison of the occurrence of antibiotics in four full-scale wastewater treatment plants with varying designs and operations. Chemosphere 68(3):428–435

Bethi B, Sonawane SH, Rohit GS et al (2016) Investigation of TiO_2 photocatalyst performance for decolorization in the presence of hydrodynamic cavitation as hybrid AOP. Ultrason Sonochem 28:150–160

Braeutigam P, Franke M, Schneider RJ et al (2012) Degradation of carbamazepine in environmentally relevant concentrations in water by hydrodynamic-acoustic-cavitation (HAC). Water Res 46(7):2469–2477

Fedorov K, Dinesh K, Sun X et al (2022) Synergistic effects of hybrid advanced oxidation processes (AOPs) based on hydrodynamic cavitation phenomenon: a review. Chem Eng J 15:432

Gao Y-Q, Gao N-Y, Deng Y et al (2012) Ultraviolet (UV) light-activated persulfate oxidation of sulfamethazine in water. Chem Eng J 195–196:248–253

Huang K-C, Zhao Z, Hoag GE et al (2005) Degradation of volatile organic compounds with thermally activated persulfate oxidation. Chemosphere 61(4):551–560

Jeong J, Song W, Cooper WJ et al (2010) Degradation of tetracycline antibiotics: mechanisms and kinetic studies for advanced oxidation/reduction processes. Chemosphere 78(5):533–540

Nihemaiti M, Miklos DB, Hübner U et al (2018) Removal of trace organic chemicals in wastewater effluent by UV/H_2O_2 and UV/PDS. Water Res 145:487–497

Ranade VV, Sarvothaman VP, Simpson A et al (2021) Scale-up of vortex based hydrodynamic cavitation devices: a case of degradation of di-chloro aniline in water. Ultrason Sonochem 70:1526

Sha H, Su X, Zhou P (2024) UV/H_2O_2 and UV/PDS treatment of an emerging cyanotoxin (aerucyclamide A): kinetics, transformation products, and toxicity. Chem Eng J 21:483

Tsitonaki A, Benjamin P, Michelle C et al (2010) In situ chemical oxidation of contaminated soil and groundwater using persulfate: a review. Crit Rev Environ Sci Technol 40(1):55–91

Wang B, Wang T, Su H (2022) Hydrodynamic cavitation (HC) degradation of tetracycline hydrochloride (TC). Separ Purif Technol 65:282

Zhang R, Yang Y, Huang C-H et al (2016) Kinetics and modeling of sulfonamide antibiotic degradation in wastewater and human urine by UV/H_2O_2 and UV/PDS. Water Res 103:283–292

Chapter 6
Study on the Washing Effectiveness of a Novel Descaling Agent for Electrochlorination Systems

Dongxia Duan, Xingyun Wang, Fengqi Xu, Qian Tang, Guofeng Yin, Mingwei Sun, and Jian Che

Abstract For sea water electrochlorination systems, scaling is an unavoidable challenge that often leads to decreased current efficiency, increased power consumption, and even poses safety risks. Therefore, it is essential to remove scale from these systems. In this study, the composition and micro-morphology of the scale in electrolytic cells were investigated. Screening and formulation experiments for washing agents were carried out. Moreover, the washing effectiveness of the solid-state descaling agent on sea water electrochlorination systems was tested. The results demonstrated that the developed solid-state washing agent significantly improved the washing efficiency. After washing, the cell voltage was reduced by ~9%, chlorine production increased by 125%, and operating costs decreased by 60%. These findings suggest that solid-state washing agents based on sulfonate and citric acid could effectively replace hydrochloric acid for descaling electrolyzers. The solid-state washing agents developed in this study offered a more efficient and cost-effective solution for electrolyzer descaling.

Keywords Electrochlorination · Electrolytic cell · Scale · Washing · Current efficiency · Sea water

6.1 Introduction

Electrochlorination is a crucial technology employed for the generation of sodium hypochlorite (Liu et al. 2023). Owing to its high oxidation–reduction potential, sodium hypochlorite serves as an effective disinfectant (Atrashkevich et al. 2024; Li

D. Duan (✉) · F. Xu · Q. Tang · G. Yin · M. Sun · J. Che
SunRui Marine Environment Engineering Company Ltd, Qingdao, Shandong, China
e-mail: dongxiaduan@hotmail.com

X. Wang
Huaneng Rizhao Power Plant, Rizhao, Shandong, China

Y. Li et al. (eds.), *16th International Conference on Environmental Science and Development*, Environmental Science and Engineering,
https://doi.org/10.1007/978-3-032-15080-6_6

et al. 2023) and bleaching agent (Scialdone et al. 2021; Brião et al. 2024). As a result, electrochlorination finds extensive applications across numerous fields, including biofouling prevention in coastal power plants and nuclear power stations (Li et al. 2023), ballast water treatment for ships (Duan et al. 2023), drinking water disinfection (Choi et al. 2023), and wastewater treatment (Scialdone et al. 2021; Brião et al. 2024).

During the electrolysis process, a large amount of OH^- is generated in the cathode region of the electrolyzer, causing an increase in the pH of the solution. As is well known, seawater contains substantial amounts of Ca^{2+}, Mg^{2+}, HCO_3^-. The pH increase within the electrolyzer promotes the formation of calcium carbonate, magnesium carbonate, and calcium hydroxide (Choj et al. 2013; Sano et al. 2018; Miyazaki et al. 2019; Tlili et al. 2003; Carré et al. 2020). These compounds have extremely low solubility and tend to deposit on the electrode and internal surfaces, forming what is commonly referred to as scale. Moreover, some of these by-products can be carried by the electrolyte to the storage tanks and dosing pipelines, where they precipitate and accumulate (Tlili et al. 2003).

In addition to calcium and magnesium scales, another common type of scale in electrolyzers is silica scale. Both calcium/magnesium deposits and silica scale exhibit low thermal conductivity and poor electrical conductivity. Consequently, scaling in electrolyzers can lead to increased power consumption, reduced chlorine production, elevated water temperatures at the outlet, and even equipment shutdown or short circuits. Scaling in the downstream storage tanks reduces the effective liquid storage volume, while scaling in the downstream pipelines results in decreased water flow and increased pump pressure (Carré et al. 2020). Therefore, controlling the formation of scale and effectively removing existing scale in the electrolyzer are crucial for ensuring the safe and stable operation of electrochlorination systems (Wang 2008; Wang et al. 2016).

Traditionally, the removal of scale from electrolyzers has relied on periodic acid washing using 10% hydrochloric acid (Wang 2008; Wang et al. 2016). However, in recent years, hydrochloric acid has been reclassified as a hazardous chemical and a precursor chemical, making it less accessible than before. Additionally, the maximum concentration of hydrochloric acid is 36% (wt%), which poses challenges in terms of transportation and storage efficiency. Moreover, hydrochloric acid is highly irritating and corrosive, posing risks to both on-site operators and equipment. Consequently, the electrochlorination industry urgently needs a new type of electrolyzer washing agent to replace hydrochloric acid, thereby ensuring the reliability and safety of seawater electrochlorination systems.

In this paper, the morphology and crystalline structure of the scale in the electrolyzer at engineering sites were analyzed. Based on the physicochemical properties of the scaling, a solid washing agent formula was developed. The practical application of this solid washing agent was also investigated at the engineering site. Our study offers a novel alternative for the maintenance of electrochlorination systems in the electrochlorination industry.

6.2 Materials and Methods

6.2.1 *Analysis of the Physicochemical Properties of Scaling*

Scale samples were collected from the seawater electrochlorination systems at Rizhao Power Plant. The samples were dried at 45 °C for 48 h. After drying, a portion of the samples was acidified and analyzed using an atomic absorption spectrometer to determine the elemental content of Ca, Mg, Fe, Ru, and Ir. The crystalline structure of the scale was analyzed by X-ray diffractometer (XRD, Rigaku SmartLab SE). Additionally, the microstructural features of the scale were examined using a scanning electron microscope (SEM, Rigaku SmartLab SE), coupled with energy-dispersive spectroscopy (EDS, Oxford X-max) (Choi et al. 2023).

6.2.2 *Screening of Components and Formulation of Solid Acid Washing Agents*

Prepare 100 mL of 5% (wt%) oxalic acid, citric acid, sulfonate solutions, and 10% hydrochloric acid solution, respectively. Add calcium carbonate, magnesium carbonate, magnesium hydroxide, calcium sulfate, and iron(III) oxide slowly to each of the above solutions, respectively, and evaluate their dissolving capabilities of these solutions for the aforementioned chemical substances.

Based on the result, select the components with the highest dissolving capabilities and mix them in various ratios from 1:1 to 6:1. Subsequently, test the dissolving capabilities of the formula on the actual scale collected from the field.

6.2.3 *Washing Performance Testing in the Field*

6.2.3.1 Test Facilities

The washing performance test of solid-state washing agent was conducted at Rizhao Power Plant. The plant is equipped a plate-mesh electrochlorination system to control biofouling in the seawater intake pipeline. The system consists of four sets of plate-mesh electrolytic cells, with seven electrolytic cells in each set, totaling 28 cells. The inflow rate of the electrolytic cells was 30 m^3/h. The operating current was 2600 A. The system also includes one set of acid washing devices, which primarily consists of an acid cleaning tank, a cleaning solution circulation pump, acid washing piping, and freshwater piping. The acid washing cycle was 15 days.

Table 6.1 Inflow water quality

Item	Content
pH	7.8–8.4
Conductivity us/cm	41,000–49,000
TSS mg/L	7.3–8.4
Ca^{2+} mg/L	376–427
Mg^{2+} mg/L	1097–1247
Cl^{-} mg/L	16,740–21,220
HCO_3^{-} mg/L	128–145

6.2.3.2 Quality of Inflow Water

For Rizhao Power Plant, the intake point of sea water electrochlorination systems was at the Rizhao port. The quality of the inflow water was listed in Table 6.1.

6.2.3.3 The Process of Washing

Poured the solid washing agent into the washing tank, then added tap water. Activated the circulation pump to thoroughly dissolve the washing agent. The concentration of the washing agent was 5% (wt%). Next, pumped the prepared washing solution into the electrolytic cells, and circulated it for 1 h. Afterward, allowed the solution to stand for 1 h, followed by another hour of circulation. Finally, drained the washing solution and thoroughly rinsed the cells with tap water.

6.2.3.4 Analysis of Chemical Parameters

TRO was measured according to the standard method of 4500 Cl–B method I (Eaton et al. 2005). The hardness for solution was analyzed by water hardness tester (Bante 932). The current efficiency of electrochlorination systems was calculated as followed:

$$\eta = \frac{G * 1000}{1.323 \times I \times n} \times 100\%$$

η—current efficiency
G—chlorine output, kg/h,
I—current, A,
N—the number of series cells.

6.3 Results and Discussion

6.3.1 The Physicochemical Properties of Scaling

The seawater electrochlorination system at Rizhao Power Plant was normally acid-washed with hydrochloric acid once every 15 days. Before washing, the shells of electrolytic cells exhibited a reddish-brown appearance. After disassembling the electrolytic cells, it was observed that the electrode plates were encrusted with scale. The scale on the electrode plates was white, with a slight reddish-brown hue. Examination of the scale under a scanning electron microscope revealed a microstructure resembling rose-like spheres (Fig. 6.1). Energy-dispersive spectroscopy (EDS) analysis showed that the scale contains high levels of oxygen, magnesium, and calcium, as well as trace amounts of sodium and chlorine (Fig. 6.2).

Elemental analysis of the scale was conducted using ICP-OES/AAS. The results indicated that the scale was primarily composed of calcium and magnesium, which was consistent with the results obtained through energy-dispersive spectroscopy (EDS). Furthermore, ICP analysis revealed the presence of a trace amount of iron within the scale, which explains the slight yellow color of the scale and the reddish-brown rust on the surface of the cell casing (Table 6.2).

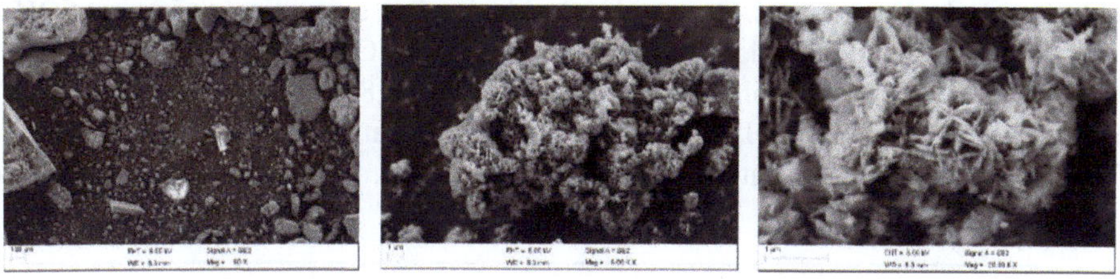

Fig. 6.1 SEM images of the scale from electrolytic cell

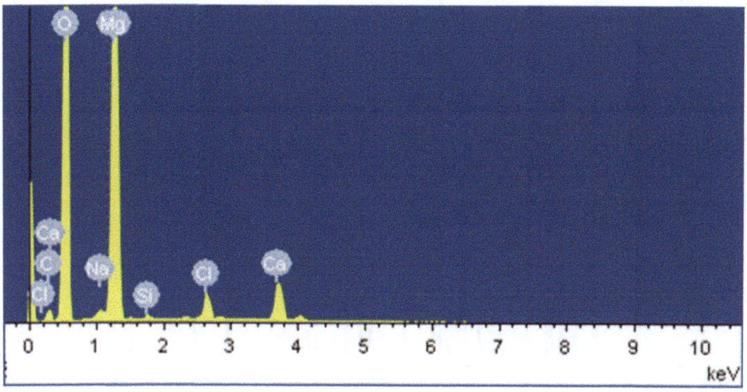

Fig. 6.2 The spectra of energy spectrum analysis

Table 6.2 The test result of ICP

Element	Weight percentage mg/g
Mg	33.355
Ca	22.989
Fe	0.032
Ru	/
Ir	/

To reveal the crystalline structure of the scale, X-ray diffraction (XRD) analysis was conducted on the sample. By comparing the diffraction patterns with standard spectra, the scale's diffraction peaks were identified as being consistent with those of brucite ($Mg(OH)_2$), aragonite ($CaCO_3$), and calcite ($CaCO_3$). Among these, the diffraction peak of brucite was the most prominent, followed by aragonite and calcite (Sano et al. 2018; Miyazaki et al. 2019; Carré et al. 2020) (Fig. 6.3). Based on the results of scanning electron microscopy (SEM), energy-dispersive spectroscopy (EDS), X-ray diffraction (XRD), and inductively coupled plasma (ICP) analysis, it could be concluded that the main components of the scale in the electrolyzer of Rizhao Power Plant were brucite and unstable aragonite and calcite, along with a minor amount of ferric oxide.

The composition of the scale could be primarily attributed to the electrolytic chlorine production process. During electrochlorination, a significant amount of OH^- was generated at the cathode of the electrolyzer. These OH^- reacted with Mg^{2+} and Ca^{2+} in the system. The solubility of $Mg(OH)_2$ was 1.8×10^{-11} mol/L (at 25 °C) and the solubility of $Ca(OH)_2$ was 5.5×10^{-6} mol/L (at 25 °C) (Eaton et al. 2005). Moreover, Mg^{2+} concentration was 1097–1247 mg/L, while the concentration of

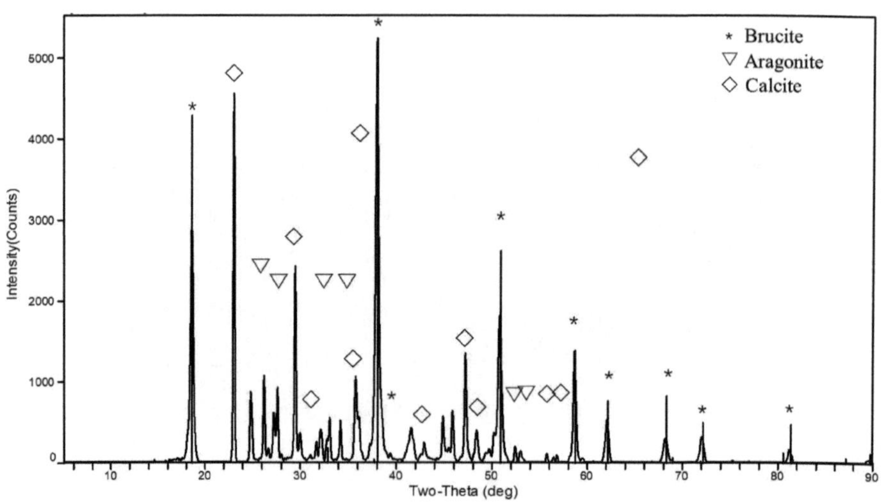

Fig. 6.3 XRD pattern of scale sample in electrolysis cell

Ca^{2+} was 376–427 mg/L in sea water. As a result, the main component of the scale was $Mg(OH)_2$ (Amrulloh et al. 2019; Neville and Morizot 2002).

Furthermore, the pH of seawater ranges from 7.8 to 8.4. It implied that the influent water of the electrolyzer contained a large amount of HCO_3^-. Because the solubility of calcium carbonate is 3.36×10^{-9} mol/L (at 25°C), the pH increase during chlorine production via electrolysis would promote the formation of calcium carbonate. Additionally, numerous studies on the relationship between electric charge and calcium-magnesium deposition have shown that the impact of electric charge on the deposition of magnesium hydroxide is more significant than that on calcium carbonate (Sano et al. 2018).

6.3.2 *Screening of Components and Formulation of Solid Acid Washing Agents*

5% (wt%) solutions of oxalic acid, citric acid, and sulfonate were prepared respectively. The dissolving capacity of these solutions for calcium carbonate, magnesium carbonate, and ferric oxide were tested. 10% hydrochloric acid solution was used as a control. The results indicated that among the tested solutions, sulfonate exhibited the best dissolution performance for calcium carbonate, magnesium carbonate, and magnesium hydroxide. Citric acid and sulfonate showed similar solubility performance for ferric oxide, while oxalic acid demonstrated relatively poor solubility for calcium carbonate, magnesium carbonate, and magnesium hydroxide (Table 6.3).

Then complex formula was then prepared by mixing sulfonate and citric acid at various ratio which the concentration of the resulting solution maintained at 5% (wt%). The dissolution performance of the complex formula for the scale of Rizhao power plant was measured. The test results indicated that the optimal dissolution performance for the scale was achieved when the ratio of sulfonate salts to citric acid was 5:1 (Table 6.4).

Table 6.3 The dissolving ability of four acids on different scaling components

Chemical	HCl	Oxalic acid	Sulfonate	Citric acid	
$CaCO_3$	Dissolution amount g/100 mL	3	0.26	2.0	0.78
	pH	5.15	2.06	5.82	4.69
$Mg(OH)_2$	Dissolution amount g/100 mL	1.6	0.30	1.32	0.40
	pH	6.37	2.66	6.63	4.71
Fe_2O_3	Dissolution amount g/100 mL	0.12	0.02	0.08	0.08
	pH	0.8	1.98	1.24	2.67

Table 6.4 The dissolving ability of compound washing agents on the scale of Rizhao power plant

Ratio	1:1	2:1	3:1	4:1	5:1	6:1
Dissolution amount of scale (g/100 mL)	1.24	1.45	1.63	1.92	2.12	1.83

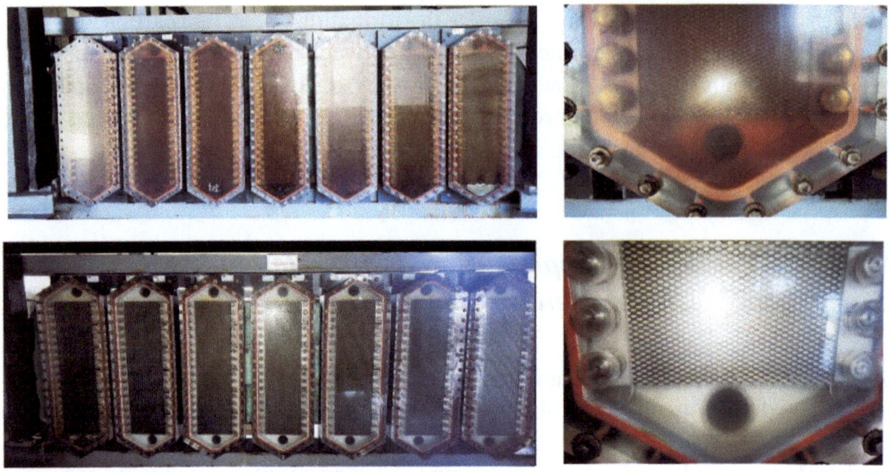

Fig. 6.4 The photo of electrolysis cell before and after cleaning (first row: before washing, second row: after washing)

6.3.3 *Washing Test in the Field*

A complex formula washing solution was prepared with a 5:1 ratio of sulfonate salts to citric acid. The concentration of washing agent was 5% too. After a 3 hwashing process, the electrolyzer shell appeared visible clear, and the reddish-brown scale completely removed. The scale layers on the electrode plates were also entirely eliminated (Fig. 6.4).

6.3.4 *The Effect of Washing on Cell Voltage*

The operating current of the electrochlorination systems was 2600 A. The cell voltage was measured. After the system was washed and resumed operation, the cell voltages were measured too. The test results showed that after washing, the cell voltage decreased from 3.202 to 2.904 V. The cell voltage decreased by 9.3% (Table 6.5).

Table 6.5 The effect of solid-state washing agent on electrolysis cell voltage

Cell voltage V	Number of cells							
	1	2	3	4	5	6	7	Average
After washing	2.801	2.808	2.968	2.984	2.867	2.921	2.985	2.904
Before washing	3.113	3.254	3.109	3.070	3.162	3.196	3.507	3.202

Table 6.6 The effect of cleaning on NaClO production

Item	TRO mg/L
After washing	372.22
Before washing	165.43

6.3.5 The Effect of Washing on Current Efficiency

After the electrolyzer resumed operation, the Total Residual Oxidants (TRO) concentration of the electrolyte was measured. The results showed that after washing, TRO production increased from 165.43 to 372.22 mg/L, an increase of ~125%. This result demonstrated that the removal of scale significantly improved the current efficiency of the electrolyzer.

The primary operating cost for electrochlorination systems is electricity power consumption which is directly related to the current efficiency of the system. Field tests showed after washing with the solid washing agent, the energy consumption of the electrolyzer was decreased by 9%, while the production of sodium hypochlorite increased by 125%. Therefore, the cost of each kilogram of sodium hypochlorite by electrochlorination system was reduced by ~60% (Table 6.6).

6.3.6 The Effect of Washing on Washing Solution

The hardness of the washing solution was measured both before and after the washing process. Before washing, the hardness of the washing solution was ~502 mg/L. This initial hardness was attributed to impurities in the tap water used for preparation and the raw chemical composition of the solid washing agent itself. After washing, the hardness value of the washing solution increased significantly. This increase was primarily due to the dissolution of the scale from the electrolyzer into the washing solution (Table 6.7).

Table 6.7 The hardness of cleaning agent

Item	Hardness (mg/L)	pH
Before washing	502	1.31
After washing	1008	2.53

6.4 Conclusion

- The scale in the electrolyzers of the Rizhao Power Plant was primarily composed of brucite and aragonite, with minor amounts of calcite and ferric oxide.
- The washing formulation compounded with sulfonate salts and citric acid could effectively dissolve the scaling layers in the electrochlorination system.
- Washing with the solid washing agent, reduced the voltage of the electrolyzer by 9%, while increasing the current efficiency by 125%. The cost of producing each kilogram of sodium hypochlorite was reduced by about 60%.

References

Amrulloh H, Simanjutak W, Situmeang RTM et al (2019) Effect of dilution and electrolysis time on recovery of Mg^{2+} As $Mg(OH)_2$ from bittern by electrochemical method. J Pure Appl Chem Res 8(1):4528

Atrashkevich A, Alum A, Stirling R et al (2024) Approaching easy water disinfection for all: can in situ electrochlorination outperform conventional chlorination under realistic conditions? Water Res 250:121014

Brião GV, da Costa TB, Antonelli R, Costa JM (2024) Electrochemical processes for the treatment of contaminant-rich wastewater: a comprehensive review. Chemosphere 355:141884

Carré C, Zanibellato A, Jeannin M et al (2020) Electrochemical calcareous deposition in seawater: a review. Environ Chem Lett 18:1193–1208

Choi E, Choi W, Lee JS et al (2023) A reflection on sustainable anode materials for electrochemical chloride oxidation. Adv Mater 35(43):e2300429

Choj J, Shim S, Yoon J (2013) Design and operating parameters affecting an electrochlorination system. J Ind Eng Chem 19(1):215–219

Duan DX, Xu FQ, Wang TY et al (2023) The effect of filtration and electrolysis on ballast water treatment. Ocean Eng 268:113301

Eaton AD, Clesceri LS, Riceand EW, Greenberg AE (2005) Standard methods for the examination of water and waste water. American public health association, Washington, pp 456–469

Li Z, Duan S, Ji XH et al (2023) Study on total residual oxidant decay in long-distance seawater intake pipeline of cooling system. Biofouling 39(7):683–690

Liu Y, Wang Y, Zhao S (2023) Journey of electrochemical chlorine production: from brine to seawater. Curr Opin Electrochem 37:101202

Miyazaki T, Arii T, Shirosaki Y (2019) Control of crystalline phase and morphology of calcium carbonate by electrolysis: effects of current and temperature. Ceram Int 45(11):14039–14044

Neville A, Morizot AP (2002) Calcareous scales formed by cathodic protection—an assessment of characteristics and kinetics. J Cryst Growth 243(3/4):490–502

Sano Y, Hao Y, Kuwahara F (2018) Development of an electrolysis based system to continuously recover magnesium from seawater. Heliyon 4:e00923

Scialdone O, Proietto F, Galia A (2021) Electrochemical production and use of chlorinated oxidants for the treatment of wastewater contaminated by organic pollutants and disinfection. Curr Opin Electrochem 27:100682

Tlili MM, Ben AM, Gabrielli C et al (2003) Influence of the interfacial pH on electrochemical $CaCO_3$ precipitation. J Electrochem Soc 150:C765

Wang XY (2008) Impact of electrolyzing sea water chlorination on sea water environment. Develop Appl Mater 23(4):32–36
Wang G, Zhang ZH, Zhang ZP et al (2016) Chemical cleaning of electrolytic sea water chlorine system. Clean World 21(9):5–11

Chapter 7
Electrochemical Detection of Amino Acids: Strategies and Environmental Applications

Xiwen Zhang and Peng Zhang

Abstract In recent years, electrochemical research has significantly advanced in the detection of substances critical to human health, with amino acids being a focal point due to their essential role in numerous biological processes. Imbalances in amino acid levels often serve as indicators of various diseases. In the context of environmental science, the detection of amino acids is also crucial for monitoring pollution, particularly in water bodies, where amino acid imbalances can signal contamination from industrial effluents or wastewater. This underscores the growing need for effective and reliable detection methods. Electrochemical technologies have demonstrated considerable promise for detecting electroactive amino acids like cysteine, tyrosine, tryptophan, methionine, and histidine. However, the detection of non-electroactive amino acids presents unique challenges because they lack intrinsic redox activity. This review focuses on recent advancements in electrochemical strategies for detecting non-electroactive amino acids such as glycine, alanine, serine, and threonine, which are of interest in environmental monitoring. Key approaches, including enzyme-based sensors, nanomaterial-modified electrodes, and indirect electrochemical methods such as molecular imprinting, are discussed. Additionally, the challenges and future directions in expanding the scope of electrochemical detection for these non-traditional electroactive amino acids are explored, with an emphasis on their potential for monitoring water quality and environmental pollution. Environmental benefits such as cost-effectiveness, low energy consumption, and high selectivity of these detection methods are also highlighted, underscoring their relevance to sustainable environmental practices.

Keywords Amino acids · Electrochemical detection · Non-electroactive amino acids · Chiral recognition · Environmental monitoring · Modified electrodes · Water quality

---

X. Zhang
School of Chemistry and Chemical Engineering, Shanghai Jiao Tong University, Shanghai, China

P. Zhang (✉)
Environmental Systems Engineering Program, University of Regina, Regina, SK, Canada
e-mail: pengzhang83@outlook.com

Y. Li et al. (eds.), *16th International Conference on Environmental Science and Development*, Environmental Science and Engineering,
https://doi.org/10.1007/978-3-032-15080-6_7

7.1 Introduction

Amino acids are the basic structural units of proteins in our body, thousands of proteins are composed of 20 different amino acids, each of which contains an amino group and a carboxyl group at each end (Jane et al. 2020), and plays an extremely important role in the life activities of organisms. They are not only indispensable in the synthesis of proteins, but also participate in a variety of metabolic pathways, regulating the synthesis of neurotransmitters, maintaining cell signaling, and regulating the acid–base balance in the body. Given its central position in physiological and pathological states (Wu 2010), accurate detection and quantitative analysis of amino acids are of great significance in many research and application fields, including clinical diagnosis (such as metabolic diseases, biomarker detection), food safety and quality control (amino acid content evaluation, adulteration detection), biochemistry and Bioengineering Research (such as cell metabolism analysis, synthetic biology process monitoring), and Environmental Science Research (soil or water body metabolite analysis).

In environmental science, detecting amino acid imbalances in natural bodies of water may offer valuable insights into contamination by agricultural runoffs or industrial effluents (Wang et al. 2022a). These imbalances could serve as early indicators of ecosystem stress, making the electrochemical detection of amino acids crucial for environmental monitoring.

In the past decades, techniques such as high performance liquid chromatography (HPLC), capillary electrophoresis (CE) and mass spectrometry (MS) have been widely used in the quantitative analysis of amino acids. With their excellent sensitivity, resolution and specificity, these methods have become the "gold standard" for amino acid analysis. However, they usually involve expensive instruments and equipment, highly specialized operators and complex sample pretreatment processes, such as derivatization, purification or dilution, which lead to time-consuming detection process and difficult to achieve portable, real-time or in situ analysis.

In contrast, electrochemical sensor technology has gradually become a popular alternative in the research of amino acid detection because of its simple operation, fast response, low cost, strong portability and adaptability to trace samples. This kind of technology can realize the direct or indirect electrochemical signal conversion of target molecules, so as to carry out quantitative analysis of samples without complex pretreatment. In addition, modern electrochemical systems can realize highly sensitive, wearable or miniaturized detection platforms by integrating microelectronic devices and nanomaterials, which are especially suitable for on-site rapid screening and instant diagnosis.

For some electroactive amino acids, such as tyrosine, tryptophan, cysteine, histidine and methionine, their molecules contain functional groups that can be oxidized or reduced (such as phenolic hydroxyl groups, indole rings, sulfhydryl groups, imidazole rings and thioethers), which can directly generate electron transfer reactions on the electrode surface and generate measurable current signals. Therefore, direct detection can be achieved without pretreatment or simple treatment. The electrochemical behavior and redox potential of these amino acids are different in different electrode materials, pH conditions and electrolyte environment, which has become the focus of electrochemical analysis research. The review published by Moulaee and Neri (2021) and Wang et al. (2022a) specifically describes the detection methods and research progress of the above five electrochemically active amino acids.

However, except for the few above, most amino acids (such as glycine, alanine, serine, threonine, etc.) are electrochemically inert molecules, that is, they do not have significant redox activity under conventional electrodes and experimental conditions. The detection of this kind of "non-electroactive amino acids" faces great challenges, and it cannot rely on the direct electron transfer path to achieve signal reading. Therefore, researchers have proposed a variety of innovative strategies to make up for this defect, including: enzyme catalyzed transformation, chemical derivatization reaction, introduction of molecular recognition elements, construction and modification of advanced nanomaterials, multifunctional integrated sensor platform and other methods.

In recent years, great progress has been made in the field of electrochemical detection of non-electroactive amino acids, which has been successfully applied to practical scenarios such as screening of amino acid disorders in plasma, environmental water monitoring, and even cancer-related metabolic marker analysis.

This review mainly focuses on and analyzes the main research progress of electrochemical detection of non-electroactive amino acids in recent years. Based on selected peer-reviewed research results, it focuses on the design ideas, sensing mechanism, performance comparison of various methods, as well as the feasibility and challenges in practical application. By summarizing the advantages and limitations of existing technologies, this paper aims to provide theoretical support and application enlightenment for subsequent research, and promote the wide application of high-performance amino acid electrochemical detection platform in clinical and industrial fields.

7.2 Electroactivity and Other Characteristics of Amino Acids

7.2.1 Electroactive Amino Acids

Amino acids can be roughly divided into electroactive and non-electroactive according to their ability to carry out redox reactions on the surface of conventional electrodes. Electroactive amino acids such as tyrosine (Tyr), tryptophan (Trp), cysteine (Cys), histidine (His) and methionine (Met) can be oxidized directly, thus achieving direct electrochemical detection. For example, the electroactivities of

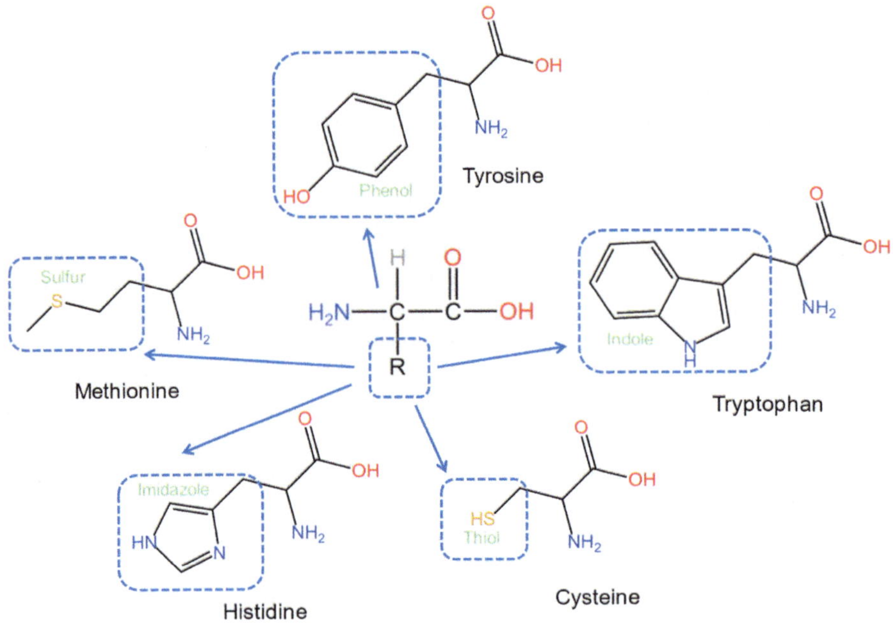

Fig. 7.1 Chemical structures and electroactive side chains of five recognized electroactive amino acids

cysteine, tyrosine, and tryptophan are assigned to thiol, phenol, and indole functions present in the side chains of these amino acids, respectively (Fig. 7.1). Methionine and histidine are other electroactive amino acids, and their electroactivities are attributed to sulfur-containing side chains and imidazole containing side chains, respectively. The mechanism of electro oxidation of amino acids is controversial among researchers.

According to the reports in the literature, some general electrooxidation pathways of cysteine (Sattarahmady and Heli 2011; Enache and Oliveira-Brett 2011), methionine (Prasad et al. 2013a), tryptophan (Paz-Zanini et al. 2015; Nguyen et al. 1986), tyrosine and histidine (Dourado et al. 2018; Ogura et al. 1999) will be listed, and the redox potential and other information will be reflected in Table 7.1. This table will help to quickly understand the electrochemical characteristics of these molecules when conducting electrochemical analysis.

The electroactive group of tyrosine (Tyr) is phenol hydroxyl (–OH), and its oxidation mechanism is that the phenol structure of tyrosine loses electrons at anodic potential to form phenol free radicals, which are then coupled or further oxidized to quinone structure. The oxidation potential is about +0.6 ~ +0.8 V, and the oxidation product can exist stably; The electroactive group of tryptophan (Trp) is the indole ring. Its oxidation mechanism is that the indole structure has a conjugated π system, which can form cationic radicals through the loss of electrons, and then rearrange or ring open. Generally, the oxidation potential is high (+0.7 ~ +1.0 V), which is

Table 7.1 Electrochemical characteristics of five electroactive amino acids

Amino acid	Main electrochemical active group	Oxidation/ reduction activity	Common oxidation potential range (versus Ag/AgCl) (V)	Activity influencing factors
Tyrosine	Phenolic hydroxyl (–OH)	Easily oxidized	+0.6 ~ +0.8	pH, modification, electrode material
Tryptophan	Indole ring	Easily oxidized	+0.7 ~ +1.0	pH, polar solvents
Cysteine	Thiol group (–SH)	Easily oxidized (forms disulfide bonds)	+0.2 ~ +0.4	pH, type of oxidant
Histidine	Imidazole ring	Weaker oxidation activity	Usually > +1.0	pH, electrode type
Methionine	Sulfur ether (–S–CH_3)	Easily oxidized	+0.6 ~ +1.0	pH, electrode modification, catalyst

greatly affected by pH and electrode materials; The electroactive group of cysteine (Cys) is thiol (–SH), and its oxidation mechanism is that the thiol group of Cys loses an electron and a proton (–SH → –S·) at anodic potential, generating sulfur radicals, further forming disulfide bonds (–S–S–), with low oxidation potential (about +0.2 ~ +0.4 V) and rapid reaction; The electroactive group of histidine (His) is the imidazole ring, and its oxidation mechanism is that the imidazole ring can form cation radicals by losing electrons (especially under basic conditions), which usually oxidize at higher potentials (often > +1.0 V), but the activity is enhanced under some modified electrodes or catalytic conditions; The electroactive group of methionine (Met) is thioether (–S–CH_3), and its oxidation mechanism is that the Thioether Group is easily oxidized to sulfoxide (MetO) or sulfone ($MetO_2$) under the action of potential. The reaction is mild, and the potential range is +0.6 ~ +1.0 v.

7.2.2 *Chirality of Amino Acids*

Amino acids are substances that widely exist in organisms. Although natural proteins are almost all composed of L-type amino acids, recent studies have found that D-amino acids (DAAS) also play important roles in the life system, such as having unique functions in neural regulation, antibacterial activity, cell signaling, etc. In addition, D-amino acid has important value in food industry, pharmaceutical synthesis and life science research. Therefore, how to accurately and efficiently detect the chiral configuration of amino acids, that is, the separation and quantitative analysis of DL isomers, has become one of the hot spots in analytical chemistry and life sciences.

In the past decade, chiral recognition methods for a variety of amino acids have emerged. As early as 2015, Martín et al. (2015) found a sensor based on graphene oxide nanobelts for simultaneous bioelectrochemical enantiomeric resolution and analysis of amino acid biomarkers. Subsequently, organic electrochemical transistors detect amino acid enantiochirality (Zhang et al. 2018), electrochemical chiral amino acid biosensors based on dopamine localized gold nanoparticles @ left-handed helical chiral carbon nanotubes (Lu et al. 2020), enantioselective recognition of amino acids based on electrochemical deposition and X-ray diffraction technology (Chong et al. 2021), and the development of chiral electrochemical sensors based on copper amino acid sulfhydryl nanorods (Pan et al. 2023) have been successively introduced. Recently, the cyclodextrin based electrochemical microporous organic network for amino acid enantiomer recognition (Zhang et al. 2024) and the chiral magneto electrochemical recognition of amino acid enantiomers (Jia et al. 2024) have been reported.

7.3 Detection Strategy of Non-Electroactive Amino Acids

Although the five amino acids (tyrosine, tryptophan, cysteine, histidine and methionine) mentioned above can directly participate in the redox reaction on the electrode surface due to their electroactive side chains, they are widely used in electrochemical detection. However, the remaining 15 natural amino acids are "non-electroactive" or "very low electroactive" in traditional electrochemical detection due to the lack of functional groups with obvious redox ability. This has brought challenges to its detection, but with the progress of material technology, nano electrode, catalyst design and other aspects, related research has also begun to make breakthroughs.

7.3.1 Analysis of Non-electroactive Amino Acids

In Table 7.2, we introduce the side chain groups of 15 common non-electroactive amino acids, among which glycine, whose side chain group contains only one hydrogen atom, has also been reported (Pérez-Ràfols et al. 2020) to have a variety of electrochemical detection methods. Based on this, we provide a series of reasons for their detection difficulties and possible breakthroughs.

Due to the lack of direct redox sites in the molecular structure of the above 15 amino acids, it is difficult to effectively identify them by traditional electrochemical detection methods. However, the current research is gradually solving this problem through a variety of means. It is worth noting that the following methods are not conflicting, but only list the most representative means of detection. Table 7.3 also lists the detection methods in chronological order of publication, limit of detection

Table 7.2 Challenges and breakthroughs for 15 non-electroactive amino acids

Amino acid	Side chain group	Detection difficulty	Detection breakthrough
Alanine (Ala)	Methyl group (–CH_3)	Simple structure; methyl group is non-electroactive	Derivatization with electroactive tags; indirect detection via metabolic intermediate (e.g., pyruvate); signal enhanced by nanomaterial-modified electrodes
Valine (Val)	Isopropyl group (–$CH(CH_3)_2$)	Branched alkyl group lacks redox-active sites	Use of signal amplification via metal ion coordination or enzyme-catalyzed deamination
Leucine (Leu)	Isobutyl group (–$CH_2CH(CH_3)_2$)	No polar or aromatic functional groups; very low electroactivity	Indirect detection via metabolic/enzyme-catalyzed products; research into sensing using weak coordination with metal nanomaterials
Isoleucine (Ile)	Sec-butyl group (–$CH(CH_3)CH_2CH_3$)	Structural isomer of leucine with similarly low electroactivity	Exploring adsorption differences on nanocomposites; combined chromatographic and electrochemical techniques
Glycine (Gly)	Hydrogen atom (–H)	Lacks any side chain; no redox-active functional group	Strong oxidation of amino/carboxyl groups as indirect signal; grafting to conductive polymers to mediate electron transfer
Glutamic acid (Glu)	γ-Carboxylic acid (–CH_2CH_2COOH)	Carboxyl group is stable and unreactive under neutral/acidic conditions	Partial electron transfer in alkaline media; detectable signal via reaction with metal oxides (e.g., MnO_2)
Aspartic acid (Asp)	β-Carboxylic acid (–CH_2COOH)	Similar to Glu; side chain is not electroactive	Selective detection via metal coordination; indirect detection through metabolic derivatives
Glutamine (Gln)	γ-Amide group (–$CH_2CH_2CONH_2$)	Amide group lacks oxidation sites; electrochemically inactive	Enzyme-assisted conversion (e.g., glutaminase) to glutamate enables indirect electrochemical detection
Asparagine (Asn)	β-Amide group (–CH_2CONH_2)	Amide side chain is not easily oxidized	Enzymatic conversion to aspartic acid using asparaginase; followed by electrochemical or chromatographic detection

(continued)

Table 7.2 (continued)

Amino acid	Side chain group	Detection difficulty	Detection breakthrough
Threonine (Thr)	Hydroxyethyl group (–CH(OH)CH_3)	Aliphatic hydroxyl group is weakly electroactive	Use of oxidants or alkaline conditions to induce oxidation; enhanced detection with graphene oxide or other catalytic materials
Serine (Ser)	Hydroxymethyl group (–CH_2OH)	Hydroxyl group difficult to oxidize in typical potential ranges	Molecularly imprinted polymers improve selectivity; electron transfer mediators enhance signal
Arginine (Arg)	Propyl-guanidino group (–$(CH_2)_3$–NHC(NH_2)NH)	Oxidation occurs at high potential; complex and prone to background interference	Interface design for signal clarity; specific catalysts (e.g., AuNPs@CNTs) enable detectable oxidation
Lysine (Lys)	Butylamino group (–$(CH_2)_4NH_2$)	Terminal amino group is protonated under physiological pH; inhibits electron transfer	Weak electrochemical signal via complexation with metal ions (e.g., Cu^{2+}); potential range extended by electrode modification
Proline (Pro)	Pyrrolidine ring (cyclic –NH$(CH_2)_3$–)	Secondary amine in a five-membered ring structure reduces reactivity	Ring-opening oxidation on nitrogen-doped carbon materials; relatively active among non-aromatic amino acids
Phenylalanine (Phe)	Benzyl group (–$CH_2C_6H_5$)	Lacks oxidizable groups such as phenolic OH; benzene ring is electrochemically inert	π–π interactions with conductive polymers (e.g., polyaniline) allow signal amplification via indirect detection mechanisms

(LOD), linear range (LR) and other information involved in some papers on electrochemical detection of non-electroactive amino acids in recent decades according to the publication time.

Enzymatic Conversion. Specific enzymes (such as amino acid oxidase, transaminase, etc.) are used to catalyze the target amino acids to produce electroactive products (such as H_2O_2, aldehydes or ketones), and then the products are detected by electrodes, so as to indirectly realize the quantification of target molecules. As shown in Article 2 (Tian et al. 2025) of the table, Electrochemical reduction-enhanced Luminol cathodic electrochemiluminescence for chiral recognition; Article 4 (Doran et al. 2024), Amino acid oxidase based electrochemical biosensor for D-serine; Article 15

Table 7.3 Records of detecting non-electroactive amino acids in recent 10 years

Num	Amino acid	Detection method	Analytical parameters	References
1	Serine (skin)	Electrochemical sensor with MIPs	LR = 48 μM–30 mM	Yuan et al. (2025)
2	L-alanine (chiral)	Electrochemical reduction of Luminol	LOD = 0.014 fM	Tian et al. (2025)
3	Serine (enantiomeric)	PVP-modified platinum electrode	LOD = 0.27 mM (L-Ser) LR = 2.5–20 mM(L-Ser) LOD = 0.16 mM (D-Ser) LR = 0.5–20 mM(D-Ser)	Amayreh and Hourani (2024)
4	Serine (brain)	Microelectrochemical biosensor	LOD = 0.33 μM	Doran et al. (2024)
5	Glycine	Modified electrode (AuCuNC@N-GQD/SPE)	LOD = 1 nM	Shamili et al. (2024)
6	Glycine (cancer)	Non-enzyme method with chitosan	LOD = 8.5 μM LR = 7–240 μM	Saini et al. (2024)
7	Lysine (chiral)	Fluorescence and circular dichroism	LOD = 16.51 nM (L-Lys), LOD = 24.38 nM (D-Lys)	Wei et al. (2024)
8	L-DOPA and L-Tyr (tryptophan)	Layer-by-layer nanodiamonds	LOD = 0.74 μM (L-Tyr)	Kaewjua and Siangproh (2023)
9	Tryptophan	CuO nanoflower-PGA modified sensor	LOD = 50 nM LR = 80 nM–100 μM	Khaleque et al. (2023)
10	Tryptophan	NiMn-LDH@PLL composites	LOD = 52.7 nM LRR = 0.1–130 μM	Qian et al. (2023)
11	D-serine	Upconversion NIR PEC sensor	LOD = 0.286 μM LR = 2–1650 μM	Huang et al. (2022)
12	Glycine (in vivo)	Wearable microneedle biosensor	LR = 25–600 μM	Wang et al. (2022c)
13	L-glutamate	Peptide aptamer based electrochemical amperometric sensor	LOD = 0.1 nM LR = 0.1 nM–1 mM	Wang et al. (2022b)
14	Aspartic acid	BSA/TiO_2 electrochemical interface	LOD = 0.37 nM (L-Asp) LOD = 0.94 nM (D-Asp) LR = 1–1000 nM (both)	Wei et al. (2021)

(continued)

Table 7.3 (continued)

Num	Amino acid	Detection method	Analytical parameters	References
15	D-serine and D-alanine	Engineered DAAO-based sensor	Not provided	Moussa et al. (2021)
16	Glycine (general)	Amperometric biosensor	LR = 25–500 μM	Wang et al. (2021)
17	Arginine	Peptide aptamer-based impedance sensor	LOD = 0.01 pM LR = 0.1 pM–0.1 mM	He et al. (2021)
18	Lysine	Graphene/In_2O_3-based electrochemical sensor	LR = 0.23–30 μM	Cheng et al. (2020)
19	Aspartic acid	Cu(II)-based molecular imprinting	Not provided	Chen et al. (2019)
20	D- and L-serine	Electrochemical sensor with layer-by-layer molecular imprinting	LOD = 0.24 ng/mL	Jaiswal et al. (2019)
21	Serine	Electrochemical Ni-NiO sensor	LOD = 0.1 μM	Dai et al. (2015)
22	D-serine	Disk-shaped amperometric sensor	LOD = 0.6 μM	Polcari et al. (2014)
23	Aspartic acid	Titanium dioxide/ CNT-based MIP	LOD = 1.73–1.79 ng/mL	Prasad et al. (2013b)
24	Methionine	Cu(II) phthalocyanine modified sensor	LOD = 2.7×10^{-8} M LR = 50–500 μM	Jeevagan and John (2012)

(Moussa et al. 2021), Electrochemical biosensor enhanced with engineered D-Amino Acid Oxidase (DAAO) for selective detection.

Introduction of Molecular Recognition Elements. These include the use of molecularly imprinted polymers (MIPs), antibody/antigen systems, aptamers, etc., which endow the sensor with highly selective binding ability to specific amino acids, and cooperate with signal transduction mechanisms (such as Faraday current or capacitance changes) to achieve quantitative detection. As shown in Article 1 (Yuan et al. 2025), Electrochemical sensor with molecularly imprinted polymer for serine; Article 13 (Wang et al. 2022b), Peptide aptamer based electrochemical amperometric sensor; Article 16 (He et al. 2021), Peptide aptamer-based impedance sensor; Article 18 (Chen et al. 2019), Cu(II)-based molecular imprinting; Article 19 (Jaiswal et al. 2019), Electrochemical sensor with layer-by-layer molecular imprinting (MIP); Article 22 (Prasad et al. 2013b), Titanium dioxide/CNT-based MIP.

Construction and Modification of Advanced Nanomaterials. By modifying the electrode surface, such as using carbon nanotubes (CNTs), graphene, metal organic framework materials (MOFs), metal nanoparticles, etc., the conductivity, surface area and reaction site density of the electrode are improved, so as to enhance the interaction between the electrode and amino acids, improve the detection sensitivity and reduce the detection limit. As shown in Article 3 (Amayreh and Hourani 2024) of the table, Enantiomeric detection using PVP-modified platinum electrodes; Article 9 (Khaleque et al. 2023) Advanced electrochemical sensor using copper oxide nanoflowers for tryptophan detection; Article 10 (Qian et al. 2023), Electrochemical sensing of tryptophan using NiMn-Layered Double Hydroxides (LDH) modified electrodes; Article 11 (Huang et al. 2022), Near-infrared light-driven photoelectrochemical biosensor for D-serine detection; Article 14 (Wei et al. 2021), Electrochemical enantioselective discrimination using BSA/TiO_2 modified electrodes; Article 17 (Cheng et al. 2020), Selective electrochemical sensor for lysine using graphene/ In_2O_3 composites; Article 20 (Dai et al. 2015), Electrochemical sensor based on electrochemical imprinting of Ni–NiO for serine detection.

Chemical Derivatization Reaction. Amino acid molecules were modified by introducing electroactive markers (such as o-benzoquinone, ferricyanide, etc.) to obtain detectable electrochemical signals. The method can be combined with HPLC or capillary electrophoresis to improve the detection specificity. As shown in Article 7 (Wei et al. 2024), Fluorescence and circular dichroism; Article 15 (Wang et al. 2021), Glycine (General) amperometric biosensor; Article 23 (Jeevagan and John 2012), Cu(II) phthalocyanine modified sensor.

Multifunctional Integrated Sensor Platform. The microfluidic chip, flexible electronics, electrochemical sensing and signal processing module are integrated to develop an amino acid detection system with multi-channel detection, fast response and data analysis capabilities, which is especially suitable for high-throughput real-time monitoring of biological samples. As shown in Article 5 (Shamili et al. 2024), Modified electrode (AuCuNC@N-GQD/SPE);Article 6 (Saini et al. 2024), Non-enzyme method with chitosan; Article 8 (Kaewjua and Siangproh 2023), Layer-by-layer nanodiamonds; Article 12 (Wang et al. 2022c), Wearable microneedle biosensor; Article 21 (Polcari et al. 2014), Disk-shaped amperometric sensor.

7.4 Conclusion: Challenges and Prospects

The combination of enzyme catalyzed reaction and electrochemical method provides a more sensitive and selective detection platform for chiral detection of amino acids. By modifying the specific enzyme to only catalyze the reaction of d- or L-type amino acids,In the identification of chiral amino acids, electrochemical methods have shown great potential in the chiral detection of amino acids with their advantages of high sensitivity, rapid response, low cost and portability. The electrochemical sensor can detect the concentration and chiral configuration of amino acids simultaneously by monitoring the redox reaction of amino acids. The integration of electrochemical sensors in environmental monitoring holds the potential to offer low-cost, real-time, and sustainable methods for assessing amino acid imbalances in ecosystems, providing a tool for early detection of environmental pollutants and contributing to efforts aimed at maintaining ecological balance.

In recent years, researchers have introduced chiral modifiers into electrochemical sensors to enhance their selectivity for specific chiral amino acids. For example, chiral molecules are used to modify the electrode surface, or nanomaterials are used to improve the responsiveness and selectivity of sensors. By modifying the specific enzyme to only catalyze the reaction of d- or L-type amino acids, and monitoring the process in real time through electrochemical signals, the isomers can be effectively distinguished. The application of nanomaterials in electrochemical amino acid detection also provides a new direction for improving sensitivity and selectivity.

Looking forward to the future, molecular imprinting technology (MIP), as a new technology, has been widely used in the chiral detection of amino acids and the concentration of complex environment in recent years. Molecular imprinting can provide a highly selective and sensitive amino acid detection platform by forming hole structures for specific amino acids in polymers. Although this technology shows good prospects in amino acid detection, it still faces some challenges. First, the preparation process of molecular imprinting method is relatively complex, and the imprinting process and hole size need to be accurately controlled to ensure its efficient recognition ability. Secondly, the stability and reusability of molecularly imprinted polymers are still a problem to be solved. Finally, how to realize the simultaneous detection of multiple amino acids is still one of the difficulties in the current research.

In the future, one of the research focuses of enzyme catalyzed detection of amino acids is to develop enzyme catalytic system with high activity, high stability and high specificity. Currently, the commonly used enzymes include amino acid oxidase, aminotransferase, dehydrogenase, etc., but these natural enzymes often show poor stability and substrate specificity in the detection of complex samples. Advances in enzyme engineering technology, such as site directed mutagenesis, structural optimization, and artificial evolution, have made it possible to construct efficient enzyme catalytic systems. For example, the improved amino acid oxidase obtained by directed evolution can maintain activity under a wider range of pH and temperature conditions, significantly improving the detection sensitivity. In addition, the development of artificial enzymes and biomimetic enzymes, such as metal organic framework (MOF)—based biomimetic catalysts, can mimic the enzyme active center and enhance the selective detection ability of specific amino acids. Quantitative detection of chiral amino acids is one of the difficulties in life sciences and pharmaceutical research. Because the properties of d- and L-type amino acids are very similar, it is difficult to accurately distinguish them by chemical analysis alone. In the future, the development of chiral selective enzymes will greatly promote the progress of chiral amino acid analysis. By means of protein engineering, improved enzymes with specific catalytic activity for D-type amino acids can be designed. Combined with chiral modified nanomaterials, the specificity of detection can be further enhanced. For example, the construction of enzyme mimicking materials with chiral nanopore structure can use the steric hindrance effect to distinguish sexual amino acids. This strategy has important application prospects in drug synthesis and chiral pollutant monitoring.

Acknowledgements The authors gratefully acknowledge the valuable comments and suggestions from the reviewers and the ICESD committees, as well as the opportunity to participate in this conference.

References

Amayreh M, Hourani MK (2024) Enantioselective electrochemical detection of a L-and D-serine at polyvinylpyrrolidone-modified platinum electrode. Bioanalysis 16(23–24):1219–1227. https://doi.org/10.1080/17576180.2024.2422209

Chen X, Zhang S, Shan X, Chen Z (2019) Derivative chiral copper(II) complexes as template of an electrochemical molecular imprinting sol–gel sensor for enantiorecognition of aspartic acid. Anal Chim Acta 1072:54–60. https://doi.org/10.1016/j.aca.2019.04.039

Cheng J, Zhong S, Wan W, Chen X, Chen A, Cheng Y (2020) Novel graphene/In(2)O(3) nanocubes preparation and selective electrochemical detection for L-lysine of camellia nitidissima chi. Materials 13(8):99. https://doi.org/10.3390/ma13081999

Chong H, Zhang G, Wu Y, Xu B, Wang G, Wang C (2021) Enantioselective recognition of amino acid based on electrochemical deposition and X-ray diffraction technology. J Inorg Biochem 218:111398. https://doi.org/10.1016/j.jinorgbio.2021.111398

Dai W, Li H, Li M, Li C, Wu X, Yang B (2015) Electrochemical imprinted polycrystalline nickel–nickel oxide half-nanotube-modified boron-doped diamond electrode for the detection of L-serine. ACS Appl Mater Interf 7(41):22858–22867. https://doi.org/10.1021/acsami.5b05642

Doran MM, Bermingham KP, Tricklebank MD, Lowry JP (2024) Characterisation of a microelectrochemical biosensor for real-time detection of brain extracellular d-serine. Talanta 278:126458. https://doi.org/10.1016/j.talanta.2024.126458

Dourado AHB, Pastrian FC, Torresi SIC (2018) The long and successful journey of electrochemically active amino acids: from fundamental adsorption studies to potential surface engineering tools. An Acad Bras Cienc 90:607–630. https://doi.org/10.1590/0001-3765201720170434

Enache TA, Oliveira-Brett AM (2011) Boron doped diamond and glassy carbon electrodes comparative study of the oxidation behaviour of cysteine and methionine. Bioelectrochemistry 81:46–52. https://doi.org/10.1016/j.bioelechem.2011.02.001

He Y, Zhou L, Deng L, Feng Z, Cao Z, Yin Y (2021) An electrochemical impedimetric sensing platform based on a peptide aptamer identified by high-throughput molecular docking for sensitive l-arginine detection. Bioelectrochemistry 137:107634. https://doi.org/10.1016/j.bioelechem.2020.107634

Huang L, Liang Z, Zhang F, Luo H, Liang R, Han F, Wu Z, Han D, Shen J, Niu L (2022) Upconversion NaYF(4):Yb/Er-TiO(2)-Ti(3)C(2) heterostructure-based near-infrared light-driven photoelectrochemical biosensor for highly sensitive and selective d-serine detection. Anal Chem 94(46):16246–16253. https://doi.org/10.1021/acs.analchem.2c04101

Jaiswal S, Singh R, Singh K, Fatma S, Prasad BB (2019) Enantioselective analysis of D- and l-serine on a layer-by-layer imprinted electrochemical sensor. Biosens Bioelectron 124–125:176–183. https://doi.org/10.1016/j.bios.2018.09.090

Jane B, Reece LAU, Cain ML, Wasserman SA, Minorsky PV, Jackson RB (2020) Campbell biology, 9th edn. Pearson, London

Jeevagan AJ, John SA (2012) Electrochemical determination of l-methionine using the electropolymerized film of non-peripheral amine substituted Cu(II) phthalocyanine on glassy carbon electrode. Bioelectrochemistry 85:50–55. https://doi.org/10.1016/j.bioelechem.2011.11.009

Jia Y, Wu W, Chen R, Wang H, Zhang C, Chen L, Yao J (2024) Magneto-electrochemical method for chiral recognition of amino acid enantiomers. Analyst 149(14):3732–3738. https://doi.org/10.1039/d4an00547c

Kaewjua K, Siangproh W (2023) Innovative electrochemical platform for the simultaneous determination of L-DOPA and L-tyrosine using layer-by-layer assembled L-proline-linked nanodiamonds on printed graphene. Mikrochim Acta 190(10):398. https://doi.org/10.1007/s00604-023-05970-1

Khaleque MA, Bacchu MS, Ali MR, Hossain MS, Mamun MRA, Hossain MI, Khan MZH (2023) Copper oxide nanoflowers/poly-l-glutamic acid modified advanced electrochemical sensor for selective detection of L-tryptophan in real samples. Heliyon 9(6):e16627. https://doi.org/10.1016/j.heliyon.2023.e16627

Lu H, Wang Z, Fan X, Wang H, Zhang Q, Fu M, Ning G, Zhang Y, Wang H (2020) Electrochemical chiral amino acid biosensor based on dopamine-localized gold nanoparticles @ left-handed spiral chiral carbon nanotubes. Anal Methods 12(31):3901–3908. https://doi.org/10.1039/d0ay00921k

Martín A, Batalla P, Hernández-Ferrer J, Martínez MT, Escarpa A (2015) Graphene oxide nanoribbon-based sensors for the simultaneous bio-electrochemical enantiomeric resolution and analysis of amino acid biomarkers. Biosens Bioelectron 68:163–167. https://doi.org/10.1016/j.bios.2014.12.030

Moulaee K, Neri G (2021) Electrochemical amino acid sensing: a review on challenges and achievements. Biosensors 11(12):502. https://doi.org/10.3390/bios11120502

Moussa S, Murtas G, Pollegioni L, Mauzeroll J (2021) Enhancing electrochemical biosensor selectivity with engineered D-amino acid oxidase enzymes for d-serine and d-alanine quantification. ACS Appl Bio Mater 4(7):5598–5604. https://doi.org/10.1021/acsabm.1c00409

Nguyen NT, Wrona MZ, Dryhurst G (1986) Electrochemical oxidation of tryptophan. J Electroanal Chem Interf Electrochem 199:101–126. https://doi.org/10.1016/0022-0728(86)87045-0

Ogura K, Kobayashi M, Nakayama M, Miho Y (1999) In-situ FTIR studies on the electrochemical oxidation of histidine and tyrosine. J Electroanal Chem 463:218–223. https://doi.org/10.1016/S0022-0728(98)00458-6

Pan Q-X, Yang Y-C, Zhao N-N, Zhang B, Cui L, Zhang C-Y (2023) Development of a chiral electrochemical sensor based on copper-amino acid mercaptide nanorods for enantioselective discrimination of tryptophan enantiomers. Anal Chim Acta 1272:341480. https://doi.org/10.1016/j.aca.2023.341480

Paz-Zanini VI, Giménez RE, Pérez OEL, de Mishima BAL, Borsarelli CD (2015) Enhancement of amperometric response to tryptophan by proton relay effect of chitosan adsorbed on glassy carbon electrode. Sens Actuat B Chem 209:391–398. https://doi.org/10.1016/j.snb.2014.11.087

Pérez-Ràfols C, Liu Y, Wang Q, Cuartero M, Crespo GA (2020) Why not glycine electrochemical biosensors? Sensors 20(14):49. https://doi.org/10.3390/s20144049

Polcari D, Kwan A, Van Horn MR, Danis L, Pollegioni L, Ruthazer ES, Mauzeroll J (2014) Disk-shaped amperometric enzymatic biosensor for in vivo detection of D-serine. Anal Chem 86(7):3501–3507. https://doi.org/10.1021/ac404111u

Prasad BB, Pandey I, Srivastava A, Kumar D, Tiwari MP (2013a) Multiwalled carbon nanotubes-based pencil graphite electrode modified with an electrosynthesized molecularly imprinted nanofilm for electrochemical sensing of methionine enantiomers. Sens Actuat B Chem 176:863–874. https://doi.org/10.1016/j.snb.2012.09.050

Prasad BB, Srivastava A, Tiwari MP (2013b) Molecularly imprinted polymer-matrix nanocomposite for enantioselective electrochemical sensing of D- and L-aspartic acid. Mater Sci Eng C Mater Biol Appl 33(7):4071–4080. https://doi.org/10.1016/j.msec.2013.05.052

Qian J, Yang J, Zhang Y, Zeng T, Wan Q, Yang N (2023) Interfacial superassembly of flower-like NiMn-LDH@poly-l-lysine composites for selective electrochemical sensing of tryptophan. Anal Chim Acta 1237:340608. https://doi.org/10.1016/j.aca.2022.340608

Saini N, Yadav D, Shirolkar M, Murugappan S, Thorat N, Kulkarni A (2024) Chitosan lecithin nanocomposite based electrochemical biosensor for glycine detection in biological matrices. Colloids Surf B Biointerf 238:113901. https://doi.org/10.1016/j.colsurfb.2024.113901

Sattarahmady N, Heli H (2011) An electrocatalytic transducer for L-cysteine detection based on cobalt hexacyanoferrate nanoparticles with a core-shell structure. Anal Biochem 409:74–80. https://doi.org/10.1016/j.ab.2010.09.032

Shamili C, Peethambharan SK, Chandran A (2024) Nanomolar level electrochemical detection of glycine on a miniaturized modified screen-printed carbon-based electrode: a comparison of performance with glassy carbon electrode system. J Mater Chem B 12(31):7557–7563. https://doi.org/10.1039/d4tb01133c

Tian T, Chen L, Li T, Wang X, Yang S, Wang H, Jiang Y, Yao X, Zhao H, Wang D, Li X (2025) Electrochemical reduction boosted luminol cathodic electrochemiluminescence for trace chiral recognition of alanine enantiomers. Bioelectrochemistry 165:108945. https://doi.org/10.1016/j.bioelechem.2025.108945

Wang Q, Liu Y, Campillo-Brocal JC, Jiménez-Quero A, Crespo GA, Cuartero M (2021) Electrochemical biosensor for glycine detection in biological fluids. Biosens Bioelectron 182:113154. https://doi.org/10.1016/j.bios.2021.113154

Wang K-L, Zhang J-X, Min D, Lv J-L, Liu D-F, Yu H-Q (2022a) Detection and quantification of antimicrobial-resistant cells in aquatic environments by bioorthogonal noncanonical amino acid tagging. Environ Sci Technol 56(22):15685–15694. https://doi.org/10.1021/acs.est.2c05024

Wang W, He Y, Gao Y, Gao H, Deng L, Gui Q, Cao Z, Yin Y, Feng Z (2022b) A peptide aptamer based electrochemical amperometric sensor for sensitive L-glutamate detection. Bioelectrochemistry 146:108165. https://doi.org/10.1016/j.bioelechem.2022.108165

Wang Q, Molinero-Fernandez A, Casanova A, Titulaer J, Campillo-Brocal JC, Konradsson-Geuken Å, Crespo GA, Cuartero M (2022c) Intradermal glycine detection with a wearable microneedle biosensor: the first in vivo assay. Anal Chem 94(34):11856–11864. https://doi.org/10.1021/acs.analchem.2c02317

Wei X, Chen Y, He S, Lian H, Cao X, Liu B (2021) Bovine serum albumin-coated titanium dioxide modified electrochemical interface for enantioselective discrimination of D/L-aspartic acid. Chirality 33(10):731–744. https://doi.org/10.1002/chir.23356

Wei S, Wang C, Wang Y, Yin X, Hu K, Liu M, Sun G, Lu L (2024) Chiral carbon dots derived from tryptophan and threonine for enantioselective sensing of L/D-lysine. J Colloid Interf Sci 662:48–57. https://doi.org/10.1016/j.jcis.2024.02.063

Wu G (2010) Functional amino acids in growth, reproduction, and health. Adv Nutr 1:31–37. https://doi.org/10.3945/an.110.1008

Yuan Y, Zhong B, Qin X, Xu H, Li Z, Li L, Wang X, Zhang W, Lou Z, Fan Y, Wang L (2025) An epidermal serine sensing system for skin healthcare. Nat Commun 16(1):2681. https://doi.org/10.1038/s41467-025-58147-0

Zhang L, Wang G, Xiong C, Zheng L, He J, Ding Y, Lu H, Zhang G, Cho K, Qiu L (2018) Chirality detection of amino acid enantiomers by organic electrochemical transistor. Biosens Bioelectron 105:121–128. https://doi.org/10.1016/j.bios.2018.01.035

Zhang X, Wang F, Chen Z (2024) Electrochemical chiral sensor for recognition of amino acid enantiomers with cyclodextrin-based microporous organic networks. Anal Chim Acta 1316:342879. https://doi.org/10.1016/j.aca.2024.342879

Chapter 8
Study of Heat Distortion Behavior and Organization of Steels for HVAC Applications

Guoan Zeng

Abstract Gleeble 3800 thermal simulation testing machine on the HVAC steel with a deformation temperature of 950–1100 °C, strain rate 0.01–5 s^{-1} thermal compression deformation, to observe the organization of different thermal deformation conditions under the evolution of the law. The results show that when the deformation temperature of 1150 °C, strain rate of 1 s^{-1} and above, HVAC steel in the isoaxial crystal grain size distribution is uniform and relatively small; when the strain rate of 1 s^{-1}, the deformation temperature of 1100 °C and above, HVAC steel softening mechanism is mainly to the dynamic recrystallization of the main; HVAC steel hot compression deformation is suitable for strain rate $\geq$1 s^{-1}, The hot compression deformation of HVAC steel is suitable for strain rate $\geq$1 s^{-1}, deformation temperature $\geq$1100 °C under the thermal deformation conditions. The deformation activation energy Q of HVAC steel is 729. 266 kJ/mol, and the rheological stress equation of HVAC steel is established by thermal compression deformation; when the strain rate is 0.01–5 s^{-1}, and the deformation temperature is 950–1150 °C, the error between the calculated and measured peak stress of the HVAC steel is within 10%, which indicates that the rheological stress equation of HVAC steel has a good performance in the actual thermal compression process. This shows that the rheological stress equation of HVAC steel has good adaptability in the actual thermal compression process.

Keyword Steel for HVAC systems · Deformation temperature · Strain rate · Organizational evolution · Constitutive equation

---

G. Zeng (✉)
Xiamen Ocean Vocational College, Xiamen, China
e-mail: zengguoan1227@163.com

Y. Li et al. (eds.), *16th International Conference on Environmental Science and Development*, Environmental Science and Engineering,
https://doi.org/10.1007/978-3-032-15080-6_8

8.1 Research Scope and Contributions

The carbon content of steel used in HVAC is between 1.0 and 2.1% (mass fraction), due to its high carbon content, HVAC steel products typically have high hardness and strength, making them suitable for manufacturing lightweight equipment and steel products with high strength and wear resistance requirements (Wei 2023). However, the high carbon content in HVAC steel can make the castings hard and brittle, making subsequent hot processing difficult. Improper selection of hot deformation process parameters during production can easily lead to defects such as cracks (Zhang et al. 2019; Wang et al. 2020; Chen et al. 2021), which to some extent limits the application of HVAC steel. Simulating the thermal deformation process of HVAC steel using laboratory rolling mills/hydraulic presses can obtain certain empirical parameters, but the simulation results often differ significantly from industrial production, and the testing cost is relatively high; The use of thermodynamic simulation methods can conveniently simulate the thermal deformation process of HVAC steel, with lower experimental costs, and can comprehensively and accurately reflect the influence of various parameters on the microstructure evolution during the thermal deformation process (Li et al. 2021; Yu et al. 2021). At present, research on HVAC steel in China is mainly focused on component design, heat treatment process optimization, and other aspects. However, there are relatively few reports on studying the thermal deformation behavior of HVAC steel through thermal simulation methods and investigating the evolution of its microstructure (Wu and Han 2018; Yang et al. 2021; Shi et al. 2021). The study of thermal simulation will help clarify the relationship between deformation parameters and stress during the thermal deformation process of HVAC steel, as well as the evolution of microstructure, and provide reference for the product design and industrial production of HVAC steel in the future.

8.2 Material and Method

Experimental steel material for Fe–1.2C–4.8Cr–0.38Mo–0.35 V heating and air conditioning steel, the main elements of the chemical composition (mass fraction): C 1.220%, Cr 4.790%, Mo 0.380%, V 0.360%, Si 0.780%, Mn 0.660%, P 0.009%, S 0.003%, the balance of Fe; matrix organization is ferrite + M2 C carbide. The matrix organization is ferrite + M2 C carbide.

The hot compression process of Fe–1.2C–4.8Cr–0.38Mo–0.35 V steel is shown in Fig. 8.1. Isothermal compression with different strain rates was carried out after holding at the deformation temperature for 60 s. The strain rate was 40% and the cooling method was water quenching. The cooling method was water quenching. During the high temperature heat deformation process, in order to avoid bonding of the samples, the sample was subjected to a high temperature heat deformation process.

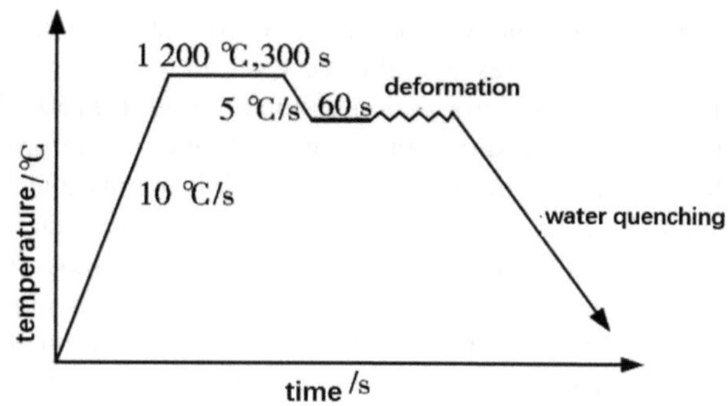

Fig. 8.1 Schematic diagram of hot compression process of ultra high carbon steel

Phenomena such as indentation and oxidization were pre-applied with lubricating oil and placed graphite flakes at both ends of the specimen and vacuumed.

The metallographic specimens were intercepted from the same part of the hot deformed specimens (along the direction of the central axis) by wire cutting method, and polished with diamond abrasive paste after cold setting, grinding with 60–1500 mesh sandpaper step by step, etching with a mixture of 10 g of potassium hydroxide +10 g of potassium ferrocyanide +100 mL of deionized water heated to a slight boil, and then blow-dried after being washed with water and alcohol; the grain shape was observed on the Leica DM6M type metallographic microscope. The grain morphology was observed on a Leica DM6M metallurgical microscope, and the grain size was measured by using the attached Image Pro plus 6.0 software, and the average grain size of 10 fields of view was taken.

8.3 Results and Analysis

8.3.1 True Stress-True Strain of High Carbon Steel Under Different Heat Distortion Conditions

Figure 8.2 shows the true stress-true strain curves of high carbon steel under different heat deformation conditions. At the initial stage of deformation, the true stress of the experimental steel at different strain rates and deformation temperatures increases rapidly with the increase of true strain, which is mainly due to the significant increase of dislocation density and the dominance of work hardening in the early stage of deformation (Guo et al. 2020). When the strain rate is 0.1 s^{-1}, the curve of deformation temperature 950–1100 °C is a dynamic recovery curve (no peak stress, in the stage of stable deformation), and the phenomenon of reduction of true stress after the peak stress occurs at the deformation temperature of 1150 °C; the rheological stresses at the strain rate of 1 s^{-1}, the deformation temperature of 1050–1150 °C and the strain rate of 5 s^{-1}, the deformation temperature of 950–1150 °C are all in the range of reaching the true strain, which is the same as that at the deformation temperature of 5 s^{-1}, the true stresses at the deformation temperature of 950–1150 °C. The rheological stresses at strain rate 1 s^{-1}, deformation temperature 1050–1150 °C and strain rate 5 s^{-1},deformation temperature 950–1150 °C all decreased after reaching the peak. At the same strain rate, the higher the deformation temperature, the lower the true stress at the same strain, which is mainly due to the fact that dislocation slip and dislocation creep in experimental steels are easier at high temperatures, and the dynamic regain and dynamic recrystallization behaviors are more pronounced (Zhang and Du 2021). During hot compression of experimental steels, work hardening and dynamic recovery due to the increase of dislocation density and dynamic softening due to dynamic recrystallization coexist, and the true stress increases when work hardening is dominant, and enters the steady-state deformation stage when both of them tend to be in equilibrium, and the true stress decreases when dynamic softening is dominant (Li et al. 2019).

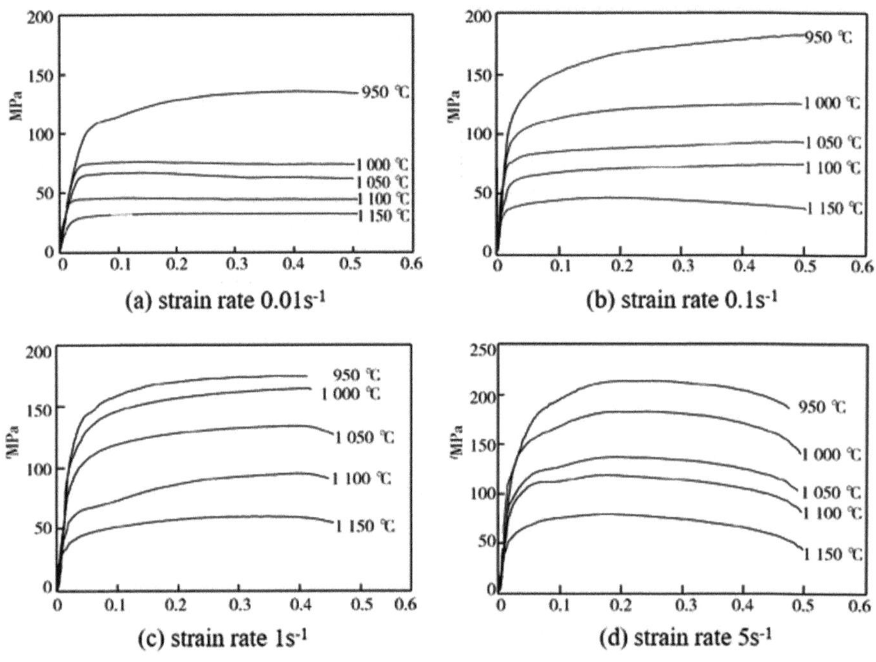

Fig. 8.2 True stress–strain curve of ultra high carbon steel

Table 8.1 Peak stress of ultra high carbon steel at different strain rates and deformation temperatures

Deflection temperature/°C	Peak stress/MPa			
	Strain rate 0.01 s^{-1}	Strain rate 0.1 s^{-1}	Strain rate 1 s^{-1}	Strain rate 5 s^{-1}
950	135.90	183.08	174.64	214.33
1000	76.40	125.50	164.18	182.97
1050	66.99	94.09	134.06	137.04
1100	45.92	75.37	95.48	119.23
1150	33.27	47.21	60.27	79.46

8.3.2 Peak Stresses at Different Strain Rates and Deformation Temperatures

Table 8.1 shows the statistical results of the peak stresses of the experimental steels at different strain rates and deformation temperatures. When the deformation temperature increases from 950 to 1150 °C, the peak stress at strain rate 0.01 s^{-1} decreases from 135.90 to 33.27 MPa; at strain rate 0.1 s^{-1} decreases from 183.08 to 47.21 MPa; at strain rate 1 s^{-1} decreases from 174.64 to 60.27 MPa; at strain rate 5 s^{-1} decreases from 214.33 to 60.27 MPa; at strain rate 5 s^{-1} decreases from 214.33 to 60.27 MPa; at strain rate 5 s^{-1} decreases from 214.33 to 60.27 MPa. 60.27 MPa; the peak stress at strain rate of 5 s^{-1} decreased from 214.33 to 79.46 MPa. When the strain rate increases from 0.01 to 5 s^{-1}, the peak stress at 950 °C increases from 135.90 to 214.33 MPa; the peak stress at 1000 °C increases from 76.40 to 182.97 MPa; the peak stress at 1050 °C increases from 66.99 to 137.04 MPa; the peak stress at 1100 °C increases from 45.92 to 137.04 MPa. The peak stress at 1050 °C increased from 66.99 to 137.04 MPa; the peak stress at 1100 °C increased from 45.92 to 119.23 MPa; the peak stress at 1150 °C increased from 33.27 to 79.46 MPa. It can be seen that when the deformation temperature from 950 to 1150 °C, the strain rate of 0.01, 0.1, 1 and 5 s^{-1} when the peak stress of the experimental steel show a gradual decrease; when the strain rate from 0.01 to 5 s^{-1}, the deformation temperature of 950, 1000, 1050, 1100 and 1150 °C, the peak stresses of the experimental steel are all decreasing. The peak stresses of the experimental steels all showed a gradual increase.

8.3.3 Grain Organisation of Experimental Steels at Different Strain Rates

Figure 8.3 shows the grain organisation of the experimental steels at different strain rates, with a deformation temperature of 1150 °C. Comparative analysis shows that there are austenitic grains of different sizes in the experimental steels at different

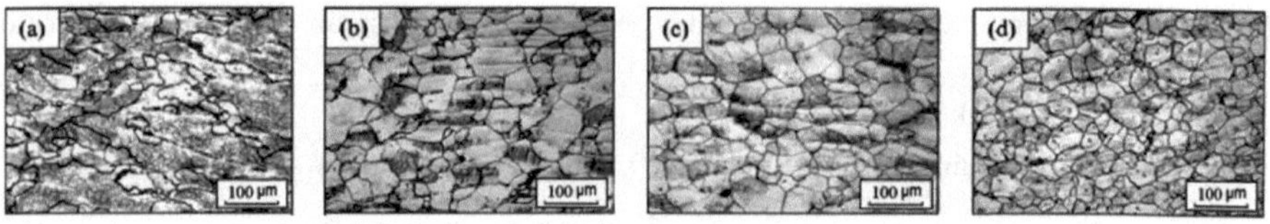

(a) strain rate 0.01s^{-1} (b) strain rate 0.1s^{-1} (c) strain rate 1s^{-1} (d) strain rate 5s^{-1}

Fig. 8.3 Grain structure of ultra high carbon steel at different strain rates

strain rates, and subcrystalline organisation can be seen in some of the austenitic grains. At a lower strain rate (0.01 s^{-1}), the experimental steel deformed at high temperature for a longer time, the merging and growth of recrystallised grains, and the subcrystalline organisation formed by dislocation cleavage can be seen inside the grains; as the strain rate increases, the experimental steel deformed at high temperature for a shorter time, and the size of recrystallised grains and the proportion of subcrystalline organisation inside the grains decrease, with the average grain size decreasing from 45 μm at 0.1–5 s^{-1}. The average grain size decreased from 45 μm at 0.1 s^{-1} to 30 μm at 5 s^{-1}. This is mainly due to the fact that with the increase of strain rate, the work hardening in the experimental steel is more significant, the nucleation time of the recrystallised grains becomes shorter, and due to the high deformation rate, these grains are preserved as fine recrystallised grains because they do not have time to grow up (Tong et al. 2019). In addition, a comparative analysis of the grain organisation at different strain rates shows that the equiaxed grains in the experimental steels are uniformly distributed and fine when the strain rate reaches 1 s^{-1} and above. Therefore, it is recommended to control the strain rate of the experimental steels at 1 s^{-1} and above in practical application.

8.3.4 Grain Organisation of Experimental Steels at Different Deformation Temperatures

Figure 8.4 shows the grain organisation of the experimental steel at different deformation temperatures with a strain rate of 1 s^{-1}. When the deformation temperature is 950 °C, the experimental steel organisation is deformed austenite grains + a small number of equiaxed dynamically recrystallised grains; when the deformation temperature is increased to 1000 °C, the experimental steel matrix organisation can be seen in the equiaxed and subcrystalline organisations, at this time, the experimental steel is still to the dynamic recovery of the main dynamic recrystallisation is not obvious; when the deformation temperature is increased to 1050 °C, the experimental steel can be seen in the grain boundaries of the continuity of the recrystallisation grains; the deformation temperature of 1150 °C the matrix organisation of the experimental steel When the deformation temperature increases to 1050 °C, the experimental steel can be seen in the grain boundary continuous recrystallisation grains; deformation

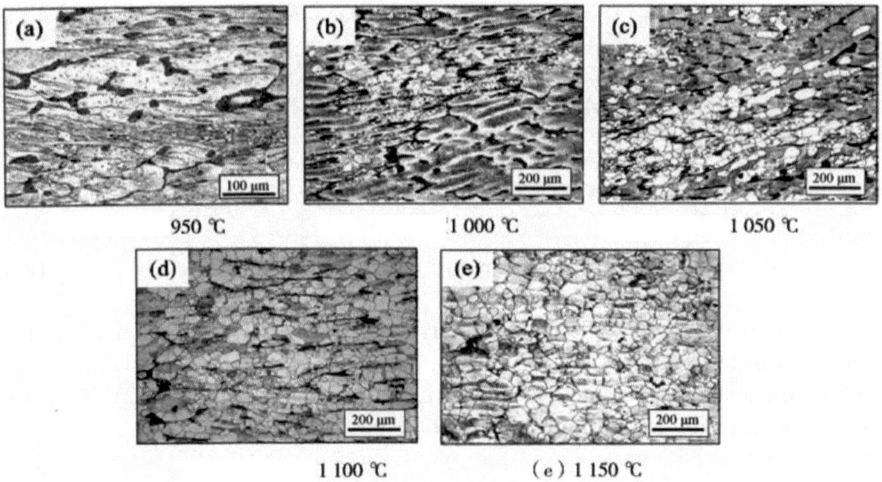

Fig. 8.4 Grain structure of ultra high carbon steel at different deformation temperatures

temperature of 1150 °C when the matrix organisation are basically uniform distribution of grain size isometric dynamic recrystallisation. When the strain rate is 1 s^{-1}, the deformation temperature increases from 950 to 1150 °C, the dislocation sliding and climbing energy in the ultra-high carbon is more, and the dynamic recrystallisation is easier (Barcellona and Palmeri 2009); at the deformation temperatures of 1100 and 1150 °C, the average size of the recrystallised grains is about 24 μm and 45 μm, respectively. In summary, at deformation temperatures of 1050 °C and below, the experimental steels were dominated by dynamic regression with a small amount of dynamic recrystallisation; at deformation temperatures of 1100 °C and above, the softening mechanism of the experimental steels was dominated by dynamic recrystallisation. According to the results of grain organisation observation at different deformation temperatures, the experimental steel is suitable for hot working and deformation at temperatures of 1100 °C and above.

8.3.5 *Rheological Stress Equation*

From Fig. 8.2 and Table 8.1, it can be seen that the stress change during hot compression deformation of experimental steels is closely related to the deformation temperature (T) and strain rate (ε), and the Zener-Hollomon parameter (Z parameter) is introduced to establish the relationship between them (Jian et al. 2020). Zener-Hollomon parameter (Z-parameter) is introduced to establish the relationship between the three (Jian et al. 2020).

$$Z = \varepsilon \exp\left(\frac{Q}{RT}\right) \tag{8.1}$$

$$\varepsilon = A1\sigma^{n_1} \exp(-Q/RT), \quad \alpha\sigma < 0.8 \tag{8.2}$$

$$\varepsilon = A2 \exp(\beta\sigma) \exp(-Q/RT), \quad \alpha\sigma > 1.2 \tag{8.3}$$

$$\varepsilon = A[\sinh(\alpha\sigma)]^n \exp(-Q/RT) \tag{8.4}$$

All stresses where: Q is the thermal activation energy of deformation, kJ/mol; R is the gas constant, 8.314 J/(mol K); A_1, A_2, A, n_1, n, α, and β ($\beta = \alpha n_1$) (Zhang et al. 2020a) are constants associated with experimental steel. Equations (8.2), (8.3), and (8.4) are obtained by logarithmic treatment, respectively:

$$\ln \varepsilon = \ln A1 + (-Q/RT) + n1 \ln \sigma \tag{8.5}$$

$$\ln \varepsilon = \ln A2 + (-Q/RT) + \beta\sigma \tag{8.6}$$

$$\ln \varepsilon = \ln A + (-Q/RT) + n \ln[\sinh \alpha(\sigma)] \tag{8.7}$$

According to the experimental results in Fig. 8.2 and Table 8.1, combined with Eqs. (8.6) and (8.7), the ln − lnσ(a) and ln − σ(b) curves of experimental steels can be obtained under different heat distortion conditions, as shown in Fig. 8.5.

The ln − σ(b) curves of the experimental steel under different heat distortion conditions can be obtained by combining the experimental results in Table 8.1 with equations. The slopes of the ln − lnσ(a) and ln − σ(b) curves are n1 and β, respectively (Qiao et al. 2017), and the slopes of the curves under different deformation conditions are analyzed in Fig. 8.2.

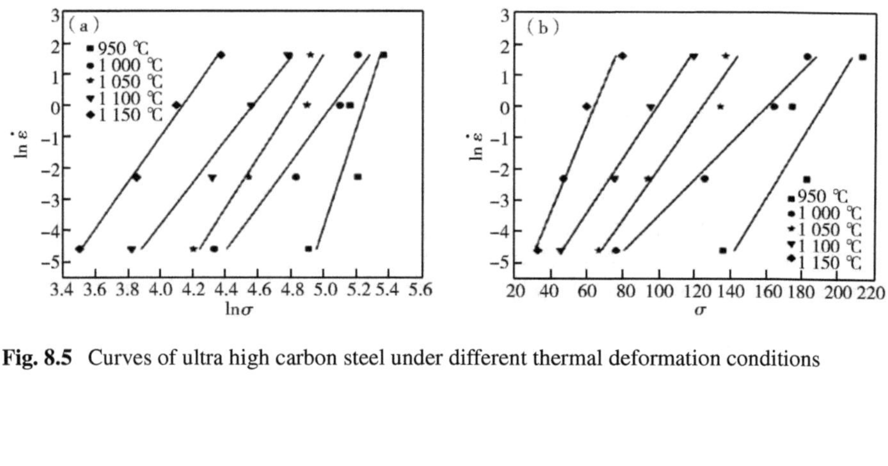

Fig. 8.5 Curves of ultra high carbon steel under different thermal deformation conditions

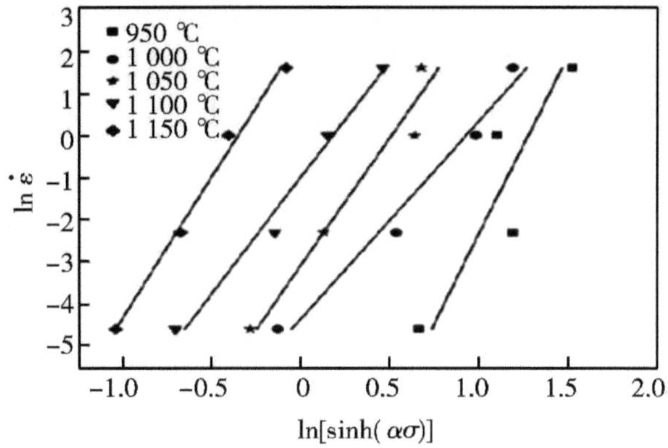

Fig. 8.6 ln·ε − ln[sinh(ασ)] relation curves of ultra high carbon steel carbon steel under different thermal deformation conditions

By fitting and averaging the slopes of the curves under different deformation conditions in Fig. 8.2, n1 and β of the experimental steel are 8.205 and 0.085, respectively, and the corresponding value of α is 0.0106.

According to Eq. (8.7), the lnε − ln[sinh(ασ)] curves of experimental steels under different heat distortion conditions can be obtained, as shown in Fig. 8.6. The lnε − ln[sinh(ασ)] curve of the experimental steel under different heat distortion conditions is shown in Fig. 8.6.

(ασ) curve is n, and n = 6.232 can be obtained by fitting and averaging the slopes of the curves at different deformation temperatures in Fig. 8.6 (Jiang et al. 2020).

From Fig. 8.6, it can be seen that there is a linear correlation between ln and ln[sinh(ασ)] of the experimental steel under different heat deformation conditions.

Further, the linear correlation between ln and ln[sinh(ασ)] for the experimental steel under different heat distortion conditions can be obtained by rationalizing equation. It can be seen that ln[sinh(ασ)] − 1/T is positively correlated, i.e., the higher the temperature, the smaller the ln[sinh(ασ)] is, and the lower the corresponding rheological stress σ is, while the lower the temperature, the larger the ln[sinh(ασ)] is, and the higher the corresponding rheological stress σ is.

The relation expression for the deformation activation energy Q can be obtained by applying 1/T to Eq. (8.7) to take the partial derivative:

$$Q = Rn\frac{\partial \ln[\sinh(\alpha\sigma)]}{\partial(1/T)} \tag{8.8}$$

From the slope of the ln[sinh(ασ)] − 1/T relationship curve (14.075) in Fig. 8.7, and substituting the values of R and n, the deformation activation energy Q = 729. 266 kJ/mol of the experimental steel can be calculated. Combined with Eq. (8.1) and Eq. (8.4), Z can be expressed as:

$$Z = \dot{\varepsilon}\exp\left(\frac{Q}{RT}\right) = A[\sinh(\alpha\sigma)]^n \tag{8.9}$$

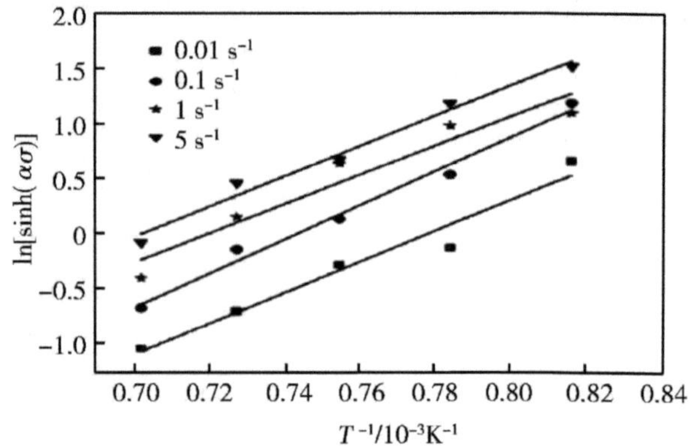

Fig. 8.7 ln[sinh(ασ)] − 1/T curves of ultra high carbon steel under different thermal deformation conditions

Substituting the Q value into Eq. (8.9), the lnZ values at different strain rates and deformation temperatures are obtained, as shown in Table 8.2. It can be seen that at the same deformation temperature, the lnZ value increases gradually as the strain rate increases from 0.01 to 5 s^{-1}, and decreases gradually as the deformation temperature increases from 950 to 1150 °C at the same strain rate.

The relationship curves of lnZ − ln[sinh(ασ)] for experimental steels under different heat deformation conditions are plotted as shown in Fig. 8.8. It can be seen that ln[sinh(ασ)] increases gradually with the increase of lnZ, and the linear correlation coefficient of the linearly fitted line of lnZ − ln[sinh(ασ)] is 0.9691, which indicates that the linear correlation of lnZ − ln[sinh(ασ)] is good (Zhang et al. 2020b). The least-squares linear regression method was used to obtain lnA = 63.35, and then the structure factor A = 3.26 × 1027.

Substituting the above calculated values of α(0.0106), n(6.232), Q(729.266 kJ/mol) and A(3.26 × 1027) into Eq. (8.4), the hyperbolic sinusoidal Arrhenius rheological stress equation for the experimental steel is given as:

$$\dot{\varepsilon} = 3.26 \times 10^{27}[\sinh(0.0106\sigma)]^{6.232}\exp\left[-\frac{729266}{RT}\right] \tag{8.10}$$

Table 8.2 lnZ values of ultra high carbon steel at different strain rates and deformation temperatures

Deflection temperature/°C	lnZ			
	Strain rate 0.01 s^{-1}	Strain rate 0.1 s^{-1}	Strain rate 1 s^{-1}	Strain rate 5 s^{-1}
950	67.121	69.423	71.726	73.335
1000	64.312	66.615	68.917	70.527
1050	61.716	64.018	66.321	67.930
1100	59.308	61.611	63.913	65.523
1150	57.070	59.732	67.561	63.284

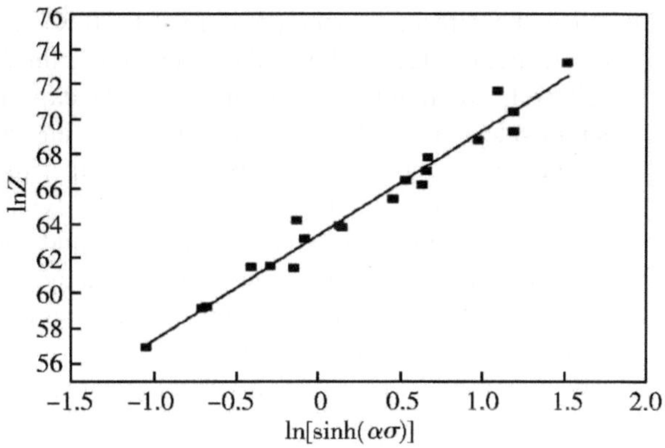

Fig. 8.8 lnZ − ln[sinh(ασ)] relation curves of ultra high carbon steel under different thermal deformation conditions

$$\sigma = \frac{1}{0.0106}\ln\left\{\left(\frac{\dot{\varepsilon}\exp(729266/RT)}{3.26\times 10^{27}}\right)^{1/6.232} + \left[\left(\left(\frac{\dot{\varepsilon}\exp(729266/RT)}{3.26\times 10^{27}}\right)^{2/6.232} + 1\right)\right]^{1/2}\right\}$$

It is worth noting that the rheological stress equation in this paper ignores the effects of strain softening and strain hardening. In order to verify the accuracy of the above rheological stress equation, the strain rate and the deformation temperature during hot compression are substituted into Eq. (8.10) and compared with Table 8.1, which shows that the peak stresses of the experimental steels at the strain rate of 0.01–5 s^{-1} and the deformation temperatures of 950–1150 °C have an error within 10% of the measured values. It is shown that the error between the calculated and measured peak stresses of the experimental steel at strain rate 0. 01–5 s^{-1} and transformation temperature 950–1150 °C is within 10%. It can be seen that the Arrhenius rheological stress equation for the experimental steel has good adaptability and application value in the hot compression process.

8.4 Conclusions

(1) When the deformation temperature rises from 950 to 1150 °C, the peak stresses of the experimental steels show a gradual decrease when the strain rate is 0.01, 0.1, 1 and 5 s^{-1}; when the strain rate rises from 0.01 to 5 s^{-1}, the peak stresses of the experimental steels show a gradual increase when the deformation temperature is 950, 1000, 1050, 1100 and 1150 °C. The peak stresses of the experimental steel at deformation temperatures of 950, 1000, 1050, 1100 and 1150 °C all showed a gradual increase.

(2) When the deformation temperature of 1150 °C, strain rate of 1 s^{-1} and above, the experimental steel in the equiaxial crystal grain size distribution is uniform and relatively small, it is recommended that the experimental steel heat deformation strain rate is suitable for control in 1 s^{-1} and above. When the deformation temperature of 1050 °C and below, the experimental steel is mainly dynamic recovery, accompanied by a small amount of dynamic recrystallization; when the deformation temperature reaches 1100 °C and above, the softening mechanism of the experimental steel is mainly based on dynamic recrystallization; when the strain rate of 1 s^{-1}, the experimental steel is suitable for hot-working deformation at a temperature of 1100 °C and above.

(3) The hyperbolic sine Arrhenius rheological stress equation of the experimental steel is

$$\dot{\varepsilon} = 3.26 \times 10^{27}[\sinh(0.0106\sigma)]^{6.232} \exp\left[-\frac{729266}{RT}\right] \tag{8.11}$$

$$\sigma = \frac{1}{0.0106} \ln\left\{\left(\frac{\dot{\varepsilon}\exp(729266/RT)}{3.26 \times 10^{27}}\right)^{1/6.232} + \left[\left(\left(\frac{\dot{\varepsilon}\exp(729266/RT)}{3.26 \times 10^{27}}\right)^{2/6.232} + 1\right)\right]^{1/2}\right\}$$

The calculated and measured peak stresses of the experimental steel at a strain rate of 0.01–5 s^{-1} and a deformation temperature of 950–1150 °C were within 10% of each other.

References

Barcellona A, Palmeri D (2009) Effect of plastic hot deformation on the hardness and continuous cooling transformations of 22MnB5microallo yed boron steel. Metall Mater Trans 40A(5):1160–1174

Chen Y, Li Y, Pang X et al (2021) Intrinsic modeling of cast 42CrMo steel considering strain compensation. For Technol 46(5):246–252

Guo WM, Ding N, Liu L et al (2020) Flow behavior and microstructure of a Mo–V–Ti microalloyed high-strength steel. Metallogr Microstruct Anal 9(2):1–9

Jian D, Guifang X, Xiaonong C et al (2020) Hot working diagram and dynamic recrystallization model of high nitrogen steel. Thermal Process Technol 49(5):7–62

Jiang Y, Wang B, Huo Y et al (2020) Heat compression deformation behavior and rheological stress principal equation of 25CrMo4 steel. J Plast Eng 27(5):167–173

Li T, Wu X, Tang BY et al (2019) Thermal deformation behavior of HG700 automobile beam steel and the establishment of rheological stress structural model. Mech Eng Mater 43(12):67–70

Li G, Wang B, He Y (2021) Thermal deformation behavior and extrudability of 12Cr13 martensitic stainless steel for nuclear power generation. J Anhui Ind Univ 38(2):125–132

Qiao GY, Cao YB, Liao B et al (2017) Effect of hot deformation and Nb precipitation on continuous cooling transformation of a high-Nb steel. Ironmaking Steelmaking 44(5):359–367

Shi R, Yu X, He W et al (2021) Study on heat distortion behavior and hot working diagram of martensitic precipitation hardening stainless steel 0Cr15Ni5Cu3Nb. J Henan Univ Sci Technol 42(4):7–15

Tong K, Han X, Cui Z (2019) Thermal deformation of new hot stamping high strength steel 22MnB5 (Nb&V). J Plast Eng 26(6):256–262

Wang R, Zhang W, Li YH et al (2020) Stress–strain behavior and microstructural evolution of ultra-high carbon Fe–C–Cr–V–Mo steel subjected to hot deformation. Mater Charact 171(10):110746–110752

Wei S (2023) Research on adhesion technology of fire water supply pipe. Adhesion 50(2):8–12

Wu C, Han S (2018) Hot deformation behavior and dynamic recrystallization characteristics in a low-alloy high-strength Ni–Cr–Mo–V steel. Acta Metall Sin 31(9):963–974

Yang D, Li W, Lin Y et al (2021) Thermal deformation behavior and intrinsic model of PH13–8Mo steel. For Technol 46(5):234–239

Yu X, Wu C, Shi R et al (2021) Experimental and simulation of microstructure evolution of 55NiCrMoV7 die steel during forging process. J Plast Eng 28(6):174–184

Zhang Y, Du S (2021) Study on the hot deformation behavior and hot working diagram of 34CrNiMo6 steel. For Equip Manufact Technol 56(3):97–105

Zhang Q, Wu T, Ma X et al (2019) Effect of tempering temperature on the organization and mechanical properties of ultra-high carbon steel. Chin Somatol Image Anal 24(2):152–159

Zhang W, Yan Z, Wang ZJ, Wang R et al (2020a) Study on hot deformation behavior and recrystallization organization of Fe–1.3C–5Cr–0.4Mo–0.4V ultra-high carbon steel. J Mech Eng 56(12):116–123

Zhang X, Liu J, He W et al (2020b) Thermal deformation behavior and high temperature plasticity of 12%Cr ultra-supercritical rotor steel. For Press Technol 45(8):184–189

Chapter 9
Nitrogen Conversion Dynamics and Removal Characterization of Anaerobic Ammonia Oxidation Process in a Zeolite Bioreactor

Dehao Zhao, Zhouzhenyang Xia, Jiajun Luo, Ze Bao, Yicheng Wu, and Haiyan Fu

Abstract This study investigates the nitrogen transformation dynamics and removal characteristics of the anaerobic ammonium oxidation (Anammox) process in a zeolite-packed bioreactor. Through 90 days continuous monitoring, the effects of zeolite packing ratios (10, 20, 30%) on concentration variations of ammonia nitrogen (NH_4^+ - N), nitrite nitrogen (NO_2^- - N), nitrate nitrogen (NO_3^- - N), and total nitrogen (TN) were systematically analyzed. Results demonstrated that with increasing packing ratios, effluent ammonia nitrogen (NH_4^+ - N) concentration decreased from 56.04 to 24.77 mg/L, accompanied by a rise in NH_4^+ - N removal efficiency from 50.49 to 79.27%. Nitrite nitrogen (NO_2^- - N) concentration peaked at 0.84 mg/L under the 20% packing ratio, while total nitrogen (TN) removal efficiency significantly improved from 20.82 to 54.64%, with effluent TN reduced to 45.90 mg/L. Nitrogen removal efficiency is synergistically enhanced by ion exchange and biofilm adsorption to improve readability, with higher packing ratios significantly enhancing the anaerobic ammonium oxidation (Anammox) reaction. This study revealed that at a 30% packing ratio, the system achieved high-efficiency nitrogen removal through the synergistic interaction of physical adsorption and biological denitrification. The findings provide critical process parameters and mechanistic insights for the application of zeolite-packed bioreactors in nitrogen-laden wastewater treatment, highlighting packing ratio optimization as a key factor in balancing nitrogen removal efficiency with system stability.

Keywords Anammox · Zeolite-packed bioreactor · Nitrogen transformation · Nitrogen removal

---

D. Zhao · Z. Xia · Z. Bao · Y. Wu · H. Fu (✉)
Key Laboratory of Environmental Biotechnology of Fujian University, Xiamen University of Technology, Xiamen, China
e-mail: fuhy@xmut.edu.cn

J. Luo
Wanjia Qingshui (Xiamen) Environmental Technology Co., LTD, Xiamen, China

Y. Li et al. (eds.), *16th International Conference on Environmental Science and Development*, Environmental Science and Engineering,
https://doi.org/10.1007/978-3-032-15080-6_9

9.1 Introduction

The persistent advancement of industrialization and urbanization has led to escalating discharge of wastewater, posing significant threats to water resources and ecological systems. Among these pollutants, nitrogenous contaminants including ammonium (NH_4^+ - N), nitrite (NO_2^- - N), and nitrate (NO_3^- - N) are prevalent detrimental constituents in wastewater, driving eutrophication and aquatic ecosystem degradation. The removal and transformation of nitrogenous pollutants constitute a critical phase in wastewater treatment processes. Biological nitrogen removal (BNR) processes, extensively applied across scales from large municipal treatment plants to laboratory-scale reactors, rely on microbial-mediated conversions of nitrogen species (He et al. 2025).

Laboratory-scale UASB reactors play a pivotal role in elucidating nitrogen removal mechanisms. These systems enable precise control of critical operational parameters including temperature, pH, dissolved oxygen (DO), and HRT, thereby facilitating systematic investigations into the transformation and removal dynamics of nitrogenous pollutants (Li et al. 2025). In a related study, zeolite was modified with NaOH and NaCl to remove ammonia and nitrogen from kimchi wastewater by modified zeolite (Dong et al. 2024). In this study, the concentrations of NH_4^+ - N, NO_2^- - N, and NO_3^- - N in influent and effluent streams of a laboratory-scale wastewater treatment reactor were subjected to continuous monitoring over a 0-day period, through controlled variations in zeolite packing ratios (Li et al. 2024a). This experimental design aimed at elucidating the dynamic interdependencies between carrier material configuration and nitrogen species transformation pathways under precisely regulated operational conditions.

9.2 Materials and Methods

9.2.1 Synthetic Wastewater and Seeding Sludge

The sludge was taken from the sedimentation tank of Xinglin Wastewater Treatment Plant (Xiamen, China).The characteristics of synthetic wastewater were as follows: 150 mg/L NH_4^+-N, 750 mg/L alkalinity (calculated by $CaCO_3$). Trace elements (solution composition: EDTA 10 mg/L, H_3BO_3 0.15 mg/L, $ZnSO_4 \cdot 7H_2O$ 0.12 mg/L, $CuSO_4 \cdot 5H_2O$ 0.03 mg/L, $FeCl_3 \cdot 6H_2O$ 1.50 mg/L, $CoCl_2 \cdot 6H_2O$ 0.15 mg/L, $MnCl_2 \cdot 4H_2O$ 0.12 mg/L, $Na_2MoO_4 \cdot 2H_2O$ 0.06 mg/L).

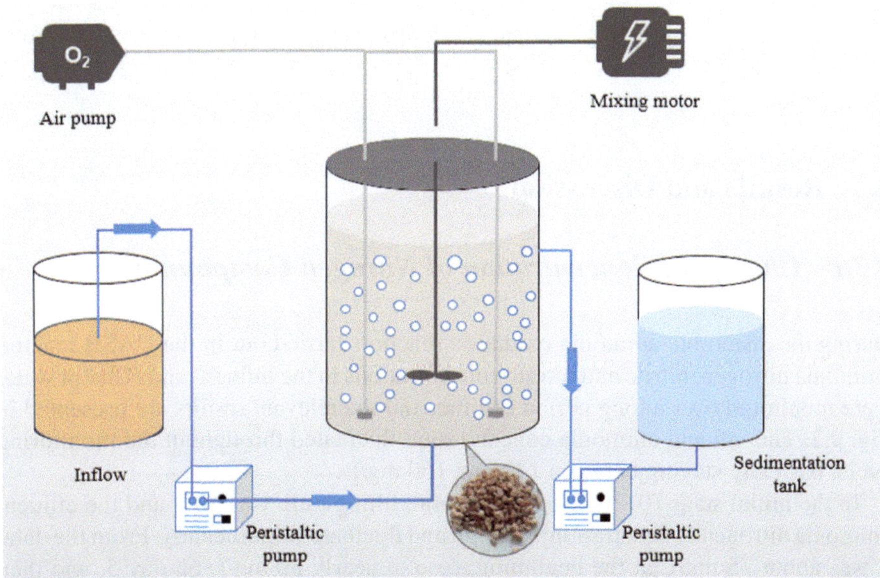

Fig. 9.1 Schematic diagram of the zeolite-packed bioreactor system

9.2.2 Reactor Set-Up

The experiment utilized an Upflow Anaerobic Sludge Blanket (UASB) reactor constructed from acrylic material, with an effective volume of 3 L (Fig. 9.1). To ensure precise thermal regulation, the reactor was externally encased in black insulating material, maintaining the internal temperature at 30 ± 1 °C throughout the operational period. Prior to experimental operation, the reactor was inoculated with anaerobic ammonium oxidation (Anammox) granular sludge at 30% of its effective volume. Zeolite was subsequently introduced at varying packing ratios in a phased manner based on operational timelines.

9.2.3 Test Items and Measurement Methods

The concentration of NH_4^+ - N was determined in accordance with the Nano reagent spectrophotometric method. The concentration of NO_3^- - N was determined by UV spectrophotometry. The concentration of chemical oxygen demand (COD) was determined by rapid extinction spectrophotometry. The pH of the samples was determined with a PHBJ-260, and the DO content was determined with a JPB-607A DO meter. Ammonia Removal Rate (ARE), Total Nitrogen Removal Rate (NRE):

$$\text{ARE}(\%) = \left(\left[NH_4^+ \text{ - N}\right]_{\text{inf}} - \left[NH_4^+ \text{ - N}\right]_{\text{eff}}\right) / \left[NH_4^+ \text{ - N}\right]_{\text{inf}} \times 100 \tag{9.1}$$

$$NRE(\%) = (TN_{inf} - TN_{eff})/TN_{inf} \times 100 \quad (9.2)$$

9.3 Results and Discussion

9.3.1 Change in Concentration of Nitrogen Compounds

During the anaerobic ammonia oxidation reaction carried out in the UASB reactor, ammonia nitrogen, nitrite and nitrate concentrations in the influent and effluent water were monitored over a long period of time, and the relevant results are presented in Fig. 9.2. The influent ammonia concentration fluctuated throughout the monitoring cycle, basically staying between 140 and 160 mg/L.

In the initial stage (0–20 days), the zeolite filling ratio was 10% and the effluent ammonia nitrogen concentration was high and fluctuated significantly. From the data, it was about 75 mg/L at the beginning, rose to nearly 80 mg/L on day 5, and then declined. This is because at the beginning of the reactor, the anaerobic ammonia oxidizing microorganisms have not been fully adapted to the environment, the population is small and the activity is low, and the ability to transform ammonia nitrogen is weak. Although the concentration of ammonia nitrogen in the influent water fluctuates but is generally at a high level, however, the microbial processing capacity is

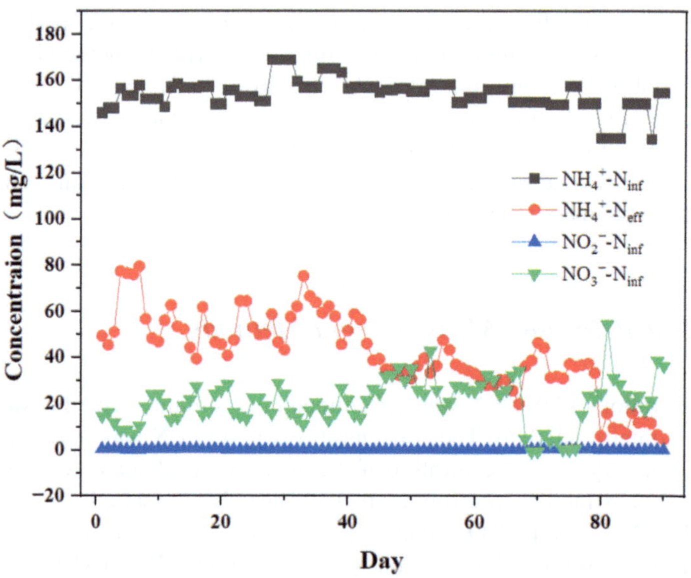

Fig. 9.2 Variations in ammonia nitrogen (NH_4^+ - N), nitrite (NO_2^- - N), and nitrate (NO_3^- - N)

insufficient, and a large amount of ammonia nitrogen is discharged without being effectively processed, resulting in a high and unstable concentration of ammonia nitrogen in the effluent water (Feng et al. 2019). As the reaction progressed to the intermediate stage (20–60 days), the effluent ammonia nitrogen concentration gradually decreased after adjusting the zeolite filling ratio to 20%. It was about 60 mg/L on day 20, to about 40 mg/L on day 40, and to about 30 mg/L on day 60. At this stage, anaerobic ammonia oxidizing bacteria are gradually enriched and their activity is increasing. They use ammonia nitrogen as the key reaction substrate, which is converted into nitrogen gas and discharged from the system through the anaerobic ammonia oxidation reaction (Jingang et al. 2021). Although the concentration of ammonia nitrogen in the influent water still fluctuates, the increase in microbial treatment capacity leads to a continuous reduction of ammonia nitrogen in the effluent water. In the late stage (60–90 days), the filling ratio of zeolite was adjusted to 30%, and the concentration of ammonia nitrogen in the effluent continued to decrease and stabilized at a lower level. On the 70th day, it was about 20 mg/L, on the 80th day, it was about 10 mg/L, and on the 90th day, it was close to 0 mg/L. At this time, the microbial ecology in the reactor was stable, and the anaerobic ammonia oxidation reaction was carried out with high efficiency, which was able to make full use of the influent ammonia nitrogen, and even if there were small fluctuations in the influent ammonia nitrogen, it could still maintain the effluent ammonia nitrogen at a very low level (Shen et al. 2022).

The effluent nitrite was stabilized at a very low level, almost close to 0 mg/L. This indicates that the microbial system in the reactor is stable, the anaerobic ammonia oxidation reaction operates stably and efficiently, and the nitrite can be utilized by the microorganisms in a timely manner, so that even if there is a slight fluctuation of the influent nitrite, the effluent nitrite can be maintained at a very low and stable level. The concentration of nitrate in the effluent was low at the initial stage, fluctuating between 10 and 20 mg/L. The nitrate production was low at this time because the anaerobic ammonia oxidation system was not perfect. At this time, the anaerobic ammonia oxidation reaction system is not perfect, and the amount of nitrate generated is small. It is possible that part of nitrite is oxidized to nitrate under some side reactions or other microbial metabolism, but the overall reaction is not intense, so the effluent nitrate concentration is low and unstable. In the intermediate stage, nitrate concentration increased and then fluctuated. It was about 20 mg/L on day 20, rose to about 30 mg/L on day 30, and then fluctuated in 20–30 mg/L. As the reaction proceeded, the environment in the reactor changed, and some autotrophic or heterotrophic microorganisms oxidized part of the nitrite to nitrate, resulting in an increase in concentration. At the same time, microbial interactions and small changes in environmental factors made nitrate production and consumption unstable, resulting in fluctuating concentrations (Deng et al. 2025).

In the later stage, the nitrate concentration still fluctuated but overall maintained in a certain range, about 10–20 mg/L. At this time, the microbial community in the reactor is stable, and the generation and consumption of nitrate reaches a relative balance. Although factors such as the entry of trace dissolved oxygen and slight changes in the composition of the influent water may cause fluctuations in concentration, the overall situation remains within a certain range, indicating that the system maintains a dynamic balance between the generation and removal of nitrate.

9.3.2 Change in Nitrogen Removal Efficiency

The dynamic changes in ammonia removal rate (ARE) and total nitrogen removal rate (NRE) visualize the performance of the reactor, as shown in Fig. 9.3. In the initial phase (0–20 days), the zeolite filling ratio was 10% and the ARE fluctuated considerably, between 40 and 70%. It starts at about 50% and then fluctuates considerably, e.g. down to about 40% on day 5 and back up to almost 60% on day 10. The NRE was lower and fluctuated sharply in this stage, which was mainly due to the fact that the anaerobic ammonia oxidation microbial community was still unstable at the early stage of reactor startup, and the population was small and not highly active. The microorganisms' ability to treat ammonia nitrogen was unstable, and at the same time, their ability to treat other nitrogenous compounds other than ammonia nitrogen in the total nitrogen was also weak (Grismer and Collison 2017).

In addition, the microbial community has poor synergistic transformation ability for various nitrogenous substrates and is greatly affected by environmental factors such as influent water quality, temperature and pH, resulting in significant fluctuations in ammonia nitrogen removal rate and total nitrogen removal rate and the total

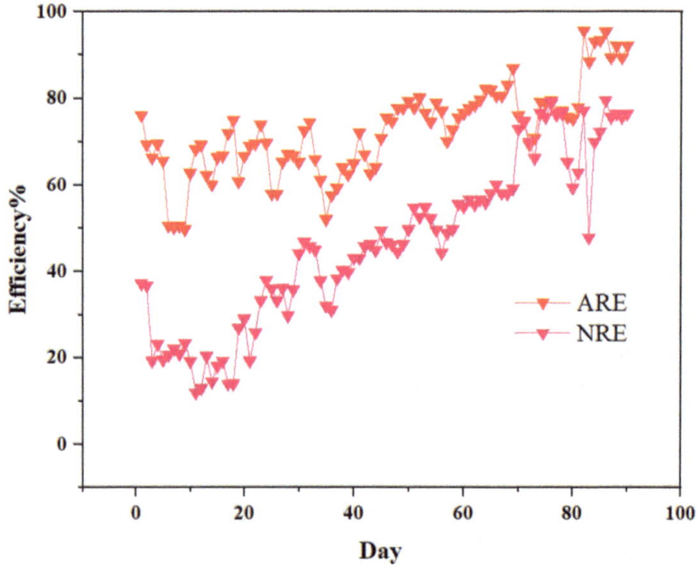

Fig. 9.3 Ammonia nitrogen removal rate and total nitrogen removal rate

nitrogen removal rate is at a low level. As time advances to the intermediate stage (20–60 days), with the increase in the zeolite filling ratio, the overall ARE tends to increase gradually from about 60% to about 80%. The NRE also increases gradually from about 10% to about 60% (Fig. 9.4). During this period, the anaerobic ammonia oxidizing bacteria gradually adapted to the reactor environment and multiplied in large quantities, and their activity also increased continuously. The microorganisms' ability to take up and transform ammonia nitrogen was greatly improved, and they could more effectively transform the influent ammonia nitrogen into nitrogen and other products to be discharged from the system, so the ammonia nitrogen removal rate continued to increase. At the same time, with the growth and enrichment of anaerobic ammonia oxidizing bacteria, its ability to transform nitrite, nitrate and other nitrogenous compounds has also improved (Lopez et al. 2008). The anaerobic ammonia oxidation reaction reacts ammonia nitrogen with nitrite to produce nitrogen, and some microorganisms may also perform reductive transformations such as nitrate, resulting in increasing total nitrogen removal rates as well.

In the later stages (60–90 days), the zeolite filling ratio is adjusted to 30%, and the ARE is maintained at a high level basically between 80 and 90%. The NRE fluctuates more, between 40 and 80%. At this time, the microbial ecosystem in the reactor tends to stabilize, the anaerobic ammonia oxidation reaction is highly efficient and stable operation, the microbial removal of ammonia nitrogen is stable, even if the influent ammonia nitrogen concentration has a certain degree of fluctuation, but also to maintain a high removal rate. Total nitrogen removal involves a variety of microorganisms and complex reaction processes (Fig. 9.5), changes in the proportion of different nitrogenous compounds in the influent water quality, and

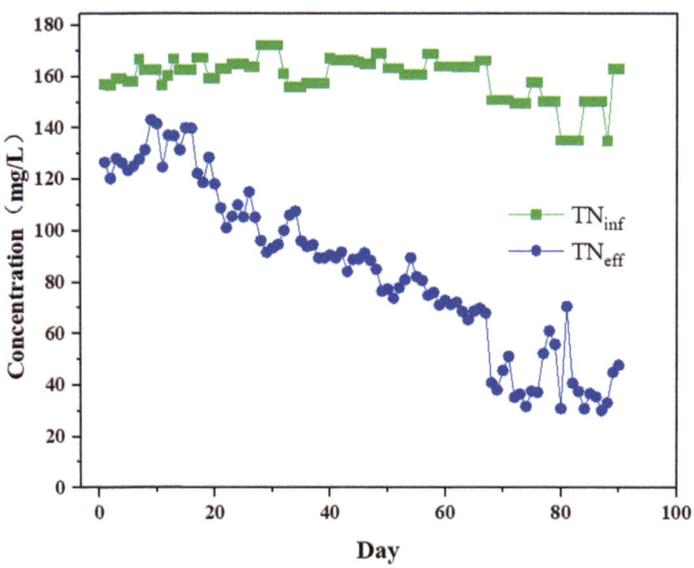

Fig. 9.4 Total nitrogen in incoming and outgoing water

small changes in environmental factors such as dissolved oxygen and acidity and alkalinity in the reactor will affect the conversion efficiency of microorganisms on various nitrogenous compounds, resulting in fluctuations in the total nitrogen removal rate (Li et al. 2024b). In summary, the microbial community of the UASB reactor was gradually stabilized during the anaerobic ammonia oxidation reaction with the increase of the operation time and the filling ratio of zeolite, and the treatment effect on ammonia nitrogen was significantly improved and stabilized. However, the total nitrogen removal rate fluctuated greatly due to various factors. Subsequent studies could focus on further optimizing the reactor operating conditions, such as precisely regulating the influent water quality and maintaining stable environmental parameters, in order to stabilize and improve the total nitrogen removal efficiency, and to promote the efficient application of anaerobic ammonia oxidation technology in the field of wastewater treatment (Luo et al. 2024). The porous structure of zeolite provides attachment surfaces for Anammox bacteria, but long-term operation may lead to an increase in biofilm thickness, blocking zeolite pores and reactor flow channels. Calcium and magnesium ions in wastewater combine with carbonate to form precipitates such as calcium carbonate ($CaCO_3$) and magnesium hydroxide ($Mg(OH)_2$), which fill zeolite pores or deposit at the bottom of the reactor. Blockage results in reduced local flow velocity in the reactor, creating dead volumes and affecting the contact efficiency between pollutants and microorganisms.

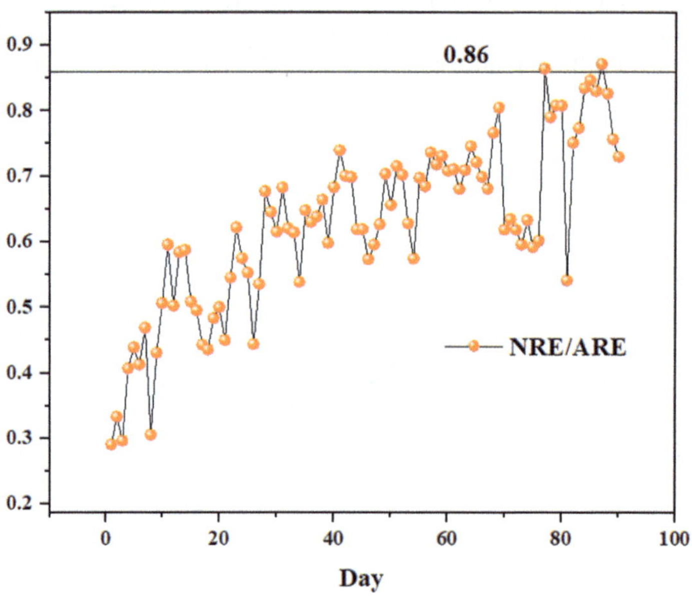

Fig. 9.5 Ratio of total nitrogen removal efficiency to ammonium removal efficiency

9.4 Conclusions

This study elucidated the impact of zeolite packing ratios on nitrogen dynamics in a UASB reactor via 90-day continuous monitoring of anaerobic ammonium oxidation (Anammox). Results demonstrated progressive adaptation of Anammox bacteria, with effluent NH_4^+ - N decreasing from 70 to 80 mg/L to near-zero levels and NH_4^+ - N removal efficiency stabilizing at 80–90%. NO_3^- - N initially increased due to nitrite oxidation but equilibrated at 10–20 mg/L post-microbial stabilization. Total nitrogen (TN) removal efficiency improved from 30 to 60%, yet exhibited significant late-stage fluctuations (40–80%), attributed to microbial-environmental synergies. A 30% zeolite packing ratio effectively stabilized NH_4^+ - N removal, validating process optimization potential. However, achieving consistent TN removal requires further investigation into microbial metabolic regulation and operational parameter tuning to enhance nitrogen transformation synergy.

Acknowledgements This research was financially supported by Fujian Provincial Industry-university Research Collaborative Innovation (2021Y4005), and Key Technology Collaborative Innovation Platform Project for Improving Urban and Rural Ecological Environment Quality in Fu-Xia-Quan National Independent Innovation Demonstration Zone (Grant No. 2022-P-024). In addition, we acknowledge support of Fujian Engineering and Research Center of Rural Sewage Treatment and Water Safety, and the Key Laboratory of Environmental Biotechnology (XMUT), Fujian Province University.

References

Deng Z, Sun C, Ma G, Zhang X, Guo H, Zhang T (2025) Anaerobic treatment of nitrogenous industrial organic wastewater by carbon–neutral processes integrated with anaerobic digestion and partial nitritation/anammox: critical review of current advances and future directions. Bioresour Technol 415:131648

Dong S, Zhang D, Zhang H, Wei S, Chen Y, Gu J (2024) Study on the removal of ammonia nitrogen by zeolite modified with NaOH and NaCl combination. Technol Water Treat 50(12):73–78

Feng Z, Wu Y, Shi Y, Xiao L, Wu G (2019) Successful startup of one-stage partial nitritation and anammox system through cascade oxygen supply and potential ecological network analysis. Sci Total Environ 696:134065

Grismer ME, Collison RS (2017) The zeolite-anammox treatment process for nitrogen removal from wastewater—a review. Water 9(11):901

He L, Dong C, Zhang L, Yan Y, Wang X, Yang H, Qian F (2025) Impacts of packed bed configuration on nitrate removal efficiency and microbial spatial distribution in solid-phase denitrification of partial nitritation/anammox granular reactor effluent. Process Saf Environ Protect 197:107075

Jingang Q, Xiyao L, Yongzhen P, Hao J (2021) Advanced nitrogen removal from landfill leachate via a two-stage combined process of partial nitrification-anammox (PNA) and partial denitrification-anammox (PDA). Sci Total Environ 810:151186

Li X, Yao J, Jia Y, Liu J, Chen Y (2024a) Initiation of anammox in an up-flow anaerobic sludge bed reactor: bacterial community structure, nitrogen removal functional genes, and antibiotic resistance genes. Water 16(23):3426

Li G, Yu Y, Li X, Jia H, Ma X, Opoku PA (2024b) Research progress of anaerobic ammonium oxidation (anammox) process based on integrated fixed-film activated sludge (IFAS). Environ Microbiol Rep 16(2):e13235

Li J, Tabassum S, Li G, Altundag H (2025) Enhanced moving bed biofilm reactor (MBBR) and an upflow anaerobic sludge blanket (UASB) reactor for landfill leachate treatment: two-stage partial nitrification-anammox process. J Water Process Eng 73:107688

Lopez H, Puig S, Ganigue R, Ruscalleda M, Balaguer MD, Colprim J (2008) Start-up and enrichment of a granular anammox SBR to treat high nitrogen load wastewaters. J Chem Technol Biotechnol 83(3):233–241

Luo J, Wu Y, Fu H, Fu M, Liu M, Guo H, Jin L, Wang S (2024) Shift in microorganism and functional gene abundance during completely autotrophic nitrogen removal over nitrite (CANON) process. J Environ Manag 359:121009

Shen H, Zhang Q, Li M, Tan X, Dong X, Wang H (2022) Research on intensive nitrogen removal of municipal sewage by mainstream anaerobic ammonia oxidation process. Chemosphere 367:143622

Part II
Watershed Hydrological Modeling and Water-Ecology-Energy-Food Nexus Management

Chapter 10
Groundwater Flow Modeling in the Wuding River Basin with MODFLOW and Environmental Response Analysis

Zhirui Yang, Ya Zhou, Guohe Huang, Yongping Li, Feng Wang, Zhipeng Xu, and Yiting Wei

Abstract Wuding River Basin is situated in a typical semi-arid region of China, where groundwater resources play a vital role in regional water resource management and ecosystem stability. A numerical groundwater model for the basin was developed using MODFLOW, incorporating high-resolution soil moisture data from the Global Land Data Assimilation System (GLDAS) to simulate groundwater dynamics and evaluate environmental response mechanisms. Results demonstrate that the model accurately captures spatial variations in the phreatic aquifer, with a correlation coefficient of 0.98 between simulated and observed water levels. The HydroSoil-CC index was employed to analyze the relationship between TSM and SA, revealing the strongest correlation (CCF = 0.81) at a two-month lag, indicative of a delayed response of groundwater recharge to soil moisture. Human activities, particularly irrigation from June to September, were found to significantly influence soil moisture, thereby driving changes in groundwater storage. These findings provide a scientific basis for water resource management in semi-arid regions.

Keywords MODFLOW · Groundwater dynamics · Agricultural irrigation · Soil moisture · Wuding River Basin

Z. Yang · G. Huang (✉) · Y. Li · F. Wang · Z. Xu · Y. Wei
State Key Laboratory of Regional Environment and Sustainability, China-Canada Center for Energy, Environment and Ecology Research, UR-BNU, School of Environment, Beijing Normal University, Beijing, China
e-mail: huang@iseis.org

Y. Zhou
School of Systems Science, Beijing Normal University, Beijing, China

Institute of Non-Equilibrium Systems, Beijing Normal University, Beijing, China

Y. Li et al. (eds.), *16th International Conference on Environmental Science and Development*, Environmental Science and Engineering,
https://doi.org/10.1007/978-3-032-15080-6_10

10.1 Introduction

Groundwater is a important resource for sustaining human life and supporting socioeconomic development, particularly in arid and semi-arid regions where surface water is scarce (Zhang et al. 2024). In the "bend area" of the Yellow River in China, groundwater plays a critical role in domestic, industrial, and agricultural activities while also supporting regional ecosystems and economic growth (Jia et al. 2006). The Wuding River Basin, located in the southeastern part of this bend area, lies on the south edge of the Mu Us Desert within the Loess Plateau of Shaanxi Province. This region is characterized by a typical temperate semi-arid continental monsoon climate, with an average annual precipitation of ~400 mm and evaporation exceeding 2000 mm (Li et al. 2016). Remarkably, increased human activities have led to a growing demand for water resources (Lin et al. 2019). Groundwater serves as a critical water source for socioeconomic development and is essential for maintaining the stability of the basin's ecosystem However, excessive groundwater extraction has resulted in declining water levels and the formation of groundwater depression cones in certain areas, threatening long-term water resource security and ecosystem stability (Xiaomei et al. 2012).

Numerical modeling provides a powerful tool for understanding groundwater dynamics and optimizing water resource management (Liu et al. 2009). The Groundwater Modeling System (GMS), developed by the United States Geological Survey, incorporates various modules, including MODFLOW, and is widely applied for groundwater flow simulations (Lai et al. 2016). Recent studies have utilized MODFLOW to assess groundwater responses to mining activities, solute transport, and irrigation (Tian et al. 2019). Research has demonstrated that soil moisture variations exhibit a lag effect on groundwater storage, which plays a crucial role in regulating recharge dynamics in semi-arid environments (Yuan et al. 2025). However, limited studies have integrated high-resolution soil moisture datasets, such as those from the Global Land Data Assimilation System (GLDAS), into MODFLOW models—especially in geologically complex regions like the Wuding River Basin (Yan et al. 2024). Although groundwater dynamics in the Wuding River Basin have been investigated previously, few studies have integrated GLDAS soil moisture data with MODFLOW to systematically evaluate the response of groundwater dynamics to soil moisture.

A groundwater model for the Wuding River Basin was developed and evaluated using MODFLOW to simulate groundwater dynamics and analyze the response mechanisms to environmental factors. The objectives of the study were as follows: (1) to construct a MODFLOW model by integrating hydrogeological data and long-term groundwater monitoring records, (2) to assess the model's performance in simulating groundwater levels and storage changes within the basin, and (3) to develop the Hydro-Soil Cross-Correlation (HydroSoil-CC) index to analyze the relationship between groundwater storage anomalies (SA) and total soil moisture (TSM), while quantifying the contributions of environmental factors, such as precipitation and human activities, to groundwater dynamics.

10.2 Methodology

10.2.1 Study Area

The study area encompasses the Wuding River Basin in the southeastern part of the Ordos Loop region, within the wind-sand region of northern Shaanxi, China (107.25°E–110.75°E, 36.75°N–39.25°N). The terrain slopes from higher elevations in the northwest to lower elevations in the southeast, with altitudes ranging from 550 to 1830 m, as illustrated in Fig. 10.1. The region is characterized by a typical temperate semi-arid continental monsoon climate, with an average annual precipitation of ~400 mm and an annual evaporation rate of about 2000 mm. Precipitation is concentrated primarily from July to September, often occurring as intense rainstorms, and exhibits significant interannual variability between wet, normal, and dry periods (Li et al. 2024). The arid climate, abundant sunlight, and frequent wind-sand activity result in strong evaporation and low vegetation cover, significantly influencing surface runoff and groundwater recharge processes (Ma et al. 2024).

The geological strata in the study area, arranged from oldest to youngest, include the Triassic, Jurassic, Lower Cretaceous, Tertiary, and Quaternary systems. The Jurassic and Cretaceous formations are the most widespread, with much of the area overlain by loose Quaternary sediments. Bedrock exposures are sparse, occurring mainly in the loess hills and along gully banks in the eastern and southeastern parts. The lithology primarily consists of interbedded sandstone and sandy mudstone (Liu et al. 2024). Quaternary loose sediments host widespread phreatic aquifers in the plains and desert regions, with groundwater depths typically ranging from 5 to 15 m. Aquifer properties are strongly influenced by lithology and fracture development, leading to significant variations in aquifer yield (Guangcai et al. 2010). Confined aquifers are predominantly found in interbedded sandstone and mudstone zones, where aquifer yield decreases with increasing depth and is heavily constrained by recharge conditions.

The region features a dense river network, with the Wuding River, a primary tributary of the Yellow River, and its tributaries (e.g., Yuxi River, Lu River, and Hailiutu River) dominating the hydrological system. River morphology is complex, with wide but unevenly distributed valleys. Shallow groundwater levels in some areas facilitate the formation of valley-type phreatic zones. Groundwater flow is controlled by topography, generally converging toward downstream rivers (Irvine et al. 2024). Recharge sources include rainfall infiltration, lateral inflow, and irrigation return flow. Human activities, particularly agricultural irrigation and mining-related energy development, have significantly impacted groundwater recharge and discharge processes (Noorbeh et al. 2024), exacerbating the depletion of groundwater resources.

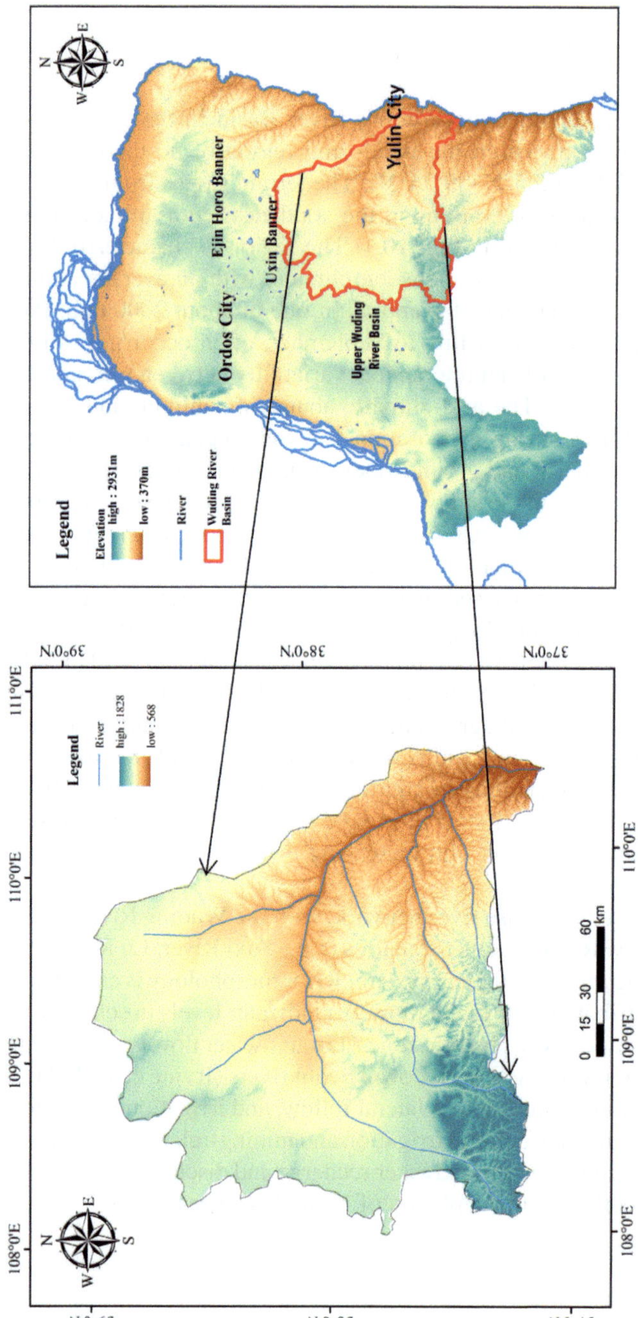

Fig. 10.1 Overview map of the Wuding River Basin

10.2.2 Data Collection

Data utilized in this study include measurements from 59 national groundwater monitoring wells, recorded at 5-day intervals throughout 2021, alongside data from 40 geological boreholes and various hydrogeological maps pertinent to the study area. These datasets were obtained from the official website of the China Geological Survey (https://geocloud.cgs.gov.cn). Meteorological data, encompassing precipitation and evapotranspiration, were sourced from the European Centre for Medium-Range Weather Forecasts (ECMWF, https://apps.ecmwf.int/datasets/). Water usage data for counties within Ordos City and Yulin City were extracted from the 2021 Water Resources Bulletin. Land use data were acquired from the National Earth System Science Data Center of China (https://www.geodata.cn/data). Total soil moisture data were derived from the NASA Global Land Data Assimilation System (GLDAS-CLSM), which provides global soil moisture dynamics at a 0.25° resolution, accessible at https://ldas.gsfc.nasa.gov/gldas.

10.2.3 Numerical Groundwater Simulation

10.2.3.1 Hydrogeological Conceptual Model

The study area, situated in the Wuding River Basin within the wind-sand region of northern Shaanxi, spans ~30,261 km^2 and includes Yuyang District, Fugu County, Hengshan District, Dingbian County, Ande County, Wubu County, Qingjian County, and parts of Wushen Banner. Guided by groundwater system theory and referencing the *China Geological Environment Atlas* published by the China Geological Survey under the Ministry of Natural Resources in 2018, the aquifer system's yield characteristics were generalized using the aquifer as the basic unit. Boundary hydraulic properties were established based on the initial groundwater flow field and topographic conditions (Zamani et al. 2022). Parameter zoning is depicted in Fig. 10.2a, while initial water level contours are presented in Fig. 10.2b. Permeability coefficient (K) and specific yield (Sy) were calculated using national monitoring well water level data and pumping test data provided by the China Geological Survey. Notably, the vertical permeability coefficient (Kz) typically decreases with depth due to the combined effects of mudstone's impervious properties, strata compaction, and cementation.

Groundwater in the study area primarily occurs in three aquifer types: Quaternary loose porous aquifers, Cretaceous sandstone aquifers, and Carboniferous to Jurassic bedrock fracture aquifers. This study focuses on unconfined aquifers, emphasizing their coupling with soil moisture. Within the regional groundwater system, confined aquifers are limited to areas where Cretaceous sandstone is underlain by interbedded mudstone and sandstone; all other major aquifers are unconfined. (1) In areas with well yield (Q) <100 m^3/d, aquifers are predominantly composed of

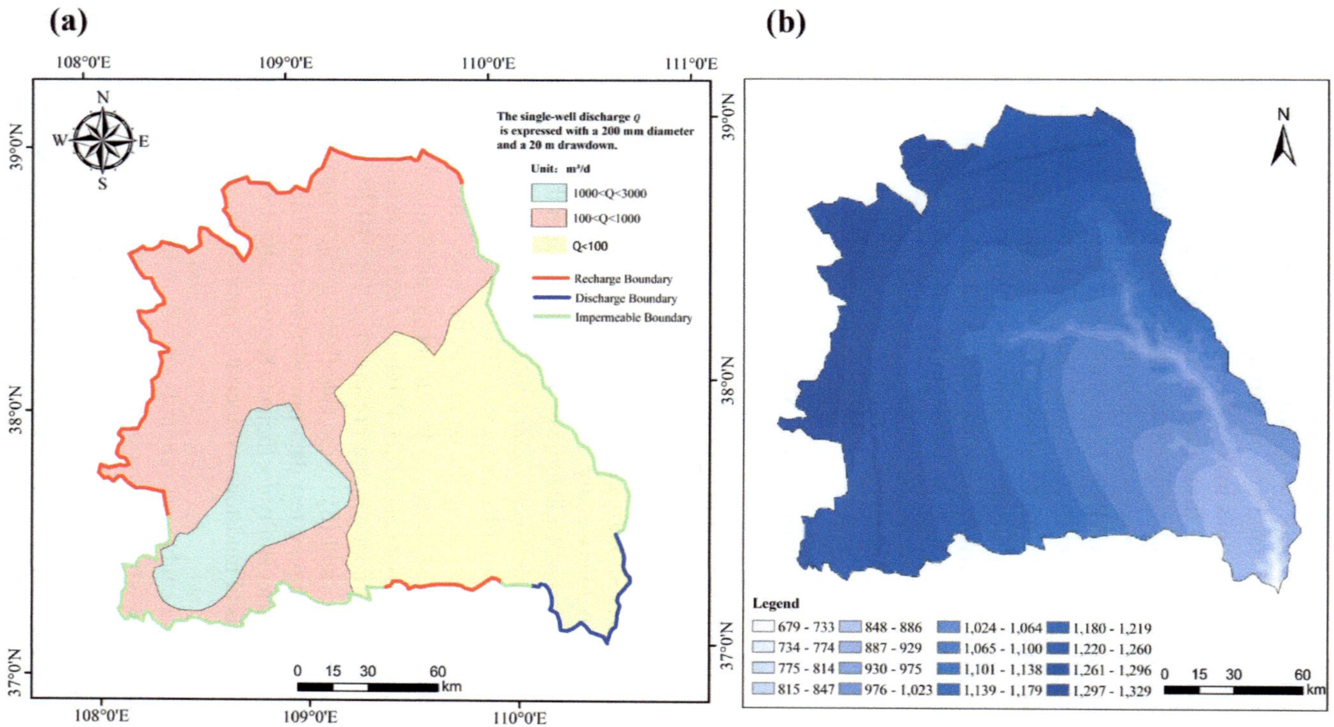

Fig. 10.2 Numerical groundwater simulation of the Wuding River Basin: **a** parameter zoning; **b** initial observed water levels

loess and Carboniferous to Jurassic sandstone–mudstone interbeds. The loess layer forms a fracture-dominated unconfined aquifer with poorly developed fractures, low permeability, and limited yield. Although the Carboniferous to Jurassic sandstone–mudstone interbeds exhibit some impervious characteristics locally, the confining layers are generally thin and discontinuous, lacking sufficient sealing capacity to form stable confined systems. Consequently, groundwater in these areas is characterized as fracture–porous unconfined with low yield. These regions, primarily agricultural, are more susceptible to fluctuations from seasonal irrigation, particularly in the loess layer compared to sandstone. (2) In areas with $100 < Q < 1000$ and $1000 < Q < 3000$ m^3/d, aquifers are mainly composed of Cretaceous sandstone and mudstone interbeds. Sandstone exhibits well-developed porosity, while mudstone provides some sealing capacity. However, due to complex geological conditions, the continuity and sealing ability of mudstone layers are limited, hindering the formation of stable confined systems. These aquifers are thus characterized as fracture–porous unconfined with moderate to high yield, offering relatively good recharge and storage capacity suitable for medium-scale groundwater development. Major industrial zones are also located in the plains of these areas.

Based on the hydrogeological conditions, the groundwater system was generalized into two aquifer layers: the first comprising Quaternary loess and Cretaceous sandstone–mudstone interbeds forming an unconfined aquifer, and the second consisting of Carboniferous to Jurassic sandstone–mudstone interbeds and deeper Cretaceous interbeds forming a fracture–porous aquifer. Integrating Darcy's law and regional hydrogeological conditions, the groundwater system was conceptualized as a three-dimensional, heterogeneous, anisotropic, and unsteady flow system.

The top of the study area is defined by the phreatic surface, which receives vertical recharge and discharge, including rainfall infiltration, irrigation return flow, pumping for human activities, and evaporation from the phreatic surface. The base is treated as an impervious boundary. Based on the initial groundwater flow field, the northwestern boundary was designated as a recharge boundary, while the southeastern area, where the Wuding River converges with the Yellow River, was set as a discharge boundary (Fig. 10.2).

10.2.3.2 Mathematical Model of Groundwater Flow

The groundwater flow in the study area was conceptualized as a three-dimensional, heterogeneous, anisotropic, and unsteady flow model, with source-sink terms exhibiting seasonal variability and hydrogeological parameters displaying significant spatial heterogeneity across different hydrogeological units. Based on Darcy's law and the principle of water balance, the governing equations were formulated as a system of differential equations to describe the well-posed problem.

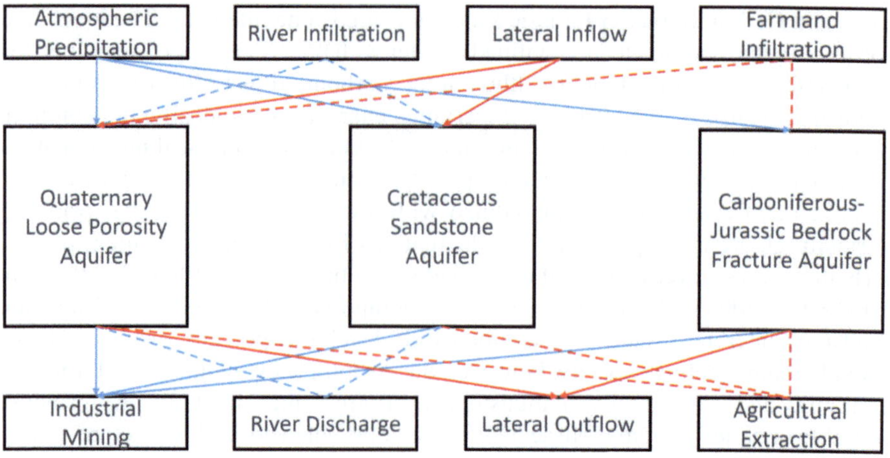

Fig. 10.3 Groundwater recharge and discharge relationships in the Wuding River Basin

$$\begin{cases} \frac{\partial}{\partial x}\left(K_x \frac{\partial h}{\partial x}\right) + \frac{\partial}{\partial y}\left(K_y \frac{\partial h}{\partial y}\right) + \frac{\partial}{\partial z}\left(K_s \frac{\partial h}{\partial z}\right) + \varepsilon = Sy\frac{\partial h}{\partial t} & x, y, z \in \mathrm{D}, t \geq 0 \\ K_x\left(\frac{\partial h}{\partial x}\right)^2 + K_y\left(\frac{\partial h}{\partial y}\right)^2 + K_s\left(\frac{\partial h}{\partial z}\right)^2 - \frac{\partial h}{\partial z}(K_s + p) + p = \mu\frac{\partial h}{\partial t} & x, y, z \in \Gamma_0, t \geq 0 \\ H(x, y, z, t)|_{t=0} = H_0 & x, y, z \in \mathrm{D}, t \geq 0 \\ K_n \frac{\partial h}{\partial \vec{n}}\Big|_{\Gamma_1} = q(x, y, z, t) & x, y, z \in \Gamma_1, t \geq 0 \end{cases} \tag{10.1}$$

Here, D represents the seepage computation domain; H denotes the total hydraulic head (m); (x, y, z) are the nodal coordinates; K_x and K_y are the horizontal permeability coefficients (m/d); K_z is the vertical permeability coefficient (m/d); ε represents the source-sink term within the seepage domain (1/d); t is time (d); p denotes the vertical water flux at the phreatic surface (m/d); S_y and μ are the specific yield (dimensionless); H_0 is the initial hydraulic head (m); H_1 represents the known hydraulic head at the boundary (m); q is the lateral flux per unit area per unit time ($m^3/(d \cdot m^2)$); Γ_0 is the upper boundary of the seepage domain (phreatic surface); Γ_1 is the second-type boundary; and n is the outward normal direction of the boundary.

Kriging interpolation was employed to estimate the elevation of the aquifer base and groundwater levels across the study area. This geostatistical method leverages the spatial autocorrelation of data to achieve optimal estimation, effectively capturing the spatial distribution characteristics of strata and water levels. The interpolation is expressed as:

$$Z(X_0) = \sum_{i=1}^{n} \lambda_i Z(x_i) \tag{10.2}$$

$$\min\{Var[Z(\hat{x}_0) - Z(x_0)]\} \tag{10.3}$$

$$\gamma(h) = \begin{cases} 0 & h = 0 \\ C_0 + C\left(1 - e^{-\frac{h}{a}}\right) & h \geq 0 \end{cases} \quad (10.4)$$

where λ_i is the weight coefficient at the i-th grid point. To minimize the variance (Var) between the predicted value $Z(X_0)$ and the actual value $Z(X_i)$, with an expected value E of zero, the variogram function $\gamma(h)$ is used, where h is the lag distance, C_0 is the nugget, C is the sill, and a is the range.

During the simulation, Darcy's law was applied to compute boundary conditions for groundwater flow. The flux values at the boundaries were derived from monitoring well data and regional permeability coefficients, expressed as:

$$q = kA\frac{H_1 - H_2}{\Delta L} = -kAi \quad (10.5)$$

where q is the flow rate (m^3); k is the permeability coefficient (m/d); A is the cross-sectional area of seepage (m^2); H_1 and H_2 are the hydraulic heads at two points, incorporating pressure, elevation, and velocity heads; ΔL is the distance between the two points (m); and i is the hydraulic gradient (dimensionless).

To evaluate the model's accuracy and reliability, multiple statistical metrics, including R^2, RMSE, and STD, were used to assess the fit between simulated and observed water levels at the end of the simulation period.

$$r = \frac{\sum_{i=1}^{n} (x_i - \overline{x})\left(y_i - \overline{Y}\right)}{\sqrt{\sum_{i=1}^{n} (x_i - \overline{x})^2}\sqrt{\sum_{i=1}^{n} \left(x_i - \overline{Y}\right)^2}} \quad (10.6)$$

In this study, x_i and y_i represent the simulated and observed water levels at the i-th grid point (Ali et al. 2024), respectively, with their respective sample means denoted accordingly.

Given that the traditional Pearson correlation coefficient may yield biased results for non-stationary or non-normally distributed data series, a nonlinear correlation coefficient, CCF, was adopted to quantify the relationship and lag effects between groundwater storage anomalies (SA) and total soil moisture (TSM) changes, providing a more accurate representation of their dynamic interactions.

$$C = \frac{\sum_{k=l+1}^{N} X_k Y_{k-1}}{\sqrt{\sum_{k=1}^{N} X_k^2 \sum_{k=1}^{N} Y_k^2}} \quad (10.7)$$

In the CCF formula, X_k and Y_k represent the sample values of SA and TSM, respectively; l denotes the time lag; and N is the total number of samples.

10.2.3.3 Development and Solution of the Numerical Groundwater Flow Model

Based on existing hydrogeological conditions and aquifer delineations, Kriging interpolation was applied to determine the elevations of the base and top of each aquifer in the model. The study area was discretized into a grid of 120 columns × 120 rows, as illustrated in Fig. 10.4. The numerical groundwater model was established using comprehensive data from January 1 to December 31, 2021.

Lateral recharge and discharge boundaries were specified as constant flux conditions. Vertical recharge, comprising monthly rainfall infiltration and irrigation return flow, was incorporated as areal recharge inputs across the study area. Given the single-crop agricultural cycle in the Wuding River Basin, irrigation activities are concentrated from June to September. Consequently, irrigation return flow was considered only during this period, with monthly allocations distributed across the relevant areas. These were combined with rainfall infiltration recharge and input into the model via the Recharge module. Evaporation was accounted for only in areas with a phreatic depth <5 m, as it is influenced by factors such as groundwater depth. Meteorological recharge to groundwater, calculated using ERA5 meteorological data and infiltration coefficients, is shown in Fig. 10.5a. Vertical discharge, including water extraction for agricultural irrigation, industrial use, domestic consumption, and ecological purposes, was represented as pumping wells distributed by administrative region. The proportion of water use types across counties is depicted in Fig. 10.5b.

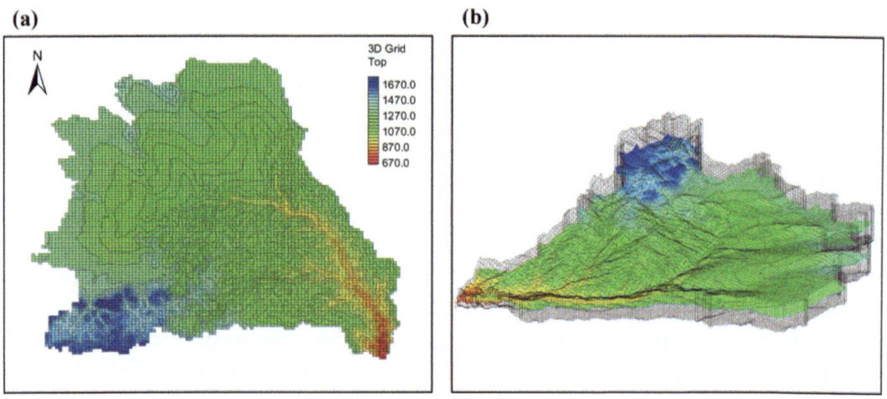

Fig. 10.4 GMS model discretization, with **a** plan view and **b** 3D view

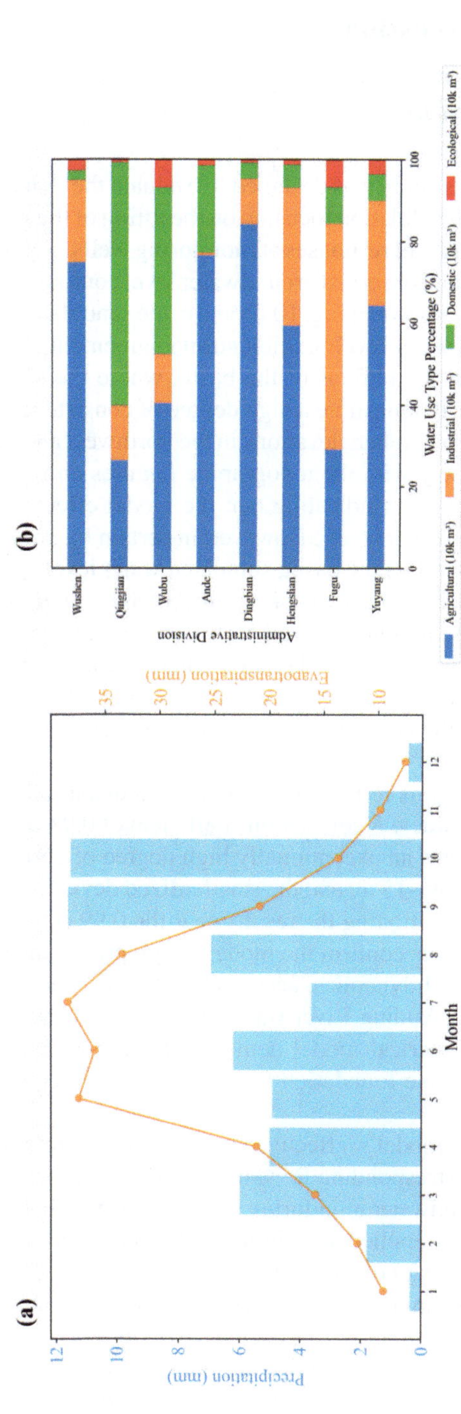

Fig. 10.5 Source-sink terms in the Wuding River Basin, with **a** meteorological recharge/discharge to groundwater and **b** human-induced recharge/discharge to groundwater

10.3 Results and Discussion

10.3.1 Model Validation

The model developed in this study was designed to evaluate the dynamics of groundwater volume changes, with validation focusing on the fitting of the groundwater flow field and the dynamic water level responses at monitoring wells. Figure 10.6 presents a comparison of simulated and observed groundwater level contours alongside Taylor diagram analysis results. As shown in Fig. 10.6b and c, the spatial distribution patterns of simulated and observed water levels exhibit strong agreement, particularly in the central and southeastern valley regions of the basin, where the shape and orientation of water level contours demonstrate a high degree of consistency. The simulated water levels, characterized by higher elevations in the northwest and lower elevations in the southeast, align closely with the topographic features and groundwater flow characteristics of the study area, indicating that the model effectively captures the spatial variability of groundwater levels. However, in certain localized areas, such as near the northeastern boundary of the basin, simulated water levels are slightly lower than observed values. This discrepancy is likely attributable to Kriging interpolation errors stemming from sparse monitoring well coverage. Insufficient well density may limit the accuracy of observed water level contours in reflecting localized variations, particularly in areas affected by industrial extraction, where drawdown cones may form, further constraining the characterization of groundwater level declines due to data scarcity.

The Taylor diagram analysis in Fig. 10.6a provides a quantitative assessment of the model's simulation accuracy. A correlation coefficient of 0.98 between simulated and observed values indicates an exceptionally high degree of correlation. The root mean square error (RMSE) is ~15 m, and the standardized deviation closely matches that of the observed data, positioning the results near the 0.99 correlation arc on the Taylor diagram. These metrics confirm the model's high precision and reliability in simulating groundwater level dynamics, accurately reflecting the spatial variability of groundwater flow in the Wuding River Basin (Khadri and Pande 2016). Overall, the MODFLOW-based numerical model demonstrates robust applicability in the study area, providing a reliable foundation for subsequent analyses of groundwater dynamics.

To further validate the model's effectiveness, Fig. 10.7 presents standardized scatter plot analyses and corresponding R^2 values for simulated and observed water levels at four national groundwater monitoring well stations within the Wuding River Basin. Standardization was applied to eliminate the influence of absolute water level differences, emphasizing consistency in fluctuation trends. The scatter plots reveal that standardized simulated and observed water levels at all four stations cluster closely around the identity line, indicating strong agreement in their temporal trends. Specifically, R^2 values for the monitoring stations are 0.8756, 0.8221, 0.9427, and 0.7914, all reflecting high levels of fit, with corresponding p-values <0.05, confirming statistically significant correlations between simulated and observed data

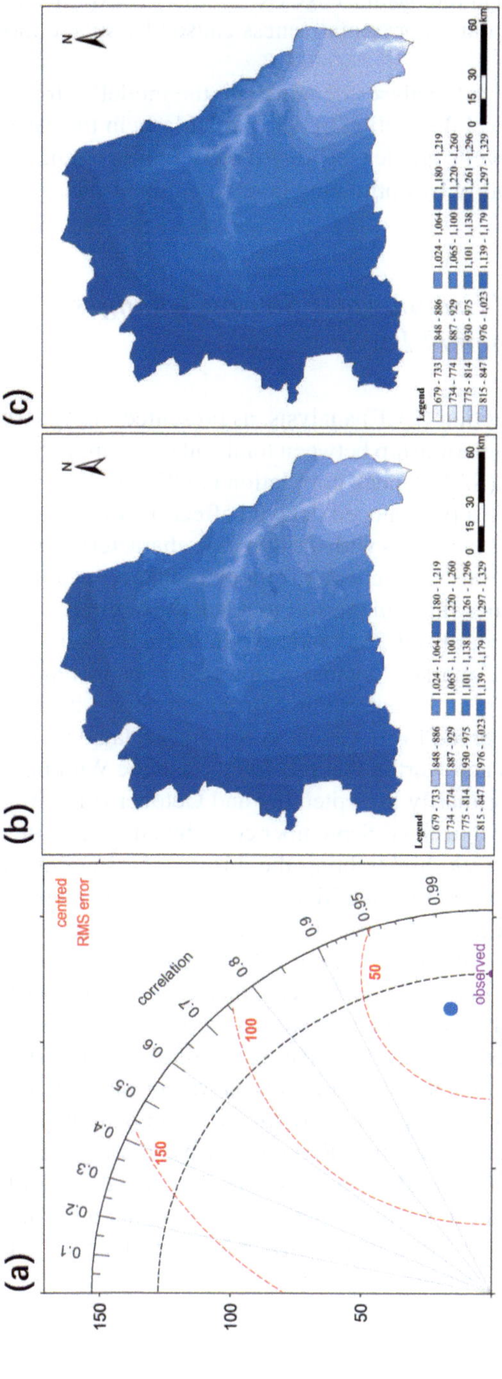

Fig. 10.6 Comparison of simulated and observed water table contour lines with Taylor diagram analysis

(Malekzadeh et al. 2019). The best fit is observed at the northeastern station (R^2 = 0.9427), while the station at the basin's edge (R^2 = 0.7914) exhibits a slightly lower fit, likely due to reduced data representativeness caused by sparse monitoring well distribution in that area.

In summary, standardized analyses demonstrate the model's strong reliability in capturing groundwater level fluctuation trends, particularly in the eastern part of the basin where monitoring wells are densely distributed. Future studies could enhance localized model applicability by optimizing boundary conditions.

10.3.2 Response of Groundwater Storage Changes to Environmental Factors

The cross-correlation function (CCF) analysis, as presented in Fig. 10.8a, was used to examine the temporal relationship between total soil moisture (TSM) and groundwater storage anomalies (SA). A peak correlation (CCF = 0.81) was observed at a two-month lag, indicating a pronounced delayed effect of soil moisture changes on groundwater storage. This lag aligns with the dynamic characteristics of groundwater recharge in semi-arid regions, where water from precipitation and human activities, such as irrigation, infiltrates through the soil to reach the aquifer.

Monthly trends of standardized TSM and SA, shown in Fig. 10.8b, reveal distinct patterns. TSM exhibits a marked decline followed by an increase from June to September, corresponding to the agricultural irrigation period in the study area. A similar trend in SA is observed from August to November, lagging TSM by approximately two months. It is noteworthy that precipitation in the Wuding River Basin in 2021 was concentrated primarily in September and October (Fig. 10.5a). However, irrigation return flow in August and September contributed to an earlier rise in TSM before the precipitation peak, highlighting the direct role of agricultural irrigation in augmenting soil moisture (Jiménez-Martínez et al. 2010). This increase in soil moisture, through infiltration, subsequently enhances groundwater storage, resulting in a significant rise in SA within two months following the irrigation season. Human water use data, as shown in Fig. 10.5b, indicate that agricultural water consumption accounts for a substantial proportion (approximately 60–80%) across administrative regions, particularly in areas with intensive agricultural activity, such as Yuyang, Hengshan, and Dingbian, where human impacts on water resources are considerable.

In conclusion, environmental factors were found to significantly influence groundwater storage changes in the Wuding River Basin by altering soil moisture. The two-month lag between TSM and SA underscores the critical role of human activities in driving groundwater dynamics in semi-arid environments. Further investigation into the mechanisms governing groundwater balance under varying scenarios is recommended to optimize water resource management strategies.

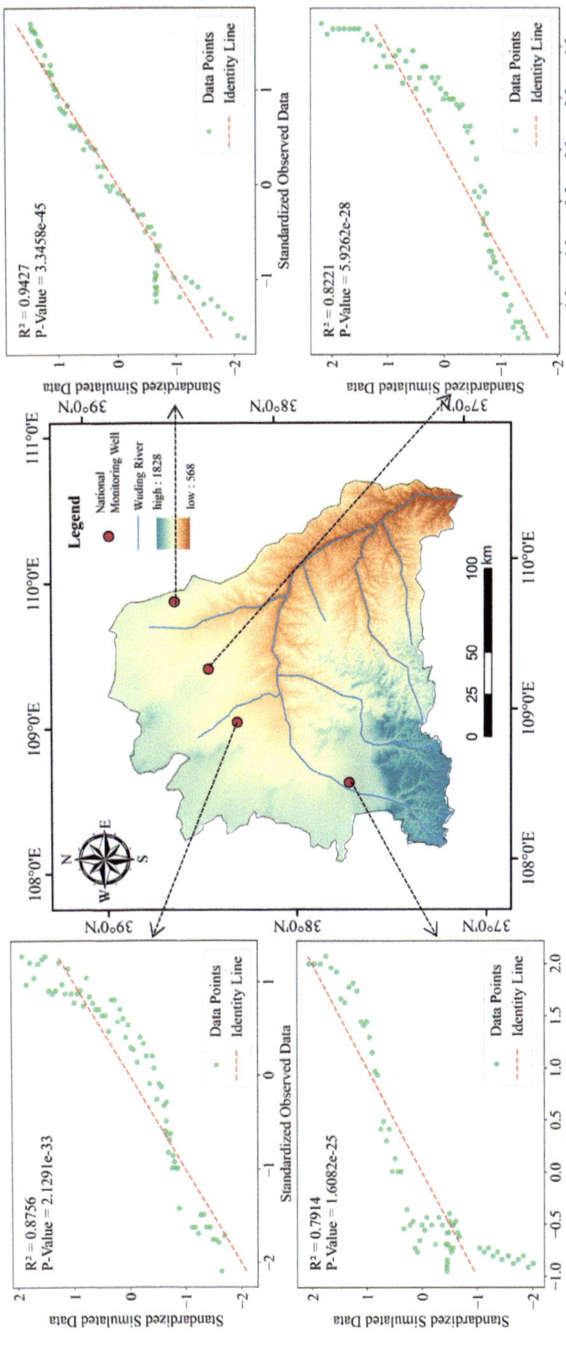

Fig. 10.7 R^2 and scatterplot analysis of simulated and observed values at four national groundwater monitoring wells

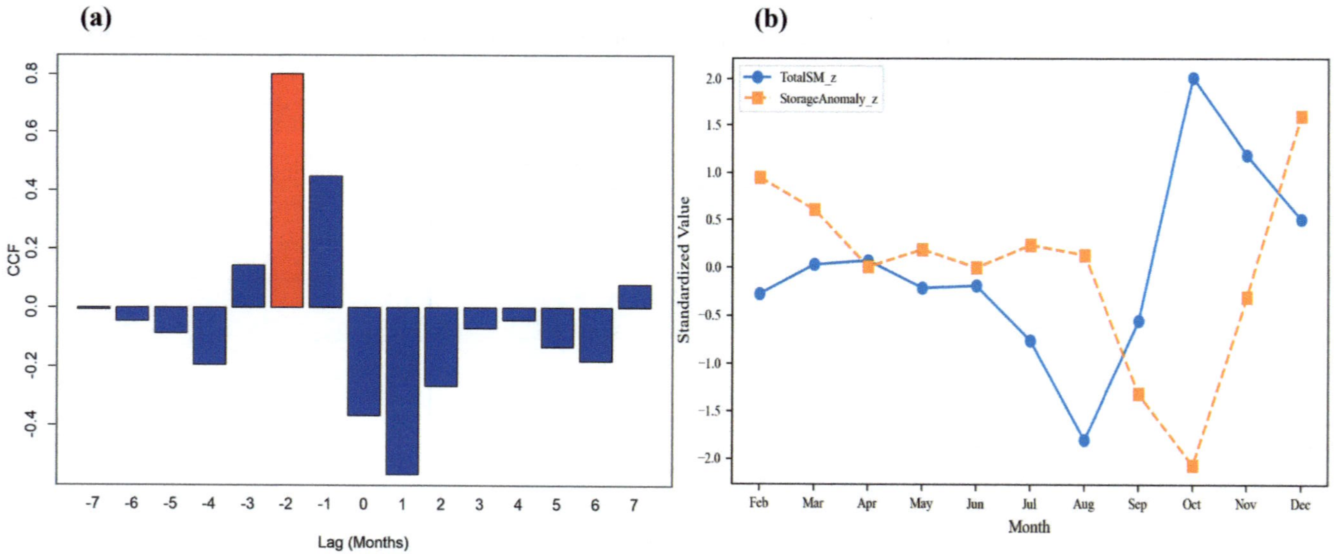

Fig. 10.8 Correlation and time-lag analysis of TSM and SA: **a** CCF analysis illustrating the lag duration with the highest correlation; **b** monthly trends of spatially averaged TSM and SA over 12 months, based on standardized Z-values

10.4 Conclusion

A groundwater model for the Wuding River Basin was developed using MODFLOW, incorporating high-resolution GLDAS soil moisture data to simulate groundwater dynamics and assess the impacts of environment responses. The model demonstrated excellent performance, with a correlation coefficient of 0.98 between simulated and observed water levels, indicating reliable representation of spatial groundwater variations. By integrating hydrogeological data, long-term monitoring records, and Kriging interpolation, the complex aquifer system of the basin, particularly the phreatic aquifer influenced by human activities, was effectively captured.

The HydroSoil-CC analysis revealed a significant lagged relationship between total soil moisture (TSM) and groundwater storage anomalies (SA), with peak correlation (CCF = 0.81) at a two-month lag. This lag underscores the delayed response of groundwater recharge to soil moisture changes in semi-arid regions (Rocha et al. 2024). Agricultural irrigation, concentrated from June to September, was identified as the primary driver of increased soil moisture, contributing substantially to groundwater recharge during and after the irrigation season. Groundwater use data further indicate that agricultural consumption, accounting for 60–80% of total extraction, dominates groundwater dynamics, particularly in agriculturally intensive areas such as Yuyang, Hengshan, and Dingbian.

Scientific support for understanding the interactions between groundwater and soil moisture in semi-arid basins was provided in this study, with the critical role of environmental factors in sustaining groundwater resources being highlighted. Numerical groundwater modeling was coupled with high-resolution soil moisture data, establishing a theoretical foundation for sustainable water resource management in semi-arid basins.

Acknowledgements This research was supported by the National Key R&D Program (2023YFC3206503) and the Natural Science Foundation (52221003, 52279002, 52279003)

References

Ali YL et al (2024) Water quality prediction based on an innovated physical and data driving hybrid model at basin scale. J Environ Inform 43(2):542

Guangcai H et al (2010) Groundwater systems and resources in the ordos basin. China. Acta Geol Sin 82:1061–1069

Irvine DJ et al (2024) Groundwater-surface water interactions research: past trends and future directions. J Hydrol 644:132061

Jia Y et al (2006) Development of the WEP-L distributed hydrological model and dynamic assessment of water resources in the Yellow River basin. J Hydrol 331(3):606–629

Jiménez-Martínez J et al (2010) Groundwater recharge in irrigated semi-arid areas: quantitative hydrological modelling and sensitivity analysis. Hydrogeol J 18(8):1811–1824

Khadri SFR, Pande C (2016) Ground water flow modeling for calibrating steady state using MODFLOW software: a case study of Mahesh River basin, India. Model Earth Syst Environ 2(1):39

Lai X et al (2016) Responses of soil water percolation to dynamic interactions among rainfall, antecedent moisture and season in a forest site. J Hydrol 540:565–573

Li S et al (2016) Response of soil moisture to hydro-meteorological variables under different precipitation gradients in the Yellow River Basin. Water Resour Manag 30(6):1867–1884

Li M et al (2024) Review: recent progress on groundwater recharge research in arid and semiarid areas of China. Hydrogeol J 32(1):9–30

Lin M, Biswas A, Bennett EM (2019) Spatio-temporal dynamics of groundwater storage changes in the Yellow River Basin. J Environ Manag 235:84–95

Liu S et al (2009) Temporal variation of soil moisture over the Wuding River basin assessed with an eco-hydrological model, in-situ observations and remote sensing. Hydrol Earth Syst Sci 13(7):1375–1398

Liu X et al (2024) Spatial variations in groundwater hydrochemistry, sources, and controls across catchments on Chinese Loess Plateau. J Hydrol Reg Stud 53:101791

Ma Z et al (2024) River–groundwater interactions in the arid and semiarid areas of northwestern China. Hydrogeol J 32(1):37–57

Malekzadeh M, Kardar S, Shabanlou S (2019) Simulation of groundwater level using MODFLOW, extreme learning machine and wavelet-extreme learning machine models. Groundwater Sustain Develop 9:100279

Noorbeh P et al (2024) Reducing carbon emission, groundwater over-exploitation and energy consumption on agricultural lands by off-farm water management practices: modernization of surface water distribution systems. J Environ Inform 44(1):59–95

Rocha VNL, Santos CAG, Silva RM (2024) Estimation of water surface reflectance and total suspended solid using MODIS images for a reservoir in the Brazilian semiarid region. J Environ Inform 44(2):1542

Tian X et al (2019) Hydrologic alteration and possible underlying causes in the Wuding River. China. Sci Total Environ 693:133556

Xiaomei F et al (2012) Soil salinity developemnt in Yellow River Delta in relation to ground water dynamics. Land Degrad Develop LAND Degrad Dev 23:175–189

Yan X et al (2024) Indirect effects of revegetation dominate groundwater recharge change at the basin scale. J Clean Product 456:142396

Yuan R, Guo S, Li Z (2025) Reforestation in the middle Yellow River has made groundwater recharge the main source of river water. J Environ Manag 381:125302

Zamani MG, Moridi A, Yazdi J (2022) Groundwater management in arid and semi-arid regions. Arab J Geosci 15(4):362

Zhang K et al (2024) Spatiotemporal interactions between soil moisture and water availability across the Yellow River Basin. China. J Hydrol Reg Stud 54:101874

Chapter 11
Research on the Application of Visualization Design Empowering IFOA-BP-AdaBoost Strong Prediction Model in Deformation Monitoring of Environmental Hydraulic Dams

Yuancheng Li and Kai Wang

Abstract Against the backdrop of sustainable development of green cities, eco-environmental security has become increasingly critical, with real-time monitoring and prediction of deformation in water conservancy dams under complex environments being of paramount importance. This study introduces visual design based on the management of original dam monitoring data to achieve real-time, dynamic, and multi-dimensional information display through line charts. An IFOA-BP-AdaBoost strong prediction model is constructed, and comparative experiments on prediction accuracy are conducted with BP and IFOA-BP algorithm models. Taking monitoring data from a specific dam as a case study, the results show that: (1) Incorporating visual design into the prediction process enhances the readability and comprehensibility of data information; (2) The root mean square error (RMSE) of the IFOA-BP-AdaBoost model is 0.3597 mm, demonstrating accuracy improvements of 33.67 and 11.19% compared to the BP and IFOA-BP models, respectively. The proposed methods are expected to provide technical support for dam monitoring and prediction in complex environments.

Keywords Environmental hydraulics · Visualization design · Improved fruit fly optimization algorithm · Deformation monitoring · IFOA-BP-AdaBoost

---

Y. Li
Guangxi Vocational Normal University, Nanning, China

K. Wang (✉)
Guangxi Construction Vocational College, Nanning, China
e-mail: 1101637063@qq.com

Y. Li et al. (eds.), *16th International Conference on Environmental Science and Development*, Environmental Science and Engineering,
https://doi.org/10.1007/978-3-032-15080-6_11

11.1 Introduction

Establishing a deformation prediction model for buildings and structures is a crucial link in ensuring the safe operation of dams (Siyuan et al. 2022). High-precision prediction of dam deformation is of vital importance for guaranteeing the safety of buildings and structures (Mengqian and Dongjian 2023).

Traditional methods such as the grey theory and time series analysis often have limited prediction accuracy when dealing with the complex issue of dam deformation (Hao and Sifa 2011), as they cannot comprehensively consider multiple influencing factors. In recent years, artificial neural networks, especially the BP neural network (Chao et al. 2015), which has the capabilities of self-adaptation, self-learning, and nonlinear fitting, have gradually attracted attention in dam deformation monitoring. Li and Zheng (2023) applied it to dam deformation prediction and achieved remarkable results. However, the arbitrary perturbation differences of the measured data still pose a challenge to the prediction accuracy of the BP neural network (Zhan et al. 2022). In order to further improve the prediction accuracy, researchers such as Schapire (1999) created the Boosting learning algorithm, a general method aimed at improving the accuracy of any given learning algorithm. The most outstanding one among them is the AdaBoost algorithm (Freund and Schapire 1995). This algorithm constructs a strong classifier by updating weak classifiers. Its core lies in finding the optimal weak classifiers, weighting them, and using the method of voting to combine multiple weak classifiers (Wang et al. 2012). At the same time, the improved fruit fly optimization algorithm is an efficient global optimization algorithm (Luo 2014). It realizes global search by simulating the foraging behavior of improved fruit flies and has the characteristics of simple calculation, easy implementation, and high precision. The IFOA-BP neural network model combines the global optimization of the improved fruit fly optimization algorithm and the local optimization of the BP neural network. It ensures that the BP neural network converges to the global optimal solution and improves the prediction accuracy. For example, Huang Junsheng and others applied the IFOA-BP neural network model to dam deformation monitoring and achieved good results. Combining the BP neural network with the characteristic of "selecting the best from the best" of AdaBoost can further enhance the prediction ability. For example, Fu Qiuxin established a BP-AdaBoost neural network and applied it to engineering deformation prediction, achieving good optimization results.

Based on the above research, this paper proposes an AdaBoost strong prediction combination model based on the BP neural network optimized by the improved fruit fly optimization algorithm—IFOA-BP-AdaBoost. In the model, the IFOA-BP neural network is designed as the "basic" prediction unit, and the improved fruit fly optimization algorithm is used for its global optimization to determine the optimal weights and thresholds. These optimized parameters are passed to the BP network, and then through its limited fine exploration ability, the best configuration of the network is found. Finally, by means of the core ideas and principles of the AdaBoost strong prediction algorithm, several such IFOA-BP neural network basic prediction units are organically combined into a powerful predictor, thus significantly improving the accuracy of inference. Practical examples show the proposed model demonstrates practical value and effectiveness in deformation monitoring and prediction for structures like dams and reservoirs (Zhou 2021).

This paper focuses on the interdisciplinary integration of the IFOA-BP-AdaBoost strong prediction model and visualization technology in visual communication design. It explores the innovative application of more intuitive, understandable and attractive visual information in enhancing the efficiency of dam deformation monitoring, providing scientific and artistic solutions for the safety management of environmental hydraulic engineering.

11.2 Theoretical Methods

11.2.1 Visualization Design

Against the backdrop of global green sustainable development and the imperative for hydraulic engineering safety management, traditional numerical reports in dam deformation monitoring face inefficiencies in information interpretation, making visualization design a critical breakthrough. This interdisciplinary practice establishes a "precision prediction-efficient communication-intelligent decision-making" closed-loop system, enhancing the recognition of key deformation features and reducing manual review time for anomalies, thus driving the transformation of engineering data from mere "visualization" to "decision-oriented intelligence." Its value lies in reconstructing the logic of data presentation through visual design, creating a new paradigm that integrates scientific rigor with artistic design for hydraulic safety management. Future research may explore deep applications of technologies such as augmented reality (AR) and digital twin.

Recent studies demonstrate that the application of visualization technologies in hydraulic engineering has formed an interdisciplinary innovation ecosystem (Jiang 2022). The Hohai University research team proposed a dynamic weighted multi-layer fusion model for full-dam sensor data, achieving adaptive multi-output forecasting of dam deformation through multi-source data mining and machine learning algorithms. Technically, the Three Gorges Dam utilized synthetic aperture radar (SAR) to detect micro-displacements as small as 0.5 mm, constructing a "point-line-plane-body" multi-level monitoring system by overlaying 3D real-scene models with historical data. This approach increased the accuracy of overall dam deformation feature recognition to 95%. These practices illustrate that visualization design is evolving from single-data display to an intelligent closed-loop integrating "data-model-decision," providing solutions that balance scientific validity and artistic intuitiveness for hydraulic engineering safety management. By transforming complex deformation data into visually encoded information—such as gradient color mappings for risk levels and interactive 3D visualizations for causal analysis—this field bridges the gap between technical precision and human cognitive efficiency, marking a transformative shift in engineering monitoring paradigms.

11.2.2 AdaBoost Strong Prediction

Since Schapire's groundbreaking innovation in the 1990s on the Boosting algorithm derived from Valiant's Probably Approximately Correct (PAC) learning model (Kearns and Vazirani 2021), this algorithm aims to construct a strong learner by combining multiple weak learners, thereby improving the accuracy of any given learning algorithm. It has had a profound impact on the field of machine learning, primarily enhancing the accuracy of randomly provided learning algorithms. Subsequently, the AdaBoost algorithm was introduced, further advancing the development of this field. Although both algorithms are comparable in efficiency, AdaBoost does not require setting a lower bound for the weak learning algorithm, thus demonstrating greater flexibility and convenience in practical applications.

As an iterative algorithm and a representative of the Boosting family, AdaBoost operates on the fundamental idea of iteratively training a series of weak learners and assigning different weights based on their performance. The final strong classifier used for decision-making is composed of these weak classifiers, which are trained using the same training dataset. During the training process, AdaBoost continuously adjusts the sample weights, enabling subsequent weak learners to focus more on the samples that were misclassified previously, thereby gradually improving the overall performance. For the detailed procedure of the AdaBoost algorithm, refer to References (Wang et al. 2012; Wang 2013).

In recent years, research on AdaBoost has focused on its generalization ability and overfitting phenomenon. Schapire et al.'s margin theory suggests that AdaBoost continues to increase the margin even after the training error reaches zero, thereby enhancing generalization ability. However, Breiman questioned this, arguing that margin maximization does not necessarily lead to better generalization performance. Subsequent studies revealed flaws in Breiman's experimental design and identified that the margin distribution, rather than mere maximization, is the key factor influencing AdaBoost's generalization performance. These findings have deepened the understanding of how AdaBoost balances training accuracy and generalization ability, providing important insights for its theoretical analysis and practical applications in complex real-world scenarios.

11.2.3 IFOA-BP

In the mid-1980s, McClelland proposed the concept of Backpropagation (BP) neural network based on the error backpropagation algorithm, demonstrating excellent nonlinear mapping, fault-tolerance, and generalization capabilities (Wan et al. 2015).

Its core mechanism involves forward propagation, backward error propagation, and final constraint adjustment.

With the deepening of research, scholar Pan Wenchao observed that improved fruit flies, when searching for food, first use olfaction to determine the general direction and then rely on vision for precise positioning. This group optimization strategy has shown practicality in solving numerous optimization problems. When detecting airborne suspended substances, improved fruit flies mainly depend on diffused odors: they are more sensitive to food with strong odor concentrations and less responsive to those with weak odors. Additionally, the distance between food and fruit flies affects their perception: greater positional differences weaken their sensitivity to low-concentration odors.

Although the FOA-BP neural network model enhances stability through ensemble learning, the fruit fly algorithm (FOA) still suffers from issues such as slow convergence speed and premature convergence in the later iteration stages, restricting further improvements in prediction accuracy. Therefore, introducing an adaptive step size to improve FOA (denoted as IFOA, improved fruit fly algorithm) can significantly enhance the parameter optimization ability, adapting to the modeling requirements of complex time-series data from dams.

In traditional FOA, the fixed step size leads to blind search behavior, making it difficult to balance global exploration and local exploitation. The adaptive step size dynamically adjusts the search range: large-scale global exploration (step size δmax) in the early stage, followed by fine-tuning in the later stage. The adjustment formula is:

$$\delta = \delta \max - (\delta \max - \delta \min) \times t/T \tag{11.1}$$

where t represents the current iteration number, and T is the total number of iterations.

The IFOA-BP neural network first employs the IFOA to explore the global solution space: starting with large-scale global exploration in the early stage and then transitioning to fine-tuning. Subsequently, the optimized parameters are fed into the BP network, which utilizes its local search capability to obtain near-optimal solutions. This combination not only ensures that the BP neural network converges to the global optimum but also improves the precision of optimization results. By integrating the global search advantage of IFOA and the local refinement ability of BP neural network, the IFOA-BP model effectively addresses the limitations of traditional FOA, providing a more robust framework for handling complex engineering problems—especially those involving time-series data with non-linear characteristics and dynamic changes, such as dam safety monitoring and structural health assessment. This hybrid approach highlights the synergy between swarm intelligence algorithms and neural networks, offering new perspectives for optimizing machine learning models in practical engineering applications.

11.2.4 IFOA-BP-Adaboost Strong Prediction Model

This study constructs the IFOA-BP-Adaboost strong prediction model using the IFOA-optimized BP neural network (IFOA-BP) as the base weak classifier. The model undergoes multiple training iterations on the same training samples, enabling the IFOA-BP neural network to compute the output of samples. In this process, various IFOA-BP weak predictors are integrated into a strong predictor through the AdaBoost algorithm, thereby achieving improved inference performance. The behavioral habits of improved fruit flies during foraging are ingeniously applied to the optimization algorithm, providing a unique framework for model construction, with the steps as follows:

Step 1: The initial positions of the IFOA swarm are randomly assigned to coordinates on the X and Y axes.

Step 2: Randomly assign foraging distances and directions to improved fruit flies.

$$\begin{cases} X_i = X_{axis} + \text{RandomValue} \\ Y_i = Y_{axis} + \text{RandomValue} \end{cases} \tag{11.2}$$

Step 3: Since the exact location of the target food is unknown, after determining the current position (Dist) of an improved fruit fly, calculate a value S—the reciprocal of the distance—as an indicator for odor concentration:

$$\begin{cases} \text{Dist}_i = \sqrt{X_i^2 + Y_i^2} \\ S_i = 1/D_i \end{cases} \tag{11.3}$$

Step 4: Based on the odor concentration indicator S obtained in Step 3, introduce a discriminant function to evaluate the (Smell) value at the fruit fly's current position:

$$\text{Smelli} = \text{Function}(\text{Si}) \tag{11.4}$$

Step 5: Identify the improved fruit fly with the highest odor concentration value:

$$[\text{bestSmell}, \text{bestindex}] = \max(\text{Smelli}) \tag{11.5}$$

Step 6: Record the highest odor concentration value and its corresponding (x, y) coordinates. The IFOA swarm then aggregates toward this optimal direction using visual information:

$$\text{Smellbest} = \text{bestSmell} \tag{11.6}$$

$$\begin{cases} X_{axis} = X(\text{bestindex}) \\ Y_{axis} = Y(\text{bestindex}) \end{cases} \tag{11.7}$$

Step 7: Perform iterative optimization by repeating Steps 2–5. After each iteration, compare whether the odor concentration in the current generation exceeds that of the previous generation. If an improvement is detected, execute the aggregation operation in Step 6. This process continuously refines the positions of the improved fruit fly swarm to explore regions with higher odor concentrations.

11.3 Experimentation

To verify the feasibility of the proposed method, this paper conducts an empirical study using the deformation monitoring values of a certain hydraulic dam as an example. Observation Point No. 1 at the crest of the river—bed dam section of the lower—reservoir dam was selected. A total of 60 non-equally spaced observations were carried out from May 2015 to November 2020. The study mainly involves influencing factors such as the temperature of the dam body, the water level, time, and the deformation amount (Y) of the dam in the Y direction, etc. (as shown in Fig. 11.1). The data is complex and voluminous, making it difficult to convert into intuitive visual information. We integrated art—based visual design into the deformation monitoring of the hydraulic dam. Line charts can clearly present the deformation trend, and comparing the inferences of different models with the actual variables can intuitively evaluate the prediction accuracy. This approach can transform complex data into intuitive visual information and enhance the efficiency of information transmission. To build the model, an interpolation method was needed to process the non-equally spaced data. Here, the Hermite interpolation method was adopted to convert the data into equally spaced deformation values. After interpolation, the deformation values clearly showed the nonlinear trend and randomness of the dam's deformation.

Three prediction schemes were thoroughly analyzed: the standard BP neural network, the IFOA-BP neural network, and the IFOA-BP-AdaBoost model optimized by the improved fruit fly algorithm (IFOA). For all three schemes, the first 50 periods of deformation observation data were used for model training, and the subsequent 10 periods were reserved for performance testing. To ensure fairness, all schemes are based on BP neural networks, with the sigmoid function set as the activation function between the input and hidden layers. The BP network structure for each scheme is consistent: an input layer with 3 nodes, a hidden layer with 7 nodes, and an output layer with 1 node. The momentum coefficient and learning rate of the models were both set to 0.01, and the maximum number of training epochs was limited to 20. For the algorithm-involved schemes (the second and third), a population size of 10 was adopted, using roulette wheel selection for parent selection. The crossover probability was set to 0.3, the mutation probability to 0.1, and the maximum number of iterations to 30. In the first scheme, network connection weights and initial thresholds were randomly generated by the model. The second and third schemes integrated the IFOA to further enhance the performance of the BP neural network. By comparing the prediction effects of these three schemes on building deformation data, this study deeply discusses their predictive capabilities

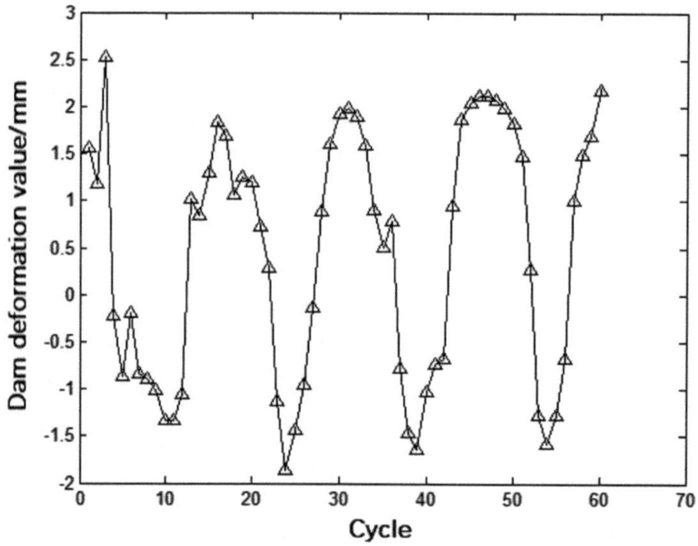

Fig. 11.1 Measure dam deformation value

and verifies the feasibility and rationality of the IFOA-BP-AdaBoost model. To evaluate the prediction accuracy of the three models, three quantitative indicators were selected: mean percentage error (MPE), mean absolute error (MAE), and root mean square error (RMSE). The specific calculation formulas for these evaluation metrics are as follows:

(1) Mean Absolute Error (MAE)

$$\mathrm{MAE} = \frac{1}{n}\sum_{t=1}^{n}\left|Y_t - \hat{Y}_t\right| \tag{11.8}$$

(2) Root Mean Square Error (RMSE)

$$\mathrm{RMSE} = \sqrt{\frac{1}{n}\sum_{t=1}^{n}\left(Y_t - \hat{Y}_t\right)^2} \tag{11.9}$$

(3) Mean Absolute Percentage Error (MAPE)

$$\mathrm{MAPE} = \frac{1}{n}\sum_{t=1}^{n}\left|\frac{Y_t - \hat{Y}_t}{Y_t}\right| \tag{11.10}$$

In the formulas, Y_t denotes the actual measured value, another Y_t represents the predicted value, n is the number of observation periods, and t indicates the t-th observation period.

The initial values from structure monitoring were used to train each model and generate calculated output values. A detailed comparison of the computational results from different scheme models is presented in Table 11.1, which systematically showcases the quantitative performance differences among the prediction approaches. This tabular summary enables a clear, standardized evaluation of their accuracy metrics, providing a critical basis for interpreting experimental outcomes and validating model effectiveness.

Based on the data in Table 11.1, Scheme 1 employs a standard BP neural network prediction model, which exhibits relatively large prediction deviations. Specifically, the absolute values of residuals exceed 0.6 mm in 4 observation periods, with a maximum residual reaching 0.7942 mm. Scheme 2 adopts the IFOA-BP model for inference, demonstrating improvements in reducing prediction residuals: the number of periods with absolute residuals exceeding 0.6 mm decreases to 3. However, its maximum residual remains at −0.7463 mm. This indicates that although the IFOA-optimized BP neural network (IFOA-BP) outperforms the standard BP model in dam deformation prediction, the enhancement in prediction accuracy is not significant, and its final training effect is still constrained by the expected output objectives. Notably, Scheme 3 utilizes the IFOA-BP-AdaBoost ensemble model, which performs exceptionally well in residual estimation. Except for the maximum residual, all other period residuals are maintained below 0.3 mm, with the maximum residual value being only −0.2512 mm. This result strongly demonstrates that compared with Schemes 1 and 2, this model achieves significantly higher precision in predicting dam deformations, highlighting the effectiveness of integrating swarm intelligence optimization (IFOA) and ensemble learning (AdaBoost) to address the limitations of single-model approaches.

In the deformation monitoring of hydraulic dams, visual and graphical representation is an intuitive and efficient method. For example, line charts clearly present deformation trends, directly reflecting the dam's dynamic changes over different time periods. By comparing the predicted variables of different models with actual measured variables, the prediction accuracy can be further evaluated. As shown in Fig. 11.2, this comparison not only visually illustrates the differences between model predictions and real data but also enables a clear assessment of each model's performance in dam deformation prediction. This chart-based presentation method offers a clear and structured approach, providing monitoring personnel with comprehensive and intuitive references to guide dam safety management and maintenance. The combination of quantitative analysis (via residual metrics) and visual interpretation (via trend charts) creates a robust framework for evaluating model reliability, especially in engineering scenarios where precise deformation monitoring is critical for structural safety. The IFOA-BP-AdaBoost model's superior performance in residuals and trend matching highlights its practical value in applications, improving dam deformation prediction accuracy and reliability.

Table 11.1 Comparison of calculation results among models/mm

Dam observation period	Measured displacement value	BP forecast value	BP residual value	IFOA-BP forecast value	IFOA-BP residual value	IFOA-BP-adaboost forecast	IFOA-BP-residual forecast
51	1.49	0.6764	0.7942	1.4568	0.0131	1.3261	0.1438
52	0.27	0.1226	0.1475	0.7786	−0.5086	0.4342	−0.1646
53	−1.27	−0.7410	−0.5293	−0.7876	−0.4823	−1.2998	0.0297
54	−1.58	−0.8177	−0.7772	−0.9518	−0.6382	−1.4179	−0.1722
55	−1.27	−0.5266	−0.7433	−0.5244	−0.7452	−1.0490	−0.2212
56	−0.69	−0.1804	−0.4994	0.0653	−0.7463	−0.4802	−0.1998
57	1.02	0.4939	0.5161	1.1525	−0.1424	1.0302	−0.0201
58	1.49	1.2412	0.2486	1.4595	0.0308	1.5493	−0.0598
59	1.67	2.2588	−0.5689	1.7424	−0.0523	1.9355	−0.2441
60	2.17	2.8934	−0.7235	2.1081	0.0614	2.4187	−0.2512

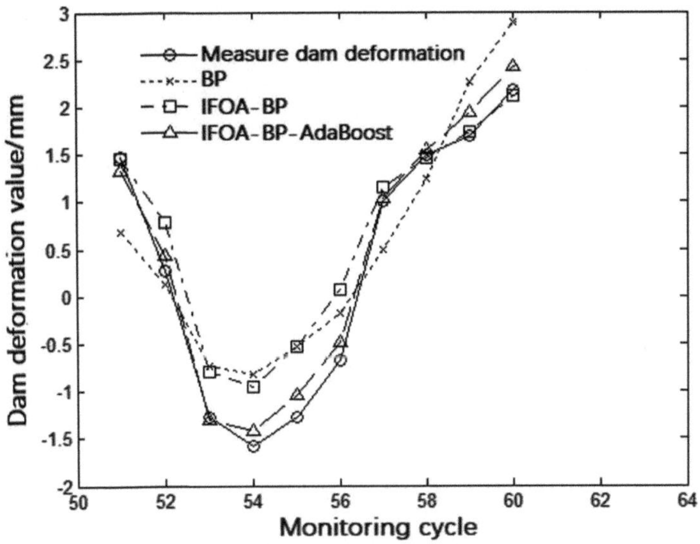

Fig. 11.2 Comparison of predicted and actual values among models

From Fig. 11.2, it can be observed that during the early and middle stages of dam deformation prediction, the deformation trends of Schemes 1 and 2 are roughly similar, with insignificant differences in predicted values. However, in the later prediction stage, the two schemes exhibit significant divergence. On the other hand, the predicted values of Scheme 3 (the IFOA-BP-AdaBoost model) maintain a high degree of consistency with the actual deformation trend of the structure throughout the entire period. Compared with the other two schemes, the deviation between Scheme 3's predictions and actual deformation values is remarkably small, indicating that the IFOA-BP-AdaBoost model performs exceptionally well in analyzing the deformation trend of buildings. This visual comparison reinforces the numerical findings in Table 11.1, demonstrating the model's superior ability to capture both short-term fluctuations and long-term trends in dam deformation, which is critical for timely safety evaluations and maintenance decisions in hydraulic engineering.

Although the comparison between the predicted values of each model and the actual values can be seen from Fig. 11.2, the colors of the graph are relatively monotonous, there is little difference between the lines, and the visual presentation is not intuitive, making it difficult to distinguish the differences between the models. In order to make the graph more intuitive and clear, combined with visualization design, different colors and clear text are now added. The graph and text are integrated for layout design to accelerate the dissemination of information among different models, so as to improve the intuitiveness and comprehensibility of the monitoring and prediction results. Please refer to Fig. 11.3.

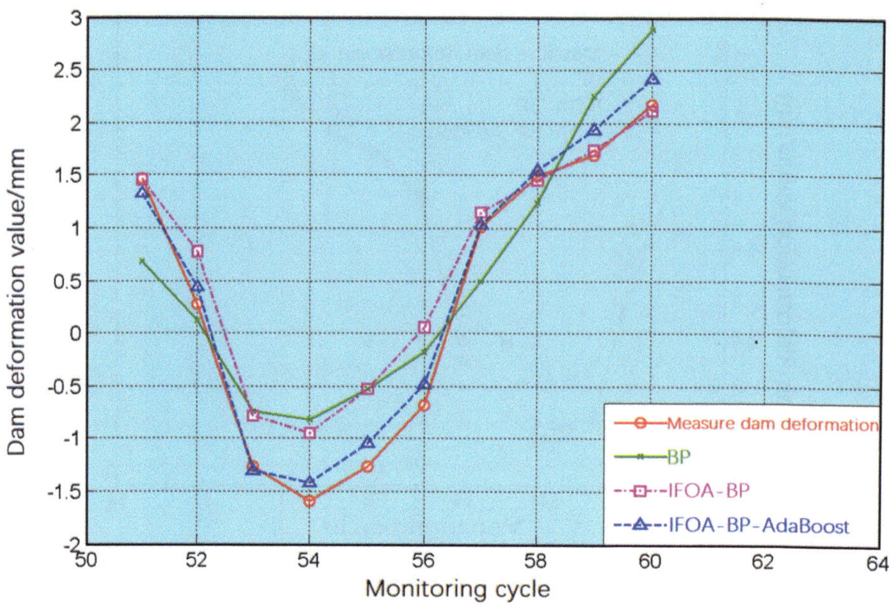

Fig. 11.3 Visualization design of the comparison between predicted values of each model and actual values

As shown in Fig. 11.4, the residuals of Scheme 1 exhibit the largest fluctuation range, while Scheme 2 reduces this range compared to Scheme 1. Notably, the residuals of Scheme 3 remain nearly stable within a range of 0.2 mm, demonstrating that the IFOA-BP-AdaBoost strong prediction model achieves high global prediction accuracy and stable predictive performance.

Similarly, although the comparison of the residuals of each model can be seen from Fig. 11.4, the visual presentation is not intuitive. In order to make the graph more intuitive and clear, visualization design is now incorporated to enhance the intuitiveness and comprehensibility of the monitoring and prediction results. Please refer to Fig. 11.5.

To comprehensively evaluate the models' performance, three quantitative evaluation indicators were adopted for detailed analysis, as presented in Table 11.2. This combined qualitative (visual residual analysis) and quantitative (metric-based assessment) approach provides a rigorous basis for comparing the models' capabilities, highlighting the IFOA-BP-AdaBoost model's superiority in both accuracy and stability for dam deformation prediction.

Based on the data in Table 11.2, when comparing the prediction accuracy of Schemes 1 and 2, Scheme 1 performs slightly worse in terms of Mean Absolute Error (MAE) and Root Mean Square Error (RMSE), with values of 0.4922 mm and 0.5423 mm, respectively. However, the two schemes exhibit comparable performance in Mean Absolute Percentage Error (MAPE), indicating that the improved fruit fly

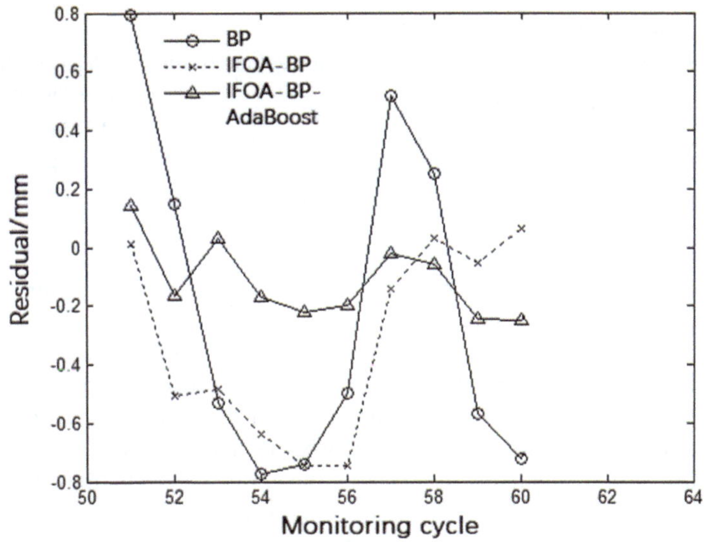

Fig. 11.4 Comparison of residuals among models

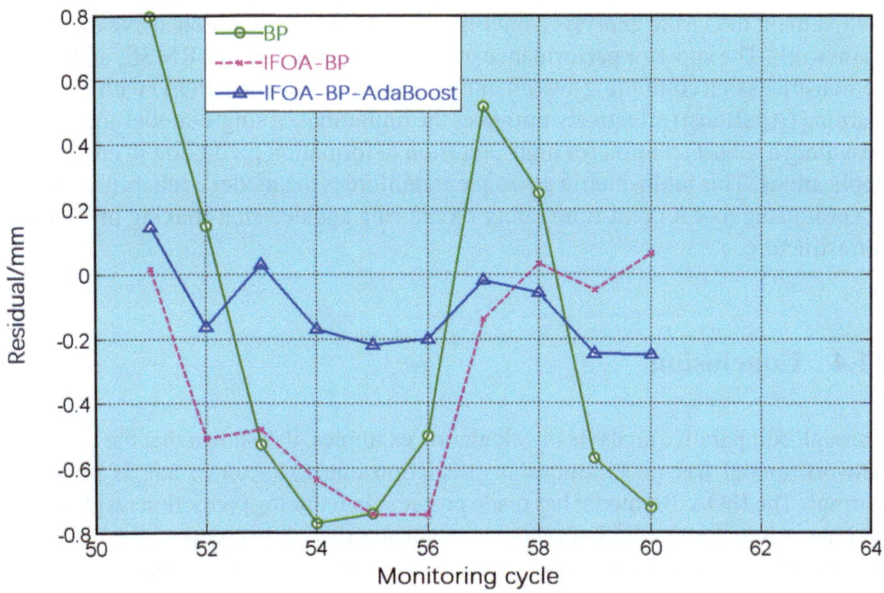

Fig. 11.5 Visualization design diagram of the comparison of residuals of each model

Table 11.2 Model accuracy comparison/mm

Prediction model	Root mean square error/m	Mean absolute error/m	Mean absolute percentage error/%
BP	0.5423	0.4922	0.3736
IFOA-BP	0.4451	0.3643	0.3774
IFOA-BP-Adaboost	0.3597	0.1576	0.1384

optimization algorithm (IFOA) only provides moderate enhancements to the BP neural network model's accuracy.

By contrast, the prediction model of Scheme 3 excels in all three accuracy metrics: its MAPE is 0.1384%, RMSE is 0.3597 mm, and MAE is 0.1576 mm—all the lowest values among the three models. The integration of the AdaBoost strong predictor allows the model to dynamically adjust the weights between weak predictors based on prediction errors, effectively aggregating the corrected predictions of the IFOA-optimized BP neural network. This "best-of-the-best" strategy not only significantly improves the model's prediction accuracy but also validates the efficiency and rationality of the proposed IFOA-BP-AdaBoost ensemble model in deformation monitoring of hydraulic dams and other structures. The quantitative results in Table 11.2 complement the visual analysis from Fig. 11.3, forming a comprehensive evaluation framework. The superior performance of Scheme 3 across MAE, RMSE, and MAPE demonstrates that combining swarm intelligence optimization (IFOA) with ensemble learning (AdaBoost) effectively mitigates the limitations of single-model approaches, providing a robust solution for high-precision deformation prediction in engineering applications. This multi-metric assessment reinforces the model's reliability, making it a promising tool for real-time safety monitoring and decision-making in hydraulic infrastructure.

11.4 Conclusion

Through comparative analysis of calculation examples, it is shown that the BP neural network model has uncertainties in prediction due to factors such as parameter settings. The IFOA-BP model has made progress in reducing prediction residuals, but its improvement in prediction accuracy is not significant. In contrast, the IFOA-BP-AdaBoost strong prediction model successfully integrates the advantages of global optimization by the improved fruit fly optimization algorithm, local optimization by BP neural network, and weighted integration by AdaBoost strong predictor, maximizing prediction accuracy. The deviation between its predicted values and actual deformation values is the smallest, with the residual fluctuation range almost stabilized within 0.2 mm. Its mean absolute percentage error (MAPE), root mean square error (RMSE), and mean absolute error (MAE) are 0.1384%, 0.3597 mm, and 0.1576 mm, respectively—all the lowest among the three models. Excelling in every prediction accuracy metric, the model confirms its feasibility and practicality in dam deformation monitoring and prediction.

This paper also explores how to integrate visualization design and artistic visual communication design into the deformation monitoring of hydraulic dams to improve the intuitiveness and comprehensibility of the monitoring results. This interdisciplinary integration not only optimizes data visualization but also provides more efficient and attractive tools for the safety management of environmental hydraulic engineering, which has important application value and provides a strong guarantee for the green, safe and sustainable development of hydraulic dams in complex environments. Although the IFOA-BP-AdaBoost strong prediction model has achieved remarkable results in the prediction of nonlinear models, more advanced prediction methods will continue to be explored in the later stage to deal with the complex challenges in dam deformation monitoring.

11.5 Fund Projects

Phased Achievements of the 2025 Guangxi Higher Education Teaching Reform Project: Research on Innovative Practice of Integrating Guangxi Ethnic Minority Cultures into Art and Design Courses in the New Media Era (Project No.: 2025JGB481). Phased Achievements of the 2025 Guangxi Vocational Education Teaching Reform Research Project: (Exploration and Research on Smart Teaching Reform of "Dual Integration, Dual Innovation, and Multi-Dimension" in "Construction Engineering Surveying" Driven by AI).

References

Chao R, Yueji L, Guangfeng P (2015) Dam deformation prediction based on grey least squares support vector machine. J Geodesy Geodyn 35(4):608–612

Freund Y, Schapire RE (1995) A decision-theoretic generalization of on-line learning and an application to boosting. Eur Confer Comput Learn Theory 21:119–139

Hao Z, Sifa X (2011) Multi-scale dam deformation prediction based on empirical mode decomposition and genetic support vector machine. Chin J Rock Mech Eng 30(S2):3681–3688

Jiang J (2022) Study on 3D visualization intelligent management methods for concrete dam safety. Doctoral Dissertation, Harbin Engineering University

Kearns M, Vazirani UF (2021) An introduction to computational learning theory. MIT Press

Li C, Zheng D (2023) NGO-XGBoost dam deformation prediction model based on multilevel data processing and its application. Water Resour Power 41(11):77–81

Luo A (2014) Reactive power optimization of power system based on improved fruit fly optimization algorithm. Technol Innov Product 5:105–107

Mengqian C, Dongjian Z (2023) Dam deformation prediction model based on FCM-WOA-LSTM and its application. Water Resour Power 41(05):71–75

Schapire RE (1999) A brief introduction to boosting. Int Joint Confer Artif Intell 15:1401–1406

Siyuan D, Lanting Z, Fei W (2022) XGBoost-LSTM variable weight combination prediction model for dam deformation and its application. J Yangtze River Sci Res Instit 39(10):72–79

Wan C, Li J, Zhao Y (2015) Dam settlement prediction based on new-dimension BP neural network-Markov chain model. J Yangtze River Sci Res Instit 32(10):23–27
Wang G, Guo L, Duan H (2012) Target threat assessment model and algorithm based on Elman-AdaBoost strong predictor. Acta Electr Sin 40(5):901–906
Wang X (2013) 43 case analyses of MATLAB neural networks. Beihang University Press
Zhan M, Chen B, Liu T (2022) Study on deformation prediction of concrete dams based on variable weight combination prediction model. Water Resour Power 40(09):115–119
Zhou X (2021) Application of ACO-BP neural network model in dam deformation interval prediction. Heilongjiang Hydr Sci Technol 49(08):200–202

Chapter 12
Assessing the Socio-Economic and Environmental Impacts of Water Supply Constraints in Inner Mongolia, China

Panpan Wang, Yongping Li, Guohe Huang, and Yanfeng Li

Abstract Previous studies have examined the effects of various policies (e.g., taxes and subsidies) on water resource management. However, limited research has systematically analyzed the socio-economic and environmental (SEE) effects of water supply constraints in water-scarce regions from an industry-wide perspective. In this study, a water-constrained computable general equilibrium (WCGE) model is developed by integrating water supply limitations as a key production factor into an extended CGE framework. The model is applied in Inner Mongolia (China) to evaluate the SEE effects of different levels of water supply reduction across industries. It is found that water supply constraints negatively impact the economy and reshapes industrial trade structures. Agriculture (AGR) and primary manufacturing (PRM) increase their value-added by investing more in labor and capital, whereas energy extraction (ENC) and services (SRV) experience declines in value-added due to their high dependence on water input. The water productivity of ENC, advanced manufacturing (ADM), and SRV improves significantly. AGR shows the largest reduction in physical water consumption but only a modest increase in water productivity (e.g., a reduction of 6.85×10^9 m^3 in physical water consumption and an increase of 12.31 CNY/m^3 in water productivity under a 50% water supply reduction scenario). Policies should focus on improving water productivity in low-efficiency sectors (e.g., AGR) and promoting the development of low-water-use industries (e.g., ADM).

Keywords Computable general equilibrium model · Inner Mongolia · Water scarcity

---

P. Wang · Y. Li (✉) · G. Huang · Y. Li
State Key Laboratory of Regional Environment and Sustainability, China-Canada Center for Energy, Environment and Ecology Research, UR-BNU, School of Environment, Beijing Normal University, Beijing, China
e-mail: yongping.li@iseis.org

Y. Li · G. Huang
Environmental Systems Engineering Program, University of Regina, Regina, SK, Canada

Y. Li et al. (eds.), *16th International Conference on Environmental Science and Development*, Environmental Science and Engineering,
https://doi.org/10.1007/978-3-032-15080-6_12

12.1 Introduction

Water resources are essential for economic activity and environmental health; however, many regions face increasing scarcity (Solovey et al. 2023; Nematian 2023). Economic systems are complex, with sectors interconnected through supply chains; thus, industries respond differently to water constraints, and shortages in one sector can affect the entire economy and environment (Wang et al. 2024). Therefore, understanding these sector-specific responses is crucial for designing effective water management strategies to ensure regional sustainable development.

The computable general equilibrium (CGE) model serves as a robust analytical framework for assessing the socio-economic and environmental (SEE) effects of water resource management policies, capturing intersectoral linkages and feedback mechanisms. For example, Zhang et al. employed a CGE-based factorial design model to evaluate the environmental and economic effects of multiple carbon and wastewater taxes in Shanxi province (Zhang et al. 2023). Li et al. developed an optimization-driven general water-management equilibrium model to facilitate integrated governance of economic and water resources, examining sectoral responses to different water policies in Guangdong province (Li et al. 2025). Inner Mongolia (located in northern China) exemplifies the challenges of water resource constraints. Its economy is heavily dependent on water-intensive industries, including agriculture and energy; however, it faces significant limitations in water availability due to climatic and geographic factors. Cai et al. analyzed the spatiotemporal characteristics of droughts in Inner Mongolia using run theory and copula methods, revealing an increasing frequency of high-severity drought events in recent decades (Cai et al. 2021). Zhou et al. assessed the water supply–demand dynamics in Inner Mongolia from 2000 to 2018 using the InVEST model and water resources bulletins, identifying substantial water shortages, particularly in the central and western regions (Zhou et al. 2023). Previous studies have examined the SEE effects of various water management policies. However, research on the impacts of water supply constraints in Inner Mongolia remains limited. In particular, the effects of reduced water availability on sectoral interactions and feedback mechanisms have yet to be comprehensively analyzed.

Therefore, the objective of this study is to develop a water-constrained CGE (WCGE) model by incorporating water supply constraints into an extended CGE framework. Specifically, this study aims to: (i) quantify the economic impacts of water scarcity, including changes in consumption, investment, and trade; (ii) evaluate sectoral adaptation in terms of patterns of production factor allocation; (iii) assess the environmental performance of sectors under varying levels of water constraints, focusing on water-saving effects and improvements in water productivity (i.e., value added per unit physical water consumption). The results will support the sustainable development of water resource management and socio-economic systems in Inner Mongolia.

12.2 Methodology

Figure 12.1 presents the framework of the WCGE model, which examines the economy-wide impacts of water scarcity. The model consists of five key modules: production, trade, income and expenditure, closure and market clearing, and scenario design.

The production module defines how goods are produced in the economy using a constant elasticity of substitution (CES) function (Nguyen 2023; Raviv et al. 2024). For example, value added consists of labor, capital, and water.

$$QVA_a = \alpha_a^{va}[\delta_a^{va} \cdot QKW_a^{\rho_a^{va}} + (1-\delta_a^{va})QLD_a^{\rho_a^{va}}]^{\frac{1}{\rho_a^{va}}} \tag{12.1}$$

$$\frac{PKW_a}{PL_a} = \frac{\delta_a^{va}}{1-\delta_a^{va}}\left(\frac{QLD_a}{QKW_a}\right)^{1-\rho_a^{va}} \tag{12.2}$$

$$PVA_a \cdot QVA_a = PKW_a \cdot QKW_a + PL_a \cdot QLD_a \tag{12.3}$$

where QVA_a represents the value added of sector a; α_a^{va} is a scale parameter; δ_a^{va} is a share parameter; QKW_a and QLD_a denote the inputs of capital-water and labor, respectively; $\rho_a^{va} = 1 - 1/\varepsilon_a^{va}$, ε_a^{va} is the elasticity substitution parameter;PVA_a, PKW_a, and PL_a are the prices of QVA_a, QKW_a, and QLD_a, respectively.

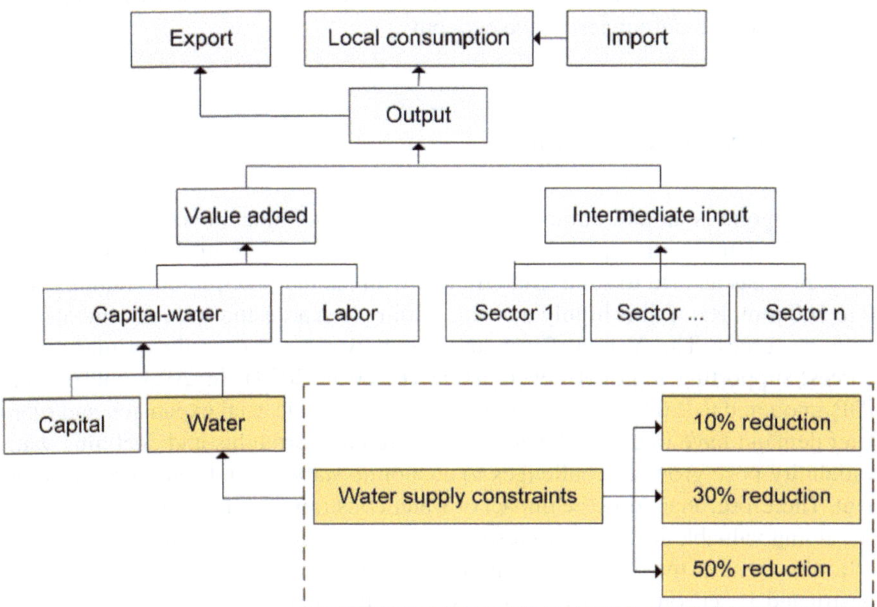

Fig. 12.1 Framework of the WCGE model

Table 12.1 Water supply and macroeconomic indicators under BAU

	BAU
Water supply ($\times 10^9$ m^3)	15.56
Final consumption ($\times 10^9$ CNY)	743.76
Investment ($\times 10^9$ CNY)	924.67
Net export ($\times 10^9$ CNY)	−185.92

The trade module governs imports and exports, employing a constant elasticity of transformation (CET) function to allocate domestic goods between local consumption and exports (Pradhan and Ghosh 2022). For imports, the model adopts the Armington assumption, which distinguishes imported goods as imperfect substitutes for domestic goods (Zeng et al. 2024). The income and expenditure module tracks financial flows among economic agents, including households, firms, the government, and foreign accounts. It ensures that income generated from production aligns with spending across the economy. The model applies a neoclassical closure to maintain market equilibrium. Under this closure, factor markets (labor, capital, and water) and product markets clear, ensuring that supply meets demand for all goods and inputs.

In this study, three scenarios are designed to assess the impact of water scarcity. Table 12.1 shows the total water supply under the business-as-usual scenario (BAU). The exogenous shock is introduced as a reduction in total water supply, set at 10% (SC1), 30% (SC2), and 50% (SC3) relative to the BAU level. These reductions are introduced into the production module, and the model subsequently evaluates their effects on sectoral output across the economy.

12.3 Study Area and Data

Inner Mongolia is an autonomous region in northern China, covering 1.18 million square kilometers with a population of approximately 23.96 million (as of 2023). The region has an arid to semi-arid climate, with annual precipitation ranging from 50 to 500 mm. Rainfall is highly uneven, leading to scarce and spatially imbalanced water resources. The Yellow River and groundwater serve as the primary water sources, supporting regional development (Li et al. 2024). In 2024, the regional GDP amounted to 2.63 trillion yuan. However, rapid industrial expansion and rising water demand have intensified water stress. Frequent droughts and declining water availability pose growing challenges to economic stability and long-term development. Therefore, analyzing the impact of water constraints on key sectors is essential for gaining valuable insights into resource management in Inner Mongolia.

In this study, the social accounting matrix (SAM) for Inner Mongolia is constructed based on the 2017 input–output table of Inner Mongolia. Table 12.2 presents the sectors included in the SAM. Socioeconomic data are sourced from the China Statistical Yearbook and the Inner Mongolia Statistical Yearbook; meanwhile,

Table 12.2 List of sectors and their abbreviations

Abbreviation	Sector
AGR	Agriculture
ENC	Energy Extraction
ENP	Energy Processing
ENT	Energy Transformation
PRM	Primary Manufacturing
ADM	Advanced Manufacturing
SRV	Services

water resource data are obtained from the China Statistical Yearbook and the Water Resources Bulletin. The elasticity parameters in the CGE model are derived from Li et al. (2025).

12.4 Result and Discussion

Table 12.1 presents the water supply and macroeconomic indicators under BAU. When the water supply was 15.56×10^9 m^3, the final consumption, investment, and net export of Inner Mongolia were 743.76×10^9 CNY, 924.67×10^9 CNY, and -185.92×10^9 CNY, respectively. The regional GDP reached 1482.51×10^9 CNY. Figure 12.2 shows sectoral exports, imports, value-added under BAU. Agriculture (AGR), energy extraction (ENC), and energy transformation (ENT) had higher exports than imports, which suggests strong export orientation and competitiveness. This trend results from resource endowments and production capacity. Primary manufacturing (PRM), advanced manufacturing (ADM), and services (SRV) had imports exceeding exports. Results indicate that PRM and SRV play a major role in trade, relying heavily on external inputs. Value-added consists of labor, capital, and water. Labor and capital accounted for 56.78% and 42.92%, respectively; however, water accounted for only 0.30%. Although water use is high, the undervaluation of water prices prevents its economic contribution from being fully reflected. SRV had the highest capital and labor input, indicating strong dependence on these resources. AGR, ENC, and PRM had high labor input, showing labor-intensive characteristics. ENC, PRM, and ENT relied heavily on capital, reflecting capital-intensive features. PRM, AGR, and SRV received the most water input, highlighting their strong dependence on water. ADM had lower inputs of labor, capital, and water, suggesting underdevelopment in Inner Mongolia.

Figure 12.3 shows sectoral physical water consumption and water productivity under BAU. AGR, PRM, and SRV accounted for 88.05, 6.92, and 3.93% of total physical water consumption, respectively. This shows that agriculture dominates water use due to high irrigation demand. Although PRM and SRV used less water than AGR, their consumption was still significant compared to other sectors, highlighting

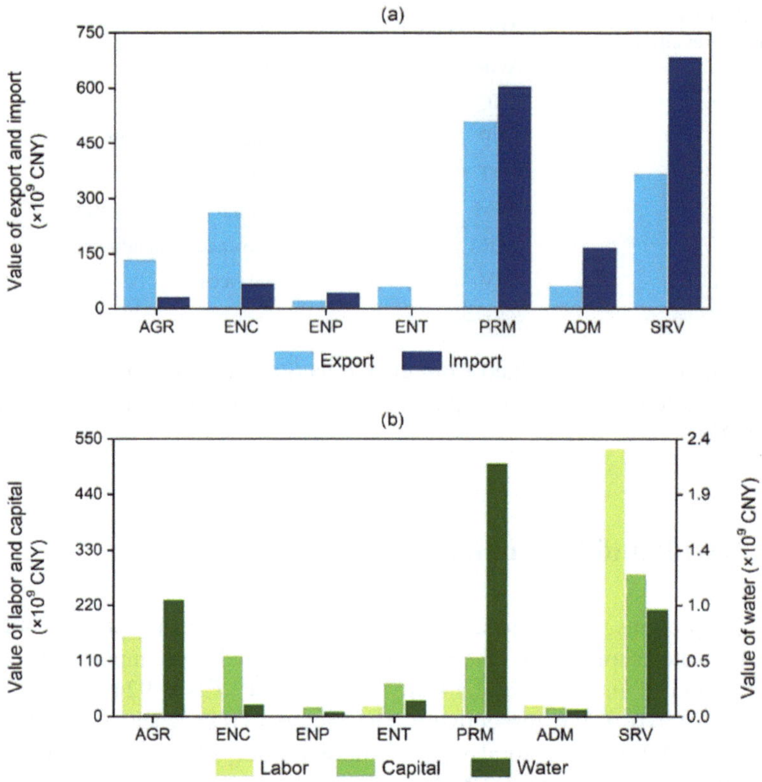

Fig. 12.2 Sectoral exports, imports, and value-added under BAU

their reliance on water for production. In sectoral water productivity, ENC achieved the highest level, followed by ADM and SRV. This indicates that ENC generates high economic output with limited water consumption, benefiting from its high value-added characteristics and production efficiency. AGR and PRM had lower water productivity due to high water consumption and lower output efficiency.

Figure 12.4 shows changes in GDP components and trade under different water supply restrictions. Compared to BAU, SC1, SC2, and SC3 reduce the water supply by 10%, 30%, and 50%, respectively. As water supply decreased, GDP declined continuously. Final consumption decreased significantly (by 0.06, 0.18, and 0.32% under SC1, SC2, and SC3, respectively), whereas investment and net exports increased slightly (by 0.02 and 0.05% in SC3, respectively). This indicates that water shortages reduce household and government consumption, directly affecting economic output. The slight increase in investment and net exports results from resource reallocation; however, this adjustment is insufficient to offset the negative impact of lower consumption. From a sectoral perspective, ENC experienced the largest reduction in exports (a decrease of 1.16% under SC3), and PRM saw the most significant

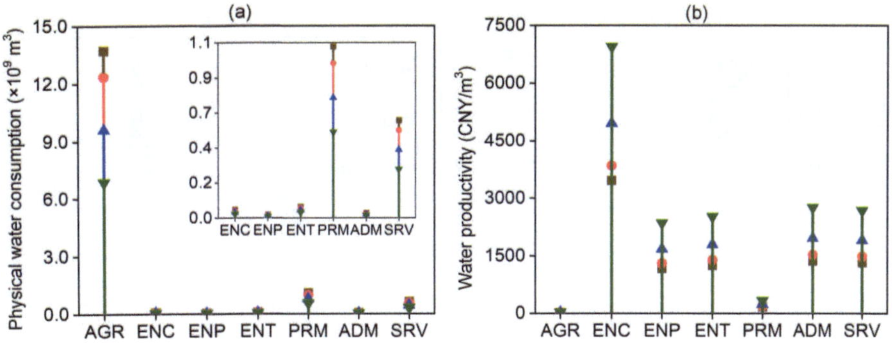

Fig. 12.3 Sectoral physical water consumption and water productivity

increase in exports (an increase of 0.64% under SC3). This is due to ENC facing direct constraints on production capacity, which weakens its export competitiveness. PRM adapts to water shortages by adjusting its production processes, thereby increasing its export share. SRV experienced the largest reduction in imports (a decrease of 0.14% under SC3), and PRM saw the largest increase in imports (an increase of 0.10% under SC3). This suggests that SRV reduces its dependence on external inputs due to lower domestic demand, and PRM increases raw material imports to support production expansion. These changes reflect the different adaptation strategies employed by sectors under water constraints.

Figure 12.5 shows the changes in sectoral value-added and its components. ENC and SRV experienced the largest declines in value-added (falling by 1758.41 and 1050.27 × 10^6 CNY under SC3, respectively), whereas AGR and PRM saw increases (rising by 462.22 and 476.72 × 10^6 CNY under SC3, respectively). Under SC3, labor input in AGR increased by 0.53%, and capital input rose by 0.63%. In PRM, labor input grew by 0.75%, and capital input increased by 0.85%, compensating for the reduction in water input. Both ENC and SRV reduced labor and capital input. Specifically, under SC3, labor input in ENC decreased by 1.02%, and capital input fell by 0.95%, while labor input in SRV declined by 0.12%. Water input decreased across all sectors, with the largest reductions occurring in PRM, AGR, and SRV, where water input fell by 42.33, 42.46, and 42.83%, respectively. These changes suggest that AGR and PRM rely more on labor and capital to increase value-added, whereas ENC and SRV struggle to substitute water with other inputs, limiting their production capacity and leading to a decline in value-added.

Figure 12.3 also shows changes in physical water consumption and water productivity under different water supply restrictions. AGR reduced water use the most, followed by PRM and SRV. ENC, ADM, and SRV showed significant improvements in water productivity, whereas AGR and PRM showed limited changes. For example, under SC3, AGR reduced water consumption by 6.85 × 10^9 m^3 and increased value-added by 462.22 × 10^6 CNY, leading to a water productivity increase of only 12.31 CNY/m^3. ENC reduced water consumption by only 0.03 × 10^9 m^3, but its value-added decreased by 1758.41 × 10^6 CNY, raising water productivity by 3491.19 CNY/

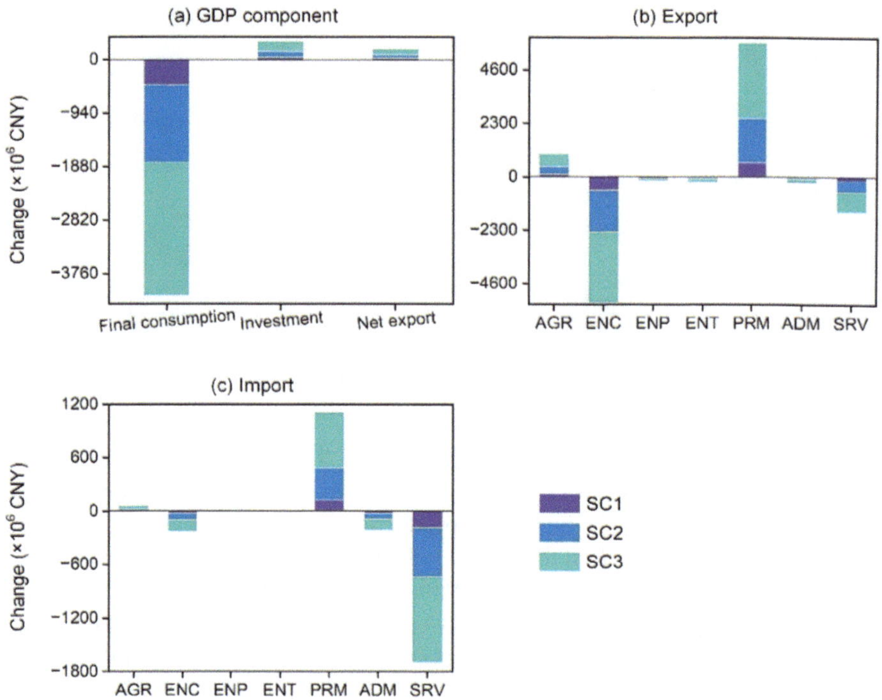

Fig. 12.4 Changes in GDP components and trade under water supply restrictions

m^3. This indicates that AGR and PRM struggle to improve economic output per unit of water due to their initially low water productivity. ENC enhances water efficiency by reducing water use and optimizing production. Although its value-added declines, its high value-added characteristics allow for strong adjustment potential under water constraints.

Generally, water shortages significantly reduce GDP and final consumption and reshape industrial trade structures. To mitigate this effect, water allocation should be optimized. Several sectors (e.g., AGR and PRM) should receive priority for production needs and improve water efficiency through technology and management. This approach helps maintain total economic output and reduces the impact of water constraints on industrial structure. AGR and PRM rely on labor and capital to compensate for water shortages. ENC and SRV have lower adaptability due to high water dependence. In Inner Mongolia, the focus on primary and service industries limits ADM, as it lacks technology or infrastructure support. Policies should promote the expansion of ADM, which has low water use and high potential. Water-saving technologies (e.g., efficient irrigation and process optimization) should also be promoted to strengthen economic resilience and improve water efficiency. The high water productivity of ENC suggests that efficiency improvements can unlock

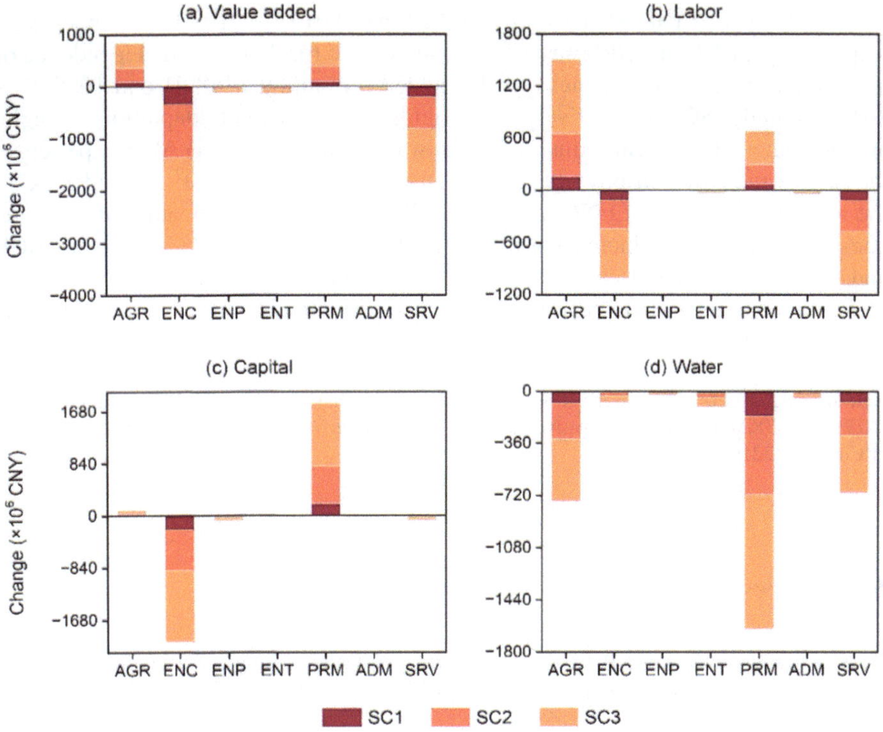

Fig. 12.5 Changes in sectoral value-added and its components under water supply restrictions

economic potential under water shortages. The low efficiency of AGR and PRM highlights room for improvement. Incentives to increase water productivity can reduce water pressure and create new growth opportunities for Inner Mongolia's sustainable development.

12.5 Conclusions

In this study, the WCGE model is developed to assess the SEE effects of water scarcity in Inner Mongolia. By incorporating water supply constraints, the model examines the impacts of reduced water availability under different scenarios. The findings are as follows: (i) as water supply decreases, Inner Mongolia's GDP continues to decline, mainly driven by a significant reduction in final consumption (a 0.32% decrease under SC3); also, ENC experiences a sharp decline in exports and SRV has a significant drop in imports; this indicates that water constraints not only reduce total economic output but also reshape industrial trade structures; (ii) despite a significant reduction in water input, AGR and PRM increase their value-added by investing more in labor and capital (rising by 462.22 and 476.72 $\times$ 10^6 CNY under SC3, respectively); ENC and SRV struggle to compensate for their strong dependence on water input, leading to declines in value-added (falling by 1758.41 and 1050.27 $\times$ 10^6 CNY under SC3, respectively); this highlights the distinct adaptation strategies and outcomes of different industries under water constraints; (iii) AGR experiences the largest reduction in physical water consumption (6.85 $\times$ 10^9 m^3 under SC3); the water productivity of ENC, ADM, and SRV improves significantly under water constraints (e.g., ENC increasing by 3491.19 CNY/m^3 under SC3), whereas AGR and PRM show limited changes (e.g., AGR increasing by 12.31 CNY/m^3 under SC3); this suggests that incentive mechanisms should be introduced to improve water productivity across all sectors, particularly in low-efficiency sectors.

Acknowledgements This research was supported by the National Key R&D Program of China (2023YFC3206503) and the Innovative Research Group of the National Natural Science Foundation of China (52221003).

References

Cai S, Song X, Hu R, Leng P, Li X, Guo D, Zhang Y, Hao Y, Wang Y (2021) Spatiotemporal characteristics of agricultural droughts based on soil moisture data in Inner Mongolia from 1981 to 2019. J Hydrol 603:127104

Li Y, Lv Y, Zhang Z, Feng X, Chen X (2024) Coupling coordination evaluation of "Three Waters" system and impulse response analysis in the Yellow River Basin. Sustain Cities Soc 102:105174

Li X, Huang G, Wang S, Li Y, Zhang X, Luo B (2025) Development of an optimization-driven equilibrium model for managing water consumption and their socio-economic effects. Sustain Cities Soc 118:106038

Nematian J (2023) A two-stage stochastic fuzzy mixed-integer linear programming approach for water resource allocation under uncertainty in Ajabshir Qaleh Chay Dam. J Environ Inf 41(1):52–66

Nguyen PT (2023) Simulating the environmental and economic effects of a carbon tax in Vietnam: a static computable general equilibrium analysis. Manag Environ Qual Int J 34(6):1647–1667

Pradhan BK, Ghosh J (2022) A computable general equilibrium (CGE) assessment of technological progress and carbon pricing in India's green energy transition via furthering its renewable capacity. Energy Econ 106:105788

Raviv O, Palatnik RR, Castellini M, Gusperti C, Vergalli S, Sirota J, Shechter M (2024) Synergies of CGE and IAM modelling for climate change implications on WEFE nexus in the Mediterranean. Clim Risk Manag 44:100608

Solovey T, Janica R, Harasymchuk V, Przychodzkal M, Yanush L (2023) Numerical modeling of transboundary groundwater flow in the Bug and San catchment areas for integrated water resource management (Poland–Ukraine). J Environ Inform 42(2):123–142

Wang P, Huang G, Li Y, Zhang Y, Cai T, Song T (2024) Unveiling key drivers of economy-water system and transforming water use pattern into sustainable development: inner-Shaan-Ning region in the Yellow River Basin. J Clean Prod 475:143651

Zeng L, Zhang W, Zhao Y, Zhang J, Jiang X (2024) Simulation analysis of Chinese new-type urbanization policy in mineral resource abundant regions based on the CGE model. Resour Policy 90:104749

Zhang J, Liu L, Xie Y, Han D, Zhang Y, Li Z, Guo H (2023) Revealing the impact of an energy–water–carbon nexus–based joint tax management policy on the environ-economic system. Appl Energy 331:120397

Zhou A, Zhao W, Pereira P (2023) Mapping and assessing freshwater ecosystem services supply and demand in Inner Mongolia (China). Landscape Ecol 38(7):1885–1902

Chapter 13
Spatial Distribution and Loss Amplification Effect of Rainstorm-Induced Cascading Hazards in the Daqing River Basin, China

Xiangxiang Jiang, Wei Fang, Qiang Huang, Quan Quan, Shengce Jia, Jie Zhang, Jianan Shang, Xin Xu, and Liangbo Zhang

Abstract In recent years, extreme precipitation events have become increasingly frequent and widespread largely due to global warming and the urban rain island effect. Intense rainstorms often trigger secondary hazards such as floods, urban waterlogging, debris flows and landslides, resulting in multi-hazard scenarios and the loss amplification effect. These rainstorm-induced cascading hazards significantly snowball the challenges of hazard risk prevention and emergency response. This study focuses on the Daqing River basin in China and initially develops a historical hazard dataset of rainstorm and its secondary hazards across 47 county-level administrative units using internet data mining and generative AI techniques. Subsequently, the spatial distribution of diverse types of rainstorm-induced cascading hazards is analyzed. The loss amplification coefficient is finally calculated from six loss dimensions—affected population, fatalities and missing persons, emergency relocations, collapsed houses, affected crop area and direct economic losses—to quantitatively examine the loss amplification effect of rainstorm-induced cascading hazards. Results show that from 2010 to 2024, a total of 404 rainstorm-induced cascading hazard events occurred in the Daqing River basin. Among them, the rainstorm-urban waterlogging cascade was the most frequent (123 times), followed by the rainstorm-landslide cascade (121 times). The frequency of rainstorm-induced cascading hazards in the central and eastern parts of the basin was significantly higher than in the west, with a marked spatial concentration in Baoding City in the central basin. The loss amplification coefficients ranged from 1.67 to 4.62 for two-hazard cascades (e.g., rainstorm-urban waterlogging, rainstorm-flood, rainstorm-landslide and rainstorm-debris flow), and from 3.06 to 6.03 for three-hazard cascades (e.g., rainstorm-flood-landslide and rainstorm-flood-debris flow). The amplification effect

X. Jiang · W. Fang (✉) · Q. Huang · Q. Quan · S. Jia · J. Zhang · J. Shang · X. Xu · L. Zhang
State Key Laboratory of Water Engineering Ecology and Environment in Arid Area, Xi'an University of Technology, Xi'an, China
e-mail: fangwei@xaut.edu.cn

Y. Li et al. (eds.), *16th International Conference on Environmental Science and Development*, Environmental Science and Engineering,
https://doi.org/10.1007/978-3-032-15080-6_13

was particularly pronounced in rainstorm-landslide among two-hazard cascades and rainstorm-flood-debris flow cascades among three-hazard cascades.

Keywords Rainstorm-induced cascading hazards · Data mining · Generative AI · Hotspot analysis loss amplification effect

13.1 Introduction

In the context of global warming, climate extremes, especially extreme precipitation events, have become frequent and widespread (Liao et al. 2019; Chen et al. 2024; Cavus et al. 2021). Their secondary hazards such as floods (Li et al. 2024), debris flows, and landslides have also shown a trend of increasing occurrence frequency and expanding impact extent (Lai et al. 2023), forming a variety of rainstorm-induced cascading hazards with sequential or simultaneous emergence of multiple hazards. According to statistics from the Emergency Events Database (EM-DAT), compared with the last century, in the past 20 years, the average annual occurrence frequencies of floods, debris flows and landslides triggered by rainstorms have increased by 83.0, 65.3 and 68.1% respectively, and the average annual economic loss has expanded by as much as 4.5 times. Rainstorm-induced cascading hazards tend to result in adverse situations of 'multiple hazards in one place' and 'loss amplification effect' (Liu et al. 2025; Pachauri et al. 2014), which pose serious threats to human life and property safety, and thus have attracted serious attention from the international community (Change 2007; Winsemius et al. 2018).

Historical records of multiple intercorrelated hazards are an important basis for conducting cascading hazard research. Internationally well-known multi-hazard datasets include Nat Cat of Munich Re (Nat Cat) (Delforge et al. 2025), Sigma of Swiss Re (Sigma) (Wirtz et al. 2014), and the Emergency Events Database (EM-DAT) of Belgium (Yi et al. 2022). Liu et al. (2018) used text mining techniques to screen 1,187 news reports on rainstorm hazards during 2008 to 2017, and disclosed spatial distribution of rainstorms, landslides, collapses, and floods using a pattern matching method. Wan et al. (2018) collected historical records of floods and droughts in the Baoji of China from 1368 to 1911 and used moving average, cumulative anomaly, and wavelet analysis methods jointly to study the spatial evolution characteristics of regional drought and flood cascading hazards in the Ming and Qing dynasties. At present, although there is a relatively large number of studies on historical datasets of rainstorms and secondary hazards both domestically and internationally, there are two major shortcomings in the extensiveness of hazard data collection and the completeness of hazard feature extraction. Firstly, hazard data sources involved previously are typically single-source, fragmented and seriously lagged, making it difficult to obtain a large amount of timely data on rainstorms and secondary hazards. Second, the feature extraction methods for massive hazard achieves from emerging social media are relatively limited, making it not easy to quickly and accurately extract key hazard features such as hazard occurrence time, hazard intensity, and casualties.

These shortages jointly reduce the completeness and accuracy of cascading hazard datasets.

The loss amplification effect is also a hot topic in cascading hazard research. Cascading hazards are a complex hazard system, and the study of the loss amplification effect requires interaction analysis of different types of hazards that occur in succession or concurrently. Wang et al. (2024) elaborated on the close associations of urban floods and their secondary hazards at different spatial scales, as well as their synergistic evolution in time. Zhang et al. (2023) developed a quantitative assessment framework for human risks under cascading hazards mainly using Bayesian networks and dynamic vulnerability models. Assessment of loss amplification effects resulting from Wenchuan earthquake-landslide-flood cascades indicated that hazard risk can increase by a factor of 4 to 5 under cascading hazard scenarios. From the perspective of system dynamics, Tang et al. (2023) revealed the loss amplification effect of typhoon cascading hazards in Shenzhen and found that infrastructure vulnerability, rapid urbanization, and emergency response lag constituted a positive feedback, leading to nonlinear superposition of loss caused by secondary hazards. At present, the study of the loss amplification effect of cascading hazards remains limited, and the understanding of its fundamental concepts, evaluation metrics and methodologies diverges across research communities.

Therefore, the current study aims to develop a historical dataset of rainstorm and its secondary hazards in county-level administrative units and to disclose the spatial distribution and loss amplification effect of diverse rainstorm-induced cascading hazards in the Daqing River basin of China, where rainstorm hazards occur frequently. The specific objectives are to: (1) compile records of rainstorms and secondary hazards and extract key hazard features via internet data mining and generative AI techniques, (2) identify hotspot and frequent types of rainstorm-induced cascading hazards through spatial pattern analysis, and (3) systematically quantify loss amplification effect of rainstorm-induced cascading hazards across six loss dimensions including affected population, fatalities and missing persons, emergency relocations, collapsed houses, affected crop area and direct economic losses. The findings are expected to support more targeted and effective risk prevention and emergency response strategies under complicated cascading hazard scenarios.

13.2 Overview of the Study Area

The Daqing River (see Fig. 13.1), with the length of 483 km, is located in the central part of Hebei Province, with numerous tributaries originating from the Taihang mountain. The Daqing River basin (38°10′N–40°12′N, 113°39′E–117°34′E) stretches from the Taihang mountain in the west to the Bohai Bay in the east, bordered by the Yongding River in the north and the Ziya River in the south. As one of the five major rivers in the Haihe River basin (Chen et al. 2022), the basin area covers 4.3×104 km^2, accounting for 13.5% of the total area of the Haihe River basin (Xiao et al. 2023; Zhang et al. 2022). The Daqing River system is primarily composed of northern and

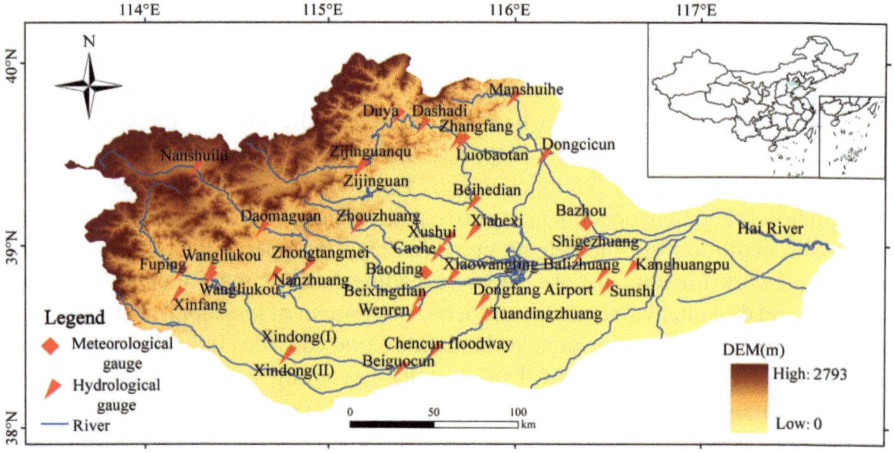

Fig. 13.1 The Daqing River basin

southern branches in the upper and middle reaches. The northern branch is the Juma River, which, after exiting the mountains at Tiesuoya, bifurcates at Luobaotan into the northern and southern Juma Rivers. The southern branch originates from the Taihang Mountains and mainly includes the Zhulong River, Tang River, Qingshui River, Fu River, Cao River, Pu River, Ping River and Xiaoyi River. Northern and southern branches all flow into Baiyangdian lake, then into the Eastern Dian lake via the New Zhaowang River, and finally reach the sea through the main stream of the Haihe River and the Duliujian River. The Daqing River basin belongs to a temperate semi-arid continental monsoon climate zone, with an average annual temperature of 11.6 °C. Precipitation varies greatly and is uneven throughout the year. The average annual precipitation ranges from 500 to 700 mm, with about 600 mm in mountainous areas and 500 mm in the plains. Heavy precipitation is concentrated in July and August, accounting for 80% of the annual total. Due to steep topography, thin soil cover, poor vegetation, and the fan-shaped distribution of many tributaries with short convergence time, floods rise and fall rapidly with concentrated water volume, making the area highly prone to rainstorm and secondary hazards such as floods, urban waterlogging, debris flows and landslides (Sun et al. 2021).

13.3 Methods

As depicted in Fig. 13.2, the study initially applies internet data mining and generative AI techniques to compile hazard records and extract key features such as occurrence time, geographic location, affected population and economic losses related to rainstorms and secondary hazards. Subsequently, the spatial distribution of rainstorm-induced cascading hazards is identified and the loss amplification coefficient is finally

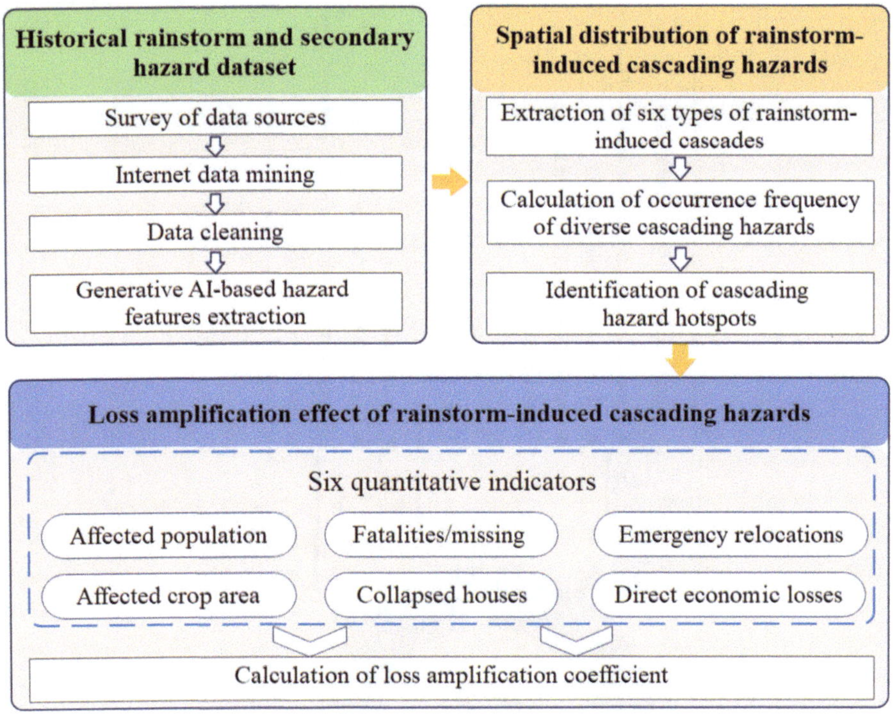

Fig. 13.2 Main procedure of cascading hazard hotspot and loss amplification effect analysis

estimated to reveal the amplification effect of cascading hazards in the Daqing River basin.

13.3.1 Four Data Sources and Rainstorm and Secondary Hazards

Using literature review, expert opinion and case studies, data sources of rainstorms and related secondary hazards were configured through literature review and expert survey and case study approaches. These sources mainly include four categories, i.e., hazard statistics published by government agencies (National Disaster Reduction Center of China, Haihe River Water Conservancy Commission of the Ministry of Water Resources, China Association for Disaster Prevention, provincial emergency management departments, flood and drought disaster prevention bulletins, etc.), international disaster databases (EM-DAT, Global Disaster Data Platform, 1981–2020 Global Concurrent Hazard Dataset), public news media (China News Service, Xinhua News, Toutiao News, Tencent News, Sohu News, etc.), and social media like Weibo. These four categories of data sources are also shown in Fig. 13.3.

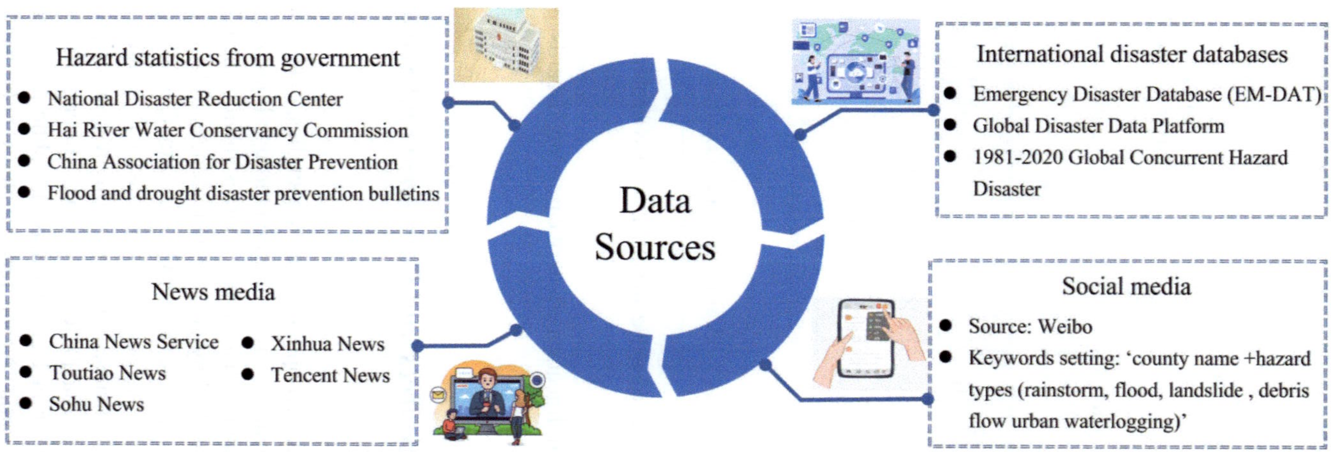

Fig. 13.3 Four categories of hazard data sources

13.3.2 Internet Data Mining Method for Rainstorm and Secondary Hazards

The compilation of hazard records through internet data mining begins with the formulation of appropriate search terms. The key search terms were set as combinations of 'county-level city name' and hazard type (rainstorm, flood, urban waterlogging, debris flow, landslide), such as 'Fangshan District Rainstorm'.

Then, an advanced search was conducted on the four major data sources (hazard statistics from government agencies, international disaster databases, news media, and social media) using Octoparse data crawler and the Python platform's Beautiful Soup, Requests, Pydantic and libraries. Keywords and the required time period were entered—for instance, 'Fangshan District Rainstorm' and '2010-01-01 00:00 to 2015-12-31 23:00'—to search and collect hazard-related records for Fangshan District within the specified time frame. Finally, based on the combination of county-level administrative unit codes and hazard categories, original hazard records related to rainstorms and secondary hazards were mined from the above data sources.

To address issues such as redundancy and lack of detailed data descriptions in the collected hazard records, data cleaning was performed. Specifically, a fuzzy string matching algorithm based on edit distance (Levenshtein Distance) was used to delete original records with similarity above 80%, reducing redundancy in hazard information.

$$Similarity = \left(1 - \frac{Levenshtein\ Distance}{\max(len(Rec\,1), len(Rec\,2))}\right) \times 100\% \tag{13.1}$$

where *Levenshtein Distance* is the edit distance measuring the degree of difference between two record strings, defined as the minimum number of single-character edit operations (insertion, deletion, substitution) required to transform one string into another, *Rec* 1 and *Rec* 2 refer to the original hazard records to be compared, and *len* represents the string length function.

13.3.3 Generative AI-Based Hazard Feature Extraction Method

After the data cleaning procedure, this study employed generative AI technology (specifically the SPARK large model developed by IFLYTEH) to extract key features of rainstorms and their secondary hazards. The specific process is as follows: First, the API interface parameters of the Spark Large Model (App_ID, App_Key, App_Secret) were configured to establish a connection channel with the model. Then, a question template was designed for hazard feature extraction. Questions are expressed as 'When and where did it happen?' 'How many people were affected?' and 'What is the total number of fatalities and missing persons?'. The response format is designed as ‘ < Occurrence time: to be filled > , < Number of people affected: to be filled > , < Fatalities and missing persons: to be filled > ’. The nine key features extracted include location, occurrence time, affected persons, fatalities persons, missing persons, emergency relocations, collapsed houses, damaged crop area, and direct economic losses. Next, multithreaded parallel processing on Python platform was carried out to extract hazard features efficiently from massive amounts of historical records collected. A dynamic delay strategy (random interval of 0.5 to 2 s) was adopted to ensure processing efficiency while effectively avoiding API call frequency limits. The extracted key hazard features were saved in a structured format into spreadsheets, and spatial visualization was carried out using geographic information system software ultimately acquiring a standardized dataset containing nine key features of rainstorm and closely related secondary hazards across 47 county-level administrative units of the Daqing River basin.

The temporal sequence of different hazards within the same county and the same hazard period were carefully analyzed to extract six types of rainstorm-induced cascading hazards (i.e., rainstorm-urban waterlogging, rainstorm-flood-landslide and so on). Then, the occurrence frequency of different types of rainstorm-induced cascading hazards in each county-level administrative unit was counted. Hotspots of rainstorm-induced cascading hazards in the Daqing River basin can be subsequently differentiated to support cascading hazard prevention and emergency response in an effective way.

13.3.4 Loss Amplification Coefficient of Rainstorm-Induced Cascading Hazards

Six hazard loss indicators—affected population, fatalities and missing persons, emergency relocations, collapsed houses, affected crop area and direct economic losses—are compared between single hazard and six types of rainstorm-induced cascading hazards to calculate the loss amplification coefficient, with the aim to quantify the loss amplification effect of cascading hazards.

Loss amplification effect can be interpreted that during the occurrence of a rainstorm-induced cascading hazards, the rainstorm hits a hazard-bearing body with initial vulnerability and causes corresponding socio-economic losses, further increasing the vulnerability of the hazard-bearing body. Then, the secondary hazards triggered by the rainstorm once again hit the increasingly vulnerable hazard-bearing body, causing further enlarged loss. When intercorrelated hazards occur successively, the vulnerability of the hazard-bearing body continues to escalate, and the hazard losses are amplified. Therefore, the losses caused by the rainstorm-induced cascading hazards are far greater than those caused by either the rainstorm alone or individual secondary hazard involved. The formula for the loss amplification coefficient of rainstorm-induced cascading hazards is as follows:

$$L = \frac{A_{loss} + B_{loss}}{A_{loss}} \tag{13.2}$$

where A_{loss} and B_{loss} refer to the various losses (number of people affected, deaths or missing, emergency relocation and resettlement numbers, crop hazard area, number of collapsed houses, direct economic losses, etc.) caused by the initial hazard A and the secondary hazard B (flood, urban waterlogging, debris flow, landslide), respectively, and L represents the loss amplification coefficient.

13.4 Results Analysis

13.4.1 Historical (2010–2024) Hazard Dataset of Rainstorms and Secondary Hazards in 47 County-Level Administrative Units of the Daqing River Basin

A large number of original hazard records on rainstorms and secondary hazards were obtained by applying internet data mining approach from four major data sources—government bulletins, international disaster databases, news information and social media. The original hazard records cover the period from January 1, 2010, to August 31, 2024, and 47 county-level administrative units of the Daqing River basin. Data mining results (Table 13.1) show that a total of 34,132 original hazard records related to rainstorms and secondary hazards in the Daqing River basin were collected during the very recent 15 years (2010–2024). Among them, social media hazard records total 32,198 entries, making it the most mined among the four data sources. The numbers of entries from government bulletins, international disaster databases, and news information are 874, 410 and 650 respectively. After data cleaning, 451 hazard entries were removed, resulting in a total of 33,681 valid hazard records. The valid hazard records from government bulletins, international disaster databases, news information, and social media are 874, 410, 636 and 31,761 entries, respectively.

From a sum of 33,681 valid hazard records, main hazard features such as occurrence time, geographic location, affected population, collapsed houses, and direct economic losses were extracted. According to the hazard category, their main features

Table 13.1 Valid records of rainstorm and secondary hazards after data cleaning

Data source	Original entries	Invalid entries	Valid entries
Government bulletins	874	0	874
International disaster databases	410	0	410
News media	650	14	636
Social media	32,198	437	31,761
Total	34,132	451	33,681

were classified and compiled to establish a four-level hazard dataset of 'Geographic location–Hazard type–Hazard features–Hazard linkage and occurrence sequence' for the 47 county-level units. Table 13.2 exemplifies the established rainstorm-related hazard dataset in Laishui County within the study area.

13.4.2 Hotspots of Rainstorm-Induced Cascading Hazards in the Daqing River Basin

The recent 15-year (2010–2024) statistics in Table 13.3 uncover a total of 404 rainstorm-induced cascading hazard events occurred in the Daqing River basin. Among them, the rainstorm-urban waterlogging, rainstorm-landslide, and rainstorm-debris flow cascading hazards occurred more frequently, with 123, 121 and 108 occurrences respectively. The rainstorm-flood, rainstorm-flood-debris flow, and rainstorm-flood-landslide cascading hazards were less frequent, which separately have occurrence frequency of 28, 19 and 5.

As depicted in Fig. 13.4, among the 47 county-level units in the Daqing River basin, 37 counties experienced rainstorm hazard events during the past 15 years. Fangshan District on the northern edge and Baoding City in the central part had the highest rainstorm cascade frequency up to 25 occurrences. Mancheng District and Anping County rank second and third, with 24 and 20 occurrences, respectively. No record was observed in ten counties such as Lingqiu County, Hunyuan County and Pingshan County, signifying the less likelihood of rainstorm cascade occurrence recently. Therefore, rainstorm-induced cascading hazards in the Daqing River basin occur more frequently in the central and eastern regions than in the west. Baoding City exhibits a notably higher spatial concentration and is identified as the primary hotspot for rainstorm-related cascading hazards within the study area.

Fu et al. (2025) developed a county-level dataset for rainstorm-flood cascading hazards in China, which revealed pronounced spatial heterogeneity of rainstorm-induced flood hazards across the country. In particular, within the Daqing River basin, county-level administrative units in the central-eastern regions experienced a notably higher frequency of rainstorm-flood cascading hazards compared to those in the western regions. This finding closely aligns with the results of the current study.

13.4.3 Loss Amplification Effect of Two- and Three-Hazard Cascades

The significant loss amplification effect of cascading hazards is quantitatively assessed by calculating the loss amplification coefficients of two-hazard cascades

Table 13.2 Historical dataset of rainstorm and secondary hazards in Laishui County

No.	Hazard category	Start date	End date	Fatalities persons	Missing persons	Affected persons	Emergency relocations	Affected crop area (thousand hectares)	Collapsed houses	Direct economic losses (million)
1	Rainstorm	2012 0725	2012 0727	33	23	/	/	/	/	/
3	Rainstorm	2012 0721	2012 0722	/	/	/	200	/	/	6.42
4	Rainstorm	2016 0710	2016 0710	/	/	/	/	/	/	5
5	Rainstorm	2017 0815	2017 0822	/	/	/	106	19.44	107	112
……	……	……	……	……	……	……	……	……	……	……
216	Flood	2013 0826	2013 0826	/	/	902	12,000	/	/	/
217	Flood	2015 0817	2015 0827	/	/	158,000	174,000	16	/	1700
218	Flood	2015 0716	2015 0720	/	/	/	8765	0.043	/	/
219	Flood	2015 0621	2015 0621	20	/	/	3000	/	/	/
220	Flood	2017 0716	2017 0718	/	/	/	3765	/	/	/

Table 13.3 Frequency of rainstorm-induced cascading hazards in the Daqing River basin

Cascading hazard category	Frequency
Rainstorm-urban waterlogging	123
Rainstorm-landslide	121
Rainstorm-debris flow	108
Rainstorm-flood	28
Rainstorm-flood-debris flow	19
Rainstorm-flood-landslide	5
Total	404

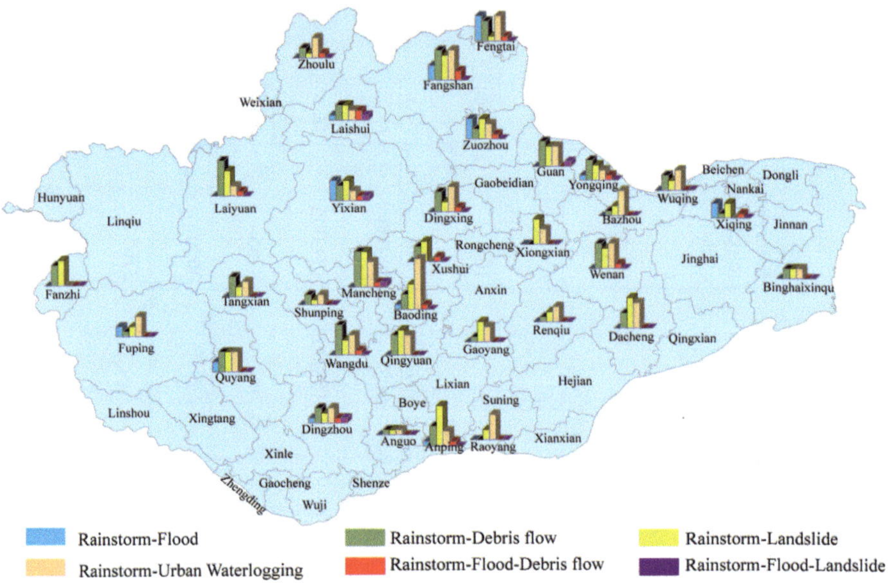

Fig. 13.4 Spatial distribution of rainstorm-induced cascading hazards at the county level

(rainstorm-urban waterlogging, rainstorm-flood, rainstorm-landslide and rainstorm-debris flow), and three-hazard cascades (rainstorm-flood-landslide, rainstorm-flood-debris flow). As shown in Table 13.4, the average loss amplification coefficients for rainstorm-flood, rainstorm-debris flow, rainstorm-urban waterlogging, rainstorm-landslide, rainstorm-flood-landslide and rainstorm-flood-debris flow cascading hazards are 2.39, 2.75, 2.76, 2.83, 3.94 and 4.40, respectively. It is noted that among two-hazard cascades, the rainstorm-landslide cascading hazard shows significantly higher loss amplification coefficients, particularly in affected Population (2.74), collapsed houses (3.06) and direct economic losses (2.08). As for three-hazard cascades, the rainstorm-flood-debris flow cascading hazard exhibits a higher average loss amplification coefficient, particularly noteworthy in terms of relocated population (6.03), collapsed houses (4.59) and direct economic losses (4.21).

Table 13.4 Loss amplification coefficients of rainstorm-induced cascading hazards

Cascading hazard category	Affected population	Fatalities/ missing	Emergency relocations	Collapsed houses	Affected crop area	Direct economic losses
Rainstorm-flood	2.24	3.33	2.23	2.42	2.06	2.03
Rainstorm-urban waterlogging	2.38	2.94	4.23	1.67	3.27	2.08
Rainstorm-debris flow	2.17	2.61	4.62	3.01	2.00	2.08
Rainstorm-landslide	2.74	3.20	3.17	3.06	2.70	2.08
Rainstorm-flood-debris flow	3.54	4.97	6.03	4.59	3.06	4.21
Rainstorm-flood-landslide	4.00	5.49	3.98	3.91	3.11	3.12

Therefore, compared with rainstorm events alone, rainstorm-induced cascading hazards show marked loss amplification effect of varying degrees across all six indicators. Moreover, the loss amplification coefficients of three-hazard cascades (3.06–6.03) are generally higher than those of two-hazard cascades (1.67–4.62). Two-hazard cascades present loss amplification coefficients approximately 168.3% higher than those of single hazards, while three-hazard cascades demonstrate an additional 55.5% increase compared to two-hazard cascades.

The present study employs a simplified method—specifically, six quantitative indicators—to assess the loss amplification effect of rainstorm-induced cascading hazards. In a recent study by Zhang et al. (2023), a five-stage probabilistic framework was developed to identify interaction mechanisms and the amplification effect of cascading hazards, which has the potential to predict losses and risks associated with various cascading hazards. Therefore, future research could benefit from sophisticated analytical techniques to enhance the reliability of the loss amplification effect and support the development of early-warning systems for mitigating cascading hazards.

13.5 Conclusion

This study investigates rainstorm-induced cascading hazards, which have been occurring with increasing frequency and attracting growing attention in the context of global warming. A historical hazard dataset covering the period from 2010 to 2024 was constructed for 47 county-level units, primarily through internet data mining and the application of generative AI techniques. In addition, spatial hotspots of rainstorm-induced cascading hazards and the consequent loss amplification effects were systematically analyzed. The main findings are summarized as follows:

(1) A total of 33,681 records on rainstorm and secondary hazards were collected, including 31,761, 874, 410 and 636 entries from social media, government bulletins, international disaster databases and news information. A four-level hazard dataset of 'Geographic location–Hazard type–Hazard features–Hazard linkage and occurrence sequence' covering 47 county-level units was established.
(2) From 2010 to 2024, a total of 404 rainstorm-induced cascading hazards occurred in the Daqing River basin. Rainstorm-urban waterlogging cascade is found to be most frequent (123 times), followed by the rainstorm-landslide cascade (121 times). The three-hazard cascades—rainstorm-flood-landslide (5 times) and rainstorm-flood-debris flow (19 times)—occurred relatively infrequently. Spatially, these cascading hazards were dominantly concentrated in the central and eastern portions, with Baoding City experiencing the highest frequency.
(3) The loss amplification coefficients ranged from 1.67 to 4.62 for two-hazard cascades and from 3.06 to 6.03 for three-hazard cascades. Two-hazard cascades result in loss roughly 168.3% higher than single hazards, while three-hazard cascades lead to an additional 55.5% increase compared with two-hazard cascades. The amplification effect was particularly pronounced in rainstorm-landslide among two-hazard cascades and rainstorm-flood-debris flow cascades among three-hazard cascades.

Acknowledgements This study was jointly funded by the National Key R&D Program of China (grant number 2023YFC3006502), the National Natural Science Foundation of China (grant number U2243216 and 52009100), China Postdoctoral Science Foundation (grant number 2021M692602) and the Open Research Fund of Yinshanbeilu Grassland Eco-hydrology National Observation and Research Station, China Institute of Water Resources and Hydropower Research (grant number YSS2022006).

References

Cavus Y, Eris E, Aksoy H et al (2021) Spatio-temporal analysis of precipitation-based drought indices in Kucuk Menderes River Basin, Turkey. EGU General Assembly Conference Abstracts, EGU21–1531

Change I P O C (2001) Climate change 2007: Impacts, adaptation and vulnerability. Genebra, Suíça

Chen L, Yang M, Liu Y et al (2022) Early warning and joint regulation of water quantity and quality in the Daqing River Basin. Water 14(19):3068

Chen X, Xu Y, Li T et al (2024) Regional rainfall damage functions to estimate direct economic losses in rainstorms: a case study of the 2016 extreme rainfall event in Hebei Province of China. Int J Disast Risk Sci 15(4):508–520

Delforge D, Wathelet V, Below R et al (2025) EM-DAT: the emergency events database. Int J Disast Risk Reduct 105509

Fu S, Schultz DM, Lyu H et al (2025) Creating a national urban flood dataset for China from news texts (2000–2022) at the county level. Hydrol Earth Syst Sci 29(3):767–783

Lai Y, Li J, Chen YD et al (2023) Compound floods in Hong Kong: hazards, triggers, and socio-economic consequences. J Hydrol Reg Stud 46:101321

Li R, Qi S, Wang Z et al (2024) Research on the heavy rainstorm-flash flood–debris flow disaster chain: a case study of the 'Haihe River '23 7'regional flood.' Remote Sensing 16(24):4802
Liao X, Xu W, Zhang J et al (2019) Global exposure to rainstorms and the contribution rates of climate change and population change. Sci Total Environ 663:644–653
Liu X, Guo H, Lin Y et al (2018) Analyzing spatial-temporal distribution of natural hazards in China by mining news sources. Nat Hazard Rev 19(3):04018006
Liu B, Guo H, Wang H (2025) An AI-driven approach to extract interrelationships between disasters. Int J Disast Risk Reduct 121:105417
Pachauri RK, Allen MR, Barros VR et al (2014) Climate change 2014: synthesis report. Contribution of Working Groups I, II and III to the fifth assessment report of the Intergovernmental Panel on Climate Change. IPCC
Sun T, Cheng W, Bo Q et al (2021) Analysis on historical flood and countermeasures in prevention and control of flood in Daqing River Basin. Environ Res 196:110895
Tang P, Zhong W, Wen J et al (2023) Developing and understanding cascading effects scenario of typhoons in coastal mega-cities from system perspectives for disaster risk reduction: a case study of shenzhen, China. Int J Disast Risk Reduct 92:103691
Wan H, Song H, Zhu C et al (2018) Spatio-temporal evolution of drought and flood disaster chains in Baoji area from 1368 to 1911. J Geog Sci 28:337–350
Wang S, Li R, Wu H (2024) Integrated expression and analysis of urban flood disaster events from the perspective of multi-spatial semantic fusion. Int J Appl Earth Obs Geoinf 132:104032
Winsemius HC, Jongman B, Veldkamp TIE et al (2018) Disaster risk, climate change, and poverty: assessing the global exposure of poor people to floods and droughts. Environ Dev Econ 23(3):328–348
Wirtz A, Kron W, Löw P et al (2014) The need for data: natural disasters and the challenges of database management. Nat Hazards 70(1):135–157
Xiao H, Zhang J, Xu M (2023) Study on temporal variability of hydrological elements in the Daqing River Basin China. J Hydrol Eng 28(11):04023032
Yi W, Yang S, Zhang L et al (2022) Comparative analysis and outlook of three global databases for meteorological disasters. Adv Clim Chang Res 18(2):253
Zhang J, Yang F, Fang H et al (2022) Study on spatial variability of rainfall-runoff and rainstorm-flood relationships in mountainous area of the Daqing River Basin. J Hydrol Eng 27(7):05022005
Zhang S, Wang B, Zhang L et al (2023) Increased human risk caused by cascading hazards–A framework. Sci Total Environ 857:159308

Chapter 14
An Input–output Analysis CNN-LSTM Model for Analyzing the Virtual Water Consumption

Shizhuo Li, Yongping Li, Guohe Huang, Panpan Wang, and Yanfeng Li

Abstract Comprehensively identifying key nodes in virtual water network and predicting variations in virtual water characteristics remains a significant challenge. In this study, an input–output analysis CNN-LSTM (IOACL) model is developed to identify the sources of virtual water consumption within the industrial chain and predict future water usage. IOACL is applied to Ningxia, one of the most water-scarce provinces in China. The major findings are as follows: (i) agriculture, food & tobacco and construction have high virtual water consumption, collectively accounting for 70% of the total virtual water consumption; in terms of physical water consumption, agriculture, construction and metal smelting are major contributors, together representing 94% of the total physical water consumption; (ii) food & tobacco, textiles and gas supply exhibit higher virtual water consumption compared to physical water consumption, which indicates that their supply chains contain large amounts of virtual water; (iii) compared to LSTM, the CNN-LSTM model demonstrates superior performance in predicting water consumption in Ningxia, achieving an R^2 value of over 0.9.

Keywords Input–output analysis · Machine learning · Virtual water

14.1 Introduction

As a fundamental natural resource essential for human survival and sustainable economic and social development, water is intricately linked to the social and economic progress. This is particularly evident in less developed regions, where an

---

S. Li · Y. Li (✉) · G. Huang · P. Wang · Y. Li
State Key Laboratory of Regional Environment and Sustainability, China-Canada Center for Energy, Environment and Ecology Research, UR-BNU, School of Environment, Beijing Normal University, Beijing, China
e-mail: yongping.li@iseis.org

Y. Li · G. Huang
Environmental Systems Engineering Program, University of Regina, Regina, SK, Canada

Y. Li et al. (eds.), *16th International Conference on Environmental Science and Development*, Environmental Science and Engineering,
https://doi.org/10.1007/978-3-032-15080-6_14

irrational industrial structure exacerbates the mismatch between the water resource endowment and production activity (Luo et al. 2024). Additionally, the increasing frequency of extreme hydrological events driven by climate change poses significant risks to water security, leading to cyclical and regional shortages (Sturla et al. 2023). Consequently, it is crucial to analyze and trace implicit water consumption within the industrial chain to effectively predict future regional water usage and mitigate potential water crises.

Input–output analysis (IOA) serves not only as a tool for examining economic interactions across sectors but also enables the tracing of material flows through supply chains (Marques et al. 2013). This methodology has been extensively applied in environmental studies to quantify embedded resources or ecological components. The ecological input–output CGE (EIOC) model developed by Wang et al. (2024a) demonstrated its utility in simulating CO_2 emission trajectories under prospective carbon taxation policies. In terms of water resources, IOA has proven effective in calculating virtual water consumption embedded within supply chains, thereby facilitating sectoral analysis of water usage (Wan et al. 2025). However, while IOA excels at deconstructing the origins of water consumption, its capacity for directly forecasting future water demand remains limited. The CNN-LSTM framework, integrating convolutional neural networks (CNN) for spatial feature extraction and long short-term memory networks (LSTM) for temporal dependency modeling, has demonstrated superior predictive accuracy in diverse domains including economics, hydrology, meteorology, and transportation (Shi et al. 2024). A notable application was presented by Cao et al. (2022), whose CNN-LSTM ensemble model provided early warnings for cyanobacterial harmful algal blooms (CyanoHABs) by predicting lake eutrophication patterns. Similarly, Li et al. (2024) leveraged CNN-LSTM to forecast new energy vehicle sales and charging infrastructure demands, identifying critical gaps in future resource allocation. Despite these advancements, the integration of IOA-derived virtual water flow analysis with CNN-LSTM architectures for regional water consumption forecasting remains underexplored.

To address this gap, an input–output analysis CNN-LSTM (IOACL) model is proposed, offering a novel methodological synthesis for water resource assessment and projection. The IOACL model integrates IOA and CNN-LSTM within an analytical structure, with empirical validation conducted in Ningxia, a hyper-arid region of China, to demonstrate its operational feasibility. Key contributions of this research include: (i) systematic tracking of virtual water flows across industrial sectors through supply chain, enabling identification of critical water-intensive nodes; (ii) strategic linkage of IOA-derived water consumption patterns to CNN-LSTM feature variables, establishing a factor-informed predictive architecture.

14.2 Methodology

Input–output analysis (IOA) can be employed to calculate the embodied water consumption flows among various sectors. Specifically, the calculation process is as follows (Zhang et al. 2014):

$$\varepsilon = \mathrm{P}[\mathrm{U} - \mathrm{H}]^{-1} \tag{14.1}$$

$$M = diag(\varepsilon)\mathrm{H} \tag{14.2}$$

$$W = diag(\varepsilon)[\mathrm{I} - \mathrm{A}]^{-1} diag(Y) \tag{14.3}$$

where **P** represents the physical water consumed by various sectors; **U** represents the economic output of various sectors; **H** represents the value flow matrix; ε represents the water consumption intensity embedded in the outputs of diverse industrial sectors; **M** represents inter-sectoral water flows triggered by intermediate input transactions within the supply chain network; **I** represents the unitary matrix; **A** represents the direct consumption coefficient matrix; **Y** represents the final consumption; **W** represents the virtual water consumed by various sectors.

The CNN-LSTM architecture synergistically integrates convolutional neural networks (CNNs) and long short-term memory networks (LSTM) for spatiotemporal feature extraction in time series analysis (Dehghani, et al. 2023; Farhangi et al. 2023). In this study, source of virtual water consumption in Ningxia is identified according to IOA analysis (Wang et al. 2024b), and then the CNN-LSTM model is used to learn the relationship between industrial water demand and various influencing factors in Ningxia, so as to build an industrial water demand prediction model of Ningxia.

14.3 Study Area and Data

The proposed IOACL model is applied to analyzing the future industrial water consumption for Ningxia province, which is one of the most water-scarce provinces in China. The Ningxia had a population of more than 7.2 million and the gross domestic product (GDP) of 531.4 billion RMB in 2023. The water usage structure in the Ningxia region is not sustainable. Agricultural irrigation accounts for 93.1% of total water consumption, compared to the national average of 68% (Xu et al. 2021). In contrast, industrial water usage constitutes only 4.5%, significantly lower than both the national average and that of the Yellow River basin (Gao et al. 2023). Thus, it is essential to develop robust methods for exploring future industrial water consumption in order to effectively manage the structure of water usage. In this study, the input–output table in 2017 (from Ningxia's Bureau of Statistics) is utilized because it is the most recent year for which input–output data from Ningxia is available. Additionally, water consumption is aggregated into 38 sectors: S1 represents the primary industry, S2 to S25 represent the secondary industry, and S26 to S38 represent the tertiary industry. The Ningxia Water Resources Bulletin offers comprehensive data on water resources, while the Ningxia Statistical Yearbook presents pertinent economic and social data. The time series distribution of machine learning training data spans from 1998 to 2022.

14.4 Result and Discussion

Figure 14.1 presents the sectoral water consumption, encompassing both physical and virtual water. From the perspective of physical water use, S1, S14 and S25 are significant contributors, collectively accounting for 94% of the total physical water consumption across all departments. In contrast, when examining virtual water use, S1, S6, S7, S12, S25, and S28 are representative departments with high virtual water consumption, accounting for 95% of the total virtual water consumption. Notably, both S1 (agriculture) and S25 (construction) are identified as high water-consuming industries across the whole industry, regardless of whether physical water or virtual water is considered. Furthermore, when accounting for hidden water consumption throughout the entire supply chain, the virtual water consumption of S25 is 17 times greater than its physical water consumption. This indicates that the water consumption in this sector is heavily reliant on the indirect water footprint associated with both upstream and downstream of the industrial chain. For instance, the supply of metal smelting, non-metallic mineral-based materials, and chemical derivatives involves substantial amounts of virtual water.

In addition, S6 (food & tobacco), S7 and S8 (textiles), S11 (oil-based, coke derivatives and nuclear fuel processing outputs), S12 (chemical derivatives), S23 (gas production and supply), S28 (accommodation & catering), S29 (information technology services), S37 (health and public administration) exhibit significantly higher virtual water consumption compared to their physical water consumption. This indicates that a substantial amount of virtual water is embedded in the supply chains of these sectors. And the hidden water associated with S29 (information technology services) is particularly significant, as servers consume an extremely high amount of power, which is indirectly linked to the virtual water used in coal-fired power generation. Notably, The virtual water consumption of S3 (petroleum and natural gas extraction), S4 and S5 (metallic and non-metallic mineral resource extraction), S9 (wood processing goods), S15 (metal products), S16 (standard and specialized machinery production), S18 (electrical systems and apparatus), S19 (telecommunications devices, computing systems and other electronic apparatus), S24 (water production and supply), S26 (wholesale & retail), S32 (leasing and business services) is negative, because this study considers the final demand for imports in the input–output analysis, indicating that the products from these sectors are primarily imported, resulting in a net inflow of virtual water.

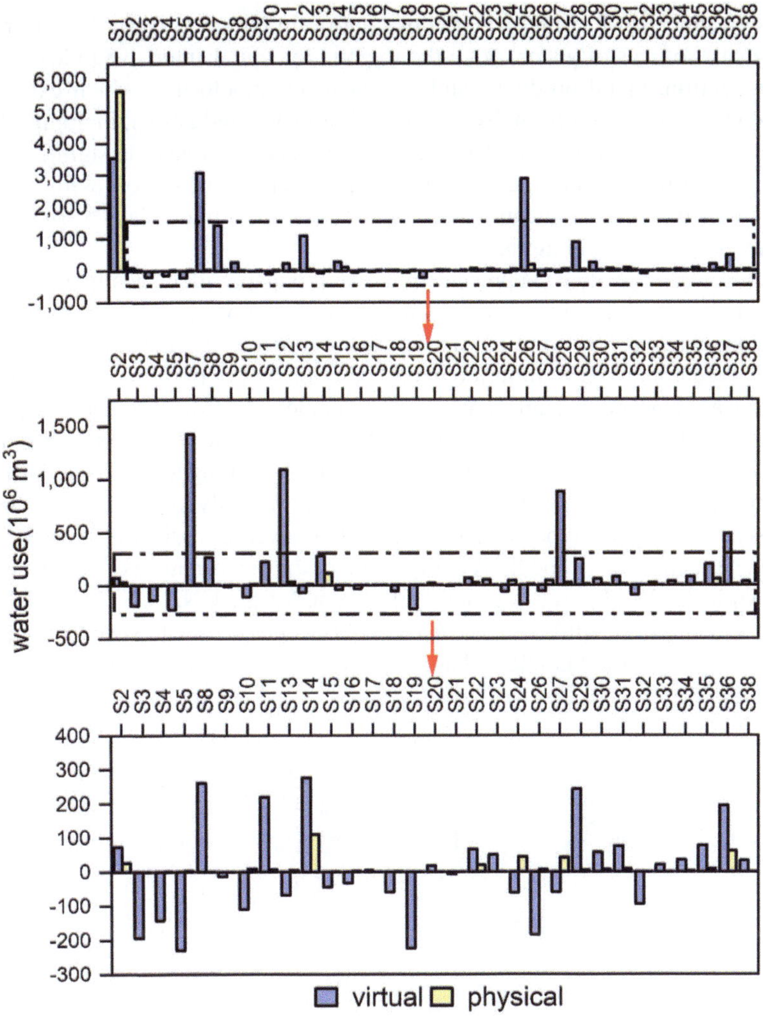

Fig. 14.1 Sectoral virtual and physical water consumption

Figure 14.2 shows the virtual water transfers among various sectors. Virtual water consumption resulting from inter-sectoral transactions exhibits an uneven distribution. For example, S1's virtual water mainly exported to S6, S7 and S28 (i.e., 2.43×10^9 m^3); S14's virtual water mainly exported to S18 and S25 (i.e., 0.12×10^9 m^3); S5's virtual water mainly exported to S25 (i.e., 0.11×10^9 m^3). The results indicate that agricultural industries supply a significant amount of virtual water-intensive products to support sectors such as food & tobacco, textiles and accommodation & catering. Additionally, the metal smelting and mining industries provide plenty of virtual water-intensive products for electrical machinery and construction. Based on the above analysis of virtual water consumption in Ningxia, it can be inferred that the primary sources of water use in the region are capital-intensive industries, such as construction, metal products, and chemical manufacturing, which exhibit high water consumption. Additionally, consumption-driven industries, including food & tobacco, textiles and accommodation & catering, also demonstrate significant water usage. Furthermore, there are numerous import-dependent sectors with a high net inflow of virtual water, indicating that import and export trade plays a crucial role in the water resources of Ningxia.

According to the analysis presented above, this paper identifies five core factors influencing water consumption in Ningxia: water consumption intensity, capital formation, resident consumption, imports and exports. The CNN-LSTM and LSTM models were employed to train the historical data, and optimal model performance for both was achieved through continuous parameter adjustments (specifically, maximizing R^2 while minimizing RMSE and MAE). The loss curve of the selected optimal model is illustrated in Fig. 14.3, which includes both training loss and validation loss. The figure demonstrates that the loss decreases continuously as the number of training increases, and the two types of loss ultimately converge, indicating that the model training has achieved a good fit. A comparison of the performance of the two models, as shown in Table 14.1, reveals that the R^2 value for the CNN-LSTM model on the test set is the highest, while its RMSE and MAE values are the lowest. Consequently, the CNN-LSTM model is selected as the predictive model for water consumption in

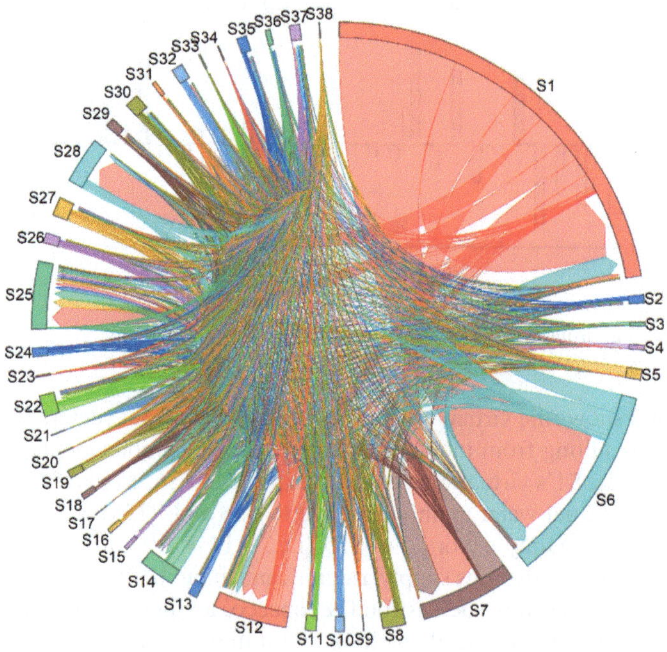

Fig. 14.2 Virtual water transfers among different sectors

(a)

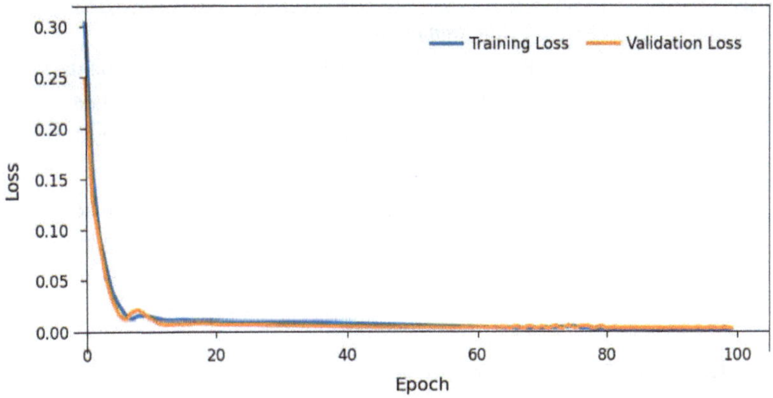

(b)

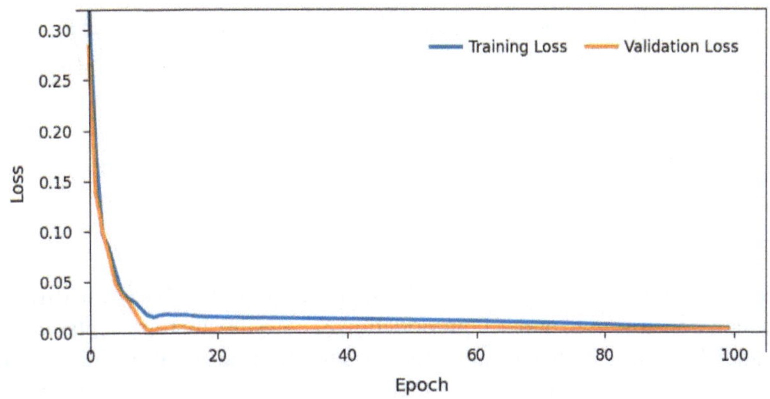

Fig. 14.3 Training and validation loss curve. *Notes* **a** CNN-LSTM model loss curve and **b** LSTM mode loss curve

Ningxia. The parameter settings for the CNN-LSTM model are detailed in Table 14.2, which includes two convolutional layers.

14.5 Conclusions

This study proposes an input–output analysis CNN-LSTM (IOACL) model to analyze the virtual water consumption system. The IOACL model excels in revealing the sources of water usage and predicting future water consumption from both mesoscopic and statistical perspectives. IOACL has been applied to the Ningxia region.

Table 14.1 Model performance

	CNN-LSTM	LSTM
Calibration period		
R^2	0.98	0.90
RMSE	1.74	3.66
MAE	1.47	2.03
Validation period		
R^2	0.90	0.86
RMSE	2.66	3.27
MAE	2.19	2.40

Table 14.2 Established CNN-LSTM architecture

Layer	Structure	Parameter
Conv1D1th	filters = 128, kernel_size = 1, activation = selu	768
Conv1D2nd	filters = 64, kernel_size = 1, activation = selu	8256
LSTM	Units = 100, activation = relu	65,600
Dense	Dense = 1	101
Optimizer	Adam (learning_rate = 0.001)	0
Epochs	100	0
batch_size	6	0

The major findings are as follows: (i) Agriculture, food & tobacco, textile, chemical product, construction, and accommodation & catering are representative departments, and they have high virtual water consumption, accounting for 95% of the total virtual water consumption; (ii) there are several sectors that exhibit higher virtual water usage compared to physical water usage, including food & tobacco, textiles, chemicals, gas supply, accommodation & catering, IT services. This indicates that their supply chains contain large amounts of virtual water; (iii) the following sectors exhibit significant negative virtual water consumption: mining, metal products, manufacturing, water supply, wholesale & retail. This phenomenon occurs because the products from these sectors are primarily imported, resulting in a net inflow of virtual water; (iv) agricultural supplied a substantial quantity of virtual water-intensive products to support sectors such as food & tobacco, textiles and accommodation & catering. Additionally, the metal smelting and mining industries provided plenty of virtual water-intensive products for electrical machinery and construction; (v) compared to LSTM, the CNN-LSTM model demonstrates superior performance in predicting water consumption in Ningxia, achieving an R^2 value of over 0.9.

Acknowledgements This research was supported by the National Key R&D Program of China (2023YFC3206503) and the Innovative Research Group of the National Natural Science Foundation of China (52221003).

References

Cao H, Han L, Li L (2022) A deep learning method for cyanobacterial harmful algae blooms prediction in Taihu Lake China. Harmful Algae 113:102189

Dehghani A, Moazam HMZH et al (2023) Comparative evaluation of LSTM, CNN, and ConvLSTM for hourly short-term streamflow forecasting using deep learning approaches. Ecol Inf 75:102119

Farhangi F, Sadeghi-Niaraki A, Safari Bazargani J et al (2023) Time-series hourly sea surface temperature prediction using deep neural network models. J Mar Sci Eng 11(6):1136

Gao H, Liu X, Wei L et al (2023) Dynamic simulation of the water-energy-food nexus (WEFN) based on a new nexus in arid zone: a case study in Ningxia China. Sci Total Environ 898:165593

Li B, Lv X, Chen J (2024) Demand and supply gap analysis of Chinese new energy vehicle charging infrastructure: based on CNN-LSTM prediction model. Renew Energy 220:119618

Luo Z, Ji L, Xie Y (2024) Multi-dimensional equilibrium optimal allocation of water resources in Beijing-Tianjin-Hebei region through spatial association network analysis. J Hydrol 642:131826

Marques A, Rodrigues J, Domingos T (2013) International trade and the geographical separation between income and enabled carbon emissions. Ecol Econ 89:162–169

Shi H, Wei A, Xu X et al (2024) A CNN-LSTM based deep learning model with high accuracy and robustness for carbon price forecasting: a case of Shenzhen's carbon market in China. J Environ Manage 352:120131

Sturla G, Ciulla L, Rocchi B (2023) Natural and social scarcity in water footprint: a multiregional input–output analysis for Italy. Ecol Ind 147:109981

Wan X, Tian G, Xia Q et al (2025) Spatial and temporal evolution of the physical-virtual water cycle and its economic coupling effects in China: a material flow analysis perspective. J Clean Product 145093

Wang PP, Huang GH, Li YP et al (2024a) An ecological input-output CGE model for unveiling CO_2 emission metabolism under China's dual carbon goals. Appl Energy 365:123277

Wang PP, Huang GH, Li YP et al (2024b) Unveiling key drivers of economy-water system and transforming water use pattern into sustainable development: Inner-Shaan-Ning region in the Yellow River Basin. J Clean Prod 475:143651

Xu J, Xiao Y, Xie G et al (2021) How to coordinate cross-regional water resource relationship by integrating water supply services flow and interregional ecological compensation. Ecol Ind 126:107595

Zhang Y, Zheng H, Fath BD et al (2014) Ecological network analysis of an urban metabolic system based on input–output tables: model development and case study for Beijing. Sci Total Environ 468:642–653

Chapter 15
Enhancing Hydrological Modeling Through an Integrated SWAT-VBMC-FA Framework: Parameter Uncertainty and Interaction Effects in the Wuding River Basin

Zhipeng Xu, Yongping Li, Guohe Huang, and Yanfeng Li

Abstract Hydrological models are essential for simulating watershed processes, yet uncertainties in parameters often lead to discrepancies between simulations and observations. This study introduced an integrated framework coupling the Soil and Water Assessment Tool (SWAT) with Variational Bayesian Monte Carlo (VBMC) and Factorial Analysis (FA) to address parameter uncertainty in hydrological modeling. The framework not only obtains the parameter posterior distributions at a low computational cost, but also quantifies the independence and interaction of the parameter factors. An application of the framework is made to the Wuding River basin in the Jiziwan region of the Yellow River to analyze the effects of parameter uncertainty. Results indicate: (i) SWAT model can accurately simulate runoff processes in the study area, with NSE values of 0.67 (calibration) and 0.63 (validation); (ii) posterior distributions showed cn2 with low uncertainty versus latq_co with higher uncertainty; (iii) factorial analysis identified cn2 as the primary driver of increased mean and peak runoff, moderated by interactions with awc and ovn, respectively. This framework enhances computational efficiency and parameter identifiability, offering a rigorous approach to uncertainty characterization and improved hydrological predictions in complex scenarios.

Keywords Factorial analysis · SWAT · Variational Bayesian Monte Carlo

---

Z. Xu · Y. Li (✉) · G. Huang · Y. Li
State Key Laboratory of Regional Environment and Sustainability, China-Canada Center for Energy, Environment and Ecology Research, UR-BNU, School of Environment, Beijing Normal University, Beijing, China
e-mail: yongping.li@iseis.org

Y. Li · G. Huang
Environmental Systems Engineering Program, University of Regina, Regina, SK, Canada

Y. Li et al. (eds.), *16th International Conference on Environmental Science and Development*, Environmental Science and Engineering,
https://doi.org/10.1007/978-3-032-15080-6_15

15.1 Introduction

Hydrological models are vital tools for studying hydrological cycles by conceptualizing and abstracting watershed processes (Edwards et al. 2015). These models use simplified mathematical formulations to represent spatiotemporal variations in water resource components and to evaluate watershed hydrological characteristics. However, no hydrological model achieves full physical accuracy due to inherent uncertainties in practical applications, resulting in discrepancies between simulated and observed data (Zhong et al. 2024). Parameters, as key model components, integrate watershed features such as vegetation cover, soil properties, and geological characteristics, significantly influencing runoff generation mechanisms. Consequently, systematic quantification of parameter uncertainties and their impacts on modeling outputs is essential (Herrera et al. 2022).

Considerable research has focused on analyzing parameter uncertainty in hydrological modeling. Advances in computational techniques have spurred the development of optimization algorithms to reduce uncertainty, including Generalized Likelihood Uncertainty Estimation (GLUE), Parameter Solution (ParaSol), and Markov Chain Monte Carlo (MCMC) methods (Rudolph et al. 2024; Zhao et al. 2024; Nikita et al. 2018; Shi et al. 2023). While informal and formal Bayesian approaches are widely used, challenges arise when applying MCMC sampling to large datasets or complex likelihood models. These limitations have prompted the adoption of Variational Inference (VI), a machine learning technique that approximates probability densities through optimization. VI offers distinct advantages over MCMC in computationally demanding models or big data contexts (Loaiza-Maya et al. 2022). Recently, Variational Bayesian Monte Carlo (VBMC) has emerged as an innovative inference framework, combining VI with Gaussian Processes (GP) and Bayesian quadrature via active sampling (Acerbi 2018). Although VBMC effectively captures complex posterior features at moderate computational cost, its use in hydrological studies remains limited. Furthermore, current methods often overlook parameter interaction effects, a critical factor in hydrological cycle simulations given the interdependent nature of watershed processes (Wang et al. 2021).

Therefore, this study develops an integrated framework coupling the Soil and Water Assessment Tool (SWAT) with VBMC and factorial analysis. This novel approach enables simultaneous achievement of two critical objectives: (i) efficient computation of parameter posterior distributions, and (ii) systematic evaluation of independent and interactive parameter effects. The framework demonstrates significant potential for enhancing parameter identifiability and improving hydrological model construction efficiency through comprehensive uncertainty characterization.

15.2 Methodology

In this study, an integrated framework coupling SWAT, VBMC, and Factorial Analysis (FA) was developed to analyze the effects of parameter uncertainty on hydrological responses and to quantify both independent and interactive parameter effects. Specifically, the SWAT model was first constructed using geographical, meteorological, and hydrological data from the study area, and sensitivity parameters were selected using the LH-OAT method. Subsequently, the VBMC method was employed to calculate the posterior distributions of the selected sensitivity parameters. Finally, factorial analysis was used to quantify the independent and interactive effects of parameters on runoff.

The SWAT model simulates long-term hydrological processes at watershed scales by integrating meteorological data, soil properties, topography, vegetation dynamics, and land management practices. Spatial heterogeneity is addressed through subdivision of the watershed into hydrologic response units (HRUs) and subbasins, ensuring computationally efficient simulations while maintaining physical realism. Model inputs were derived from geographic datasets, including DEM, soil maps, land use/cover classifications, and meteorological observations. Parameter sensitivity was evaluated using the Latin Hypercube One-factor-At-a-Time (LH-OAT) method, which combines Latin hypercube sampling with local sensitivity analysis. Sensitivity rankings were determined based on t-statistics and p-values, with parameters exhibiting higher absolute t-values and lower p-values identified as dominant controls on runoff generation. The top nine sensitive parameters were selected for subsequent uncertainty quantification.

VBMC, an advanced Bayesian inference framework, was employed to approximate parameter posterior distributions by synergizing variational inference (VI) with Gaussian process (GP) surrogates and Bayesian quadrature. This approach leverages GP regression to approximate the likelihood function and employs active sampling to strategically select evaluation points, significantly reducing computational costs while maintaining accuracy. The likelihood function $\mathcal{L}(\theta)$ is defined as:

$$\mathcal{L}(\theta) \sim \mathrm{GP}\big(\mu(\theta), k(\theta, \theta') + \sigma_n^2\big) \tag{15.1}$$

where $\mu(\theta)$ represents the mean function, $k(\theta, \theta')$ is a Matérn covariance kernel, and σ_n^2 accounts for observational noise. Unlike traditional Markov Chain Monte Carlo (MCMC) methods that rely on iterative sampling, VBMC constructs a parameterized variational distribution through optimization, reducing the Kullback–Leibler (KL) divergence from the true posterior.. The KL function is as follows:

$$KL\big[q(\theta)\|p(\theta|y)\big] = \mathbb{E}_{q(\theta)}\left[\log\frac{q(\theta)}{p(\theta)p(y|\theta)}\right] \tag{15.2}$$

where $p(\theta)$ denotes the prior distribution, and $p(y|\theta)$ is the likelihood. VBMC iteratively refines the GP surrogate through active sampling, balancing exploration and exploitation to achieve computational efficiency. In this study, VBMC was implemented using 10,000 iterations with a GP prior, enabling efficient identification of posterior modes and uncertainty bounds for the selected parameters.

A three-level full factorial design was adopted to systematically quantify independent and interactive effects of the five selected parameters on runoff simulations. Each parameter was assigned three levels representing low, medium, and high values within their physically plausible ranges. The experimental matrix comprised 243 combinations, with SWAT simulations executed for each scenario. The response Q was modeled as a polynomial function:

$$Q = \beta_0 + \sum_{i=1}^{5} \beta_i x_i + \sum_{1 \le i < j \le 5} \beta_{ij} x_i x_j + \in \tag{15.3}$$

where $x_i \in \{-1, 0, +1\}$ denotes normalized parameter levels (low, medium, high), β_i represents main effects, β_{ij} quantifies two-way interactions, and $\in \sim N(0, \sigma^2)$ is Gaussian noise. Regression coefficients (β) were estimated via least squares, with effect magnitudes ranked by standardized coefficients.

15.3 Study Area and Data

The Wuding River is located in northern Shaanxi Province, China, and is a major tributary of the Yellow River (Fig. 15.1). The river flows through the Loess Plateau, covering a drainage area of approximately 30,261 km^2. The basin exhibits a temperate semi-arid continental climate. Its average yearly temperature falls between 7.6–9.9 °C, with annual precipitation spanning 350–430 mm (Tian et al. 2019). Precipitation is unevenly distributed throughout the year, with the majority occurring during the summer months (June to September). The river experiences significant seasonal variations in flow, with peak runoff occurring during the summer and reduced flow during the winter. Historical records show that the river's runoff has exhibited significant interannual variability, with a notable decreasing trend from 1975 to 2020. In recent years, the basin has faced challenges such as water resource shortages and ecological degradation, driven by climate change and human activities.

The DEM data were sourced from the Geospatial Data Cloud platform, utilizing the publicly available SRTM (Shuttle Radar Topography Mission) terrain product with a spatial resolution of 90 × 90 m. Meteorological data were obtained from the China Meteorological Data Service Center, covering the period from 2004 to 2019 at a daily temporal resolution. Land use data came from the IGSNRR, part of the Chinese Academy of Sciences, with a spatial resolution of 30 m for the year 2015. Soil data were derived from the Harmonized World Soil Database (HWSD), which provides global soil information at a standardized spatial resolution.

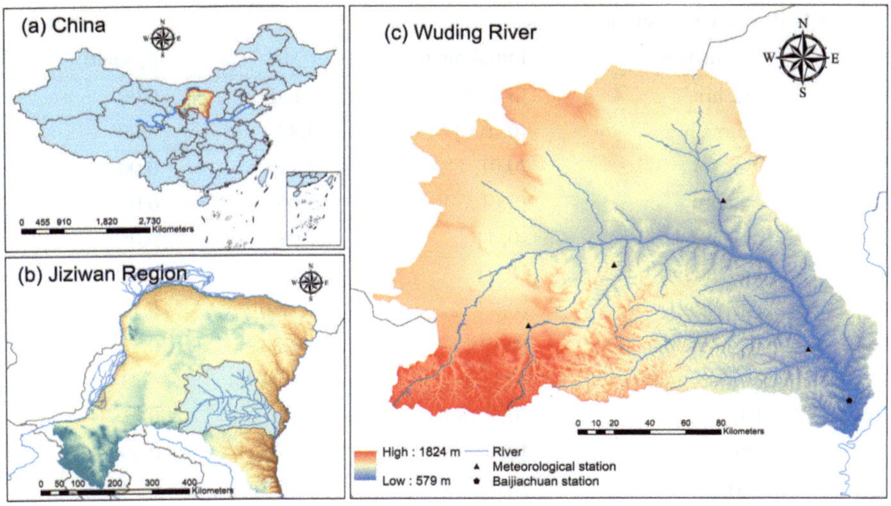

Fig. 15.1 Study area

15.4 Result and Discussion

The SWAT model was constructed using geographic, meteorological, and hydrological data from the Wuding River basin. Parameter sensitivity analysis was performed with the LH-OAT (Latin Hypercube One-factor-At-a-Time) method, with results presented in Table 15.1. Among the 21 parameters analyzed, the top nine (cn2, canmx., awc, ovn.hru, alpha, latq_co, epco, flo_min, and bf_max) exhibited high sensitivity, indicated by large absolute t-values and p-values less than 0.25. These findings highlight their significant influence on runoff generation. For instance, cn2.hru (SCS curve number) showed a t-value of −6.07 and a p-value of 0.00, suggesting strong control over surface runoff simulation. Similarly, canmx (maximum canopy storage) yielded a t-value of 4.82 and a p-value of 0.00, underscoring the regulatory role of vegetation in hydrological processes. Additionally, the AR1 parameter was incorporated to account for potential first-order autocorrelation in residuals, enhancing the statistical reliability of the model.

Posterior distributions of the selected nine sensitive parameters (bolded parameters in Table 15.1) and the AR1 parameter were calculated using the VBMC method, with 10,000 iterations performed. Convergence of the evidence lower bound (ELBO) was achieved after approximately 3,800 iterations (Fig. 15.2b). Results, including means, standard deviations, coefficients of variation, and confidence intervals (2.5 to 97.5%), are summarized in Table 15.2 and illustrated in Fig. 15.2a. For example, the mean of cn2 was estimated at 78.79, with a SD of 6.70 and a CV of 0.09, spanning a posterior range of 67.19 to 89.96. This indicates a relatively concentrated distribution with low uncertainty. In contrast, latq_co (lateral flow coefficient) had a mean of 0.31, a SD of 0.22, and a CV of 0.72, with a posterior range of 0.03 to

Table 15.1 Parameter sensitivity analysis

No.	Parameters	Initial range	*t*_stat	*p*_value
1	cn2.hru	[35, 95]	−6.07	0.00
2	canmx.hru	[0, 100]	4.82	0.00
3	awc.sol	[0.01, 1]	3.12	0.00
4	ovn.hru	[0.01, 30]	2.06	0.04
5	alpha.aqu	[0, 1]	1.98	0.05
6	latq_co.hru	[0, 1]	1.73	0.08
7	epco.hru	[0, 1]	−1.66	0.10
8	flo_min.aqu	[0, 50]	−1.36	0.17
9	bf_max.aqu	[0.1, 2]	−1.21	0.23
10	lat_ttime.hru	[0.5, 180]	1.14	0.25
11	slope.hru	[0, 0.9]	−1.06	0.29
12	surlag.bsn	[0.05, 24]	0.96	0.34
13	chk.rte	[0, 500]	−0.79	0.43
14	revap_co.aqu	[0.02, 0.2]	−0.79	0.43
15	cherod.rte	[0, 0.6]	0.65	0.51
16	k.sol	[0, 2000]	−0.52	0.60
17	cn3_swf.hru	[0, 1]	−0.43	0.67
18	esco.hru	[0, 1]	0.30	0.76
19	revap_min.aqu	[0, 50]	−0.20	0.84
20	lat_len.hru	[1, 150]	−0.14	0.89
21	perco.hru	[0, 1]	−0.07	0.98

0.83, reflecting higher uncertainty. Such differences likely stem from the physical processes represented by each parameter and their spatial variability within the basin.

Runoff simulation results from the SWAT model, integrated with VBMC-derived posterior parameter distributions, are shown in Fig. 15.3a. During the calibration period, the Nash–Sutcliffe efficiency coefficient (NSE) reached 0.67, and the coefficient of determination (R^2) was 0.68. In the validation period, NSE was 0.63, and R^2 was 0.62. These metrics demonstrate satisfactory model performance in capturing runoff dynamics in the Wuding River basin. Residual analysis further validated statistical assumptions: the normal probability plot of transformed residuals (Fig. 15.3b) confirmed adherence to a normal distribution, while the autocorrelation function plot (Fig. 15.3c) revealed low autocorrelation beyond the first order, indicating no significant violation of the independent and identically distributed assumption. These outcomes affirm the robustness of the VBMC approach in enhancing model reliability.

Figure 15.4 displays the individual and interactive effects of five sensitive parameters (cn2, canmx, awc, ovn, and alpha) on mean runoff, derived from a three-level full factorial design with 243 SWAT simulations. The parameter cn2 (SCS curve

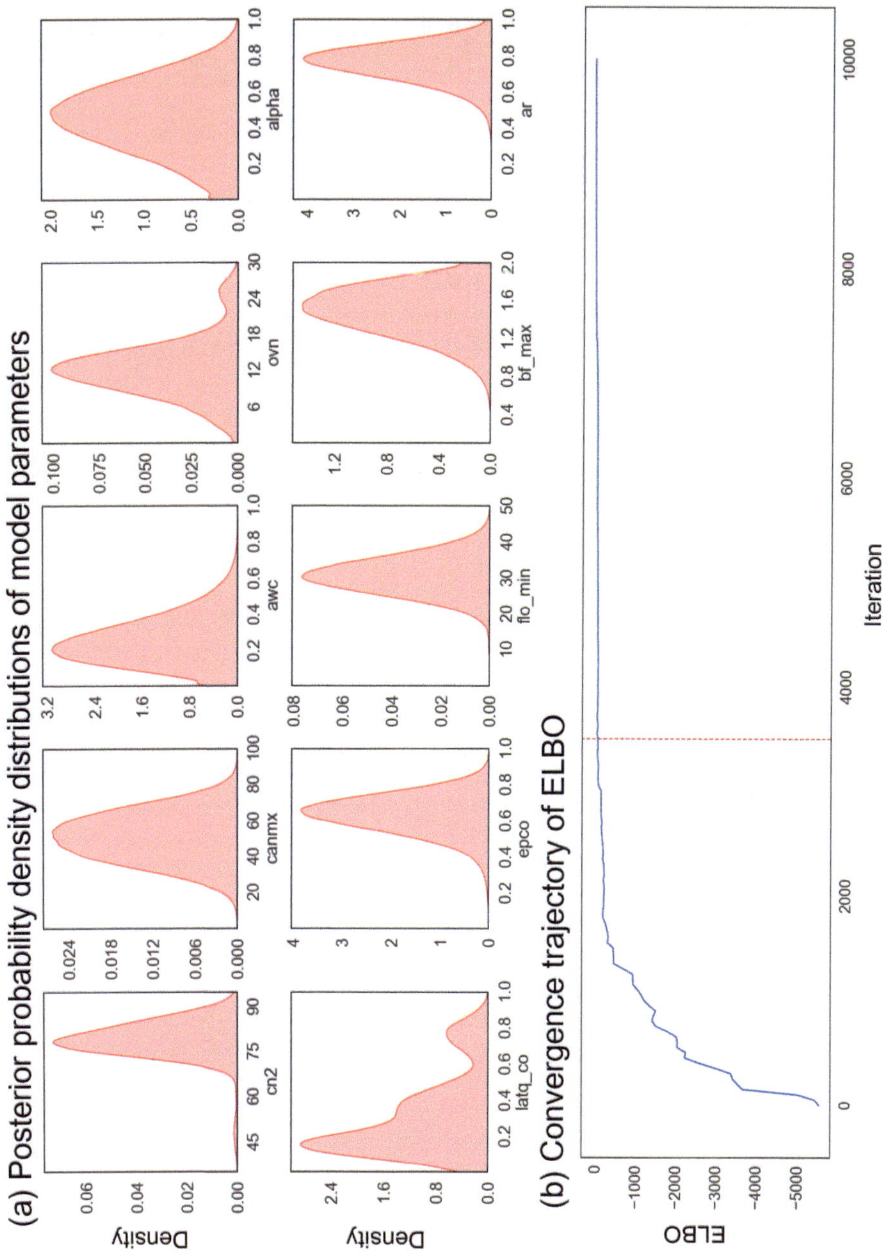

Fig. 15.2 **a** Posterior probability density distributions of model parameters and **b** convergence trajectory of ELBO

Table 15.2 Posterior distribution statistics of sensitive parameters

No.	Mean	Standard deviation	Coefficient of variation	2.50%	Median	97.50%
cn2	78.79	6.70	0.09	67.19	78.92	89.96
canmx	50.57	13.65	0.27	24.30	50.79	75.87
awc	0.26	0.13	0.52	0.04	0.24	0.56
ovn	12.37	4.87	0.39	3.44	12.10	25.20
alpha	0.45	0.19	0.42	0.07	0.46	0.81
latq_co	0.31	0.22	0.72	0.03	0.24	0.83
epco	0.63	0.11	0.18	0.39	0.64	0.83
snomelt_lag	30.45	5.30	0.17	20.20	30.41	40.88
flo_min	1.48	0.27	0.18	0.90	1.50	1.92
ar	0.75	0.10	0.14	0.53	0.76	0.93

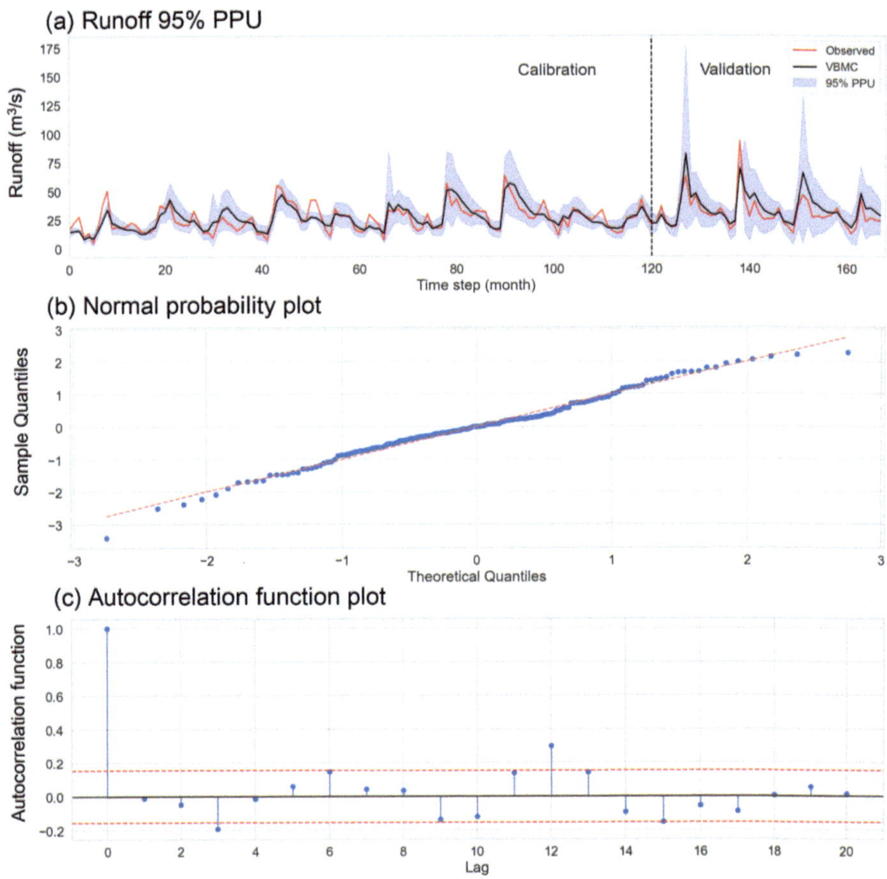

Fig. 15.3 **a** Runoff simulation results, **b** normal probability and **c** autocorrelation function plot of transformed residuals

number) showed the strongest individual effect, significantly increasing mean runoff by reducing infiltration capacity, consistent with its high sensitivity. The parameter canmx (maximum canopy storage) reduced mean runoff by enhancing interception, reflecting its regulatory role. A notable interaction emerged between cn2 and awc (available water capacity), where higher awc values moderated runoff increases under elevated cn2 conditions, suggesting soil moisture storage buffers surface runoff. This interplay underscores the value of factorial analysis in capturing parameter dependencies, enhancing model interpretability for the Wuding River basin.

Figure 15.5 examines the effects on peak runoff, with cn2 again exerting the dominant individual influence, amplifying peak flows during intense rainfall due to its control over runoff responsiveness. The parameter ovn (Manning's roughness

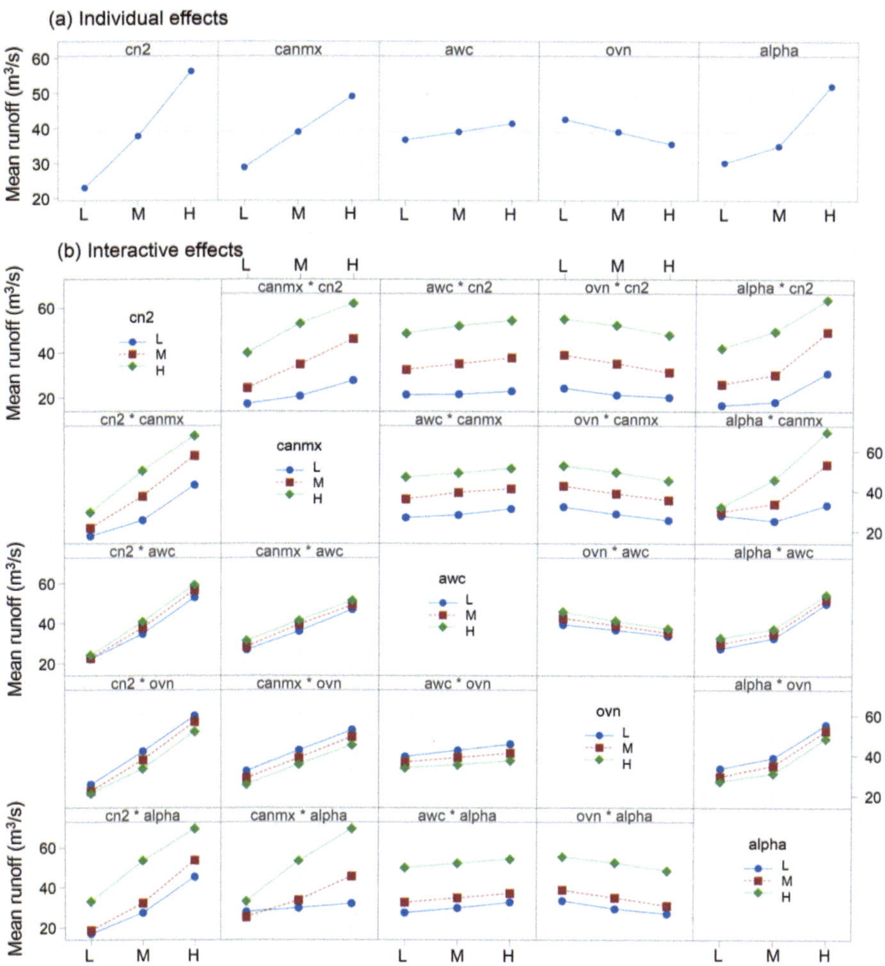

Fig. 15.4 Individual and interactive effects of parameters on mean runoff

coefficient) significantly reduced peak runoff by slowing overland flow velocity, aligning with its sensitivity ranking. A key interaction between cn2 and ovn revealed that increased roughness mitigated peak runoff escalation under high cn2 scenarios, highlighting their combined impact on flow dynamics. The parameter alpha (baseflow recession constant) also interacted with cn2, slightly reducing peak magnitudes by enhancing baseflow contributions. These findings emphasize the framework's ability to quantify interactive effects, which are critical for understanding peak flow behavior in the Wuding River basin and improving hydrological predictions under extreme conditions.

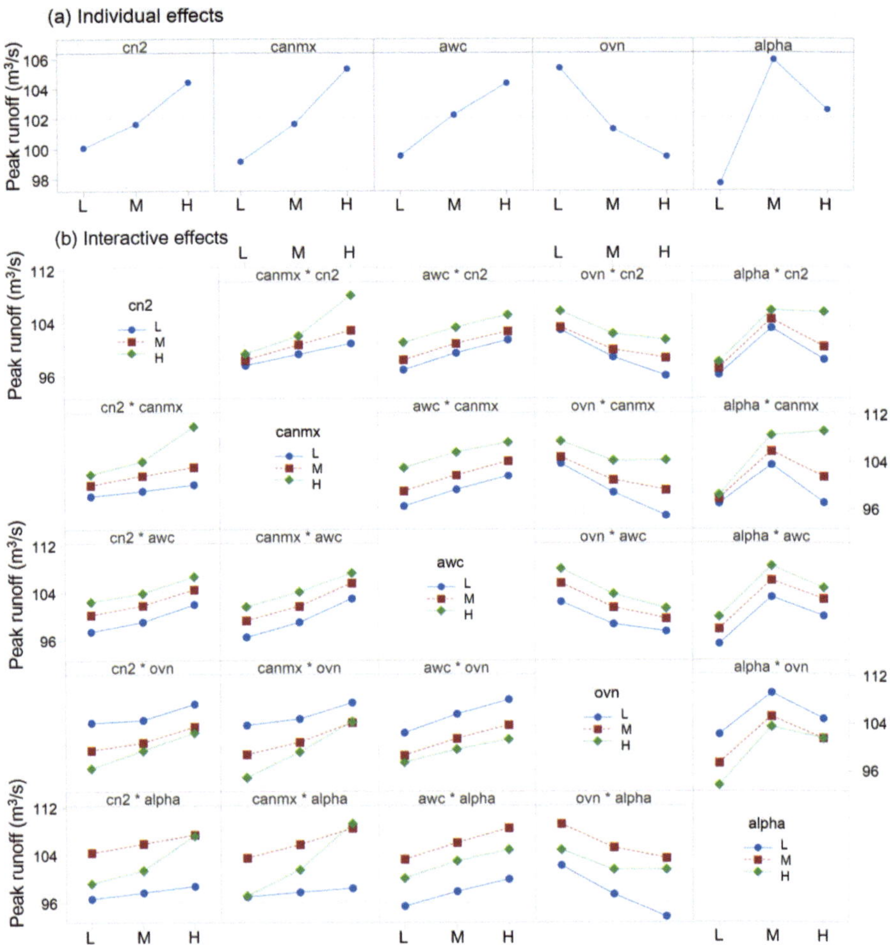

Fig. 15.5 Individual and interactive effects of parameters on peak runoff

15.5 Conclusions

This study developed an integrated SWAT-VBMC-FA framework to quantify parameter uncertainty and interaction effects in hydrological modeling. The approach efficiently computes posterior distributions, evaluates parameter interdependencies, and enhances model reliability, offering advantages over traditional methods in computational cost and scalability. Applied to the Wuding River basin, the framework yielded the following results: (i) identification of nine sensitive parameters via LH-OAT, with cn2, canmx, and ovn most influential, and VBMC achieving convergence after approximately 3,800 of 10,000 iterations, showing low uncertainty for cn2 (mean = 78.79, CV = 0.09) and higher variability for latq_co (mean = 0.31, CV = 0.72); (ii) robust runoff simulation performance, with NSE values of 0.67 (calibration) and 0.63 (validation), and residuals adhering to normality and low autocorrelation beyond first-order; (iii) factorial analysis revealing cn2 as the dominant driver of increased mean runoff, moderated by awc interaction, and peak runoff, tempered by ovn and alpha interactions, highlighting critical parameter interdependencies. This approach enhances parameter identifiability and modeling efficiency, providing a valuable tool for hydrological research and watershed management in complex regions.

Acknowledgements This research was supported by the National Key R&D Program of China (2023YFC3206503) and the Innovative Research Group of the National Natural Science Foundation of China (52221003).

References

Acerbi L (2018) Variational Bayesian Monte Carlo. Adv Neural Inform Process Syst 31

Edwards PJ, Williard KW, Schoonover JE (2015) Fundamentals of watershed hydrology. J Contemp Water Res Educ 154(1):3–20

Herrera PA, Marazuela MA, Hofmann T (2022) Parameter estimation and uncertainty analysis in hydrological modeling. Wiley Interdiscip Rev Water 9(1):e1569

Loaiza-Maya R, Smith MS, Nott DJ, Danaher PJ (2022) Fast and accurate variational inference for models with many latent variables. J Econ 230(2):339–362

Nikita S, Prabhat Kumar Singh D, Shyam B (2018) A comparison of SWAT model calibration techniques for hydrological modeling in the Ganga River watershed. Eng 4(5):643–652

Rudolph MG, Wöhling T, Wagener T, Hartmann A (2024) Extending GLUE with multi-level methods to accelerate statistical inversion of hydrological models. Water Resour Res 60(10):e2024WR037735

Shi P, Yang T, Yong B, Xu CY, Li Z, Wang X, Qin Y, Zhou X (2023) Some statistical inferences of parameter in MCMC approach and the application in uncertainty analysis of hydrological simulation. J Hydrol 617:128767

Tian X, Zhao G, Mu X, Zhang P, Tian P, Gao P, Sun W (2019) Hydrologic alteration and possible underlying causes in the Wuding River China. Sci Total Environ 693:133556

Wang F, Huang G, Cheng G, Li Y (2021) Multi-level factorial analysis for ensemble data-driven hydrological prediction. Adv Water Resour 153:103948

Zhao L, Li H, Li C, Zhao Y, Du X, Ye X, Li F (2024) Enhanced SWAT calibration through intelligent range-based parameter optimization. J Environ Manage 367:121933

Zhong L, Lei H, Yang J (2024) Development of a distributed physics-informed deep learning hydrological model for data-scarce regions. Water Resour Res 60(6):e2023WR036333

Chapter 16
A Semi-infinite Trapezoidal Fuzzy Chance-Constrained Programming Coupled Multi-criteria Decision Analysis (MCDA) Approach for Water-Energy Nexus Management in Coastal Cities Under Uncertainties

Zhiyi Peng and Lei Jin

Abstract Integrated water resources management from a water-energy nexus perspective in coastal cities is essential for promoting sustainable urban development. However, the complex interactions and uncertainties in such nexus systems present substantial challenges for decision-makers. To address these challenges, this study proposes a novel semi-infinite trapezoidal fuzzy programming approach integrated with multi-criteria decision analysis (MCDA) to support robust decision-making for water-energy nexus management under uncertainties. The proposed framework is capable of handling infinite constraint sets characterized by both fuzziness and variability, thereby capturing a wide range of potential system responses. A real-world application in a coastal city is conducted to demonstrate the method's practicality and effectiveness. Within this context, multiple water supply strategies are evaluated considering decision-makers' preferences, evolving energy consumption patterns, and varying levels of water resource availability. The results indicate that surface and transferred water sources consistently contribute approximately 70% of the total supply. Concurrently, the share of unconventional sources—including recycled and desalinated water—increased from 15.1 to 20.9% across the two planning periods, reflecting a significant shift toward diversified and resilient water sourcing under tightening supply conditions. Moreover, the embodied electricity consumption associated with different supply sources is assessed, revealing potential trade-offs between energy demand and water security. These insights underscore the importance of adopting flexible, risk-informed resource management approaches in response to uncertain and interdependent water-energy systems. The developed method provides

Z. Peng · L. Jin (✉)
College of Environmental Science and Engineering, Xiamen University of Technology, Xiamen City, China
e-mail: jinlei777@outlook.com

Y. Li et al. (eds.), *16th International Conference on Environmental Science and Development*, Environmental Science and Engineering,
https://doi.org/10.1007/978-3-032-15080-6_16

a valuable tool for decision-makers seeking to enhance sustainability, resilience, and efficiency in coastal urban water-energy planning.

Keywords Semi-infinite programming · Trapezoidal fuzzy sets · Water-energy nexus · Multi-criteria decision analysis · Uncertainties

16.1 Introduction

The dual challenge of global climate change coupled with rapid urbanization is putting coastal cities at dire risk of water shortages. This comes alongside an increase in the energy consumption of water supply systems. Achieving an optimization balance within the water-energy systems poses a challenge for researchers (Lv 2020). Taking Xiamen city as an example, in 2022, the region's per capita water resources stood at a meager 34.6% of the national average. Furthermore, the percentage of energy-intensive seawater desalination surpassed 23% at that time. This situation creates a glaring imbalance between the resources and energy consumption, which requires a more systematic approach. Traditional methods of single objective planning focus on either water balance or energy singularly, which rarely fulfill practical management needs (Ji et al. 2020). There is a growing demand for new optimization frameworks that could address dynamic, uncertain, multi-objective trading over time.

Even though some attempts have been made to integrate fuzzy logic and multi-criteria decision-making techniques into the optimization of water resources, major gaps still exist concerning the problems of water resources in coastal cities: Zhang et al.'s stochastic planning method (2025) dealing with some parameter uncertainties, is still not able to capture the dramatic seasonal and asymmetric fluctuations in the supply and demand of water resources. Tan's team's interval fuzzy planning method (2019) does improve to some degree the system's performance but lacks temporal adaptability. Planning developed by Mu's research team (2020), while somewhat increasing the system's robustness, simultaneously decreases its sufficiency in adapting to dynamics over time. Most notably, even though the two-level fuzzy planning model developed by Bai et al. does accomplish the optimal intersectoral allocation of water resources (Bai et al. 2023), the static fuzzy affiliation function that is used tends to underestimate the dynamics of water level changes during dry and rainy seasons in coastal areas (Yang et al. 2024). In addition, existing studies generally ignore the rigid requirements of energy consumption constraints for energy-intensive technologies such as desalination (Lin et al. 2021; Im 2020), and there is often a lack of effective articulation between the optimization results and the local government's water planning policies, leading to a significant reduction in the value of decision support.

This research tackles this overarching problem by an innovative approach that applies semi-infinite trapezoidal fuzzy chance-constrained programming along with multi-criteria decision making for optimization. By employing the time-varying trapezoidal fuzzy number (TV-TFN) to describe the uncertainty of hydrological parameters and implementing the 12-month phase affiliation function to describe intra-annual distributions of elements such as rainfall, we propose a two-level mapping approach utilizing alpha-intercept mappings combined with fuzzy opportunity deductions to dynamically balance the decision maker"s risk aversion and total system benefits. Moreover, combining economic, environmental, and energy efficiency enables us to analyze water allocation policies using a multi-criteria evaluation system (Panagopoulos 2021). A multi-criteria assessment integrating economy, environmental protection, and energy efficiency provides additional dimensions to evaluate water allocation plans with quantitative precision (Lei et al. 2014; Yang et al. 2024). The methodology's innovation stems from: (1) For the first time, introducing semi-infinite planning theory into the area of water-energy coupled system optimization greatly enhanced computational efficiency through the discretization of crucial time points; (2) building a fuzzy opportunity constraint processing mechanism that effectively takes the dynamic features of time resolved the applicability of conventional static fuzzy planning in seasonal water resource allocation; (3) achieving a breakthrough in directly converting regional water policies into model constraints; and (4) establishing regional water policy as model constraints boosted the policy guidance value of the optimization outcomes. This study presented new theoretical concepts for addressing the water-energy contradiction in coastal cities and demonstrated considerable application value in the empirical study of Xiamen City. When the ratio of externally transferred water exceeded 50%, the system's energy consumption exhibited an inflection point growth, which provided a quantitative conclusion serving as a crucial foundation for designing the scale of inter-basin water transfer projects in national water network construction planning. Theoretically, the study broadened the range of applications of fuzzy mathematics in time-varying system optimization. Practically, by comparing the model results with the Xiamen 14th Five-Year Plan, it was confirmed that the proposed methodology delivered more precise decision-making support for the integrated management of regional water-energy resources.

16.2 Methodology

The TV-FCCP (Time-Varying Trapezoidal Fuzzy Chance Constrained Programming) framework presented in this research was aimed at resolving the problem of balancing uncertainty and energy-efficiency during water resource allocation in coastal cities, incorporated a discrete node selection mechanism of semi-infinite optimization rooted in conventional fuzzy planning and linked the construction approach of a dynamic fuzzy affiliation function, which notably enhanced the optimization dependability and computational effectiveness of the water-energy system.

16.2.1 Time-Varying Trapezoidal Fuzzy Number

Traditional fuzzy planning methods have limitations when it comes to dealing with seasonal hydrological fluctuations, so this study aimed at these limitations and put forward a dynamic uncertainty modeling method based on Time-Varying Trapezoidal Fuzzy Number (TV-TFN), which is different from the fixed affiliation function in traditional fuzzy set theory because TV-TFN can achieve a precise description of the annual cyclical variation of crucial parameters like rainfall utilizing the dynamic adjustment of 12 phases and its mathematical expression form is as follows:

$$\tilde{A}(t) = (a(t), b(t), c(t), d(t)) \tag{16.1}$$

The boundary parameters a(t), b(t), c(t), and c(t) are functions of the time variable t, and their particular values were dynamically determined through the analysis of the statistical features of the historical rainfall series, taking into account the seasonal nature of the precipitation process, so a broad trapezoidal form was utilized during the flood—prone period from April to September because the intervals of b(t)− a(t) and d(t)− c(t) were widened to accommodate the intense randomness of extreme precipitation occurrences like typhoons, whereas a slender trapezoidal form was adopted during the dry season from October to March to decrease the uncertainty range and enhance the model's responsiveness to low precipitation levels, and the calibration of the dynamics of each boundary parameter was carried out by fitting the statistical properties (for example, mean, standard deviation, and extreme value) of the historical rainfall data and then optimized using an enhanced genetic algorithm, with the Kolmogorov–Smirnov test serving as the evaluation standard for the goodness-of-fit to guarantee that the fuzzy distributions had a high degree of conformity with the measured data, and its membership function conformed to:

$$\mu_{\tilde{A}(t)}(x) = \begin{cases} 0, & x < a(t), \\ \frac{x-a(t)}{b(t)-a(t)}, & a(t) \le x < b(t), \\ 1, & b(t) \le x \le c(t), \\ \frac{d(t)-x}{d(t)-c(t)}, & c(t) < x \le d(t), \\ 0 & x > d(t). \end{cases} \tag{16.2}$$

where $\mu_{\tilde{A}(t)}(x)$ denotes the high or low eligibility of the element x to belong to the fuzzy subset $\tilde{A}$. The closer the value of $\mu_{\overline{A}}(x)$ is to 1, the lower the eligibility of x to belong to $\tilde{A}$; and when $\mu_{\overline{A}}(x)$ is 1 or 0, the fuzzy set degenerates into the classical set, a property that makes the trapezoidal fuzzy number both flexible and rigorous in dealing with fuzzy information. By plotting the curve of the affiliation function $\mu_{\overline{A}}(x)$ (see Fig. 16.1), the distribution characteristics of the trapezoidal fuzzy numbers can be understood more intuitively, which provides the basis for the subsequent model construction and analysis.

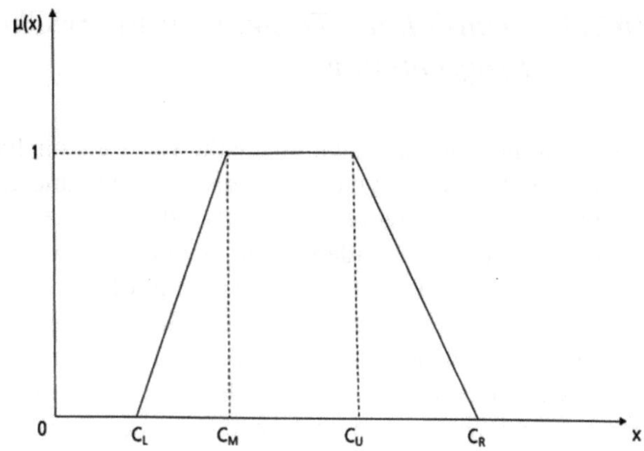

Fig. 16.1 The curve of the affiliation function $\mu_{\overline{A}}(x)$

16.2.2 Trapezoidal Fuzzy Linear Programming

$$\max\{\tilde{z} = \tilde{\mathbf{c}}^T x\} \\ s.t.\begin{cases} \tilde{\mathbf{A}}x \le \tilde{\mathbf{b}} \\ x \ge 0 \end{cases} \tag{16.3}$$

where $C = (c_1, c_2, \cdots, c_n)$ is an n-dimensional clear type row vector, and the elements of $\underset{\sim}{\mathbf{A}}$ and $\underset{\sim}{\mathbf{b}}$ are all or partially trapezoidal fuzzy numbers. $\underset{\sim}{\mathrm{A}} = (\tilde{a}_{ij})_{m\times n}$ is the matrix of technical coefficients, $\underset{\sim}{\mathbf{b}} = (\tilde{b}_1, \tilde{b}_2, \ldots, \tilde{b}_m)^T$ is the resource vector, $\tilde{c} = (\tilde{c}_1, \tilde{c}_2, \ldots, \tilde{c}_n)$ is objective coefficient, $x = (x_1, x_2, \ldots, x_n)^T$ is decision variable coefficients.

Equation (16.3) is equivalent to the following interval objective programming model:

$$\max\{\tilde{z} = \mathbf{E}(\tilde{\mathbf{c}}^T)x\} \\ s.t.\begin{cases} E(\tilde{\mathbf{A}})x \le E(\tilde{\mathbf{b}}) \\ x \ge 0 \end{cases} \tag{16.4}$$

where $E(\tilde{\mathbf{c}}) = (E(\tilde{c}_1), E(\tilde{c}_2), \ldots, E(\tilde{c}_n))^T, E(\tilde{\mathbf{b}}) = (E(\tilde{b}_1), E(\tilde{b}_2), \ldots, E(\tilde{b}_m))^T$, and $E(\tilde{\mathbf{A}}) = (E(\tilde{a}_{ij}))_{m\times n}$ are interval expectation vector and matrix of corresponding trapezoidal fuzzy numbers (TrFNs).

16.2.3 Semi-infinite Trapezoidal Fuzzy Chance-Constrained Programming

Conventional interval planning methods based on fixed boundaries often fail to capture the temporal dynamics of uncertain parameters in water resource systems. To address this limitation, this study adopts a semi-infinite programming framework that models time-dependent parameters as continuous functions over the planning horizon T, requiring constraints of the form $\mathbf{A(t)x} \leq \mathbf{b}(t)$ to hold for all $t \in T$. However, solving such infinite-dimensional problems directly poses computational challenges (Hua et al. 2020). To ensure tractability, an approximation strategy is applied by discretizing the time domain into a finite set of representative points—such as seasonal peaks and extreme events—identified based on historical extremal patterns, thereby transforming the original model into a finite-dimensional optimization problem.

$$(t) \in \left[\underline{A}(t), \overline{A}(t)\right], t \in T = [t_L, t_U] \tag{16.5}$$

Further, to balance the conservatism of the solution with the computational efficiency, the relaxation variable \eta is introduced to moderately relax the constraint boundaries to form a modified optimization objective with explicit physical meaning $min_x\left\{c^T x + \lambda\eta\right\}$, its' constraints are rewritten as $\max_{1\leq k\leq m}[A(t_k)x - b(t_k)] \leq \eta$. This approach was theoretically akin to constructing robust feasible domains at chosen time points but in practice it substantially reduced computational complexity.

16.2.4 Model Solution

The model solution process integrates the representation of fuzziness via trapezoidal fuzzy numbers and the treatment of semi-infinite chance constraints under uncertainties. To address imprecise coefficients in the objective function and constraints, fuzzy parameters are first expressed using trapezoidal fuzzy numbers of the form $\tilde{c} = (c_1, c_2, c_3, c_4)$, where $[c_2, c_3]$ denotes the core and $[c_1, c_4]$ defines the support.

Using the α-cut technique, the fuzzy coefficients are transformed into interval values at a given confidence level $\alpha \in [0, 1]$. The corresponding interval is defined as:

$$\mathrm{c}^{\alpha} = [\mathrm{c}_2 + (1-\alpha)(\mathrm{c}_1 - \mathrm{c}_2), c_3 + (1-\alpha)(c_4 - c_3)] \tag{16.6}$$

This allows the fuzzy linear programming model to be reformulated as an interval linear programming model (Ong and Wan 2018).

$$(E(\tilde{c}))^T x = \sum_{j=1}^{n} E(\tilde{c}_j)x_j = \left[\frac{1}{2}\sum_{j=1}^{n}(c_{j1} + c_{j2}\right)x_j, \frac{1}{2}\sum_{j=1}^{n}(c_{j3} + c_{j4})x_j]$$

$$\sum_{j=1}^{n} E(\tilde{a}_{ij})x_j = \left[\frac{1}{2}\sum_{j=1}^{n}(a_{ij1} + a_{ij2})x_j, \frac{1}{2}\sum_{j=1}^{n}(a_{ij3} + a_{ij4})x_j\right]$$

$$\mathrm{E}(\tilde{b}_i) = \left[\frac{1}{2}(b_{i1} + b_{i2}), \frac{1}{2}(b_{i3} + b_{i4})\right]. \tag{16.7}$$

Equation (16.4) is equivalent to the bi-objective linear programming model as follows:

$$\max\left\{Z_1 = \frac{1}{2}\sum_{j=1}^{n}(c_{j1} + c_{j2})x_j\right\}$$

$$\max\left\{Z_2 = \frac{1}{4}\sum_{j=1}^{n}(c_{j1} + c_{j2} + c_{j3} + c_{j4})x_j\right\}$$

$$\text{s.t.}\begin{cases} \sum_{j=1}^{n}(a_{ij3} + a_{ij4})x_j \le (b_{i3} + b_{i4})(i = 1, 2, \cdots, m) \\ (1-\alpha)\sum_{j=1}^{n}(a_{ij1} + a_{ij2})x_j + \alpha\sum_{j=1}^{n}(a_{ij3} + a_{ij4})x_j \\ \le (1-\alpha)(b_{i1} + b_{i2}) + \alpha(b_{i3} + b_{i4})(i = 1, 2, \cdots, m) \\ \boldsymbol{x} \ge 0 \end{cases} \tag{16.8}$$

Let θ_1 be the goal of the objective $Z_{i'}$, δ_i^- be the underachievement of goal, δ_i^+ be the overachievement of goal, P_i be the preemptive priority assigned to the objective $Z_{i'}$ (i = 1, 2). Then, by using the goal programming approach, Eq. (8) can be changed into a linear goal program as follows:

$$\min\{Z = P_1\delta_1^+ + P_2\delta_2^+\}$$

$$s.t.\begin{cases} -\frac{1}{2}\sum_{j=1}^{n}(c_{j1} + c_{j2})x_j + \delta_1^- - \delta_1^+ = \theta_1 \\ -\frac{1}{4}\sum_{j=1}^{n}(c_{j1} + c_{j2} + c_{j3} + c_{j4})x_j + \delta_2^- - \delta_2^+ = \theta_2 \\ \sum_{j=1}^{n}(a_{ij3} + a_{ij4})x_j \le (b_{i3} + b_{i4})(i = 1, 2, \cdots, m) \\ (1-\alpha)\sum_{j=1}^{n}(a_{ij1} + a_{ij2})x_j + \alpha\sum_{j=1}^{n}(a_{ij3} + a_{ij4})x_j \\ \le (1-\alpha)(b_{i1} + b_{i2}) + \alpha(b_{i3} + b_{i4})(i = 1, 2, \cdots, m) \\ \boldsymbol{x} \ge 0 \end{cases} \tag{16.9}$$

After undergoing this transformation, the intricate linear programming problem that had trapezoidal fuzzy numbers was turned into a solvable bi-objective linear programming format, presenting an effective means for dealing with the ambiguity issue in real-world water distribution.

16.3 Applications

16.3.1 Overview of the Study Area

Located on the southeast coast of China, Xiamen (latitude 24°23'-24°54'N, longitude 117°53'-118°26'E) is a typical island-type city, as shown in Fig. 16.2. It has a total land area of 1,700.61 square kilometers and a sea area of 390 square kilometers. By 2022, its population had reached 5.31 million with an urbanization rate of 90.1% (Xiamen Statistics Bureau, 2023). The city experiences a subtropical monsoon climate with an average annual precipitation of 1,387 mm, over 78% of which is concentrated between April and September, leading to extreme temporal and spatial unevenness in water availability (Xiamen Water Resources Bulletin, 2022).

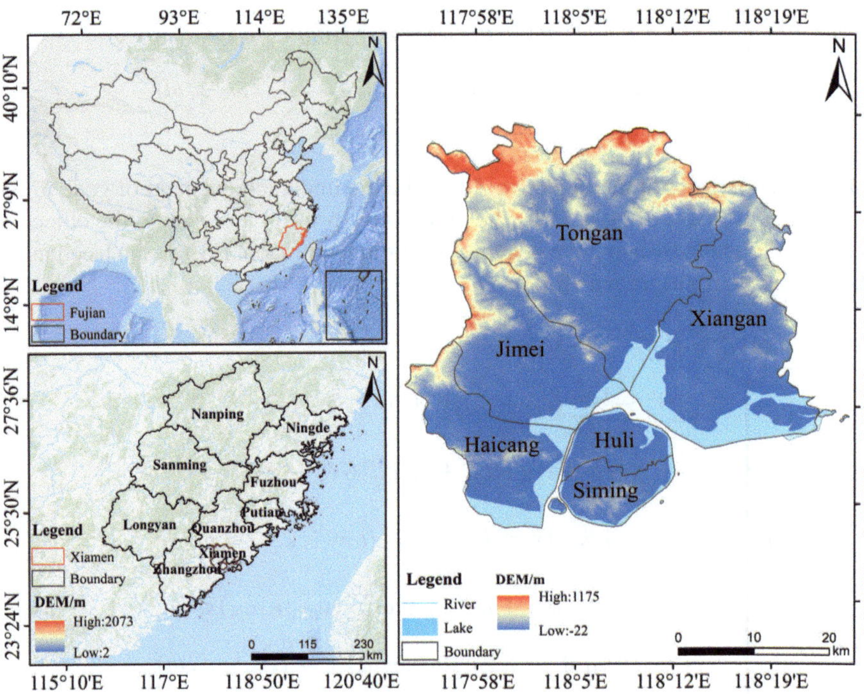

Fig. 16.2 Study area

Despite relatively high precipitation, Xiamen's local water resources are limited to 1.23 billion m^3 annually—less than one-fourth of the national per capita average—classifying it as a severely water-scarce region (water scarcity index = 0.62). Its water supply system relies mainly on three sources: surface water, groundwater, and non-conventional sources. Surface water, primarily sourced from the Beixi Diversion Project, accounts for about 52% of the total supply, though its availability has fluctuated considerably in recent years (CV = 0.31). Groundwater is minimal (0.25 billion m^3) and suffers from seawater intrusion, with 17% of wells exceeding national chloride limits. Non-conventional sources include reclaimed wastewater (110 million m^3) and desalinated water (0.15 billion m^3), with desalination being highly energy-intensive (4.8 kWh/m^3).

In 2022, the city's water use structure was dominated by municipal (41.2%) and industrial (34.7%) consumption, with ecological and agricultural uses at 18.6% and 5.5%, respectively. Rapid economic growth and industrial upgrading have driven a 6.8% annual increase in water demand. Simultaneously, rising sea levels and increasing drought frequency due to climate change have posed challenges to system stability.

Xiamen is selected as a representative case due to its high external water dependency, strong water-energy linkage (1.24 kg CO_2/m^3), and status as an early pilot in China's "sponge city" program, offering both implementation capacity and practical experience for integrated water-energy management research.

16.3.2 Data Collection

This study backed the decision—making analysis regarding water resources management through the integration of data from various sources, getting the fundamental data from the Xiamen Water Resources Bulletin, the Xiamen Statistical Yearbook (Xiamen municipal bureau of statistics 2012), and public government documents like the Five-Year Plan for National Economic and Social Development while also referring to crucial technical parameters from domestic and international literature; taking into account the uncertainty of future water demand, the study established three water demand scenarios, namely low, medium, and high ones, whose probability distributions and values were figured out by fitting historical statistical data and local government report evaluation approaches; the essential economic and technical parameters of the water system (such as the economic efficiency coefficient, water resource cost range, fluctuation extent of wastewater treatment cost, and water conservation investment threshold) were taken from related literature, and the specific parameter values could be seen in Table 16.1.

Moreover, to evaluate the effects of dynamic alterations in water accessibility, this study developed three scenarios grounded in historical hydrological information, forecasts of extreme climates, and environmental policy limitations: the base—line scenario (H0) depicted traditional water resource situations, while the H1 and H2 scenarios simulated extreme circumstances of a 4% and 8% decrease in the quantity

Table 16.1 The main economic and technical parameters in the case study

Parameter	Period	
	T = 1(2021–2025)	T = 2(2025–2030)
Water consumption benefit coefficient, $\tilde{NB}_{jt}$(RMB/m^3)		
Agricultural (j = 1)	(25,29,36,40)	(35,37,45,48)
Industrial (j = 2)	(310,340,380,410)	(330,360,410,440)
Municipal (j = 3)	(5,6,7,8)	(6,7,8,9)
Environmental and ecological (j = 4)	(6,7,9,10)	(8,10,12,14)
Water supply cost coefficient, CWS_{it}(RMB/m^3)		
Agricultural (j = 1)	0.008	0.007
Industrial (j = 2)	0.011	0.010
Municipal (j = 3)	0.006	0.005
Environmental and ecological (j = 4)	0.009	0.008
Water conversation cost coefficient, CSW_{jt}(RMB/m^3)		
Agricultural (j = 1)	0.012	0.010
Industrial (j = 2)	0.015	0.013
Municipal (j = 3)	0.021	0.017
Environmental and ecological (j = 4)	0.013	0.011
Maximum available water supply, AW_{it}(10^{6}m^3)		
Surface water (i = 1)	420	440
Groundwater (i = 2)	250	250
Transferred water (i = 3)	500	525
Surface water (i = 1)	420	440
Electricity consumption coefficient for water treatment process, $\widetilde{ECW}_{jt}$ (kWh/m^3)		
Agricultural (j = 1)	(0.20,0.22,0.24,0.26)	(0.18,0.20,0.22,0.24)
Industrial (j = 2)	(0.70,0.74,0.78,0.82)	(0.65,0.68,0.72,0.75)
Municipal (j = 3)	(0.33,0.35,0.39,0.42)	(0.30,0.32,0.35,0.38)
Environmental and ecological (j = 4)	(0.10,0.11,0.13,0.14)	(0.09,0.10,0.12,0.13)
Sewage discharge coefficient, γ_{jt}		
Agricultural (j = 1)	0.3	0.25
Industrial (j = 2)	0.7	0.65
Municipal (j = 3)	0.85	0.88
Environmental and ecological (j = 4)	0.1	0.15
Conservation water, SW_{jt}		
Agricultural (j = 1)	300	650
Industrial (j = 2)	1000	1750
Municipal (j = 3)	700	1250
Environmental and ecological (j = 4)	80	275

of scarce water resources under H0 conditions respectively, and the methodological framework offered strong optimization backing for multi-criterion decision-making by linking interval parameters, probabilistic scenarios, and fuzzy set theory.

H0: Conventional.

H1: H0 reduced by 4% of scarce water conditions.

H2: H0 reduced by 8% of scarce water conditions.

16.3.3 Modeling Formulation

In this paper, a regional water allocation management problem taking into account water–energy connections was formulated aiming to achieve the maximum desirable net benefits and a 10-year planning horizon was taken into account with the initial period from 2021–2025 and the subsequent period from 2026–2035; in this research, the advantages of water utilization in diverse sectors, as shown in Fig. 16.3, along with the expenses related to water provision, wastewater treatment, and water preservation were taken into consideration while ignoring the procedures of water purification, distribution, and wastewater collection which simplified the flow patterns, lessened the modeling complexity, facilitated visual modeling, and strengthened the system integration. Therefore, with the optimization objective of maximizing the total net benefit, the water use benefit (f_1) is calculated minus the cost of water supply (f_2),, the cost of wastewater treatment (f_3), and the cost of water conservation (f_4). In addition, in this study, the interactions in the coupled water-energy system are assumed to be linear for simplicity. Uncertainties in economic parameters, technical coefficients, and future water demand were considered in the decision-making. Finally, a fuzzy opportunity-constrained planning model for regional water allocation is developed:

$$Maxf = f_1 - f_2 - f_3 - f_4 \tag{16.10}$$

1. Water benefit

$$f_1 = \sum_{t=1}^{2}\sum_{j=1}^{4} \widetilde{NB}_{jt} \cdot Q_{jt} \tag{16.11}$$

t denotes the planning period, with the entire planning period divided into two periods, t = 1 for 2021–2025 and t = 2 for 2026–2030, and j denotes the water use sector (j = 1 for the agricultural sector, j = 2 for the industrial sector, j = 3 for the municipal sector, and j = 4 for the ecological sector), $\widetilde{NB}_{jt}$ is the water use efficiency coefficient of user j in period t, and x_{jt} is the water consumption of user j in period t.

2. Cost of water supply

$$f_2 = \sum_{t=1}^{2}\sum_{i=1}^{3}\sum_{j=1}^{4} CWS_{jt} \cdot W_{jit} \tag{16.12}$$

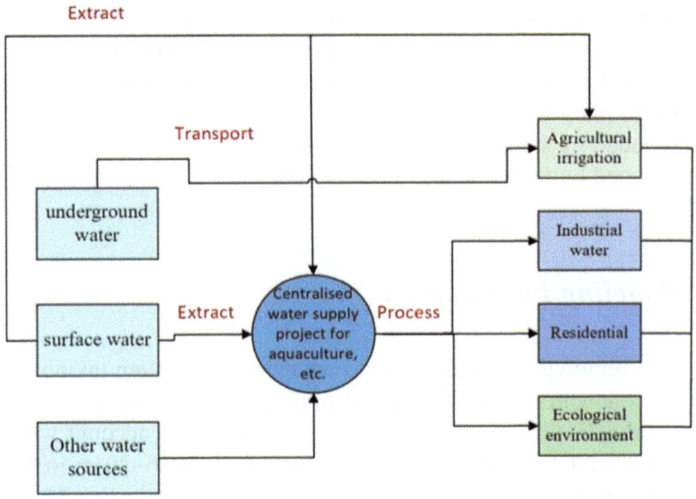

Fig. 16.3 Water supply system construction

where i denotes the source of water supply (i = 1 denotes surface water, i = 2 denotes groundwater, and i = 3 denotes externally transferred water), CWS_{jit} is the cost coefficient of water supply for source i, and W_{jit} denotes the amount of water supplied from source i to end-user j.

3. Sewage treatment costs

$$f_3 = \sum_{t=1}^{2} \sum_{j=1}^{4} \gamma_{jt} \cdot \eta_{jt} \cdot CWT_{jt} \cdot W_{jt} \tag{16.13}$$

4. Saving water costs

$$f_4 = \sum_{t=1}^{2} \sum_{j=1}^{4} CSW_{jt} \cdot SW_{jt} \tag{16.14}$$

For the above objective functions, the relevant constraints are established.

(1) Energy consumption constraints

$$\sum_{t=1}^{2} \sum_{j=1}^{4} \widetilde{ECW_{jt}} Q_{jt} \leq \tilde{E}\max \tag{16.15}$$

$\tilde{E}max$ denotes the maximum permissible upper limit of electrical energy consumption, taken as (1000,1200,1500,1800).

(2) Water supply constraints

$$\sum_{j=1}^{4} W_{jit} \leq AW_{it}, \forall i \in 1, 2, 3 \tag{16.16}$$

(3) Water conservation control constraints

$$CrSW_{jt} \leq e_{jt} \cdot \widetilde{ED}_{jt} \geq \lambda, \forall j, t \tag{16.17}$$

where e_{jt} denotes the percentage of available water savings and $\widetilde{ED}_{jt}$ denotes the expected water use by end-user j.

(4) Water supply constraints

$$Cr\{W_{jit} + SW_{jt} \geq \widetilde{ED}_{jt}\} \geq \lambda, \forall j, t \tag{16.18}$$

(5) Water consumption control limits

$$\sum_{i=1}^{4}\sum_{j=1}^{4} W_{jit} \leq TWA_t, \forall t \tag{16.19}$$

TWA_t is the upper limit of total water consumption in period t.

(6) Non-negative constraints on variables

$$Q_{jt} \geq 0 \tag{16.20}$$

16.4 Results and Discussion

16.4.1 Results Analysis

To assess the efficacy and practicality of the suggested semi-infinite time-varying trapezoidal fuzzy chance-constrained planning (SI-TFCCP) linked with multi-criteria decision analysis (MCDA), three water resource availability situations (H0: standard, H1: 4% decrease, H2: 8% decrease) and four confidence degrees ($\alpha = 0$, 0.3, 0.6, 0.9) were devised according to Xiamen's traits and simulations under these pairings were carried out and examined.

Results show that system net benefits increase with higher α, rising from 3.072 to 4.231 billion CNY under H0, a 37.8% improvement ((as presented in Fig. 16.4), while stricter water scenarios (H1, H2) reduce net benefits by 5.8 and 12.6%, respectively. Higher confidence levels enhance system robustness and reduce resource waste, with water utilization improving from 78.6 to 92.4%. However, flexibility against extreme

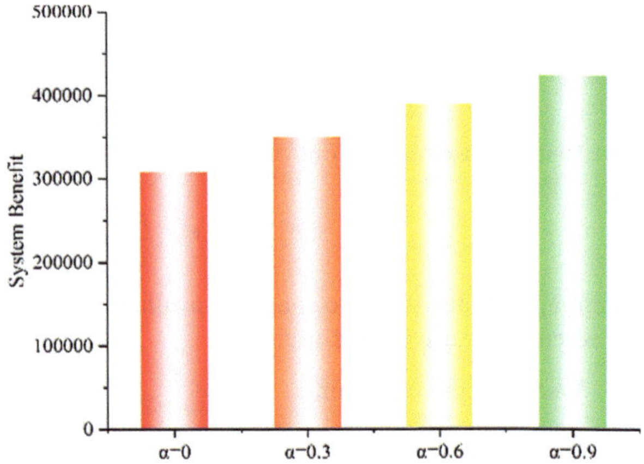

Fig. 16.4 Changes in total system net benefits at different α levels

events decreases, suggesting that adaptive adjustment of α based on climate forecasts and urban development is necessary.

Sectoral analysis reveals that the municipal sector contributes the most (47.2–52.3%), followed by industry ((as presented in Fig. 16.5), whereas agriculture and the ecological aspect had smaller portions and were subjected to more significant reductions when there were constraints, which underlined the necessity for protective measures for agriculture and ecosystems during times of scarcity.

The sensitivity analysis carried out in H2 showed that the optimization of water resources—an increase of 21% in surface water, 18.5 per cent in wastewater reuse, and 25 per cent in storage capacity—still led to an increase of 17.3 per cent in net benefits, demonstrating that the proposed optimal framework is adaptable and feasible under extreme resource pressure.

16.4.2 Discussion

In order to comprehensively evaluate water allocation schemes in various scenarios, this study uses the Multiple Criteria Decision Analysis (MCDA), which is weighted in terms of economic benefits (45%), environment (35%) and energy efficiency (20%) (Li et al. 2020). Under normal conditions (H0), results show that higher confidence levels ($\alpha = 0.9$) lead to improved system performance, increasing composite scores from 0.745 to 0.862. This optimization achieves the largest net benefit of 4.23 billion CNY, with a reduction in carbon intensity of 1.24 kwh/m ^{3}, and a cap on energy consumption of less than 1.03 billion kWh, showing a balanced economic and environmental performance.

However, water scarcity (H1 and H2) shows trade-offs in resource management, with the overall score falling by 3.2 to 7.5% as a result of increased dependence on energy intensive unconventional resources such as desalination. Even at high

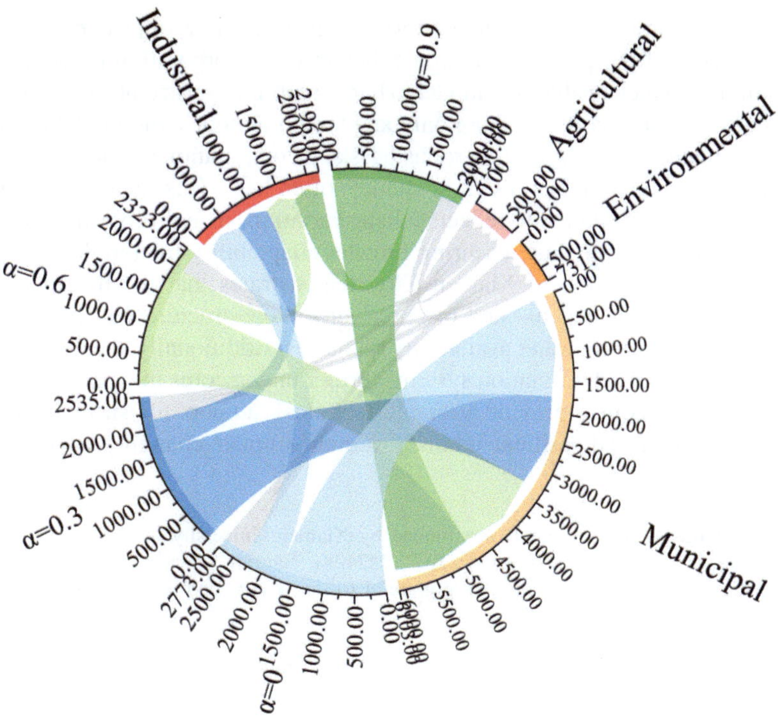

Fig. 16.5 Water allocated to each sector at different α levels

confidence levels (α = 0.9), carbon emissions in H2 remain 6.5% above baseline values, indicating persistent sustainability challenges under resource constraints.

Additional analysis highlights that the strongest response to the adjustment of confidence levels is the economic performance, whereas the environmental objectives call for more stringent management of the expansion of unconventional water sources. In addition, sensitivity tests indicate that changing policy priorities towards environmental protection changes the ranking of the system, increasing the need for flexible MCDA frameworks for coastal water management. All of these results provide valuable information for the balancing of competing goals in the planning of urban water and energy systems.

16.5 Conclusion

In this research, a semi-infinite time-varying trapezoidal fuzzy chance-constrained programming (SI-TFCCP) model linked with multi-criteria decision analysis (MCDA) was formulated to optimize the management of urban water resources amidst uncertainty. The model effectively addresses dynamic uncertainties while balancing economic, environmental, and energy efficiency objectives. A significant advantage of the proposed approach lies in the incorporation of semi-infinite programming, which enables the model to handle infinite constraint sets over continuous time domains. By discretizing only key temporal nodes, the model achieves a practical balance between solution robustness and computational efficiency—something traditional interval or fully finite models often lack. This feature substantially enhances the model's ability to reflect realistic variations in hydrological and energy-related parameters over time, ensuring broader adaptability to complex real-world systems. The application to Xiamen City demonstrates that the model not only improves system net benefits and resource allocation structure but also enhances resilience under varying water availability scenarios. Additionally, the dynamic optimization of seawater desalination systems further underscores the model's practical utility. These outcomes provide valuable technical support for sustainable water resource planning and resilient urban governance under changing environmental conditions.

Acknowledgements This research was funded by Xiamen University of Technology National Natural Science Foundation Key Project (XPYM2408). Natural Science Foundation of Fujian Province (2025J011275). The authors are grateful to the editors and the anonymous reviewers for their insightful comments and suggestions.

Declarations Conflicts of interest/Competing interests Authors declare no conflict of interest/competing interests.

References

Bai R, Jin L, Zhuo B et al (2023) Development of a type-2 fuzzy bi-level programming model coupling MCDA analysis for water resources optimization under uncertainty[M/OL]//YANG Z. Environmental Science and Technology: Sustainable Development. Cham: Springer International Publishing, pp 47–59. https://doi.org/10.1007/978-3-031-27431-2_4

Hua L et al (2020) Floodwater utilization based on reservoir pre-release strategy considering the worst-case scenario. Water 12(3):892. https://doi.org/10.3390/w12030892

Im S et al (2020) Techno-economic evaluation of an element-scale forward osmosis-reverse osmosis hybrid process for seawater desalination. Desalination 476:114240. https://doi.org/10.1016/j.desal.2019.114240

Ji L, Wu T, Xie Y et al (2020) A novel two-stage fuzzy stochastic model for water supply management from a water-energy nexus perspective[J/OL]. J Clean Prod 277:123386. https://doi.org/10.1016/j.jclepro.2020.123386

Lei J, Guohe H, Yurui F, Li W, Tiffney W (2015) A pseudo-optimal inexact stochastic interval t2 fuzzy sets approach for energy and environmental systems planning under uncertainty: a case study for Xiamen City of China. Appl Energy 138: 71–90. https://doi.org/10.1016/j.apenergy.2014.10.024

Li X et al (2020) Multi-water resources optimal allocation based on multi-objective uncertain chance-constrained programming model. Water Resour Manag 34(13):4167–4181. https://doi.org/10.1007/s11269-020-02697-z

Lin S, Zhao H, Zhu L, et al (2021) Seawater desalination technology and engineering in China: a review. Desalination 498:114728. https://doi.org/10.1016/j.desal.2020.114728

Lv J, Li Y P, Huang G H, Suo C, Mei H, Li Y (2020) Quantifying the impact of water availability on China's energy system under uncertainties: a perceptive of energy-water nexus. Renew Sustain Energy Rev 134:110321. https://doi.org/10.1016/j.rser.2020.110321

Musa A (2021) Goal programming model for optimal water allocation of limited resources under increasing demands. Environ Develop Sustain 23:5956–5984. https://doi.org/10.1007/s10668-020-00856-1

Ong J, Wan SP (2018) A new trapezoidal fuzzy linear programming method considering the acceptance degree of fuzzy constraints violated[J/OL]. Knowl Based Syst 148:100–114. https://doi.org/10.1016/j.knosys.2018.02.030

Panagopoulos A (2021) Energetic, economic and environmental assessment of zero liquid discharge (ZLD) brackish water and seawater desalination systems. Energy Convers Manag 235:113957. https://doi.org/10.1016/j.enconman.2021.113957

Tan Y, Lu Y, Zhao H, et al (2019) An optimal allocation model for large complex water resources system considering water supply and ecological needs. Water 11(4):843. https://doi.org/10.3390/w11040843

Xiamen Municipal Bureau of Statistics (2012) Xiamen special economic zone in 2012 yearbook [EB/OL]. https://tjj.xm.gov.cn/zfxxgk/zfxxgkml/tjsjzl/tjnj/

Yang C, Lei J, Haiyan F, Yurui F, Ruolin B, Yi W (2024a) A novel fuzzy best-worst multicriteria decision-making method based on the dual interval algorithm. J Environ Inform 46(2):155–175. https://doi.org/10.3808/jei.202400523

Yang C, Lei J, Haiyan F, Yurui F, Ruolin B, Yi W (2024b) Investigating the impact of climate change and policy orientation on energy–carbon–water nexus under multi-criteria analysis. Renew Sustain Energy Rev 189:114032. https://doi.org/10.1016/j.rser.2023.114032

Zhang X, Jin L, Fan Y et al (2025) Improved semi-infinite interval type-2 fuzzy multi-objective programming with integrated multi-criteria decision analysis for sustainable water resource management[J/OL]. J Water Process Eng 73:107687. https://doi.org/10.1016/j.jwpe.2025.107687

Chapter 17
Spatial–Temporal Variations in Blue and Green Water Resources in the Jiziwan Region of the Yellow River Basin

Yue Qiu, Yongping Li, Guohe Huang, Zhipeng Xu, and Yanfeng Li

Abstract This study develops and applies an integrated framework for analyzing spatiotemporal patterns of blue-green water. The framework combines Latin Hypercube sampling with one-factor-at-a-time (LH-OAT) method, SWAT model, and blue-green water accounting methodology, enabling simultaneous simulation of streamflow processes and quantification of blue-green water distribution across spatiotemporal dimensions. Applied to the Jiziwan Region of the Yellow River Basin (YRB), the framework reveals: (i) the SWAT model was demonstrated to be well-suited for the study area, the parameters (CN2, TRNSRCH and SLSUBBSN)are the most sensitive parameters to streamflow output; (ii) the spatial distribution of blue and green water resources in the basin is uneven, the average annual amount of blue water in the subbasins of the study area ranged from about 0 to 85 mm, and the average annual amount of green water ranged from about 140 mm to 680 mm. (iii) interannual variability of blue and green water fluxes during 2006–2019 exhibited strong precipitation dependence, with green water demonstrating greater temporal stability compared to the more dynamic blue water response.

Keywords Blue and green water · SWAT model · Yellow River Basin

Y. Qiu · Y. Li (✉) · G. Huang · Z. Xu · Y. Li
State Key Laboratory of Regional Environment and Sustainability, China-Canada Center for Energy, Environment and Ecology Research, UR-BNU, School of Environment, Beijing Normal University, Beijing, China
e-mail: yongping.li@iseis.org

Y. Li · G. Huang
Environmental Systems Engineering Program, University of Regina, Regina, SK, Canada

Y. Li et al. (eds.), *16th International Conference on Environmental Science and Development*, Environmental Science and Engineering,
https://doi.org/10.1007/978-3-032-15080-6_17

17.1 Introduction

Water serves as an essential natural resource that underpins biological existence, socioeconomic development, and ecological equilibrium (Yu et al. 2024). Understanding the water cycle is essential for sustainable societal development. Given population growth and climate change impacts, assessing regional water resource distribution is critical for effective water management. Blue water denotes surface and groundwater resources (lakes, rivers, aquifers) directly available for human use, whereas green water represents precipitation-stored soil moisture ultimately transpired by vegetation (Falkenmark 1995). The blue-green water paradigm facilitates systematic quantification of water resources, thereby improving governance strategies across local and regional scales. Hydrological modeling serves as a critical tool in this approach (Zhang et al. 2024a). The application of the blue-green water paradigm enables rigorous hydrological evaluation, providing a scientific basis for optimizing water governance at local and regional levels (Ding et al. 2024).

The Soil and Water Assessment Tool (SWAT) model, a widely adopted semi-distributed parametric hydrologic model, has emerged as a powerful computational framework for simulating regional hydrological processes. Its distinctive capability to explicitly quantify both blue and green water fluxes makes it particularly valuable for integrated water resource assessments. Numerous studies have demonstrated the model's effectiveness in streamflow prediction and comprehensive water balance analysis, validating its utility in hydrological research (Xu et al. 2023). Research on blue-green water assessment proliferated during the 2010s, with significant methodological advancements in hydrological modeling. The SWAT model, initially designed for watershed-scale hydrological simulation, demonstrates robust capability in quantifying spatiotemporal dynamics of both blue and green water fluxes. Subsequent studies have extensively validated SWAT's application for streamflow modeling and its unique capacity to simulate the partitioning of water resources between blue and green water components (Chen et al. 2024).

The Yellow River, China's second-longest river system, ranks among the world's most sediment-laden waterways. The basin exhibits pronounced spatial heterogeneity in water resource distribution, resulting in acute water scarcity across specific regions (F and Z 2022). Numerous scholars have conducted water shortage assessments for the Yellow River Basin (YRB), but as the Jiziwan Region of the YRB, where water shortage and soil erosion are the most serious (Wang et al. 2024), fewer studies have been conducted to directly target this region.

This study develops an integrated modeling framework for systematic evaluation of spatiotemporal patterns in blue and green water fluxes across heterogeneous watersheds. The methodology synergistically combines Latin Hypercube Sampling with one-factor-at-a-time (LH-OAT), the SWAT model, and blue-green water accounting methods, enabling simultaneous simulation of streamflow processes and quantification of blue-green water fluxes across spatiotemporal scales. This study has two primary research objectives: (i) identification of basin-wide sensitive parameters through systematic sensitivity analysis; (ii) comprehensive characterization of spatiotemporal patterns in blue and green water fluxes across the watershed. The findings provide a scientific foundation for evidence-based water resource management in the Jiziwan Region of the YRB, offering actionable insights for policymakers.

17.2 Methodology

In this study, a research framework combining the SWAT model, Latin Hypercube sampling with one-factor-at-a-time (LH-OAT) method, and the blue-green water quantification method was developed for complex watershed systems. The SWAT model was established based on the basic data of the study area, and the main parameters affecting the hydrological model were determined using the LH-OAT method. The model outputs enabled systematic evaluation of spatiotemporal blue-green water flux partitioning patterns across the study area.

17.2.1 SWAT Model

The Soil and Water Assessment Tool (SWAT), developed by the USDA Agricultural Research Service (ARS), is a semi-distributed, physically-based hydrological model designed for watershed-scale simulations (Li et al. 2024). It is particularly well suited to the simulation of complex, large subbasins and long-term hydrologic processes (Zhang et al. 2024b).

In this study, the SWAT simulation was divided into three distinct periods: the warm-up period (2004–2006), the calibration period (2006–2015), and the validation period (2016–2019). The SWAT model was calibrated and validated using the Sequential Uncertainty Fitting algorithm version 2 (SUFI-2), which is applicable to a wide range of hydrologic models and complex hydrologic processes. Through iterative optimization and uncertainty quantification, it can significantly improve the simulation accuracy and reliability of hydrologic models.

The SWAT model contains over 210 hydrologic parameters. In SWAT-CUP, the LH-OAT methods to compute objective functions through multiple regression analysis, thereby identifying the most sensitive parameters. Parameter sensitivity was evaluated through t-test analysis, with significance determined by larger absolute t-statistics ($|t| > 2.0$) and smaller p-values ($p < 0.05$).

17.2.2 Blue and Green Water Assessments

Blue-green water assessment is a crucial concept in hydrological cycle research, primarily used to quantify different components of water resources. In this study, the SWAT model was employed to evaluate green and blue water resources at both the

HRU (Hydrological Response Unit) and subbasin scales. Green water (*Gw*) includes evapotranspiration (*ET*, mm) and soil water content (*SW*, mm), respectively. Blue water (*Bw*) is computed as the combined contribution of surface water yield (*WYLD*) and deep aquifer recharge (*DA_RCHG*). The equations for the calculations are as follows:

$$Gw = ET + SW \tag{17.1}$$

$$Bw = WYLD + DA\_RCHG \tag{17.2}$$

17.2.3 Model Simulation Metrics

In this study, the simulation accuracy of the SWAT model is evaluated using three statistical metrics: the coefficient of determination (R^2), the Nash–Sutcliffe efficiency coefficient (*NSE*), and the percent bias (*PBIAS*). The coefficient of determination (R^2) quantifies the linear relationship between the model results and the observed data.

The NSE assesses the predictive accuracy of hydrologic models by comparing the model's goodness-of-fit to the variance of the observed data. Values of R^2 and *NSE* greater than 0.50, along with *PBIAS* values within the range of $\pm$ 15 to $\pm$ 20%, indicate that the SWAT model was acceptably calibrated. To assess the agreement between observed and simulated values, the following equations were used:

$$PBIAS = 100\hat{A} \cdot \left[\frac{\sum_{i=1}^{n} Q_{i'} - Q_i}{\sum_{i=1}^{n} Q_i}\right] \tag{17.3}$$

$$R^2 = \frac{\left[\sum_{i=1}^{n}\left(Q_i - \underset{i}{Q}\right)\left(Q_{i'} - \underset{i}{Q'}\right)\right]^2}{\sum_{i=1}^{n}(Q_i - \underset{i}{Q})^2 \sum_{i=1}^{n}\left(Q_{i'} - \underset{i}{Q'}\right)^2} \tag{17.4}$$

$$NSE = 1 - \frac{\sum_{i=1}^{n}(Q_i - Q_{i'})^2}{\sum_{i=1}^{n}\left(Q_i - \underset{i}{Q}\right)^2} \tag{17.5}$$

Here, Q_i and $Q_{i'}$ represent the observed and simulated data, respectively; Q_i and $Q_{i'}$ represent the mean observed and simulated data, respectively; n is the total number of observations.

17.3 Study Area and Data

The YRB is a vital water resource region in China, and its water resources status significantly influences the ecological environment, economic development, and social life within the basin. The YRB exhibits significant spatial heterogeneity in water availability, with relative abundance in the upper reaches but severe scarcity in the middle-lower sections. The study area of this paper is selected as the Jiziwan Region of the YRB. Due to the combined effect of topography and climate, the distribution of precipitation in the study area is inhomogeneous, with more precipitation in the mountainous areas and relatively less in the river valleys and plains. The streamflow of the Yellow River in this area also shows significant seasonal variations, with abundant water during the summer flood season and a sharp decrease during the winter dry season. Figure 17.1 illustrates the study area's geographical setting and the distribution of hydrometeorological monitoring infrastructure.

This study employed the SWAT model to simulate blue and green water fluxes within the watershed. Required inputs comprised a 90 m SRTM digital elevation model (DEM) for stream network and subbasin delineation, land use and soil maps for HRU generation, along with meteorological observations and reservoir operation records. The SWAT model simulation and validation incorporated observational data from 36 meteorological stations, 2 hydrological gauging stations and 1 reservoir, the

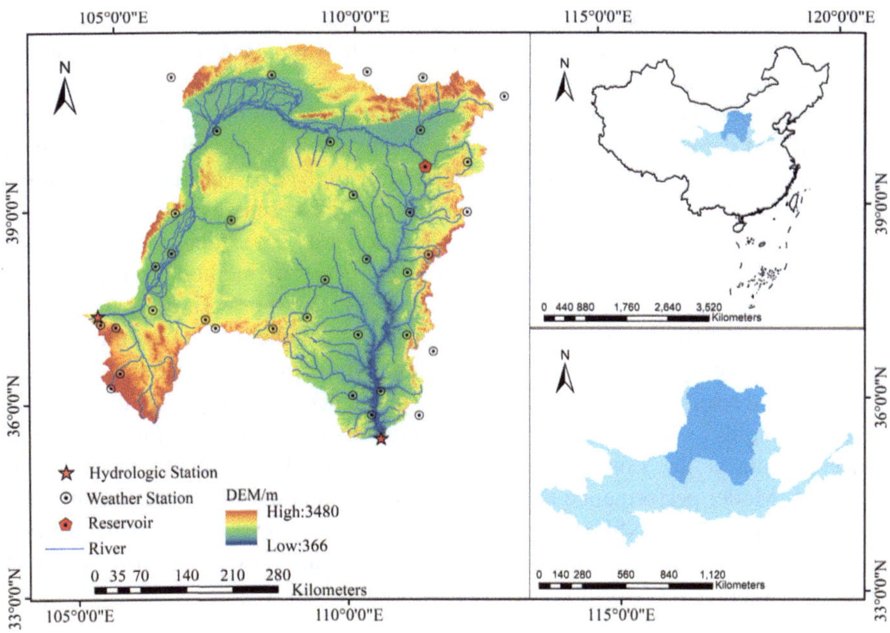

Fig. 17.1 Overview of the study area

data are on a daily scale and span 2006–2019. Upstream boundary conditions were defined using discharge records from hydrologic stations at the study area entrance.

17.4 Results and Discussion

17.4.1 Model Setup and Evaluation

This study developed a SWAT model for streamflow simulation, with the watershed partitioned into 40 sub-subbasins comprising 476 HRUs. The spatial distributions of sub-subbasin boundaries, soil types, land use classifications, and slope characteristics are presented in Fig. 17.2. Watershed delineation was based on topographic analysis using a 90-m digital elevation model. Land use soil type also has a significant impact on subbasin hydrology. The study area consists of six land use types: 48.6% grassland (PAST), 23.9% cropland (AGRL), 14.0% barren land (BARR), 8.7% forested land (FRST), 2.8% townships (URBN), and 2.0% water bodies (WR) in 2015. The study area contains 72 soil forms, which were categorized into 17 soil groups based on their typological characteristics, with more than 50% of the soil groups being dominated by compact soils (CMc) and argillaceous soils (Arc). The slope of the study area was classified based on different numerical intervals, and it can be seen that most of the study area has a gentle slope.

Parameter sensitivity analysis was conducted through 1000 model simulations, evaluating 25 parameters representing key hydrological processes. Monthly prediction results identified 3 statistically significant parameters (CN2, TRNSRCH and SLSUBBSN) demonstrating both high sensitivity ($p < 0.05$) and strong influence on streamflow (large t-statistics), as illustrated in Fig. 17.3. The remaining parameters were systematically ranked based on their relative sensitivities. CN2 denotes the curve number for runoff estimation under moderate antecedent moisture conditions (AMC II) in the Soil Conservation Service (SCS) method. A higher CN2 value indicates greater runoff potential, making it the most critical parameter in the model. TRNSRCH represents the proportion of channel water loss through bed seepage, simulating the common downstream flow depletion phenomenon in arid and semi-arid rivers. It ranked as the second most sensitive parameter in the analysis.

The monthly streamflow simulations demonstrated strong agreement with observed data at the Longmen hydrological station (basin outlet), as shown in Fig. 17.4. Model performance metrics exceeded evaluation thresholds during both calibration ($R^2 = 0.79$, NSE = 0.62) and validation ($R^2 = 0.90$, NSE = 0.76) periods, with PBIAS values also meeting acceptable criteria. These results confirm the SWAT model's reliability for hydrological simulation in the study area.

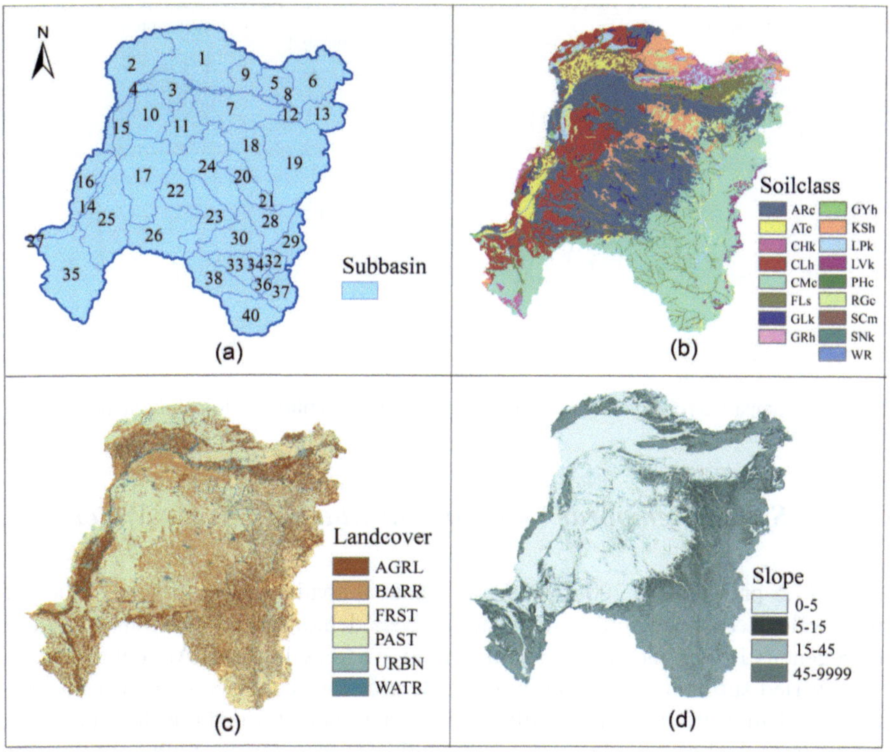

Fig. 17.2 Subbasin delineation **a**, soil class **b**, land cover **c** and slope profile **d** of the study area

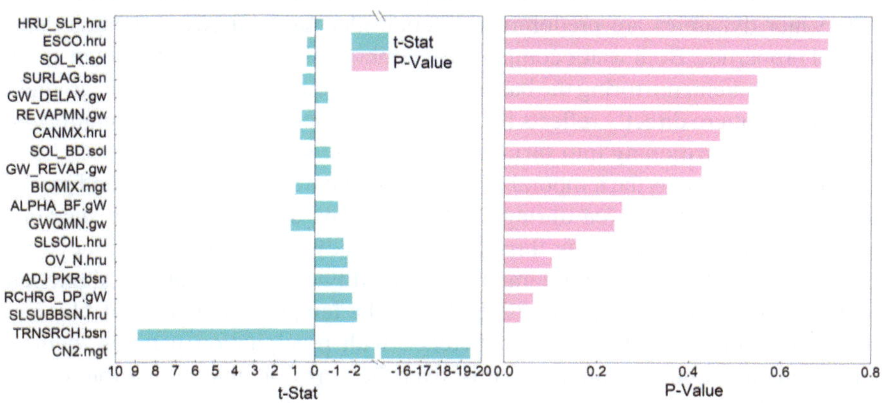

Fig. 17.3 P-value and t-test of parameter sensitivity

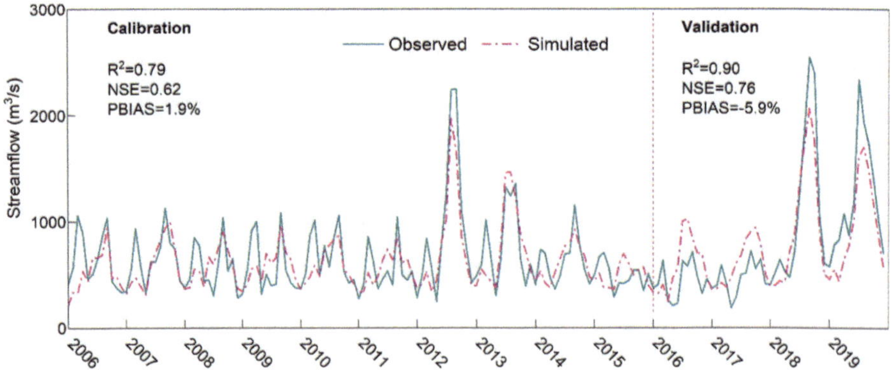

Fig. 17.4 Comparison of simulated and observed monthly streamflow hydrographs at the Longmen hydrological station

17.4.2 Spatial–temporal Variations in Blue and Green Water

Figure 17.5 presents the 2006–2019 mean annual spatial distributions of blue and green water across the study area. The analysis demonstrated pronounced spatial heterogeneity in blue and green water distribution across the basin. Annual blue water fluxes varied substantially (0–85 mm), while green water fluxes ranged from 140–680 mm, demonstrating consistently higher magnitudes throughout the study area. This hydrological partitioning reflects the region's characteristic high evapotranspiration rates and severe soil erosion, particularly across the Loess Plateau's central reaches. Precipitation constitutes the primary driver of blue water variability, whereas green water dynamics are predominantly controlled by land cover characteristics and soil properties. Both water components exhibit a distinct upstream–downstream increasing gradient, mirroring the basin's northwest-southeast precipitation gradient. This spatial pattern directly reflects the orographic enhancement of precipitation and its subsequent partitioning into surface flows and soil storage.

Figure 17.6 presents the temporal dynamics of blue water, green water, and precipitation across the study area during 2006–2019. The analysis reveals a sustained increase in green water fluxes throughout the study period, whereas blue water resources exhibited marked interannual variability with particularly significant increases during 2012–2013 and 2015–2016. The temporal dynamics of blue and green water fluxes exhibit strong coherence with precipitation patterns, demonstrating precipitation's dominant control over their spatiotemporal distribution. As the primary hydrological driver, precipitation governs both surface runoff generation (blue water) and soil moisture retention (green water), ultimately determining basin-scale water availability. More precipitation creates surface streamflow and groundwater recharge, increasing river and groundwater storage. Concentrated precipitation may lead to flooding, significantly increasing blue water resources in the short term.

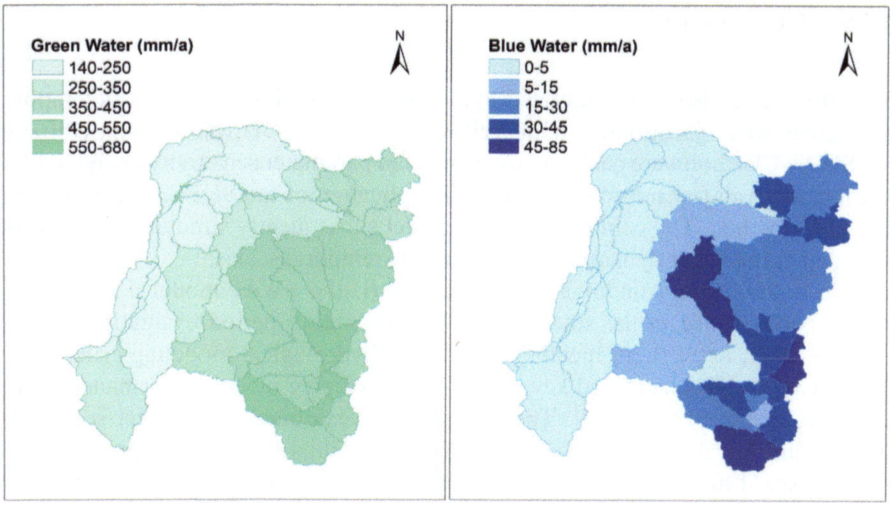

Fig. 17.5 Spatial distribution of annual average values of blue and green water resources

More precipitation is absorbed by the soil, increasing soil water content for utilization by plants. Vegetation grows vigorously, increasing evapotranspiration and green water resources.

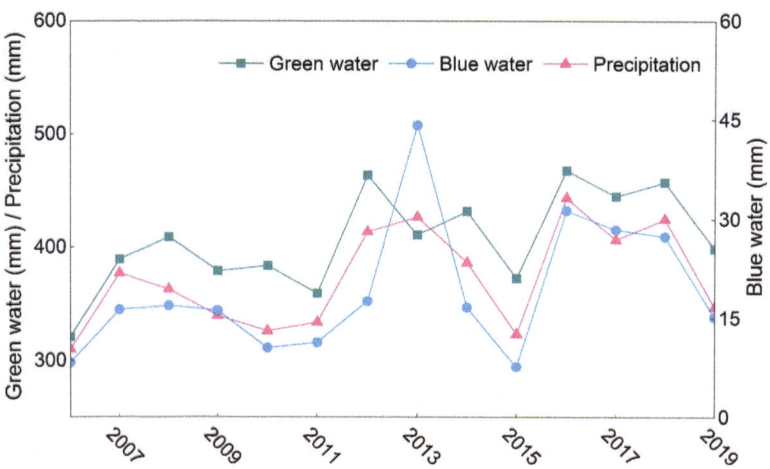

Fig. 17.6 Annual variations of blue and green water resources and precipitation from 2006 to 2019

17.5 Conclusions

This study establishes an integrated modeling framework to assess spatiotemporal blue-green water partitioning in the Yellow River Basin's Jiziwan Region. Utilizing the SWAT-CUP software package, we conducted parameter sensitivity analysis, calibration, and validation through combined application of the SUFI-2 algorithm and LH-OAT method. The calibrated SWAT model subsequently simulated basin-scale blue and green water dynamics across multiple temporal scales.

The following key findings were revealed: (i) the SWAT model demonstrated robust performance in the study area, with calibration period evaluation metrics (NSE = 0.62, R^2 = 0.79) meeting established hydrological modeling criteria; (ii) Sensitivity analysis identified key parameters governing model performance: CN2, TRNSRCH and SLSUBBSN. These parameters exhibited the highest sensitivity indices during calibration; (iii) The study area exhibits significantly greater green water fluxes (140–680 mm) compared to blue water fluxes (0–85 mm), attributable to the region's characteristic high evapotranspiration rates and pronounced soil erosion; (iv) during 2006–2019, both blue and green water fluxes exhibited precipitation-dependent interannual variability, with distinct response patterns. Green water demonstrated greater temporal stability, showing a steady increasing trend, whereas blue water displayed marked interannual fluctuations, reflecting its stronger sensitivity to precipitation extremes. Overall, this study advances the understanding of hydrological processes in the Yellow River Basin's Jiziwan Region through comprehensive runoff modeling and water availability assessment.

Acknowledgements This research was supported by the National Key R&D Program of China (2023YFC3206503) and the Innovative Research Group of the National Natural Science Foundation of China (52221003).

References

Chen G, Zuo D, Xu Z (2024) Changes in water conservation and possible causes in the Yellow River Basin of China during the recent four decades. J Hydrol 637:131314

Ding BB, Zhang JM, Zheng PF, Li ZD, Wang YS, Jia GD, Yu XX (2024) Water security assessment for effective water resource management based on multi-temporal blue and green water footprints. J Hydrol 632:130761

F YJ, Z A (2022) Spatiotemporal differentiation and driving patterns of water utilization intensity in Yellow River Basin of China: comprehensive perspective on the water quantity and quality. J Clean Product 369:133395

Falkenmark M (1995) Land-water linkages: a synopsis

Li JQ, Yang N, Shen ZY (2024) Evaluation of the water quality monitoring network layout based on driving-pressure-state-response framework and entropy weight TOPSIS model: a case study of Liao River China. J Environ Manag 361:121267

Wang PP, Huang GH, Li YP, Zhang YF, Cai TC, Song TN, Liu YY, Xu ZP, Shen ZY (2024) Unveiling key drivers of economy-water system and transforming water use pattern into sustainable development: Inner-Shaan-Ning region in the Yellow River Basin. J Clean Prod 475:14365

Xu ZP, Li YP, Huang GH, Shen ZY (2023) Developing an integrated PCE-ANOVA-RF method for uncertainty quantification of hydrological model–The Amu Darya River Basin in Central Asia. J Hydrol 624:129941

Yu ZB, Yin S, Bai JH, Wang C, Chen GZ, Wang W, Wang YQ, Cui BS, Liu XH, Li XW (2024) Suaeda salsa in relation to hydrological connectivity: from the view of plant trait networks. J Environ Inform 43(1):41–49

Zhang W, Liang W, Gao X, Li JY, Zhao X (2024a) Trajectory in water scarcity and potential water savings benefits in the Yellow River basin. J Hydrol 633:130998

Zhang X, Chen P, Dai S (2024b) A study on drought assessment in the Yi Luo River Basin based on SWAT. Ecol Ind 169:112852–112852

Chapter 18
Chance-Constrained Fuzzy Programming-Based Water-Ecology-Energy-Food Nexus Model for the City of Ordos

Jinxian Sun, Yongping Li, Guohe Huang, and Yanxiao Zhou

Abstract In this study, a chance-constrained fuzzy programming (CCFP) method is proposed to synergistically coordinate the water-ecology-energy-food (WEEF) nexus management. CCFP can address both probabilistic randomness and ambiguous fuzziness within the decision-making framework. Then a CCFP-WEEF model is formulated for Ordos. The main research results are: (i) groundwater remains the main water source in Ordos (accounting for an average of 43.7%) and the share of reclaimed water would rise significantly during 2046–2050; (ii) the water allocation of agriculture would be the largest among different users (reaching 4.8×10^9 m^3) and the water allocation of the ecology and energy would increase significantly during 2046–2050; (iii) the water resource allocation in each county would generally increase and the water allocation in Dongsheng would be the largest (reaching an average of 578.2 $\times$ 10^6 m^3 per year); (iv) systematic uncertainties significantly influence water allocation schemes, policymakers should incorporate the impacts of uncertainties into the WEEF nexus management, and formulate appropriate management strategies according to different risk tolerance levels.

Keywords Collaborative management · System optimization · Uncertainties

J. Sun · G. Huang
China-Canada Center of Energy, Environment and Sustainability Research, UR-SDU, School of Environmental Science and Engineering, Shandong University, Qingdao, China

Y. Li (✉) · Y. Zhou
State Key Laboratory of Regional Environment and Sustainability, China-Canada Center for Energy, Environment and Ecology Research, UR-BNU, School of Environment, Beijing Normal University, Beijing, China
e-mail: yongping.li@iseis.org

Y. Li · G. Huang
Environmental Systems Engineering Program, University of Regina, Regina, SK, Canada

Y. Li et al. (eds.), *16th International Conference on Environmental Science and Development*, Environmental Science and Engineering,
https://doi.org/10.1007/978-3-032-15080-6_18

18.1 Introduction

The compounding pressures of population expansion and intensified urbanization intensify the coupling between water scarcity, ecological deterioration, energy insecurity, and food system vulnerabilities (Lv et al. 2024). Against this backdrop, coordinated governance strategies are necessary for the water-ecology-energy-food (WEEF) nexus system to resolve interconnected resource conflicts. In addition, due to the impacts of climate-induced fluctuations and anthropogenic perturbations, there are various uncertainties in the system (Kujala et al. 2013). Therefore, it is important to incorporate the uncertainty method into the optimization model.

Fuzzy programming (FP) can effectively address the problem of incomplete information or linguistic ambiguity (Zhong et al. 1994). By employing membership functions to quantify subjective knowledge and qualitative descriptions, it avoids the excessive reliance on precise data inherent in traditional deterministic models (Kundu and Islam 2018). Hu et al. (2011) developed a feasibility-based imprecise fuzzy programming method to address uncertainties expressed as fuzzy and intervals. Zhang et al. (2023) proposed a decision-making framework based on spatial simulation and fuzzy multi-objective optimization to achieve sustainable land-use patterns. Belamkar et al. (2023) developed a multi-objective mixed-integer transportation model, where FP was applied to tackle uncertain economic parameters including revenue and purchase cost. However, the membership function in FP is subjective, and there are deficiencies in risk quantification. Chance-constrained programming (CCP) enhances the resilience of complex systems to random disturbances by incorporating probabilistic uncertainties into the optimization framework. Integrating chance-constrained and fuzzy programming can synergistically address fuzzy uncertainties and probabilistic uncertainties inherent in complex systems.

Therefore, the purpose of this study is to propose a chance-constrained fuzzy programming (CCFP) model to synergistically manage the WEEF nexus. By integrating fuzzy programming and chance-constrained programming, the model provides quantitative decision-making support for Ordos to alleviate water scarcity and balance energy, food and ecological demands under complex resource constraints.

18.2 Methodology

Fuzzy programming can effectively address uncertainty with imprecise information. However, fuzzy sets heavily rely on the subjective definition of membership functions, and fuzzy programming lacks probabilistic interpretation, making it challenging to quantify and assess risks. Therefore, chance-constrained fuzzy programming (CCFP) can be proposed to tackle imprecise information and probabilistic uncertainty by integrating FP and CCP. The CCFP can be formulated as follows (Ma et al. 2024):

$$Maxf = \sum_{j=1}^{J} c_j x_j \tag{18.1}$$

subject to:

$$\Pr\left[\sum_{j=1}^{J} e_{kj} x_j \le h_k\right] \ge 1 - p, \forall k \tag{18.2}$$

$$Cr\left[\sum_{j=1}^{J} \underset{kj}{\tilde{a}} x_j \ge b_k\right] \ge \alpha, \forall k \tag{18.3}$$

$$x_j \ge 0, \forall j \tag{18.4}$$

where x_j is decision variables, c_j are coefficients for decision variable x_j; e_{kj} and a_{kj} are technical coefficients for decision variable; a_{kj} is fuzzy parameter; h_k and b_k are right-side coefficients; p represents the default probability level. The Cr constraint is set according to the credibility programming, which means that the credibility level of satisfying the corresponding constraint should be greater than or equal to α. Suppose a_{kj} is a fuzzy parameter. Then, CCFP can be converted as follows:

$$Maxf = \sum_{j=1}^{J} c_j x_j \tag{18.5}$$

subject to:

$$\Pr\left[\sum_{j=1}^{J} e_{kj} x_j \le h_k\right] \ge 1 - p, \forall k \tag{18.6}$$

$$2(1-\alpha)\sum_{j=1}^{J} a_{kj} x_j + (2\alpha - 1)\sum_{j=1}^{J} \underset{-kj}{a} x_j \ge b_k, \forall k \tag{18.7}$$

$$x_j \ge 0, \forall j \tag{18.8}$$

18.3 Case Study

18.3.1 Study Area

Ordos faces severe challenges in balancing water resources shortage, ecosystem degradation, energy supply–demand imbalances, and food security (Hu et al. 2024). The water availability in Ordos in 2023 was approximately 1.8×10^9 m^3. The per capita water resource availability is about 810.7 m^3. This value is lower than the international water shortage threshold (1000 m^3 per capita) (Falkenmark 1989). Among these water resources, as groundwater dominates (accounting for 57.2%), the exploitation of groundwater seriously endangers ecological security. In addition, Ordos is a major energy city, but agriculture is the main water consumer. Therefore, strengthening the comprehensive management of water-ecology-energy-food is crucial for alleviating the water resource conflicts.

18.3.2 Modeling Formulation

Based on the WEEF model, the CCFP method is applied in Ordos with the aim of maximizing the system benefits. The framework incorporates six water sources, 19 water users and five planning periods (2026–2050). To address the uncertainties of fuzziness and randomness in the WEEF system, three p level (i.e., 0.05, 0.10 and 0.15) and three α level (i.e., 0.5, 0.75 and 1) are designated. Two energy, food and ecology demand levels (i.e., low and high) are also considered to quantitatively assess the impacts of energy, food and ecology demand on WEEF system. Therefore, the CCFP-WEFE model is expressed as:

$$Maxf = PROENE + PROAGR + PROIND + PRORES + PROECO \tag{18.9}$$

(1) Benefit for energy:

$$\begin{aligned}
PROENE = {} & \\
& \sum_{t}^{T}\sum_{j=1}^{7}\sum_{i=1}^{9}(GW_{ijt} + SW_{\mathrm{ijt}} + OW_{\mathrm{ijt}} + RW_{\mathrm{ijt}} + MW_{\mathrm{ijt}} + UW_{\mathrm{ijt}})/EWEX_{\mathrm{ijt}} \times PE_{\mathrm{ijt}} \\
& -\sum_{t}^{T}\sum_{j=1}^{7}\sum_{i=1}^{9}(GW_{ijt} + SW_{\mathrm{ijt}} + OW_{\mathrm{ijt}} + RW_{\mathrm{ijt}} + MW_{\mathrm{ijt}} + UW_{\mathrm{ijt}})/EWEX_{\mathrm{ijt}} \times UEC_{\mathrm{ijt}} \\
& -\sum_{t}^{T}\sum_{j=1}^{7}\sum_{i=1}^{9}(GW_{ijt} + SW_{\mathrm{ijt}} + OW_{\mathrm{ijt}} + RW_{\mathrm{ijt}} + MW_{\mathrm{ijt}} + UW_{\mathrm{ijt}}) \times UEU_{\mathrm{ijt}}
\end{aligned} \tag{18.10}$$

(2) Benefit for agriculture:

$$PROAGR = \sum_{t}^{T}\sum_{j=13}^{19}\sum_{i=1}^{9}(GW_{ijt} + SW_{\mathrm{ijt}} + OW_{\mathrm{ijt}} + RW_{\mathrm{ijt}} + MW_{\mathrm{ijt}} + UW_{\mathrm{ijt}}) \times BW_{\mathrm{ijt}} - \sum_{t}^{T}\sum_{j=13}^{19}\sum_{i=1}^{9}(GW_{ijt} + SW_{\mathrm{ijt}} + OW_{\mathrm{ijt}} + RW_{\mathrm{ijt}} + MW_{\mathrm{ijt}} + UW_{\mathrm{ijt}}) \times UEU_{\mathrm{ijt}} \tag{18.11}$$

(3) Benefit for industry (non-energy):

$$PROIND = \sum_{t}^{T}\sum_{j=8}^{8}\sum_{i=1}^{9}(GW_{ijt} + SW_{\mathrm{ijt}} + OW_{\mathrm{ijt}} + RW_{\mathrm{ijt}} + MW_{\mathrm{ijt}} + UW_{\mathrm{ijt}}) \times BW_{\mathrm{ijt}} - \sum_{t}^{T}\sum_{j=8}^{8}\sum_{i=1}^{9}(GW_{ijt} + SW_{\mathrm{ijt}} + OW_{\mathrm{ijt}} + RW_{\mathrm{ijt}} + MW_{\mathrm{ijt}} + UW_{\mathrm{ijt}}) \times UEU_{\mathrm{ijt}} \tag{18.12}$$

(4) Benefit for resident:

$$PRORES = -\sum_{t}^{T}\sum_{j=9}^{9}\sum_{i=1}^{9}(GW_{ijt} + SW_{\mathrm{ijt}} + OW_{\mathrm{ijt}} + RW_{\mathrm{ijt}} + MW_{\mathrm{ijt}} + UW_{\mathrm{ijt}}) \times UEU_{\mathrm{ijt}} \tag{18.13}$$

(5) Benefit for ecology:

$$PROECO = \sum_{t}^{T}\sum_{j=10}^{12}\sum_{i=1}^{9}(GW_{ijt} + SW_{\mathrm{ijt}} + OW_{\mathrm{ijt}} + RW_{\mathrm{ijt}} + MW_{\mathrm{ijt}} + UW_{\mathrm{ijt}})/GWD_{\mathrm{jt}} \times EE_{jt} - \sum_{t}^{T}\sum_{j=10}^{12}\sum_{i=1}^{9}(GW_{ijt} + SW_{\mathrm{ijt}} + OW_{\mathrm{ijt}} + RW_{\mathrm{ijt}} + MW_{\mathrm{ijt}} + UW_{\mathrm{ijt}}) \times UEU_{\mathrm{ijt}} \tag{18.14}$$

subject to:

(1) Water demand constraint

$$SW_{ijt} + GW_{ijt} + OW_{ijt} + RW_{ijt} + MW_{ijt} + UW_{ijt} \geq DW_{ijt}, \ \forall i, j, t \tag{18.15}$$

(2) Water availability constraint

$$\Pr\left[\sum_{j}^{J}\sum_{i}^{I} GW_{ijt} \times \gamma_{ijt} \leq TGA_t \times PGE_t\right] \geq 1 - p, \ \forall t \tag{18.16}$$

(3) Sewage reuse constraint

$$Cr\left[\sum_{j}^{J}\sum_{i}^{I}\left(SW_{ijt} + GW_{ijt} + OW_{ijt} + RW_{ijt} + MW_{ijt} + UW_{ijt}\right) \times \tilde{\theta}_{ijt} \geq \sum_{j}^{J}\sum_{i}^{I} RW_{ijt}\right] \geq \alpha, \forall t \tag{18.17}$$

(4) Energy extraction constraint

$$\sum_{i}^{I}\left(SW_{ijt} + GW_{ijt} + OW_{ijt} + RW_{ijt} + MW_{ijt} + UW_{ijt}\right)/EWEX_{ijt} \leq AE_{jt}^{\max} \tag{18.18}$$

(5) Food security constraint

$$\sum_{i}^{I}\left(SW_{ijt} + GW_{ijt} + OW_{ijt} + RW_{ijt} + MW_{ijt} + UW_{ijt}\right)/WPC_{ijt} \times YA_{ijt} \geq FD_{jt}, \forall \text{j, t} \tag{18.19}$$

(6) Ecological security constraint

$$\sum_{i}^{I}\sum_{j=10}^{13}(SW_{ijt} + GW_{ijt} + OW_{ijt} + RW_{ijt} + MW_{ijt} + UW_{ijt})/GWD_{jt} \geq TDE_t, \ \forall t \tag{18.20}$$

18.4 Result and Discussion

Figure 18.1 shows the system benefits and water allocation under 72 scenarios. According to the results, the water resource allocation in Ordos would range from 53.3×10^9 m^3 to 55.1×10^9 m^3, S50 corresponds to the smallest water allocation, and S69 represents the largest. The system benefits would be 25.8×10^{12}–30.9×10^{12} yuan, S17 yields the highest system benefits, and S54 yields the lowest. Water resource allocation under different scenarios would have variations, indicating that total water allocation adjusts according to the demand levels. The trend of system benefits aligns with water allocation. Increasing available amount of water resource

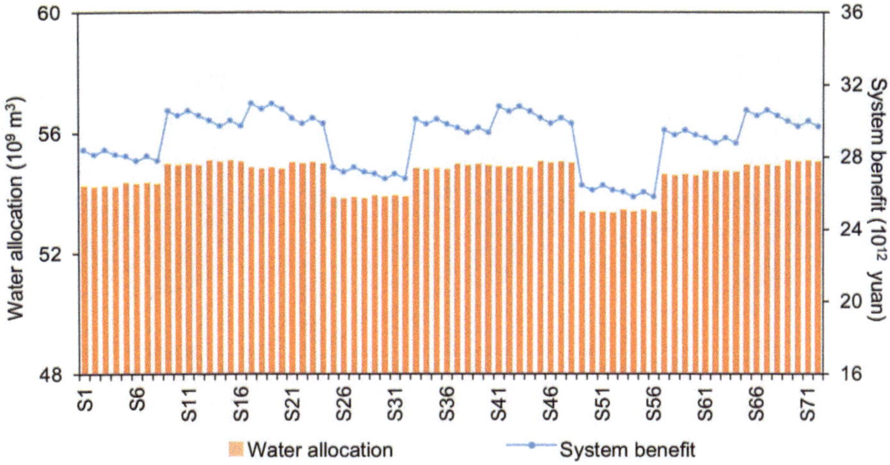

Fig. 18.1 The water allocation and system benefit under all scenarios

by enhancing the utilization of transferred water and non-conventional water would increase system benefits. However, as water allocation remains at a high level, the growth in benefits would show a decelerating trend, suggesting a potential diminishing marginal utility effect. Result implies that once water allocation reaches a certain scale, further increases in water input yield progressively smaller gains in system benefits. It is necessary to optimize the water utilization structure and improve water-use efficiency. To achieve sustainable water resource management in Ordos, decision-makers should ensure the rational allocation of water resources to maintain optimal system benefits, but also avoid excessive dependence on water input to prevent inefficiency and unsustainable resource consumption. Moreover, higher p level (from 0.05 to 0.15) corresponds to a 2.1% increase in water allocation, indicating that a higher risk tolerance level can provide more water resources, but at the cost of increased systemic risk.

Figure 18.2 illustrates the proportions of different water sources. The results indicate that groundwater would dominate at an average of 43.7%, followed by transferred water at 33.8%, and other sources contribute minimally, indicating Ordos relies heavily on groundwater and transferred water. Over planning periods (from $t = 1$ to $t = 5$), the proportion of non-conventional water such as reclaimed water and mine water would increase, reflecting strategic adjustments toward alternative sources. Ordos remains dependent on conventional sources, necessitating balanced planning to avoid over-reliance on single sources and advancing non-conventional water development. Under high-risk scenarios (i.e., $p = 0.15$), surface water and groundwater would increase, suggesting greater risk tolerance enables higher utilization of these water sources. As p rises from 0.05 to 0.15, surface water would increase by 0.3% and groundwater would increase by 4.2%. As α increases from 0.5 to 1, reclaimed water would decrease by 1.3%. The changes in the p level and the α level affect

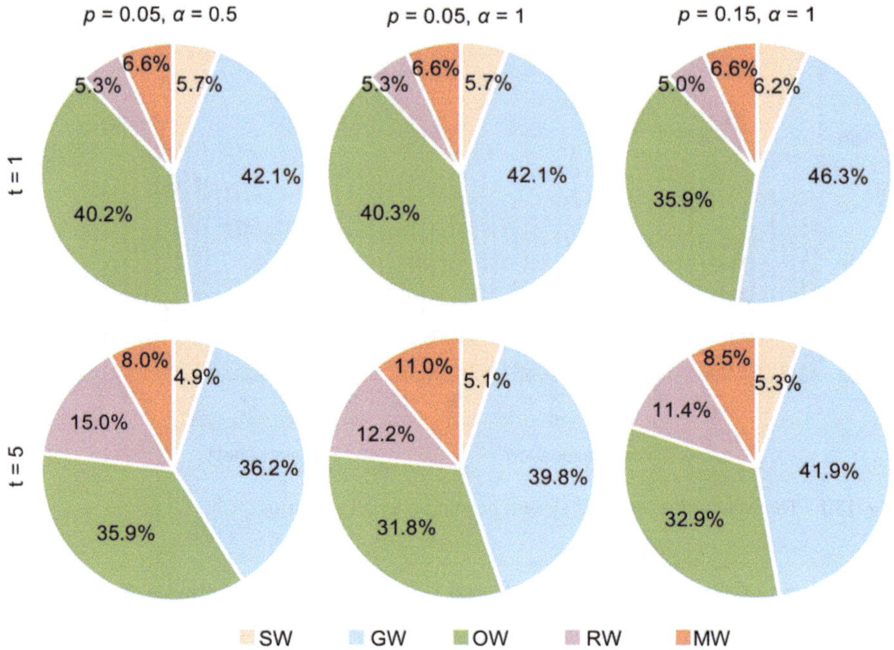

Fig. 18.2 The proportion of different water sources

the changes in the results, indicating that the uncertainties represented by randomness and fuzziness will influence the proportion results of various water sources. Policy-makers need to consider the impact of uncertainties when making plans.

Figure 18.3 presents water allocation across sectors under different scenarios. Results demonstrate that food consistently receives the highest water allocation (reaching 4.8×10^9 m^3 by the planning horizon), followed by energy (2.6×10^9 m^3), highlighting their dominant position in Ordos' economic structure and substantial water demands for production processes. Temporal analysis reveals a general upward trend in water allocation from $t = 1$ to $t = 5$, reflecting growing urban water demand. Besides, ecology and energy exhibit the most significant increases (56.2% and 131.2% growth respectively by the planning horizon compared to initial levels). It shows that the ecology and energy can obtain relatively large benefits per unit of water. Policy-makers should give priority to allocating the surplus water to the ecological and energy. Strategic water reallocation among sectors proves to be essential for optimizing system benefits.

Figure 18.4 shows the water resource allocation in different counties. According to the results, the water allocation in Dongsheng would be the highest, reaching an average of 578.2×10^6 m^3 per year by the end of the planning period. It is followed by Hanggin and Wushen, which can reach 420.1×10^6 m^3 and 359.0×10^6 m^3 respectively. The water allocation in Otog Front would be the lowest (101.0×10^6 m^3). Under different scenarios, there is a similarity in the changing

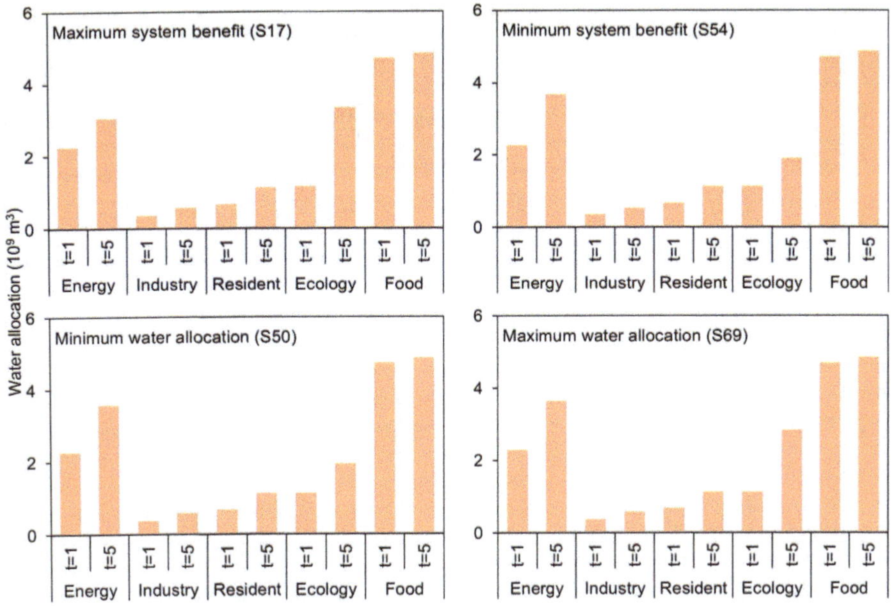

Fig. 18.3 Water allocation among different users

trends of water resource allocation in different counties. From t = 1 to t = 5, the water resource allocation would generally increase, which reflects the growth of water demand brought about by the overall development of Ordos. Among them, due to the combined effects of industrial concentration, population urbanization, and the implementation of high-water-consuming projects, Dongsheng would have the largest increase, with an average increase of about 1.9 times. It is necessary to precisely allocate water resources according to the functional positioning of the region and its development needs. For core areas with a large water consumption like Dongsheng, it is necessary to strengthen water conservation management and the recycling of water resources. At the same time, the system benefits should be comprehensively considered, and the efficiency of water resource utilization should be improved by optimizing the water resource allocation among different counties.

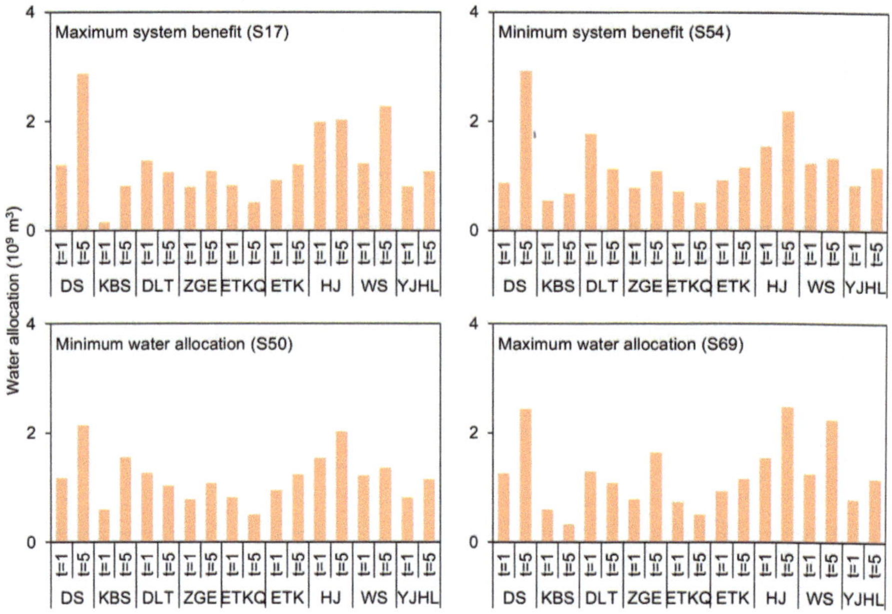

Fig. 18.4 Water allocation for different counties

18.5 Conclusion

In this study, a chance-constrained fuzzy programming (CCFP) method is proposed. This method has advantages in dealing with fuzziness and probabilistic uncertainties simultaneously. Then, this method is applied to Ordos, and 72 scenarios are formulated to manage the water-ecology-energy-food (WEEF) nexus framework.

The major findings are as follows: (i) Groundwater remains the main water source in Ordos during the planning period, accounting for an average of 43.7%, followed by transferred water (accounting for 33.8%). Reclaimed water would have a larger proportion than 2023. The rational utilization of various water sources is crucial for alleviating water shortage in Ordos. (ii) During the planning period, agriculture would have the highest water allocation, and the water allocation of ecology and energy would increase significantly over time. (iii) Water allocations across all counties would show an upward trend during the planning period, and Dongsheng would have the highest allocation, followed by Hanggin. It is necessary to rationally allocate water resources according to the development needs of different regions. (iv) As the p-value increase, the total water resource allocation would rise 2.1%, and the system benefits would increase by 11.9%. The proportions of surface water and groundwater would increase by 0.3 and 4.2% respectively. As the α level increases, the proportion of reclaimed water would decrease by 1.3%. In the management of the WEEF system, policymakers should formulate suitable water resource allocation strategies by considering uncertainties and risk perceptions.

Acknowledgements This research was supported by the National Key R&D Program of China (2023YFC3206503), the National Natural Science Foundation of China (52279003) and the Innovative Research Group of the National Natural Science Foundation of China (52221003).

Appendix

i	Counties in Ordos; i = 1, 2, 3, …, 9 (for Dongsheng, Kangbashi, Dalad, Jungar, Otog Front, Otog, Hangjing, Wushen and Ejin Horo)
j	Water users; j = 1, 2, 3, …, 19 (for coal, oil, gas, biomass, wind, hydro, solar, industry (non-energy), residents, woodland, grassland, waterbody, wheat, corn, rice, beans, beets, sunflower and animal husbandry)
t	Planning period, t = 1, …, 5 (2026–2050)
$SW_{i,j,t}$	The amount of surface water allocation (m^3)
$GW_{i,j,t}$	The amount of groundwater allocation (m^3)
$OW_{i,j,t}$	The amount of transferred water allocation (m^3)
$RW_{i,j,t}$	The amount of reclaimed water allocation (m^3)
$MW_{i,j,t}$	The amount of mine water allocation (m^3)
$UW_{i,j,t}$	The amount of other water allocation (m^3)
$EWEX_{i,j,t}$	Water consumption per unit of energy j (m^3/kg standard coal)
$PE_{i,j,t}$	Local price of energy j (yuan/kg standard coal)
$UEC_{i,j,t}$	Production cost per unit of energy (yuan/kg)
$UEU_{i,j,t}$	Unit water distribution cost (yuan/m^3)
$BW_{i,j,t}$	Benefit of water allocation (yuan/m^3)
$GWD_{j,t}$	Water consumption per unit of ecological land (m^3/ha)
$EE_{j,t}$	Ecological value per unit (yuan/ha)
$DW_{i,j,t}$	The water demand of users j in county i under planning period t (m^3)
$\gamma_{i,j,t}$	Efficiency of water resources allocation (%)
TGA_t	Total groundwater availability under planning period t (m^3)
PGE_t	Maximum over-extraction coefficient
$\theta_{i,j,t}$	Reclaimed water reuse rate (%)
$AE_{jt}^{\max}$	maximum extraction of energy resources (kg)
$FD_{j,t}$	Food demand of users j under planning period t (kg)
TDE_t	Total food demand under planning period t (kg)

References

Belamkar P, Biswas S, Baidya A, Majumder P, Bera UK (2023) Multi-objective optimization of agro-food supply chain networking problem integrating economic viability and environmental sustainability through type-2 fuzzy-based decision making. J Clean Prod 421:138294

Falkenmark M (1989) The massive water scarcity now threatening Africa: why isn't it being addressed? Ambio 18(2):112–118

Hu Q, Huang G, Cai Y, Huang Y (2011) Feasibility-based inexact fuzzy programming for electric power generation systems planning under dual uncertainties. Appl Energy 88(12):4642–4654

Hu Z, Tang S, Mo S, Shi X, Yin X, Sun Y, Liu X, Duan L, Miao P, Liu T, Wu J (2024) Water storage changes (2003–2020) in the Ordos Basin, China, explained by GRACE data and interpretable deep learning. Hydrogeol J 32(1):307–320

Kujala H, Burgman MA, Moilanen A (2013) Treatment of uncertainty in conservation under climate change. Conserv Lett 6(2):73–85

Kundu T, Islam S (2018) Neutrosophic goal geometric programming problem and its application to multi-objective reliability optimization model. Int J Fuzzy Syst 20(6):1986–1994

Lv C, Hu Y, Ling M, Luo A, Yan D (2024) Comprehensive evaluation and obstacle factors of coordinated development of regional water–ecology–energy–food nexus. Environ Dev Sustain 26(8):20001–20025

Ma Y, Li Y, Huang G, Liu Y, Zhang Y (2024) Collaborative management of water-energy-food-ecosystems nexus in Central Asia under uncertainty. Water Resourc Res 60(3):e2023WR035166

Zhang T, Xin X, He F, Wang X, Chen K (2023) How to promote sustainable land use in Hangzhou Bay, China? a decision framework based on fuzzy multiobjective optimization and spatial simulation. J Clean Prod 414:137576

Zhong Q, Yue Z, Guangyuan W (1994) On fuzzy random linear programming. Fuzzy Sets Syst 65(1):31–49

Chapter 19
A Scenario-Based Optimization Model for Water-Ecology-Energy-Food Nexus Management

Yanxiao Zhou, Yongping Li, Guohe Huang, and Yanfeng Li

Abstract The uncertainty of water resource availability poses a significant challenge to the sustainable development. In this study, a scenario-based optimization model is developed for the integrated management of the water-ecology-energy-food (WEEF) nexus. The model is applied to the Jiziwan Area of the Yellow River, where 27 scenarios are designed to address uncertainty of water resource availability. Results indicate that: (i) compared to the current situation, the proportion of surface water would decrease from 68.7 to 48.4%, and the proportion of groundwater would decline from 25 to 16.4% following WEEF optimization, which underscores the urgent need to diversify water sources to mitigate over-exploitation of traditional resources; (ii) with increased water availability, the water allocation for agriculture and resident would rise, reflecting the influence of resource uncertainty on water allocation pattern; (iii) among all water users, agriculture would remain the primary consumer, experiencing a decline in water allocation ranging from 16.4 to 20.3% compared to the current situation; ecology would be the second-largest water user and its water allocation would increase by 11.4–14.1%. The results indicate a positive shift toward environmental restoration and the sustainable development of the WEEF nexus, which can help decision-makers make informed decisions.

Keywords Jiziwan area · Scenario analysis · Water-ecology-energy-food nexus

Y. Zhou · Y. Li (✉) · G. Huang · Y. Li
State Key Laboratory of Regional Environment and Sustainability, China-Canada Center for Energy, Environment and Ecology Research, UR-BNU, School of Environment, Beijing Normal University, Beijing, China
e-mail: yongping.li@iseis.org

Y. Li · G. Huang
Environmental Systems Engineering Program, University of Regina, Regina, SK, Canada

Y. Li et al. (eds.), *16th International Conference on Environmental Science and Development*, Environmental Science and Engineering,
https://doi.org/10.1007/978-3-032-15080-6_19

19.1 Introduction

Water scarcity has become a global challenge, which can intensify ecosystem degradation and pose significant threats to energy security and food production (Fernandes et al. 2023). To recognize the complex interconnections among water, ecology, energy and food systems, the water-ecology-energy-food (WEEF) nexus has garnered significant attention (Tan et al. 2025). However, due to the randomness of natural processes (e.g., climate change and precipitation), incomplete regional information, and the inherent complexity of the water resource system, many parameters within the nexus system are subject to uncertainty, such as the availability of water resources (Nematian 2023). These uncertainties pose significant challenges to the modeling, optimization and management of the WEEF nexus. Therefore, it is necessary to explore the interplay within the WEEF nexus and analyze the impact of uncertainty factors for sustainable management and decision-making.

Numerous studies have been carried out on the interconnections and integrated management among water, food as well as energy systems. For example, Lee and Jung (2018) estimated the efficiency of agricultural water within the water-food nexus; the results indicated that crop types and irrigation techniques can significantly enhance water use efficiency and economic productivity (Lee and Jung 2018). Li et al. (2023) developed a water-energy-food nexus model to maximize the performance of agriculture, forestry and animal husbandry systems (Li et al. 2023). Radmehr et al. (2024) formulated a multi-objective method for the water-food-energy nexus, which aimed to improve the productivity of energy and water resources, reduce environmental damage and guarantee food security (Radmehr et al. 2024). However, in certain regions, the water-energy-food nexus is also closely related to the ecological environment. Currently, research on the interactive relationships and holistic optimization of water, ecology, food as well as energy, remains limited.

Therefore, the purpose of this research is to present a scenario-based optimized model for water-ecology-energy-food (SO-WEEF) nexus system. The key innovations and contributions include: (i) multi-scenario analysis is used to address uncertainties in the availability of multiple water sources; (ii) an SO-WEEF model is proposed for the Jiziwan Area, where different inflow levels are examined to capture the interactions among different water sources; (iii) the results can provide decision support for managers to effectively mitigate water shortage, promote ecological restoration and ensure food security.

19.2 Application

19.2.1 Problem Statement

The Jiziwan area is situated in the central of the Yellow River, encompassing the cities of Zhongwei, Wuzhong and Yinchuan in Ningxia, Yulin and Yan'an in Shaanxi, as well as Ordos in Inner Mongolia (Ren et al. 2022). This region experiences an arid climate, with an annual rainfall varying between 250 and 650 mm, but the evaporation reaches 800–1000 mm (Yin et al. 2023). However, constrained by the natural conditions, the imbalance between water supply and demand is particularly prominent (Zhang et al. 2023). Therefore, collaborative management of the WEEF nexus in the Jiziwan Area is essential to alleviate water shortage and achieve the synergetic development of water, energy, food and ecological systems.

19.2.2 SO-WEEF Modeling Formulation

Based on multi-scenario analysis, a water-ecology-energy-food (WEEF) nexus management model is developed for the Jiziwan Area of the Yellow River. The model considers eight water sources, aiming to maximize water allocation benefit over the planning period from 2026 to 2040. Additionally, in order to address the uncertainty in the WEEF nexus system, three availability levels of surface water, groundwater and unconventional water are considered, respectively. The SO-WEEF model is proposed as follows:

$$
\begin{aligned}
&\mathrm{Max}\, f = \\
&\sum_{i=1}^{I}\sum_{j=1}^{10}\sum_{t=1}^{T} PRW_{ijt} \times \begin{pmatrix} SW_{ijt} + GW_{ijt} + OW_{ijt} + IRW_{ijt} + DRW_{ijt} \\ +HW_{ijt} + PW_{ijt} + BW_{ijt} + MW_{ijt} \end{pmatrix} \\
&+\sum_{t=1}^{T}\sum_{i=1}^{I}\sum_{j=11}^{14} \begin{pmatrix} SW_{ijt} + GW_{ijt} + OW_{ijt} + \\ HW_{ijt} + IRW_{ijt} + DRW_{ijt} + \\ ERW_{ijt} + PW_{ijt} + BW_{ijt} + MW_{ijt} \end{pmatrix} \times \begin{pmatrix} CR_{jt} + ST_{jt} + RM_{jt} \\ +WR_{jt} + BD_{jt} + RA_{jt} \\ +SC_{jt} + FP_{jt} \end{pmatrix} \\
&/GWD_{jt} \\
&+\sum_{t=1}^{T}\left\{\left[TYA_t - \sum_{i=1}^{I}\sum_{j=11}^{14} SW_{ijt}/\gamma\gamma_{jt} - \sum_{i=1}^{I} SEW_{ikt}/\gamma e_{kt} - WYL_t\right] \times \overline{bUlan_t}\right\} \\
&+\sum_{i=1}^{I}\sum_{k=1}^{K}\sum_{t=1}^{T} RED_{kt} \times \begin{pmatrix} SEW_{ikt} + GEW_{ikt} + OEW_{ikt} + IREW_{ikt} + DREW_{ikt} \\ +EREW_{ikt} + PEW_{ikt} + BEW_{ikt} + MEW_{ikt} \end{pmatrix}
\end{aligned}
\tag{19.1}
$$

subject to:

(1) Water resource availability constraints

$$\sum_{i=1}^{I}\sum_{j=1}^{J} SW_{ijt} \Big/ \gamma\gamma_{jt} + \sum_{i=1}^{I}\sum_{k=1}^{K} SEW_{ikt}/(\gamma e_{kt}) + WYL_t \leq TYA_t + \sum_{i=1}^{I}\sum_{j=1}^{7} SW_{ijt} \times RWR_t \tag{19.2}$$

$$\sum_{i=1}^{I}\sum_{j=1}^{J} GW_{ijt} \Big/ \gamma\gamma_{jt} + \sum_{i=1}^{I}\sum_{k=1}^{K} GEW_{ikt}/(\gamma e_{kt}) \leq TGA_t \times PGE_t + \sum_{i=1}^{I}\sum_{j=1}^{7} GW_{ijt} \times RWR_t \tag{19.3}$$

(2) Transferred water constraint

$$\sum_{i=1}^{I}\sum_{j=1}^{J} OW_{ijt} \Big/ \gamma\gamma_{jt} + \sum_{i=1}^{I}\sum_{k=1}^{K} OEW_{ikt}/(\gamma e_{kt}) \leq TOA_t \tag{19.4}$$

(3) Water demand constraints

$$\begin{aligned} &SW_{ijt} + GW_{ijt} + OW_{ijt} + IRW_{ijt} + DRW_{ijt} \\ &+ERW_{ijt} + HW_{ijt} + PW_{ijt}+BW_{ijt} + MW_{ijt} \geq DW_{ijt} \end{aligned} \tag{19.5}$$

$$\begin{aligned} &SEW_{ikt} + GEW_{ikt} + OEW_{ikt} + PEW_{ikt} + BEW_{ikt} \\ &+IREW_{ikt} + DREW_{ikt} + EREW_{ikt} + MEW_{ikt} \geq ELEW_{ikt} \end{aligned} \tag{19.6}$$

(4) Food demand constraint

$$\sum_{i}^{I}\sum_{j=1}^{7} \begin{pmatrix} SW_{ijt} + GW_{ijt} + OW_{ijt} + DRW_{ijt} \\ +HW_{ijt} + PW_{ijt} + BW_{ijt} + MW_{ijt} \end{pmatrix} \Big/ WPC_{ijt} \times YA_{ijt} \geq \sum_{i=1}^{I} \begin{matrix} FD_t \\ \times PO_{it} \end{matrix} \tag{19.7}$$

(5) Energy consumption constraint

$$\sum_{i}^{I}\sum_{j=1}^{7} \begin{pmatrix} SW_{ijt} + GW_{ijt} + OW_{ijt} + DRW_{ijt} \\ +HW_{ijt} + PW_{ijt}+BW_{ijt} + MW_{ijt} \end{pmatrix} /(1 + LF_{ijt})/WPC_{ijt} \times KLD_{jt} \leq TED_t^{\max} \tag{19.8}$$

(6) Cooling water constraint

$$\sum_{i}^{I}\sum_{k}^{K}\text{rec}_{kt}\times\begin{pmatrix}SEW_{ikt}+GEW_{ikt}+OEW_{ikt}\\+DREW_{ikt}+IREW_{ikt}+PEW_{ikt}\\+BEW_{ikt}+MEW_{ikt}\end{pmatrix}\geq\sum_{i}^{I}\sum_{j}^{J}ERW_{ijt}+\sum_{i}^{I}\sum_{k}^{K}EREW_{ijt}\tag{19.9}$$

(7) Ecological area constraint

$$\sum_{i}^{I}\begin{pmatrix}SW_{ijt}+GW_{ijt}+OW_{ijt}+IRW_{ijt}+DRW_{ijt}\\+ERW_{ijt}+PW_{ijt}+BW_{ijt}+MW_{ijt}+HW_{ijt}\end{pmatrix}/GWD_{jt}\geq TDE_{jt},\quad j=11,12,13,14\tag{19.10}$$

19.3 Result and Discussion

Figure 19.1 visualizes the optimal water allocation of multiple water departments under 27 scenarios. Results indicate that the total water allocation for Jiziwan in 2026–2040 would be 63.5×10^9 m^3 to 76.3×10^9 m^3; the system benefit would be 14.8×10^{12} RMB¥ to 14.9×10^{12} RMB¥. Figure 19.1b indicates the water allocation proportion of six cities. Ordos would have the highest proportion of water resource allocation, accounting for 34% under LLL scenario, which indicates that Ordos would be the primary water-consuming region. This is attributed to Ordos's role as a national energy base and an important agricultural and animal husbandry area. The proportions of surface water, groundwater, transferred water and unconventional water resources under the LLL scenario would be 48.4%, 16.4%, 16.2% and 18.9%, respectively (as shown in Fig. 19.1c). The optimization results indicate that the proportions of surface water and groundwater would decrease by 20.3% and 8.5%, respectively, compared to the current situation.

Figure 19.2 illustrates the water allocation proportions of multiple users under different scenarios. Agriculture and ecology would be the primary water users in Jiziwan Area, accounting for 40.6–44.5% and 21.9–24.5% of the total water allocation, respectively. Compared with the current data, the water allocation for agriculture would decrease by 16.4–20.3%; the water allocation for ecology would increase by 11.4–14.1%. The uncertainty of water availability would affect the water distribution structure. For example, as available surface water and groundwater resources increase (i.e., from low to high levels), the proportions of agricultural and residential water allocation would increase; the share for other sectors would reduce. Compared

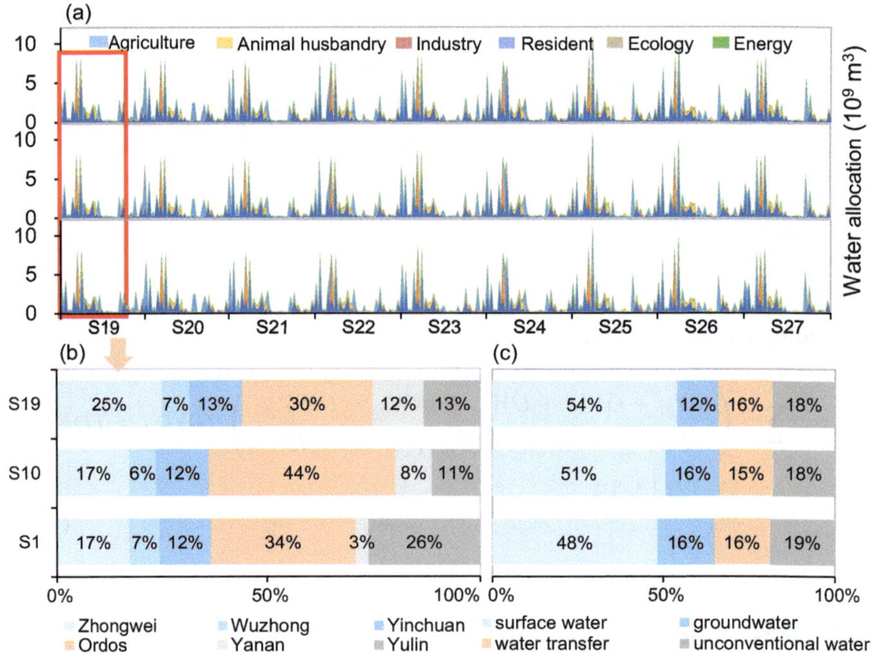

Fig. 19.1 Water allocation schemes under all planning schemes

to the LLL scenario, the water allocation for agriculture under HLL scenario would rise by 3.6×10^9 m^3, and the water allocation for resident would increase by 1.5×10^9 m^3 over the planning period.

Figure 19.3 presents the agricultural planting structures in 43 counties under LLL scenario. From the county-level perspective, Dongsheng, Dalate, Haiyuan, Otog Front Banner, Tongxin, Hangjin and Yuyang would be the primary crop cultivation areas, and the cultivated area of these seven counties would account for 59.3% of the total cultivated area. At the municipal level, Ordos would have the largest cultivated area in Jiziwan Area, comprising 34–56% across different periods, followed by Yan'an. Figure 19.3c illustrates the agricultural cropping pattern in the Jiziwan Area. The results indicate that the cultivated area of various crops would rank as follows: corn (30.9%) > sunflower (25.3%) > wheat (16.7%) > rice (13.6%) > soybean (9.7%) > sugar beet (4.6%). The selection of these crops is closely related to local climatic and soil conditions. High-yield and drought-tolerant crops (e.g., corn) would be particularly well-suited to the arid climate of the Jiziwan Area, maintaining high yield under limited water availability. In contrast, water-intensive crops such as rice would have a small cultivated area, especially in regions with scarce water resources, which reflects the significant impact of local water resources on agricultural planting structure.

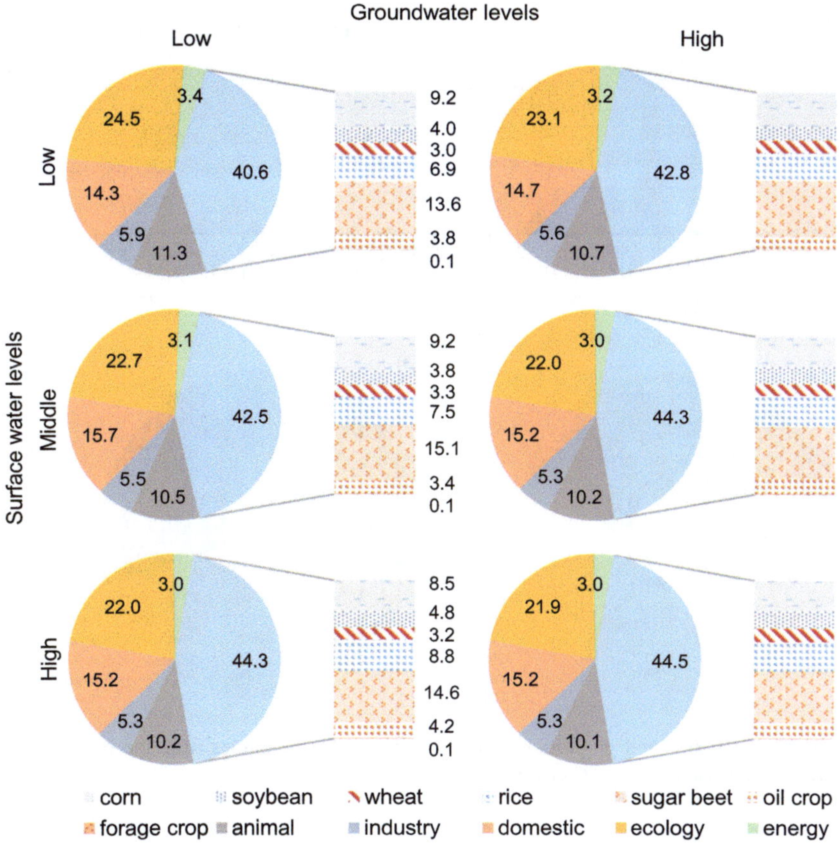

Fig. 19.2 The water allocation proportion of each user under different scenarios

19.4 Conclusions

In this study, a scenario-based optimized model for the water-ecology-energy-food (SO-WEEF) nexus has been developed. Scenario analysis has been employed to generate WEEF management schemes under uncertainty and assess the robustness of the model. Specifically, different levels have been set for surface water, groundwater and unconventional water availability to address uncertainties in water resources caused by climate change and human activity. The proposed model has been applied to WEEF management in the Jiziwan Area with the objective of maximizing system benefit.

Some findings and policy recommendations can be derived from the optimized results. Firstly, surface water, groundwater, transferred water and unconventional water resources would constitute 48.4%, 16.4%, 16.2% and 18.9% of total water

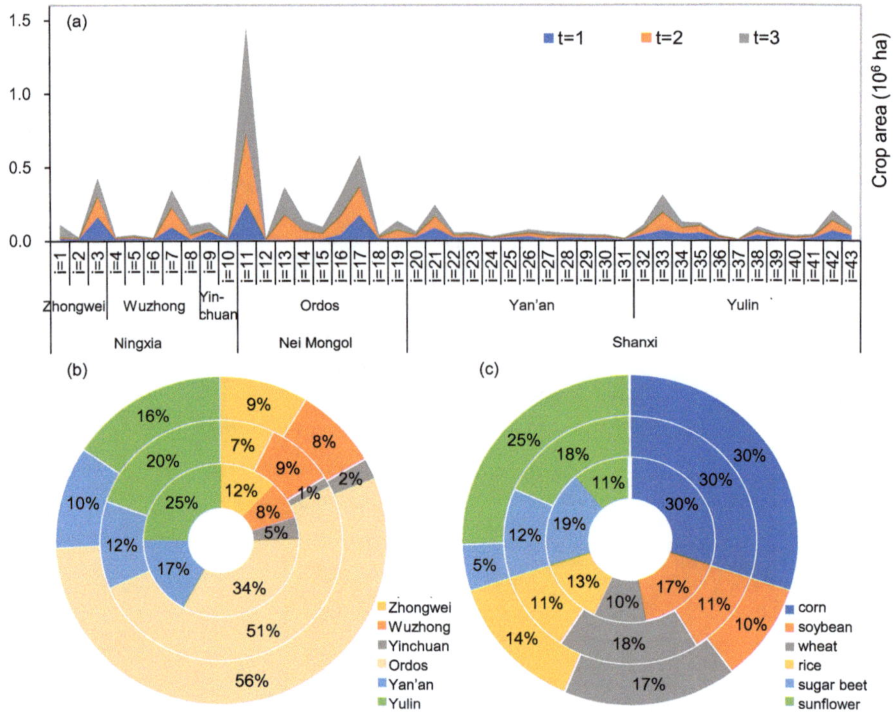

Fig. 19.3 The crop area and proportion under scenario LLL (the circles from the inside to the outside represent periods 1–3, respectively)

allocation. Compared to the present situation, the share of surface water and groundwater would reduce by 20.3% and 8.5%, respectively. Governments should incentivize the utilization of unconventional water resources, such as desalinated seawater, rainwater harvesting and treated wastewater recycling, to reduce the dependence on overexploited surface and groundwater systems. Secondly, as water availability increases, water allocation for agriculture as well as resident would rise, and that for other sectors would decrease, reflecting the response of water distribution to resource fluctuations. Finally, among all water-using sectors, agriculture and ecology would account for the largest proportions. Compared to the current situation, the proportion of water allocated to crops would reduce from 70.9 to 44.5%, and that of ecology would increase from 10.5 to 24.5%. Therefore, decision-makers should promote sustainable agricultural water use and enhance water-use efficiency. Measures such as encouraging water-saving and precision irrigation technologies should be implemented to offset the reduction in agricultural water allocation.

Acknowledgements This research was supported by the National Key R&D Program of China (2023YFC3206503), the National Natural Science Foundation of China (52279003) and the Innovative Research Group of the National Natural Science Foundation of China (52221003).

Appendix: Nomenclatures

i counties in Jiziwan Area, i = 1, 2, …, 43.
j water users, j = 1, 2, …, 14.
k energy users, k = 1, 2, …, 13.
t planning periods, t = 1, 2, …, 3 (2026 - 2040).
SW_{ijt} water allocation from surface water for user j in region i (m^3).
GW_{ijt} water allocation from groundwater for user j in region i (m^3).
OW_{ijt} transferred water allocation for user j in region i (m^3).
HW_{ijt} water allocation from reservoir for user j in region i (m^3).
IRW_{ijt} industrial reclaimed water allocation for user j in region i (m^3).
DRW_{ijt} domestic reclaimed water allocation for user j in region i (m^3).
ERW_{ijt} energy reclaimed water allocation for user j in region i (m^3).
PW_{ijt} rainwater allocation for user j in region i (m^3).
BW_{ijt} water allocation from brackish water for user j in region i (m^3).
MW_{ijt} water allocation from mine water for user j in region i (m^3).
SEW_{ikt} surface water allocation for energy sector k in region i (m^3).
GEW_{ikt} groundwater allocation for energy sector k in region i (m^3).
OEW_{ikt} transferred water allocation for energy sector k in region i (m^3).
REW_{ikt} reclaimed water allocation for energy sector k in region i (m^3).
PEW_{ikt} rainwater water allocation for energy sector k in region i (m^3).
MEW_{ikt} mine water allocation for energy sector k in region i (m^3).
BEW_{ikt} brackish water allocation for energy sector k in region i (m^3).
PRW_{ijt} net benefit of water allocation (¥/m^3).
GWD_{jt} water consumption per unit of ecological land (m^3/ha).
TYA_t inflow from the upper reaches of the Yellow River (m^3).
WYL_t downstream ecological flow (m^3).
$\gamma\gamma_{jt}$ water distribution efficiency in the production and ecological sectors.
γe_{kt} water distribution efficiency in the energy sectors.
$bUlan_t$ ecological value coefficient (¥/m^3).
RED_{kt} benefit of energy sectors (¥/kg).
FED_{ikt} energy demand of sector k in region i (kWh).
CEE_{kt} cost of energy extracted (¥/kg).
AEE_{ikt} amount of energy extracted (kg).
CEP_{kt} processing cost of energy (¥/kg).
AEP_{ikt} amount of energy conversion (kWh).

AEC_{ikt} energy processing capacity (kWh).
$VCEC_{kt}$ energy conversion cost (¥/kg).
LF_{ijt} irrigation leaching share (%).
WPC_{ijt} water demand of crop (m^3/ha).
FC_{ijt} fixed cost per unit planted area of crop (¥/ha).
TGA_t availability of groundwater (m^3).
PGE_t maximum extraction coefficient of groundwater.
RWR_t water resource reflux rate (%).
TOA_t availability of transferred water (m^3).
YA_{ijt} crop yield of per unit area (kg/ha).
FD_t per capita food demand (kg).
PO_{it} population.
TED_t^{max} maximum available energy in agricultural process (kg).
TDE_{jt} minimum ecological land (ha).

References

Fernandes LS, Galvão A, Santos R, Monteiro S (2023) Impact of water reuse on agricultural practices and human health. Environ Res 216:114762

Lee SO, Jung Y (2018) Efficiency of water use and its implications for a water-food nexus in the Aral Sea Basin. Agric Water Manag 207:80–90

Li HY, Li M, Fu Q, Singh VP, Liu D, Xu YW (2023) An optimization approach of water-food-energy nexus in agro-forestry-livestock system under uncertain water supply. J Clean Prod 407:137116

Nematian J (2023) A two-stage stochastic fuzzy mixed-integer linear programming approach for water resource allocation under uncertainty in Ajabshir Qaleh Chay Dam. J Environ Inf 41(1):52–66

Radmehr R, Brorsen BW, Shayanmehr S (2024) Adapting to climate change in arid agricultural systems: an optimization model for water-energy-food nexus sustainability. Agric Water Manag 303:109052

Ren HR, Liu B, Zhang ZR, Li FX, Pan K, Zhou ZL, Xu XS (2022) A water-energy-food-carbon nexus optimization model for sustainable agricultural development in the Yellow River Basin under uncertainty. Appl Energy 326:120008

Tan Z, Li H, Song Q, Wang Z, Cao Y (2025) Synergistic optimization and interaction evaluation of water-energy-food-ecology nexus under uncertainty from the perspective of urban agglomeration. Sustain Cities Soc 124:106291

Yin DY, Yu HC, Shi YY, Zhao MY, Zhang J, Li XS (2023) Matching supply and demand for ecosystem services in the Yellow River Basin, China: a perspective of the water-energy-food nexus. J Clean Prod 384:135469

Zhang YT, Yang PG, Liu J, Zhang XC, Zhao Y, Zhang Q, Li L (2023) Sustainable agricultural water management in the Yellow River Basin China. Agric Water Manag 288:108473

Part III
Low-Carbon Technology and Carbon Management Innovation

Chapter 20
Land Use Planning for Carbon Emission Mitigation Based on the BLP-PLUS Model: A Case Study of Fujian Province

Hejie Zhou, Pangpang Gao, Zhihao Liu, and Shiqi Li

Abstract Carbon emissions from land use are a major driver of global climate change, and their accurate characterization and forecastion are essential for regional sustainability. This study integrates a bi-layer programming (BLP) framework with the Patch-based Land-Use Simulation (PLUS) model to establish a BLP-PLUS system for Fujian Province, China. Economic benefit and carbon emission are treated as dual optimization objectives, while cultivated land protection and other policy constraints are imposed. A stage-wise optimization strategy is applied under three economic growth scenarios, including low, medium, and high, to generate alternative land allocation schemes. The PLUS model is then used to simulate the spatial evolution of land use and the associated carbon emissions for each scenario. Results show that the proposed framework effectively reconciles economic growth with emission-reduction goals, offering robust policy support and practical guidance for low-carbon land-space planning in Fujian Province.

Keywords Land use change · Bi-layer programming · PLUS model · Scenario design · Low carbon development

20.1 Introduction

Land use optimization is pivotal for easing the twin pressures of rapid urban expansion and climate change. As cities grow, competition for land intensifies, thereby threatening ecosystem integrity, increasing carbon emissions, and underming long-term sustainability. Because land use decisions directly shape carbon sequestration, biodiversity, and other essential ecosystem services, effective land-use management is indispensable for any pathway toward sustainable development and carbon neutrality.

H. Zhou · P. Gao (✉) · Z. Liu · S. Li
School of Environmental Science and Engineering, Xiamen University of Technology, Xiamen, China
e-mail: Gaopp@xmut.edu.cn

Y. Li et al. (eds.), *16th International Conference on Environmental Science and Development*, Environmental Science and Engineering,
https://doi.org/10.1007/978-3-032-15080-6_20

Balancing urban growth, food security, and ecological protection is inherently complex. Main-stream optimization algorithms, including genetic algorithms (GA), particle-swarm optimization (PSO), simulated annealing, typically maximize land use efficiency under narrowly defined constraints, often overlooking broader environmental and social dimensions (Guo et al. 2023; Song et al. 2025). System-dynamics (SD) models capture long-term trends but struggle to represent the feedback loops linking land, population, and economic change (Xu and Li 2024; Liu et al. 2022). Spatially explicit frameworks, such as conversion of land use and its effects at small regional extent (CLUE-S), cellular automata (CA), and future land use simulation model (FLUS) simulate land-use transitions over time, yet they rarely integrate economic, ecological, and social objectives simultaneously (Zhang and Zhang 2023). These limitations underscore the need for integrated, multi-objective approaches that embed spatial and temporal dynamics, which are prerequisites for achieving both carbon neutrality and sustainable development (Wang and Han 2021; Zhang et al. 2023; Chen et al. 2023).

To address these gaps, this study couples bi-layer programming (BLP) framework with the Patch-based Land-Use Simulation (PLUS) model, creating a BLP-PLUS system for land use optimization (LUO) in Fujian Province. The framework simultaneously minimizes carbon emissions and maximizes economic benefits under explicit spatial and temporal constraints. Through a stage-wise optimization strategy and three economic-growth scenarios (including low, medium, and high), the model produces flexible, policy-relevant land-allocation schemes that reconcile competing objectives. The proposed BLP-PLUS approach is capable of offering a robust decision-support platform that aligns land-use planning with carbon–neutral and sustainability goals.

20.2 The Research Area and Data Sources

20.2.1 Overview of the Research Area

Fujian Province, located in southeastern China between 23°33’ and 28°20′ north latitude and 115°50′ and 120°40′ east longitude, covers an area of 124,000 square kilometers in Fig. 20.1. The province features diverse topography, including the Wuyi Mountains in the north, along with plains, valleys, and coastal wetlands. Its landscape is composed of forests, agricultural lands, and other ecosystems. The rapid urbanization and industrialization have led to the encroachment of forests, grasslands, and water bodies, resulting in reduced ecological carbon sink capacity and environmental degradation. With the region’s economic growth, the challenge lies in balancing economic development with ecological preservation, making land use optimization crucial for sustainable future growth.

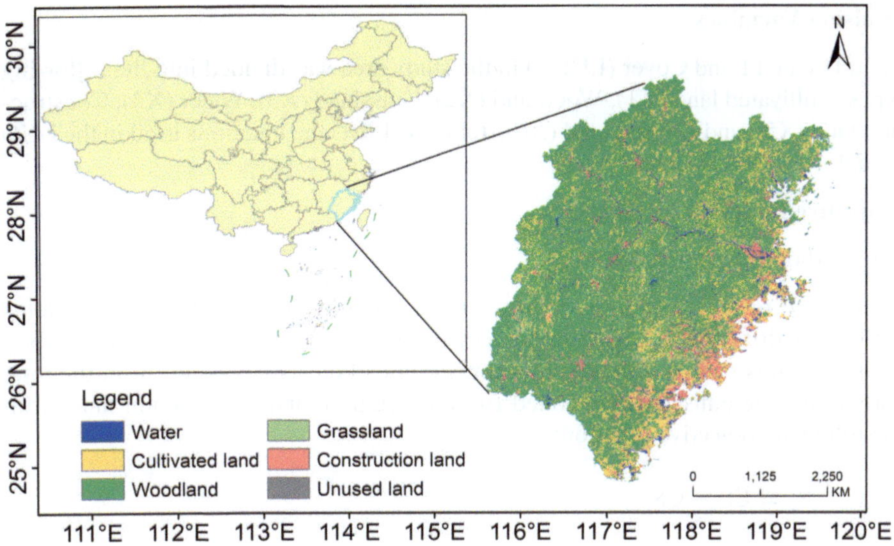

Fig. 20.1 Study area overview map

20.2.2 Data Sources

The modeling framework in this study requires quantitative data, including the GDP from agriculture, forestry, animal husbandry, fisheries, and the combined secondary and tertiary industries, as well as population, fertilizer usage, agricultural film and pesticide usage, and energy consumption. These data are taken from the data in the Statistical Yearbook of Fujian Province 2020. Carbon emission coefficients for agricultural activities and energy use, as well as carbon absorption coefficients for land types like forests, are derived from studies (Cao et al. 2024; IPCC 2006). Data on forest coverage and the cultivated land red line come from the "Fujian Provincial Land and Spatial Planning (2021–2035)."

20.3 Model Formulation

20.3.1 BLP Model

In this study, a bi-layer optimization model is developed to coordinate regional economic development and ecological protection. The BLP model comprises three parts: decision variables, objective functions and constraints.

Decision Variables

Land Use and Land Cover (LULC) in the study area was divided into the following types: Cultivated land (X1), Woodland (X2), Grassland (X3), Water (X4), Construction land (X5), and Unused land (X6). Table 20.1 lists the notations used in the LUO model.

Objective Function

(1) Carbon emission reduction

Energy and resource consumption by human activities in socio-economic systems is the main driver of increased carbon emissions. In order to reduce regional carbon emissions, this can be achieved by subtracting carbon sequestration from the total carbon dioxide emissions generated from agricultural and construction land, with the following objective function:

$$
\begin{aligned}
F_1 &= CE - CS \\
&= [(CE^{cultivated} + CE^{construction})] - \sum_{k=2}^{4} X_k \times S_k \\
&= [CE^{fertilizer} + CE^{machinery} + CE^{irrigation} + CE^{film} + CE^{pesticide}] \\
&+ [CE^{energy} + CE^{human})] - \sum_{k=2}^{4} X_k \times S_k \\
&= X_1 \times [kf \times Q^{fertilizer} + (km \times Q^{electricity} + kc \times Q^{machinery}) \\
&+ kr \times \alpha + ka \times Q^{a} + kp \times Q^{p}] - \sum_{n=1}^{N} X_5 \times (Q_n \times K_n) \\
&+ PO \times f_h] - \sum_{k=2}^{4} X_k \times S_k
\end{aligned}
\tag{20.1}
$$

(2) Economic benefit

LUO should be based on ensuring regional economic benefits. This study aims to quantify the economic dimension by maximizing economic benefits, as represented in the following equation:

$$
F_2(X) = Max \sum_{k=1}^{6} D_k X_k \tag{20.2}
$$

Constraints

(1) Total area constraint

Table 20.1 Definitions of symbols

Indices	Definition
CE, CS	Carbon emissions and sequestration (kg C)
$CE^{cultivated}$	Carbon emissions from agriculture (kg C)
$CE^{machinery}$	Carbon emissions from machinery (kg C)
$CE^{irrigation}$	Carbon emissions from irrigation (kg C)
CE^{film}	Carbon emissions from agricultural films (kg C)
$CE^{pesticide}$	Carbon emissions from pesticides (kg C)
CE^{enery}	Carbon emissions from energy consumption (kg C)
CE^{human}	Carbon emissions from human respiration (kg C)
S_k	The carbon sequestration coefficient for kind k land (kg C/km^2)
kf	The carbon emission factors for fertilizers (kg C/kg)
km	The carbon emission factors for electricity for agricultural machinery (kg C/kw)
kc	Carbon emission factors from agricultural machinery (kg C/kw)
kr	Irrigation coefficient (kg C/km^2)
ka	Carbon emission factor for agricultural films (kg C/kg)
kp	Carbon emission factors for pesticides (kg C/kg)
$Q^{fertilizer}$	Amount of fertilizer used per unit area of cultivated land (kg/km^2)
$Q^{electricity}$	Total electricity consumption of agricultural machinery per unit area of cultivated land (kw/km^2)
$Q^{machinery}$	Fossil fuel energy carbon emissions from agricultural machinery per unit area of cultivated land (kg C/km^2)
α	The irrigation ratio
Q^a	Amount of agricultural film used per unit area (kg/km^2)
Q^p	Amount of agricultural film used per unit area (kg/km^2)
Q_n	The amount of energy n used per unit area of construction land (kg/km^2)
K_n	The carbon emission factor for energy n (kg C/kg)
PO	Population of construction land (persons)
f_h	The carbon emission factors for human breathing (kg C/person·yr^{-1})
D_k	Economic coefficient for this category per unit area (Million/km^2)
P	Yield per unit area (kg/km^2)
TLA	Total area (km^2)
N	The total population of the region (persons)
d	Per capita food demand (kg /person)
W_k	Land area in category k in base year 2020

The sum of the areas of each land use type shall be equal to the total area of Fujian Province:

$$X_1 + X_2 + X_3 + X_4 + X_5 + X_6 = 124000\,\text{km}^2 \tag{20.3}$$

(2) Forest cover constraint

According to the Fujian Provincial Land Spatial Plan (2021–2035), Fujian Province has set a target for forest cover: by 2035, the province's forest cover will reach 65.37%.

$$X_2/\text{TLA} \geq 65.37\% \tag{20.4}$$

(3) Food security constraint

Ensure food security by ensuring that total food production meets the food needs of the total population.

$$P \cdot X_1 \geq N \cdot d \tag{20.5}$$

(4) Cultivated land red line

The minimum arable-land area must not fall below Fujian's cropland-protection "red line." According to the Land Spatial Plan of Fujian Province (2021–2035), this baseline is set at 8940 km^2 of cultivated land.

(5) Landscape diversity constraint

In order to maintain biodiversity and reduce ecological vulnerability, the combined percentage of farmland, grassland and water area in the optimization scenario is maintained equal to or greater than that observed in the ND scenario.

$$\frac{X_1 + X_3 + X_4}{\text{TLA}} \geq \frac{W_1 + W_3 + W_4}{\text{TLA}} \tag{20.6}$$

(6) Economic growth rate constraint

According to the Work Report of the People's Government of Fujian Province in 2024, the expected growth target of Fujian Province's gross domestic product (GDP) in 2024 is about 5.5%, and let a = 5.5%.

$$\sum_{k=1}^{6} D_{kt}X_{kt} \cdot (1 + a) \leq \sum_{k=1}^{6} D_{k(t+1)}X_{k(t+1)} \tag{20.7}$$

(7) Construction land constraint

The construction land will increase in the future due to economic growth. The maximum increase in area should be 20% compared to 2020.

$$80\% \times W_5 \leq X_5 \leq 120\% \times W_5 \tag{20.8}$$

(8) Area change constraint

From a practical application point of view, to ensure that the results of the optimization can be applied to real-life planning, the area of each optimized LULC type should not vary too much from that of the previous year. Therefore, the change in area for each LULC type was set to be less than 10%.

$$90\% \times X_{k,t-1} \leq X_{k,t} \leq 110\% \times X_{k,t-1} \tag{20.9}$$

(9) Nonnegative constraint

$$X_k \geq 0, \forall k \tag{20.10}$$

20.3.2 PLUS Model

Integration of the PLUS model. Using land use data for Fujian Province in 2020, we compiled 16 key drivers of land-use change: elevation, slope, mean air temperature, precipitation, aridity index, soil type, soil-erosion class, population density, GDP, night-time light intensity, livestock density, and Euclidean distances to highways, other urban roads, railways, rivers, and settlements (Xu et al. 2024). All layers were resampled to a common resolution and fed into the PLUS model to simulate the spatial evolution of land use across the province. Land use demand for 2020–2060—calculated under three economic-growth scenarios (low, medium, and high)—was subsequently input to forecast alternative land use patterns. The simulations reveal the spatial distribution and trajectories of future land use under each scenario, providing a robust scientific basis for regional land spatial planning and policy-making.

20.3.3 Design of Optimization Proceedings

This study develops a multi-objective fuzzy dynamic programming model to optimize land use from 2020 to 2060 by balancing carbon emissions and economic benefits. The model considers six land types across five periods, with objectives to minimize carbon emissions and maximize economic benefits (Liu et al. 2019). The model integrates three economic growth scenarios, including low-speed growth (5%), medium-speed growth (5.5%), and high-speed growth (6%), to assess the impact of varying economic development rates on land use and carbon emissions. An improved fuzzy satisfaction algorithm solves the trade-off between carbon reduction and economic growth by deriving ideal/anti-ideal solutions, defining linear membership functions, and applying a max–min principle for compromise optimization (Zhang et al. 2025). Dynamic constraints, including resource limits, policy requirements, and land transition rules, are considered across different growth scenarios (Wang et al. 2024). This approach provides a comprehensive analysis of land use optimization under multiple economic conditions, supporting sustainable development and carbon neutrality efforts.

Bi-Layer Optimization Strategy

This study adopts a two-tier optimization strategy. The first layer takes minimizing carbon emissions as the core objective, aiming to reduce carbon emissions and achieve the goal of low-carbon development through optimizing land use allocation (Yuan et al. 2024). The optimization objective of the second layer is to maximize economic benefits, aiming to promote the improvement of land use efficiency and support the growth of the regional economy. In order to comprehensively consider the trade-off between these two objectives, the model introduces the concept of global satisfaction for fuzzy optimization (Rong et al. 2022). At each stage, based on global satisfaction, the model will perform multi-objective optimization based on the conflict between carbon emissions and economic benefits to ensure that the optimal balance between carbon emission reduction and economic development is found.

Multi-Scenario Optimization Strategy

In order to cope with different economic development trends in the future, this paper sets up three economic growth scenarios: low economic growth, medium economic growth and high economic growth (Fan et al. 2023; Wang et al. 2024). Three different economic growth rates are set according to the policy direction of Fujian Province: 5%, 5.5% and 6%. Each scenario reflects different macroeconomic development conditions, and the model optimizes the land use accordingly to these scenarios. In the low-growth scenario, economic growth is slower and land use optimization will focus on resource conservation and ecological protection; in the medium-growth scenario, economic development and ecological protection are balanced, and land resources are allocated rationally; in the high-growth scenario, emphasis is placed on efficient land use and productivity enhancement, while carbon emissions need to be controlled. Through multi-scenario analysis, the impacts of different economic development paths on land use patterns and carbon emission control strategies can be assessed to provide a basis for policy formulation (Wang et al. 2023).

20.4 Result and Discussion

20.4.1 The Changes of Land Allocation Under Different Scenarios

The analysis of land use changes under different development speeds (*low-speed, medium-speed, and high-speed*) from 2020 to 2060 reveals distinct trends for each land type. For example, cultivated land shows a gradual increase under all scenarios, with the highest growth rate observed in the low-speed development scenario (*green*), followed by medium-speed development (*blue*), and the slowest increase under high-speed development (*red*) in Fig. 20.2. This suggests that slower economic expansion allows for more preservation and gradual expansion of agricultural land, whereas rapid growth leads to greater pressure on land resources. In contrast, woodland exhibits a consistent decline across all scenarios, with the steepest reduction in the low-speed development scenario. This unexpected trend may be due to the earlier conversion of woodland to meet gradual but sustained land demands. In faster development scenarios, woodland loss occurs but at a relatively slower cumulative rate, possibly due to more stringent conservation measures accompanying rapid urban expansion. Grassland shows a notable pattern where it initially increases, especially under medium- speed and high-speed development, and then stabilizes or slightly declines after 2040. In the low-speed development scenario, grassland fluctuates but ultimately shows a slight recovery towards the later periods. This trend suggests that grassland is influenced by both land degradation and subsequent ecological restoration efforts, which vary across different economic conditions.

The water area decreases steadily under all scenarios, with similar reduction rates across all development speeds. This indicates that the loss of water bodies is more related to overall land-use pressures rather than the specific speed of economic development. In contrast, construction land increases significantly in all scenarios, with the fastest expansion observed under high-speed development, indicating strong urbanization and infrastructure growth. This aligns with the pressures of rapid economic growth, which lead to more conversion of other land types into construction land areas. Finally, unused land exhibits minimal change in the low-speed development scenario, a slight decline under medium-speed development, and remains at the lowest level throughout under high-speed development. This suggests that rapid urban expansion leads to the more complete utilization of previously unused lands, intensifying land-type transformations. Overall, the results reflect the varying impacts of economic growth rates on land use and the environment, with high-speed development leading to the most significant and intense changes in land use patterns.

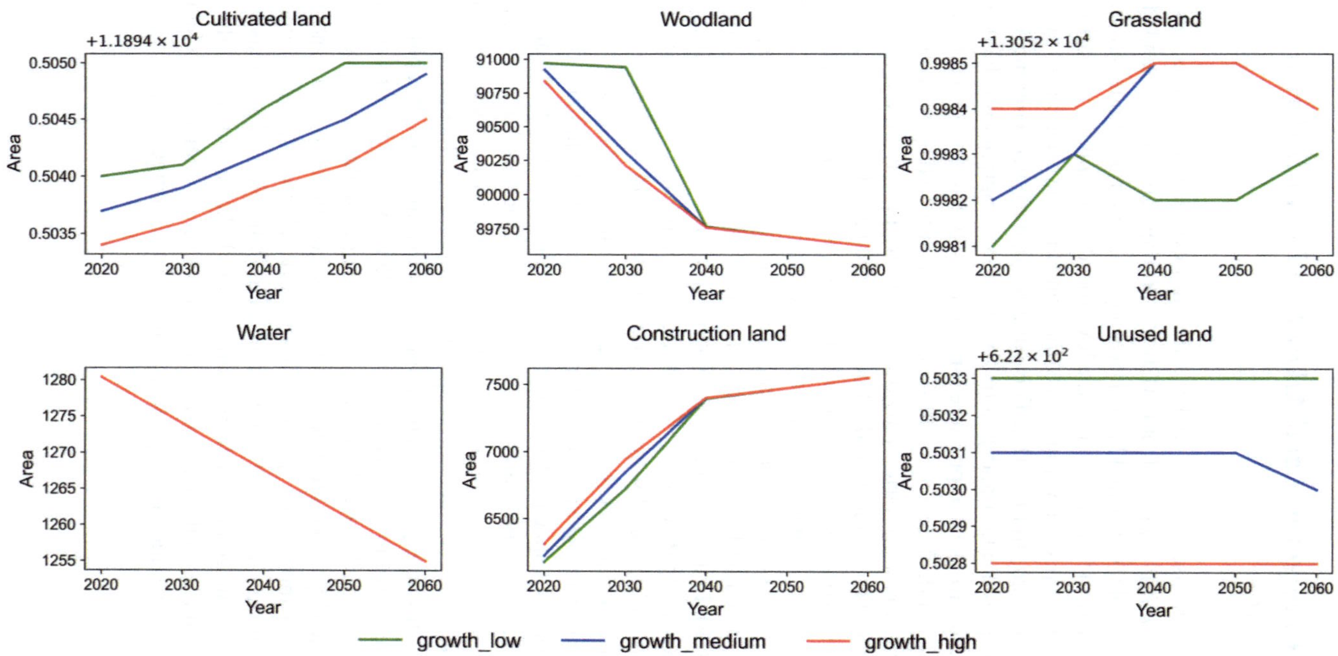

Fig. 20.2 Trajectory of land-use demand under three economic growth rate scenarios

20.4.2 Spatial Changes in Land Use Under Different Scenarios

The spatial distribution maps of land use in Fujian Province were generated by inputting the actual land use data in 2020 and the projected data in 2060 under three different economic growth scenarios into the PLUS-CA model based on multi-class stochastic patch seeding (CARS) in Fig. 20.3. These maps show the spatial trends of various land types in Fujian Province under three different scenarios (*low growth, medium growth and high growth*).Under the low growth scenario (*growth_low_2060*), the increase of arable land and construction land is relatively flat, while the areas of forest land and grassland remain relatively stable. The area of unused land also does not show significant changes. This trend reflects the impact of lower urbanization and slower economic development on land use. Under the medium growth scenario (*growth_medium_2060*), the areas of arable land and construction land expand more notably, particularly around urban areas. The decrease in forest land becomes more pronounced compared to the low-growth scenario, while grassland areas remain relatively stable. The change in water area is not significant. This pattern reflects the accelerated land conversion driven by moderate economic growth.

In the high growth scenario (*growth_high_2060*), construction land expands significantly, especially in the eastern region, accompanied by a reduction in forest land and grassland. Arable land continues to expand to meet food and urban demands, rather than declining. The change in water area remains minimal. Meanwhile, unused land shows a marked decline, indicating intensified land exploitation and utilization with rapid economic development. These spatial distribution maps enable visual observation of the changes in land use types under different scenarios, especially the transitions among arable land, construction land, and ecological land. This analysis

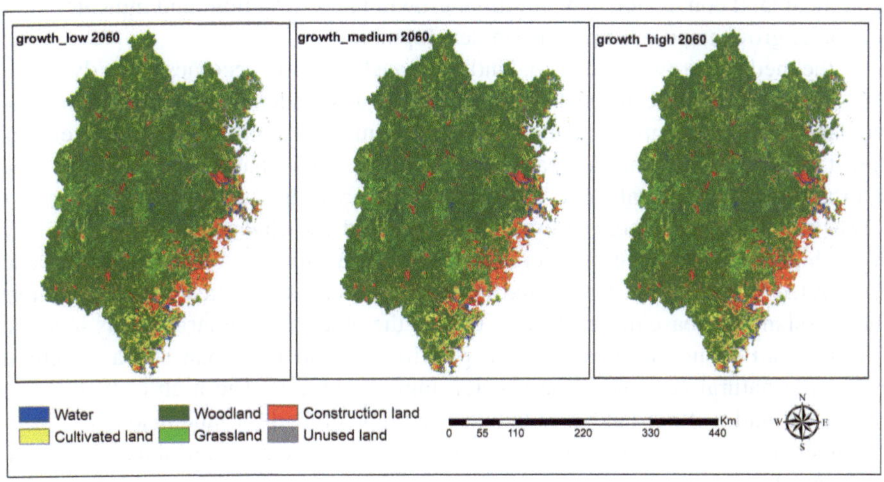

Fig. 20.3 Patterns of land type distribution in 2060 under three scenarios

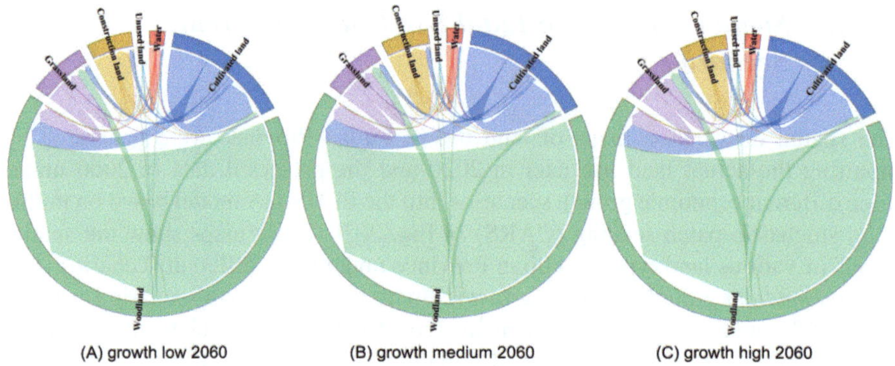

Fig. 20.4 Conversion of land types under three scenarios

helps to understand the spatial patterns of future land use changes and provides a valuable reference for the optimal allocation of land resources and ecological protection efforts.

20.4.3 Land Resource Optimization Under Different Scenarios

Land use transformation is relatively modest in the low-growth scenario. The conversion of unused land to cropland is significant, suggesting that agricultural expansion remains a priority even in a low-growth scenario in Fig. 20.4. However, changes in woodland and grassland are relatively small, reflecting limited urbanization and ecological land conversion. The minimal growth in construction land indicates slow economic growth and inhibited urban development.

In the medium growth scenario, land use transformation becomes more dynamic. The significant increase in construction land indicates moderate urban expansion, and cropland continues to grow, but at a slower rate compared to the high growth scenario. Forest land and grassland decrease significantly as more natural land is converted to urban and agricultural land. Unutilized land continues to decrease, albeit at a slower rate than in the high growth scenario. Land use changes are most dramatic in the high growth scenario. There is a rapid expansion of construction land and accelerated urbanization. Cultivated land also increases significantly, while forest and grassland decrease more severely than in the other two scenarios. This scenario reflects rapid economic development, prioritizing land for urban and agricultural use, while natural ecosystems are increasingly displaced. The high conversion of unutilized land to other land uses indicates a high demand for land resources.

Under all three scenarios, unutilized land is converted to cropland and construction land, but the extent of this conversion depends on the growth rate. More land is retained for ecological functions in the low growth scenario, while the loss of natural habitat is most severe in the high growth scenario. The balance between economic development (urbanization) and ecological preservation (woodland, grassland) is crucial for long-term sustainability. The modelling results clearly show how different growth rates affect land use patterns and can guide future land resource planning to achieve a sustainable balance between development and environmental protection.

20.5 Conclusions

In this study, a bi-layer optimization model was developed and integrated with the PLUS model to predict land use changes and carbon emission trends in Fujian Province from 2020 to 2060. The model simultaneously balances multiple objectives—economic growth and carbon emission reduction—under varying economic-growth scenarios (low, medium, and high). By incorporating dynamic constraints such as resource limitations, policy requirements, and land transition rules, the bi-layer fuzzy optimization approach effectively resolves trade-offs between competing goals, providing a flexible and accurate framework for sustainable land use planning.

The results indicate that under the low-growth scenario, forest and grassland areas remain relatively stable (approximately 89,626 km^2 and 13,053 km^2 by 2060), with expansion primarily occurring in less urbanized regions to support ecological preservation and carbon neutrality. In contrast, the high-growth scenario is characterized by rapid expansion of construction land (from 6312 km^2 to 7548 km^2), which encroaches upon cultivated and natural lands, leading to significant habitat loss. The fuzzy optimization framework successfully balances economic development with emission reduction, underscoring the critical importance of strategic land use management. Consequently, prioritizing low-growth pathways combined with rational construction land planning and stronger protection of ecological lands is vital for achieving sustainable development and carbon neutrality.

Despite these contributions, several limitations warrant further investigation. The current model does not incorporate non-economic and social factors, such as demographic changes and policy enforcement, which can substantially impact land use decisions. Moreover, the study is geographically focused on Fujian Province; therefore, validating the model's applicability across regions with diverse socio-economic and environmental contexts remains a necessary future step. Lastly, advancing the optimization algorithm to enhance model robustness and adaptability will be essential for broader practical implementation.

References

Cao X, Wang H, Song Y et al (2024) An optimization-simulation modeling framework for exploring the dynamic correlation between the coordinated development of the human-nature nexus and land use under uncertainty. Sustain Cities Soc 112:105613

Chen N, Xin C, Zhang B et al (2023) Contribution of multi-objective land use optimization to carbon neutrality: a case study of Northwest China. Ecol Ind 157:111219

Fan H, Si Q, Dong W et al (2023) Land use change and landscape ecological risk prediction in urumqi under the shared socio-economic pathways and the representative concentration pathways (SSP-RCP) scenarios. Sustainability 15(19):14214

Guo P, Wang H, Qin F et al (2023) Coupled MOP and PLUS-SA model research on land use scenario simulations in Zhengzhou metropolitan area, central China. Remote Sens 15(15):3762

IPCC (2006) 2006 IPCC guidelines for national greenhouse gas inventories, vol 4. Agriculture, Forestry and Other Land Use, Geneva, Switzerland

Liu HX, Li YP, Yu L (2019) Urban agglomeration (Guangzhou-Foshan-Zhaoqing) ecosystem management under uncertainty: a factorial fuzzy chance-constrained programming method. Environ Res 173:97–111

Liu Y, You K, Jiang Y et al (2022) Multi-objective optimal scheduling of automated construction equipment using non-dominated sorting genetic algorithm (NSGA-III). Autom Constr 143:104587

Rong Q, Zeng J, Su M et al (2022) Prediction and optimization of regional land-use patterns considering nonpoint-source pollution control under conditions of uncertainty. J Environ Manage 306:114432

Song M, An M, He W et al (2025) Research on land use optimization based on PSO-GA model with the goals of increasing economic benefits and ecosystem services value. Sustain Cities Soc 119:106072

Wang G, Han Q (2021) The multi-objective spatial optimization of urban land use based on low-carbon city planning. Ecol Ind 125:107540

Wang Q, Guan Q, Sun Y et al (2023) Simulation of future land use/cover change (LUCC) in typical watersheds of arid regions under multiple scenarios. J Environ Manage 335:117543

Wang Y, Li M, Jin G (2024) Exploring the optimization of spatial patterns for carbon sequestration services based on multi-scenario land use/cover changes in the changchun-Jilin-Tumen region, China. J Clean Prod 438:140788

Xu Q, Li K (2024) Multi-scenario simulation of land use and assessment of carbon stocks in terrestrial ecosystems based on SD-PLUS-invest coupled modeling in Nanjing city. Forests 15(10):1824

Xu X, Kong W, Wang L et al (2024) A novel and dynamic land use/cover change research framework based on an improved PLUS model and a fuzzy multiobjective programming model. Eco Inform 80:102460

Yuan Z, Xu J, Xu D et al (2024) The carbon emissions risk evolution and low-carbon optimization in a typical mountainous region on the western edge of the Sichuan Basin, China. Ecol Ind 162:112064

Zhang X, Zhang D (2023) Urban carbon emission scenario prediction and multi-objective land use optimization strategy under carbon emission constraints. J Clean Prod 430:139684

Zhang Z, Hu B, Jiang W et al (2023) Spatial and temporal variation and prediction of ecological carrying capacity based on machine learning and PLUS model. Ecol Ind 154:110611

Zhang H, Xi Q, Chen L et al (2025) Robust fuzzy dynamic integrated environmental-economic-social scheduling considering demand response and user's satisfaction with electricity under multiple uncertainties. Sci Rep 15(1):6881

Chapter 21
Development of a Two-Stage Stochastic Low-Carbon Recirculating Aquaculture System Planning Model

Xiao Li, Jiawei Li, Yi Jin, and Shubo Xu

Abstract Driven by the "dual-carbon" goals, China's aquaculture systems urgently need to transform towards low energy consumption and low carbon emissions. From the perspective of optimizing aquaculture systems, this study utilizes the two-stage stochastic programming method to develop a low-carbon recirculating aquaculture system planning model. It explores the low-carbon development path of the aquaculture industry under the influence of stochastic uncertainties, providing scientific and technological support for decision-makers to formulate relevant plans and policies. Taking Fujian Province as the research area, this study verifies the model by comparing the actual aquaculture data with the model's prediction results, and deeply analyzes its performance in terms of accuracy and reliability. Additionally, it summarizes the potential problems that may occur in the practical application of the optimized low-carbon recirculating aquaculture system planning model, and then puts forward targeted decision-making suggestions to address these issues. This study has the following findings: To achieve the dual carbon goals and maximize economic benefits, the aquaculture industry in Fujian Province should (i) gradually replace traditional feed with low-carbon feed such as microalgae; (ii) extensively utilize wind and solar resources for energy supply to aquaculture equipment.

Keywords Aquaculture · Low-carbon · Energy structure optimization

---

X. Li (✉)
School of Film Television and Communication, Xiamen University of Technology, Xiamen, China
e-mail: lixiaonick@163.com

J. Li · Y. Jin · S. Xu
School of Environmental Science and Engineering, Xiamen University of Technology, Xiamen, Fujian, China

Y. Li et al. (eds.), *16th International Conference on Environmental Science and Development*, Environmental Science and Engineering,
https://doi.org/10.1007/978-3-032-15080-6_21

21.1 Introduction

21.1.1 Background

It is widely agreed that all countries' economic sectors should detect low-carbon development pathways to limit average global increased temperature below 1.5–2.0 °C by 2100 (Qiao et al. 2024). China, as a great carbon emission country, has promised that its social-economic sectors would gradually be green and low-carbon. Aquaculture is an important sector in the national economy. According to the Ministry of Agriculture, the total amount of aquatic products was 68.6591 million tons in 2022, increasing by 2.62% year-on-year. Among them, the amount of aquaculture products was 55.6546 million tons (the ratio of marine aquaculture products to freshwater aquaculture products was 50.4:49.6), increasing by 3.17% year-on-year (NFES 2023). It is indicated that aquaculture is the main production type of aquatic products in China. The aquaculture process generally involves multiple procedures such as lighting, oxygen supply, water treatment and etc. Energy is an essential resource to support each procedure. For instance, in the oxygenation procedure, the oxygen production of the aerator is approximately 0.5–0.8 kg/kWh, and the flow rate of the water pump is about 20–30 m^3/kW. It can be inferred that the energy (mainly electricity) consumption of aquaculture is great. The low-carbon development of aquaculture can consider renewable energy conversion technologies (e.g., construction of photovoltaic panel, wind turbine generator), aquaculture equipment improvement, aquaculture process upgrading and etc. The adoptions of them should consider investment cost, carbon emission mitigation efficiency, energy supply stability and etc. How to balance these perspectives is a challenge.

21.2 Literature Review

Based on system optimization technique, several researchers proposed low-carbon aquaculture system planning model to handle the challenge. Nguyen et al. (2021) proposed a hybrid energy isolated-island aquaculture system planning model to explore the effects of adopting supplementary energy (i.e., solar and wind) sources on costs and carbon emissions. Erdemir and Dincer (2024) constructed a green hydrogen-oxygen production and power generation planning system based on solar energy and applied it in five cities in the United States, demonstrating that this system could promote sustainable aquaculture. Xu et al. (2024) optimized the energy structure of the Zhoushan city's aquaculture system, as well as proposed planning schemes about energy substitution, breeding scale and energy transmission efficiency to promote carbon emission reduction under three scenarios. Zhang et al. (2024) developed a renewable energy based recirculating aquaculture system planning model to improve the sustainability (e.g., reduce cost, mitigate carbon emissions). Dewi et al. (2025) proposed a solar-hydro energy integrated aquaponics system in Indonesia, results indicated that the system effectively improve resource consumption efficiency and reduce power-grid electricity supply. These studies show that adopting renewable energy to replace power-grid electricity is a major low-carbon development measure for aquaculture system. Building a low-carbon aquaculture system optimization model with hybrid energy is meaningful.

In a real-world system planning problem, decision makers may often need to make plans before the reality come true. Future aquatic product demands may have stochastic features; the previous studies have difficulties in providing decision making support under such uncertainty and may result in great losses.

Two-stage stochastic programming (TSP) method is an appropriate technique to deal with stochastic information in the systems. The concept of TSP is to set a pre-regulated target (e.g., breeding numbers) in the first-stage, and quantifying stochastic information (e.g., market demands). When the planning period comes, the market demands may be lower than, equal to, or higher than the target. For the "lower" scenario, managers need to take actions to reduce potential economic losses; for the "equal" scenario, managers may obtain the satisfactory benefit; for the "higher" scenario, managers cannot obtain the maximum benefit, these are called the second-stage compensation decisions. A number of researchers applied the TSP method in the field of water resources allocation (WRA), energy system planning (ESP) and etc. Nematian (2023) proposed a TSP-WRA model to balance a river dam's water allocation target and stochastic water availability. Yang et al. (2024) developed a TSP-ESP model to balance the newly-built capacity and stochastic wind speed and solar radiation. Wang et al. (2025) developed a TSP-WRA model to weigh the water use efficiency and water inflow probability. Liu and Zhang (2025) developed a TSP-ESP model to weigh the capacity configuration and stochastic unit unavailable. Tu et al. (2025) developed a TSP-WAR model to balance water market allocation and stochastic water demand. The TSP method has not been introduced in an aquaculture system optimization model.

21.3 Innovations

To fill the research gap, this study aims to proposes a two-stage stochastic low-carbon recirculating aquaculture system planning model (TLRAM). The model has advantages of: (i) gaining insight into the tradeoff between system benefit and probable economic losses generated by pre-regulated target; (ii) quantifying the complex interrelationships among multiple system components. The model is then applied to Fujian province (China) as a real-world case study. Results can contribute to generating optimal plans relating to maximized system benefit, renewable energy mix, number of fish fry, water utilization, pollutants, CO_2 reductions and etc. during 2026–2030. Decision makers can make aquaculture plans or policies according to the model results.

21.4 Model Development

An TLRAM is formulated as follows (Parameters are listed in Appendix A):

Objective Functions

$$\text{Max f} = (1) - (2) - (3) - (4) - (5) - (6) - (7) - (8) - (9)$$

Incomes of selling aquatic product:

$$\sum_{h=1}^{3}\sum_{i=1}^{3}\sum_{k=1}^{5}(PF_{i,k} \times QS_{i,k}+\Delta PF_{i,k} \times \Delta QS_{i,k,h} \times p_h) \tag{21.1}$$

Costs of purchasing fish fry:

$$\sum_{h=1}^{3}\sum_{i=1}^{3}\sum_{k=1}^{5}(CF_{i,k} \times QF_{i,k} + \Delta CF_{i,k} \times \Delta QF_{i,k,h} \times p_h) \tag{21.2}$$

Costs of purchasing feed:

$$\sum_{k=1}^{5}\sum_{l=1}^{3}(CFD_{k,l} \times QFD_{k,l}) \tag{21.3}$$

Costs of purchasing aquatic equipment:

$$\sum_{k=1}^{5}(CE_k \times NEE_k) \tag{21.4}$$

Costs of equipment operation power generation:

$$\sum_{k=1}^{5}(NE_k \times ELC_k \times CBE_k) \tag{21.5}$$

Costs of purchasing medicine:

$$\sum_{k=1}^{5}(CD_k \times MD_k) \tag{21.6}$$

Costs of water utilization:

$$\sum_{k=1}^{5} (CWU_k \times WU_k) \tag{21.7}$$

Costs of pollutant treatment:

$$\sum_{k=1}^{5} \sum_{t=1}^{3} (P_{k,t} \times LL_{k,t}) \tag{21.8}$$

Costs of carbon emission mitigation:

$$\sum_{k=1}^{5} (FCS_k \times CFC_k + DAC_k \times CDA_k) \tag{21.9}$$

Constraints

(a) Product demand:

$$QS_{i,k} + \Delta QS_{i,k,h} \geq QFF_{i,k,h}, \forall i, k, h \tag{21.10}$$

$$QFF_{i,k,h} \leq PRE_{i,k,h} \leq QFF_{i,k,h} \times 1.1, \forall i, k, h \tag{21.11}$$

(b) Aquatic equipment capability:

$$R_i \geq PRE_{i,k=1,h}, \forall i \tag{21.12}$$

$$PRE_{i,k-1,h} + QF_{\mathrm{i},k} + \Delta QF_{i,k,h} - QC_{i,k-1,h} \geq PRE_{i,k,h}, k \geq 2, \forall i, h \tag{21.13}$$

$$PRE_{i,k,h} \leq SEC_{i,k} \times NE_k, \forall i, k, h \tag{21.14}$$

$$NE_{\mathrm{k}} = NER + NEE_k, k = 1 \tag{21.15}$$

$$NE_{\mathrm{k}} = NE_{k-1} + NEE_k, k \geq 2 \tag{21.16}$$

(c) Fishing-selling balance:

$$QS_{i,k} + \Delta QS_{i,k,h} \leq QC_{i,k,h}, \forall i, k, h \tag{21.17}$$

(d) Fry Number-fishing balance:

$$QC_{i,k,h} \leq PRE_{i,k,h}, \forall i, k, h \tag{21.18}$$

(e) Water availability:

$$\sum_{i=1}^{3} WW_{i,k} \times PRE_{i,k,h} \leq WU_k, \forall k \tag{21.19}$$

$$WU_k \leq WA_k, \forall k \tag{21.20}$$

(f) Pollutant treatment:

$$\alpha_{k,t} \times (\sum_{i=1}^{3} (PRE_{i,k,h} \times EP_{i,k,t}) + \sum_{l=1}^{3} (QFD_{k,l} \times EN_{k,l,t})) \geq P_{k,t}, \forall k, t \tag{21.21}$$

$$\sum_{i=1}^{3} (PRE_{i,k,h} \times EP_{i,k,t}) + \sum_{l=1}^{3} (QFD_{k,l} \times EN_{k,l,t}) - P_{k,t} \leq PL_{k,t}, \forall k, t \tag{21.22}$$

(g) Nutrient supply:

$$\sum_{\mathrm{i}=1}^{3} NMI_{i,k,l} \times PRE_{i,k,h} \leq QFD_{\mathrm{k,l}} \leq \sum_{i=1}^{3} NMA_{i,k,l} \times PRE_{i,k,h}, \forall k, l \tag{21.23}$$

(h) Medicine supply:

$$MD_{\mathrm{k}} \leq MMD_k, \forall k \tag{21.24}$$

(i) Carbon emission:

$$(NE_k \times ELC_k \times CEE_k) + \sum_{l=1}^{3} (QFD_{k,l} \times CEFD_{l,k}) + \sum_{i=1}^{3} (PRE_{i,k,h} \times CEF_{i,k}) - FCS_k - DAC_k \leq CEL_k, \forall k \tag{21.25}$$

$$FCS_k \leq MFCS_k, \forall k \tag{21.26}$$

$$DAC_k \leq MDAC_k, \forall k \tag{21.27}$$

(j) Non-negative variable:

$$\Delta QS_{k,i}, \Delta QF_i, QFD_{k,l}, NE_k, MD_k, PRE_{i,k,h}, QC_{i,k,h}, WU_k, P_k \geq 0 \quad (21.28)$$

21.5 Study Area

21.5.1 Study Area

Fujian Province is located in the eastern coastal region of China, with a total mainland coastline length of 3752 km and a straight-line length of about 635 km, ranking first in the country for coastline curvature. The maritime area is vast, covering 136 $\times$ 10^3 km^2. Fujian Province has extensive shallow sea and tidal flats, with an area of about 14.77 $\times$ 10^3 km^2 within the 30-m isobath, and about 2000 of intertidal flats. Additionally, the province features a dense network of inland rivers, streams, reservoirs, and lakes, providing more area for aquaculture than the land used for agriculture, making it exceptionally well-suited for the aquaculture industry. However, the development of aquaculture leads to significant electricity consumption, which in turn increases carbon emissions. Therefore, reducing carbon emissions while ensuring the growth of the aquaculture industry has become a major challenge for the aquaculture sector in Fujian Province.

21.6 Data Source

The main data sources are yearbook, statistical communiqué, authoritative literature, etc. Tables 21.1, 21.2 and 21.3 contain collected historical data, specifically the selling prices of mature fish, the emission limits for aquaculture in Fujian Province, and the annual fish production in Fujian Province, which can provide data support for this model. Table 21.4 presents the demand for fish in the market under 3 different levels of market demand during the planning period of this model.

Table 21.1 Price list for adult fish sales (CNY/kg)

Types	2021	2022	2023
Large yellow croaker	32.06	37.17	37.50
Common eel	74.38	80.57	84.00
Tilapia	12.98	13.46	11.55

Table 21.2 Effluent discharge limits for aquaculture tailwater in marine receiving waters of Fujian Province (mg/L)

Item	First level	Second level
COD	≤ 10	≤ 20
Total nitrogen	≤ 5.0	≤ 7.0
Total phosphorus	≤ 0.5	≤ 1.0

Table 21.3 Fish production in Fujian Province (10^3 t)

Types	2021	2022	2023
Large yellow croaker	207.01	215.61	238.68
Common eel	56.38	58.73	55.94
Tilapia	122.50	128.70	132.50

Table 21.4 Market demand (10^3 t)

Types	Level	2026	2027	2028	2029	2030
Large yellow croaker	h = 1	267.04	286.87	308.17	331.05	355.63
Common eel	h = 2	281.10	301.97	324.39	348.48	341.35
Tilapia	h = 3	295.89	317.86	341.46	366.82	394.05
Large yellow croaker	h = 1	46.15	44.79	43.46	42.18	40.94
Common eel	h = 2	48.56	47.14	45.75	44.40	43.09
Tilapia	h = 3	51.13	49.62	48.16	46.74	45.36
Large yellow croaker	h = 1	134.53	139.92	145.52	151.35	157.41
Common eel	h = 2	141.61	147.28	153.18	159.32	165.70
Tilapia	h = 3	149.07	155.04	161.25	167.70	174.12

21.7 Results and Discussion

This study is based on the real data collected about aquaculture in the research area. We set key parameters in advance and programs the model on the Lingo platform. By inputting the parameter data, it calculates the optimal solution.

21.8 Aquaculture Scale Planning

Figures 21.1. and 21.2. shows that over the 5-year planning period, the total fish release and sales volume of planned fish species show an upward trend. Among them, the fish fry release and sales volume of large yellow croaker exhibit a significant annual increase, with average annual growth rates of [11.21, 12.27, 12.55] % and [6.67, 7.42, 7.43] %, respectively; tilapia shows a slight increase, with average annual growth rates of [9.72, 10.02, 10.07] % and 4.00%; while eel displays a stable farming trend. This is because, compared to the other two species, large yellow croaker has higher economic value and lower carbon emission coefficients. From a market demand perspective, the largest fluctuation influenced by market demand is observed in large yellow croaker, with the highest uncertainty occurring in 2027, where the fish fry release volume and sales volume vary by 34.09×10^3 t between the highest and lowest market demands. Common eel is the least sensitive to market demand, with its fish fry release volume and sales volume in the next 5-year plan not exceeding 5.32×10^3 t under 3 market demands.

Figure 21.3. shows the three types of fish feed used during the planning period. The usage of microalgae grew the fastest, with an annual growth rate of 12.06%; insect protein and soybean meal had annual growth rates of 8.11% and 5.65%, respectively. This is due to the lower carbon emissions of microalgae. To achieve the goal of reducing carbon emissions annually, aquaculture needs to select feed with lower

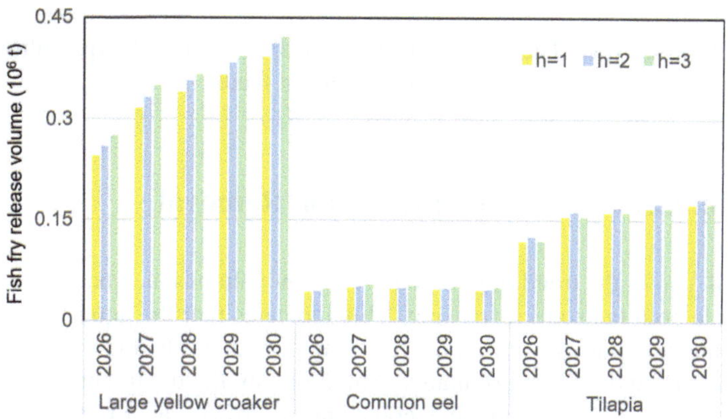

Fig. 21.1 Fish fry release volume under 3 market demand levels

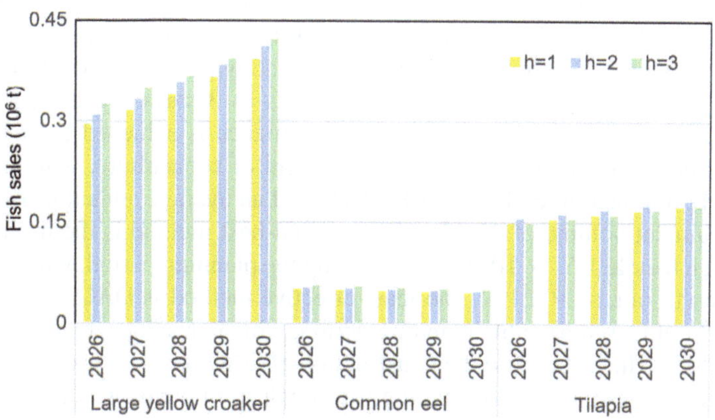

Fig. 21.2 Fish sales under 3 market demand levels

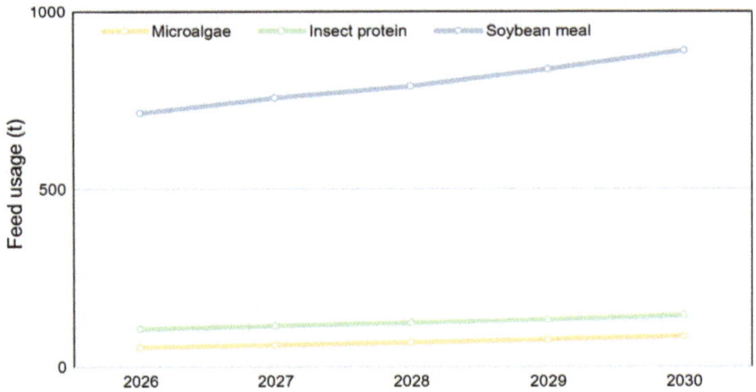

Fig. 21.3 Feed usage

carbon emission levels. Therefore, during the planning period, aquaculture should use more microalgae to replace other high-carbon feeds.

21.9 Pollutant and Carbon Emission Removal

Figure 21.4. shows the trend of pollutant treatment volumes in aquaculture, showing that all three pollutants are increasing. This indicates that the aquaculture industry is developing continuously during the planning period, partly due to stricter permissible emission standards for pollutants. Among them, the treatment volume of total nitrogen increases rapidly, with an average annual growth rate of 26.34%, followed by total phosphorus, with an average annual growth rate of 16.33%, while COD has the slowest growth, with an average annual increase of 7.41%. This corresponds to the previously mentioned need to gradually reduce the use of soybean meal as aquaculture feed. Replacing soybean meal with microalgae and insect protein can effectively reduce COD production, thereby achieving pollution reduction.

Figure 21.5. shows the CO_2 reduction effects of 3 carbon treatment methods in the model. The best carbon reduction performance is achieved through energy structure adjustments (EAD) in the power system of aquaculture, with an average annual carbon reduction of 1.41×10^6 t CO_2 and an average annual growth rate of carbon treatment volume of 25.95%. The second-best method is the use of carbon sinks, which handle 1.67×10^6 t of CO_2 during the planning period, with an average annual growth rate of 10.21%. The least effective method is DAC, which handles 0.20×10^6 t of CO_2 during the planning period, with an average annual growth rate of only 5.00%. This is because the construction of DAC facilities is challenging and involves high investment costs, making it difficult to implement. Therefore, future carbon reduction efforts in aquaculture should focus on energy structure adjustments, utilizing wind and solar energy for energy supply to achieve carbon reduction goals.

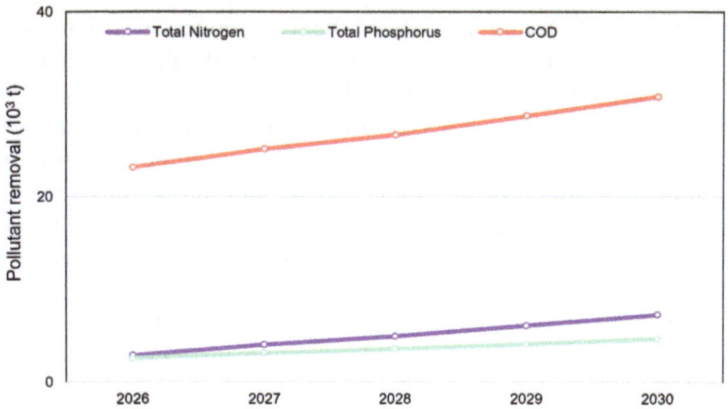

Fig. 21.4 Pollutant removal

Fig. 21.5 Carbon removal

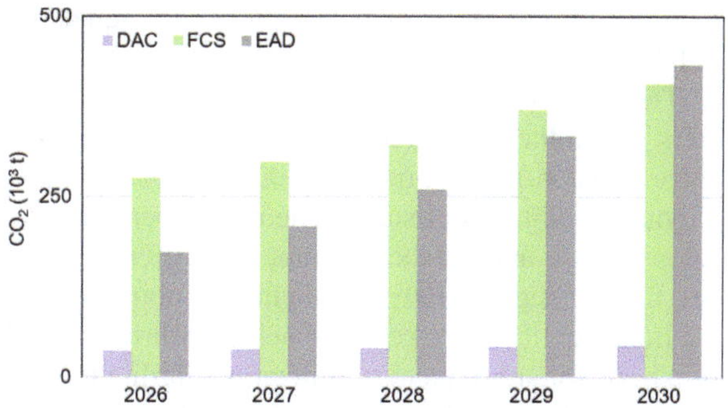

21.10 Conclusion

Under the backdrop of the dual carbon goals, this study constructs and validates a two-stage low-carbon aquaculture system design model, which has practical value and provides methodological basis for the green transformation of the aquaculture industry. Based on the theory of uncertainty scenario analysis, by utilizing two-stage stochastic programming methods, this study integrates carbon emission monitoring technology with recirculating water process upgrades and feed formula improvements to address the issue of isolated handling of single factors in previous research. Additionally, the model enhances its ability to cope with complex environmental changes by setting buffer modules to address market demand fluctuations. The results indicate that compared to traditional models, this model offers lower carbon emission intensity, higher production efficiency, and higher resource recycling rates.

Through the operation and analysis of the TSP model, it is demonstrated that this study has achieved reasonable conclusions. Specifically, the conclusions can be summarized as follows: (i) The model shows certain advantages in resource management and ecological benefits. By substituting microalgae and insect protein for conventional feed ingredients, it controls feed costs while reducing unit carbon emissions in the breeding process. This approach not only reduces feed waste but also further improves resource utilization, thereby decreasing pollutant emissions and effectively lowering the risk of water pollution from aquaculture. (ii) Based on the RAS (Recirculating Aquaculture System), this study reduces the dependence of aquaculture on fresh water, enhancing water resource utilization efficiency, thus providing crucial support for the sustainable development of the aquaculture industry. It was found that the use of wind and solar energy significantly contributes to the carbon reduction effects in aquaculture, followed by carbon sink measures, with direct air capture technology being the costliest and least practical. Therefore, in the absence of better negative carbon technologies, utilizing wind and solar energy to replace fossil fuels for powering fishery aquaculture is the most reliable carbon reduction method.

The model demonstrates excellent economic performance. Analysis of key parameters within the model indicates strong robustness and adaptability. Through a two-stage decision-making mechanism, the process is divided into a trial sales stage and a dynamic adjustment stage. The model performs well in controlling error rates in sales volume and catches volume for different fish species (large yellow croaker, eel, tilapia), significantly enhancing economic returns. During the trial sales stage, the model collects market feedback data through small-scale trials, providing scientific basis for subsequent optimization; during the dynamic adjustment stage, based on market demand fluctuations, the model precisely adjusts key parameters such as feed formulas and catch volumes, effectively balancing the proportion of high-risk high-yield fish species and low-risk stable-yield fish species. This provides important support for economic optimization and low-carbon transformation in the aquaculture industry.

Acknowledgements This work was supported by Fujian Provincial Social Science Foundation Project (FJ2024B082).

Appendix A: Parameter list

Subscription:

h	h Is market demand level; $h = 1$ to 3 means low, medium, high
i	i Is type of fish; $i = 1$ to 3 means large yellow croaker, common eel, tilapia
k	k Is planning period,$k = 1$ to 5 means 2026 to 2060
l	l Is type of feed,$l = 1$ to 3 means microalgae, insect protein, soybean meal

t t Is type of pollutant, $t = 1$ to 3 means TN, TP, COD

Decision variables:

DAC_k Carbon removal by direct air capture (t)
FCS_k Carbon removal by carbon sink (t)
MD_k Amount of medicine (t)
NE_k Number of aquaculture equipment (set)
$P_{k,t}$ Pollutant treatment (t)
$PRE_{i,k,h}$ Number of adult fish
$\Delta QS_{i,k,h}$ Selling amounts of fish in the second-stage (t)
$\Delta QF_{i,k,h}$ Number of fish fry in the second-stage (t)
$QFD_{k,l}$ Amount of breed (t)
$QC_{i,k,h}$ Number of fish caught
WU_k Water utilization (m^3)

Other parameters:

CBE_k Unit cost of equipment operation power generation (CNY/MWh)
CD_k Unit cost of medicine (CNY/t)
CDA_k Unit cost of direct air capture (CNY/t)
CE_k Unit cost of purchasing equipment (CNY/set)
CEE_k Carbon emission efficiency of electricity generation (t/MWh)
$CEF_{i,k}$ Carbon emission efficiency (t /t fish)
$CEFD_{k,l}$ Carbon emission efficiency (t/t feed)
CEL_k Maximum carbon emission allowance (t)
$CF_{i,k}$ Unit cost of purchasing fish fry in the first-stage (CNY/t)
$\Delta CF_{i,k}$ Unit cost of purchasing fish fry in the second-stage (CNY/t)
CFC_k Unit cost of carbon sink (CNY/t)
$CFD_{k,l}$ Unit cost of purchasing feed (CNY/t)
CWU_k Unit cost of water utilization (CNY/m^3)
$EP_{i,k,t}$ Pollutant generation efficiency (t /t fish)
$EN_{k,l,t}$ Pollutant generation efficiency (t/t feed)
ELC_k Electricity power consumption per equipment (MWh/set)
$LL_{k,t}$ Unit cost of pollutant treatment (t/CNY)
$MFCS_k$ Maximum amount of carbon sink (t)
$MDAC_k$ Maximum amount of direct air capture (t)
MMD_k Maximum medicine allowance (t)
$NMA_{i,k,l}$ Maximum feed supply (t/t)
$NMI_{i,k,l}$ Minimum feed supply (t/t)
NEE_k Number of additional equipment (set)

NER Number of equipment at beginning (set)
$PF_{i,k}$ Unit cost of selling fish in the first-stage (CNY/t)
$\Delta PF_{i,k}$ Unit cost of selling fish in the second-stage (CNY/t)
p_h $p_{\text{Low}} = 0.2, p_{\text{medium}} = 0.6, p_{\text{high}} = 0.2$, probability of demand
$PL_{k,t}$ Pollutant discharge allowance (t)
$QF_{i,k}$ Targeted number of fish fry in the first-stage (t)
$QFF_{i,k,h}$ Fish demand (t)
$QS_{i,k}$ Targeted number of selling fish in the first-stage (t)
R_i Initial number of fish (t)
$SEC_{i,k}$ Equipment's aquatic capability (t/set)
WA_k Water availability (m^3)
$WW_{i,k}$ Water utilization coefficient (m^3/t)
$\alpha_{k,t}$ Treatment efficiency (%)

References

Dewi T, Risma P, Oktarina Y, Dwijayanti S, Mardiyati EN, Sianipar AB, Hibrizi DR, Azhar MS, Linarti D (2025) Smart integrated aquaponics system: hybrid solar-hydro energy with deep learning forecasting for optimized energy management in aquaculture and hydroponics. Energy Sustain Dev 85:101683

Erdemir D, Dincer I (2024) Development of renewable energy based green hydrogen and oxygen production and electricity generation systems for sustainable aquaculture. J Clean Prod 434:140081

Liu JF, Zhang SJ (2025) Two-stage stochastic optimal design and operation of distributed energy resource systems considering unit unavailability. Energy Rep 13:3279–3290

Nematian J (2023) A two-stage stochastic fuzzy mixed-integer linear programming approach for water resource allocation under uncertainty in Ajabshir Qaleh Chay Dam. J Environ Inform 41(1):52–66

NFES: National Fisheries Economic Statistical Bulletin (2023) China fisheries (in Chinese) (08):7–8

Nguyen NT, Matsuhashi R, Chau Vo TTB (2021) A design on sustainable hybrid energy systems by multi-objective optimization for aquaculture industry. Renew Energy 163:1878–1894

Qiao YN, Wang YX, Zhang SY, Stoner AMK, Santos J (2024) Effects of 1.5 °C global warming on pavement climatic factors and performance. Transp Res Part D-Transp Environ 136:104393

Tu Y, Lu YZ, Lev B (2025) An interval two-stage robust stochastic programming under a bi-level multi-objective framework toward river basin water resources allocation. Comput Oper Res 180:107045

Wang S, Lan Z, Tan Q, Zhang T, Cao Y (2025) Optimizing water use efficiency in industrial parks under uncertainty: an inexact two-stage stochastic partial fractional programming method. J Water Process Eng 71:107351

Xu X, Gui F, Zhao S (2024) Simulation of fishery energy consumption and carbon emissions in Zhoushan city based on the long-range energy alternatives planning system model. Front Energy Res 10:1319915

Yang KY, Wang QP, Zhao L (2024) Two-stage stochastic programming for multi-objective optimization of sustainable utility systems integrating with combined heat and power units. J Clean Prod 451:142143

Zhang RS, Chen T, Wang Y, Short M (2024) An optimisation approach for the design and operation of recirculating aquaculture systems integrated with sustainable hybrid energy systems 477:143860

Chapter 22
Multi-Area Distributed Low-Carbon Optimal Scheduling Considering Carbon Capture Power Plant

Tengteng Jia and Haoyong Chen

Abstract The power industry, as the core carbon emission source, is facing the pressure of low carbon transformation. This paper focuses on the collaborative carbon reduction effect of carbon capture technology in multi-regional interconnected systems, and constructs a system operation architecture and modeling framework for low-carbon collaborative requirements. Based on this, this paper constructs a multi-regional distributed low-carbon optimal scheduling model for carbon capture power plants. The upper model aims at minimizing the cross-regional power deviation, and the superior dispatching agency coordinates the power exchange between regions. The lower model aims at minimizing the operating cost of the system, and the regional power grids make decisions independently and coordinate with the main problems of the superior dispatching institutions. Finally, the promotion effect of carbon trading mechanism and carbon capture technology on multi-regional collaborative carbon reduction is verified by simulation, and the balance between economy and low carbon is realized. In addition, this paper quantifies the impact of carbon trading mechanism and carbon capture power plants on the scheduling results through sensitivity analysis, which provides theoretical support and practical guidance for the low-carbon operation of multi-area interconnected systems.

Keywords Carbon capture technology · Carbon trading · Low carbon power dispatching

22.1 Introduction

Global climate change has become a major challenge facing mankind. Reducing greenhouse gas emissions and developing a low-carbon economy have become a global consensus. As one of the main sources of carbon emissions, the low-carbon

T. Jia · H. Chen (✉)
South China University of Technology, Guangzhou, China
e-mail: eehychen@scut.edu.cn

Y. Li et al. (eds.), *16th International Conference on Environmental Science and Development*, Environmental Science and Engineering,
https://doi.org/10.1007/978-3-032-15080-6_22

transformation of the power industry is crucial to achieving global climate goals. According to the National Energy Agency, global energy-related carbon dioxide emissions in 2023 reached 37.4 billion tons, a record high. As the main industry of China 's carbon emission source, the power industry accounts for more than 40% of carbon emissions, which is the key area for China to achieve the double carbon goal.

The carbon market mechanism is the basis of carbon market research and has significant potential in promoting energy conservation and emission reduction. Relevant research mainly focuses on carbon quota allocation, carbon price prediction, trading mechanism and so on. In terms of carbon quota allocation, the national carbon market is in its infancy. Reference (Li et al. 2022) established a quota allocation model for the power generation industry, and clarified the total amount of quota issuance and the responsibilities and processes of each participant. In Reference (Xiao et al. 2022), a carbon quota model was established based on the entropy method to achieve a more reasonable allocation considering the electricity forecast and carbon emission reduction targets. The literature (Liu 2020) analyzes the impact of regional differences, carbon content, and corporate performance on the allocation of quotas in the power industry.

Carbon capture and storage (CCS) technology, as a key technology to realize low-carbon flexible transformation of thermal power units, has attracted great attention from the international community. The fifth assessment report of the Intergovernmental Panel on Climate Change (IPCC) shows that carbon capture technology plays a key role in climate change and is critical to achieving zero carbon emissions and even negative carbon emissions targets. At present, the research on carbon capture and storage technology mainly focuses on the technical path and optimal operation decision of carbon capture power plants. The literature (Mikulčić et al. 2019) analyzes the basic principles and operating characteristics of carbon capture power plants, and sorts out the research status of its development in the power industry; reference (Li et al. 2025) studied the operating characteristics of flexible carbon capture power plants and quantified the environmental and economic benefits of their coordinated operation with renewable energy systems. In (Yuan et al. 2024; Bao et al. 2025), the operation characteristics and peaking capacity of carbon capture power plants are analyzed, and the conclusion that carbon capture power plants have greater peaking depth and better peaking response speed is obtained.

Carbon capture power plants have the characteristics of flexible regulation. A large number of studies have combined them with renewable energy power generation to promote the consumption of renewable energy and reduce the carbon emissions of the system. In (Akbari-Dibavar et al. 2021; An et al. 2024), considering the trade-off between cost and carbon emissions, the carbon capture power plant is combined with conventional power generation to verify the rationality and economy of the proposed model. In Reference (Li et al. 2024), the carbon capture power plant is combined with wind power and photovoltaic to establish a low-carbon economic dispatch model with the minimum comprehensive cost as the objective function, so that the power grid can obtain more significant economic and low-carbon benefits.

22.2 Operation Characteristics and Modeling of Carbon Capture Power Plant

Carbon capture power plant is a new type of power generation system which combines traditional fossil fuel power plant with carbon capture technology. The system is modified by installing carbon capture devices in traditional thermal power units (such as coal-fired power plants) to capture, compress and seal the carbon dioxide generated during combustion, thereby greatly reducing carbon emissions and having good low-carbon performance. According to the time point and mode of carbon dioxide capture, the technical routes of carbon capture power plants are mainly divided into three categories: pre-combustion, oxyfuel and post-combustion, as shown in Table 22.1.

Among them, the post-combustion capture technology does not change the original coal-fired power generation system, and only needs to add carbon capture equipment at the back end, which is suitable for the transformation of existing power plants. Therefore, the technology is relatively mature and widely used. In view of the maturity of post-combustion capture technology and the ability to transform existing power plants, this paper uses post-combustion capture technology for research. In this mode, the operation mode of carbon capture power plant mainly includes basic operation mode and flexible operation mode. Under the basic operation mode, each operation unit of the carbon capture power plant is in the rated working condition, and each part operates independently, which cannot fully reflect the flexibility of its operation. The flue gas split operation mode is one of the core technologies of the flexible operation mode. It accurately controls the amount of carbon dioxide entering

Table 22.1 Technical route of carbon capture power plant

Type	Characteristics	Advantages	Disadvantages	Applicable scenarios
Post-combustion	The flue gas is treated after combustion to separate and recover carbon dioxide	High technical maturity	The energy consumption is large, usually accounting for 15–30% of the total output of the power plant	Retrofit of existing power plants
Pre-combustion	Convert fuel into syngas and separate carbon dioxide before combustion	The higher the concentration of CO_2, the higher the capture efficiency	The technology is complex and the investment cost is high	New power plant
Oxyfuel	Combustion in a pure oxygen environment produces a high concentration of carbon dioxide	The concentration of CO_2 is extremely high, and the capture process is simple	It needs to be equipped with an air separation device to provide pure oxygen, and the energy consumption is large	New power plant or large-scale transformation

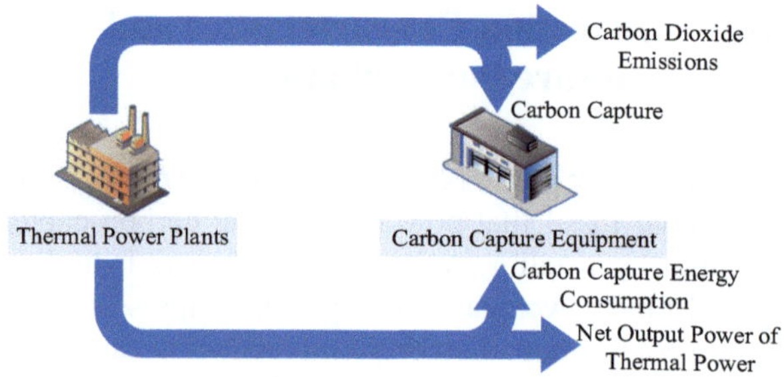

Fig. 22.1 Power flow of carbon capture power plant

the carbon capture equipment by adjusting the flue gas split ratio, thereby improving the economy and low carbon of the overall operation of the system.

The total output power of the thermal power plant consists of two parts: the net output power of the thermal power plant and the carbon capture energy consumption, as shown in Fig. 22.1. The energy consumption of the carbon capture system consists of fixed energy consumption P_M and operating energy consumption P_Y. P_M is the power lost due to the change of the operating state of the unit caused by the transformation into a carbon capture power plant. This part of energy consumption accounts for a small proportion in the carbon capture system and is usually negligible. Specifically, it can be expressed as:

$$P_G = P_M + P_Y + P_J \tag{22.1}$$

where, the energy consumption of carbon capture equipment P_Y can be expressed as:

$$P_Y = \lambda_Y E_{YCO2} \tag{22.2}$$

where, λ_Y is the energy consumption required to capture unit carbon dioxide; E_{YCO2} is the amount of carbon dioxide captured can be further expressed as:

$$E_{YCO2} = \theta_Y \delta_Y E_G \tag{22.3}$$

where, θ_Y is the carbon capture efficiency; δ_Y is the flue gas split ratio, and $0 \leq \delta_Y \leq 1$; E_G is the actual carbon dioxide emissions of thermal power plants can be expressed as:

$$E_G = \varphi_G P_G \tag{22.4}$$

where, φ_G is the carbon emission intensity, that is, the amount of carbon dioxide emitted by the production unit of electric energy.

The actual carbon dioxide emissions of thermal power plants can also be expressed as the sum of net carbon dioxide emissions E_J and the amount of carbon dioxide captured by carbon capture equipment E_{YCO2}, that is

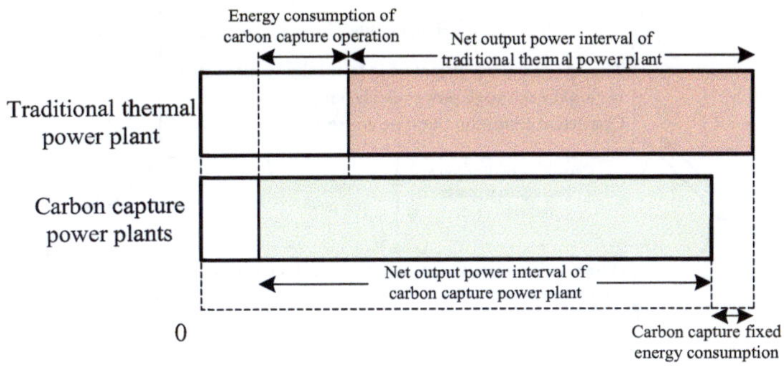

Fig. 22.2 The output operation interval of carbon capture power plant

$$E_G = E_J + E_{YCO2} \tag{22.5}$$

The carbon capture system consumes a certain amount of electric energy during operation. This part of the power is allocated to the carbon capture system to ensure its normal operation, thus affecting the net output of the thermal power unit. Based on the mathematical model formula (22.2–22.5)–(22.2–22.9) of carbon capture power plants, the net output power of carbon capture power plants can be derived:

$$P_J = (1 - \varphi_G \lambda_Y \theta_Y \delta_Y) P_G - P_M \tag{22.6}$$

Therefore, the net output power operation interval of the carbon capture power plant can be expressed as:

$$(1 - \varphi_G \lambda_Y \theta_Y \delta_Y) P_G^{\min} - P_M \le P_J \le P_G^{\max} - P_M \tag{22.7}$$

The output operation range of the carbon capture power plant is shown in Fig. 22.2.

Due to the existence of carbon capture energy consumption, the net output power of carbon capture power plants breaks through the lower limit of the net output power of traditional thermal power plants. Therefore, the carbon capture power plant has stronger down-regulation ability, wider output regulation range than the traditional thermal power unit, and the flexibility of the system operation is improved. This feature makes carbon capture power plants have good peak shaving, backup and load tracking capabilities in the power system, and can better adapt to changes in power demand and provide support for the stable operation of the power system.

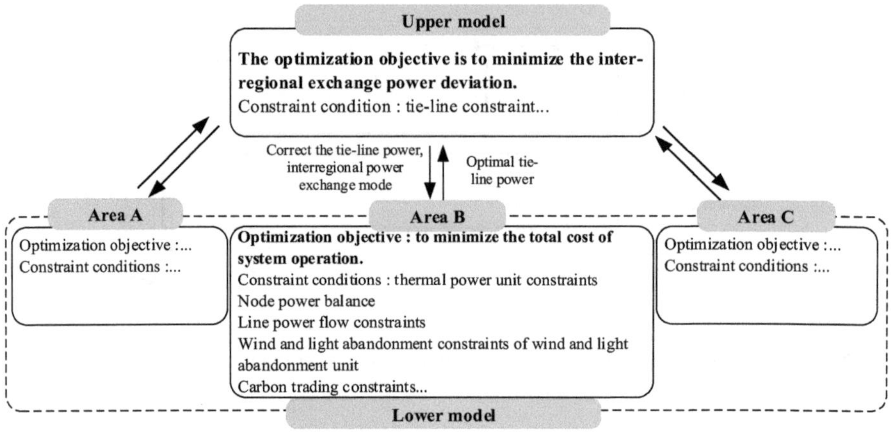

Fig. 22.3 Decentralized scheduling framework for multi-area interconnected systems

22.3 Multi-Regional Distributed Low-Carbon Optimal Scheduling Model for Carbon Capture Power Plants

22.3.1 Multi-Area Interconnected System Modeling Framework

This chapter builds a distributed scheduling framework for multi-area interconnected systems based on the target cascading analysis technique, as shown in Fig. 22.3. In this framework, the superior dispatching agency, as the coordinator of the whole system, is responsible for formulating the initial inter-regional power exchange plan and dynamically adjusting the inter-regional power allocation strategy through the distributed coordination mechanism. While independently scheduling the regional problem, each regional dispatching agency optimizes the solution with the goal of minimizing the system operating cost, and feeds the updated tie-line power information back to the superior dispatching agency. Based on the feedback information of the region, the superior scheduling agency conducts global optimization scheduling with the goal of minimizing the inter-regional exchange power deviation, generates inter-regional power exchange plans, and coordinates the operation strategies of each region.

22.3.2 Upper Model

The upper model assumes the responsibility of global coordination. The objective function is as follows:

$$\begin{aligned}\min\sum_{A\in\Omega}\sum_{\tau\in N_\tau}\sum_{t\in T}\left[\psi_{\tau,t}\left(P^{a,0}_{\tau,t}-P^{a}_{\tau,t}\right)+\left\|\zeta_{\tau,t}\circ\left(P^{a,0}_{\tau,t}-P^{a}_{\tau,t}\right)\right\|_2^2\right]\\+\sum_{B\in\Omega}\sum_{\tau\in N_\tau}\sum_{t\in T}\left[\psi_{\tau,t}\left(P^{b,0}_{\tau,t}-P^{b}_{\tau,t}\right)+\left\|\zeta_{\tau,t}\circ\left(P^{b,0}_{\tau,t}-P^{b}_{\tau,t}\right)\right\|_2^2\right]\end{aligned}\tag{22.8}$$

where, Ω is the number of regions; N_τ is the total number of contact lines in the region; $\psi_{\tau,t}$ and $\zeta_{\tau,t}$ are the penalty weights related to the linear term and the quadratic term, respectively; $P^{a,0}_{\tau,t}$ and $P^{b,0}_{\tau,t}$ are the reference values of DC tie-line power fed back to regions A and B by the superior dispatching agency respectively; $P^{a}_{\tau,t}$, and $P^{b}_{\tau,t}$ are the DC tie-line power transmitted from regions A and B to the superior dispatching agency.

(1) Tie-line consistency constraint

$$P^{a,0}_{\tau,t}=P^{b,0}_{\tau,t}\tag{22.9}$$

let $P^{a,0}_{\tau,t}=P^{b,0}_{\tau,t}=P^{0}_{\tau,t}$, other constraints are introduced according to the operating characteristics of the DC tie line.

(2) Tie-line transmission limit constraint

$$\underline{P}_\tau\le P^{0}_{\tau,t}\le\overline{P}_\tau\tag{22.10}$$

where, $\overline{P}_\tau$ and $\underline{P}_\tau$ are the upper and lower limits of the transmission power of the tie line, respectively.

(3) Tie-line transmission capacity constraints

$$\underline{Q}_\tau\le\sum_{t\in T}P^{0}_{\tau,t}\le\overline{Q}_\tau\tag{22.11}$$

where, $\overline{Q}_\tau$ and $\underline{Q}_\tau$ are the upper and lower limits of the power transmission of the tie line, respectively, depending on the power transmission plan.

(4) Constraints on the number of tie-line adjustments

$$\sum_{t\in T}\left(\alpha^{+}_{\tau,t}+\alpha^{-}_{\tau,t}\right)\le\overline{N}_\tau\tag{22.12}$$

where, $\overline{N}_\tau$ is the maximum adjustment number of tie lines in a scheduling cycle.

(5) Power direction continuity constraint

$$\begin{cases}\alpha^{+}_{\tau,t}+\alpha^{-}_{\tau,t}\le 1\\ \alpha^{+}_{\tau,t-1}+\beta^{-}_{\tau,t}\le 1\\ \alpha^{+}_{\tau,t}+\beta^{+}_{\tau,t-1}\le 1\end{cases}\tag{22.13}$$

where, $\alpha_{\tau,t}^{+}$ and $\alpha_{\tau,t}^{-}$ are binary variables, which represent the upward and downward adjustment power of the tie line, respectively.

(6) Tie-line power adjustment constraint

$$-S_{\tau} \leq P_{\tau,t}^{0} - P_{\tau,t-1}^{0} \leq S_{\tau} \tag{22.14}$$

where, S_{τ} is the maximum adjustment rate of the tie line.

22.3.3 Lower Model

The solution goal of low-carbon optimal dispatching of regional power grid is to minimize the total operation cost of each region and construct the low-carbon optimal dispatching model of each sub-region under the premise of satisfying the safe operation constraints of regional power grid. Taking regional power grid A as an example, its objective function is expressed as:

$$\begin{aligned}
\min C &= C_Y + C_K + C_{Carbon}^{1} + C_{ED} \\
&= \sum_{t\in T}\sum_{i\in N_I} u_{i,t}\left(a_i P_{i,t}^{G^2} + b_i P_{i,t}^{G} + c_i\right) + \sum_{t\in T}\sum_{j\in N_J} u_{j,t}\left(a_j P_{j,t}^{G^2} + b_j P_{j,t}^{G} + c_j\right) \\
&+ \sum_{t\in T}\sum_{i\in N_I}\left(x_{i,t}SU_i + v_{i,t}SD_i\right) + \sum_{t\in T}\sum_{j\in N_J}\left(x_{j,t}SU_j + v_{j,t}SD_j\right) \\
&+ \sum_{t\in T}\kappa_{wt}P_{WT,t}^{shed} + \sum_{t\in T}\kappa_{pv}P_{PV,t}^{shed} + \sum_{t\in T}\kappa_{load}P_{Load}^{loss} + \gamma\left(Q_p - Q_q\right) + C_{ED} \\
&+ \sum_{\tau\in N_\tau}\sum_{t\in T}\left[\psi_{\tau,t}\left(P_{\tau,t}^{0} - P_{\tau,t}\right) + \left\|\zeta_{\tau,t}\circ\left(P_{\tau,t}^{0} - P_{\tau,t}\right)\right\|_2^2\right]
\end{aligned} \tag{22.15}$$

where, C_{ED} is the depreciation cost of carbon capture equipment; T is the scheduling period; $\sum_{\tau\in N_\tau}\sum_{t\in T}\left[\psi_{\tau,t}\left(P_{\tau,t}^{0} - P_{\tau,t}\right) + \left\|\zeta_{\tau,t}\circ\left(P_{\tau,t}^{0} - P_{\tau,t}\right)\right\|_2^2\right]$ is the deviation correction term of DC tie-line power; $P_{\tau,t}^{0}$ is the reference value of DC tie-line power fed back to the lower level by the higher level dispatching institution.

(1) Node balance constraint

$$\mathrm{A}\left(\mathrm{P}^{G} - \mathrm{P}_{shed}^{WT,PV}\right) + \mathrm{A}_{\tau}\mathrm{P}_{\tau} = \mathrm{P}^{Load} + \mathrm{XP}_{L} \tag{22.16}$$

where, A is the generator-node correlation matrix;. P^{G}. is the unit output matrix; $\mathrm{P}_{shed}^{WT,PV}$ is the abandoned wind and abandoned light power matrix; A_{τ} is the tie-line-node correlation matrix; P_{τ} is the tie-line power matrix; P^{Load} is the node load matrix; X is the node-branch incidence matrix; P_{L} is a branch correlation matrix.

(2) Branch power flow power

$$-\overline{P}_m^l \leq P_{m,t}^l \leq \overline{P}_m^l \tag{22.17}$$

where, $P_{m,t}^l$ and $\overline{P}_m^l$ are the transmission power and the maximum transmission power of the branch at the moment.

(3) Thermal power unit constraints

$$x_{i,t}\underline{P}_i^G \leq P_{i,t}^G \leq x_{i,t}\overline{P}_i^G \tag{22.18}$$

$$\begin{cases} P_{i,t}^G - P_{i,t-1}^G \leq x_{i,t-1}\left(R_i^U - S_i^U\right) + S_i^U \\ P_{i,t-1}^G - P_{i,t}^G \leq x_{i,t}\left(R_i^D - S_i^D\right) + S_i^D \end{cases} \tag{22.19}$$

$$S_i^U = S_i^D = \frac{1}{2}\left(\underline{P}_i^G + \overline{P}_i^G\right) \tag{22.20}$$

$$\begin{cases} \sum\limits_{k=t}^{t+T_i^{off}-1} \left(1 - x_{i,k}\right) \geq T_i^{off}\left(x_{i,t-1} - x_{i,t}\right) \\ \sum\limits_{k=t}^{t+T_i^{on}-1} x_{i,k} \geq T_i^{on}\left(x_{i,t} - x_{i,t-1}\right) \end{cases} \tag{22.21}$$

$$\begin{cases} \sum\limits_{t=1}^{T} \left(x_{i,t} - x_{i,t-1}\right) \leq \overline{T}_i^{on} \\ \sum\limits_{t=1}^{T} \left(x_{i,t-1} - x_{i,t}\right) \leq \overline{T}_i^{off} \end{cases} \tag{22.22}$$

where, $\underline{P}_i^G$ and $\overline{P}_i^G$ are the minimum and maximum technical output of thermal power units respectively; R_i^U and R_i^D are respectively the upward and downward climbing rate limits of thermal power units i; when $\underline{P}_i^G > R$, the minimum output of the unit start-up exceeds the ramp rate, and the ramp constraint will cause the unit to fail to meet the start-up requirements. In order to simplify the model, the maximum rise rate of the unit start-up and the maximum drop rate of the shutdown are expressed by the formula (22.20); T_i^{off} and T_i^{on} are the minimum shutdown and minimum start-up time of the thermal power unit, respectively.

(4) Carbon trading constraints

$$Q_p = \eta \sum_{t=1}^{T} P_t^G \tag{22.23}$$

$$Q_q = \sigma \sum_{t=1}^{T} P_t^G \tag{22.24}$$

$$C_{Carbon} = \gamma \left(Q_p - Q_q \right) \tag{22.25}$$

where, η is the carbon emission intensity of the unit; P_t^G is the output of the thermal power unit during the period t; σ is the free allocation proportion of carbon emission quota per unit of unit power; γ is the unit carbon trading price.

(5) Carbon capture power plant constraints

The constraint of carbon capture power plant is expressed by (1–5).

(6) Wind power and photovoltaic power curtailment constraints

$$0 \leq \frac{P_{WT,t} - P_{WT,t}^{shed}}{P_{WT,t}} \leq \overline{\kappa}_{wt} \tag{22.26}$$

$$0 \leq \frac{P_{PV,t} - P_{PV,t}^{shed}}{P_{PV,t}} \leq \overline{\kappa}_{pv} \tag{22.27}$$

where, $P_{WT,t}$ and $P_{PV,t}$ are respectively the output of wind power and photovoltaic units at the moment; $P_{WT,t}^{shed}$ and $P_{PV,t}^{shed}$ are the amount of light abandoned by wind power and photovoltaic units at the moment; $\overline{\kappa}_{wt}$ and $\overline{\kappa}_{pv}$ are the maximum allowable curtailment rate of wind power and photovoltaic units respectively.

22.3.4 Model Solution

The convergence criterion of the model only needs to check whether the relative error between the reference value of the DC tie-line power issued by the superior dispatching institution and the DC tie-line power uploaded by the lower regions meets the convergence conditions. The specific convergence basis is as follows:

$$\begin{cases} \dfrac{\left| P_{\tau,t}^{a,0} - P_{\tau,t}^{a} \right|}{P_{\tau,t}^{a,0}} \leq \varepsilon_1 \\ \dfrac{\left| P_{\tau,t}^{b,0} - P_{\tau,t}^{b} \right|}{P_{\tau,t}^{b,0}} \leq \varepsilon_2 \end{cases} \tag{22.28}$$

where, ε_1 and ε_2 are the convergence threshold, and the value is 0.1%.

If the above formula does not satisfy the accuracy in the l th iteration, the penalty coefficient $\psi_{\tau,t}$ and $\zeta_{\tau,t}$ are updated by the following formula.

$$\begin{cases} \psi_{\tau,t}(l) = \psi_{\tau,t}(l-1) + 2\zeta_{\tau,t}(l-1)^2\left[P^{a,0}_{\tau,t}(l-1) - P^{a}_{\tau,t}(l-1)\right] \\ \zeta_{\tau,t}(l) = \alpha\zeta_{\tau,t}(l-1) \end{cases} \quad (22.29)$$

where, α is the update coefficient, take 1.5. When $l = 1$, $\psi_{\tau,t}(1) = 1$ and $\zeta_{\tau,t}(1) = 0.5$.

22.3.5 *Example Analysis*

22.3.5.1 Data

In order to verify the effectiveness of the proposed multi-area distributed low-carbon scheduling model, this chapter uses a three-area 117-node interconnected power system model for simulation verification. The model is composed of three improved IEEE39 node systems. If the carbon capture power plant is considered, the conventional thermal power units G1 and G9 in each region are transformed into carbon capture power plants. A typhoon power plant is connected at nodes 8 and 33, and a photovoltaic power plant is connected at nodes 8 and 39, as shown in Fig. 22.4. Among them, area A has the smallest load demand and the most abundant wind and light resources. Area B has the largest load demand and the most scarce wind and light resources.

This section sets up two scenarios for comparative analysis, as follows:

Scenario 1: Optimal scheduling of multi-area interconnected systems (excluding carbon trading and carbon capture power plants), as a benchmark scenario;
Scenario 2: Low-carbon optimal scheduling of multi-area interconnected systems with carbon trading and carbon capture power plants.

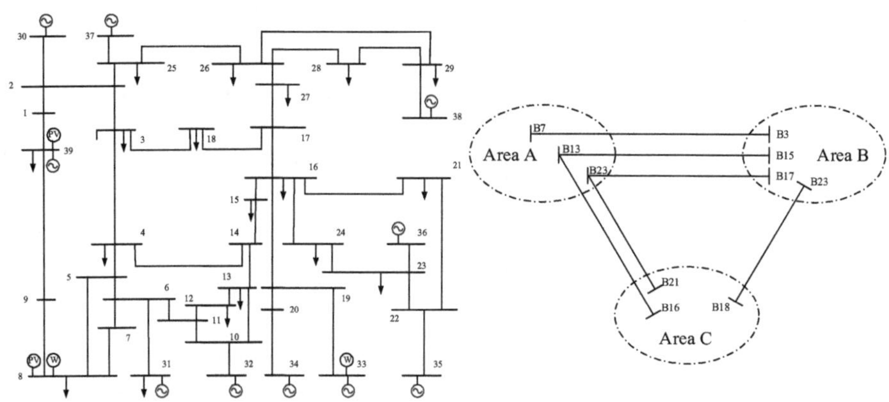

Fig. 22.4 System structure diagram

22.3.6 Carbon Capture Power Plant Strength Analysis

Due to the high carbon emission intensity of the thermal power unit G9 and the significant output in each scenario, the output of the unit G9 is analyzed here, and then the output characteristics of the carbon capture power plant are studied. Figure 22.5 shows the output of the thermal power unit G9 under Scenario 1 and Scenario 2, and Fig. 22.6 shows the carbon emission of the carbon capture unit G9 under Scenario 2.

It can be seen from the above figure that the net emissions of carbon capture power plant G9 are always maintained at a low level. The reason is that during the power generation process, the carbon capture power plant allocates part of the power generation to the carbon capture device to ensure stable carbon capture intensity.

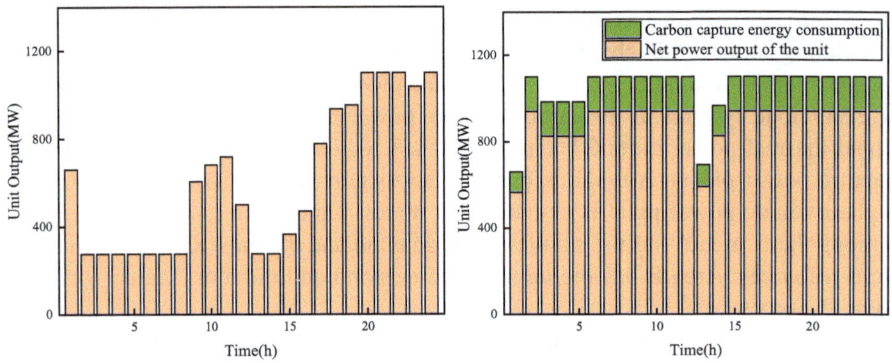

(a) The output diagram of unit G9 under scenario 1.

(b) The output diagram of carbon capture power plant G9 under scenario 2.

Fig. 22.5 Unit G9 output diagram

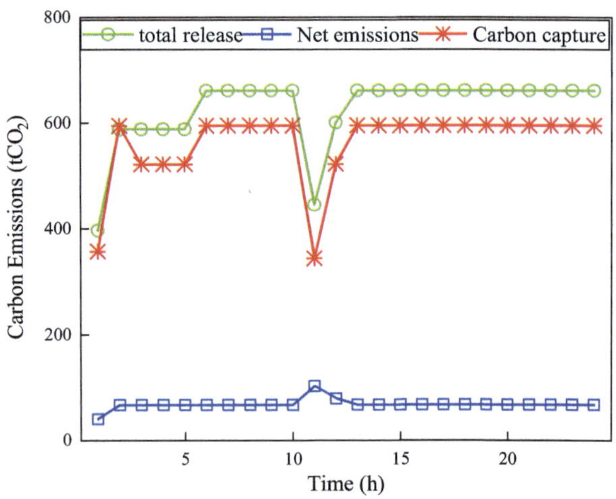

Fig. 22.6 Carbon capture unit G9 carbon emissions figure

However, due to the limitation of carbon capture technology efficiency, the system can only capture 90% of the total carbon emissions, so the remaining 10%. Carbon emissions constitute the main net emission source of the power plant.

Carbon capture power plants have unique operating characteristics in power dispatching. In the period of low load, by reducing its own net output, it creates space for the consumption of renewable energy and relieves the pressure of power grid. During the peak load period, the challenge of wind and light curtailment and load loss can be effectively dealt with by flexibly adjusting the energy consumption of carbon capture.

During the high wind power period of 0–5, the system load is at a low ebb, and the wind power output increases significantly. In order to improve the new energy consumption capacity, the traditional thermal power unit actively reduces the output, and the carbon capture unit G9 also adopts the low output operation strategy. This coordinated regulation method effectively alleviates the peaking pressure of the power grid and promotes the full utilization of wind power. At 11am, this time is the peak load period, and the net emission of the unit G9 shows an obvious peak. This is because the system increases the use of high-carbon emission units in response to peak loads, resulting in a simultaneous increase in carbon capture energy consumption, thereby reducing the net output of the power plant.

22.3.7 Carbon Trading Mechanism Parameters

The carbon trading price not only profoundly affects the economy of power dispatching operation, but also forms a key constraint on the dispatching strategy of carbon capture power plants. In order to further explore the impact of carbon price on the scheduling decision of carbon capture power plants, this study compares and analyzes the carbon price levels of 30 yuan / t and 80 yuan / t under scenario three. The scheduling results of thermal power units with a carbon trading price of 30 yuan/t and 80 yuan/t are shown in Fig. 22.7.

It can be found that the carbon capture unit G1 always maintains high utilization rate operation based on its power generation characteristics. In contrast, the carbon capture unit G9 is greatly affected by the carbon price. At low carbon prices, its utilization rate is relatively low. When the carbon price rises, the utilization rate of the unit is significantly improved. This shows that when the carbon price is at a low level, the utilization rate of carbon capture units is not high. The reason is that the lower carbon price at this time makes the cost of carbon emissions account for a relatively small proportion of the total cost, and the penalty cost brought by carbon emissions is lower than the coal-fired cost and start-stop cost of thermal power units and the benefits obtained by reducing the carbon capture device to capture carbon dioxide. Therefore, the system preferentially calls high-carbon units, which is not conducive to the system's constraints on carbon emissions.

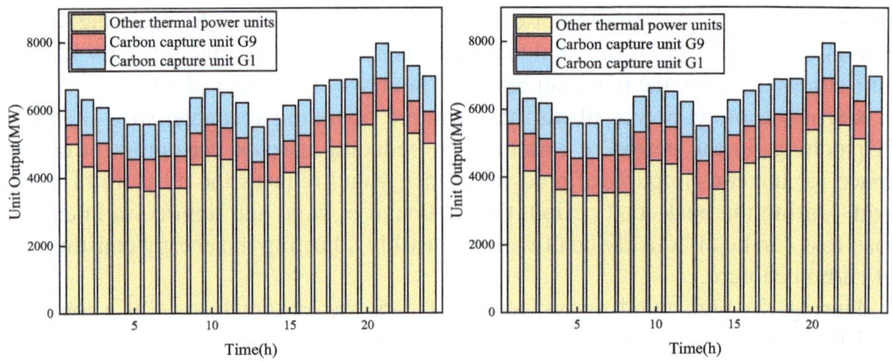

Fig. 22.7 Thermal power unit scheduling results with carbon trading prices of 30 yuan/t and 80 yuan/t

22.3.8 Flue Gas Split Ratio of Carbon Capture Power Plant

In order to further analyze the influence of the flue gas split ratio on the carbon emissions of the system, this section sets different flue gas split ratio limits for Scenario 2, and obtains the relationship curve between the flue gas split ratio and the carbon emissions of the system, as shown in Fig. 22.8.

It can be seen from the trend of the curve in the figure that with the increase of the flue gas split ratio from 0.75 to 1, the carbon emissions of the system show a downward trend. When the flue gas split ratio is increased to more than 0.95, the downward trend of carbon emissions is obviously slowed down. Due to the upper limit of the performance of the carbon capture system, when the flue gas split ratio is close to the limit value 1, the capture device reaches full load operation state. At this time, even if the flue gas split ratio continues to increase, the space for improving the carbon capture efficiency is extremely limited. This shows that the flue gas split ratio

Fig. 22.8 Relationship curve between flue gas split ratio and carbon emission

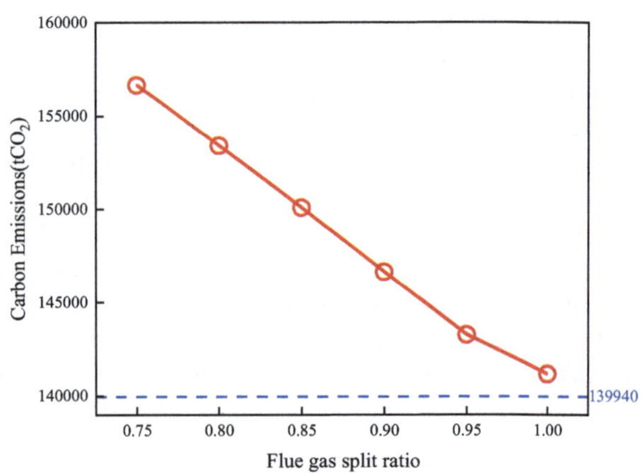

determines the operating intensity of the carbon capture system, that is, the higher the flue gas split ratio, the greater the proportion of flue gas entering the capture device, and the stronger the capture capacity of the power plant to carbon dioxide, thereby significantly reducing the carbon emissions of the system. It is worth noting that even if the flue gas split ratio is set to a limit value of 1, the corresponding system carbon emissions are still significantly higher than the carbon emission level when the flue gas split ratio is used as a decision variable. This shows that when the flue gas split ratio is forced to be limited to a fixed parameter, the system cannot achieve global optimization by dynamically adjusting other operating parameters such as unit output. In contrast, when the flue gas split ratio is set as the decision variable, the system can break through the physical limitation of a single variable through multi-parameter collaborative optimization, so as to achieve the optimal scheduling decision at a better balance point. Therefore, in actual operation, it is necessary to balance the relationship between carbon capture efficiency and energy consumption cost, so as to avoid the economic deterioration caused by blind pursuit of limit parameters.

22.4 Conclusion

This paper focuses on the core issue of low-carbon dispatching in power system, and constructs a multi-regional distributed low-carbon optimal dispatching model with carbon capture power plants, which provides theoretical support and technical path for the low-carbon development of power system. The main conclusions are as follows:

(1) Carbon capture power plant has a deeper peak regulation performance than conventional thermal power units. Making full use of the advantages of flexible regulation of carbon capture power plants can effectively suppress the fluctuation of wind and solar output, improve low-carbon benefits, and ensure the safe operation of power systems in a wider range.
(2) The sensitivity analysis of the flue gas split ratio of the carbon capture power plant shows that the increase of the flue gas split ratio can reduce the carbon emission of the system. When it is close to the limit value, the efficiency improvement of the carbon capture power plant is limited.

This paper is supported by National Key Research and Development Program of China (Key technologies on intelligent dispatch of power grid under 20% new energy integration scenario, 2022YFB2403500) and Research on the Cooperative Development Mechanism of New Energy Industry in South China Region (ZDPY202407).

References

Akbari-Dibavar A, Mohammadi-Ivatloo B, Zare K et al (2021) Economic-emission dispatch problem in power systems with carbon capture power plants. IEEE Trans Ind Appl 57(4):3341–3351

An Y, Jia CH, Ning XY et al (2024) Low-carbon economic scheduling considering wind-PV correlation and flexible carbon capture in complementary system. Acta Energiae Solaris Sinica 45(11):40–49

Bao G, Zhang YH, Peng X et al (2025) Optimal dispatching of virtual power plant considering information gap decision theory and oxy-fuel combustion carbon capture technology. Electr Meas Instrum 1–12

Li XR, Fu XQ, Li GD (2022) Carbon quota mechanism design of provincial power generation industry under the background of carbon peaking. Electr Power Constr 43(12):74–82

Li YT, Li K, Xing XM et al (2024) Optimal scheduling of the integrated energy system considering oxy-fuel combustion technology. Power Syst Clean Energy 40(08):1–10+17

Li J, Zhang C, Davidson RM et al (2025) Assessing the synergies of flexibly-operated carbon capture power plants with variable renewable energy in large-scale power systems. Applied Energy 377(PB):124459

Liu QL (2020) Thoughts and suggestions on carbon emission quota allocation in power generation industry. Environ Sci Surv 39(S1):13–15

Mikulčić H, Skov IR, Dominković DF et al (2019) Flexible carbon capture and utilization technologies in future energy systems and the utilization pathways of captured CO_2. Renew Sustain Energy Rev 114:109338

Xiao Y, Pei F, Hu W et al (2022) Grey prediction model of optimal electricity quantity of carbon quota based on entropy method. Sci Technol Eng 22(19):8355–8362

Yuan X, Liu J, Chen H et al (2024) Effect of carbon capture technology application on peak shaving capacity of coal-fired units. Power Gener Technol 45(3):373–381

Chapter 23
Driving Forces of Carbon Dioxide Emissions from the Standpoint of Industrial Characteristic: A Case Study in the Yangtze River Delta

Zhechen Zhou, Su Xu, Jun Wang, Yuting Huang, and Yarong Cao

Abstract Urban agglomerations are of utmost importance in facilitating the progress of urbanization and industrialization. The study of the factors influencing CO_2 emissions, particularly when considering the cities' specialized industrial economy structure within urban agglomerations, holds significant importance in developing effective strategies to reduce CO_2 emissions. Using the cross-sectional dataset, this paper initially employed cluster analysis to classify the sample cities according to the proportion of output values. Subsequently, based on the obtained classification results, the driving forces of CO_2 emissions at the municipal level were examined by utilizing quantile regression. The findings indicate that: (1) The sample cities can be classified into five groups according to their pillar industries, with significant disparities in CO_2 emissions among these five types of cities. (2) Population growth, the rise in the share of secondary industry, and land urbanization actually lead to an escalation of CO_2 emissions. The latter two factors have a more pronounced impact on cities with both higher and lower CO_2 emissions compared to cities with

---

Zhechen Zhou and Su Xu—Co-first authors.

Z. Zhou
Department of Building and Real Estate, The Hong Kong Polytechnic University, Hong Kong, China

S. Xu · J. Wang (✉)
Department of Landscape Planning and Design, East China University of Science and Technology, Shanghai, China
e-mail: wangjun81@ecust.edu.cn

Y. Huang
Department of Urban Planning, College of Architecture and Urban Planning, Tongji University, Shanghai, China

Y. Cao
Department of Landscape Architecture, Faculty of Architecture, South China University of Technology, Guangzhou, China

Y. Li et al. (eds.), *16th International Conference on Environmental Science and Development*, Environmental Science and Engineering,
https://doi.org/10.1007/978-3-032-15080-6_23

moderate emissions. (3) The adoption of a polycentric and highly connected development model can facilitate the reduction of CO_2 emissions, particularly in areas with high levels of emissions. (4) The case of the YRDUA does not support the traditional Environmental Kuznets Curve hypothesis. Our findings offer a foundation for the low-carbon construction in urban agglomerations from the perspectives of socioeconomic progress and urban structure organization.

Keywords Carbon dioxide emissions · Industrial characteristic · Driving forces · Quantile regression · Yangtze River Delta urban agglomeration

23.1 Introduction

Global climate change is currently manifested as global warming, which is dominated by rising average temperatures and extreme weather events (Newman and Noy 2023; Li et al. 2022a). This presents a substantial peril to the sustainable development of humanity. The interim assessment from the World Meteorological Organization regarding the global climate indicates that the worldwide average temperature in 2021 exceeded the pre-industrial baseline by 1.1 °C, resulting in significant climate hazards (IPCC 2022). The increase in atmospheric carbon dioxide (CO_2) concentration is the primary driver of global warming, with the principal sources of CO_2 being fossil fuel combustion and changes in land use (Wang et al. 2021, 2023). Recent studies predict that cities will account for 85% of CO_2 emissions by 2030 (Shan et al. 2019). Urban agglomerations, as crucial conduits of urbanization and human activities, frequently serve as the primary contributors to CO_2 emissions. Effective governance of CO_2 emissions and the promotion of environmentally friendly development paths in urban agglomerations are crucial components of global initiatives to reduce and adapt to the consequences of global warming.

The environmental consequences resulting from the advancement of the economy and society are a critical factor in global climate change. Substantial advancements have been achieved in studying the effects of human activities (Rosa and Dietz 2012). Based on the famous IPAT and STIRPAT equation (environmental stress equals to population, affluence, technology, and other factors in a multiplicative form), multiple studies have consistently demonstrated that population growth is the primary driver of CO_2 emissions (Zhou and Wang 2018; Mohmmed et al. 2019; Ou et al. 2019; Cui et al. 2020; Shi et al. 2020). Nevertheless, there is a lack of consensus among scholars about the effects of trade, urbanization, consumption, and spatial structure. One prominent theory in this regard is the Environmental Kuznets Curve (EKC) hypothesis. Put simply, the correlation between environmental stress and economic growth follows an inverted-U pattern. Once a certain threshold is reached, economic growth can actually facilitate the decrease of environmental stress (Mohammed et al. 2024; Bekun et al. 2021; Wang et al. 2022a). However, a significant amount of empirical research demonstrates that the aforementioned inverted-U pattern mainly occurs in relation to the influence of water or air toxins on ecosystems, and greenhouse gas emissions do not seem to follow this pattern (Rosa and Dietz 2012; Zhou and Wang 2018). Moreover, multiple researchers contend that urbanization can result in regional economic growth at the cost of higher energy consumption and the consequent emission of greenhouse gases (Zhu et al. 2022; Wang et al. 2017). Nevertheless, the fact that the production and services needed for urban development will be satisfied by other regions, which leads to a rise in CO_2 emissions in the producing area, this process is commonly referred to as "the metabolic rift" (VandeWeghe and Kennedy 2007; Foster et al. 2010). Likewise, environmentally harmful and energy-intensive industrial operations in developed areas tend to relocate to less developed areas. This approach, aimed at achieving their own eco-friendly objectives, will result in an increase in CO_2 emissions from the less developed regions they are associated with (Cole 2004; Jorgenson 2009). The phenomenon has transformed from the early phases of international trade to the spatial spillover effect observed between urban agglomerations within countries (Shen et al. 2021; Yu et al. 2020). At the same time, researchers have extensively examined the effects of urban structure. Compact cities exhibit greater land-use efficiency, resulting in reduced CO_2 emissions at several levels (Li et al. 2022b; Zhou et al. 2025). Nevertheless, empirical evidence indicates that an excessive concentration of population in first-tier cities resulted in a corresponding rise in CO_2 emissions (Ou et al. 2019). The disparities in the findings of these studies can be linked to variations in the geographical and time scope of the research conducted. Furthermore, the previous research mostly relies on mean regression models to analyze the average effect of contributing factors and illustrates the trend of concentration. However, it fails to include the potential variations in emission characteristics among cities with varying levels of emissions, thus cannot support in-depth analysis of CO_2 emission characteristics at the municipal level and provide tailored emission reduction suggestions.

China suggests utilizing urban agglomerations as key platforms for population and economic concentration, creating interconnected urban clusters to achieve collaborative advantages and promote high-quality development. Several studies indicate that the economic growth resulting from the demographic dividend in developing countries often surpasses the increase in CO_2 emissions attributed to population growth. In the meantime, urban agglomerations tend to exhibit significant clustering characteristics, which is also reflected in the concentration of CO_2 emissions (Cui et al. 2020). Additionally, all sorts of cities are interconnected and interdependent, and there is an unavoidable industrial division of labor among cities within the urban agglomerations. For instance, the primary energy supply in the Yangtze River Delta region is derived from the Huainan-Huaibei coal mining base (Shan et al. 2018), emission intensive operations of both heavy and light manufacturing cities contribute to the consumption and production activities of other cities within urban agglomerations (Feng et al. 2013; Mi et al. 2016). Therefore, urban agglomerations are characterized not only by their physical proximity, but also by their emphasis on internal specialization, collaboration, and complementary roles. Currently, certain academics have categorized cities based on their GDP share or urban form (Ramaswami et al. 2018; Shi et al. 2022). However, there is still a need for further research on the variations in CO_2 emission characteristics among different types of cities, and isolate plausibly exogenous variation of pillar industries when studying the driving mechanisms of CO_2 emissions in urban agglomerations.

This study seeks to comprehensively assess the factors that influence CO_2 emissions within urban agglomerations. To accomplish this objective, we chose the Yangtze River Delta urban agglomeration (YRDUA) as the subject of our research and initially calculated the landscape index of this urban agglomeration. Furthermore, cities are classified according to their GDP share of industrial output in the study area. Ultimately, quantile regression was utilized to assess the correlation between CO_2 emissions and other indicators, the outcomes of city categorization were employed as the clustering robust standard error for clustering variables to isolate plausibly exogenous variation. This paper has three main contributions. Firstly, this research thoroughly examines the allocation of roles across urban agglomerations and the subsequent influence of emission characteristics from cities with distinct main sectors, ensuring that the findings are more closely aligned with actuality. Additionally, we employed quantile regression to further examine the impact of each variable on cities with varying levels of emissions. Ultimately, this research examines both linear and nonlinear associations to determine if the CO_2 emission patterns of the YRDUA align with the EKC hypothesis.

23.2 Data and Methods

23.2.1 Study Area

The focus of this study is the Pan-YRDUA, which encompasses a large territory comprising Jiangsu, Zhejiang, Anhui, and Shanghai (Fig. 23.1). In total, there are 41 municipal-level administrative regions within this region, covering a land area of 350,000 km^2. The YRDUA is a crucial strategic area in China, possessing significant comprehensive capabilities. The southern portion of the Yangtze River Delta is predominantly characterized by mountainous terrain with karst landforms, while the northern part consists of a flat plain. Additionally, the estuary located on the eastern side of the region is in close proximity to Shanghai and Jiangsu. The region boasts abundant natural resources, encompassing both subtropical monsoon climate zones and temperate monsoon temperature zones. By 2022, the urbanization rate has reached 71.65%, with a permanent population of 235 million, or 16.64% of China's total population at that time. The gross domestic product (GDP) had achieved a value of 24.47 trillion yuan, representing 24.26% of the total national GDP over the corresponding time period. Being a significant testbed for development in China, the dual carbon goal has introduced fresh development objectives and requirements to the YRDUA. Consequently, conducting research on reducing CO_2 emissions in the YRDUA holds immense practical importance in spearheading the nation's low-carbon development.

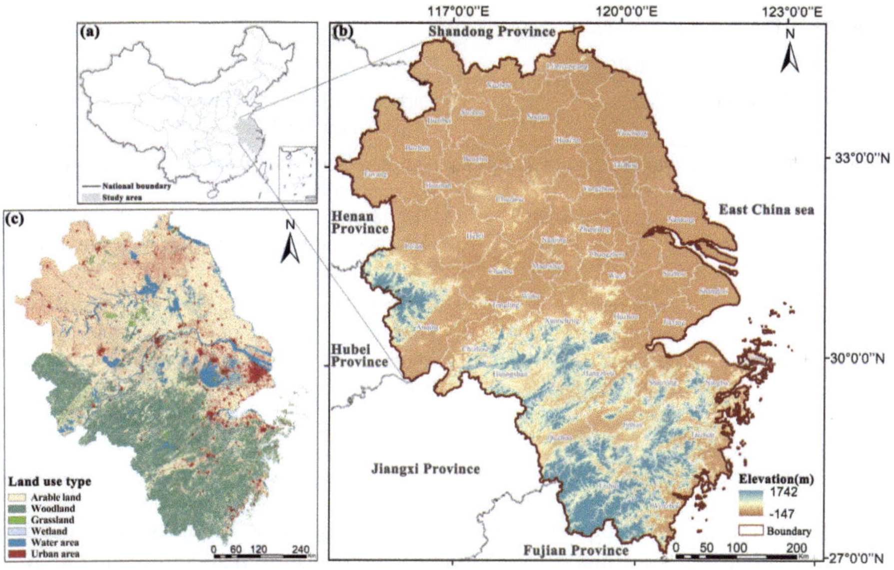

Fig. 23.1 **a, b** Location of the YRDUA **c** land use type of the YRDUA

23.2.2 Data Set and Pre-processing

The study incorporates three distinct categories of data: CO_2 emission data, geography data, and socio-economic data (Table 23.1). It is worth noting that China's national population census is conducted every ten years, and it is the only one that includes detailed population data like "residential population living in the urban area", which are used to accurately calculate the urbanization rate of the population, as well as cross-sectional data on various population characteristics, economic development, urban construction, housing, living consumption, and income. The latest census was conducted in 2020. However, during that year, countries worldwide enforced a range of stringent measures, such as company and factory closures, travel restrictions, and remote work, in order to contain the transmission of COVID-19 (Guan et al. 2020). The implementation of these steps has resulted in a substantial decrease in the need for fossil fuels, consequently contributing to a decrease in CO_2 emissions (Zheng et al. 2020). Given the significant upheaval caused by the COVID-19 pandemic on the socio-economic system, the year 2010 was chosen as the research timeframe for our research.

The CO_2 emission data were obtained from the study conducted by Shan et al. (2017, 2018, 2019), utilizing the IPCC methodology and integrating the province emission account system provided by the National Development and Reform Commission of China. The emissions inventory is composed of 46 socioeconomic sectors. In the CO_2 emission calculation in 2010, industrial energy consumption was also divided into sub industries according to the types of industrial enterprises, thus

Table 23.1 Data used in this study

Data type	Data name	Data source
CO_2 emission data	CEADs	Carbon Emission Accounts and Datasets (https://www.ceads.net.cn/)
Geographic data	Land use data of Yangtze River Delta Administrative divisions	National Geomatics Center of China (http://www.globallandcover.com/GLC30Download/index.aspx) Resource and Environment Science and Data Center (http://www.resdc.cn/)
Socio-economic data	Demographic characteristics Economic development Urban construction and housing Living and income	Chinese Population Census Yearbook 2010, Statistical Yearbooks of Provinces and Municipalities and Statistical Communiqué on Socio-Economic Development

forming the energy consumption table corresponding to the industry classification of Chinese National Administration of Quality Supervision, Inspection and Quarantine (Shan et al. 2017, 2018, 2019, 2022). The land cover raster data has a spatial resolution of 30 m × 30 m, which provides benefits in terms of both spatiotemporal features and data accuracy. The socio-economic data were sourced from several reliable publications, including the census year books, statistical yearbooks, and statistical communiqué on socio-economic development.

23.2.3 Research Methods

Selection and Calculation of Landscape Pattern Index

Land use change is a clear indication of how human activities affect the environment (Mundaca and Markandya 2016). Altering the type of land cover in a region can have significant effects on the structure, processes, and function of ecosystems (Priess et al. 2007; Ricketts et al. 2008; Zhao et al. 2004; Zhou and Wang 2022). It can also impact transportation, infrastructure construction, and ultimately the resource use efficiency of cities in the region. Multiple studies have demonstrated a distinct connection between urban structure and regional CO_2 emissions (Shi et al. 2022; Wang et al. 2022b). We utilized four variables Percentage of construction Landscape (PLAND), Mean construction patch Shape Index (SHAPE_MN), Mean construction patch Radius of Gyration (GYRATE_MN), and construction Aggregation Index (AI) to denote the rates of land urbanization, urban form complexity, concentration and diffusion, and the presence of multi-center and single-center cities within the study area, respectively. These four landscape indices were calculated at the municipal level using Fragstats4.2.

Cluster Analysis

Cluster analysis is a statistical method used to classify research samples based on their similarity, with the goal of grouping similar objects together (Edwards and Cavalli-Sforza 1965). Given the non-uniqueness of the indicators in the sample data, cluster analysis was employed to group the urban samples. This grouping was based on similarities in features, resulting in samples within the same group being closer in geometric distance, while the distance between other groups is greater.

K-means clustering, also referred to as quick clustering, efficiently addresses the issue of poor efficiency in hierarchical clustering (Bouguettaya et al. 2015). Additionally, K-means clustering allows for pre-determining the number of classification outcomes and can capture more complex relationship among data points in flat structure (Ikotun et al. 2023). Based on a prior study comparing several distance metrics employed in the K-means algorithm, it was shown that the Euclidean distance metric outperforms the others (Singh et al. 2013). For these reasons, we employed the K-means algorithm, utilizing the Euclidean distance metric, to categorize cities in our investigation. Moreover, as the industry sector accounts for the majority of CO_2 emissions, comprising 87% of the overall emissions, cities' pillar industry and their specialized industrial economy structure could strongly affect cities' CO_2 emissions characteristic. According to the Chinese National Administration of Quality Supervision, Inspection and Quarantine, the industrial sector contains 39 subcategories, which can be further categorized into high-tech industries, light manufacturing, heavy manufacturing, and energy production. As outlined above, we collected the industrial outputs of 39 sectors, clustered cities based on the GDP share of the four sectors to the industrial outputs.

Extended STIRPAT Model and Quantile Regression

The IPAT equation, introduced by Ehrlich and Holdren in 1971 (Ehrlich and Holdren 1971), is currently extensively employed for analyzing the factors that contribute to environmental impact. The independent variable is the environmental impact, whereas the influencing factors consist of population, affluence, and technology. Nevertheless, the IPAT identity is limited to only three driving forces and cannot be employed for hypothesis testing. In order to address this issue, Dietz and Rosa devised the STIRPAT model (Dietz and Rosa 1997), which is a stochastic version of the IPAT model:

$$I_i = aP_i^b A_i^c e_i \tag{23.1}$$

The overall environmental impact (I) can be determined by the population size (P), affluence (A), and other influencing factors (e). The constant of the scaling model is represented by "a", while the net effect of population and affluence is denoted by "b" and "c" respectively. These influencing factors encompass not only technology in the traditional IPAT equation, but also urbanization, social organization, trade, society, and all other elements that influence human impact on the environment, in addition to population and income. Empirical research typically adopts the logarithmic form for both sides:

$$\ln I_i = \ln a + b \ln P_i + c \ln A_i + \ln e_i \tag{23.2}$$

Furthermore, other explanatory variables can be integrated into the STIRPAT model to examine various factors that contribute to environmental impacts. Previous studies commonly employed GDP per capita as a measure of affluence. However, the bivariate test in this study revealed a strong correlation between GDP per capita and the urbanization rate. This phenomenon occurs because of the correlation between increasing affluence in societies and the expansion of metropolitan areas, resulting in a proportional increase in consumption levels (Rosa and Dietz 2012). Considering the model's multicollinearity and parsimony, we utilized the urbanization rate as a representation of both population urbanization and affluence.

In summary, this study used the STIRPAT model to analyze the diverse factors that contribute to CO_2 emissions in prefecture-level cities. The permanent resident population (POP) was used to represent the population, while the urbanization rate (UR) represents population urbanization and affluence. The proportion of secondary production (SEC) was used to indicate the degree of industrialization, and the Percentage of construction Landscape (PLAND) represents the extent of land urbanization. Additionally, Mean construction patch Shape Index (SHAPE_MN) was used to measure the complexity of urban boundaries, while Mean construction patch Radius of Gyration (GYRATE_MN) and construction Aggregation Index (AI) were used to assess the centrality and concentration of urban structure. The ultimate STIRPAT model is as follows:

$$\begin{aligned} \ln CO_{2i} = {} & \alpha + \beta_1 \ln POP_i + \beta_2 \ln UR_i + \beta_3 \ln SEC_i + \beta_4 \ln PLAND_i + \beta_5 \ln SHAPE\_MN_i \\ & + \beta_6 \ln GYRATE\_MN_i + \beta_7 \ln AI_i + e_i \end{aligned} \tag{23.3}$$

The traditional regression model uses the sum of squared residuals as the objective function of minimization, which represents mean regression and illustrates the trend of concentration. However, this method is vulnerable to the influence of extreme values. In order to address this issue, Koenker and Bassett introduced Quantile Regression (QR) as a solution (Roger and Bassett 1978). This method utilizes the weighted average of the absolute value of the residual as the minimized objective function. By doing so, it becomes less influenced by extreme values and offers greater robustness compared to mean regression. Additionally, QR has the capability to furnish extensive data regarding the distribution of circumstances (Das et al. 2024). Suppose the population qth quantile of conditional expectation $E(x|y)$ is y_q, $y_q(x)$ is a linear function of x:

$$q = P(Y \le y_q) = F_y(y_q) \tag{23.4}$$

$$y_q(x_i) = x_i' \beta_q \tag{23.5}$$

where $F_y(y)$ is the distribution function of y, β_q is the q quantile regression coefficient, and its estimator $\widehat{\beta_q}$ is defined by the following formula:

$$\min_{\beta_q} \sum_{i:y_i \geq x_i'\beta_q}^{n} q\left|y_i - x_i'\beta_q\right| + \sum_{i:y_i < x_i'\beta_q}^{n} (1-q)\left|y_i - x_i'\beta_q\right| \tag{23.6}$$

23.3 Result and Discussion

23.3.1 Overview of Urban Structure Characteristics of the YRDUA

Figure 23.2 displays the spatial distribution of the urban area within the YRDUA in 2010. Generally speaking, the urban area in Shanghai was the most extensive and had a high level of connectivity. Many secondary cores interconnected extensive urban land patches in the area, demonstrating a significant level of aggregation. The scale of urban area in Jiangsu Province was large, but the structure had obvious differentiation between north and south. The core patch of urban land was large, and the aggregation degree was high in the south-developed areas such as Suzhou, Wuxi, and Changzhou, while the patches were more abundant, dense, and dispersed in the northern areas. The overall development intensity of urban land in Zhejiang Province was low, so the shape of the construction patch had strong regularity, and the patch density was small. Ningbo, Jiaxing, Jinhua, Shaoxing, and other cities showed an obvious multi-center structure. The structure of Hangzhou tended to be unicentric, while Quzhou and Lishui were more dispersed. Urban land in Anhui Province was characterized by scattered development, high patch density, and low cohesion. The urban structure of Huangshan, Chizhou, and Xuancheng, which are near Zhejiang Province, was similar to that of other cities in Zhejiang Province.

23.3.2 Cluster Analysis Results and Analysis of Variance

The sample cities were classified using cluster analysis. The K-means algorithm in SPSS was utilized, with a predetermined number of four clusters. The classification results of pillar industries yielded four groupings of cities: 19 high-tech cities, 9 light manufacturing cities, 10 heavy manufacturing cities, and 2 energy production cities. We conducted further robustness checks comparing different distance metrics and using a hierarchical cluster method to classify the cities, finding that Euclidean distance metric outperforms the others and the result of the K-means clustering showed more consistence with practical situation. As urbanization and industrialization progress, high-tech cities will transition into serve-oriented cities. High-emission

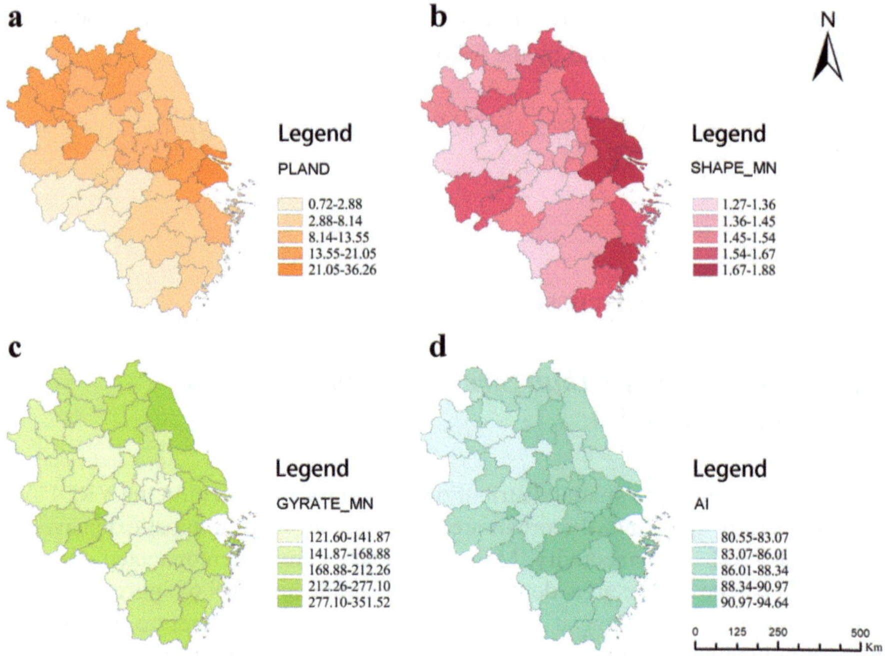

Fig. 23.2 The spatial structure of urban area in the YRDUA

industries will be relocated to the surrounding cities of urban agglomerations, while pillar industries will shift towards low-emission service sectors (Ramaswami et al. 2018). Shanghai and Nanjing, two of the cities being considered, are categorized as service-oriented cities due to the service industry contributing over 50.63% of the GDP output value. The results are shown in Fig. 23.3.

Significant disparities exist in the social and economic progress within urban agglomerations. Cities have distinct roles and responsibilities, and their emission profiles are influenced to some extent by their unique industrial structures. As an illustration, the industrial output value varied between 3.2 billion yuan in Bozhou and 301.1 billion yuan in Shanghai. Similarly, the CO_2 emissions per 1000 yuan of GDP ranged from 0.16 tons for high-tech cities to 0.86 tons for energy production cities. Although service-oriented cities have the highest CO_2 emissions, the high proportion of high-tech industry's share in industry output makes them have relatively low emission intensity. The analysis of variance was conducted to evaluate the impact of cities' pillar industries on CO_2 emissions. The results revealed significant variations in CO_2 emissions among different city groups ($F = 10.395$, $\text{Sig} = 0.000$). The results of the homogeneity of variance indicate that the variance homogeneity test for CO_2 emissions among cities with different major industry categories yielded an observed value of 2.569 with a corresponding p-value of 0.055. Based on a 5% significance level, we can conclude that there is no statistically significant variation in the overall

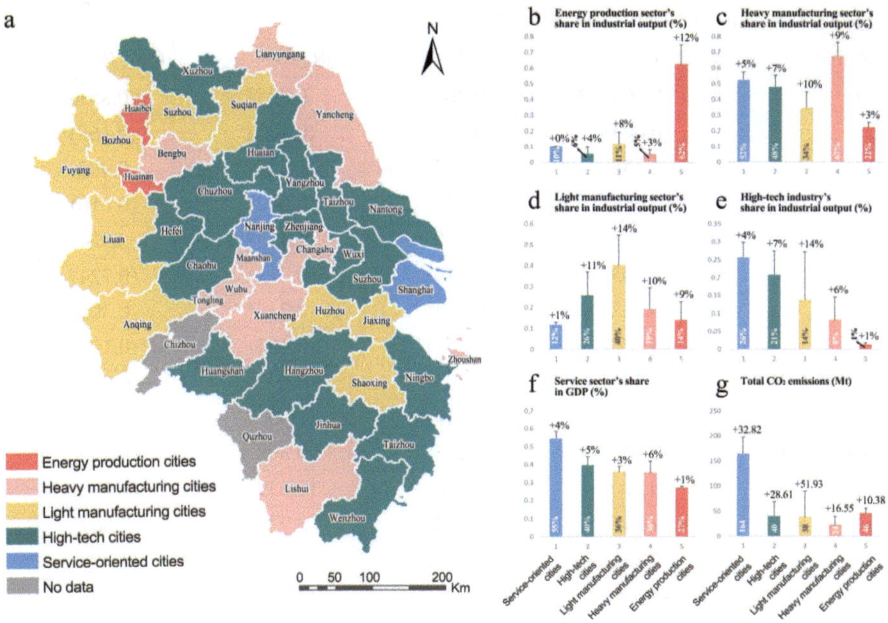

Fig. 23.3 **a** Five city groups, **b–g** mean and standard deviation of economic structure and CO_2 emissions

variance of CO_2 emissions among cities. This satisfies the requirement of variance homogeneity in the analysis of variance.

23.3.3 Driving Forces of CO_2 Emissions in the YRDUA

Before conducting parameter estimation, it is imperative to evaluate the existence of multicollinearity among the variables. The variance inflation factor (VIF) is 3.04, suggesting that there is no significant multicollinearity among the variables. To overcome the problem of misspecification caused by the disparities in pillar industries of each city in the YRDUA, we utilized the results of urban classification in 3.2 as the clustering variables for the cluster robust standard error. Based on the previous studies (Zhou and Wang 2018; Kang et al. 2016; Zhang et al. 2017), we examined potential nonlinear relationships in the model by Link test, the results indicated that there was insufficient evidence to support the EKC hypothesis. The linear regression results of the STIRPAT model and QR are presented in the Table 23.2.

POP showed significant positive correlation with CO_2 emissions in OLS and QR. Population growth is a significant measure of socio-economic progress in emerging nations, and it is frequently accompanied with a rapid increase in the need for energy for both daily living and commercial activities. With the growth of the population,

Table 23.2 Results of STIRPAT model and QR

Variables	OLS	QR_25	QR_50	QR_75
POP	0.905***	0.864***	0.895***	0.851***
	(0.094)	(0.242)	(0.166)	(0.168)
UR	0.858	0.975	1.233**	0.902
	(0.554)	(0.878)	(0.602)	(0.610)
SEC	2.575***	2.890***	2.437***	2.924***
	(0.503)	(1.042)	(0.715)	(0.724)
PLAND	0.268**	0.373**	0.217**	0.303***
	(0.070)	(0.154)	(0.105)	(0.107)
SHAPE_MN	– 0.585	0.264	– 0.690	– 0.0320
	(0.831)	(2.049)	(1.406)	(1.423)
GYRATE_MN	– 0.621**	– 0.213	– 0.308	– 0.930*
	(0.210)	(0.786)	(0.539)	(0.546)
AI	7.045**	2.755	3.729	7.567*
	(2.465)	(5.852)	(4.016)	(4.065)
Cons	– 44.34***	– 29.47	– 31.87**	– 46.39***
	(10.09)	(22.18)	(15.22)	(15.41)

Note We performed QR for 25%, 50%, and 75% quantiles. * $P < 0.1$, ** $P < 0.05$, *** $P < 0.01$, All variables underwent natural logarithm process. Standard errors in parentheses

societal norms and behaviors may undergo transformation. For instance, individuals may experience a rise in their standard of living, characterized by increased affluence. This often coincides with the adoption of high-consumption lifestyles, which in turn result in greater energy consumption. Moreover, the surge in consumer demand necessitates expanded production and alterations in land use, ultimately contributing to a rise in CO_2 emissions. Shanghai and Suzhou, as examined in this study, possess a substantial population within the YRDUA and exhibit the greatest levels of CO_2 emissions. Furthermore, this study also examined the quadratic term of POP, the findings indicate that the nonlinear relationship were not statistically significant. This aligns with the outcomes reported by Dietz et al. when they excluded the outlier values (Dietz and Rosa 1997). The population distribution in the YRDUA has not reached the same extreme disparity as the transnational panel. As a result, developing countries still exhibit a linear correlation between population and CO_2 emissions.

The coefficients of UR showed significant only in the 50% quantiles ($P<0.01$). Contradict with prior research which has indicated that urbanization is highly probable to lead to a rise in CO_2 emissions (Zhu et al. 2022) or follow an inverted-U pattern with CO_2 emissions in cross-country panel data (Zhang et al. 2017). An escalation of urbanization rate in developing countries will lead to a corresponding surge in the need for transportation, construction, industrial and commercial operations, as well as waste management in developing countries. Furthermore, living in densely populated urban areas may foster a contemporary lifestyle characterized by substantial energy consumption, thereby directly contributing to an upsurge in CO_2 emissions. However, during the urbanization process, the population may transition from primary industries to energy production manufacturing with high emission intensity or high-tech industries with low emission intensity. An example of a low CO_2 emission intensity city with a thriving tourism industry in the YRDUA is Huangshan. This could be one of the reasons why UR does not follow the inverted-U pattern of the traditional EKC hypothesis and does not influence CO_2 emissions in the OLS, 25% quantile, and 75% quantile.

The coefficients of SEC revealed that this indicator has a substantial driving effect on CO_2 emissions. The secondary industry, particularly heavy industry, releases a significant quantity of carbon dioxide through production processes, transportation and logistics requirements, and waste management (Zheng et al. 2018). Consequently, the CO_2 emissions of cities have shown a statistical increase in tandem with the growth of the secondary industry. As an illustration, Huainan, Huaibei, and Ma'anshan, being the primary energy providers and steel manufacturers in the Yangtze River Delta, exhibit higher levels of CO_2 emissions compared to other cities of comparable size.

PLAND showed a significant driving effect on CO_2 emissions, which quantifies the expansion of the percentage of urban area, in response to the need for land to support the development of socio-economic activities. Following the reform and opening up policies, the YRDUA has experienced substantial urban expansion. This expansion has directly influenced the land use changes in the YRDUA, typically accompanied by extensive infrastructure development, leading to the release of CO_2. Prior research has also identified comparable coefficients for urban sprawl, thereby corroborating the conclusion that urban sprawl has a profound positive impact on CO_2 emission (Ou et al. 2019; Zhu et al. 2022). Simultaneously, the transformation of carbon sink areas, such as wetlands and woodlands, into areas for construction purposes directly diminishes the capacity of the region to store carbon and leads to irreversible harm to the region's ability to absorb CO_2. This worsens CO_2 emissions and initiates regional environmental alterations, such as the heat island effect.

The SHAPE_MN did not have a substantial effect on CO_2 emissions in this research. SHAPE_MN is a concise and direct indicator of the overall complexity of the constructed land. A higher SHAPE_MN value corresponds to a greater overall shape complexity. The presence of intricate land boundaries in construction leads to a potential increase in the time and distance required for the movement of elements within the area, thereby resulting in higher CO_2 emissions (Liu et al. 2020). Nevertheless, in the YRDUA, the intricate structure of the city is influenced by the stage of urban development, climate conditions, and land topography. In highly developed regions, there is a tendency for high-emission heavy industry or energy sectors to expand into the surrounding areas in a dispersed or isolated manner, as a result of spatial spillover effects (Shen et al. 2021). Consequently, this expansion leads to a rise in emissions in the neighboring regions. Moreover, in Zhejiang Province and certain regions of Anhui Province, situated south of the Qinling Mountains-Huaihe River Line, the growth of urban areas is significantly constrained by the local topography.

As a result, the expansion of construction land in these areas is frequently characterized by intricate and irregular boundaries. Furthermore, these regions tend to prioritize the development of environmentally sustainable primary industries. Hence, the intricacy of the boundary of the developed land has minimal impact on the CO_2 emissions.

As shown, there is a substantial negative correlation (at a 95% confidence level) between GYRATE_MN and CO_2 emissions. Conversely, we found a significant positive association (at a 95% confidence level) between AI and CO_2 emissions. Quantile Regression revealed significant levels just at the upper quantiles, while the impact of the remaining quantiles was not statistically significant. GYRATE_MN quantifies the mean distance between patches of construction areas, with higher GYRATE_MN values indicating greater separation between urban areas. Aggregation index (AI) quantifies the level of clustering of construction land plots within the city. A higher AI value indicates a more pronounced tendency of concentrated distribution of urban area. The correlation analysis between the two variables and carbon emissions indicates that implementing a growth plan with multiple agglomeration centers can effectively mitigate CO_2 emissions in high-emission regions. In theory, if the average distance between urban land patches increases, it will result in a decline in the connection between urban plots, leading to a higher transportation demand and increased CO_2 emissions (Chen et al. 2011). Nevertheless, urbanized regions in developing countries often concentrate an excessive population and resources, leading to the depletion of the natural environment's ability to sustain them. This undermines the benefits of urban agglomeration and contributes to a rise in CO_2 emissions. In the multi-center expansion mode, there is a significant distance between each agglomeration center. However, each agglomeration region maintains a reasonably high level of urban operation and land use efficiency without exceeding the environmental carrying capacity. The organization possesses a comprehensive infrastructure that enables the execution of various social and economic activities. Additionally, it helps minimize the unnecessary movement of production and living elements across different regions, as well as the phenomenon of commuting during peak hours. Simultaneously, the central city has the potential to influence atmospheric circulation, thus mitigating the adverse effects of the spatial dispersion of CO_2 (Liu et al. 2020). This can expedite the CO_2 cycle process, ultimately achieving environmental sustainability.

23.4 Conclusion and Policy Implications

This research analyzes the influence of socioeconomic and urban structure factors on CO_2 emissions from the standpoint of industrial characteristic, using cross-sectional data from the YRDUA in 2010, addressing the gap in current studies concerning the cities' specialized industrial economy structure. Initially, the urban structure of the YRDUA was analyzed by quantifying the prefecture-level landscape pattern. It was observed that there was a clear spatial variation in the urban structure. Furthermore, employing cluster analysis, the cities in the YRDUA were classified into five separate groups based on the GDP share of industrial sectors. Cities with different pillar industry types exhibited notable disparities in CO_2 emission characteristics. Finally, we focused on analyzing the critical elements that influence CO_2 emissions in the YRDUA. This was achieved by employing the STRIRPAT model in conjunction with quantile regression, and the results of city classification were taken as the clustering variable of cluster-robust standard error in regression model to isolate plausibly exogenous variation, which provides a robust lower bound on the extent of unobserved variables within the same city group due to manufacturing structure. The findings indicate that the case of the YRDUA does not support the traditional EKC hypothesis, affluence only exerts a favorable influence on CO_2 emissions in cities with moderate emissions, the impact of affluence on emissions tend to weak rather than reverse with the pace of development. The increase in population, the escalation of secondary production, and the trend of land urbanization significantly impact the rise in CO_2 emissions. Notably, the latter two factors exert a more substantial effect on emissions in both high and low emission cities than in regions characterized by moderate emissions. As for the urban structure, exclusively concentrating land development mode will result in a increase in CO_2 emissions. On the other hand, a decentralized urban agglomeration structure with strong interconnections is a successful approach to reducing CO_2 emissions, particularly in areas with high levels of emissions.

To meet China's ambitious goal of decreasing carbon emissions intensity by 65% under the Paris Agreement and attaining carbon neutrality by 2060, policymakers must formulate more precise and localized CO_2 reduction policies. We provide policymakers the chance to mitigate CO_2 emissions by implementing spatial planning strategies at the city level, while maintaining the city's existing specialized industrial economy structure and energy infrastructure. Developed regions in urban agglomerations rely on support from other cities for energy, raw materials, and industrial goods. Implementing excessively extreme tactics ignoring the resource endowment and advantage differences may lead to substantial modifications in the industrial structure, hence undermining the economic core of developing cities and causing adverse impacts on the economy and employment. Cities with a strong emphasis on high-tech and service-oriented sectors should give top priority to the establishment of a low-carbon economy. Conversely, cities that primarily focus on energy production and heavy manufacturing should concentrate their efforts on clean technology and enhancing energy efficiency.

Although cities with low CO_2 emissions are mainly affected by socioeconomic factors, it is crucial to additionally consider the spatial pattern of future urban development in order to effectively optimize or decouple economic growth from carbon emissions. For still-developing cities, degree of urban development concentration should be limited to keep away from the "old" development paths of developed cities. Underdeveloped cities possess constrained resources for public services and the capability to sustain urban areas. An extremely dense urban structure is not conducive to CO_2 emissions reduction. Taking into account urban connectedness while improving urban compactness will aid in mitigating avoidable carbon emissions resulting from the transportation of goods and individuals across cities. For developed cities, the adoption of a polycentric and highly connected development model of anti-urbanization can effectively promote environmental sustainability by utilizing the advanced infrastructure and technical efficiency of small and medium-sized urban areas. Moreover, optimizing the urban land use pattern in the YRDUA is crucial. Prioritizing the administration of the green spaces that effectively capture carbon dioxide in the area to maintain ecological sustainability. To achieve sustainable development and address emission reduction, it is necessary to carefully manage the balance between regional expansion and the environmental capacity of urban areas.

23.5 Limitations and Future Work

This work has some limitations, primarily concerning the timeliness of the data. Further research should prioritize the following two aspects to outline the cities' industrialization process, and extend the emission reduction pathways from the perspective of social division of labor.

First, due to data availability, our study utilized cross-sectional data in 2010 to characterize the industrial features of the cities, which could not reveal the dynamic evolutions of the cities' pillar industries. Future efforts should be focused on developing panel data to demonstrate the cities' dynamic of pillar industries. Moreover, our research focused on city's specialized industrial economy structure, which could potentially bias the results toward cities that have a higher concentration of industry and power generation. Future studies may benefit from a greater focus on the mitigation potential of industrial and nonindustrial sectors. Our future plans involve conducting a comprehensive study of over 100 cities in China's five major urban agglomerations, taking into account the emission reduction potential of the agricultural, industrial and service sectors.

Acknowledgements This research is supported by Shanghai Philosophy and Social Science Project (NO.2023BCK009).

References

Bekun FV, Gyamfi BA, Onifade ST, Agboola MO (2021) Beyond the environmental Kuznets Curve in E7 economies: accounting for the combined impacts of institutional quality and renewables. J Clean Prod 314:127924. https://doi.org/10.1016/j.jclepro.2021.127924

Bouguettaya A, Yu Q, Liu X, Zhou X, Song A (2015) Efficient agglomerative hierarchical clustering. Expert Syst Appl 42(5):2785–2797. https://doi.org/10.1016/j.eswa.2014.09.054

Chen Y, Li X, Zheng Y, Guan Y, Liu X (2011) Estimating the relationship between urban forms and energy consumption: a case study in the Pearl River Delta, 2005–2008. Landsc Urban Plan 102(1):33–42. https://doi.org/10.1016/j.landurbplan.2011.03.007

Cole MA (2004) Trade, the pollution haven hypothesis and the environmental Kuznets curve: examining the linkages. Ecol Econ 48(1):71–81. https://doi.org/10.1016/j.ecolecon.2003.09.007

Cui C, Cai B, Bin G, Wang Z (2020) Decennary spatial pattern changes and scaling effects of CO_2 emissions of urban agglomerations in China. Cities 105:102818. https://doi.org/10.1016/j.cities.2020.102818

Das K, Krzywinski M, Altman N (2024) Quantile regression. Nat Methods. Accessed 14 Jan 2024 [Online]. Available: http://www-nature-com-s.webvpn.njmu.edu.cn:8118/articles/s41592-019-0406-y

Dietz T, Rosa EA (1997) Effects of population and affluence on CO_2 emissions. Proc Natl Acad Sci 94(1):175–179. https://doi.org/10.1073/pnas.94.1.175

Edwards AWF, Cavalli-Sforza LL (1965) A method for cluster analysis. Biometrics 21(2):362–375. https://doi.org/10.2307/2528096

Ehrlich PR, Holdren JP (1971) Impact of population growth. Science 171(3977):1212–1217. https://doi.org/10.1126/science.171.3977.1212

Feng K et al (2013) Outsourcing CO_2 within China. Proc Natl Acad Sci U S A 110(28):11654–11659. https://doi.org/10.1073/pnas.1219918110

Foster JB, Clark B, York R (2023) The ecological rift: capitalisms war on the earth. NYU Press, 2010. Accessed 30 Nov 2023 [Online]. Available: https://www.jstor.org/stable/j.ctt9qg075

Guan D et al (2020) Global supply-chain effects of COVID-19 control measures. Nat Hum Behav 4(6), Art. No. 6. https://doi.org/10.1038/s41562-020-0896-8

Ikotun AM, Ezugwu AE, Abualigah L, Abuhaija B, Heming J (2023) K-means clustering algorithms: a comprehensive review, variants analysis, and advances in the era of big data. Inf Sci 622:178–210. https://doi.org/10.1016/j.ins.2022.11.139

IPCC (2022) Climate change 2022: mitigation of climate change. In: Contribution of working Group III to the sixth assessment report of the Intergovernmental Panel on Climate Change [Shukla PR, Skea J, Slade R, Al Khourdajie A, van Diemen R, McCollum D, Pathak M, Some S, Vyas P, Fradera R, Belkacemi M, Hasija A, Lisboa G, Luz S, Malley J (eds)]. Cambridge University Press, Cambridge, UK and New York, NY, USA

Jorgenson AK (2009) The transnational organization of production, the scale of degradation, and ecoefficiency: a study of carbon dioxide emissions in less-developed countries. Hum Ecol Rev 16(1):64–74

Kang Y-Q, Zhao T, Yang Y-Y (2016) Environmental Kuznets curve for CO_2 emissions in China: a spatial panel data approach. Ecol Indic 63:231–239. https://doi.org/10.1016/j.ecolind.2015.12.011

Li G et al (2022a) Global impacts of future urban expansion on terrestrial vertebrate diversity. Nat Commun 13(1), Art. No. 1. https://doi.org/10.1038/s41467-022-29324-2

Li Z et al (2022b) Exploring differentiated impacts of socioeconomic factors and urban forms on city-level CO_2 emissions in China: spatial heterogeneity and varying importance levels. Sustain Cities Soc 84:104028. https://doi.org/10.1016/j.scs.2022.104028

Liu K, Xue M, Peng M, Wang C (2020) Impact of spatial structure of urban agglomeration on carbon emissions: an analysis of the Shandong Peninsula, China. Technol Forecast Soc Change 161:120313. https://doi.org/10.1016/j.techfore.2020.120313

Mi Z et al (2016) Consumption-based emission accounting for Chinese cities. Appl Energy 184:1073–1081. https://doi.org/10.1016/j.apenergy.2016.06.094

Mohmmed A et al (2019) Driving factors of CO_2 emissions and nexus with economic growth, development and human health in the top ten emitting countries. Resour Conserv Recycl 148:157–169. https://doi.org/10.1016/j.resconrec.2019.03.048

Mohammed S et al (2024) Assessment of the environmental kuznets curve within EU-27: steps toward environmental sustainability (1990–2019). Environ Sci Ecotechnol 18:100312. https://doi.org/10.1016/j.ese.2023.100312

Mundaca L, Markandya A (2016) Assessing regional progress towards a 'green energy economy.' Appl Energy 179:1372–1394. https://doi.org/10.1016/j.apenergy.2015.10.098
Newman R, Noy I (2023) The global costs of extreme weather that are attributable to climate change. Nat Commun 14(1), Art. No. 1. https://doi.org/10.1038/s41467-023-41888-1
Ou J, Liu X, Wang S, Xie R, Li X (2019) Investigating the differentiated impacts of socioeconomic factors and urban forms on CO_2 emissions: empirical evidence from Chinese cities of different developmental levels. J Clean Prod 226:601–614. https://doi.org/10.1016/j.jclepro.2019.04.123
Priess JA, Mimler M, Klein A-M, Schwarze S, Tscharntke T, Steffan-Dewenter I (2007) Linking deforestation scenarios to pollination services and economic returns in coffee agroforestry systems. Ecol Appl 17(2):407–417. https://doi.org/10.1890/05-1795
Ramaswami A, Jiang D, Tong K, Zhao J (2018) Impact of the economic structure of cities on urban scaling factors: implications for urban material and energy flows in China. J Ind Ecol 22(2):392–405. https://doi.org/10.1111/jiec.12563
Ricketts TH et al (2008) Landscape effects on crop pollination services: are there general patterns? Ecol Lett 11(5):499–515. https://doi.org/10.1111/j.1461-0248.2008.01157.x
Roger K, Bassett G (1978) Regression quantiles. Econometrica 46(1):33–50. https://doi.org/10.2307/1913643
Rosa EA, Dietz T (2012) Human drivers of national greenhouse-gas emissions. Nat Clim Change 2(8), Art. No. 8. https://doi.org/10.1038/nclimate1506
Shan Y et al (2017) Methodology and applications of city level CO_2 emission accounts in China. J Clean Prod 161:1215–1225. https://doi.org/10.1016/j.jclepro.2017.06.075
Shan Y et al (2018) City-level climate change mitigation in China. Sci Adv 4(6):eaaq0390. https://doi.org/10.1126/sciadv.aaq0390
Shan Y, Liu J, Liu Z, Shao S, Guan D (2019) An emissions-socioeconomic inventory of Chinese cities. Sci Data 6(1), Art. No. 1. https://doi.org/10.1038/sdata.2019.27
Shan Y et al (2022) City-level emission peak and drivers in China. Sci Bull 67(18):1910–1920. https://doi.org/10.1016/j.scib.2022.08.024
Shen W, Liang H, Dong L, Ren J, Wang G (2021) Synergistic CO_2 reduction effects in Chinese urban agglomerations: perspectives from social network analysis. Sci Total Environ 798:149352. https://doi.org/10.1016/j.scitotenv.2021.149352
Shi K et al (2020) Effects of urban forms on CO_2 emissions in China from a multi-perspective analysis. J Environ Manage 262:110300. https://doi.org/10.1016/j.jenvman.2020.110300
Shi F, Liao X, Shen L, Meng C, Lai Y (2022) Exploring the spatiotemporal impacts of urban form on CO_2 emissions: evidence and implications from 256 Chinese cities. Environ Impact Assess Rev 96:106850. https://doi.org/10.1016/j.eiar.2022.106850
Singh A, Yadav A, Rana A (2013) K-means with three different distance metrics. Int J Comput Appl 67:13–17. https://doi.org/10.5120/11430-6785
VandeWeghe JR, Kennedy C (2007) A spatial analysis of residential greenhouse gas emissions in the Toronto Census metropolitan area. J Ind Ecol 11(2):133–144. https://doi.org/10.1162/jie.2007.1220
Wang S, Liu X, Zhou C, Hu J, Ou J (2017) Examining the impacts of socioeconomic factors, urban form, and transportation networks on CO_2 emissions in China's megacities. Appl Energy 185:189–200. https://doi.org/10.1016/j.apenergy.2016.10.052
Wang J et al (2021) Anthropogenic emissions and urbanization increase risk of compound hot extremes in cities. Nat Clim Change 11(12), Art. No. 12. https://doi.org/10.1038/s41558-021-01196-2
Wang Q, Wang X, Li R (2022a) Does urbanization redefine the environmental Kuznets curve? An empirical analysis of 134 countries. Sustain Cities Soc 76:103382. https://doi.org/10.1016/j.scs.2021.103382
Wang Y, Niu Y, Li M, Yu Q, Chen W (2022b) Spatial structure and carbon emission of urban agglomerations: spatiotemporal characteristics and driving forces. Sustain Cities Soc 78:103600. https://doi.org/10.1016/j.scs.2021.103600

Wang C et al (2023) Impacts of climate change, population growth, and power sector decarbonization on urban building energy use. Nat Commun 14(1), Art. No. 1. https://doi.org/10.1038/s41467-023-41458-5

Yu X, Wu Z, Zheng H, Li M, Tan T (2020) How urban agglomeration improve the emission efficiency? A spatial econometric analysis of the Yangtze River Delta urban agglomeration in China. J Environ Manage 260:110061. https://doi.org/10.1016/j.jenvman.2019.110061

Zhang N, Yu K, Chen Z (2017) How does urbanization affect carbon dioxide emissions? A cross-country panel data analysis. Energy Policy 107:678–687. https://doi.org/10.1016/j.enpol.2017.03.072

Zhao B, Kreuter U, Li B, Ma Z, Chen J, Nakagoshi N (2004) An ecosystem service value assessment of land-use change on Chongming Island, China. Land Use Policy 21(2):139–148. https://doi.org/10.1016/j.landusepol.2003.10.003

Zheng B et al (2018) Trends in China's anthropogenic emissions since 2010 as the consequence of clean air actions. Atmos Chem Phys 18(19):14095–14111. https://doi.org/10.5194/acp-18-14095-2018

Zheng B et al (2020) Satellite-based estimates of decline and rebound in China's CO_2 emissions during COVID-19 pandemic. Sci Adv 6(49):eabd4998. https://doi.org/10.1126/sciadv.abd4998

Zhou C, Wang S (2018) Examining the determinants and the spatial nexus of city-level CO_2 emissions in China: a dynamic spatial panel analysis of China's cities. J Clean Prod 171:917–926. https://doi.org/10.1016/j.jclepro.2017.10.096

Zhou ZC, Wang J (2022) Evolution of landscape dynamics in the Yangtze River Delta from 2000 to 2020. J Water Clim Change 13(3):1241–1256. https://doi.org/10.2166/wcc.2022.307

Zhou Z, Xu S, Wang J, Shen H (2025) Spatiotemporal impacts of urban structure and socioeconomic factors on carbon dioxide emissions: a case study of the Yangtze River Delta urban agglomeration. Ecosyst Health Sustain 11:0290. https://doi.org/10.34133/ehs.0290

Zhu E, Qi Q, Chen L, Wu X (2022) The spatial-temporal patterns and multiple driving mechanisms of carbon emissions in the process of urbanization: a case study in Zhejiang, China. J Clean Prod 358:131954. https://doi.org/10.1016/j.jclepro.2022.131954

Chapter 24
Optimizing Negative Emission Technology Portfolios in the Northwest Region with Consideration of Carbon Price Risk

Juanjuan Yu, Qianqian Gao, and Cai Suo

Abstract To effectively address climate change and mitigate global warming, negative emission technologies (NETs) have become the key to achieving deep decarbonization and climate goals. However, in the face of the diversity and complexity of NETs, how to conduct the optimal technology portfolio planning to achieve their large-scale, high-efficiency and sustainable deployment is a key issue that urgently needs to be solved at present. In this study, an optimization model is developed based on modern portfolio theory for identifying optimal NETs portfolio. A real case study of Northwest China is conducted to analyze the optimal NETs portfolio with consideration of carbon price risk. The results show that the total negative emissions in the northwest region would reach 9.23 Gt CO_2, and the total economic benefit would be USD 467 billion. Among all NETs, enhanced weathering would make the greatest contribution and would dominate the portfolio of NETs.

Keywords Negative emission technologies · Mixed integer nonlinear programming · Optimization · Northwest region

24.1 Introduction

Global climate change has posed severe challenges to human society. The Sixth Assessment Report by the IPCC indicated that to achieve the Paris Agreement's goal of limiting temperature increase to 1.5 °C, the world must reach "net-zero emissions" by around 2050 and transition to "net-negative emissions" thereafter (IPCC 2023). Currently, mainstream emission reduction strategies can be divided into two categories: conventional mitigation approaches and NETs. Conventional

J. Yu · Q. Gao · C. Suo (✉)
School of Management Science and Engineering, Nanjing University of Information Science and Technology, Nanjing, China
e-mail: sc7481823@163.com

Y. Li et al. (eds.), *16th International Conference on Environmental Science and Development*, Environmental Science and Engineering,
https://doi.org/10.1007/978-3-032-15080-6_24

approaches focus on reducing carbon emissions at their source, such as phasing out coal-fired power plants, promoting electric vehicles, and electrifying industrial processes. While these measures can significantly reduce emissions, they struggle to fully eliminate "hard-to-abate" emissions in sectors like industry, agriculture, and transportation (IPCC 2022). In contrast, NETs address this critical gap by directly removing CO_2 from the atmosphere and achieving long-term sequestration. To meet mid-century net-zero targets, the large-scale deployment of these technologies has become indispensable (IPCC 2023).

As the world's largest carbon emitter, China has proposed the "dual carbon" goals, committing to peak carbon emissions by 2030 and achieve carbon neutrality by 2060 (Zhang et al. 2023). China's emission reduction pathway holds global significance for meeting the goals of the Paris Agreement. As a national energy and chemical industry base, the northwest region (Shaanxi, Gansu, Qinghai, Ningxia and Xinjiang) plays a unique dual role. The northwest region is an area with high carbon emissions. The total carbon emissions in 2023 reached 1.49 Gt CO_2, accounting for 11.81% of the national total. Meanwhile, this area has a unique potential for negative emissions. The northwest region possesses vast unused land, abundant wind and solar resources, and superior geological storage conditions, providing a robust foundation for deploying NETs (Liu et al. 2022; Sun et al. 2022; Wang et al. 2023). Therefore, the optimal deployment of NETs in the northwest region is of great significance for the realization of the "dual carbon" goals and national climate governance.

24.2 Case Study

24.2.1 Objective Function

A mixed integer nonlinear programming model based on modern portfolio theory is developed for optimizing NET portfolios. The objective function of the model is to maximize economic returns by optimizing the deployment of NETs. The detailed objective function can be formulated as follows:

$$\sum_{t=1}^{T}\sum_{m=1}^{M}\sum_{r=1}^{R} x_{tmr}(P_t - C_{tmr})\left[\left(\frac{(1+r_E)^{\Delta T}-1}{r_E}\right)(1+r_E)^{-t\Delta T}\right] \tag{24.1}$$

where t represents planning period; m represents negative emission technology (as shown in Table 24.1); r indicates province; r_E is economic discount rate. P_t is carbon price; C_{tmr} is deployment cost of NET and x_{tmr} is the negative emission allocated to NET.

Table 24.1 Description of six NETs

NET	Description
Bioenergy with carbon capture and storage (BECCS)	BECCS is a technology that involves generating energy through the combustion of bioenergy sources and capturing the CO_2 emitted, then storing it underground
Afforestation/reforestation (AR)	Increasing carbon stocks by establishing new forests on lands where they previously did not exist or by replanting trees in areas that have been deforested
Soil carbon sequestration (SCS)	Soil carbon sequestration refers to the process of increasing soil organic matter content through improved agricultural and land management practice
Biochar (BC)	Biochar is a carbon-rich material produced by heating biomass materials in a low-oxygen environment
Enhanced weathering (EW)	Enhanced weathering is the process of spreading natural or engineered minerals over large areas of land to accelerate the Earth's natural weathering processes through chemical reactions between the minerals and atmospheric CO_2
Direct air carbon capture and storage (DACCS)	Direct air carbon capture and storage is a process where CO_2 is captured directly from the atmosphere

24.2.2 *Constraints*

A number of constraints are set to ensure the feasibility of the deployment of NETs. These constraints can be described as follows:

(1) Constraint for carbon price risk

$$\sum_{t=1}^{T}\sum_{m=1}^{M}\sum_{n=1}^{M}\sum_{r=1}^{R} x_{tmr}x_{tnr}\rho_{mnt}\sigma_{mt}\sigma_{nt} \le p \tag{24.2}$$

$$\sigma_{mt} = \sqrt{\frac{1}{T=1}\sum_{t=1}^{T}\left(P_{mt} - \overline{P_t}\right)^2} \tag{24.3}$$

$$\rho_{mnt} = \frac{\sum_{t=1}^{T}(P_{mt} - \overline{P}_m)(P_{nt} - \overline{P}_n)}{\sqrt{\sum_{t=1}^{T}(P_{mt} - \overline{P}_m)^2}\sqrt{\sum_{t=1}^{T}(P_{nt} - \overline{P}_n)^2}} \tag{24.4}$$

where σ_{mt} is the degree of carbon price volatility and ρ_{mnt} is the correlation among six NETs.

(2) Constraint for negative emission target

$$\sum_{t=1}^{T}\sum_{m=1}^{M}\sum_{r=1}^{R} x_{tmr}\left[\left(\frac{(1+r_C)^{\Delta T}-1}{r_C}\right)(1+r_C)^{-t\Delta T}\right] \geq H \quad (24.5)$$

where H is the negative emission target and r_C is the carbon value discount rate.

(3) Constraint for negative emission potential

$$0 \leq x_{tmr} \leq x_{tmr}^{U} b_{tmr} \quad (24.6)$$

$$b_{tmr} \leq b_{(t+1)mr} \quad (24.7)$$

$$b_{tmr} \in \{0, 1\} \quad (24.8)$$

where x_{tmr}^{U} is the maximum negative emission potential of each NET and b_{tmr} is 0–1 binary variable to control whether each NET is activated or deactivated.

(4) Constraint for resource availability

$$\sum_{m=1}^{M} x_{tmr} E_{mer} \leq S_{ter} \quad \forall t, e, r \quad (24.9)$$

where E_{mer} is environmental footprint of NETs and S_{ter} is the maximum resources that can be used by NETs.

(5) Constraint for geological storage

$$\sum_{t=1}^{T}\sum_{m=1}^{M} x_{tmr} y_{mr} \Delta t \leq W_r \quad \forall r \quad (24.10)$$

$$y_{mr} \in \{0, 1\} \quad (24.11)$$

24.2.3 Data Collection

The planning horizon (2021–2060) is divided into eight periods with each having an interval of 5 years. The research data are mainly come from the National Bureau of Statistics of China (https://www.stats.gov.cn/), CSMAR Database (https://data.csmar.com/), relevant reports and published papers (Chen et al. 2016, 2023; Fuss et al. 2018; Guillemette and Turner 2018; Shen et al. 2021; Migo-Sumagang et al. 2022, 2023; China Carbon Market Long-Term Price Forecast Report 2023; Deng et al. 2024; He et al. 2024).

24.3 Results Analysis

24.3.1 NETs Portfolio Analysis

In this study, six NETs are considered, including biochar (BC), bioenergy with carbon capture and storage (BECCS), direct air carbon capture and storage (DACCS), enhanced weathering (EW), afforestation/reforestation (AR) and soil carbon sequestration (SCS). Figure 24.1 presents the optimal portfolio of NETs in each province at different time periods. During the planning period, the total negative emission in the five northwestern provinces would be 9.23 Gt CO_2. Among them, Shaanxi would achieve the highest total negative emissions, reaching 4.1 Gt CO_2, while Qinghai would record the lowest at only 0.41 Gt CO_2. Qinghai and Ningxia would exhibit a continuous upward trend in negative emissions, whereas Gansu, Shaanxi, and Ningxia are projected to peak between 2045 and 2050, followed by a gradual decline. Among all NETs, EW would have the highest proportion, reaching 34.53%. It has low resource consumption and great potential for emission reduction, providing favorable conditions for large-scale deployment. The contribution of BECCS would be the smallest, accounting for only 3.6%, and it would only be deployed in Xinjiang. The limited deployment of BECCS stems from its high costs and large resource consumption, which diminish its competitiveness compared to other NETs.

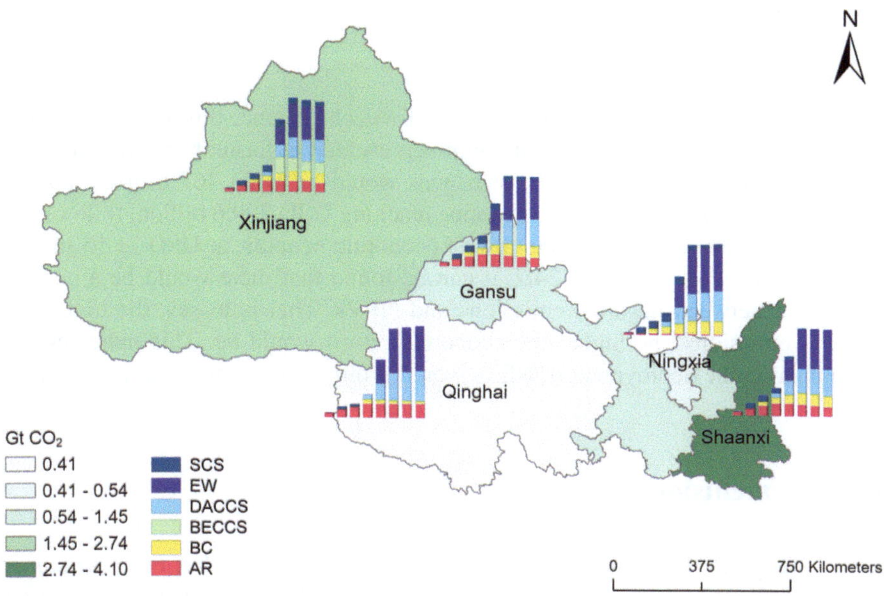

Fig. 24.1 Negative emissions achieved by each province throughout the entire planning period. The color gradient reflects the sum of the negative emissions of six NETs in each province. Each bar represents the total negative emissions in a five-year period. All bar charts use a uniform vertical scale

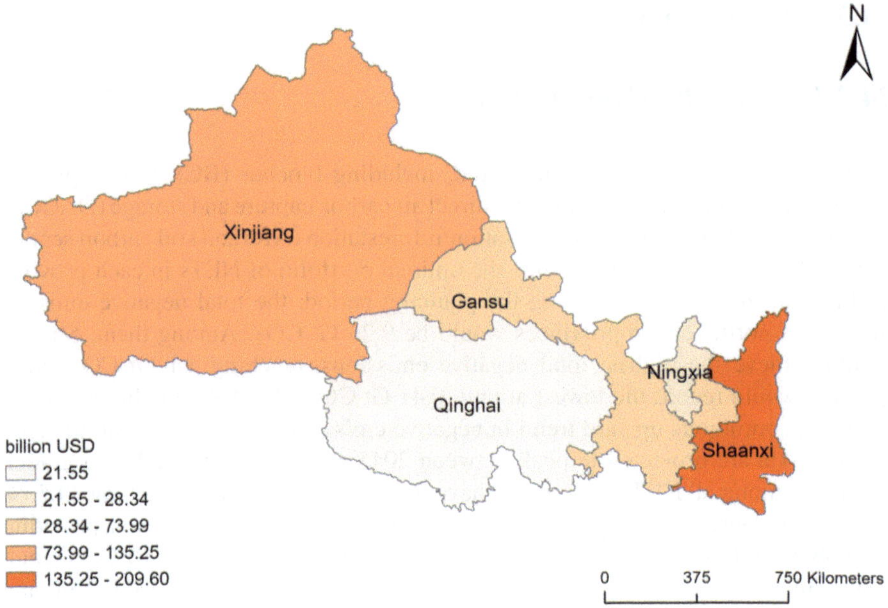

Fig. 24.2 Economic returns from NETs for each province throughout the entire planning period. The color gradient reflects the economic benefits of each province

24.3.2 Economic Returns Analysis

Figure 24.2 illustrates the economic benefits for each province throughout the whole planning period. The total economic returns presented in form of net present values (NPVs) for the five northwestern provinces would be USD 467 billion. Shaanxi would have the highest economic benefits, reaching USD 209.6 billion, followed by Xinjiang. Qinghai would have the lowest economic benefits, at USD 21.55 billion. By comparing Figs. 24.1 and 24.2, it can be found that there would be a positive correlation between negative emissions and NPVs. That is to say, the higher the negative emissions, the higher the economic return would be. This indicates that strengthening the deployment of NETs would promote regional economic growth.

24.4 Conclusion

In this study, a mixed integer nonlinear programming model has been developed for optimizing NET portfolios. The model considers the negative emissions target, negative emission potential and resource limits. The model also accounts for the uncertainty of carbon price fluctuations. The five northwestern provinces are taken as case studies. Results show that the total negative emissions in the northwest region would reach 9.23 Gt CO_2. Among all NETs, enhanced weathering would make the greatest contribution and would dominate the portfolio of NETs. The total economic benefit would be USD 467 billion. Negative emissions would be positively correlated with economic benefits. Strengthening the deployment of NETs not only contributes to environmental protection but may also become an important driving force for promoting regional economic growth.

References

2023 China Carbon Market Long-Term Price Forecast Report. Refinitiv Carbon Group; 2023. Available at: https://energylaw.chinalaw.org.cn/portal/article/index/id/3356.html

Chen DH, Pan YC, Wu CY (2016) Marginal abatement costs of CO_2 emission in China and its regional differences. China Popul Resour Environ 26(10):86–93

Chen W, Luo Y, Cao L, Dan L, Huang L, Deng XZ (2023) Research progress and review on carbon dioxide removal technology. Clim Change Res 19(05):672–682

Deng X, Teng F, Chen MP, Du ZL, Wang B, Li RQ, Wang P (2024) Exploring negative emission potential of biochar to achieve carbon neutrality goal in China. Nat Commun 15:1085

Fuss S, Lamb WF, Callaghan MW, Hilaire J, Creutzig F, Amann T, Beringer T, De Oliveira Garcia W, Hartmann J, Khanna T, Luderer G, Nemet GF, Rogelj J, Smith P, Vicente JV, Wilcox J, Del Mar Zamora Dominguez M, Minx JC (2018) Negative emissions—part 2: costs, potentials and side effects. Environ Res Lett 13:063002

Guillemette Y, Turner D (2018) The long view: scenarios for the world economy to 2060. OECD Publishing, Paris, 22

He LS, Wan JH, Zhang JQ, Liang JL, Liu ZJ, Li WP (2024) CO_2 geological sequestration and eevaluation of CO_2 geological sequestration potential in China. Gansu Geol 33(01):59–71

IPCC (2022) Summary for policymakers. In: climate change 2022: impacts, adaptation, and vulnerability. Contribution of Working Group II to the sixth assessment report of the intergovernmental panel on climate change. Cambridge University Press (in press)

IPCC (2023) Summary for policymakers. In: Climate change 2023: synthesis report. Contribution of Working Groups I, II and III to the sixth assessment report of the intergovernmental panel on climate change [Core Writing Team, Lee H, Romero J (eds)]. IPCC, Geneva, Switzerland, pp 1–34

Liu H, Niu T, Yu Q, Yang L, Ma J, Qiu S (2022) Evaluation of the spatiotemporal evolution of China's ecological spatial network function-structure and its pattern optimization. Remote Sens 14(18)

Migo-Sumagang MV, Tan RR, Tapia JFD, Aviso KB (2022) Fuzzy mixed-integer linear and quadratic programming models for planning negative emissions technologies portfolios with synergistic interactions. Clean Eng Technol 9:100507

Migo-Sumagang MV, Tan RR, Aviso KB (2023) A multi-period model for optimizing negative emission technology portfolios with economic and carbon value discount rates. Energy 275:127445

Shen XB, Chen Y, Lin BQ (2021) The impacts of technological progress and industrial structure distortion on China's energy intensity. Econ Res J 56(02):157–173

Sun YK, Luo X, Wang DJ, Liu YF, Liu JP (2022) Energy production and consumption characteristics and energy development path in Northwest China. Chin J Eng Sci 24(6)

Wang GQ, Li T, Chen JL, Sun JM, Zhu T, Bu T, Huang BT (2023) Assessment of carbon dioxide sequestration potential of ultramafic rocks in Northwest China. Northwest Geol 56(01):186–193

Zhang C, Zou X, Lin C (2013) Carbon footprint prediction of thermal power industry under the dual-carbon target: a case study of Zhejiang Province, China. Sustainability 15(4)

Chapter 25
Technical Pathway Planning and Key Issues Consideration for CCUS Projects in Ordos City

Li Ding, Linfeng Wu, Di Tang, Yue Zeng, Jia Liu, and Tao Wei

Abstract To address the carbon reduction challenges during the energy transition in Ordos City, an integrated full-process CCUS (Carbon Capture, Utilization, and Storage) technical framework is proposed. This study establishes a comprehensive CO_2 source-sink database and a screening system for storage site selection, and develops a dynamic source-sink matching model optimized for economic benefits. Low-energy absorbents and an integrated combined cooling, heating, and power (CCHP) capture system are developed and demonstrated. Additionally, low-energy solid waste leaching agent formulations are designed to enable solid waste mineralization technologies, including the production of nano-calcium carbonate for carbon sequestration. A multi-physics numerical simulation platform is constructed to elucidate CO_2 transport mechanisms in geological formations, guiding injection parameter optimization and facilitating storage pilot tests. This framework aims to achieve efficient, low-energy, and cost-effective capture of low-concentration CO_2 from coal-fired power plants by implementing a multi-site source-sink matching and screening system. It further seeks to identify feasible technological pathways and scale-up potential for CO_2 mineralization and storage utilizing coal-fired power plant fly ash. The study reveals mechanisms of geological CO_2 utilization and storage, completes site screening, safety assessment, and evaluates storage capacity and resource potential, thereby integrating a full-process CCUS technology system. The proposed solution is expected to provide an effective pathway to support Ordos City's carbon neutrality goals.

Keywords Integrated CCUS full-chain · Pathway planning · Carbon reduction in Ordos · Source-sink matching optimization

L. Ding · L. Wu (✉) · D. Tang · Y. Zeng · J. Liu · T. Wei
Peking University Ordos Institute of Energy, Ordos, China
e-mail: wulinfeng@ordos.pku.edu.cn

Y. Li et al. (eds.), *16th International Conference on Environmental Science and Development*, Environmental Science and Engineering,
https://doi.org/10.1007/978-3-032-15080-6_25

25.1 Introduction

With the intensification of global climate change, promoting the low-carbon transition of energy systems has become a strategic consensus among nations to achieve sustainable development. CCUS technology, as an indispensable pathway for deep decarbonization—particularly in hard-to-abate sectors such as coal-fired power, steel, and cement—is entering the stage of large-scale deployment. In recent years, several million-tonne-scale CCUS demonstration projects have been constructed and put into operation worldwide, including major initiatives in the United States, Canada, and China (Rubin et al. 2015). However, the large-scale deployment of CCUS still faces a series of critical scientific and engineering challenges, particularly the lack of mature technical pathways adapted to the industrial structures and geological conditions of resource-based regions (Bui et al. 2018).

Located in western China, Ordos City is a typical resource-based city, with a long-standing reliance on high-carbon industries such as coal mining, power generation, and coal chemical production, resulting in formidable carbon reduction challenges. In recent years, driven by China's "dual-carbon" strategy, Ordos has been designated as a key demonstration area for the large-scale application of CCUS technologies (Guo et al. 2022). Nevertheless, the current promotion of integrated CCUS systems in Ordos faces multiple bottlenecks, including the absence of a top-level strategic planning framework, high energy consumption in low-concentration CO_2 capture processes, insufficient technologies for the resource utilization of bulk industrial solid wastes, and complex subsurface responses induced by large-scale supercritical CO_2 injection.

This study focuses on addressing the core challenges associated with the large-scale deployment of CCUS in Ordos City and proposes an integrated CCUS technological pathway tailored to the region's industrial structure and geological characteristics. The research encompasses four main aspects: constructing a CCUS pathway planning and dynamic source-sink matching system. developing high-efficiency, low-energy CO_2 capture technologies and demonstration systems for low-concentration flue gas from coal-fired power plants. exploring CO_2 mineralization and synergistic sequestration pathways utilizing industrial solid wastes and establishing a high-precision CO_2 geological storage and dynamic monitoring platform to support the optimization of large-scale injection parameters and safety evaluations. This study aims to provide technological support for achieving carbon neutrality in Ordos City and to offer replicable demonstration experiences for the deployment of CCUS in other resource-based regions.

25.2 Key Technological Advances in CCUS

25.2.1 CO_2 Source-Sink Mapping and Cluster Deployment

Source-sink matching serves as a core component in CCUS pathway planning, acting as a crucial bridge between carbon emission sources and storage/utilization sites, and is key to achieving emission reduction targets and optimal resource allocation (Benson and Orr 2008). CCUS pathway planning is a complex, multi-integrated process that spans several stages, including storage site selection, source-sink coupling, industrial symbiosis, and pathway optimization. Under the dual drivers of China's "dual-carbon" goals and industrial restructuring, constructing an efficient and orderly CCUS pathway system has become a strategic choice for Ordos City to realize a green and low-carbon transformation. Although China has made preliminary progress in CO_2 storage suitability assessments, utilization technologies, and project deployment (Mao et al. 2021), source-sink matching remains a relatively underdeveloped foundation, especially at the regional scale where system integration and multi-objective coordination face significant challenges.

Source-sink matching constitutes a multi-objective, multi-constraint combinatorial optimization problem, involving systematic integration and trade-offs across factors such as carbon source distribution, emission intensity, capture costs, storage capacity, transport pathways, and utilization potential. Internationally, research on this topic began earlier and has formed relatively mature systems. For instance, as early as 2008, the Massachusetts Institute of Technology proposed a CO_2 source-sink matching method based on GIS and mathematical modeling, pioneering systematic research in this area (Smyth et al. 2007). In recent years, initiatives such as the STRATEGY CCUS project in Europe (Vulin et al. 2023) and the Net Zero Teesside project in the United Kingdom (Perry 2023) have significantly enhanced the economic viability and feasibility of cluster-based CCUS deployment through deep integration of spatial layout optimization and pipeline network planning. China is also actively advancing source-sink matching research. For example, Sinopec's "Qilu-Dongying" project employed a collaborative optimization strategy to enhance carbon flow pathway efficiency (Zheng et al. 2024), while PetroChina explored scenario-based matching methods adapted to complex geological structures in the Changqing Oilfield, providing feasible pathways for source-sink integration (Chen et al. 2024).

However, Ordos City has yet to establish a systematic CO_2 storage suitability evaluation mechanism tailored to different reservoir types (e.g., saline aquifers, depleted oil and gas reservoirs, coal seams), and carbon source data remain insufficiently refined at the regional scale. These gaps constrain the efficiency and scientific rigor of collaborative planning for CO_2 resources and storage sites. Given Ordos' advantages—dense high-concentration carbon sources, abundant storage potential, and a comprehensive industrial system—it represents an ideal location for deploying CCUS clusters. Moving forward, source-sink matching should be leveraged as a technological anchor to build an intelligent platform for pathway optimization and industrial symbiosis scheduling. This would drive a systemic transition from isolated demonstrations to regional integration, and from project-level practices to full-chain coordination, thereby providing a solid foundation for exploring replicable and scalable CCUS development pathways.

25.2.2 Low-Energy CO_2 Capture System Development

CO_2 capture remains the most costly segment within the CCUS chain, accounting for over 60% of the total cost, and serves as a critical determinant of overall economic feasibility (Rodrigues et al. 2022). Developed regions such as Europe and North America have vigorously promoted CO_2 capture technologies through policy incentives and financial support, including mechanisms like the U.S. tax credit schemes and the EU Emissions Trading System (Quemin 2022). In terms of technological pathways, chemical absorption has become the mainstream method for decarbonization in sectors such as power generation, steelmaking, and cement production, owing to its strong adaptability and high capture efficiency. Notable examples include Canada's Boundary Dam project and the Petra Nova project in the United States, which utilize Shell's CANSOLV and Mitsubishi's KS-1 processes, respectively, both achieving million-ton-per-year CO_2 capture capacities (Liu et al. 2018; Jenkins 2015). Although China entered the field later, it has made rapid progress in recent years through demonstration projects (Liu et al. 2017), successfully commissioning a series of CO_2 capture facilities ranging from 100,000 to 1 million tons per year, with plans underway to construct the world's largest 1.5-million-ton-per-year capture project.

Chemical absorption technology has undergone three generations of evolution, primarily distinguished by reductions in regeneration energy consumption. The first generation, dominated by single-amine solvents such as MEA, exhibited high energy demands and strong corrosivity. The second generation introduced blended amine absorbents, achieving enhanced performance and lowering energy consumption to 2.5–3.4 GJ per ton of CO_2 (Wang et al. 2011). hird-generation technologies aim to further reduce energy consumption to 2.0–2.4 GJ per ton of CO_2, although they currently remain at laboratory or pilot-scale stages.

Overall, while chemical absorption provides a solid foundation for large-scale deployment, its high energy intensity and operational costs continue to present significant barriers. Future efforts must focus on achieving technological breakthroughs and optimizing engineering integration to lower the energy consumption and costs associated with capture systems. This will be essential for facilitating deep emissions reductions in coal-fired power plants and advancing the clustered development of CCUS in regions like Ordos.

25.2.3 CO_2 Mineralization via Solid Waste Utilization

Alkaline industrial wastes possess exceptionally high carbonation potential, making them ideal materials for CO_2 mineralization utilization. Through reactions with CO_2 to form stable carbonates, these wastes enable long-term greenhouse gas sequestration while producing high-value-added products applicable across chemical engineering, construction materials, soil remediation, and other fields. Consequently, CO_2 mineralization has emerged as a key technological pathway for carbon reduction and resource utilization within the industrial sector. As a major hub for industry, energy, and raw materials, Ordos City generates massive annual quantities of alkaline solid wastes such as fly ash, coal gangue, carbide slag, and steel slag. However, the region's utilization rates remain low, and long-term accumulation of these wastes imposes severe ecological and environmental pressures, creating an urgent need for the development of green and efficient solid waste valorization technologies.

CO_2 mineralization technologies can be broadly categorized into traditional direct carbonation, indirect carbonation, and integrated absorption-mineralization (IAM) processes (Sanna et al. 2014). Among them, direct carbonation involves the direct reaction of gaseous CO_2 with reactive components in solid waste (Liu et al. 2021), typically proceeding via gas–solid or gas–liquid-solid multiphase reactions. This pathway features a simple process flow and is suitable for large-scale treatment of low-value wastes. However, it suffers from limitations such as low reaction rates, stringent requirements for temperature and pressure, and poor economic viability. In particular, the difficulty of recovering additives used in gas–liquid-solid systems further escalates operational costs. Consequently, direct carbonation is primarily applied to the production of mid- to low-end products such as carbon-negative building materials. For example, the Institute of Organic Functional Materials and Applications, in collaboration with Tongji University, developed an industrial-scale (10,000-ton) process for CO_2 mineralization of solid wastes into carbon-negative building materials, marking significant progress toward industrial deployment of direct carbonation technology.

In contrast, indirect carbonation extracts alkaline metal ions from solid wastes using acids, bases, or salt solutions, followed by CO_2 reaction to generate high-purity carbonates (Luo and He 2021). This approach offers greater product flexibility and high value-added potential, with tunable product properties including particle size, crystal form, and purity—ideal for applications in nanomaterials, electronics, and pharmaceutical excipients. However, conventional leaching agents are often costly, difficult to recycle, and environmentally hazardous. For instance, early pilot tests in Finland employing NH_4Cl for steel slag carbonation faced setbacks due to the agent's high volatility and low selectivity. Future breakthroughs must therefore focus on developing renewable, highly selective, and environmentally benign leaching agents to enable sustainable, high-value indirect carbonation.

IAM technologies integrate the CO_2 capture and mineralization processes, sharing reaction media or intermediates to significantly reduce energy consumption and equipment load, thereby realizing a seamless "capture–mineralization–resource utilization" process. IAM currently represents the most promising development pathway. An illustrative case is the chemical looping mineralization CCUS pilot project constructed by China Energy Investment Corporation in Datong, Shanxi Province, where a proprietary cyclic medium captures CO_2 and subsequently reacts with carbide slag to produce calcium carbonate, creating a closed-loop "CO_2–solid waste co-origin" system. This approach simultaneously achieves emission reduction, waste minimization, and efficiency enhancement, offering strong replicability and scalability for coal chemical clusters in Inner Mongolia and similar regions.

Moreover, geological mineralization and biological mineralization are emerging as important complementary pathways for CO_2 mineralization utilization. Basalt, peridotite, and other alkaline rocks possess natural potential for CO_2 mineral trapping and can be integrated with artificial mineralization to establish multi-scale carbon sequestration systems. In the realm of biological mineralization, technologies based on microbially induced carbonate precipitation (MICP) have shown unique advantages in soil stabilization, mine site rehabilitation, and ecological restoration. These processes enable green CO_2 fixation under mild conditions and are particularly suitable for integration with ecological initiatives such as desertification control and agropastoral development, thus offering innovative pathways for regional sustainable development.

25.2.4 CO_2 Mineralization via Solid Waste Utilization

The target reservoirs for CO_2 geological storage primarily include depleted oil and gas fields and deep saline aquifers, offering both carbon reduction benefits and the potential for enhanced oil and gas recovery, thus demonstrating promising economic and deployment prospects. To date, approximately 200 CO_2 storage projects have been initiated worldwide (Canle 2023), with around 70% incorporating Carbon Dioxide Enhanced Oil Recovery (CO_2-EOR) technologies that promote the synergistic development of storage and utilization. The United States began exploring CO_2-EOR in the 1950s, gradually establishing a mature commercial system (Novak Mavar et al. 2021). In China, CO_2 flooding experiments commenced in the 1960s in fields such as Daqing and Shengli, and in recent years, demonstration projects have been implemented in regions including Jiangsu, Jilin, Shaanxi, and Shandong. However, large-scale commercial application remains in its early stages, transitioning from engineering demonstrations. In comparison, deep saline aquifers offer greater storage capacity and broader applicability, and are regarded as the primary pathway for achieving large-scale carbon reductions in the future.

The Ordos region, characterized by diverse geological formations and favorable storage conditions, possesses an estimated deep saline aquifer storage potential of approximately 13.3 billion tonnes (Yang et al. 2019). This provides a solid foundation for developing CCUS projects exceeding one million tonnes per year. Nonetheless, key technical challenges must be addressed, including reservoir evaluation, large-scale CO_2 injection management, and integrated optimization strategies, to establish a stable and efficient storage system. Ensuring the safety of CO_2 geological storage hinges on comprehensive monitoring technologies, critical for maintaining operational security and long-term system stability. Storage operations must proactively mitigate risks such as fault reactivation, caprock failure, CO_2 leakage, groundwater contamination, and induced seismicity. Internationally, countries like the United States, Canada, and Norway have developed relatively sophisticated multidisciplinary monitoring systems integrating geophysical, geochemical, and microseismic techniques to track CO_2 migration and assess reservoir integrity in real time.

In China, CCUS demonstration projects in Ordos and Guangdong have deployed a combination of seismic monitoring, downhole chemical sampling, and fiber-optic sensing technologies. Furthermore, artificial intelligence, the Internet of Things, and high-resolution equipment are gradually being integrated to enhance monitoring intelligence and real-time capabilities. However, challenges remain, including the lack of unified national technical standards and regulations, the limited domestic manufacturing of key monitoring equipment, insufficient sensitivity for detecting minor leakages, and the need for long-term monitoring expertise. Future efforts should prioritize the development of advanced monitoring equipment, the establishment of standardized technical frameworks, and the construction of multi-scale, multi-dimensional, and integrated CO_2 storage safety monitoring networks to safeguard the secure operation and environmental integrity of large-scale CCUS deployments.

25.3 Key Technological Advances in CCUS

To address the carbon reduction challenges faced by resource-based cities and to promote the large-scale deployment of CCUS in Ordos, this study proposes an integrated CCUS technological pathway encompassing the full chain of capture–utilization–storage–monitoring, as illustrated in Fig. 25.1. This pathway integrates multidisciplinary approaches including geological resource assessment, energy consumption optimization, synergistic utilization of industrial solid waste, and high-precision storage monitoring. It specifically targets the systemic and integrative bottlenecks that currently constrain the large-scale development of CCUS, aiming to establish a comprehensive and efficient deployment framework.

25.3.1 CO_2 Source-Sink Mapping and Cluster Deployment

To enable large-scale CCUS deployment in Ordos, the primary task is to clarify the spatial distribution and dynamic coupling relationships between regional CO_2 emission sources and geological storage sinks. A comprehensive source-sink database covering the full life cycle will be established, and an optimized cluster deployment

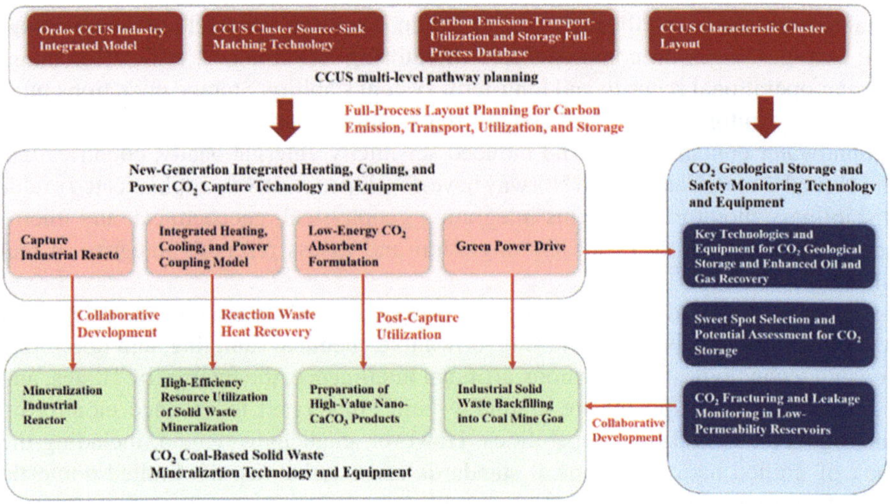

Fig. 25.1 CCUS integrated technology roadmap in Ordos City

model will be developed based on economic viability and engineering adaptability, providing a scientific basis for subsequent capture, transport, and storage stages.

Construction of Source-Sink Database and Spatial Distribution Analysis. Based on emission data from major high-carbon industries in Ordos, including coal-fired power plants, coal chemical industries, and cement production, key parameters such as CO_2 emission intensity, emission type, and emission stability will be compiled to build a standardized CO_2 emission source database. Simultaneously, geological storage sinks such as deep saline aquifers and depleted oil and gas reservoirs will be identified based on regional geological surveys and previous storage research. GIS spatial analysis techniques will be employed to visualize the spatial coupling between sources and sinks, revealing characteristics of concentration and belt-like distribution, thereby providing foundational support for cluster-based deployment.

Technical Route for Geological Storage Potential Assessment. For typical storage targets such as deep saline aquifers and depleted oil and gas fields in Ordos, a multi-scale, multi-parameter suitability evaluation system will be established. Geological factors including reservoir thickness, porosity, permeability, caprock integrity, structural stability, and fault development will be comprehensively considered. A suitability index model will be constructed using a hierarchical weighting method. Fine-scale geological modeling and storage resource estimation will be conducted based on drilling, logging, seismic, and regional structural data to identify the most promising storage blocks. A preliminary assessment of CO_2 geological storage suitability in Ordos has been carried out based on this methodology, as shown in Fig. 25.2.

Source-Sink Matching Optimization Model Based on Economic Factors. To overcome the limitations of traditional distance-based source-sink matching

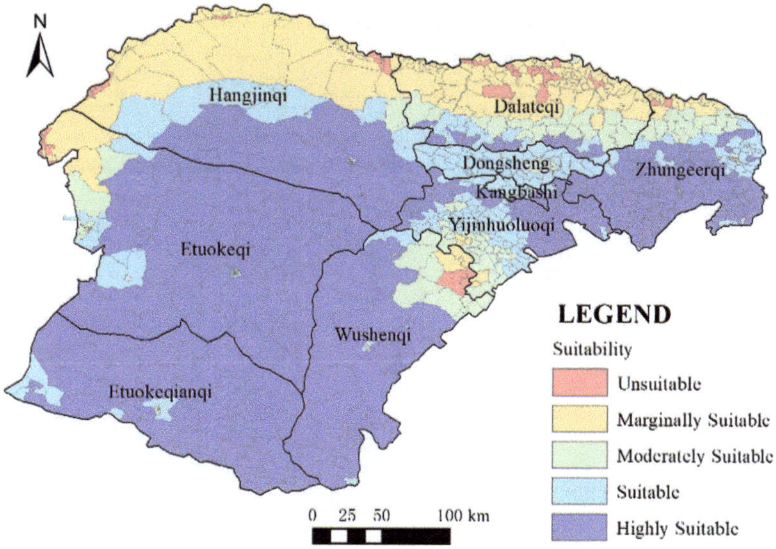

Fig. 25.2 Distribution map of CO_2 geological storage suitability in Ordos City

approaches, an economic-driven matching model is developed with the objective of minimizing unit storage costs. This model comprehensively considers capture costs, transportation distance and routing, injection costs, construction conditions, and economies of scale. Optimal one-to-many matching solutions for multi-source–multi-sink systems are obtained through a hybrid solution method combining integer programming and genetic algorithms. Time constraints and dynamic carbon emission factors are introduced into the modeling process to enhance the engineering applicability and long-term adaptability of the matching results.

Design of Spatiotemporal Dynamic Deployment Strategies. Based on the optimized matching results, phased and region-specific CCUS cluster deployment strategies are proposed. System dynamics and scenario simulation methods are employed to predict carbon emission scenarios and industrial restructuring pathways over different time periods, enabling dynamic adjustment of the capture–transport–storage network layout. A cluster deployment pathway characterized by "proximity priority, resource condition priority, and economic efficiency priority" is developed, balancing carbon reduction efficiency with storage safety, and providing a scientific basis for the steady advancement of large-scale CCUS deployment. The overall technical route is illustrated in Fig. 25.3.

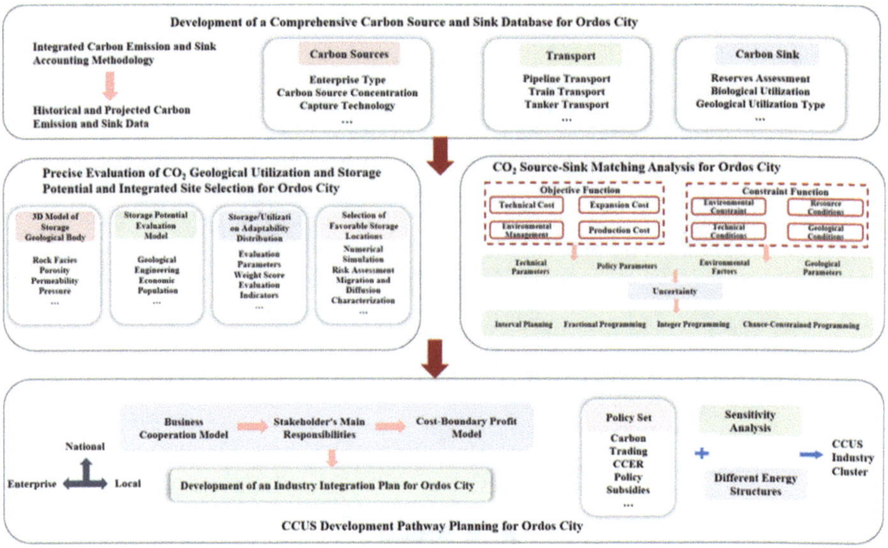

Fig. 25.3 Path map of integrated CCUS technology with a capacity of ten million tons in Ordos City

25.3.2 *Low-Energy CO_2 Capture System Development*

The CO_2 volume fraction in flue gas from coal-fired power plants typically ranges from 10 to 15%, accompanied by high humidity and various impurities such as SO_2, NO_x, and particulates, which impose higher demands on traditional monoethanolamine (MEA) absorption systems. Under these conditions, the capture energy consumption is high, corrosion is severe, and operational stability is poor, which are the primary limiting factors for large-scale engineering deployment. Therefore, there is an urgent need to improve CO_2 capture efficiency and significantly reduce specific energy consumption by optimizing both absorbent performance and the overall process system.

Optimization of High-Efficiency, Low-Energy Absorbent Formulation. To address the aforementioned issues, this study develops a novel blended amine-based absorbent system that balances high absorption capacity, rapid mass transfer performance, and low regeneration heat demand. Through molecular simulation and experimental screening techniques, the amine structure combinations were optimized to enhance CO_2 selectivity and reaction rate, while significantly reducing the heat requirements for regeneration. Under experimental conditions, the absorption rate of the developed absorbent was approximately 30% higher than that of MEA, and the regeneration energy consumption was reduced by more than 15%, laying a solid foundation for optimizing system energy efficiency.

System Integration Design and Integrated Heating, Cooling, and Power (CHP) CO_2 Capture Optimization. To further enhance the overall energy utilization efficiency of the system, this study proposes an integrated heating, cooling, and power (CHP) CO_2 capture scheme based on the coupling utilization of power plant waste heat. By integrating low-temperature waste heat from the boiler flue gas of coal-fired power plants, the structure and operational parameters of the solution regeneration heat exchanger are optimized to achieve hierarchical utilization of thermal energy during the absorption–desorption process. Additionally, the power supply system and thermal storage devices of the power plant are combined to enable dynamic energy allocation under unstable operating conditions, effectively mitigating the impact of fluctuations in peak and off-peak electricity prices on system operating costs.

Pilot Testing and Demonstration Project Technical Parameters. Based on the aforementioned technological system, the new absorbent formulation has been applied in the 10,000-ton-level bio-boiler CO_2 capture demonstration project in Guigang, Guangxi. The system achieves a capture rate of over 90%, with CO_2 purity exceeding 99%, and a thermal energy consumption of 2.0 GJ/ton CO_2. The typical amine escape is less than 5 mg/Nm^3. A pilot-scale integrated heating, cooling, and power (CHP) CO_2 capture device is currently under construction in Ordos, Inner Mongolia. Building upon the pilot validation, the design for scaling up to the hundred-ton level is underway, with plans for demonstration applications at typical coal-fired power plants in Ordos. The technical roadmap is shown in Fig. 25.4. Moving forward, the process flow and equipment selection will be continuously optimized in response to CCUS project requirements, enhancing the system's modularity and intelligence to support large-scale deployment.

25.3.3 *CO_2 Mineralization via Solid Waste Utilization*

Characterization of Typical Solid Waste Resources in Ordos. The fly ash and coal gangue generated from coal-fired power plants in the Ordos region are rich in alkaline metal elements such as Ca and Mg, which exhibit potential reactivity for CO_2 mineralization, as illustrated in Fig. 25.5. However, due to challenges such as compositional heterogeneity, low intrinsic reactivity, and poor leachability, traditional mineralization processes often suffer from low sequestration efficiency and high energy consumption, hindering their large-scale application. In this study, the physical and chemical properties of representative solid wastes were systematically analyzed to identify the reactive components suitable for mineralization reactions. Furthermore, appropriate pretreatment methods were proposed to enhance mineralization efficiency, providing essential baseline data for subsequent process development.

Optimization of Low-Energy Mineralization Processes and Leaching Agent Systems. To enhance the efficiency of mineralization reactions while reducing energy consumption, a novel low-energy leaching agent system was developed to selectively extract mineralization-active ions such as Ca^{2+} and Mg^{2+}. By analyzing reaction

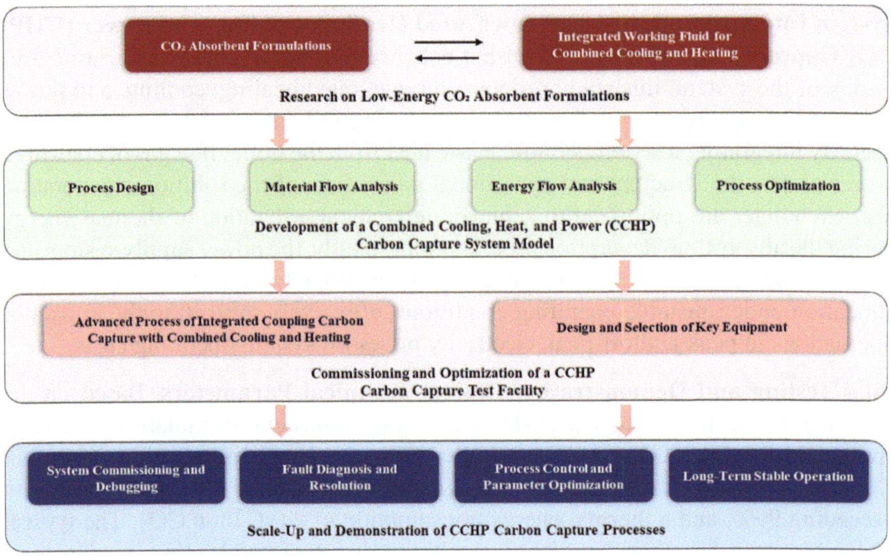

Fig. 25.4 Technical roadmap for CO_2 capture

kinetics and mass transfer mechanisms, key process parameters—including leaching temperature, liquid-to-solid ratio, and reaction time—were systematically optimized to maximize leaching efficiency and ion concentration. Additionally, alternative carbonation pathways under ambient pressure and low-temperature conditions were explored to minimize the overall thermal load of the reaction, thereby improving the economic viability of the mineralization process.

Preparation of Nano-Scale Calcium Carbonate and Pathways for Resource Utilization. In parallel with the CO_2 mineralization process, this study explores the development of high-value-added products. A crystal growth control mechanism for calcium carbonate was established, enabling precise regulation of crystal morphology and particle size by adjusting the nucleation environment and reaction parameters. Preliminary experiments successfully synthesized nano-sized $CaCO_3$ particles with spherical and calcite-type morphologies, exhibiting potential applications in industries such as coatings, plastics, and rubber. This approach provides an economic driver for CO_2 mineralization-based storage. In the future, a three-dimensional synergistic pathway integrating "carbon reduction–waste minimization–value addition" will be constructed to enhance the market adaptability of carbon mineralization technologies.

Construction of a Multi-pathway Synergistic Solid Waste-Based CO_2 Storage System. Beyond the utilization of fly ash, this study systematically reviews multiple CO_2 mineralization approaches, including the thermal activation of coal gangue, co-mineralization with mining waste, and the use of alkaline minerals from construction waste. Based on the diverse industrial scenarios in Ordos, an adaptive and flexibly deployable multi-pathway synergistic solid waste-based CO_2 storage strategy

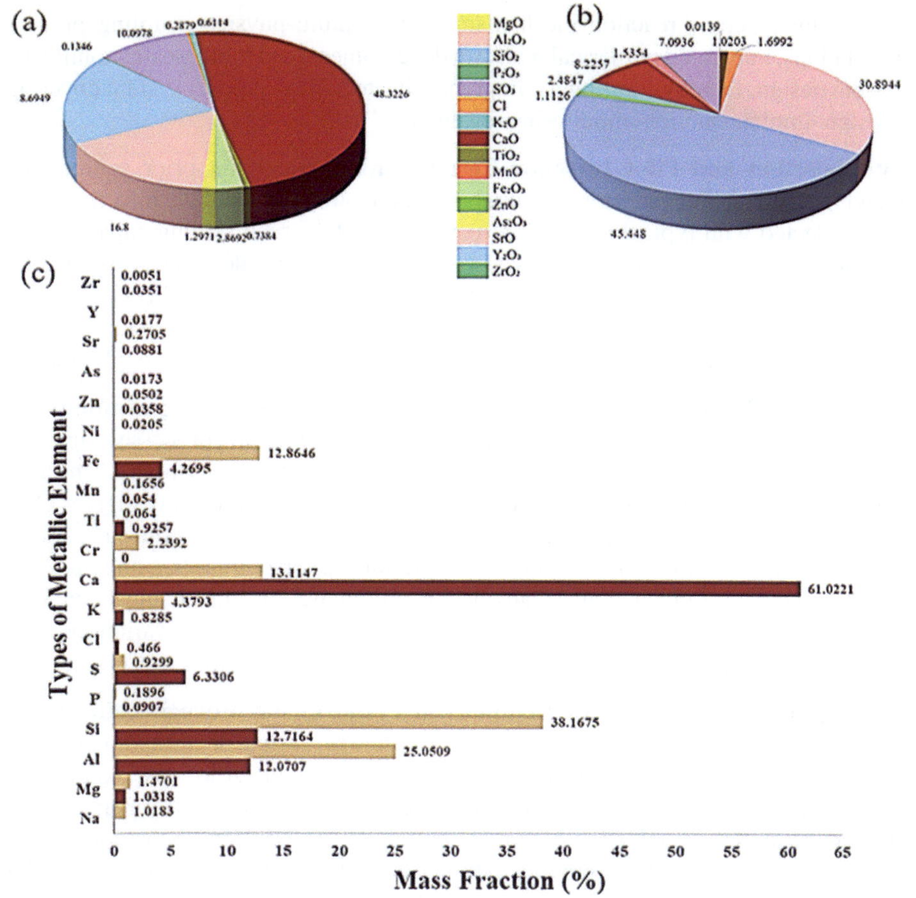

Fig. 25.5 **a** Oxide composition of fly ash from a plant in Etuoke Qi, **b** oxide content of fly ash from a plant in Etok Front Qi, **c** metal element content of fly as

is proposed. Through process integration and site-coupling optimization, a regionalized, on-site CO_2–solid waste co-utilization model is established, laying a solid foundation for the future demonstration of million-ton-scale resource-based carbon storage projects.

25.3.4 *Geological Storage and Monitoring Platform*

To ensure the long-term security and engineering feasibility of large-scale CO_2 geological storage, an integrated technical framework combining storage and monitoring was established. This framework covers site selection and pilot injection testing, microscopic reaction mechanism studies, multi-physics coupling process simulations, integrated numerical platform development, and multi-scale monitoring system design, thereby forming a comprehensive technical support system covering "design–implementation–monitoring–evaluation."

Site Selection and Pilot Injection Test Planning. Based on prior source-sink matching and geological storage suitability evaluations, candidate injection sites were selected with typical reservoir structures, good caprock sealing capacity, and complete baseline data. Pilot-scale CO_2 injection tests were designed, targeting an injection rate of 50,000–100,000 tonnes per year. These tests aim to collect geological response data throughout the injection–migration–storage process, providing model validation and boundary conditions for future large-scale engineering deployments.

Microscopic Reaction Mechanisms and Multi-physics Coupling Simulations. The thermal–hydraulic–mechanical–chemical (THMC) coupling processes triggered by supercritical CO_2 injection are critical to storage stability assessments. Through core flooding experiments and X-CT imaging analyses, this study investigates CO_2 migration pathways and interfacial reaction mechanisms within the pore structure of the reservoir. The temporal evolution of dissolution, capillary trapping, and mineralization processes is elucidated. Meanwhile, a multiphase flow–mechanical coupling model is developed to simulate fracture propagation, in-situ stress variations, and caprock risk during injection, enabling quantitative risk evaluations of CO_2 storage.

Development of an Integrated Numerical Platform for Storage–Utilization–Monitoring. To enhance the intelligence of storage design, a three-dimensional numerical simulation platform tailored to various types of Ordos reservoirs (e.g., sandstone, coal seams, saline aquifers) was constructed. This platform integrates geological modeling, injection parameter optimization, long-term behavior prediction, and monitoring data inversion modules. It enables dynamic visualization of the CO_2 injection process and, by coupling with real-time monitoring data, identifies abnormal response zones and provides proactive early warnings for storage security.

Design of a Multi-scale Monitoring System for CO_2 Storage. Safety monitoring is fundamental to guaranteeing long-term storage stability. Based on the geological features and engineering layout of the Ordos storage sites, a multi-scale, multi-method combined monitoring system was designed, incorporating "downhole monitoring–surface observation–distributed fiber optics–remote sensing inversion." Multi-parameter data, including temperature, pressure, microseismicity, and trace gases, are collected. Real-time strain detection based on distributed optical fiber technology enhances the ability to identify CO_2 migration pathways and abnormal leakage events. A dynamic monitoring system characterized by "visualization, controllability, and early warning capability" is thus established, as illustrated in Fig. 25.6.

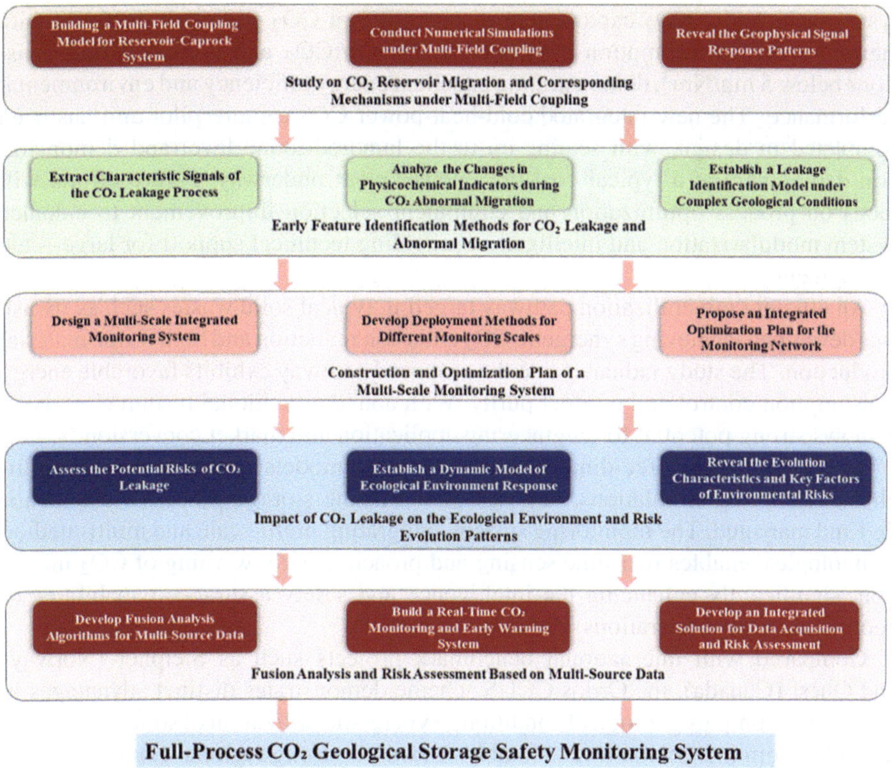

Fig. 25.6 Technical roadmap of the full-process CO_2 geological storage safety monitoring system

25.4 Conclusions

This study establishes an integrated CCUS technological pathway tailored for Ordos City, covering key aspects such as source-sink layout optimization, low-energy consumption capture systems, solid waste mineralization utilization, and geological storage and monitoring platforms. Based on this framework, a systematic evaluation and result analysis were conducted from the perspectives of energy efficiency, technical feasibility, and regional adaptability. The main conclusions are as follows.

A spatial coupling atlas between major emission sources and high-quality storage sites was developed based on a multi-source-driven CO_2 source-sink database and storage suitability evaluation system. An economic constraint model was introduced to optimize the matching pathways, enhancing the overall synergy of the system. Furthermore, a "multi-point distribution–clustered injection–dynamic adjustment" deployment strategy was proposed, significantly improving the adaptability and operational flexibility of the CCUS system in response to emission fluctuations.

The newly developed absorption system has been applied to a 10,000-tonne-level biomass boiler CO_2 capture demonstration project in Guigang, Guangxi. The system achieved a CO_2 capture rate of over 90%, a CO_2 purity above 99%, with thermal energy consumption controlled at 2.0 GJ/tCO_2 and typical amine emissions below 5 mg/Nm^3, demonstrating excellent energy efficiency and environmental performance. The new integrated cold-heat-power CO_2 capture pilot unit has been completed in design, with scaling-up to the hundred-tonne level and demonstration deployment at a typical coal-fired power plant underway. Future efforts will focus on process optimization and equipment selection improvement to enhance system modularization and intelligence, providing technical support for large-scale deployment.

An efficient mineralization pathway targeting typical solid wastes such as fly ash was developed, achieving synergetic CO_2 emission reduction and high-value material production. The study indicates that the proposed pathway exhibits favorable energy consumption control and product purity. With abundant regional resource reserves, it shows strong potential for engineering application and market conversion.

By constructing three-dimensional geological models and conducting multi-physics coupling simulations, key risk points in the storage process were identified and managed. The monitoring system, integrating multi-scale and multi-method technologies, enables real-time sensing and proactive early warning of CO_2 migration, significantly enhancing the intelligence and safety assurance capabilities of geological storage operations.

Compared with international benchmark projects such as Sleipner (Norway) and Quest (Canada), the Ordos CCUS scheme demonstrates distinct advantages in adapting to complex geological conditions, synergistic utilization of solid wastes, and overall system integration level. Notably, it exhibits strong regional compatibility in low-concentration CO_2 capture and mineralization utilization, offering demonstrative significance. The analysis suggests that this scheme could serve as a replicable and scalable CCUS technological pathway model for other resource-oriented cities.

To facilitate the deployment of such projects, it is imperative to establish a unified regulatory framework covering the entire CCUS chain—capture, transportation, utilization, and storage. This framework should clearly define legal responsibilities at each stage and set environmental safety standards. Effective integration of CCUS with policy instruments such as carbon markets and carbon taxes is also essential. It is recommended that investors prioritize regional projects characterized by high-concentration emission sources and high-quality storage sites, with a focus on cluster-based demonstration projects located in areas with favorable policy environments and strong industrial synergy. As the green finance system continues to evolve, the commercial viability of CCUS will gradually improve. The coordinated advancement of policy and capital is expected to become a key driving force for large-scale deployment.

References

Benson SM, Orr FM Jr (2008) Carbon dioxide capture and storage. MRS Bull 33(4):303–305

Bui M, Adjiman CS, Bardow A, Anthony EJ, Boston A, Brown S, Mac Dowell N et al (2018) Carbon capture and storage (CCS): the way forward. Energy Environ Sci 11(5):1062–1176

Canle C (2023) Global status of CCS 2023

Chen C, Ma S, Wang X, Shen J, Qin Y, Ling Z, Song Y (2024) CCUS source-sink matching model based on sink well placement optimization. Fuel 377:132812

Guo H, Lyu X, Meng E, Xu Y, Zhang M, Fu H, Song K et al (2022) CCUS in China: challenges and opportunities. In: SPE improved oil recovery conference? SPE, p D031S031R002

Jenkins J (2015) A case study of the Petra Nova carbon capture project. The Paulson Institute

Liu HJ, Were P, Li Q, Gou Y, Hou Z (2017) Worldwide status of CCUS technologies and their development and challenges in China. Geofluids 2017(1):6126505

Liu W, Teng L, Rohani S, Qin Z, Zhao B, Xu CC, Liang B et al (2021) CO_2 mineral carbonation using industrial solid wastes: a review of recent developments. Chem Eng J 416:129093

Liu H, Consoli C, Zapantis A (2018) Overview of carbon capture and storage (CCS) facilities globally. In 14th Greenhouse gas control technologies conference Melbourne, pp 21–26

Luo Y, He D (2021) Research status and future challenge for CO_2 sequestration by mineral carbonation strategy using iron and steel slag. Environ Sci Pollut Res 28:49383–49409

Mao X, Xian Z, Jingli FAN, Lin G, Dong X (2021) Status quo, opportunities and challenges of integrated application of coal-to-hydrogen and CCUS technology in China. J Min Sci Technol 6(6):659–666

Novak Mavar K, Gaurina-Međimurec N, Hrnčević L (2021) Significance of enhanced oil recovery in carbon dioxide emission reduction. Sustainability 13(4):1800

Perry AVM (2023) Net zero, net profit: an analysis of offshore UK carbon capture and storage

Quemin S (2022) Raising climate ambition in emissions trading systems: the case of the EU ETS and the 2021 review. Resour Energy Econ 68:101300

Rodrigues C, Pinheiro H, De Sousa ML (2022) Clean energy transition challenge: the contributions of geology. Affordable Clean Energy 47

Rubin ES, Davison JE, Herzog HJ (2015) The cost of CO_2 capture and storage. Int J Greenhouse Gas Control 40:378–400

Sanna A, Uibu M, Caramanna G, Kuusik R, Maroto-Valer MM (2014) A review of mineral carbonation technologies to sequester CO_2. Chem Soc Rev 43(23):8049–8080

Smyth RC, Hovorka SD, Meckel T, Breton C, Paine JG, Hill GR, Center GCC et al (2007) Potential sinks for geologic storage of carbon dioxide generated in the Carolinas. US Bureau of Economic Geology, summary report, 1–14

Vulin D, Močilac IK, Jukić L, Arnaut M, Vodopić F, Saftić B, Cvetković M et al (2023) Development of CCUS clusters in Croatia. Int J Greenhouse Gas Control 124:103857

Wang M, Lawal A, Stephenson P, Sidders J, Ramshaw C (2011) Post-combustion CO_2 capture with chemical absorption: a state-of-the-art review. Chem Eng Res des 89(9):1609–1624

Yang H, Zhao X, Kang Y, Chen L, Huang C, Wang H (2019) CO_2 geological storage suitability and potential evaluation in the Ordos Basin. Prog Clim Change Res 15(1):95

Zheng B, Niu J, Zhang K (2024) Current status and outlook of CCUS industry in China. In: annual report on China's petroleum, gas and new energy industry (2022–2023), pp 289–306

Chapter 26
Optimizing Carbon Reduction Pathways in Aviation Through Carbon Emission Trading and Sustainable Aviation Fuels

Shuwei Jin, Xingyu Zhu, and Li Wang

Abstract Aviation carbon emissions mostly happen at high altitudes, where CO_2 has a much stronger impact on climate. Cutting down these emissions is tough because of technical barriers. Sustainable aviation fuel and carbon trading mechanisms are important measures to address carbon reduction in the aviation industry. The application of sustainable aviation fuel and carbon trading mechanisms in aviation is related to different sectors and multiple complexities. In this study, a multi-level programming model was developed to optimize carbon emissions in the aviation industry through carbon trading and sustainable aviation fuel. An interactive solution mechanism is developed to resolve the complex models with a multilevel decision-making framework. Then the optimized results of fleet planning, carbon allocation, and sustainable aviation fuel blending ratio can be obtained, which can provide decision support for the government to formulate aviation carbon abatement policies and the transformation of airline operations.

Keywords Aviation · Carbon abatement · Carbon trading · Multi-level programming · Sustainable aviation fuel

26.1 Introduction

The aviation industry is facing multi-dimensional pressures from international carbon emission targets, technology upgrade bottlenecks and biofuel development, and decarbonization is becoming increasingly urgent with fleet expansion and net-zero targets (An et al. 2025). The aviation sector witnessed a 10.4% global passenger traffic surge in 2024 compared to 2023, with China's ascension as the world's second-largest air transport market accelerating environmental pressure (Wang et al. 2023). The civil aviation industry is experiencing restorative growth (as shown in

S. Jin (✉) · X. Zhu · L. Wang
School of Transportation Science and Engineering, Civil Aviation University of China, Tianjin, China
e-mail: swjin@cauc.edu.cn

Y. Li et al. (eds.), *16th International Conference on Environmental Science and Development*, Environmental Science and Engineering,
https://doi.org/10.1007/978-3-032-15080-6_26

Fig. 26.1). Despite significant short-term fluctuations, the overall trend is gradually moving toward recovery. The 2023 rebound not only confirms the resilience of market demand but also highlights the urgency of addressing carbon emissions in the future. Current projections indicate a sixfold increase in civil aviation carbon emissions, potentially reaching 690 million tons by 2060 (Gössling and Humpe 2020; Wong et al. 2024). To curb projected carbon emission growth, the aviation sector has adopted ICAO's binding 2% annual fuel efficiency mandate (2021–2050), ratified at its 41st Assembly (Hu et al. 2022). This transformation aligns with the Paris Agreement's 1.5 °C threshold, operationalized through IATA-ATAG's 2050 net-zero roadmap (Delbecq et al. 2023). The carbon emissions associated with the aviation industry in China, the world's second largest air transport market, have increased significantly. This has led to pressure from the international community to regulate emissions and reduce climate change.

There are many approaches that governments and airlines have researched in order to reduce carbon emissions to meet their reduction targets, including sustainable aviation fuel, carbon trading mechanisms, efficiency improvement, hydrogen-powered electric aircraft (Zhang et al. 2025) and electric-powered aircraft (Hopkins and Tavares 2025; Viswanathan et al. 2022). Among them, sustainable aviation fuel and carbon trading mechanism demonstrate better potential for emission reductions (Shahriar and Khanal 2022; Lau et al. 2024). Studies indicate that the costs of conventional jet fuel are currently 2.2 to 8 times lower than those of sustainable aviation fuel (Watson et al. 2024; Hamdan et al. 2022). In comparison, carbon trading mechanisms, which incorporate carbon trading benefits into the aviation market, not only help to alleviate climate change, but also facilitate the economic operation of airlines (Tanveer et al. 2024; Qiu et al. 2017). Consequently, combining sustainable aviation fuel with carbon trading mechanism is an effective way to establish optimal emission reduction strategies.

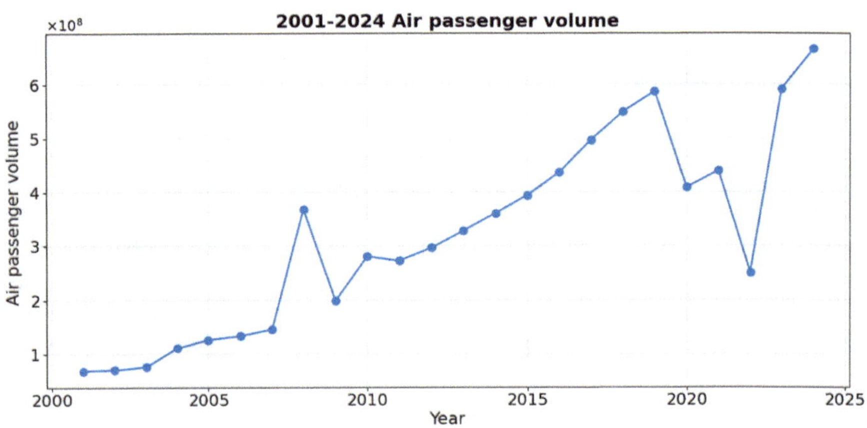

Fig. 26.1 Air passenger volume in 2001–2024

26.2 Case Study

The case study concentrates on airlines operating on a single route. It encompasses six airlines: Air China (CA), Sichuan Airlines (3U), Chengdu Airlines (EU), Okay Airways (BK), China United Airlines (8L), and Tianjin Airlines (GS). Various aircraft types are involved in the study, such as A319, A320, A321, A330, 32A, 32B, and B737. Combining carbon emission trading and sustainable aviation fuels offers an effective way to establish optimal emission reduction strategies.

26.2.1 Objective

The objective of this study is to develop a multi-level programming model for optimizing carbon reduction pathways in aviation through carbon emission trading and sustainable aviation fuels. The three-level decision-making framework comprises upper-level (government) aims to minimize environmental costs through sustainable aviation fuel and carbon trading mechanism strategies (including carbon reduction benefits and environmental costs), middle-level (airline) seek to maximize economic benefits composed of fixed and variable costs, and lower-level (consumer) strive to maximize utility. The detailed objective function can be formulated as follows:

$$f_U = -\sum_{i \in i_{total}} \sum_{k \in k_{total}} \underbrace{X_{i,k} \times T_{CO_2}}_{carbon\,abatement} + + \underbrace{SW + Air + WAT + SPE}_{ecological\ effect} \tag{26.1}$$

$$\begin{aligned} f_M = & \underbrace{\sum_{k \in k_{total}} X_{i,k} \times H_k \times P_{i,k}}_{airfare} - \underbrace{\sum_{k \in k_{total}} X_{i,k} \times H_k \times L \times FIX_{i,k}}_{fixed\ cost} \\ & - \underbrace{\sum_{k \in k_{total}} X_{i,k} \times AP \times L \times FU_k \times (P_{SAF} \times \theta_{i,k} + P_{jet} \times (1 - \theta_{i,k})) - C_{CO_2}}_{variable\ cost} \end{aligned} \tag{26.2}$$

$$\begin{aligned} f_L &= \max_{P_{i,k}} \left\{ \min[U_1(P_{i,k}), U_2(P_{i,k}), U_3(P_{i,k}), \ldots, U_i(P_{i,k})] \right\} \\ U_i(P_{i,k}) &= TR_i + PT_i \end{aligned} \tag{26.3}$$

where i is a set of airlines; k is a set of aircraft types; $X_{i,k}$ represents integer variables used to determine how many type k aircrafts are selected for airline i, with these variables being determined by the airlines; T_{CO_2} is the parameter related to carbon emissions; WAT is the sustainable aviation fuel production water consumption effect; SW is the sustainable aviation fuel production sewage effect; Air is the sustainable aviation fuel production air pollutant effect; SPE is the sustainable aviation fuel production species damage effect; $P_{i,k}$ is the airfare of airline i on aircraft k; $FIX_{i,k}$ is the parameter related to the fixed costs of the airline i; $\Theta_{i,k}$ is the blending ratio of sustainable aviation fuel; H_k is the economic parameter related to airlines; FU_k is the fuel efficiency of different aircraft k; C_{CO_2} is the carbon cost; AP is air passenger volume; TR_i is the utility of airline punctuality; PT_i is the utility of airfares.

26.2.2 Constraints

The constraints include carbon allowance trading, environmental regulation, airline capacity and passenger demand.

(1) *Constraints for carbon allocation*: The carbon emissions cap comprises both free allocated allowances and tradable allowances subject to market-driven pricing. The Government allocates the free allowances in a ratio based on the historical allocation method.

$$JE_{i,k} = \sum_{k \in k_{total}} X_{i,k} \times FU_k \times L \times H_k \times \alpha_{jet} \tag{26.4}$$

$$0 \leq JE_{i,k} \leq Q_{i,free} + Q_{i,trade} \tag{26.5}$$

(2) *Constraints for carbon trading*: Airline's carbon allowance does not surpass the government-allocated free carbon allowances, and the tradable allowances for airlines must maintain non-negative values. Total allowance trading volume must remain within the government-allocated tradable allowances ceiling.

$$\sum_{i \in i_{total}} (Q_{i,+} - Q_{i,-}) \leq Q_{i,trade} \tag{26.6}$$

$$C_{i,CO_2} = Q_{i,-} \times P_{CO_2} \tag{26.7}$$

$$0 \leq \sum_{i \in i_{total}} Q_{i,free} + Q_{i,trade} \leq Q_{i,\max} \tag{26.8}$$

(3) *Constraints on environmental regulation*: Environmental impact of SAF includes water consumption, solid waste, and sewage discharge.

$$U_{SAF} \times 1000 \times ET \big/ Pd_{SAF} \leq WT_{\max} \tag{26.9}$$

$$SOW \times Pd_{SAF} \times (1 - RUR) \leq TES \tag{26.10}$$

$$SWD \times Pd_{SAF} - SV \times BR \leq SEDI \tag{26.11}$$

(4) *Airline capacity and passenger demand constraints*: Airline fleet allocation adheres to its operational capacity constraints while ensuring compliance with minimum demand requirements.

$$\sum_{k \in k_{total}} X_{i,k} \leq M_{i,k} \tag{26.12}$$

$$\sum_{k \in k_{total}} D_{demand} \leq \sum_{k \in k_{total}} x_{i,k} \times H_k \tag{26.13}$$

(5) *SAF blending ratio constraint*: SAF blending ratios vary depending on the SAF production technology, and currently SAF is blended with conventional jet fuel at a maximum of 50% (Khalifa et al. 2022).

$$0 < \theta_{i,k} \leq 0.5 \tag{26.14}$$

where JE_{ik} represents CO_2 released from the conventional jet fuel; a_{jet} is the carbon emission factor for conventional jet fuel; $Q_{i,free}$ is the amount of freely allocated carbon allowances; $Q_{i,trade}$ is the amount of carbon allowances traded through the carbon market; Q_{total} is the total amount of carbon allowances; $Q_{i,+}$ is the amount of carbon allowances that need to be purchased through carbon market trading; $Q_{i,-}$ is the amount of carbon allowances that need to be sold through carbon market trading; P_{CO_2} is the carbon price; C_{i,CO_2} is the carbon related cost; $Q_{i,\max}$ is the maximum amount of carbon allowances allowed to be allocated; U_{SAF} is the sustainable aviation fuel transformation ratio; ET is the seasonal evapotranspiration; Pd_{SAF} is the sustainable aviation fuel production from the bio-refinery; $WT_{\max}$ is the maximum water consumption; SOW is the solid waste emissions; RUR is the recycling ratio for solid waste; TES is the environmental capacity of solid waste; SWD is the amount of wastewater discharged by unit of sustainable aviation fuel; SV is the amount of sewage treated per unit area; BR is the floorspace of the biorefinery; $SEDI$ is the sewage discharge standard; $M_{i,k}$ is the maximum amount of flights that airline i can provide; D_{demand} is the airline's future passenger demand.

26.2.3 Data Collection

The economic and technical data in the model are derived from systematic analyses of representative government reports and relevant academic literature. Data on carbon emissions ratio and classifications for aircraft are shown in Table 26.1. The study of airline economic cost and technical data is based on the Civil Aviation Statistics Yearbook of China, the Ecological Environment Statistics Annual Report, Civil Aviation Network of China (CANC), China New Energy Network, China Civil Airport Network, Flight manager DAST, OAG analytic, government reports, official

Table 26.1 Aircraft type and carbon emission ratio

Major types of aircraft	Aircrafts	Carbon emission ratio ($tCO2/10^4 \cdot km$)
Wide-body passenger aircraft	B787	10. 40
	A330-200	
	A330-300	
	A350	
Narrow-body passenger aircraft	B737-600 B737-700 B737-800 B737-900	10.82
	A319	
	A320/A320neo	
	A321	

website newsletters, expert consultations and literature analyses. These parameters are mainly derived from the Civil Aviation Administration of China (CAAC) website and national carbon market trading data.

26.3 Results Analysis

Figure 26.2 shows the optimized results of sustainable aviation fuel utilization and carbon allocation in different seasons (see Fig. 26.2). The free carbon allowances would have a decreased tendency in different seasons. For example, the free allowance for 3U would decrease. This can be attributed to the high demand for air passenger transport during the summer and autumn seasons, which leads to an increase in carbon emissions. As the season changes, sustainable aviation fuel blending ratio decreases from 12.36 to 10.73%. Meanwhile, seasonal declines from summer-autumn seasons to winter-spring seasons can be obtained due to market share compression and higher production costs for biofuels. Thus, the implementation of seasonally adjusted carbon allowance allocations and sustainable aviation fuel blending ratios by the government and airlines can effectively achieve carbon abatement and economic efficiency objectives.

This dynamic mechanism also influences the results of airline fleet planning. Airlines would choose different aircraft types based on the hybrid mechanism in different seasons. The results demonstrate that big carriers (e.g., Air China (CA), Sichuan Airlines (3U)) can reduce emissions by selecting aircraft types according to seasonal needs.

The carbon emission flow for different airlines is shown in Fig. 26.3, which demonstrates a comparison between carbon emissions and the flow of carbon emissions in different seasons (see Fig. 26.3). Carbon emissions are predicted to increase from 103×10^3 tons to 108×10^3 tons. The blending ratio of sustainable aviation fuel utilization would vary with different airlines and aircraft. For example, CA

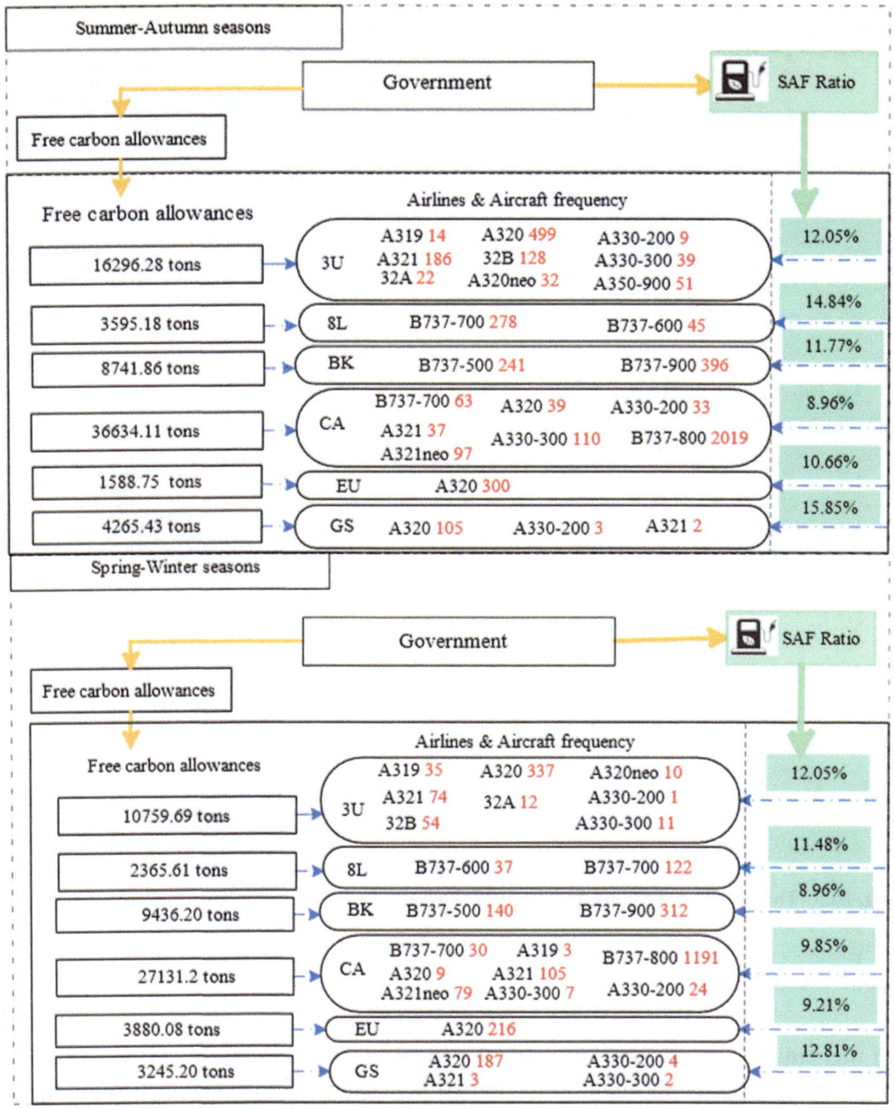

Fig. 26.2 The optimized results of carbon trading and sustainable aviation fuel utilization

carriers can blend 4.69% SAF to A320, while 3U carriers blend ratio less than 1%. In addition, the proportion of sustainable aviation fuel blending ratio would vary depending on the scale of the airline and the type of aircraft. Especially, compared to smaller carriers, larger airlines are better positioned to achieve significant emission reductions through sustainable aviation fuels.

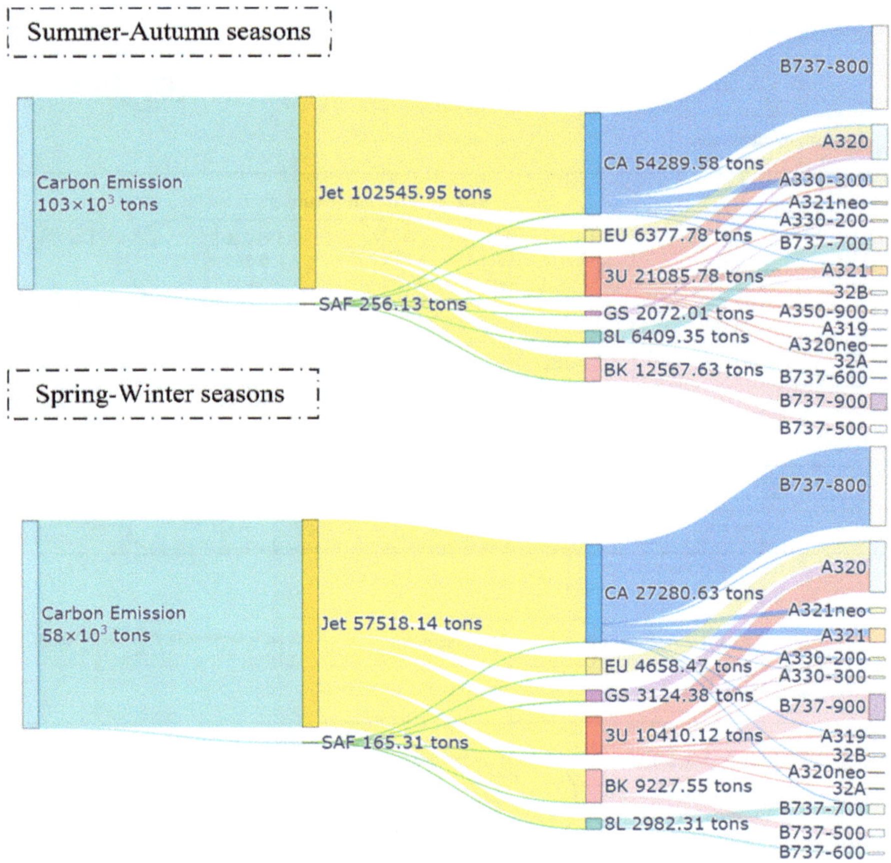

Fig. 26.3 The results of carbon emission (unit: tons)

26.4 Conclusion

In this study, a multi-level model was proposed for optimizing carbon reduction pathways in aviation through carbon emission trading and sustainable aviation fuels. In addition, the modelling process involves air transport demand, sustainable aviation fuel environmental costs and carbon allowance allocations. The model is then transformed into three deterministic sub-models by interaction algorithm, and desired results can be obtained. Results show that the combination of sustainable aviation fuel and carbon trading mechanism is an effective way to achieve carbon abatement and airline fleet planning. The potential for carbon abatement varies for different aircraft types and different carriers. These constructive findings can provide an efficient carbon abatement strategy for different airlines.

References

An HW, Wang Y, Wang YX, Liu J, Tang XL, Yi HH (2025) Civil aviation emissions in China in 2019: characteristics and abatement potential. J Environ Sci 51:225–237

Delbecq S, Fontane J, Gourdain N, Planès T, Simatos F (2023) Sustainable aviation in the context of the Paris Agreement: a review of prospective scenarios and their technological mitigation levers. Prog Aerosp Sci 141:100920

Gössling S, Humpe A (2020) The global scale, distribution and growth of aviation: implications for climate change. Glob Environ Chang 65:102194

Hamdan S, Jouini O, Cheaitou A, Jemai Z, Granberg TA, Josefsson B (2022) Air traffic flow management under emission policies: analyzing the impact of sustainable aviation fuel and different carbon prices. Transp Res Part a: Policy Pract 166:14–40

Hopkins BJ, Tavares FT (2025) Calculating the impact of replacing legacy batteries with standardized lithium-ion versions in defense aircraft. Next Energy 6:100208

Hu YJ, Yang LS, Cui HF, Wang HL, Li CJ, Tang BJ (2022) Strategies to mitigate carbon emissions for sustainable aviation: a critical review from a life-cycle perspective. Sustain Prod Consumption 33:788–808

Khalifa R, Alherbawi M, Elomri A, Al-Ansari T (2022) Alternative fuels' blending model to facilitate the implementation of carbon offsetting and reduction Scheme for International Aviation. Fuel 326:124974

Lau JC, Wang YS, Ang T, Seo JCF, Khadaroo S, Chew JJ, Lup AK, Sunarso J (2024) Emerging technologies, policies and challenges toward implementing sustainable aviation fuel (SAF). Biomass Bioenerg 186:107277

Qiu R, Xu J, Zeng Z (2017) Carbon emission allowance allocation with a mixed mechanism in air passenger transport. J Environ Manage 200:204–216

Shahriar MF, Khanal A (2022) The current techno-economic, environmental, policy status and perspectives of sustainable aviation fuel (SAF). Fuel 325:124905

Tanveer U, Ishaq S, Hoang TG (2024) Enhancing carbon trading mechanisms through innovative collaboration: case studies from developing nations. J Clean Prod 482:144122

Viswanathan V, Epstein AH, Chiang YM, Takeuchi E, Bradley M, Langford J, Winter M (2022) The challenges and opportunities of battery-powered flight. Nature 601:519–525

Wang K, Zhang YY, Wei YM (2023) China's aviation passenger transport can reduce CO_2 emissions by 2.9 billion tons by 2050 if certain abatement options are implemented. One Earth 6(8):1050–1065

Watson MJ, Machado PG, Silva AV, da Silva AV, Saltar Y, Ribeiro CO, Nascimento CAO, Dowling AW (2024) Sustainable aviation fuel technologies, costs, emissions, policies, and markets: a critical review. J Clean Prod 449:141472

Wong FH, Kez DA, Rio DD, Foley A, Rooney D, Abai M (2024) Decarbonizing and offsetting emissions in the airline industry: current perspectives and strategies. Energy 313:133809

Zhang J, Shen CF, Qin YP, Chen JY (2025) Forecasting of natural gas based on a novel discrete grey seasonal prediction model with a time power term. Energ Strat Rev 58:101677

Part IV
Assessment and Control of Air Pollution

Chapter 27
The Air Pollution Assessment of Waste Incineration Power Plant—A Case Study in Chongqing, China

Zhenhao Zha and David Holland

Abstract This study assesses the atmospheric impact of a waste incineration power plant in Chongqing, China, using the AERSCREEN model. Emission data from March–September 2024 were analyzed to simulate dispersion of SO_2, NO_x, TSP, CO, and HCl at three nearby receptor points. Results showed average proportion concentrations (P) of all pollutants were below 100% of China's environmental quality standards (GB 3095-2012), confirming regulatory compliance. Chronic health risks, evaluated via the U.S. EPA's hazard index (HI), yielded values (HI < 1) significantly below the threshold, indicating negligible long-term impacts on residents. The framework provides a replicable approach for environmental assessments of similar facilities in densely populated regions.

Keywords AERSCREEN model · Air pollutant · Health risks · Hazard index · Average concentration (C) · Average proportion concentration (P)

27.1 Introduction

Rapid urbanization and rising living standards have led to unprecedented growth in municipal solid waste generation, with global annual production reaching 2.01 billion tons (The World Bank 2018). Inadequate waste management poses significant environmental and public health risks, including soil contamination, groundwater pollution, and the proliferation of pathogens (e.g., mold, parasites). Conventional disposal methods such as landfills are increasingly unsustainable in densely populated regions due to land scarcity and greenhouse gas emissions. Consequently, waste-to-energy (WtE) technologies, particularly incineration, have emerged as critical solutions

Z. Zha (✉) · D. Holland
New York University, New York, NY, USA
e-mail: zz4036@nyu.edu

D. Holland
e-mail: dmh4@nyu.edu

Y. Li et al. (eds.), *16th International Conference on Environmental Science and Development*, Environmental Science and Engineering,
https://doi.org/10.1007/978-3-032-15080-6_27

for reducing waste volume while generating renewable energy. Chongqing City, a megacity in southwestern China has a mountainous terrain, limits landfill capacity, and rapid industrialization exacerbates pollution risks. Waste-to-energy (WtE) incineration has emerged as a critical solution, reducing waste volume while generating energy. However, incineration also emits pollutants such as SO_2, NO_x, particulate matters, and dioxins, raising concerns about air quality and public health.

Chongqing City, a megacity in southwestern China, faces unique challenges: mountainous terrain limits landfill capacity, and rapid industrialization exacerbates pollution risks. While WtE power plants mitigate landfill dependence and enhance energy security, their atmospheric impacts in complex terrains remain understudied. This study evaluates a waste incineration power plant in Chongqing, addressing three objectives: (1) simulating atmospheric dispersion of key pollutants (SO_2, NO_x, TSP, CO, HCl) using the AERSCREEN model; (2) assessing chronic health risks to nearby populations via the U.S. EPA's hazard index (HI); and (3) proposing policy implications for WtE projects in mountainous urban settings. The findings provide a framework for balancing waste management and air quality in densely populated regions with geographical constraints.

27.2 Method

27.2.1 Human Health Risks

To evaluate pollutant dispersion and health risks, this study employs the AERSCREEN online model developed by Shijiazhuang Huan'an Technology Co. Ltd., a platform optimized for compliance with China's air quality standards. This tool integrates terrain elevation data, meteorological extremes, and source parameters (e.g., stack height, emission rates) to simulate maximum ground-level concentrations of pollutants under conservative scenarios.

Non-carcinogenic health risks were assessed using the Hazard Index (HI), a U.S. EPA-endorsed metric for chronic exposure (Rovira et al. 2010; Goumenou and Tsatsakis 2019). The HI quantifies cumulative risks from multiple pollutants using the following equations:

$$HQ = \frac{C_i}{RfC_i} * \frac{IR * ET * EF * ED}{BW * AT * 365 * 24} \tag{27.1}$$

$$HI = \sum_{i}^{n} (HQ)i \tag{27.2}$$

In the formulas above, *HQ* (Hazard Quotient) is the health risks of inhaling one pollutant; *HI* (Hazard Index) is the cumulative health risks of all pollutant(s); *n* is the total number of pollutant(s); C_i is the average concentration of the chemical in environment (μg/m^3); RfC_i is the reference concentration (μg/m^3); IR is the inhalation rate (m^3/day), takes 20; ET is exposure time (h/day), takes 8; EF is exposure frequency (day/year), takes 320; ED is the exposure duration (year), takes 30; BW is the body weight (kg), takes 70; AT is the average exposure time (year), takes 25. The values taken are based on multiple relevant literatures (Du et al. 2014; Wang et al. 2009).

27.2.2 Modelling

This atmospheric simulation model also includes topographical conditions. The satellite map of the area around the power plant is shown in Fig. 27.1.

All five kinds of pollutant data from the pollution source and the related environmental parameters are shown in Tables 27.1 and 27.2, respectively. The smoke flow rate, smoke temperature, and each pollutant emission rate data take the daily average data of 214 days (March 1st, 2024 ~ September 30th, 2024) of "Historical data table of daily average values of waste gas emission outlets of Chongqing Qianjiang Sanfeng Environmental Protection Industry Development Co., Ltd. (domestic waste incineration power plant) in 2024" (March ~ September) published by Chongqing Qianjiang District Ecological Environment Bureau (Qianjiang District Ecological Environment Bureau 2024). Pollutant emission rate data is determined by daily emission data of each pollutant divided by 24 h. The environmental parameters are automatically given by the AERSCREEN model.

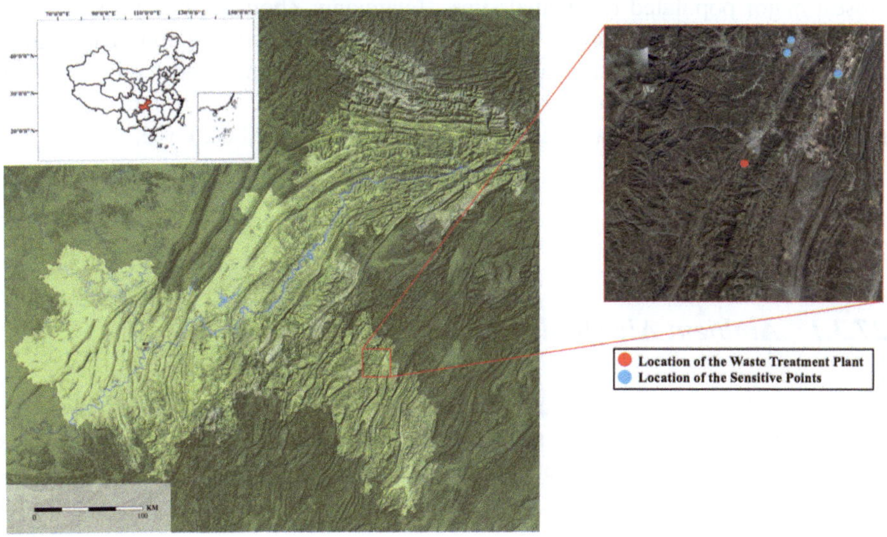

Fig. 27.1 The terrain of region around the pollution source

Table 27.1 The related data of the pollution source

Altitude (m)	Chimney height (m)	Chimney diameter (m)	Smoke flow rate (m/s)	Smoke temperature (K)	Pollutant emission rate (kg/h)				
					SO_2	NO_x	TSP	CO	HCl
631	110	1.18	16.06	427.01	3.65	15.35	0.33	1.05	2.08

Table 27.2 The related data of the environment

Min. temperature (°C)	Max. temperature (°C)	Min. wind speed (m/s)	Land use	Humidity	Land type	TIBL	Assessment region
− 6.5	39.5	0.5	Broadleaf forest	Humid	Rural	No	50 km × 50 km

Table 27.3 The information of 3 simulated sensitive points

	Altitude (m)	Longitude	Latitude	Distance to source (km)
Qianjiang	787	108.7621	29.5244	11.962
Zhoubai	555	108.8151	29.5081	13.268
Xishan	600	108.7661	29.5368	13.394

A preliminary geographical survey confirmed no significant human activities or habitation within a 10 km radius of the incineration plant. Therefore, the three closest major populated residential sites—Qianjiang, Zhoubai, and Xishan—were selected as simulated sensitive points to assess pollutant dispersion. These sites, located 11.962 km, 13.268 km, and 13.394 km from the plant respectively, as shown in Table 27.3, represent the nearest inhabited areas and were prioritized as critical receptors for evaluating potential environmental and public health risks.

27.3 Results

27.3.1 Ambient Air Quality Standard

The scope of the waste incineration power plant project's simulation involves the second-class ambient air functional area according to Chinese laws, so the pollutants in the ambient air of the assessment area shall comply with the second-class standard limit in the "Ambient Air Quality Standard" (GB 3095-2012) (Environmental Development Center 2016).

27.3.2 Simulation Analysis

The simulation results of the average concentration (C), the average proportion concentration (P), and the ambient air quality standard of five pollutants are shown in Table 27.4.

From the simulation results, it can be obtained that the average proportion concentration (P) of all five pollutants (SO_2, NO_x, TSP, CO, HCl) doesn't exceed the standard limit ($P < 100\%$), which complies with the national ambient air quality standard. The influence on the nearby residential sites is low during the operation of this waste-to-energy plant. The Hazard Quotient (HQ) of each pollutant at each site and their cumulative Hazard Index (HI) can be calculated using the simulation results, as shown in Table 27.5.

The RfC of each pollutant follows the primary standards (provide public health protection, including protecting the health of "sensitive" populations such as asthmatics, children, and the elderly) in the annual mean of the National Ambient Air Quality Standards (NAAQS) (Environmental Protection Agency 2024). The RfD (Reference Dose) for NO_2 and SO_2 is 0.053 ppm and 0.03 ppm, respectively (Hesami Arani et al. 2021), which is roughly 99.7 $\mu g/m^3$ for NO_2 and 78.6 $\mu g/m^3$ for SO_2 under 1 STP. The RfC for TSP is 75.0 $\mu g/m^3$. Although HCl is not regulated under the NAAQS, the "California Air Pollution Control Officers Association Air Toxics 'Hot Spots' Program, Revised 1992 Risk Assessment Guidelines" has provided a reference concentration for chronic inhalation (RfC) for HCl at 7.0 $\mu g/m^3$ (Hydrochloric

Table 27.4 Simulation results of five pollutants at three places

Pollutant	Sensitive point	Ambient air quality standard ($\mu g/m^3$)	C ($\mu g/m^3$)	P (%)	Compliance
SO_2	Qianjiang	500	5.6736	1.1347	Yes
	Zhoubai	500	4.7410	0.9482	Yes
	Xishan	500	4.9502	0.9900	Yes
NO_x	Qianjiang	250	23.8602	9.5441	Yes
	Zhoubai	250	19.9382	7.9753	Yes
	Xishan	250	20.8180	8.3272	Yes
TSP	Qianjiang	300	0.5130	0.1710	Yes
	Zhoubai	300	0.4286	0.1429	Yes
	Xishan	300	0.4476	0.1492	Yes
CO	Qianjiang	10,000	1.6321	0.0163	Yes
	Zhoubai	10,000	1.3638	0.0136	Yes
	Xishan	10,000	1.4240	0.0142	Yes
HCl	Qianjiang	50	3.2332	6.4664	Yes
	Zhoubai	50	2.7017	5.4034	Yes
	Xishan	50	2.8209	5.6418	Yes

Table 27.5 Hazard index (HI) at three places

Sensitive points	Pollutants	RfC ($\mu g/m^3$)	C ($\mu g/m^3$)	HQ	HI
Qianjiang	SO_2	78.6	5.6736	0.0072	0.0782
	NO_x	99.7	23.8602	0.0240	
	TSP	75	0.5130	0.0007	
	HCl	7	3.2332	0.0463	
Zhoubai	SO_2	78.6	4.7410	0.0060	0.0653
	NO_x	99.7	19.9382	0.0200	
	TSP	75	0.4286	0.0006	
	HCl	7	2.7017	0.0387	
Xishan	SO_2	78.6	4.9502	0.0063	0.0682
	NO_x	99.7	20.8180	0.0209	
	TSP	75	0.4476	0.0006	
	HCl	7	2.8209	0.0404	

acid is a federal hazardous air pollutant … n.d.). Pollutant CO lacks related standard (averaging time: 1 year) data, thereby being unable to calculate here.

The calculated cumulative Hazard Index (HI) values at the three sensitive sites—Qianjiang (HI = 0.0782), Zhoubai (HI = 0.0653), and Xishan (HI = 0.0682)—are significantly below the U.S. EPA's acceptable risk threshold (HI < 1). These results indicate that the combined chronic health risks from exposure to multiple pollutants (SO_2, NO_x, TSP, and HCl) emitted by the waste incineration plant are negligible for nearby residents, including vulnerable populations such as children, the elderly, and individuals with pre-existing respiratory conditions. The low HI values reflect the effectiveness of emission control measures in minimizing pollutant concentrations, as well as the plant's compliance with air quality standards. Notably, HCl, despite contributing the highest individual Hazard Quotient (HQ = 0.04–0.046) due to its lower Reference Concentration (RfC = 7.0 $\mu g/m^3$), remains well within safe limits. The exclusion of CO from the HI calculation, owing to a lack of annual RfC data, introduces a minor uncertainty; however, given its minimal simulated concentrations (~ 1.4–1.6 $\mu g/m^3$) and the absence of evidence suggesting significant synergistic effects with other pollutants, its omission is unlikely to alter the overall conclusion. These findings align with global studies emphasizing that modern waste-to-energy facilities, when properly regulated and equipped with advanced pollution control technologies, pose minimal chronic health risks. For example, a 2024 study shows that waste incineration plants with advanced disposal technologies (such as ultra-low emission technology and various kinds of scrubber) and scientific classification methods can significantly reduce air pollutants emission (Wei et al. 2024).

27.4 Discussion

The U.S. Environmental Protection Agency (EPA) provides many numerical models of atmospheric pollutant diffusion, most of which are based on the Gaussian plume model, including the AERSCREEN model. The Gaussian plume model mainly requires: (1) the distribution of pollutant concentration on the Y and Z axes conforms to the normal distribution; (2) the wind speed is uniform and stable in the entire space; (3) the pollutant emission is continuous and uniform; (4) the mass of pollutants is conserved during the diffusion process, that is, the pollutants do not settle, decompose or combine, and the ground reflects them completely without absorption or adsorption. The Gaussian model is the most widely used diffusion model because of its intuitive physical meaning, simple mathematical expression, convenient relationship between physical quantities and mathematical deduction, ease of mastering and calculating, and has been verified in a large number of small-scale diffusion experiments, especially in experiments on plains and flat grasslands. However, the Gaussian model is difficult to adapt to the changes in wind fields and cannot handle local circulation caused by terrain. It does not consider the removal of pollutants by chemical oxidation and dry and wet deposition. When the Gaussian model is used to predict the air quality in complex terrain, it often brings large errors. In order to solve the atmospheric transport and diffusion problems under complex terrain conditions and improve the application effect of the Gaussian model on complex terrain, many derived models based on the Gaussian model have emerged.

In March 2011, The U.S. Environmental Protection Agency (EPA) officially released AERSCREEN as a new generation of estimation model, which replaced SCREEN3 as the new estimation model for U.S. air quality. The AERSCREEN model is a valuable tool for preliminary air quality assessments and environmental planning, helping researchers identify regions at risk of exceeding air quality standards. It coupled related kernels of the AERMOD model (AERMOD, AERMAP, BPIPPRM), which can calculate the maximum ground concentration of point sources, flare sources, surface sources, and volume sources, as well as the maximum ground concentration of pollutants under special conditions such as building downwash and shore fumigation (TIBL), which makes a significant improvement compared with SCREEN3 model. The estimation model embeds a variety of meteorological combination conditions, and it can also conservatively (compared with AERMOD) estimate the maximum impact of a pollution source on ambient air quality without inputting meteorological conditions, giving the maximum ground concentration at different downwind distances (Xin et al. 2011).

The pollution source in the study is located in hilly terrain. Under such complex terrain conditions, the degree of influence on the diffusion of pollutants is closely related to the surrounding environmental condition. One of the important factors is the wind. Overall, places in the downwind direction of the pollution source usually have higher pollutant concentration, and higher wind speed is generally conducive to most pollutants' dispersion. The wind rose plot at the coordinates of the pollution source in 2023 (measured by hour) is shown in Fig. 27.2. The wind direction frequency data comes from the European Centre for Medium-Range Weather Forecasts (ECMWF). As shown in the picture, influenced by the East Asian Monsoon, which is shown as the red arrows in Fig. 27.3 (Ye et al. 2022), the frequency of wind mainly comes from ESE and SE directions, which is 21% and 16%, respectively. Consequently, the wind direction most of the time is perpendicular to the relative direction (NNE) of 3 sensitive points selected above, indicating that the wind at the pollution source doesn't contribute to the dispersion of emitted pollutants towards the direction of several nearby residential sites. This deduction also aligns with the simulation results that the average proportion concentration (P) of all five pollutants is far lower than the standard limit.

Generally, under complex terrain conditions, the diffusion of pollutants mainly changes along the valley. When encountering high mountains, pollutants diffuse along the mountain walls to the depths of the valley. At the same time, the concentration contour line distribution diagram shows a curved diffusion trend in the area affected by obstacles, while in the plains, because there are no obstacles, the concentration contour line distribution diagram shows a circular arc line and diffuses evenly. In the study, the maximum concentration of all pollutants occurs at the place 875 m downwind from the pollution source, where there's a mountain range presence, which aligns with the general pattern.

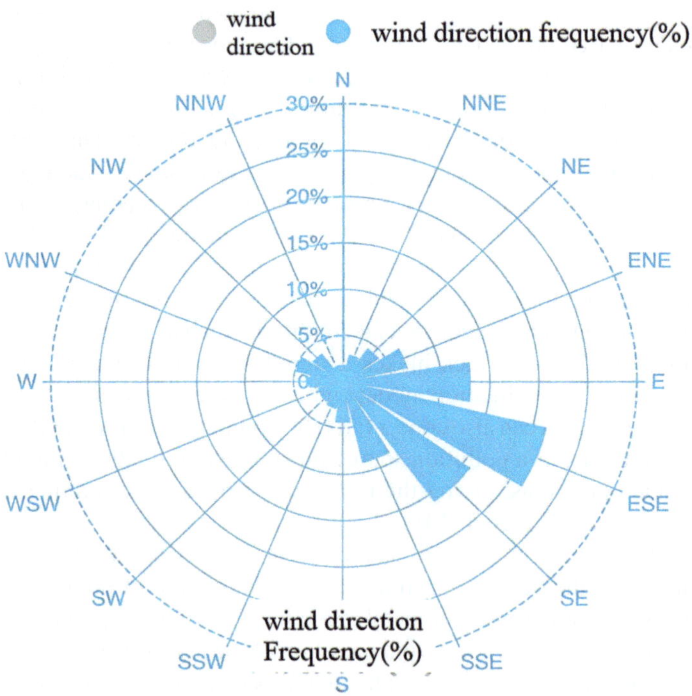

Fig. 27.2 The wind rose plot of the pollution source in 2023

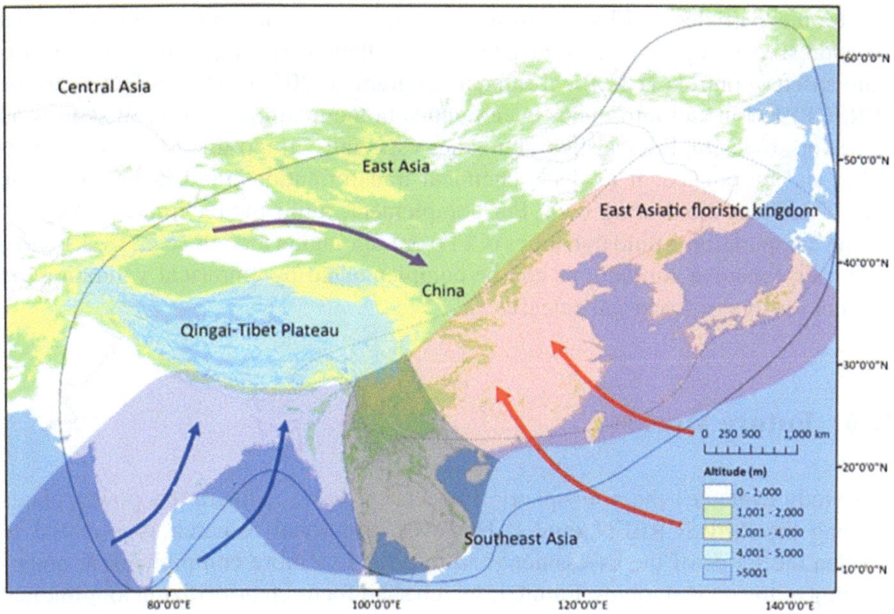

Fig. 27.3 The East Asian Monsoon (red arrows)

The findings of this study also align with broader discussions on the environmental and health impacts of waste-to-energy (WtE) technologies, which are increasingly adopted as sustainable solutions for municipal solid waste management. While incineration reduces landfill dependency and generates energy, concerns persist about air pollutant emissions, particularly in densely populated regions. In general, waste incineration contributes to resource recovery but must coexist with pollution prevention. Public acceptance of WtE facilities, however, hinges on transparent risk communication and participatory governance, ensuring communities trust scientific assurances of safety.

27.5 Limitations

Several limitations and uncertainties exist in the assessment of environmental and health risks associated with pollutant dispersion from the waste incineration power plant in the study due to various reasons. Like most studies that evaluate the health impacts caused by air pollution, the results of all such studies should be interpreted with caution.

The absence of carcinogenic pollutant data such as dioxins and furans precluded a Lifetime Cancer Risk (LCR) assessment, potentially underestimating long-term health risks for residents (Buekens and Huang 1998). Emerging pollutants, such as ultrafine particles (UFPs) or dioxins—common in incinerator emissions but not addressed here—also warrant attention, as their toxicity profiles may necessitate stricter controls (UN Environment Programme 2023). Also, reliance on the AERSCREEN model introduces uncertainties in mountainous terrain: its Gaussian framework oversimplifies airflow dynamics especially in terrains like valley channeling and turbulence. It neglects chemical transformations, and lacks validation against ambient monitoring data. For complement, multi-model frameworks integrating computational fluid dynamics (CFD) or Lagrangian models (e.g., CALPUFF) for terrain-sensitive simulations can be complemented by empirical validation via ambient sensors or tracer gas releases.

27.6 Future Works

This study evaluates the atmospheric impacts of waste incineration power plants utilizing the AERSCREEN model. However, additional research is essential to broaden the scope of the assessment and to provide a more comprehensive understanding of the related topics. Future research shall focus on prioritizing longitudinal health studies, expanding pollutant inventories and optimizing modelling to address the insufficiencies of current study.

Future explorations shall integrate carcinogenic into health risk assessments if related pollutant data are available. It would fill the current vacancy in Lifetime Cancer Risk (LCR) evaluation, which is important particularly for vulnerable populations such as the elderly people and children. Also, expanding pollutant inventories to include background emissions such as regional industries and traffic would clarify cumulative exposure risks and synergistic effects, which further enhancing the accuracy of environmental assessments for waste-to-energy plants in geographically complex urban areas. Refining dispersion modeling in mountainous terrain requires multi-model frameworks to account for valley channeling and pollutant deposition, complemented by ambient monitoring data for validation.

27.7 Conclusions

The study uses the AERSCREEN online model calculation platform developed by Shijiazhuang Huan'an Technology Co. Ltd. for estimation, simulates the impact of a waste incineration power plant project in Chongqing City, China on the surrounding environment and the health risks to residents in several nearby sensitive points.

The results of model simulation turn out that the simulated average proportion concentration (P) of all five kinds of pollutants: sulfur dioxide (SO_2), nitrogen oxides (NO_x), total suspended particulate matter (TSP), carbon monoxide (CO), and hydrogen chloride (HCl) in three nearby sensitive residential sites (Qianjiang, Zhoubai, Xishan) are all less than 100%, which complied with China's Class II air quality standards. The Hazard Index (HI) at three sensitive points around the project is far less than the acceptable level (HI < 1) introduced by the U.S. EPA, representing that the pollutants discharged by the power plant basically will not cause chronic health risks to residents living in the surrounding sensitive points. Dominant wind patterns (ESE/SE), influenced by the East Asian Monsoon, and mountainous terrain further limited pollutant dispersion toward residential areas, aligning with simulated low exposure levels. These findings suggest that modern waste-to-energy facilities, when properly equipped with advanced emission control technologies and strategically sited with consideration of local meteorological and geographical factors, can effectively operate within established environmental and health protection frameworks.

While these results are specific to the studied facility in Chongqing's unique geographical context, the methodology and assessment framework developed in this study may be cautiously applied to similar waste-to-energy projects in other mountainous urban areas, provided that site-specific adjustments are made for local emission profiles, terrain characteristics, and meteorological conditions. The study demonstrates the importance of comprehensive environmental impact assessments that integrate dispersion modeling with health risk evaluation, offering a valuable approach for supporting sustainable urban waste management strategies without compromising public health.

Data Availability Simulation results and wind direction frequency data: Accessibility restricted due to privacy concerns. Access requests can be submitted to the corresponding author at zz4036@nyu.edu, subject to a data use agreement. The online modelling simulation platform developed by Shijiazhuang Huan'an Technology Co. Ltd. is available at: https://www.ihamodel.com.

Declaration of Competing Interest The authors declare that they have no known competing financial interests or personal relationships that could have appeared to influence the work reported in this paper.

References

Buekens A, Huang H (1998) Comparative evaluation of techniques for controlling the formation and emission of chlorinated dioxins/furans in municipal waste incineration. J Hazard Mater 62(1):1–33. https://doi.org/10.1016/s0304-3894(98)00153-8

Du Z, Mo J, Zhang Y (2014) Risk assessment of population inhalation exposure to volatile organic compounds and carbonyls in urban China. Environ Int 73:33–45. https://doi.org/10.1016/j.envint.2014.06.014

Environmental Development Center (trans) (2016) GB 3095-2012. Ambient air quality standards, 1 Jan 2016. https://english.mee.gov.cn/Resources/standards/Air_Environment/quality_standard1/201605/t20160511_337502.shtml

Environmental Protection Agency (2024) EPA, 16 Dec 2024. https://www.epa.gov/criteria-air-pollutants/naaqs-table

Goumenou M, Tsatsakis A (2019) Proposing new approaches for the risk characterisation of single chemicals and chemical mixtures: the source related hazard quotient (HQS) and hazard index (HIS) and the adversity specific hazard index (HIA). Toxicol Rep 6:632–636. https://doi.org/10.1016/j.toxrep.2019.06.010

Hesami Arani M, Jaafarzadeh N, Moslemzadeh M, Rezvani Ghalhari M, Bagheri Arani S, Mohammadzadeh M (2021) Dispersion of NO_2 and SO_2 pollutants in the rolling industry with AERMOD model: a case study to assess human health risk. J Environ Health Sci Eng 19(2):1287–1298. https://doi.org/10.1007/s40201-021-00686-x

Hydrochloric acid is a federal hazardous air pollutant … (n.d.). https://ww2.arb.ca.gov/sites/default/files/classic//toxics/tac/factshts1997/hydchlor.pdf

Qianjiang District Ecological Environment Bureau (2024) Historical data table of daily average values of waste gas emission outlets of Chongqing Qianjiang Sanfeng Environmental Protection Industry Development Co., Ltd. (domestic waste incineration power plant) in March–September 2024. Qianjiang District People's Government of Chongqing. https://www.qianjiang.gov.cn/bmjd/xzfgzbm/qsthjj_49066/zwgk_49068/gkml/hjgl/hjjc/

Rovira J, Mari M, Nadal M, Schuhmacher M, Domingo JL (2010) Environmental monitoring of metals, PCDD/Fs and PCBS as a complementary tool of biological surveillance to assess human health risks. Chemosphere 80(10):1183–1189. https://doi.org/10.1016/j.chemosphere.2010.06.016

The World Bank (2018) What a waste 2.0. In: Trends in solid waste management. https://datatopics.worldbank.org/what-a-waste/trends_in_solid_waste_management.html

UN Environment Programme (2023) Annual report 2022, 6 Feb 2023. UNEP. https://www.unep.org/resources/annual-report-2022

Wang Z, Duan X, Liu P, Nie J, Huang N, Zhang J (2009) Human exposure factors of Chinese people in environmental health risk assessment. Res Environ Sci 22(10):1164–1170

Wei J, Zhang L, Li H (2024) Pollutants control strategies for modern waste incinerators: status quo in China in the context of municipal solid waste classification. In: Solid waste management for resource-efficient systems, pp 205–231. https://doi.org/10.1016/b978-0-443-23775-1.00010-2

Xin B, YinYin F, Meng L, Feng D (2011) New prediction method for atmospheric environmental impact based on AERSCREEN. Procedia Environ Sci 10:1308–1314. https://doi.org/10.1016/j.proenv.2011.09.209

Ye J-W, Tian B, Li D-Z (2022) Monsoon intensification in East Asia triggered the evolution of its flora. Front Plant Sci 13. https://doi.org/10.3389/fpls.2022.1046538

Chapter 28
Development and Performance Study of Composite Dust Suppressant Based on Polymer Blending for Industrial Solid Waste Landfill

Lisong Zhu, Yu Wang, Fujun He, and Yang Meng

Abstract Industrial solid waste landfills are prone to generate dust, which adversely affects the health of operators and the environment. The objective of this study was to develop an environmentally friendly composite dust suppressant for landfill dust. Sodium dodecyl benzene sulfonate (SDBS) was used as a surfactant with excellent performance, glycerol as a water-retaining agent, and carboxymethyl cellulose (CMC) as a binder to prepare an environmentally friendly composite dust suppressant (EFCDS) based on polymer blending with the effects of wettability, bonding, wind erosion resistance, degradability, and freeze–thaw resistance. The optimal formulation of the composite dust suppressant was determined using response surface methodology. The test results show that the optimal mixing concentration of the composite dust suppressant was 0.538% SDBS, 0.448% glycerol, and 0.359% CMC. The composite dust suppressant is non-toxic, low metal corrosive, degradable and its main raw materials (SDBS and CMC) are from nature. The above results show that the composite dust suppressant can effectively and long-term inhibit the dust in the landfill, and it is a kind of high-efficiency dust suppressant material in line with the concept of environmental protection.

Keywords General industrial solid waste landfill · Dust suppressant · Orthogonal experiment · Performance testing

28.1 Introduction

General industrial solid waste landfills face dust pollution during disposal (Buelvas et al. 2025). Dust from transport, loading/unloading, and stockpiling is widespread, concentrated, and has diverse sources. It degrades air quality, visibility, and endangers

L. Zhu (✉) · Y. Wang · F. He · Y. Meng
Shaanxi Deqi Environmental Protection Co., Yulin, China
e-mail: 6306217@qq.com

Y. Li et al. (eds.), *16th International Conference on Environmental Science and Development*, Environmental Science and Engineering,
https://doi.org/10.1007/978-3-032-15080-6_28

human health, especially the respiratory system. It may contain harmful substances like formaldehyde and accelerates equipment wear. In open environments, the unimpeded spread of fine dust amplifies both public health risks and environmental contamination (Mohammad et al. 2023). So, effective dust control is crucial to protect surrounding ecosystems, soil fertility, and air quality.

Current dust control mainly uses water spraying and chemical suppressants. Water spraying consumes much water and evaporates quickly, being ineffective in arid/semi-arid areas (Koda et al. 2023). Chemical suppressants can cluster dust or form crusts, but have limitations such as high cost, toxicity, corrosiveness, and poor biodegradability, restricting their use (Mahandra et al. 2021; Omane et al. 2018). Thus, developing biodegradable, non-secondary-polluting, effective, and low-cost dust suppression techniques is urgent.

Dust suppressants are trending towards composite, multi-functional, polymer, and eco-friendly types (Wang et al. 2024a; Wei et al. 2021; Zhang et al. 2022; Zhou et al. 2018). Composite suppressants combine components to achieve wetting, bonding, water absorption, etc., enhancing dust suppression (Ha et al. 2024; Yu et al. 2025a). They often use natural polymer materials, which are biodegradable and non-toxic (Wang et al. 2024b; Wu et al. 2020; Zhou et al. 2023; Yu et al. 2025b).

In this study, carboxymethyl cellulose (CMC), a biodegradable and non-toxic natural polymer, was chosen as the main raw material for dust suppressants. Sodium dodecyl benzene sulfonate (SDBS) and glycerol were used as auxiliary raw materials. SDBS, a common anionic surfactant, has wetting and other properties and is biodegradable. Glycerol, a natural, eco-friendly, and biodegradable material, is abundant in fats and oils (Li et al. 2024).

The study formulated an environmentally friendly composite dust suppressant (EFCDS) using SDBS as surfactant, glycerol as water-retaining agent, and CMC as binder. EFCDS has bonding, water-retaining, and wetting properties. It penetrates landfill particulate matter, retains moisture, forms an adhesive film to consolidate waste, and seal dust sources. With most raw materials being eco-friendly, EFCDS shows great potential for cleaner production.

28.2 Materials

The dust soil samples used in this paper were taken from the Jinjie landfill in Shaanxi, China. Samples were taken from the surface layer of the dust heap in five representative areas of the landfill in accordance with the manual sampling tubes specified in the national standard (GB/T.12573 2008). The collected dust samples were put in plastic bags, numbered and recorded with information such as sampling date, sampling method, sampling location, ambient temperature, and humidity. Further, dust samples collected from the site are first sieved using an 80-mesh standard dividing sieve to remove debris. They are then dried in an 80 °C blast drying oven until their weight stabilizes. Finally, the dried samples are stored in a silica gel desiccator for use as test dust samples (see Fig. 28.1).

Fig. 28.1 Dust sample specimen after sieving

The dust sample is a composite structural mixture of crystals, vitreous, and a small amount of unburnt charcoal, and the scanning electron microscope (SEM) (Razmjooei and O'Regan 2024) test results are shown in Fig. 28.2. The microstructure of the dust sample is mainly composed of spherical particles with smooth surfaces and porous particles with loose and irregular structures. These smooth-surface spherical particles, with particle sizes between 10 and 100 μm, are glass beads. The formation of glass beads occurs in a boiler at high temperatures, approximately 1400 °C. Within the combustion chamber, ash melts due to the intense heat. Upon exiting the boiler and entering the flue, the molten ash undergoes rapid cooling. This sudden temperature change, coupled with surface tension, causes the ash particles to contract and form spherical glass beads. In the dust sample, the presence of irregular particles is notably higher. These irregular granular materials primarily consist of hard gypsum and unburned charcoal particles. Their particle size distributions vary, and the surfaces of the dust particles are characterized by roughness. Additionally, the dust particles exhibit loose packing and low cohesion.

The sodium dodecyl benzene sulfonate (SDBS) and glycerol were obtained from Sinopharm Chemical Reagent Co., Ltd. (Shanghai, China). The carboxymethyl cellulose (CMC) was obtained from Shanghai Aladdin Biochemical Technology Co., Ltd. (Shanghai, China). All the chemicals were analytical reagents. EFCDS is an aqueous solution of SDBS, glycerol, CMC, and water mixed homogeneously in specific proportions at room temperature. The reactor used for preparing the EFCDS is constructed from stainless steel or plastic.

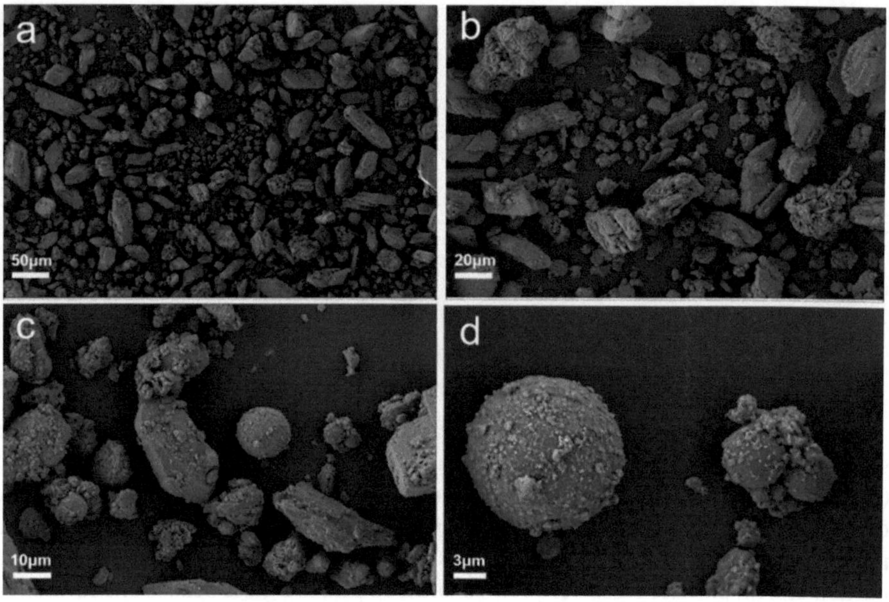

Fig. 28.2 Scanning electron microscope of dust samples

28.3 Experiments

28.3.1 Experimental Flowchart

First, the optimal mixing ratio of SDBS, glycerol, and CMC was determined using response surface methodology (RSM) (Al Qudah et al. 2025). The surface tension, penetration rate, water retention, and adhesion were selected as response variables for the RSM analysis. Subsequently, the dust suppression efficiency and performance of the EFCDS were assessed. The detailed experimental step for this study is depicted in Fig. 28.3.

28.3.2 Experimental Steps

In this study, SDBS, glycerol, and CMC were utilized as surfactants, water-retaining agents, and binders, respectively, for the EFCDS. SDBS enhances the wetting of dust particles by diminishing the surface tension of water, which in turn augments the penetration depth of the dust suppressant into the dust matrix. Glycerol serves to mitigate water evaporation from the dust, thereby maintaining the dust in a moist state for an extended period. CMC, on the other hand, elevates the viscosity of the EFCDS, facilitating the formation of a cohesive layer that binds the dust particles together.

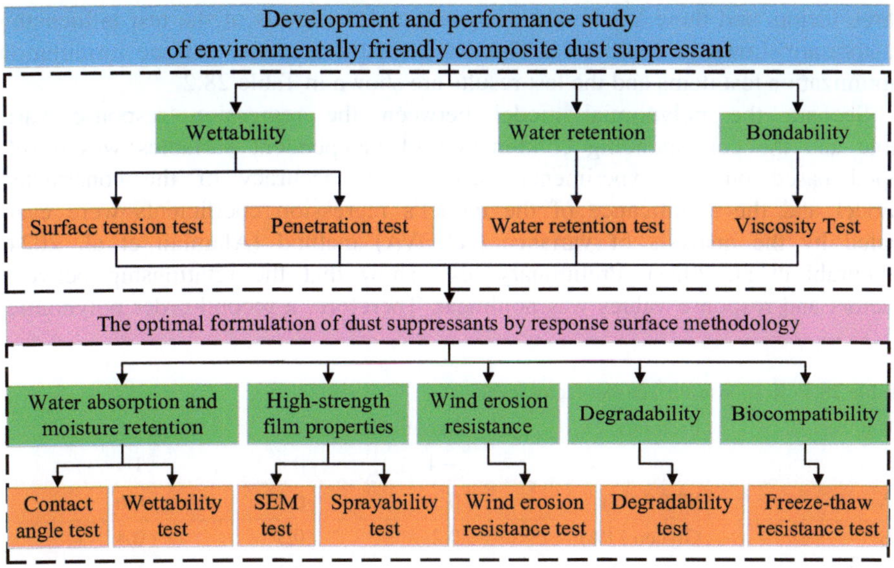

Fig. 28.3 Experimental flowchart

Collectively, these three components enable the EFCDS to achieve simultaneous bonding (via CMC), water retention (via glycerol), and wetting (via SDBS).

To ensure that all the effects of EFCDS meet the requirements for general industrial solid waste landfill dust applications, it is necessary to determine the optimum formula of the three ingredients of EFCDS. The RSM refers to a set of experimental designs based on the theory of experimental design. It is used to explore the relationship between influencing factors and the response variable within a specified experimental space. RSM aims to achieve high modeling accuracy while minimizing the number of experimental runs. By doing so, it can derive a mathematical model of the objective function that predicts the response values at non-sampled points. This methodology is a combination of mathematical modeling and statistical analysis, and it is commonly employed for process optimization or the refinement of formulation ratios. In this study, a three-level, three-factor RSM was used to determine the optimum formula of EFCDS. The pH value, water retention, and viscosity were selected as the three responses. Data processing for the RSM was done by Design-Expert 11.0 software. The RSM analysis was conducted in a three-step process as follows.

First, the experimental design was conducted using the Box-Behnken design (BBD), a type of RSM, to optimize the experimental conditions. In this study, Design-Expert 11 software was used to design a response surface test to analyze the effects of three factors, SDBS mass fraction (A), glycerol mass fraction (B), and CMC mass fraction (C), on the pH (Y_1), water retention (Y_2), and viscosity (Y_3) of the dust suppressant, and − 1, 0, and 1 were used as the high, medium, and low levels of the three factors, respectively. The response surface synthesis optimization test with three factors and three levels was designed, and the levels of the test influencing factors are shown in Table 28.1. The design of the response surface formulation optimization test items and the test results are shown in Table 28.2.

Second, the polynomial model between the regression response variable and the corresponding combination of independent variables was developed based on the experimental results. The accuracy of the constructed model and the significance of the model's regression coefficients were evaluated by the analysis of variance (ANOVA) method (Alamian et al. 2024; Shourabi et al. 2024). Preliminary tests show that the relationship between factors and response values was nonlinear. Therefore, a second-order polynomial

Table 28.1 Response surface test factor levels

Number	Factors	Variable level		
		− 1	0	1
A	SDBS (%)	0.3	0.45	0.6
B	Glycerol (%)	0.2	0.3	0.4
C	CMC (%)	0.2	0.3	0.4

Table 28.2 Response surface formulation optimization test project design and results

Number	Variables			Responses		
	A	B	C	Y_1	Y_2	Y_3
	SDBS (%)	Glycerol (%)	CMC (%)	pH value	Water retention (%)	Viscosity (Pa s)
1	0.3	0.2	0.3	10.5	0.345	14.73
2	0.6	0.2	0.3	10.09	0.323	11.29
3	0.3	0.4	0.3	10.01	0.412	13.88
4	0.6	0.4	0.3	10.03	0.414	11.97
5	0.3	0.3	0.2	10.06	0.375	9.00
6	0.6	0.3	0.2	10.12	0.369	6.84
7	0.3	0.3	0.4	10.02	0.379	26.32
8	0.6	0.3	0.4	10.1	0.381	20.46
9	0.45	0.2	0.2	10.1	0.285	7.73
10	0.45	0.4	0.2	10.03	0.405	8.83
11	0.45	0.2	0.4	10.1	0.34	20.08
12	0.45	0.4	0.4	10.03	0.42	19.87
13	0.45	0.3	0.3	10.05	0.442	12.44
14	0.45	0.3	0.3	10.07	0.449	13.45
15	0.45	0.3	0.3	10.1	0.453	13.41
16	0.45	0.3	0.3	10.08	0.455	13.13
17	0.45	0.3	0.3	10.05	0.44	13.03

equation was used in the RSM analysis in the conventional form of Eq. (28.1).

$$y = b_0 + \sum_{i=1}^{k} b_i x_i + \sum_{i=1}^{k} b_{ii} x_i^2 + \sum_{i=1}^{k-1} \sum_{j=1}^{k} b_{ii} x_i x_j + \varepsilon \tag{28.1}$$

where y is the joint each factor level combination measurement of the impact value, k is the number of variables, b_0 is a constant term, b_i represents the linear impact coefficient, b_{ij} is the linear and linear interaction impact coefficients, x_i represents the independent variables, and ε is the residuals of the constructed model.

The significance of the fitted model and its constituent items was evaluated using ANOVA and the Fisher test (F-test), with the F-value and p-value serving as specific indicators of this assessment (Caporale and Gil-Alana 2019; Liew et al. 2009). In addition, some other metrics were used to analyze the experimental data with descriptive statistics such as sum of squares (SS), degrees of freedom (DF), squared variance (R^2), adjusted coefficient of determination (R^2 adj.), coefficient of variation (c.v.) and adequate precision (AP) (Pouladi et al. 2019).

Finally, the optimum formula of EFCDS was determined based on application requirements, such as pH, water retention, and viscosity. Under the optimized conditions obtained, the reliability and predictive ability of the constructed model is further evaluated by experimentally comparing the error between the predicted value of the model and the experimental mean value.

28.4 Results and Discussion

28.4.1 Response Surface Methodology Analysis

Table 28.2 shows the pH value, water retention, and viscosity of 17 groups of EFCDS with different dust suppression solution concentrations.

Based on the experimental results shown in Table 28.2, the second-order polynomial regression model were developed to describe the relationship between the mass fractions of the three ingredients and the corresponding response variables: pH value (Y_1), water retention (Y_2), and viscosity (Y_3). These models are represented by Eqs. (28.2)–(28.3). A, B, and C denoted the coded numbers of SDBS, glycerol, and CMC, respectively.

$$\begin{aligned} Y_2 = {} & 0.4478 - 0.003A + 0.04475B + 0.01075C + 0.006AB + 0.002AC \\ & - 0.01BC - 0.0304A^2 - 0.0439B^2 - 0.0414C^2 \end{aligned} \tag{28.2}$$

$$\begin{aligned} Y_3 = {} & 20.684 - 1.54625A + 0.09B + 6.66625C + 0.3825AB - 0.675AC \\ & - 0.3275BC - 3.2195A^2 - 4.497B^2 - 2.0595C^2 \end{aligned} \tag{28.3}$$

As can be seen from Table 28.2, the pH of the dust suppressant solution tends to be stable and does not fluctuate significantly with changes in the concentration of each component solution.

28.4.2 Response Surface Result Analysis

The shape of the contour lines and the density of the curves were used to judge the magnitude of the interaction effect of the factors on the response surface. The elliptical shape of the contour lines and the high density of the curves reflect a strong interaction between the two factors, while a round shape and a gentle curve reflect a weak interaction. The contour lines and response surfaces for the effects of each factor on surface drying time are shown.

The contours and response surfaces of the effect of each factor on water retention are shown in Fig. 28.4. Analyzing Fig. 28.4, it can be seen that the contour plots of $Y_2 = f(A, B)$, $Y_2 = f(A, C)$, and $Y_2 = f(B, C)$ are elliptical, indicating that the effects of A and B, A and C, and B and C act significantly. From the $Y_2 = f(A, B)$ response surface plot, when the mass fraction of SDBS is certain, with the increase of glycerol mass fraction, the water retention rate of the dust suppressant gradually increases, and the growth gradually becomes smaller. From the $Y_2 = f(A, C)$ response surface plot, we can get that, when the mass fraction of SDBS is kept certain, the increase in the mass fraction of CMC makes the water retention rate show the first increase and then decrease. The $Y_3 = f(B, C)$ response surface plot shows that when the mass fraction of CMC is certain, the water retention rate of the dust suppressant gradually becomes larger with the increase of the mass fraction of glycerol.

The contour and response surface of the effect of each factor on viscosity is shown in Fig. 28.5. Analyzing Fig. 28.5, it can be seen that the contour plots of $Y_3 = f(A, B)$ and $Y_3 = f(A, C)$ are circular with a gentle curve, indicating that the interaction of A and B, A and C on the viscosity of dust suppressant is not significant. The contour plot of $Y_3 = f(B, C)$ is elliptical, indicating that the interaction of B and C on the viscosity of the dust suppressant is significant. From Fig. 28.5d, it can be seen that when the CMC mass fraction is 0.3%, the dust suppressant viscosity increases and then decreases with the increase of glycerol mass fraction. When the glycerol amount fraction is 0.4%, the dust suppressant viscosity increases with the increase of CMC mass fraction.

28.4.3 Optimization

The optimum ratios of SDBS, glycerol, and CMC mass fractions of the dust suppressant were obtained by response surface optimization as 0.538%, 0.448%, and 0.359%, respectively. The theoretical values of water retention and viscosity of the dust

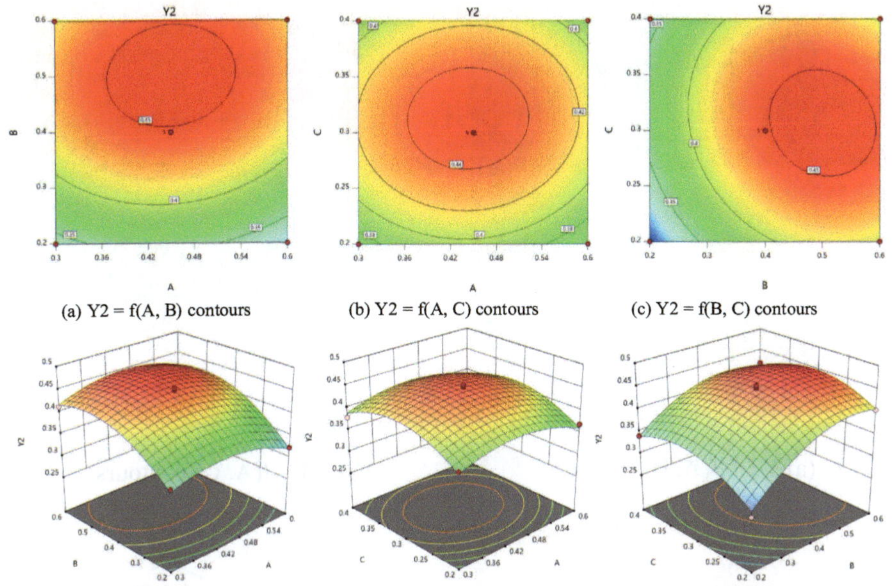

(a) Y2 = f(A, B) contours (b) Y2 = f(A, C) contours (c) Y2 = f(B, C) contours

(d) Y2 = f(A, B) response surface (e) Y2 = f(A, C) response surface (f) Y2 = f(B, C) response surface

Fig. 28.4 Contours and response surfaces of the effect of each factor on water retention rate

suppressant under the optimum conditions were 0.436% and 21.386 Pa s, respectively. Taking into consideration the convenience of the operation, the optimum ratios of SDBS, glycerol and CMC mass fractions were 0.54%, 0.45%, and 0.36%, respectively. Considering the convenience of practical operation, the optimal ratios of SDBS, glycerol, and CMC mass fractions were 0.54%, 0.45% and 0.36%, respectively. In order to test the accuracy and reliability of the model, five parallel tests were carried out under the conditions (see Table 28.3), and the actual verified water retention rate and viscosity of the dust suppressant were 0.55% and 21.62 Pa s, with an error of 2.2 and 1.08%, which was basically the same as that of the predicted results of the regression model. The results are basically consistent with the regression model prediction, indicating that the optimal parameters obtained by BBD response surface have certain accuracy, reliability, and practicality, which can effectively guide the design of dust suppressant ratio.

28.4.4 Environmentally Friendly Composite Dust Suppressant Preparation

Figure 28.6 is a schematic diagram of the preparation of the composite dust suppressant. First of all, the wetting agent, binder, water retaining agent monomer according to a certain group distribution ratio dissolved in water, and then placed in a water

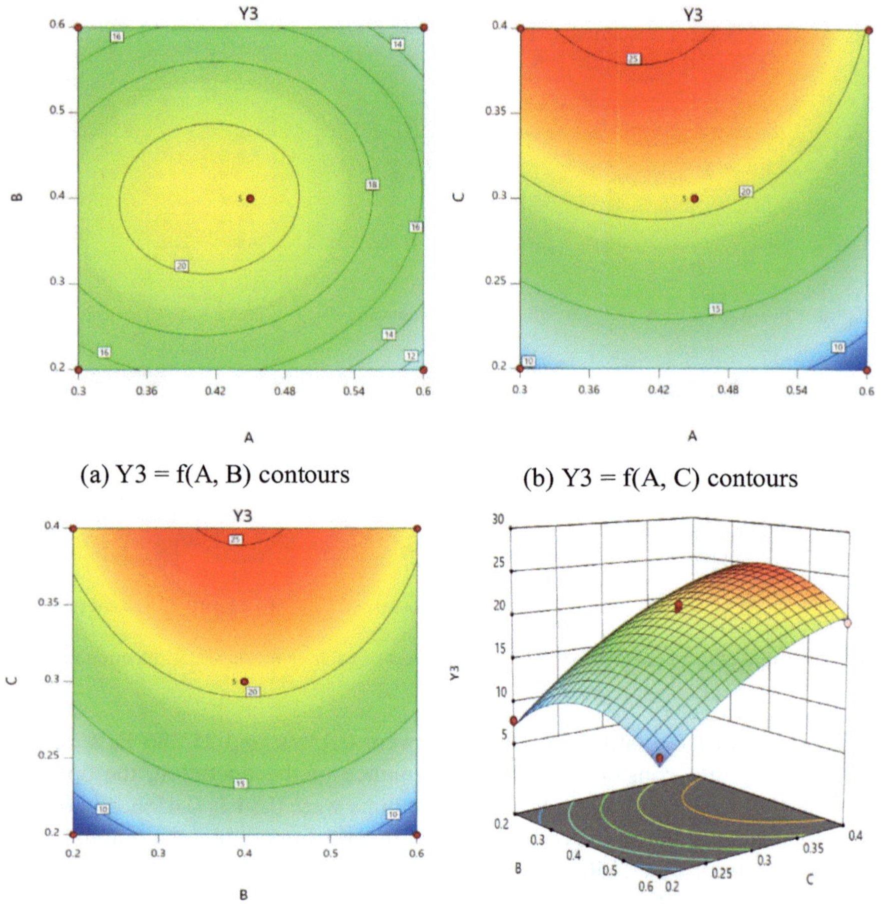

(a) Y3 = f(A, B) contours

(b) Y3 = f(A, C) contours

(c) Y3 = f(B, C) contours

(d) Y3 = f(B, C) response surface

Fig. 28.5 Contours and response surfaces of the effect of each factor on viscosity

Table 28.3 Test results of water retention rate and viscosity of the best formulated dust suppressant

	1	2	3	4	5	Average value
Water retention (%)	0.57	0.55	0.56	0.52	0.55	0.55
Viscosity (Pa s)	22.11	21.78	21.37	21.19	21.65	21.62

bath stirring, to obtain a highly concentrated composite dust suppressant solution, and then after drying, grinding treatment, to obtain the dust suppressant powder; dust suppressant powder dissolved in water after the dust suppressant solution obtained after the dust samples can be effectively inhibited by spraying on the dust.

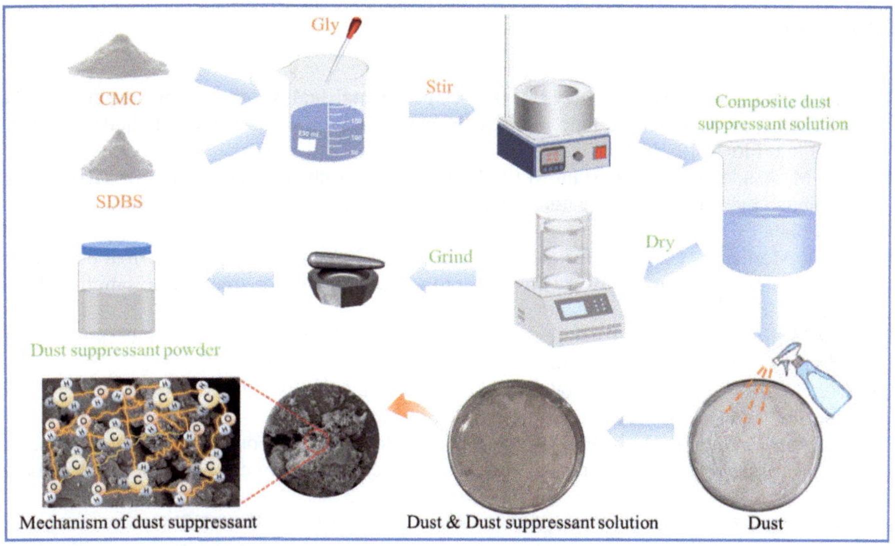

Fig. 28.6 Schematic diagram of the preparation of composite dust suppressant

28.5 Conclusion

In this study, an EFCDS suitable for general industrial solid waste landfills was developed and the property of EFCDS was evaluated. The following conclusions were drawn:

(1) The RSM models for pH, water retention and viscosity of EFCDS were developed with high significance, accuracy and reliability. The optimum formula of EFCDS was determined as 0.54% SDBS, 0.45% glycerol, and 0.36% CMC.
(2) EFCDS, composed of SDBS, glycerin, and CMC, is made from environmentally friendly materials that promote the efficient use of natural resources and support sustainable development goals. Therefore, EFCDS is an environmentally friendly dust suppression material with a good dust suppression effect.

Acknowledgements The authors acknowledge the financial support provided by the 2023 Shaanxi Provincial State-owned Capital Operation Budget Special Funds for Science and Technology Innovation Project and the Key Research and Development Program of Shaanxi Province of China (Program No. 2025CY-YBXM-602).

References

Al Qudah S, Radwan A, El-Sharkawy II (2025) Experimental analysis of heat transfer in a circular tube fitted with new twisted tape insert with rings using response surface methodology. Int J Heat Fluid Flow 112:109713

Alamian M, Asadollahi S, Dehestani M (2024) Optimum mechanical properties of repair mortar containing hybrid fibers using the Taguchi method and analysis of variance. J Build Eng 98:111049

Buelvas V, Morhart L, Casavant T, Fleming I, Mahler C (2025) Evaluation of thermal parameters with geothermal energy recovery from a cold climate municipal solid waste landfill. Waste Manage 193:11–22

Caporale GM, Gil-Alana L (2019) Testing the Fisher hypothesis in the G-7 countries using I(d) techniques. Int Econ 159:140–150

GB/T.12573 (2008) Sampling method for cement. General Administration of Quality Supervision, Inspection and Quarantine, People's Republic of China

Ha M-X, Liu Y, Feng Y, Hu X-M, Zhao Y-Y, Liu J-D et al (2024) Force mechanism analysis of composite microbial dust suppressants based on extracellular polymeric substances (EPS) mode components. J Environ Manage 370:122926

Koda E, Osiński P, Podlasek A, Markiewicz A, Winkler J, Vaverková MD (2023) Geoenvironmental approaches in an old municipal waste landfill reclamation process: expectations vs reality. Soils Found 63(1):101273

Li J, Tang Z, Ahmadi M, Yang H, Gerlach D, Kooi BJ et al (2024) One-pot synthesis of alanine from glycerol without additional H_2 over a rutile-supported Ru catalyst. Appl Catal B Environ Energy 124960

Liew AW-C, Law N-F, Cao X-Q, Yan H (2009) Statistical power of Fisher test for the detection of short periodic gene expression profiles. Pattern Recogn 42(4):549–556

Mahandra H, Wu C, Ghahreman A (2021) Leaching characteristics and stability assessment of sequestered arsenic in flue dust based glass. Chemosphere 276:130173

Mohammad A, Paleologos EK, Ogrodnik P, Koda E, Osiński P, Podlasek A et al (2023) Occurrence and ecotoxicological effects of fires at municipal solid waste landfills. Environ Geotech 40:1–14

Omane D, Liu WV, Pourrahimian YJAPR (2018) Comparison of chemical suppressants under different atmospheric temperatures for the control of fugitive dust emission on mine hauls roads. Atmos Pollut Res 9(3):561–568

Pouladi B, Fanaei MA, Baghmisheh G (2019) Optimization of oxidative desulfurization of gas condensate via response surface methodology approach. J Clean Prod 209:965–977

Razmjooei MJ, O'Regan M (2024) Improved paired light and scanning electron microscope imaging technique for identifying nannofossils in Arctic sediments. Geobios

Shourabi FN, Mirzabozorg H, Zhuge Y (2024) Optimization of UHPC based on reactive powder concrete: Box-Behnken design, analysis of variance, components interaction effects and sensitivity analysis. J Build Eng 98:111291

Wang R, Yu H, Xie Y, Li W, Qi H (2024a) Study on protein-polysaccharide environmental foam dust suppressant based on Maillard reaction. J Mol Liq 410:125644

Wang X, Chen H, Shi Y, Lin Y, Yang J, Li X (2024b) Development of an eco-friendly dust suppressant based on modified guar gum: performance and mechanism analysis. Powder Technol 439:119681

Wei J, Zhao Y, Yu S, Du J, Hu X, Bai G et al (2021) Environment-friendly dual-network hydrogel dust suppressant based on xanthan gum, polyvinyl alcohol and acrylic acid. J Environ Manage 295:113139

Wu M, Hu X, Zhang Q, Zhao Y, Sun J, Cheng W et al (2020) Preparation and performance evaluation of environment-friendly biological dust suppressant. J Clean Prod 273:123162

Yu R, Huang Z, Zeng M, Li W, Liu Q, Ding H et al (2025a) Study on the performance of a composite dust suppressant in open-pit mines. Atmos Pollut Res 16(2):102379

Yu K, Zhou L, Peng S, Tian Y, Huang H, Chen X et al (2025b) One-pot preparation of carboxymethyl cellulose-based multifunctional active film embedded with Cu-BTC (benzenetricarboxylic acid) metal-organic framework for mango preservation. Food Packag Shelf Life 47:101416

Zhang X-N, Sun H, Jiang P, Cui X-Z, Wang J-R (2022) Development and performance characterization of a composite dust suppressant for Yellow River alluvial silt using response surface methodology. J Clean Prod 376:134293

Zhou G, Ma Y, Fan T, Wang G (2018) Preparation and characteristics of a multifunctional dust suppressant with agglomeration and wettability performance used in coal mine. Chem Eng Res Des 132:729–742

Zhou G, Xu Y, Wang Q, Jiang B, Ren B, Zhang X et al (2023) Wetting-consolidation type dust suppressant based on sugarcane bagasse as an environmental material: preparation, characterization and dust suppression mechanism. J Environ Manage 330:117097

Chapter 29
Particulate Matter Deposition on the Interior Surfaces of Louver-Shaded Building with Cross Ventilation

Jianwen Zheng, Qiuhua Tao, Zhixian Hong, and Han Wang

Abstract In hot and humid climate regions, exterior shading devices are widely employed to reduce cooling loads and improve daylighting in buildings. While their effects on thermal comfort and energy performance have been extensively studied, limited research has addressed how these devices influence natural ventilation and indoor particulate matter (PM) deposition. This study investigates PM deposition patterns in a multi-story building equipped with horizontal louvered shading under natural cross ventilation. The building is divided into zones based on floors, row units, and corridors. Using computational fluid dynamics (CFD) simulations based on the realizable k-ε turbulence model, the study evaluates the airflow field and particle deposition velocity for PM sizes ranging from 0.1 to 2.5 μm, under various wind directions and louver angles. Results show that as particle size increases, deposition velocity also increases, with 0.1 μm particles being most sensitive to airflow variation. Deposition tends to concentrate around doors, windows, and corners. Notably, different wind directions significantly affect spatial deposition patterns for 0.1 μm particles, with greater non-uniformity observed when louvers are located on the leeward side. These findings offer new insights into the interactions between shading design, ventilation, and indoor pollutant behavior, providing a reference for optimizing shading systems to balance energy efficiency and indoor air quality.

Keywords Natural ventilation · Shading louvers · Particulate matter deposition · CFD simulation

J. Zheng · Q. Tao (✉) · Z. Hong · H. Wang
School of Marine Equipment and Mechanical Engineering, Jimei University, Xiamen, China
e-mail: qhtao@jmu.edu.cn

J. Zheng · Q. Tao
Fujian Province Key Laboratory of Energy Cleaning Utilization and Development, Xiamen, China

J. Zheng
Department of Electronics, Information and Bioengineering, Polytechnic University of Milan, Milan, Italy

Y. Li et al. (eds.), *16th International Conference on Environmental Science and Development*, Environmental Science and Engineering,
https://doi.org/10.1007/978-3-032-15080-6_29

29.1 Introduction

Building energy consumption accounts for approximately one-third of the total societal energy use, and this proportion continues to rise (Building Energy Research Center of Tsinghua University 2016). In response to the growing demand for energy conservation in buildings, natural ventilation and architectural shading have emerged as effective strategies to reduce air conditioning energy consumption. During summer, a significant portion of building energy use is attributed to cooling loads. Exterior shading devices can mitigate incoming solar radiation and thereby effectively lower cooling energy demands. Additionally, such devices help improve the uniformity of daylight distribution, contributing to reduced lighting energy consumption. As a result, exterior shading systems have been widely adopted in various types of buildings. However, existing research on architectural shading has primarily focused on its effects on indoor thermal and daylighting conditions as well as its energy-saving potential, while limited attention has been paid to how changes in natural ventilation caused by exterior shading devices may influence indoor pollutant deposition.

Natural ventilation can be categorized into three types based on the driving mechanism: buoyancy-driven ventilation, wind-driven ventilation, and combined buoyancy-wind-driven ventilation. In studies focusing on the ventilation performance of buildings with shading devices, Sun et al. (2018) found through simulations that the installation of horizontal shading louvers increased the non-uniformity of the indoor airflow field and the air age. Jiang et al. (2022) conducted wind tunnel experiments and numerical simulations to analyze airflow patterns around shading louvers and indoor air age distribution in a cubic building under different louver rotation angles. Zheng et al. (2022) used numerical simulations to assess the impact of fixed shading louvers on the inter-room spread of viral aerosols. However, in the studies of natural ventilation and pollutant dispersion in shaded buildings, limited attention has been paid to the division of indoor ventilated spaces by floors and row units with interior corridors. As a result, existing research lacks data across different building zones, making it difficult to provide effective guidance for the design and regulation of shading components.

PM is a major component of urban air pollution due to industrialization and urbanization, and it exists in various forms such as dust, smoke, haze, and smog, with particle diameters ranging from approximately 0.001 to 100 μm (Li 2017; Jiang et al. 2018). Inhalable particles (PM_{10}, less than 10 μm) and respirable particles ($PM_{2.5}$, less than 2.5 μm) pose significant health risks, and while their deposition can reduce long-term accumulation and human exposure, it may also lead to contamination and corrosion of indoor surfaces and equipment, facilitate the spread of pollutants between rooms under varying airflow conditions, and ultimately affect indoor air quality. The deposition velocity of particulate matter represents the rate at which particles pass through the concentration boundary layer near surfaces and settle. However, current studies have mainly focused on the dispersion of gaseous pollutants, with little investigation into the behavior of particulate matter, including its transport and deposition on indoor surfaces.

This study divides a multi-story building equipped with single-sided horizontal shading louvers into zones based on floors, row units, and internal corridors. Numerical simulations are conducted to analyze the indoor airflow field under natural ventilation conditions. The distribution of indoor particulate matter deposition velocity is investigated across different louver rotation angles, aiming to provide a reference for the design of ventilation strategies and the regulation of louver angles in buildings with exterior shading.

29.2 Numerical Model

The multi-story building selected for this study is located in Xiamen, China. which is in a hot summer and warm winter climate zone, where strong solar radiation makes shading louvers widely used. As a coastal city, Xiamen experiences frequent land-sea breezes, with an annual average wind speed of up to 3.41 m/s. Therefore, the effective utilization of natural ventilation can significantly reduce the ventilation energy consumption.

29.2.1 Turbulence Model

To simulate airflow around buildings using CFD, Reynolds-averaged Navier–Stokes (RANS) models are widely adopted due to their balance between accuracy and computational efficiency. Among them, the realizable k-ε model has shown good agreement with experimental data and performs well across various wind attack angles, making it suitable for urban environments. Therefore, the realizable k-ε model is selected in this study to capture the airflow characteristics around the louver-shaded building.

29.2.2 Reference Building

The wind tunnel experiment conducted by Jiang et al. was used as a reference for the outside numerical simulation, as shown in Fig. 29.1a (Jiang et al. 2019). In the reference experiment, wind velocity fields around a cubic building were measured using a geometric scale of 1:30. The simplified geometric model of the louver-shaded building is shown in Fig. 29.1. The test section of the wind tunnel measures 12 m $\times$ 1.8 m $\times$ 1.8 m, with a blockage ratio of less than 8%. The scaled dimensions of the test building are 0.5 m in length, width, and height. A total of 26 rows of shading louvers were installed on one side of the windward facade. Each row consists of five

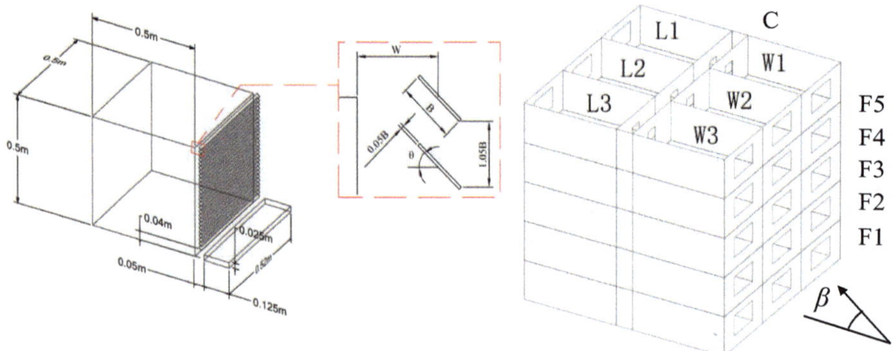

(a) Outside geometry (b) Inside geometry

Fig. 29.1 Geometric model of cross-ventilation building with shading louvers

louvers, collectively covering the entire windward surface. Each individual louver measures 93 mm (length) × 17.5 mm (width) × 1.0 mm (thickness), and the vertical spacing between adjacent louver rows is 18.3 mm. In the simplified model, the five louvers in each row are treated as a single louver unit with a length equal to the total length of the five louvers. As shown in Fig. 29.1, all shading louvers are set at a rotation angle of $\theta = 45°$, with $W = 0.025$ m and $B = 0.0175$ m. The inside feature was then added in Fig. 29.1b for five floors (F1–5). Each floor had six units (L1–3 and W1–3) with one corridor (C). The wind attack angle β is set as 0, 90 and 180° in different cases.

29.2.3 Computational Domain

In the computational domain for the external flow field of the louver-shaded building, the distance from the rooftop to the top boundary of the domain is set as $5h_0$, the distance from the windward facade to the inlet boundary is $5h_0$, and the distance from the leeward facade to the outlet boundary is $10h_0$. The distance from the building side-walls to the lateral boundaries of the domain is also $5h_0$. These sizes of computational domain are recommended by Franke et al. (2011).

In the final mesh, the blockage ratio is less than 3%, meeting the standard requirements for external flow simulations. Near-wall mesh elements on building walls, the ground, and the shading louvers have a minimum height of 0.0003 m ($0.0006h_0$), ensuring that the dimensionless wall distance y^+ remains below 5 and approximately equals 1. This allows the use of enhanced wall treatment for resolving the boundary layer near all surfaces. The final mesh consists of approximately 2 million cells.

29.2.4 Boundary Conditions

In the computational domain, the inlet boundary is defined as a velocity inlet, while the outlet is set as a pressure outlet with free outflow. The top and side boundaries of the domain are specified as symmetry boundaries. The ground surface, building walls, and shading louver surfaces are defined as no-slip walls. The incoming flow is treated as the velocity inlet. Based on measurements from the reference wind tunnel experiment, the inlet wind velocity U and turbulence intensity I follow specific vertical distribution profiles. The velocity profile is represented by a power-law function, as described in Eq. (29.1):

$$U(z) = U_{ref}\left(\frac{z}{h_0}\right)^{\alpha} \tag{29.1}$$

$$I(z) = 0.1\left(\frac{z}{h_0}\right)^{-\alpha-0.05} \tag{29.2}$$

$$k(z) = [U(z)I(z)]^2 \tag{29.3}$$

$$\varepsilon(z) = C_{\mu}^{0.5}k(z)\frac{U(z)}{h_0}\alpha\left(\frac{z}{h_0}\right)^{\alpha-1} \tag{29.4}$$

where $U(z)$ is the incoming wind speed at height z (m/s), and U_{ref} is the reference velocity at the building height h_0. U_{ref} is 3.41 m/s at $h_0 = 0.5$ m. The exponent α represents the power-law index associated with surface roughness. For suburban terrain, α is taken as 0.27. At the inlet boundary, in addition to the wind speed U, the turbulence intensity I, turbulent kinetic energy k, and turbulence dissipation rate ε are defined using Eqs. (29.2), (29.3), and (29.4), with the model constant C_{μ} set to 0.09.

29.2.5 Particle Matter Deposition

The deposition velocity can be calculated using the following formula (Lai and Nazaroff 2000).

Deposition velocity on indoor vertical surfaces:

$$u_v = \frac{u^*}{I} \tag{29.5}$$

Deposition velocity on upward-facing horizontal surfaces:

$$u_{h_1} = \frac{v_s}{1 - \exp\left(-\frac{v_s I}{u^*}\right)} \quad (29.6)$$

Deposition velocity on downward-facing horizontal surfaces:

$$u_{h_1} = \frac{v_s}{\exp\left(\frac{v_s I}{u^*}\right) - 1} \quad (29.7)$$

In these equations, u^* is the friction velocity (m/s), and v_s is the settling velocity (m/s). The formulas for other related parameters (Lai and Nazaroff 2000) are as follows:

$$I = 3.64 Sc^{2/3}\left(a^* - b^*\right) + 39 \quad (29.8)$$

$$a^* = \frac{1}{2}\ln\left[\frac{\left(10.92 Sc^{-1/3} + 4.3\right)^3}{Sc^{-1} + 0.0609}\right] + \sqrt{3}\tan^{-1}\left(\frac{8.6 - 10.92 Sc^{-1/3}}{\sqrt{3} \times 10.92 Sc^{-1/3}}\right) \quad (29.9)$$

$$b^* = \frac{1}{2}\ln\left[\frac{\left(10.92 Sc^{-1/3} + r^+\right)^3}{Sc^{-1} + 7.669 \times 10^{-4}(r^+)^3}\right] + \sqrt{3}\tan^{-1}\left(\frac{2r^+ - 10.92 Sc^{-1/3}}{\sqrt{3} \times 10.92 Sc^{-1/3}}\right) \quad (29.10)$$

$$u^* = \sqrt{k\sqrt{0.09}} \quad (29.11)$$

$$Sc = \frac{v}{D} \quad (29.12)$$

$$r^+ = \frac{d_p u^*}{2v} \quad (29.13)$$

where Sc is the Schmidt number, v is the kinematic viscosity of air (m^2/s), and D is the Brownian diffusion coefficient of the particle (m^2/s). When $y^+ = r^+$, it indicates that the particle concentration at the wall surface is zero. d_p is the particle diameter (m).

For aerosol particles, the settling velocity v is calculated using the following equation (Zhou 2017):

$$v_s = \frac{d_p^2 g C_c\left(\rho_p - \rho\right)}{18\mu} \quad (29.14)$$

where μ is the dynamic viscosity of air (Pa s), and ρ_p is the particle density (kg/m^3). For most particles, the density of air is negligible compared to the particle density, and the equation can be simplified as:

$$v_s = \frac{\rho_p g C_c d_p^2}{18\mu} \tag{29.15}$$

where C_c is the Cunningham slip correction factor associated with non-continuum effects. The expression for C_c for gases is given by (Lai and Nazaroff 2000):

$$C_c = 1 + \frac{2l}{d_p}\left[A_1 + A_2\exp\left(\frac{-A_3 d_p}{l}\right)\right] \tag{29.16}$$

where $A_1 = 2.514$, $A_2 = 0.8$, and $A_3 = 1.0$. l is the mean free path of gas molecules, taken as 0.065 μm. For particles with a diameter d_p, the Brownian diffusion coefficient D can be calculated based on the Cunningham slip correction factor (Zhou 2017) using the following formula:

$$D_p = \frac{k_b T_K C_c}{3\pi \mu d_p} \tag{29.17}$$

where k_b is the Boltzmann constant, taken as 1.380649×10^{-23} J/K, and T_K is the absolute temperature, taken as 288.15 K.

29.3 Results and Discussion

This study firstly investigates particle surface deposition speed for different particle sizes in different building floors to find the horizontal and vertical differences of surface deposition. Then three wind directions are considered to evaluate the impact of various airflow patterns on particle deposition in specific floors. The validation of numerical simulation and airflow characterization in this study refer to our previous research (Zheng et al. 2020; Jiang et al. 2022).

29.3.1 Cases for Different Particle Sizes

Table 29.1 lists the average deposition velocities of particles with particle sizes of 0.1, 0.5, 1.0, and 2.5 μm on the surfaces of various rooms in a cross flow ventilated building when the rotation angle is 45° at a wind direction angle of 0°. Overall, as the particle size increases, the magnitude of sedimentation velocity tends to increase, and the values range from 7.99×10^{-7} to 5.13×10^{-5} m/s in different rooms. For particles with a diameter of 0.1 μm, there is a significant difference in the average sedimentation rate between different rooms. Except for particles with a diameter of 0.1 μm, the sedimentation velocity fluctuation of particles larger than 0.1 μm in the same direction in different rooms is less than 5%. In these rooms, the sedimentation of larger particles is not greatly affected by indoor wind fields, but mainly by gravity.

This is because for small-scale particles with $d_p \leq 0.1$ μm, sedimentation is mainly controlled by Brownian diffusion, and the smaller the particle size, the faster the sedimentation rate. For mesoscale particles with $0.1 < d_p < 1$ μm, sedimentation is mainly controlled by diffusion collision, and the larger the particle size, the faster the sedimentation rate. For large-scale particles with $d_p \geq 1$ μm, sedimentation is mainly controlled by inertia, and the larger the particle size, the faster the sedimentation rate (Jiang et al. 2018). For a specific layer, particles with a diameter of 0.1 μm have a higher average sedimentation rate in the inner corridor than in other rooms, while particles larger than 0.1 μm have a lower average sedimentation rate in the inner corridor than in other rooms. The interior corridor is more prone to depositing small-scale particles than other rooms, because the vertical surfaces with door edges of the inner corridor accounts for a larger proportion of the interior corridor surface area. For rooms, due to a larger horizontal ground area than the interior corridor, large and medium-sized particles are more likely to deposit. Based on the relationship between sedimentation rate and particle size, 0.1 μm particles were mainly selected for analysis in the following study.

29.3.2 *Cases for Different Wind Directions*

Figure 29.2 shows the distribution contour maps of deposition velocity for 0.1 μm particles on the first and fifth floors of the cross-ventilated building at a louver rotation angle of 45°, under wind direction angle β of 0, 90, and 180°. As shown in Fig. 29.2, regardless of the wind direction angle, the highest deposition velocities within both the corridor and the rooms are consistently observed on the surfaces near doors and windows. In addition to these areas, deposition velocity is slightly higher at the corners of the inner surface of the windward exterior wall near the window, while in leeward rooms, higher deposition occurs on the floor area near the interior door. Comparing the 0 and 180° wind directions, when the louvers are located on the leeward facade rather than the windward side, more contour lines appear in the plots, indicating increased non-uniformity in particle deposition. Under the 90° wind direction, the non-uniformity of deposition velocity is slightly higher in rooms near the leeward side compared to those near the windward side, but both the magnitude and the non-uniformity are significantly lower than those observed under the 0 and 180° conditions.

29.4 Summary

This study used CFD simulations to analyze the deposition of particulate matter in a cross-ventilated building equipped with horizontal shading louvers. Results show that as particle size increases from 0.1 to 2.5 μm, the average deposition velocity increases significantly, with values ranging from 7.99×10^{-7} to 5.13×10^{-5} m/s

Table 29.1 Average surface deposition velocity of various particle sizes in a cross-ventilated building with louvers (unit: m/s)

Room ID	$d_p = 0.1$ μm	$d_p = 0.5$ μm	$d_p = 1.0$ μm	$d_p = 2.5$ μm
F1-C (corridor)	7.99E−07	1.59E−06	5.24E−06	2.97E−05
F1-W1	8.20E−07	2.65E−06	9.02E−06	5.13E−05
F1-W2	5.39E−07	2.60E−06	9.00E−06	5.14E−05
F1-W3	8.20E−07	2.65E−06	9.02E−06	5.13E−05
F1-L1	5.81E−07	2.61E−06	8.99E−06	5.13E−05
F1-L2	6.28E−07	2.61E−06	9.00E−06	5.14E−05
F1-L3	5.82E−07	2.61E−06	8.99E−06	5.13E−05
F2-C (corridor)	9.44E−07	1.61E−06	5.25E−06	2.97E−05
F2-W1	8.72E−07	2.66E−06	9.02E−06	5.13E−05
F2-W2	5.44E−07	2.61E−06	9.00E−06	5.14E−05
F2-W3	8.72E−07	2.66E−06	9.02E−06	5.13E−05
F2-L1	8.50E−07	2.65E−06	9.01E−06	5.13E−05
F2-L2	7.43E−07	2.63E−06	9.01E−06	5.14E−05
F2-L3	8.53E−07	2.65E−06	9.01E−06	5.13E−05
F3-C (corridor)	1.20E−06	1.65E−06	5.28E−06	2.97E−05
F3-W1	9.77E−07	2.67E−06	9.03E−06	5.13E−05
F3-W2	5.90E−07	2.62E−06	9.01E−06	5.14E−05
F3-W3	9.76E−07	2.67E−06	9.03E−06	5.13E−05
F3-L1	1.03E−06	2.67E−06	9.03E−06	5.13E−05
F3-L2	8.60E−07	2.64E−06	9.02E−06	5.14E−05
F3-L3	1.03E−06	2.67E−06	9.03E−06	5.13E−05
F4-C (corridor)	1.45E−06	1.69E−06	5.30E−06	2.97E−05
F4-W1	1.08E−06	2.69E−06	9.04E−06	5.13E−05
F4-W2	6.79E−07	2.63E−06	9.01E−06	5.14E−05
F4-W3	1.08E−06	2.69E−06	9.04E−06	5.13E−05
F4-L1	1.19E−06	2.69E−06	9.04E−06	5.13E−05
F4-L2	1.02E−06	2.66E−06	9.03E−06	5.14E−05
F4-L3	1.19E−06	2.69E−06	9.04E−06	5.13E−05
F5-C (corridor)	1.57E−06	1.69E−06	5.30E−06	2.97E−05
F5-W1	1.27E−06	2.71E−06	9.05E−06	5.13E−05
F5-W2	1.00E−06	2.64E−06	9.02E−06	5.14E−05
F5-W3	1.30E−06	2.71E−06	9.05E−06	5.13E−05
F5-L1	1.16E−06	2.67E−06	9.03E−06	5.13E−05
F5-L2	1.15E−06	2.65E−06	9.03E−06	5.14E−05
F5-L3	1.17E−06	2.67E−06	9.03E−06	5.13E−05

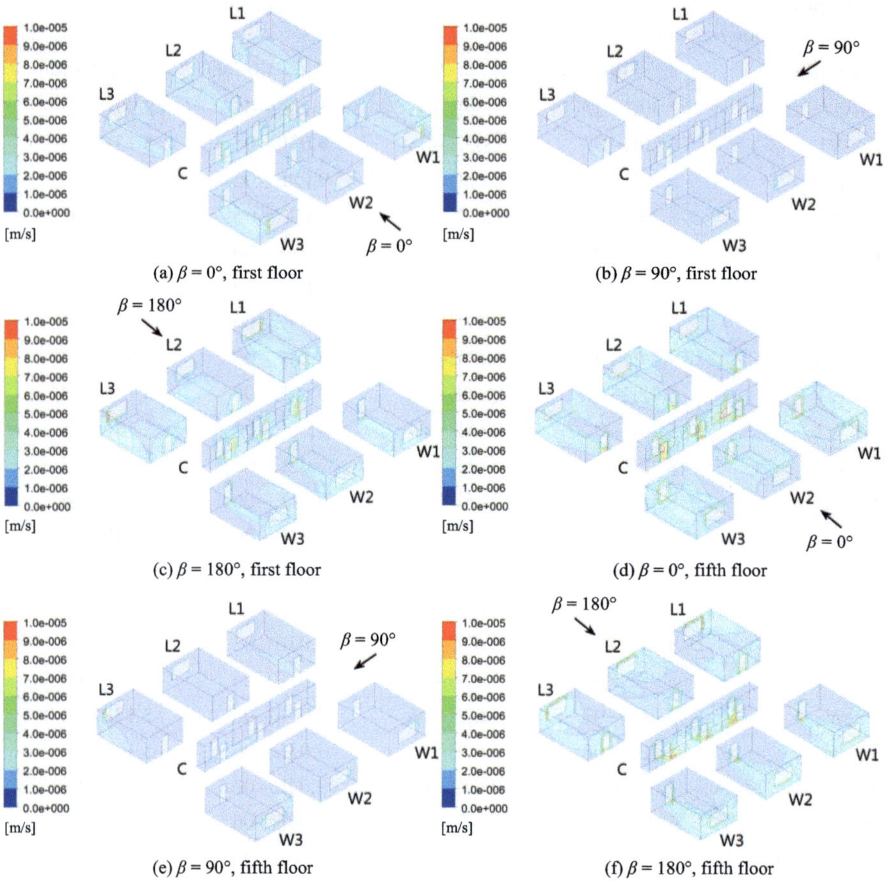

Fig. 29.2 Deposition velocity distribution of particulate matter in a cross-ventilated building with louver angle at 45° under different wind directions

in typical rooms. For 0.1 μm particles, deposition is more sensitive to airflow variations, especially near doors and windows. Wind direction affects spatial deposition uniformity, with the 180° case showing more contour lines and higher non-uniformity than the 0° case. These findings provide a reference for optimizing louver design to improve both ventilation performance and indoor air quality.

Acknowledgements This research work was supported by the National Natural Science Foundation of China (No. 52308096), the Natural Science Foundation of Fujian Province (No. 2024J01110), and the Xiamen Science and Technology Subsidy Project (No. 2024CXY0309).

References

Building Energy Research Center of Tsinghua University (2016) Annual development report on China building energy efficiency. China Architecture & Building Press, Beijing (in Chinese)
Franke J, Hellsten A, Schlunzen KH et al (2011) The COST 732 best practice guideline for CFD simulation of flows in the urban environment: a summary. Int J Environ Pollut 44(1–4):419–427
Jiang Z, Chen J, Wen H (2018) Aerosol mechanics and applications. Metallurgical Industry Press, Beijing (in Chinese)
Jiang F, Li Z, Zhao Q et al (2019) Flow field around a surface-mounted cubic building with louver blinds. Build Simul 12(1):141–151
Jiang F, Tao S, Tao Q, Yuan Y, Zheng J (2022) The effect of louver blinds on the wind-driven cross ventilation of multi-storey buildings. J Build Eng 54:104614
Lai ACK, Nazaroff WW (2000) Modeling indoor particle deposition from turbulent flow onto smooth surfaces. J Aerosol Sci 31(4):463–476
Li L (2017) Environmental chemistry. Harbin Institute of Technology Press, Harbin (in Chinese)
Sun N, Cui Y, Jiang Y (2018) Lighting and ventilation-based building sun-shading design and simulation case in cold regions. Energy Procedia 152:462–469
Zheng J, Tao Q, Li L (2020) Wind pressure coefficient on a multi-storey building with external shading louvers. Appl Sci 10(3):1128
Zheng J, Tao Q, Chen Y (2022) Airborne infection risk of inter-unit dispersion through semi-shaded openings: a case study of a multi-storey building with external louvers. Build Environ 225:109586
Zhou Z (2017) Fundamentals of electro-optical countermeasure materials. Beijing Institute of Technology Press, Beijing (in Chinese)

Part V
Wastewater Treatment and Life Cycle Assessment

Chapter 30
Life Cycle Assessment of Environmental Performance in a Municipal Wastewater Treatment Plant: A Dynamic Comparison (2022–2024)

Wenbo Huang, Haiyan Fu, Zhouzhenyang Xia, Jiajun Luo, Ze Bao, Yicheng Wu, and Jiangxue Long

Abstract Significant changes in operational conditions have impacted the environmental performance of sewage treatment plants. This study employed a life cycle assessment to compare the environmental impacts of a Xiamen sewage treatment plant during 2022 and 2024, focusing on its sewage treatment phase. Using SimaPro software with the ReCiPe 2016 Midpoint method and defining the treatment of 1 cubic meter of sewage as the functional unit, key parameters including greenhouse gas emissions, energy consumption, and chemical usage were quantified. The 2022 emission profile identified direct greenhouse gas releases, specifically methane and nitrous oxide, as the dominant climate forcing agents, accounting for 4.26 kg CO_2-equivalent (CO_2eq) per functional unit. Secondary contributions arose from grid electricity consumption (0.381 kg CO_2eq), reflecting energy-intensive operational processes. By 2024, improvements in operational efficiency and the restoration of maintenance protocols reduced direct gas emission-related carbon emissions to 3.85 kg CO_2eq. Increased disinfectant use and microplastic presence during the study period were found to inhibit microbial activity, thereby exacerbating energy demands. The study highlights that optimizing direct emission management, enhancing energy efficiency, and strengthening equipment maintenance are critical strategies for mitigating environmental impacts. Integrating renewable energy, refining treatment processes, and addressing emerging pollutants are essential for promoting sustainable development in the sewage treatment industry. These findings provide theoretical and practical insights for low-carbon operation and environmental risk management amid operational changes and recovery.

W. Huang · H. Fu · Z. Xia · Z. Bao · Y. Wu · J. Long (✉)
Key Laboratory of Environmental Biotechnology of Fujian University, Xiamen University of Technology, Xiamen, PR China
e-mail: 2011121601@xmut.edu.cn

J. Luo
Wanjia Qingshui (Xiamen) Environmental Technology Co., Ltd., Xiamen, China

Y. Li et al. (eds.), *16th International Conference on Environmental Science and Development*, Environmental Science and Engineering,
https://doi.org/10.1007/978-3-032-15080-6_30

Keywords Wastewater treatment · Life cycle assessment · Greenhouse gases · SimaPro

30.1 Introduction

Modern wastewater infrastructure has undergone significant operational transformations in response to emerging global challenges. As critical environmental protection systems, wastewater treatment plants (WWTPs) provide essential contaminant removal services that maintain hydrological ecosystem integrity and prevent anthropogenic pollutant dispersion (Hospido et al. 2004). Prior to the pandemic, WWTPs were designed and operated based on conventional sewage generation patterns, with influent primarily originating from domestic, industrial, and commercial sources. Established treatment processes—such as the activated sludge method, anaerobic digestion, and advanced oxidation—were optimized to reduce key pollutants (biochemical oxygen demand, chemical oxygen demand, nutrients, and suspended solids) to meet regulatory discharge standards (Corominas et al. 2013).

Increased use of personal protective equipment like masks and gloves in households and healthcare facilities introduced microplastics and novel contaminants into wastewater streams. Furthermore, the widespread adoption of chlorine-based disinfectants significantly altered influent composition. These shifts potentially compromise the efficacy of biological treatment processes, as elevated disinfectant concentrations may inhibit microbial activity critical for pollutant degradation (Dang et al. 2023). Pandemic-related operational challenges, such as workforce shortages due to isolation protocols and disrupted equipment maintenance, threatened treatment efficiency and environmental performance. Given these systemic changes, conducting a comprehensive Life Cycle Assessment (LCA) of WWTPs is imperative to evaluate their environmental impacts before and after the COVID-19 pandemic. LCA provides a holistic framework to quantify environmental burdens across all operational phases, from raw material extraction and energy consumption to sludge management and effluent discharge (Li et al. 2021).

This study focuses on the sewage treatment phase of wastewater treatment plants in Xiamen and did not include the assessment of construction, sludge treatment, and decommissioning phases. This assumption is based on the following considerations: the sewage treatment phase is the direct impact link for changes in influent composition and process operation challenges, and its environmental impacts are highly relevant to the research objectives. However, the infrastructure energy consumption in the construction phase, pollutant migration risks in sludge treatment, and resource recovery potential in the decommissioning phase may all significantly affect the overall sustainability of wastewater treatment plants. By comparing LCA results from 2022 and 2024, study identify critical environmental impact drivers, assess the adaptability of current treatment processes to emerging challenges, and propose sustainable strategies for post-pandemic wastewater management. The findings hold significant implications for advancing the sustainability of the wastewater treatment industry while ensuring long-term environmental protection and public health resilience.

30.2 Materials and Methods

30.2.1 Definition of the Scope and Objectives of LCA

The primary objective of conducting a life cycle assessment (LCA) for wastewater treatment plants (WWTPs) is to holistically evaluate their environmental, economic, and social impacts across their entire lifecycle—from construction and operation to decommissioning or abandonment. This study focuses specifically on the sewage treatment phase of a WWTP in Xiamen, China. A functional unit of "treatment of one cubic meter of sewage" is defined (Samsami et al. 2020). Key pollutants emitted during this phase—including Chemical Oxygen Demand (COD), ammonia nitrogen, and greenhouse gases (GHGs)—are quantified across air, water, and soil media. Their contributions to critical environmental challenges, such as global warming, stratospheric ozone depletion, aquatic eutrophication, and ecotoxicity, are systematically assessed (Al-Anbar et al. 2022). By quantifying these impacts, the study identifies environmental hotspots in the treatment process, establishes a foundation for targeted mitigation strategies (process optimization and emission reduction), and enables comparative analysis of environmental performance across different operational periods.

30.2.2 Life Cycle Inventory

Life Cycle Inventory (LCI) is a critical component of Life Cycle Assessment, involving the systematic collection and quantitative analysis of inputs (materials, energy) and outputs (gaseous emissions, wastewater, solid waste) across the entire life cycle of a product, process, or service. Table 30.1 presents the influent and effluent water quality characteristics of WWTPs across different years, along with their discharge into water bodies. Table 30.2 details annual power consumption metrics, including energy usage for pumping, water treatment, and sludge management. Table 30.3 summarizes the chemical agents utilized in water treatment processes during the same periods. Direct emissions during the treatment phase were calculated using formulas (30.1) and (30.2) to WWTPs (IPCC 2006). Methane emissions per cubic meter of treated sewage were 0.1196 kg in 2022 and 0.1068 kg in 2024, while nitrous oxide emissions were 0.000644 kg (2022) and 0.000734 kg (2024), respectively.

$$E(CH_4) = (COD_{IN} - COD_{OUT}) \times \rho_S \times B_0 \tag{30.1}$$

$$E(N_2O) = (TN_{IN} - TN_{OUT}) \times EF_{N_2O} \tag{30.2}$$

Selecting an appropriate Life Cycle Impact Assessment (LCIA) method is essential to align with environmental analysis goals and derive actionable insights. This study employed SimaPro software with the ReCiPe 2016 Midpoint (H) method. The Midpoint approach offers advantages in data accessibility and indicator quantification, as its methodology is well-established and straightforward. For instance, ReCiPe 2016 Midpoint (H) quantifies climate change impacts by applying predefined global warming potential (GWP) factors to greenhouse gas emissions (Rebello et al. 2021). Its eighteen impact categories, including climate change, terrestrial acidification, and freshwater eutrophication—provide granular insights into distinct environmental issues, enabling researchers and policymakers to pinpoint critical areas for intervention. In contrast, the Endpoint method aggregates results into broader categories (human health, ecosystem quality). While this approach provides system-level assessment capabilities, it fails to resolve critical sector-specific environmental externalities. The Midpoint (H) framework addresses this limitation through its taxonomic classification system, enabling comparative analysis across studies sharing common impact pathways and facilitating precision mitigation planning.

Table 30.1 Qualitative characteristics in the effluent

	Biochemical oxygen demand (mg/L)		COD (mg/L)		Total phosphorus (mg/L)		Total nitrogen (mg/L)		Ammonia nitrogen content index (mg/L)		Volume (m^3)
	In	Out	In	Out	In	Out	In	Out	In	Out	
2022	179.00	1.60	358.00	14.00	5.18	0.15	47.20	6.880	36.90	0.03	105,004.000
2024	166.00	0.90	308.00	12.00	5.11	0.17	52.80	6.96	43.10	0.06	112,896.000

Table 30.2 The power consumption of sewage treatment plants

	Increase energy consumption	Energy consumption in water treatment	Energy consumption for sludge treatment
2022	0.037	0.296	0.013
2024	0.039	0.299	0.011

Table 30.3 The chemical consumption of sewage treatment plants

	Sodium acetate (t)	Polymerized ferrous sulfate (t)
2022	7.810	7.430
2024	3.760	7.490

30.3 Results and Discussion

30.3.1 2022 Life Cycle Assessment Results

The LCI analysis quantifies critical environmental parameters associated with sewage treatment phases during the operational years 2022 and 2024. As illustrated in Fig. 30.1, direct greenhouse gas (GHG) emissions and electricity consumption emerged as the predominant contributors to global warming potential (GWP), accounting for 4.26 kg CO_2-equivalent (CO_2eq) and 1.201 kg CO_2eq per functional unit, respectively. Operational emissions from the treatment facility, predominantly methane and nitrous oxide, combined with indirect emissions from grid electricity usage, constitute the primary drivers of this environmental impact category. The analysis further identified measurable ionizing radiation effects, with water inflow/outflow processes and grit removal operations contributing 0.0166 kBq Co-60 equivalent (Co-60-eq) and 0.00616 kBq Co-60-eq, respectively. These values suggest the presence of radionuclides or radiation-associated compounds during these operational stages, though their specific sources remain uncharacterized. Targeted investigations are required to elucidate the origin, speciation, and transport mechanisms of radioactive species in influent/effluent streams and sedimentation processes.

For both Ozone Formation (Human Health) and Ozone Formation (Terrestrial), the total impacts are relatively minor. Notably, Electricity Consumption contributes to these categories, suggesting that nitrogen oxide precursors emitted during power usage influence ozone formation. The negative value for Freshwater Eutrophication indicates that, using a positive baseline reference, the treatment process reduces eutrophication impacts on water bodies rather than enhancing them.

In 2022, the sewage treatment plant exhibited substantial environmental impacts in Global Warming, Resource Scarcity, and Ecological Toxicity. Electricity consumption and chemical usage emerged as critical contributors across multiple impact categories, highlighting their role as key focus areas for improving energy efficiency, reducing emissions, and mitigating environmental harm (Başkurt et al. 2017). Direct greenhouse gas emissions also remain a pivotal factor in driving impacts like global warming.

Using the IPCC 2013 GWP 100a (based on the 100-year global warming potential framework), the contribution of different stages to global warming was quantified. Figure 30.2 further clarifies that direct greenhouse gas emissions and electricity consumption are the dominant factors driving the sewage treatment plant's global warming impact.

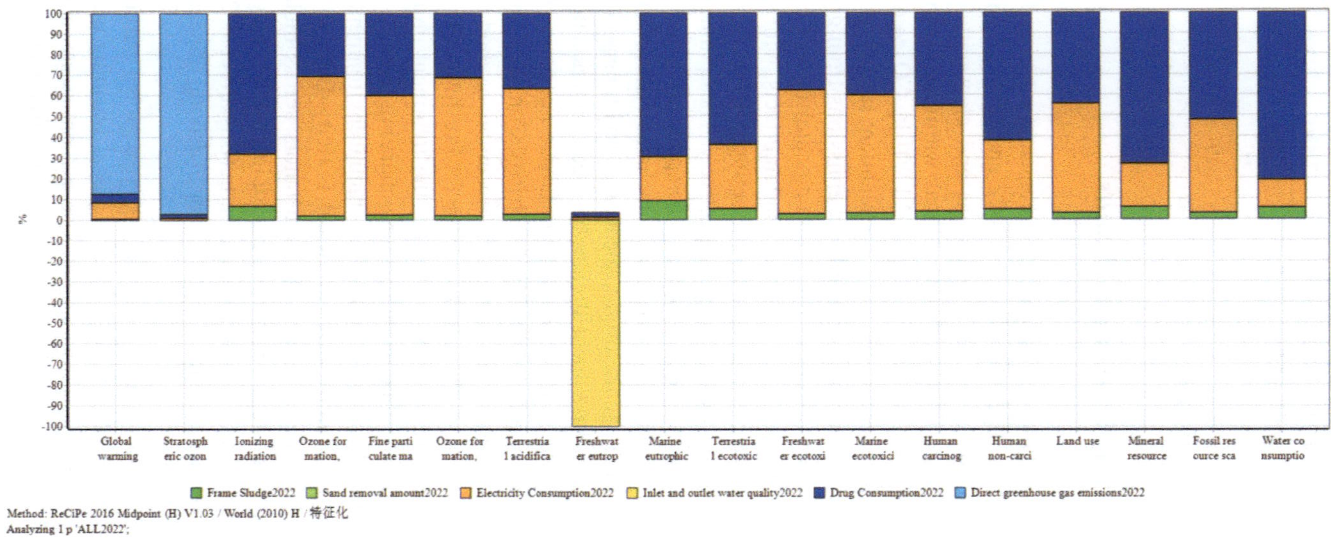

Fig. 30.1 Life cycle assessment using ReCiPe method on January 1, 2022

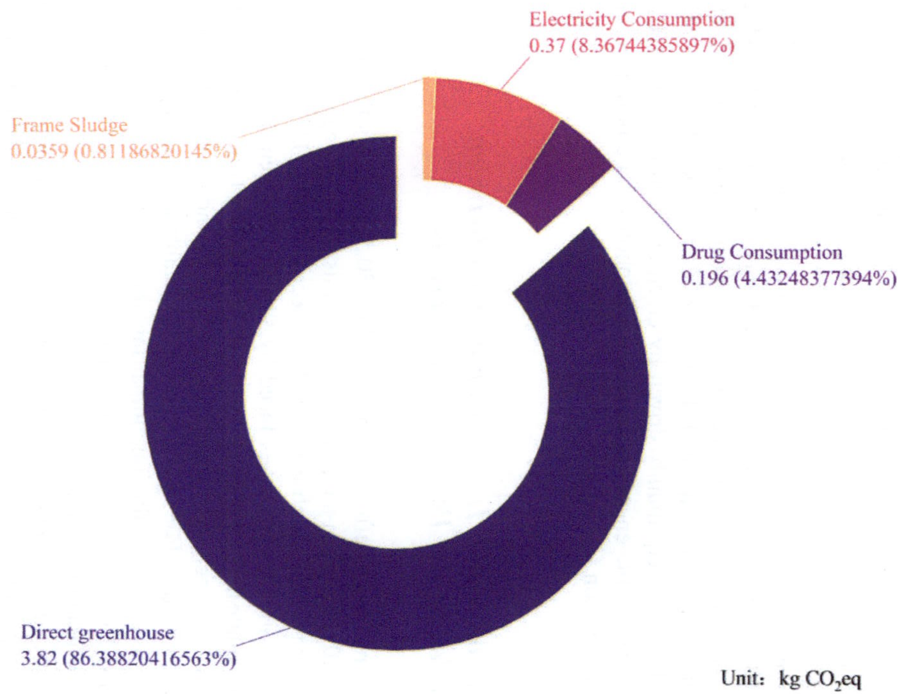

Fig. 30.2 Carbon emissions under the IPCC method in 2022

30.3.2 2024 Life Cycle Assessment Results

Figure 30.3 demonstrates that Direct Greenhouse Gas Emissions and Electricity Consumption contribute significantly to the Global Warming impact, accounting for 3.85 kg CO_2eq and 0.384 kg CO_2eq, respectively. The proportional impacts of other stages on environmental categories remain consistent with those observed in 2022. The climate impacts presented in Fig. 30.4 were quantified following the standardized methodology outlined in the IPCC Fifth Assessment Report, employing the 100-year global warming potential (GWP_{100}) metric for equivalent CO_2 conversion. This analytical continuity ensures cross-comparability with the emission patterns demonstrated in Fig. 30.3. The value of Total GWP is 3.97 kg CO_2eq, mainly from Direct Greenhouse (3.45 kg CO_2eq, accounting for 86.9%) and Electricity Consumption (0.373 kg CO_2eq, accounting for 9.4%). Drug Consumption contributes 0.114 kg CO_2eq (2.9%), while the impact of other processes (such as sand and gravel removal, inlet and outlet water) can be ignored. In the ReCiPe method, the value of Fossil Resource is 0.127 kg oileq, mainly driven by electricity consumption (0.0747 kg oileq). Promote renewable energy sources, such as photovoltaic power generation, to reduce dependence on fossil fuels.

30.3.3 Environmental Impact Comparison Between Pandemic and Post-pandemic Periods

Regarding the greenhouse effect comparison, Fig. 30.5 illustrates the proportional distribution of different green-shaded areas within this impact category for 2022 and 2024. The higher proportion observed in 2022 decreased by 2024, a trend closely linked to the operational challenges faced by sewage treatment plants during the pandemic. Inadequate and delayed equipment upkeep reduced operational efficiency, leading to increased energy consumption and corresponding rises in electricity-related carbon emissions. For instance, improper maintenance of aeration systems caused uneven oxygen distribution, impairing the activity of sewage-degrading microorganisms. To compensate, extended aeration times were required, escalating power usage (Delre et al. 2019).

The systematic restoration of equipment maintenance and monitoring protocols improved mechanical efficiency, reduced energy waste, and consequently decreased carbon emissions linked to electricity consumption (Fang et al. 2016). A comparative analysis of the relative contributions (represented by green-shaded areas) across environmental impact categories between 2022 and 2024 clearly demonstrates the evolving environmental footprint of the wastewater treatment plant. The pandemic caused multifaceted operational disruptions in 2022, which elevated emissions and exacerbated impacts in multiple categories (Tran et al. 2024). By 2024, systematic improvements in three operational domains—automated process controls, resilient resource allocation systems, and predictive maintenance frameworks—demonstrated

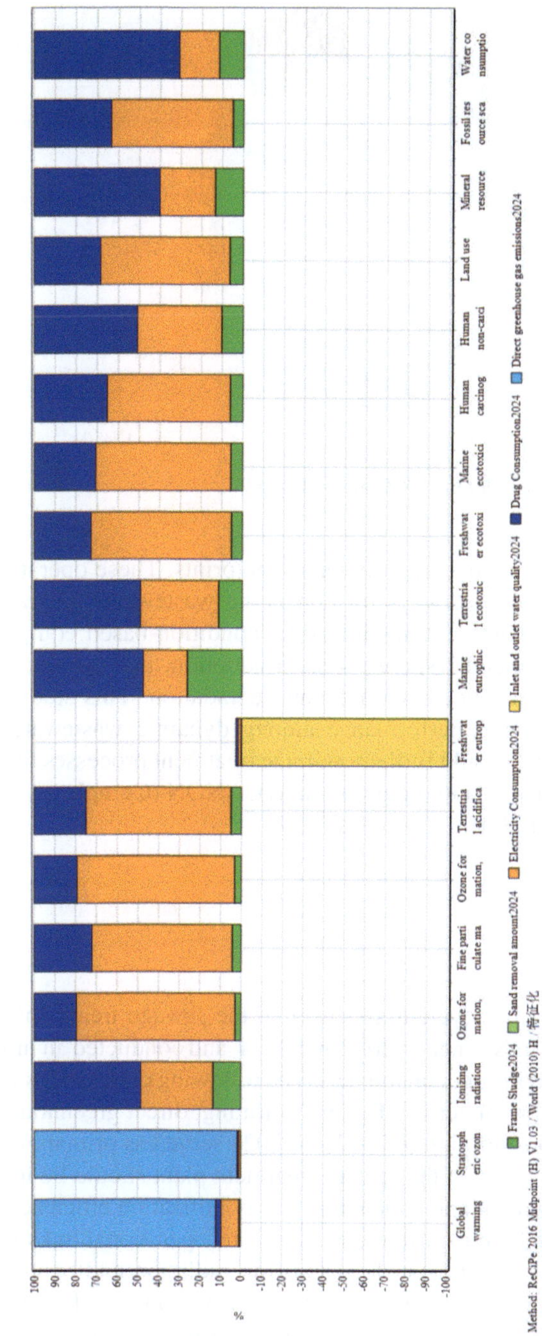

Fig. 30.3 Life cycle assessment using ReCiPe method on January 1, 2024

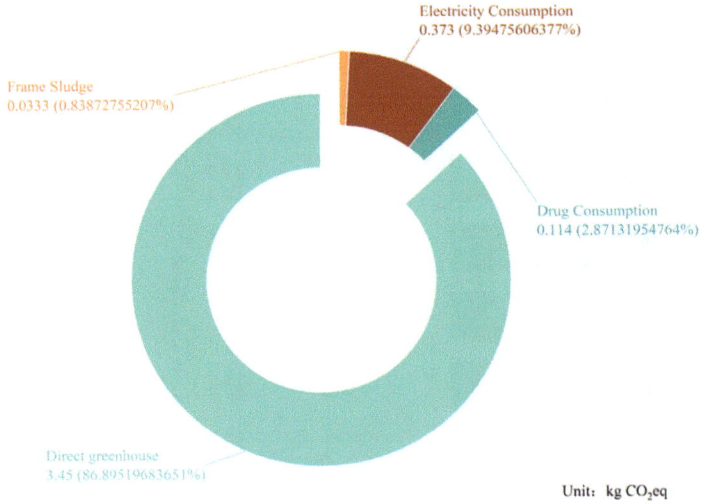

Fig. 30.4 Carbon emissions under the IPCC method in 2024

measurable success in reducing environmental footprints. These operational refinements establish actionable pathways for enhancing wastewater treatment sustainability through targeted interventions including condition-based equipment monitoring, smart energy demand-response system adoption, and advanced oxidation process integration for enhanced contaminant degradation. This study reveals the disparities in environmental performance among dynamic wastewater treatment plants, and future research could further optimize treatment processes by integrating emerging nitrogen removal technologies such as CANON (Completely Autotrophic Nitrogen Removal Over Nitrite) (Luo et al. 2024).

30.4 Conclusions

This study conducted a life cycle assessment of the sewage treatment stages of a sewage treatment plant in Xiamen in 2022 and 2024, and conducted an in-depth analysis of differences in environmental impacts. The following conclusions were drawn: key environmental factors: in terms of global warming, direct greenhouse gas emissions and electricity consumption in 2022 and 2024 served as primary contributors, the emission reduction focus of the sewage treatment plant should lie in optimizing direct emission management and enhancing energy utilization efficiency. Additionally, the study quantified the contributions of various operational links within other environmental impact categories such as ionizing radiation and ozone formation, offering targeted improvement directions. The impact and trend: leading to heightened environmental impacts across multiple categories in 2022, and equipment drove

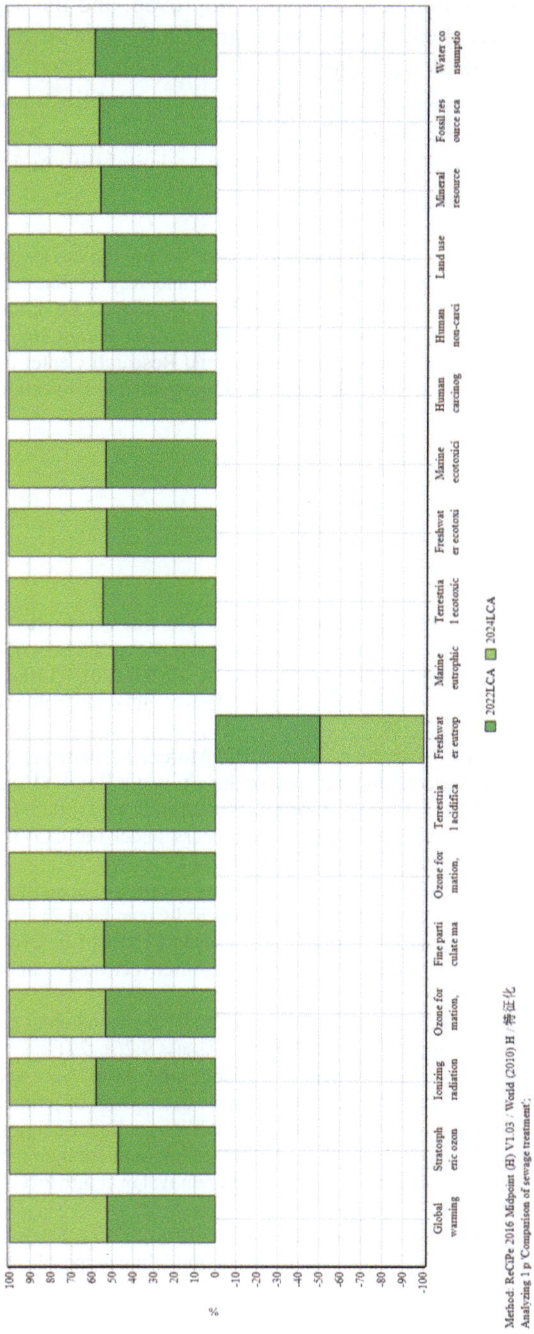

Fig. 30.5 Comparison of life cycle assessments using ReCiPe method in different periods

up electricity consumption, thereby exacerbating carbon emissions. In 2024, operations returned to normal, equipment efficiency improved, and the negative effects across various environmental categories were significantly alleviated. It can further align with China's dual carbon goals and the EU's Green New Deal, integrate smart water management digital monitoring with sludge resource utilization technologies, and simultaneously participate in the formulation of international wastewater treatment resilience standards, so as to strengthen policy adaptability and technological forward-looking.

Acknowledgements This research was financially supported by Fujian Provincial Industry-university Research Collaborative Innovation (2021Y4005), and Key Technology Collaborative Innovation Platform Project for Improving Urban and Rural Ecological Environment Quality in Fu-Xia-Quan National Independent Innovation Demonstration Zone (Grant No. 2022-P-024). In addition, we acknowledge support of Fujian Engineering and Research Center of Rural Sewage Treatment and Water Safety, and the Key Laboratory of Environmental Biotechnology (XMUT), Fujian Province University.

References

Al-Anbar MA, Altaee S, Kareem S (2022) Using life cycle assessment (LCA) in appraisal sustainability indicators of Najaf wastewater treatment plant. Egypt J Chem 65(9):513–519

Başkurt M, Kocababuç I, Binici E, Dulekgurgen E, Özgün ÖK, Taşli R (2017) Life cycle assessment as a decision support tool in wastewater treatment plant design with renewable energy utilization. Desalin Water Treatment 93:229–238

Corominas L, Foley J, Guest JS, Hospido A, Larsen HF, Morera S, Shaw A (2013) Life cycle assessment applied to wastewater treatment: state of the art. Water Res 47(15):5480–5492

Dang C, Zhang Y, Zheng M, Meng Q, Wang J, Zhong Y, Wu Z, Liu B, Fu J (2023) Effect of chlorine disinfectant influx on biological sewage treatment process under the COVID-19 pandemic: performance, mechanisms and implications. Water Res 244:120453

Delre A, ten Hoeve M, Scheutz C (2019) Site-specific carbon footprints of Scandinavian wastewater treatment plants, using the life cycle assessment approach. J Clean Prod 211:1001–1014

Fang LL, Valverde-Pérez B, Damgaard A, Plósz BG, Rygaard M (2016) Life cycle assessment as development and decision support tool for wastewater resource recovery technology. Water Res 88:538–549

Hospido A, Moreira MT, Fernández-Couto M, Feijoo G (2004) Environmental performance of a municipal wastewater treatment plant. Int J Life Cycle Assess 9:261–271

IPCC (2006) https://www.ipcc-nggip.iges.or.jp/public/2006gl/index.html. Accessed 28 May 2022

Li Y, Luo X, Huang X, Wang D, Zhang W (2021) Life cycle assessment of municipal wastewater treatment processes regarding energy production from the sludge line. Energies 14(2):356

Luo J, Wu Y, Fu H, Fu M, Liu M, Guo H, Jin L, Wang S (2024) Shift in microorganism and functional gene abundance during completely autotrophic nitrogen removal over nitrite (CANON) process. J Environ Manage 359:121009

Rebello TA, Roque RP, Gonçalves RF, Calmon JL, Queiroz LM (2021) Life cycle assessment of urban wastewater treatment plants: a critical analysis and guideline proposal. Water Sci Technol 83(3):501–514

Samsami Z, Jafari AJ, Delnavaz M, Naderi A, Rashvanlou RB, Dehghanifard E (2020) Environmental assessment of a wastewater treatment plant using life cycle assessment (LCA) approach: case of Ain Taoujdate Morocco. Int J Adv Res Eng Technol (IJARET) 11(5)

Tran HV, Phan HA, Bui HM (2024) Life cycle assessment of industry wastewater treatment plant: a case study in Vietnam. RSC Sustain 3(3):1415–1423

Chapter 31
Research on Functional Verification Test Methods for Water and Waste System

Caixia Liu, Yifeng Zhang, and Yudi Wang

Abstract This paper introduces a civil aircraft water and waste system function verification test platform and test method, which can be used to design and test the core function and performance of water and waste system. Through the study of the system bench test, the comprehensive performance of the water and waste system can be subjectively and objectively evaluated for continuous improvement. The test bench can provide a reference for the water and waste system principle prototype and the real machine system design scheme and design method selection and improvement.

Keywords Water and waste system · Test-bed · Function verification

31.1 Introduction

Civil aircraft water and waste system includes two sub-systems, drinking water system and wastewater treatment system, and its main users are the crew members and passengers on board and ground service staff. The main function of the potable water system is to provide water for the on-board galley taps, lavatory taps, toilet areas, etc. by storing a certain amount of water and using pressure water supply and distribution piping to meet the needs of passengers' on-board life. The main function of the wastewater treatment system is to provide and maintain a clean environment on board by treating on-board waste water including wash water from the bathroom and galley and wastewater from the toilet, and discharging the wastewater during ground maintenance.

The development of aircraft water supply system has gone through four stages: the first stage is the use of independent water storage tanks for water supply in the cabin; the second stage is the use of gravity water supply; the third stage is

C. Liu (✉) · Y. Zhang · Y. Wang
State Key Laboratory of Airliner Integration Technology and Flight Simulation, Shanghai Aircraft Design and Research Institute, Shanghai, China
e-mail: liucaixia@comac.cc

Y. Li et al. (eds.), *16th International Conference on Environmental Science and Development*, Environmental Science and Engineering,
https://doi.org/10.1007/978-3-032-15080-6_31

the use of pressure water supply; the fourth stage is a high degree of integration and information exchange with on-board system based on pressure water supply to achieve mechatronic integration. Comparing gravity water supply and pressure water supply, the former is simple in structure, easy to maintain and operate. Its product has fewer accessories and is lighter in weight. However, its water tank needs to be placed on the roof of passenger cabin, which is mostly a small space, so the installation position of the water tank will be limited. Especially for large airliners this type of water supply does more harm than good. The water tank, using pressure water supply, can be installed in the larger cargo hold under the floor of the passenger cabin, where there are no space limitations. Pressure water supply adds a set of pressurization equipment and is more complicated to arrange than gravity water supply, but works more reliably. In terms of ground maintenance, gravity water system relies on gravity to discharge standing water in the system, but pressure water supplies can be discharged under pressure, which can speed up the discharge rate. Therefore, most airliners use pressure water supply. The water system consists of a water tank, a water quantity indication system, a water tank pressurization system, a water service panel, a distribution line and a discharge system (Long et al. 2016; Lei 2012).

The aircraft wastewater treatment system is described separately for washing effluent and toilet effluent. There are two forms of washing wastewater treatment system, one type that discharges directly into the atmosphere outside the aircraft, and another type that discharges first into the aircraft's internal sewage tanks and then treats it after reaching the ground. The latter, accompanying the creation of vacuum wastewater systems, is used in many aircraft types. To simplify the system, the washing wastewater and toilet wastewater usually share a common wastewater tank, a set of vacuum generating units, ground service units, etc. Toilet sewage has gone through the "dry toilet" type, the self-circulating type and the electronically controlled vacuum type. The self-circulating type utilizes sewage stored in a sewage tank to flush the toilet. Press the water electric door after using the toilet and the water in the sewage tank, filtered and pressurized, is injected into the toilet by the flushing pump for flushing. After flushing, all dirt is stored as it enters the sewage tank. Compared with self-circulating system, vacuum wastewater system, taking the vacuum principle to draw feces and other wastewater into the wastewater tank, is more hygienic and water-efficient. The water used to flush the toilet uses clean water from the water system, which greatly improves the hygienic condition for toilet flushing. In addition, this system is less polluting to the cockpit environment, more flexible when modifying the cockpit interior, more maintainable and more advanced and so on. The principle of the system is to make use of the pressure difference generated between the passenger compartment and the wastewater tank so that the toilet excreta is transported to the wastewater tank (Zhang et al. 2009; Xiao 2013).

31.2 Principle of Operation

31.2.1 Analysis of General Requirements

The test platform should be able to verify the core functions and performance of the water and waste system, and provide the overhead environment simulation device, tank pressurization simulation device, filling/discharging simulation device, etc., which are required for its operation. The overall scheme layout is shown in Fig. 31.1.

The pressurized water supply system test platform is mainly composed of a water filling/supply system and a pressurization system. The water filling system provides sufficient and clean water for the whole system. The pressurization system provides the power for the water supply system to deliver water to the user units. The water supply system provides users with the required water. Specifically, it includes water tank, variable frequency water pump, water supply tank, air compressor, filter, sensor, valve, pipeline, controller and so on.

The wastewater system test platform mainly consists of a wastewater treatment system, an overhead environmental simulation system and a wastewater discharge system. The wastewater treatment system drains washing sewage and toilet sewage into the sewage tank. The overhead environment simulation system provides the power for the sewage to enter the sewage tank, and the wastewater discharge system discharges the water from the sewage tank into the sewage collection tank. Specifically, it includes a hand washing sink assembly, vacuum toilet assembly, sewage tank, sewage storage tank, sensors, vacuum pump, valve, pipeline, controllers, and ground service panels, etc.

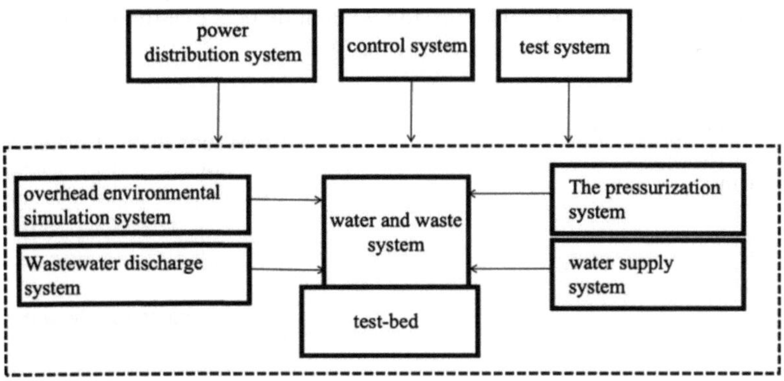

Fig. 31.1 The overall scheme layout

31.2.2 Working Principle of Drinking Water System

The working principle of the pressurized water supply system test platform is shown in Fig. 31.2. Through the water filling system and the pressurization device, the water supply tank can normally deliver a fixed amount of water to the water points in the kitchen and washroom.

The principle of water supply tank filling as shown in Fig. 31.3, the water supply system is mainly through the water tank and frequency conversion pump for water filling. As the water system also meets the flushing system water requirements, set up a tee at the outlet end of the pump, and set up water valves and cleaning valves. In order to meet the water supply tank water pressure regulation, in the water pump and the tee to increase the pressure sensor. Water tank through the hose and tap water faucet connection, and a water level observation tube is arranged on the top of the water tank. In order to facilitate the cleaning of the water tank itself, there is a tee connector and a mechanical switch installed at the connection between the inlet of the water pump and the water filling tank.

Figure 31.4 shows the pressurization principle diagram of the water supply tank, the pressure booster device is mainly to simulate the aircraft water supply tank pressure environment, to ensure that the pressurization water supply system can reach the normal water supply pressure. The simulation of the environment is mainly achieved through the air compressor, when the pressure in the water supply tank decreases, the control system starts the air compressor. By detecting the pressure sensor and controlling the state of the proportional valve, the output pressure is adjusted. In order to avoid high pressure beyond the capacity of the pressurised water supply system, a tee is connected between the air compressor and the water tank, and an evacuation valve is set at the end, so that the control system will start the evacuation valve to reduce the pressure when it detects that the water pressure is too high.

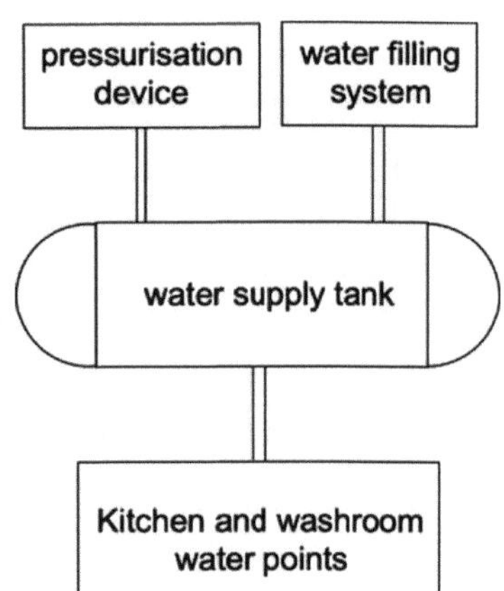

Fig. 31.2 Working principle of the pressurized water supply system

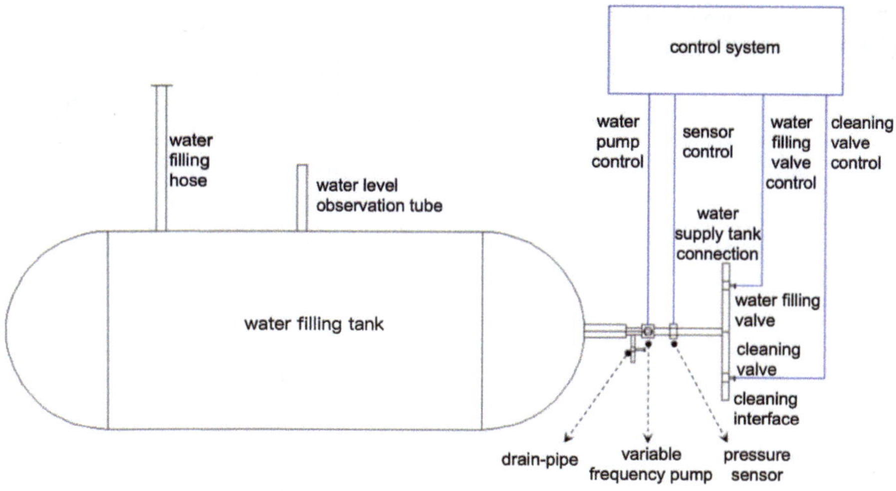

Fig. 31.3 Principle of water supply tank filling

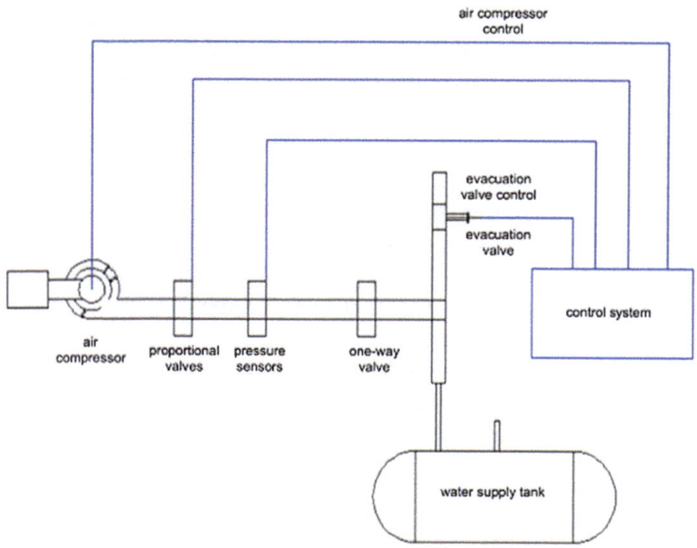

Fig. 31.4 Principle of pressurization of water supply tanks

31.2.3 Working Principle of Wastewater System

Wastewater system test platform working principle shown in Fig. 31.5. The high-altitude environment simulation device makes the pressure difference between the kitchen launching point and the lavatory launching point and the sewage tank, and the pressure difference is the driving force to discharge all the generated wastewater into

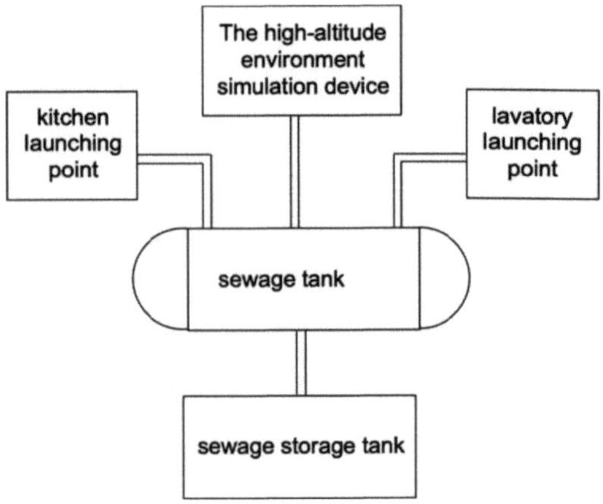

Fig. 31.5 Working principle of wastewater system

the sewage tank, and the sewage in the sewage tank is discharged into the sewage storage tank as needed. The general sewage storage tank is fixed on the transfer vehicle, which is easy to move and dispose of sewage.

The working principle of the high-altitude environment simulation system is shown in Fig. 31.6 which simulates the pressure difference at different altitudes through the pressure adjustment of the vacuum tank. Sensors set in the vacuum tank monitor the pressure in the vacuum tank in real time, and control the vacuum pump and air supplement valve to work as needed. The vacuum tank is connected to the vent pipe of the sewage tank.

The wastewater discharge system mainly contains two functions: residual water of the water supply system, waste water collection and cleaning of the sewage tank. The working principle is shown in Fig. 31.7. Above the sewage storage tank is set up with a hose to connect with the sewage tank or water supply tank, and the connection interface is a quick-disconnecting connector. A sewage pipe is arranged at the bottom of the side of the sewage storage tank, and a mechanical valve is arranged on the sewage pipe to control sewage discharge. When the cleaning function is used, the cleaning interface in the water refilling equipment is connected to the sewage tank, the control system issues a command, the water refilling valve is closed, the cleaning

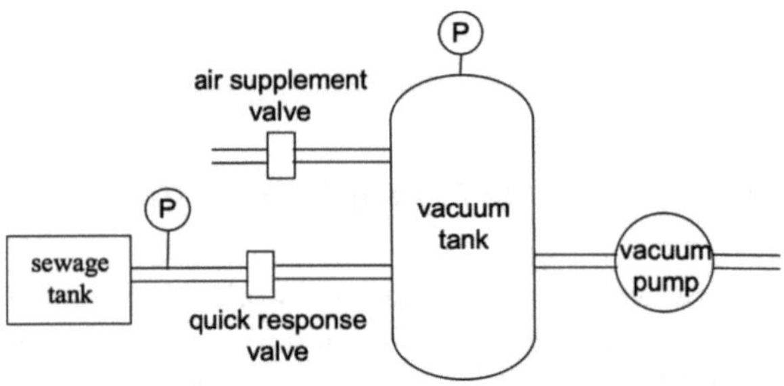

Fig. 31.6 Working principle of the high-altitude environment simulation system

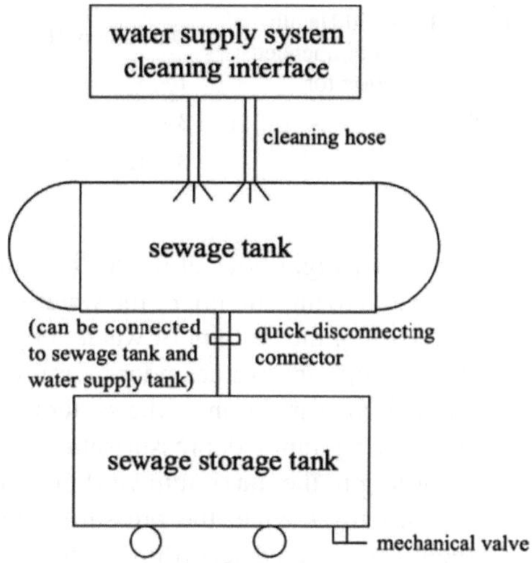

Fig. 31.7 Schematic diagram of wastewater discharge system

valve is opened, and the pump works after the command is issued to carry out the cleaning function.

31.3 Test Verification Project

At least water supply system water filling function test, drainage function test, pressure boosting performance test, water supply function and performance test; wastewater vacuum system performance test, wastewater system toilet flushing function test, sewage discharge function test, washing function test, washing sewage drainage function test and other system tests, check the system working condition, in order to verify whether the system meets the relevant requirements (HB 8488-2014 2014; HB 8489-2014 2014; CCAR25-R4 2011; Radio Technical Commission for Aeronautics 2010).

31.3.1 Key Performance Design Requirements

(1) The volume of the water supply tank is determined by the formula "number of crew members × maximum flight hours × water consumption (per person per hour) + margin". Table 31.1 is the water consumption of passengers recommended by the World Health Organization. Table 31.2 is the water storage capacity of foreign civil aircraft (General Editorial Board 2002; He 2017). According to the general requirements of China's civil aircraft water supply system is generally appropriate for the crew to provide at least 0.24 L per

Table 31.1 World Health Organization recommended water consumption for travellers

Duration of flight (h)	Water consumption per passenger (L)
1 ~ 3	1.7 or 1.36
3 ~ 5	3.10 or 2.5
5 ~ 12	4.54 or 3.64

person per flight hour of domestic water. Water supply tank volume should be compared to the amount of the water supply to leave a certain margin, generally take 10% of the amount of water supply.

(2) When supplying water under pressure, the tank must withstand the maximum internal pressurisation of the system as well as the maximum differential pressure during rapid depressurisation. Positive pressure load can be designed according to the maximum working pressure of the system 1.5 times.

(3) Pressurizing device, the pressure value of the pressure water supply system should be selected according to the size of the water supply, the distance of the water supply and the height of the water supply point. Under normal flight and normal pressure, with the switch open 50%, the recommended water flow rate of the water outlet in the cockpit is 3.785 L/min.

(4) The volume of the sewage tank is determined by the formula "number of crew members × maximum flight hours × amount of waste water (per person per hour) + margin". According to the general requirements of China's civil aircraft sewage discharge system, when the general flight time is less than 3 h, it is necessary to provide at least 0.96 ~ 1.12 L per person sewage tank volume for the crew.

(5) The strength of the sewage tank should be enough to withstand the negative pressure caused by the vacuum inside the tank. It should be able to withstand a vacuum of 1.5 times the maximum pressure difference between inside and outside the cabin.

Table 31.2 Water storage capacity of foreign civil aircraft

Aircraft type	Number of passengers	Total water storage (L)	Per capita water consumption (L)
A320	150 ~ 180	200	1.1 ~ 1.3
B707	189 ~ 219	390	1.78 ~ 2
B737	130 ~ 149	128	0.85 ~ 1
B787	228	270	1.18
Trident	115 ~ 149	158	1.06 ~ 1.37
MD-82	155 ~ 172	177.89	1.03 ~ 1.15

31.3.2 Key Performance Design Requirements

The water and waste system test verification method is as follows:

(1) Water supply system water filling function test. In order to minimize the ground maintenance time before the aircraft takes off, the design of pipeline and water filling connector should ensure that when the pressure of water added is a certain value (varying with different types of aircraft due to different working pressures), the time of water tank filling should not be more than 10 min. The key collection parameters are water tank filling pressure and flow, water tank level and effective filling time.
(2) Water supply system drainage function test. Drainage should be smooth and exhaustive. In order to minimize the ground maintenance time after landing of the aircraft, the discharge time of the water supply system should not exceed 10 min. The key parameters to be collected are the amount of water discharged and the time of discharging.
(3) Pressure boosting performance test of the water supply system. The test conditions should at least determine the tank volume, tank pressure adjustment range, tank positive pressure load, system water supply pressure, tank water volume and pressure booster pre-set value, etc. The time to pressurize the water supply tank to the target pressure should be as small as possible. Pressurization to the target pressure should be as small as possible. The key acquisition parameters are the supply gas pressure used for tank pressurization and the effective pressurization time.
(4) Water supply function and performance test. The water flow is smooth in the water supply process, and the flow rate of a single faucet outlet is much higher than the minimum set value. The key collection parameters are the flow of faucet outlet at a fixed time, the instantaneous water supply pressure and flow of water supply pipeline.
(5) Wastewater vacuum system performance test to test the time it takes for a sewage tank to go from atmospheric pressure to a permissible pressure at which a toilet can be used, and to test the change in pressure in the sewage tank when the toilet is flushed. The key acquisition parameter is the pressure in the wastewater tank.
(6) Wastewater system toilet flushing function test, should be able to flush clean sewage in the toilet, each flushing cycle should be no more than 15 s. Toilet without residual single flush water volume is not greater than 0.25 L. The key acquisition parameter is the flushing cycle time and flushing water volume.
(7) Sewage discharge function test. It should be able to rely on gravity for sewage discharge, and the sewage tank should be drained within 10 min. The key acquisition parameter is the effective time of complete discharge of sewage.
(8) Washing function test to test whether the flushing flow rate of the sewage tank meets the requirements. The key acquisition parameters are the instantaneous water supply flow and water supply pressure of sewage tank cleaning.
(9) Washing sewage drainage function test. Drainage should be smooth and exhaustive.

31.4 Summary

As one of the important aircraft systems, the water and waste system is responsible for providing safe and stable water for passengers and crew and treating wastewater generated from daily life. In order to better master the technology related to the water and waste system, it is necessary to build a ground test platform that provides the required high-altitude environment, pressurized environment, power supply environment, water filling, water draining, and aircraft bumpiness and other environmental simulations of the system through peripheral facilities. This paper introduces a test platform and verification method that can be used for the functional verification of aircraft water and waste systems. Through the study of the system bench test, it can provide reference for the selection and improvement of the design scheme and design method of the principle prototype and real system of water and wastewater system. The engineering practice proves that the test bench and the test method are feasible for assessing the water and waste system.

References

CCAR25-R4 (2011) Civil aviation regulations of China. Part 25—airworthiness standards for transport aircraft. Civil Aviation Administration of China

General Editorial Board of Aircraft Design Manual (2002) Aircraft design manual (No. 11): internal facilities of civil aircraft. Aviation Industry Press, Beijing

HB 8488-2014 (2014) General requirements for civil aircraft water supply systems

HB 8489-2014 (2014) General requirements for civil aircraft sewage disposal systems

He Y (2017) Analysis of Airbus A320 aircraft water system composition and daily service work. Tibet's Sci Technol 6:77–80

Lei M (2012) Research on types of water system for new civil aircraft. Adv Aeronaut Sci Eng 3(2):229–234

Long J, Liu F, Zhang Z (2016) Modern aircraft structures and systems. Northwestern Polytechnical University Press, Xi'an

Radio Technical Commission for Aeronautics (2010) DO-160G environmental conditions and test procedures for airborne equipment. Washington

Xiao S (2013) Introduction of vacuum toilet sewage treatment system for civil aircraft. Sci Technol Inf 16:405–406

Zhang W, Xiao S, Lei M (2009) Civil aircraft galleys wastewater treatment technology research. Civ Aircr Des Res 3:44–47

Chapter 32
Research on Flammable Liquid Drainage Path Design and Analysis of Civil Aircraft

Wenjie Xu and Yadi Liu

Abstract The design of civil aircraft flammable liquid drainage path is an important part of aircraft fire safety. The reasonable liquid drainage path design can improve the corrosion resistance of aircraft structure parts and ensure the aircraft operation safety. The airworthiness requirements and design indexes of civil aircraft liquid drainage path are defined in this paper, the relationship between liquid drainage outlet size and liquid drainage rate is studied, the mathematical model of liquid drainage rate is established, and the design method of liquid drainage path is proposed. According to the liquid drainage path of the fuselage, the evaluation method and process of the liquid drainage path on the ground and in the flight stage are summarized, which provides technical guidance for the design and analysis of the drainage path for civil aircraft.

Keywords Aircraft fire safety · Flammable liquid · Drainage path · Drainage rate · Liquid drainage hole

32.1 Introduction

The fire safety of flammable liquid is an important part of the fire safety design and verification of civil aircraft. CCAR 25.863 stipulates the fire safety design requirements for the flammable liquid leakage area of the whole civil aircraft, and in the flammable liquid leakage area, reasonable structural internal drainage paths and external drainage outlets need to be designed to ensure that the flammable liquid leakage can be discharged out of the aircraft in a timely manner without the accumulation of hazardous quantities, and that it can be prevented from discharging out of the airplane. Flammable liquid can be prevented from re-entering the fuselage or other areas (ventilation in-lets and outlets, hot surfaces, etc.) that may pose a hazard,

W. Xu (✉) · Y. Liu
COMAC Shanghai Aircraft Design and Research Institute, Shanghai, China
e-mail: xuwenjie@comac.cc

Y. Li et al. (eds.), *16th International Conference on Environmental Science and Development*, Environmental Science and Engineering,
https://doi.org/10.1007/978-3-032-15080-6_32

thus reducing the risk of fire. During the airworthiness certification process, ground and in-flight liquid drainage tests of flammable liquid spill areas are required to verify the aircraft's liquid drainage capability (Yin et al. 2014; Zhu et al. 2022).

Based on the airworthiness requirements, this paper studies the liquid drainage design indicator, liquid drainage capacity calculation method, proposes a common liquid drainage hole design method, liquid drainage path analysis method, and provides effective guidance for the design of civil aircraft liquid drainage.

32.2 Civil Aircraft Liquid Drainage Design Requirements

The design of civil aircraft liquid drainage needs to consider the liquid drainage outlet, liquid drainage rate, liquid accumulation area and liquid drainage path, etc. The corresponding design guidelines are summarized as follows (FAA 2002).

(1) Liquid drainage measures should enable flammable liquid leaks to be discharged out of the aircraft in a timely manner, in the case of a leakage discharge rate of 3.8 L/min at the point of leakage and a discharge volume of 15 L, more than 90% of the liquid can be discharged within 10 min;
(2) The liquid reservoir storage capacity of the drainage area needs to match the discharge capacity of the drainage pipe to prevent liquid from spilling out of the reservoir;
(3) No more than 44 ml (1.5 oz) of fluid accumulates in a single effusion site in each drainage area;
(4) The location of the drainage outlet should avoid the area affected by high temperature jets;
(5) The drainage path should be designed to avoid contact with ignition sources, reducing fire hazards.
(6) The external drainage path is capable of preventing the drained fluid from reentering the aircraft's internal danger zones during the flight phase.

32.3 Civil Aircraft Liquid Drainage Hole Design Methods

In order to meet the regulation of CCAR25.863 for the verification of liquid drainage during ground test and flight test, it is necessary to calculate and analyze the rate of liquid drainage and the amount of liquid accumulation to determine the form, size and location of the liquid drainage hole for the aircraft (Ke et al. 2022; Liu and Wang 2011; Hu 2022).

32.3.1 Mathematical Modeling of Discharge Rate

To ensure sufficient drainage capacity, there is a basic principle for the design of drainage holes on civil aircraft: drainage holes should be designed in front of each bulkhead, with the cross-sectional area of the drainage outlet not <65 mm^2, and the cross-sectional area of the internal drainage channel in the structure not <70 mm^2.

For areas where liquid storage is present, it is necessary to consider the relationship between the liquid storage capacity of the liquid storage tank and the drainage capacity of the drainage pipe, and establish a calculation model for the size of the drainage outlet in special liquid storage areas.

According to the Reynolds Transport Theorem as shown in Fig. 32.1, v_1, v_2 are uniformly distributed in the inlet and outlet regions, the continuity equation for the flow in a constant, incompressible, flow tube is obtained:

$$v_2 A_2 - v_1 A_1 = 0 \tag{32.1}$$

where v_1, A_1 are the inlet flow rate and inlet area respectively; v_2, A_2 are the outlet flow rate and outlet area respectively.

Considering equal flow parameters at each point on the inlet and outlet cross sections, and the fluid continuity equation, the Bernoulli equation is obtained:

$$gz_1 + \frac{v_1^2}{2} + \frac{p_1}{\rho} = gz_2 + \frac{v_2^2}{2} + \frac{p_2}{\rho} \tag{32.2}$$

For the drainage rate problem, it can be abstracted as a mode, which is shown in Fig. 32.2, there is a large container with a bottom area of A_1, which contains a liquid height of z, the liquid is discharged by opening a drainage hole of area of A_2 below the large container. The upper surface of the container is subjected to the pressure of p_1, the rate of the container liquid surface falling for v_1, the pressure of the exit of the small holes for p_2, the rate of the liquid drainage from the small holes for v_2 (Chen et al. 2017; Weng et al. 2018; Zhang 2015).

Establish the Bernoulli equation for the vessel level and outlet, where $z_1 - z_2 = z$, $p_1 = p_2$ are obtained:

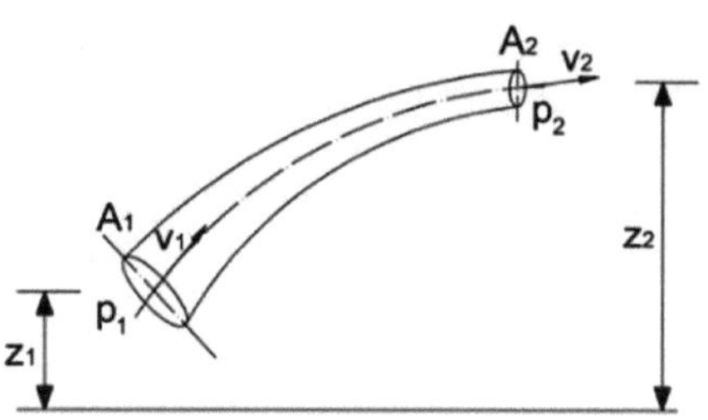

Fig. 32.1 Schematic of flow tube calculation

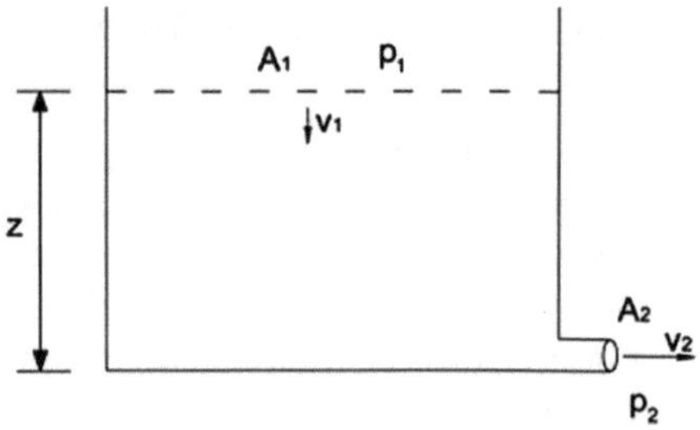

Fig. 32.2 Schematic of liquid discharge

$$gz + \frac{v_1^2}{2} = \frac{v_2^2}{2} \tag{32.3}$$

In the liquid drainage test process, in the leakage source leakage and drainage hole discharge liquid synchronization, in order to ensure that there is no large amount of liquid diffusion out of the liquid tank, it is necessary to ensure that the leakage flow rate is equal to the flow rate of discharge, the liquid level of the liquid tank z remains unchanged, so it is necessary to design the dimensions of the liquid drainage hole area A_2 in the above situation.

Based on the flow continuity equation and Bernoulli's equation, the formula for the drainage hole area A_2 was obtained:

$$A_2 = \frac{v_1 \cdot A_1}{\sqrt{2gz + v_1^2}} = \frac{Q}{\sqrt{2gz + \left(\frac{Q}{A_1}\right)^2}} \tag{32.4}$$

where Q is the drainage flow rate (volume of liquid drainage per unit of time), according to AC 25.863 draft the discharge flow rate is usually set to 3.8 L/min.

From the above equation, it can be seen that the size of the drainage hole area is directly proportional to the bottom area of the liquid accumulation and inversely proportional to the quadratic root of the height.

While calculating the area of the drainage hole, in order to meet the demand for liquid to be discharged out of the body within a specified time, the corresponding drainage time needs to be calculated, and if it is not possible to meet the requirements, the hole diameter needs to be enlarged.

After the leakage source has finished leaking, liquid is discharged only from the discharge port of the accumulation tank, and the corresponding discharge time is calculated as follows:

Since $A_1 >> A_2$, according to the flow continuity equation, $v_1 << v_2$, it can be obtained by neglecting the $v_1^2/2$ term:

$$v_2 = \sqrt{2gz} \tag{32.5}$$

Calculate the height of the liquid level in a container as a function of time:

$$\frac{dz}{dt} = -v_1 = -\frac{v_2 \cdot A_2}{A_1} = -\frac{A_2}{A_1} \cdot \sqrt{2gz} \tag{32.6}$$

The time for the liquid to be completely discharged from the liquid level z can be calculated by integration:

$$t = \frac{A_1}{A_2} \cdot \sqrt{\frac{2z}{g}} \tag{32.7}$$

As can be seen from the above equation, the time for all the liquid in the accumulation tank to flow out is proportional to the quadratic root of the accumulation bottom area and the initial height of the liquid surface and inversely proportional to the area of the drainage hole.

32.3.2 Design of Drainage Hole Location

The location of the liquid drainage port should be arranged at the lowest point of the liquid drainage, and at the same time if the angle between the axis of the liquid drainage pipe and the horizontal plane, it is recommended to be >5° in order to ensure that the liquid accumulated inside the hanging can be discharged smoothly under the action of gravity.

As the liquid discharged from the liquid drainage port are flammable liquid, it is necessary to consider the temperature limit of the engine jet region under the temperature limit operating conditions (including normal operating conditions and engine failure, engine performance degradation conditions) the temperature field distribution of the engine inner/outer connotation tail nozzle jet, a model of the jet temperature field as shown in Fig. 32.3, the liquid drainage port location should be arranged to avoid high-temperature spray area, arranged in the lower temperature zone.

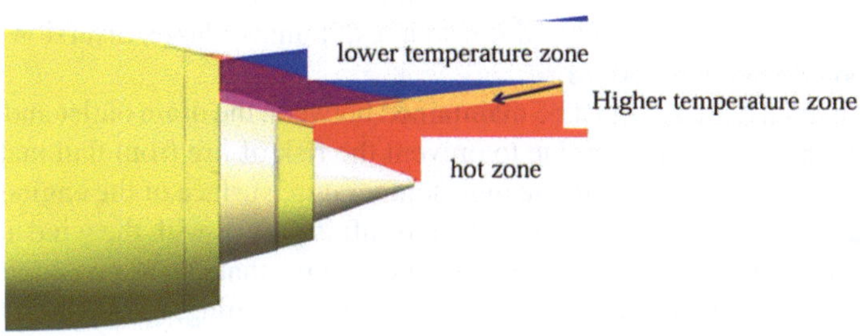

Fig. 32.3 Schematic of jet temperature field

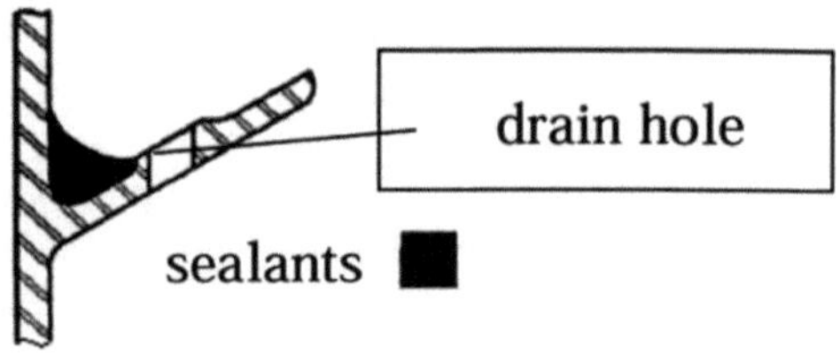

Fig. 32.4 Schematic diagram of sealant filling

32.3.3 Design of Drainage Hole Location

According to AC 25.863 draft, for areas where are susceptible to the accumulation of flammable liquid, the volume of liquid accumulated in a single tank should be <44 ml. If there is a notch in the part structure on the discharge path where the volume of liquid accumulated is >44 ml, the area can be filled in by means of a new discharge port, discharge path, or the use of a sealant as shown in Fig. 32.4.

32.4 Methods for Analyzing Drainage Paths in Civil Aircraft

After specifying the form, location and size of the aircraft drainage holes, it is necessary to analyze and demonstrate liquid drainage path to ensure the safety of liquid drainage.

32.4.1 Methods for Analyzing Drainage Paths on the Ground Stage

The specific discharge path of the flammable liquid from the leak source needs to be analyzed during the ground stage, with the main design requirements being.

Drainage paths should avoid ignition sources, if unavoidable, diversion measures are required (Zheng et al. 2018).

A certain axial distance must be maintained between the drain outlet and the high-temperature tail cone of the engine to prevent the risk of fire from flammable liquid dripping from the drain rod onto the high-temperature surface of the engine tail cone under the action of the tail wind when the aircraft is parked with the wind at its back.

In response to Design Requirement (1), the designs that can be considered are the addition of a deflector strip or the addition of a drain bushing, as shown in Fig. 32.5.

For the design requirements (2) need to consider the trajectory of the liquid discharged from the liquid drainage port in the aircraft shutdown state, the operating conditions of operational wind speed. The integral method can be used to calculate the angle of deflection of the droplets under the action of wind speed, where the force

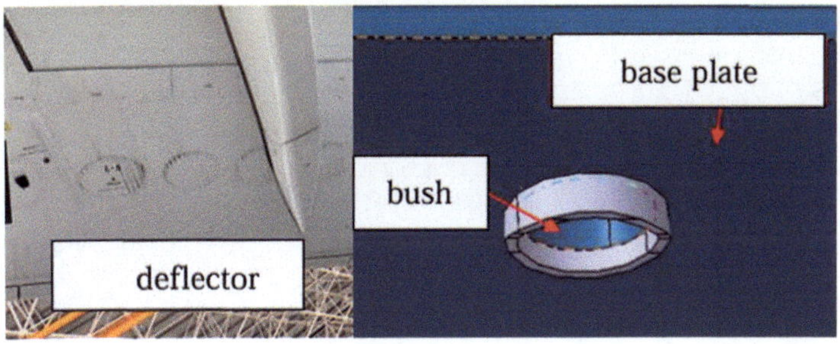

Fig. 32.5 Schematic diagram of drainage measures

coefficient of the droplets is calculated according to the following formula:

$$C_d = \frac{24}{\text{Re}}\left(1.0 + 0.197\text{Re}^{0.68} + 2.6 \times 10^{-4}\text{Re}^{1.38}\right) \tag{32.8}$$

Where Re is the Reynolds number.

Based on the liquid trailing force, buoyancy and gravity, the displacement of the droplet at different moments is calculated separately:

$$F_{force} = \frac{A}{2} \cdot \rho \cdot (v_{air} - v_{droplet})^2 \cdot C_d \tag{32.9}$$

$$u_d = u_{d0} + \frac{F_{forcecomponent}}{m} \cdot \mathrm{d}t \tag{32.10}$$

$$v_d = v_{d0} + \left(\frac{F_{buoyancy} + F_{forcecomponent}}{m} - g\right) \cdot \mathrm{d}t \tag{32.11}$$

The droplet trajectories under the influence of different wind speeds of 15 and 10 m/s are calculated and shown in Fig. 32.6.

According to the research results, it can be seen that usually wind speed below grade 5 has no effect on the takeoff of the aircraft, grade 6–8 gusts need to depend on the direction of the wind and judge whether to take off or not at their discretion, and all the aircraft are grounded in case of gusts of grade 9 or above. The location of the liquid drainage outlet should be able to prevent liquid droplets from falling onto the surface of the tailbone under the influence of a wind speed of 10 m/s (force 5 wind).

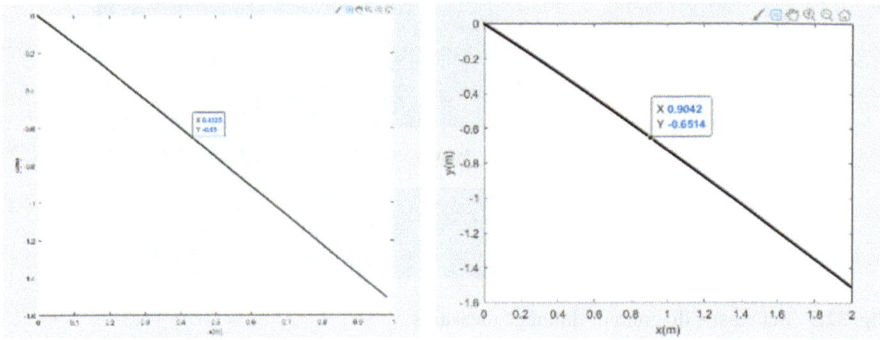

(a) 10 m/s wind speed (b) 15 m/s wind velocity

Fig. 32.6 Schematic of the fluid trajectory

32.4.2 Methods for Analyzing Drainage Paths in the Flight Stage

In the aircraft flight phase, the external flammable liquid drainage path needs to be analyzed, to make sure there is capability to prevent the drained fluid from re-enter the internal danger zones of aircraft. The flow steps are shown in Fig. 32.7.

The main factors affecting the external drainage path of an airplane are the airplane configuration, the incoming flow Mach number, the angle of approach and the sideslip angle. Therefore, considering the typical flight scenarios of the airplane, different jamming configurations are selected as the computational working conditions, and each configuration is considered to have four states, namely, high-speed sideslip, high-speed no sideslip, low-speed sideslip, and low-speed no sideslip.

Grid division of aircraft configurations with different card positions and encrypted control at the wing and jet regions, the meshing of an civil aircraft is shown in Fig. 32.8. In order to investigate the influence of the grid size and the thickness of the boundary layer on the calculation results, the calculation is carried out by controlling the grid base size, the number of grid layers in the boundary layer, the growth rate, and the calculation is divided from sparse to dense, and the grid-independence verification is carried out, and the mesh with the appropriate accuracy is selected.

The CFD calculation method is used to discretize the RANS equations using the finite volume method, and the SST $k - \omega$ two-equation model is selected for the turbulence model. Based on the following assumptions: the drainage flammable liquid is atomized into very small droplets, the gravity of the small droplets themselves is negligible relative to the aerodynamic force of the air on the droplets, and the droplets move in the air and on the surface of the fuselage with the movement of the airflow. Detailed boundary condition settings is shown in Table 32.1.

In the analysis, the flow trajectory of the flammable liquid after it is discharged from the body is characterized by the flow line released from the location of the drainage outlet as a means of determining the contamination of the flammable liquid

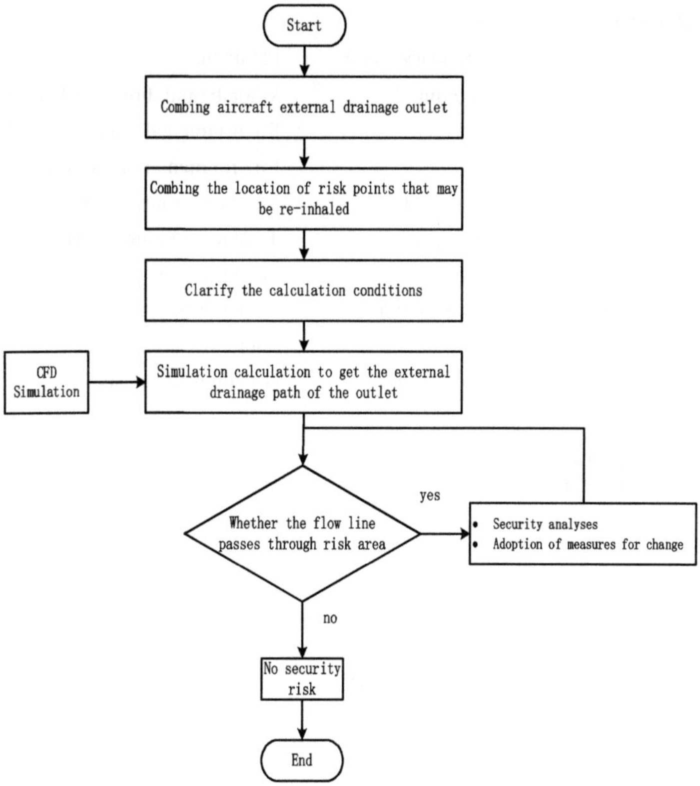

Fig. 32.7 Flow of external drainage path analysis

Fig. 32.8 Grid diagram

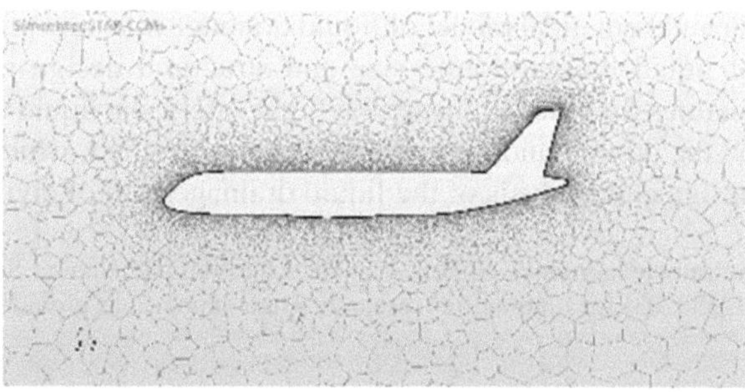

on the body, and the trajectory of the external drainage path for a particular model is shown schematically in Fig. 32.9.

Table 32.1 Boundary condition settings

Region	Boundary name	Parameter
Far field	Opening	Velocity and direction *Ma* (m/s)
		Environmental temperature T_0 (K)
		Environmental pressure P_0 (Pa)
Engine inlet duct entrance	Flow outlet	Mass flow rate $\dot{m}$ (kg/s)
Engine outer duct inlet	Pressure inlet	Total temperature T_1 (K)
		Total pressure P_1 (Pa)
Engine inner duct outlet	Pressure inlet	Total temperature T_2 (K)
		Total pressure P_2 (Pa)
Airframe surface	Wall surface	No-slip wall surface

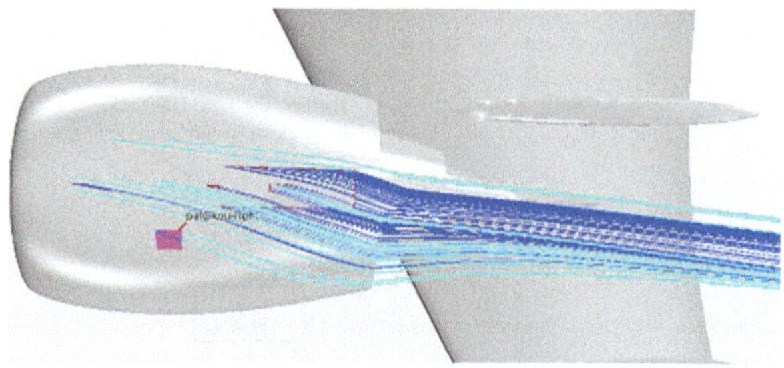

Fig. 32.9 Schematic of the drainage path trajectory of a model

32.5 Conclusion

This paper discusses the design requirements of civil aircraft liquid drainage, studies the mathematical model of liquid drainage rate, and proposes the calculation method of liquid drainage hole size and structural design means for different structural features. In order to ensure the flight safety, the liquid drainage path analysis method on the ground and in the flight stage is clarified, which provides a reference for the design and analysis of the liquid drainage path of civil aircraft.

Acknowledgements Sincere thanks to the National Natural Science of China (NSFC, Grant No. U2333203) for funding this research. Their support has provided a foundation for this study.

References

Chen A, Xie J, Wen X (2017) Draining scheme of aircraft under influence of multiple factors. Aeronaut Soc China J Chengdu Aircraft des Instit 4:5428

FAA (2002) Flammable fluid fire safety: AC 25.863 draft. FAA, Washington

Hu RQ (2022) Numerical study on flow coefficient of aero-engine nacelle discharge hole. D Nanjing University of Aeronautics and Astronautics, Beijing

Ke J, Wang M, Tao W et al (2022) Analysis of liquid drainage from local structure of APU hatch. J Civil Aircraft des Res 02:62–66

Liu L, Wang D (2011) Preliminary method for determining the size of APU cabin drain hole. J Civil Aircraft des Res 04:21–23

Weng C, Cui W, Lian W et al (2018) Design and research on civil aircraft slide sleeve fluid discharge path. J Civil Aircraft des Res 01:62–66

Yin W, Yu S, Tang H (2014) Study on fire safety design requirements for civil aircraft. J Civil Aircraft des Res 02:11–30

Zhang J (2015) Design of APU hatch liquid drainage for civil aircraft. J Civil Aircraft des Res 04:45–47

Zheng Q, Sun Y, Wang X et al (2018) Effect of particle size distribution of flammable liquid droplets on their ignitability. J J Saf Environ 05:1779–1783

Zhu R, Xia W, Hao Y et al (2022) Technical analysis of airworthiness certification for flammable liquid fire safety. J J Civil Aviat 04(6):96–98

Chapter 33
Identification and Analysis of Key Environmental Impact Factors in the End-of-Life Treatment of Household Appliances in China

Leixu Huang, Mengyu Zhai, Yingfang Zhu, Weiwei Zhai, Chao Wang, Shuo Shi, Yingao Gao, Gang Dong, Fengling He, Canyan Chen, and Shuai Su

Abstract The rapid growth of the home appliance industry has caused a sharp rise in discarded appliances, making making their proper disposal is a vital environmental issue. This study employed the life cycle assessment (LCA) approach to systematically evaluate the environmental impacts of end - of - life management for refrigerators, air conditioners, and washing machines. The research scope covers transportation, disassembly, and reuse of disassembly products. Through collecting real - world and background data for inventory analysis, six typical environmental impact indicators were selected for evaluation.Results show that drum washing machines have relatively high environmental costs in waste treatment. Their ADPe is 1.71E-06 kg Sb eq and ADPf is 2.78E+02 MJ, higher than those of refrigerators and indoor air conditioners. Contribution analysis indicates that the iron recycling stage of refrigerators has the greatest impact on the six environmental aspects; the disassembly and recycling of four elements of indoor air conditioners significantly affect ADPe; and

L. Huang · M. Zhai · Y. Zhu · C. Chen
Institute of Circular Economy, Beijing University of Technology, Beijing, China

College of Material Science and Engineering, Beijing University of Technology, Beijing, China

C. Chen
e-mail: ccyseven@gmail.com

W. Zhai · C. Wang
Qingdao Penghai Software Co., Ltd., Qingdao, Shandong, China

S. Shi · Y. Gao
Faculty of Information Science and Engineering, Ocean University of China, Qingdao, China

G. Dong · F. He
Qingdao Hailuyuan Recycling Technology Co., Ltd., Qingdao, Shandong, China

S. Su (✉)
State Grid Integrated Energy Service Group Co., Ltd., Beijing, China
e-mail: 18810803173@163.com

Y. Li et al. (eds.), *16th International Conference on Environmental Science and Development*, Environmental Science and Engineering,
https://doi.org/10.1007/978-3-032-15080-6_33

the iron recycling stage of drum washing machines has the most significant environmental impacts.In the assessment of recycling benefits, except for copper, recycled aluminum, iron, and plastic consume far less fossil energy than primary materials. For example, recycled aluminum saves 4.63 MJ compared to primary aluminum. Except for iron, recycling copper, aluminum, and plastic brings significant carbon reduction benefits. This study provides a basis for optimizing home appliance waste disposal and promoting industry sustainability.

Keywords Typical home appliances · Environmental impact assessment · Resource recycling benefits · Life cycle assessment

33.1 Introduction

33.1.1 Background

The burgeoning global home appliance industry has significantly improved quality of life but concurrently presents formidable challenges in managing Waste Electrical and Electronic Equipment (WEEE), commonly known as e-waste. The accelerated pace of product replacement, driven by the integration of technologies like the Internet of Things (IoT) and Artificial Intelligence (AI), further exacerbates the generation of obsolete appliances. According to the "Global E-waste Monitor" report jointly released by the International Telecommunication Union and the United Nations Institute for Training and Research, in 2022, a total of 62 million tons of electronic waste was generated globally. The report predicts that, based on the current development trend, by 2030, the generation of electronic waste will increase by 33% compared to 2022, reaching 82 million tons (Tsinghua Media, XXX). Waste household appliances are an important component of electronic waste. Taking China as an example, according to the data published by Web Consulting, in 2024, a total of 92 waste electrical appliance recycling and processing enterprises nationwide standardized the recycling and processing of "four appliances and one computer", with the quantity reaching 95,597,862 units/sets (Web Consulting, xxx) However, the current state of waste appliance disposal is concerning. A large volume is processed through informal channels, leading not only to inefficient recovery of valuable resources (such as metals and plastics) and significant waste but also posing grave threats to soil, water, air, and human health due to the improper handling and release of hazardous substances such as heavy metals and persistent organic pollutants during disassembly. Consequently, establishing efficient and environmentally sound collection and treatment systems for waste appliances, maximizing resource recovery, and controlling environmental risks has become an imperative for promoting sustainable industry development and advancing ecological civilization.

33.1.2 Background

Life Cycle Assessment (LCA), a standardized environmental management tool (ISO 14040/14044), has been increasingly applied within the home appliance sector to evaluate product environmental footprints, inform eco-design, and benchmark against environmental performance. Regarding refrigerators, a typical large home appliance, several studies have conducted LCA evaluations from various perspectives. For instance, Zhang et al. (2010) pioneered an evaluation system based on enterprise ERP data, focusing primarily on the environmental performance of the refrigerator production phase. Hou (2015) broadened the scope to encompass four key stages: raw material production, refrigerator manufacturing, transportation, and use, providing a relatively comprehensive life cycle environmental assessment. Yang et al. (2016), evaluating a large sample (10,000 units), quantitatively demonstrated through comparison of energy consumption and environmental impacts across stages that the use phase contributes most significantly to the life cycle environmental impact, followed by the raw material acquisition phase, while the disposal phase's impact was considered relatively minimal. Similarly, LCA research on air conditioning (AC) products has progressed. Zhou (2018) analyzed the main environmental impacts across the entire life cycle of a household air conditioner, from raw material extraction, production, and use to disposal, also concluding that the use phase is the largest contributor to environmental impact. Jiang et al. (2016), utilizing specialized LCA software (GaBi), conducted a detailed full life cycle assessment of a household air conditioner. Beyond calculating the total carbon emissions and confirming the high contribution of the use phase, they further employed uncertainty and sensitivity analyses to identify key factors influencing the product's carbon footprint, subsequently proposing targeted improvement measures. Maggie et al. (2022) from a broader perspective, emphasized the significant negative impact of AC systems on climate change due to their energy consumption and greenhouse gas emissions, highlighting the necessity of comparing different technologies and methods to enhance sustainability.

Beyond evaluating individual product life cycles, some research has begun to focus on the disposal or recycling systems themselves. Anne-Sophie et al. (2016), adhering to the ISO 14040 framework, conducted an LCA of Brazil's refrigerator recycling system. They systematically analyzed its environmental impacts and potential advantages, exploring its feasibility as an energy-saving measure. This highlights the practical value of in-depth studies on specific regional or dedicated recycling systems. Furthermore, other research has examined broader industry practices. Gabriela et al. (2017), through case studies involving five manufacturers and two industry associations in Brazil and other emerging economies (covering internal environmental management, reverse logistics, green procurement, eco-design, LCA, waste management, and green manufacturing), found that while waste management is a relatively common environmental practice, the implementation of green procurement and Life Cycle Assessment remains comparatively limited. This indirectly suggests that while LCA is applied in academic research, its adoption and depth of application within industry, particularly in emerging markets, still have considerable room for expansion.

In summary, existing research has laid a crucial foundation for understanding the environmental impacts of home appliances, consistently revealing the dominance of the use phase in their life cycle environmental burden. However, this has often led to relatively less attention being paid to the EoL phase. EoL assessments are frequently conducted at a macro level, or their impact is considered secondary and thus simplified in some studies. With improvements in product energy efficiency, the deepening emphasis on circular economy principles, and advancements in EoL treatment technologies, the environmental impact of the EoL phase and its optimization potential are gaining increasing importance. Specifically, the current body of research exhibits the following limitations:

(i) Lack of systematic, detailed comparative analysis of the EoL phase for typical large home appliances: Few studies have systematically compared the EoL environmental profiles of multiple major appliances like refrigerators, air conditioners, and washing machines within a unified methodological framework and boundary, covering the entire process from collection and transport through disassembly to the recycling of major materials.
(ii) Insufficiently clear attribution of environmental contributions within the EoL phase: Detailed quantitative analysis is lacking regarding the specific contributions of internal EoL stages (e.g., transport, disassembly, recycling/regeneration of different materials like scrap steel, copper, aluminum, plastics) to the overall environmental impact, and the identification of key environmentally significant stages (hotspots).
(iii) Need for refinement in boundaries and data for assessing recycling benefits: When comparing the environmental impacts of recycled versus virgin materials, the definition of assessment boundaries and the accuracy and representativeness of recycling process data are crucial for robust results. This area requires further investigation, particularly assessments based on localized, operational data.

33.1.3 Research Objectives and Scope

Therefore, this study aims to address the aforementioned research gaps. Selecting three representative large home appliances in China – a refrigerator (121L-220L, a HFC-free), an air conditioner (split wall-mounted indoor unit), and a front-load washing machine—we employ the Life Cycle Assessment (LCA) methodology to systematically evaluate the environmental impacts associated with their EoL phase. The system boundary for this study is explicitly defined as a "gate-to-gate" extension, commencing from the transportation of the waste appliance from collection points, proceeding through disassembly and treatment, and concluding with the regeneration of major recovered materials (scrap steel, copper, aluminum, plastics) into secondary raw materials. The primary objectives are:

(i) Quantify Environmental Impacts: To calculate the overall contribution of the EoL phase for each of the three appliance types across six environmental impact categories: Abiotic Depletion Potential (ADPe elements / ADPf fossil fuels), Global Warming Potential (GWP), Acidification Potential (AP), Eutrophication Potential (EP), and Human Toxicity Potential (HTP).
(ii) Identify Key Stages and Impact Categories: To elucidate the contribution ratios of different stages—including transport, disassembly, and the recycling/regeneration of major materials—to each environmental impact category through contribution analysis and normalization. This will identify the key EoL stages (hotspots) driving environmental impacts and determine the relatively most significant environmental impact categories.
(iii) Evaluate Resource Recovery Benefits: To compare the environmental impacts associated with producing major materials (copper, aluminum, iron, plastics) via the studied recycling pathways versus virgin material production routes, thereby quantifying the pollution reduction and carbon mitigation benefits derived from the resource recovery of typical EoL home appliances.
(iv) The findings of this research are intended to provide a scientific basis and data support for optimizing EoL treatment processes for large home appliances in China, informing relevant environmental management policies, and promoting the development of a circular economy within the appliance sector.

33.2 Method

This study employs the Life Cycle Assessment (LCA) method and, in accordance with the framework specified by the ISO 14040/14044 series standards of the International Organization for Standardization (Zuo and Nie 2003)(as shown in Fig. 33.1), conducts a systematic assessment of the environmental impacts during the End-of-Life (EoL) stage of typical home appliance products. LCA mainly encompasses four interrelated stages: goal and scope definition, life cycle inventory analysis, life cycle impact assessment, and life cycle interpretation.

33.2.1 Goal and Scope Definition

The core objective of this study is to utilize the Life Cycle Assessment (LCA) method to systematically evaluate and compare the environmental impacts of the End-of-Life (EoL) stage of three typical large—scale home appliances (refrigerators, indoor units of air conditioners, and washing machines) in the context of China. Specifically, it aims to identify the key environmental impact links and major environmental problem categories, and quantitatively assess the environmental benefits of the recycling and utilization of major recovered materials compared to the production of virgin materials.

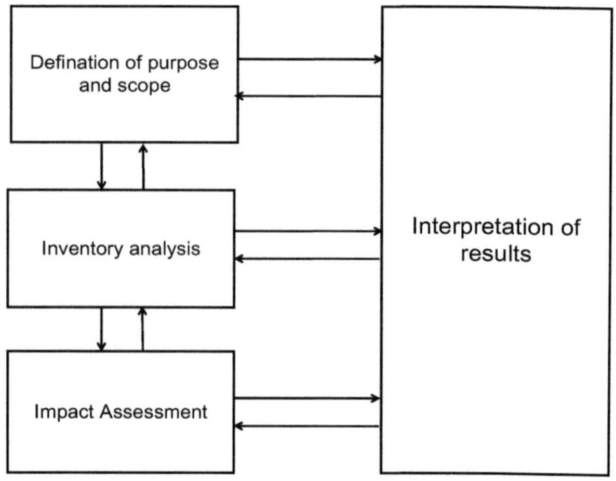

Fig. 33.1 Technical framework of life cycle assessment

To enable comparable assessments, this study defines the Functional Unit (FU) as "processing one discarded typical household appliance", which specifically includes: a 121L–220L CFC—free household refrigerator (EMR); an indoor unit of a split—wall—mounted air conditioner (EMR); and a front—load household washing machine (EMR).

The system boundary of the study is defined as an extended form of "Gate–to–Gate", as shown in Fig. 33.2. This boundary commences from the transportation of waste home appliances from regional collection points to the centralized treatment center. It encompasses a series of pretreatment operations within the treatment center, such as disassembly, crushing, and sorting, and further extends to the specific recycling and regeneration processes of major disassembly products (scrap steel, scrap copper, scrap aluminum, and waste plastics), until secondary raw materials available for market reuse (such as recycled metal ingots and recycled plastic pellets) are formed. This boundary clearly includes the energy (electricity, diesel) required in the process, the necessary input of auxiliary materials, and various environmental emissions (exhaust gas, wastewater, solid waste, etc.) generated during the process. It should be noted that the system boundary does not include: the collection activities of waste home appliances before they reach the user—end or collection points; the final disposal of ultimate waste (such as residues) generated during the treatment process that exceeds the definition of the recycling process (for example, being sent to landfills or incinerators); and the use stage of recycled materials in subsequent product manufacturing. The key assumptions include using a diesel—powered vehicle for transportation with an average transportation distance of 100 km based on actual investigations, calculating the impact of electricity consumption based on the Chinese mixed power grid, and setting the recovery rates of major materials according to the recycling and treatment statistical data provided by enterprises.

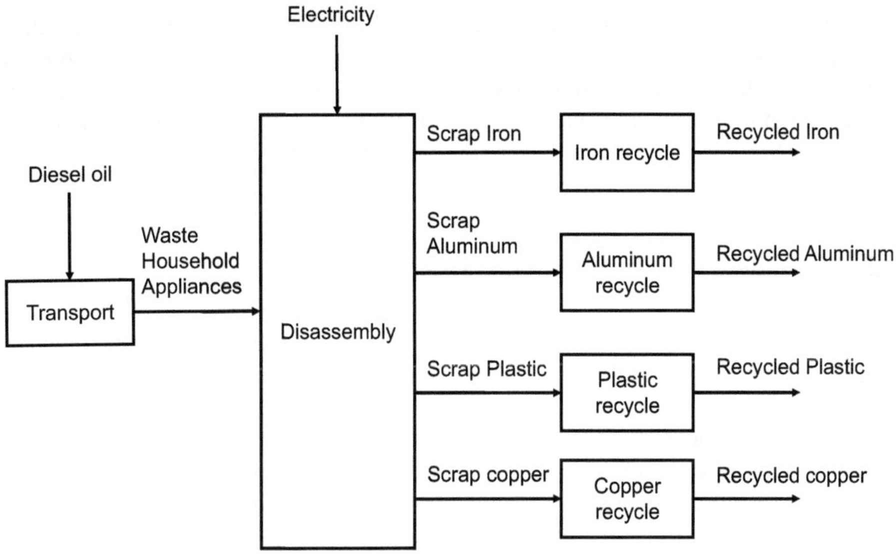

Fig. 33.2 System boundary diagram of the waste treatment stage of typical home appliances

33.3 Life Cycle Inventory Analysis, LCI

The life cycle inventory analysis stage aims to collect and quantify the data of inputs (resources, energy) and outputs (products, emissions, waste) for all processes within the system boundary. The data sources of this study are divided into two categories:

(i) Foreground Data: Foreground data mainly involves the operation data of the transportation and disassembly processes of waste home appliances. This part of the data is provided by Qingdao Hailuyuan Recycling Technology Co., Ltd. It includes first—hand or enterprise—calculated data (in 2023) such as the power consumption, diesel consumption required for processing a unit functional unit (e.g., each refrigerator), the mass balance of disassembly products (various metals, plastics, other wastes), and the average transportation distance.

(ii) Background Data: Background data mainly involves energy production (such as electricity and diesel), primary material production (for comparison), and the recycling processes of disassembly products (such as scrap steel smelting, recycled aluminum production, pyrometallurgical or hydrometallurgical recycling of scrap copper, and waste plastic pelletizing). This part of the data is mainly sourced from professional LCA databases. (MLC Databases 2025.1), The data inventories for the copper and aluminum recycling processes were obtained by referring to relevant literature (Chen et al. 2019; Zhang 2017). All background data were adjusted and matched according to the functional unit of this study. The environmental impact data for primary material production adopted the "Cradle–to–Gate" boundary.

In this study, LCA for Experts (10.9.1) is used for LCA modeling and data calculation. Data with geographical and technical representativeness consistent with the situation in China are adopted as much as possible. The foreground data reflect the actual operations of specific enterprises, while the background data are selected from widely recognized databases to ensure the reliability of the results.

33.3.1 Life Cycle Impact Assessment, LCIA

After completing the life cycle inventory analysis and collecting the data of material and energy exchanges within the system boundary, the Life Cycle Impact Assessment (LCIA) stage aims to transform these inventory data into an understanding of potential environmental impacts. To this end, this study selects the CML2001—Aug. 2016 methodology, which is widely applied in international LCA practices, for the quantitative assessment of environmental impacts. This methodology is integrated into the GaBi software platform used in this study.

Considering that the treatment process of waste electrical and electronic equipment (WEEE) generally involves resource consumption, energy use, potential release of harmful substances, and related regional and global environmental issues, and also referring to the environmental focus areas in related research fields (such as power battery recycling), within the framework of the CML—IA method, this study focuses on selecting the following six environmental impact categories that can reflect these core issues for in—depth analysis:

- Abiotic Depletion Potential (elements) (ADPe)
- Abiotic Depletion Potential (fossil fuels) (ADPf)
- Global Warming Potential (GWP, over a 100 - year time horizon)
- Acidification Potential (AP)
- Eutrophication Potential (EP)
- Human Toxicity Potential (HTP).

In the assessment process, characterization is carried out first. That is, according to Eq. (33.1), the emission amount of each substance or resource consumption amount (Uik) obtained from the inventory analysis is multiplied by the corresponding characterization factor (CFk) provided by the CML—IA method, and the total potential impact value (EI) of each environmental impact category (k) is obtained by accumulation.

$$\mathrm{EI} = \sum (CFk \cdot Uik) \tag{33.1}$$

Subsequently, in order to compare the relative importance among different environmental impact categories, normalization is implemented. According to Eq. (33.2), the characterization results (Ek) of each impact category are divided by the corresponding normalization reference value (Ik, the specific values and sources of which

Table 33.1 Classification of environmental impacts and standardized baseline

Values types of environmental impact	Main substances	Standardization factor
ADPe	Cu, Ni, Fe, Mn	3.61E+08
ADPf	Coal Crude oil Natural gas	3.80E+14
AP	NOx, SO_2, HCl, SOx, NH_3	2.39E+11
EP	COD, BOD, NH_3, NOx	1.58E+11
GWP	CO_2, CH_4, N_2O, CCl_4	4.22E+13
HTP	Er Yuying	2.58E+12

are based on Table 33.1 and represent the total background load of a specific region and year). The normalization result (*Sk*) provides a dimensionless indicator, which is convenient for identifying the areas of relatively more prominent environmental problems within the system of this study.

$$Sk = Ek/Ik \tag{33.2}$$

33.3.2 *Calculation of Environmental Benefits*

To clearly quantify the improvement in environmental performance brought about by replacing virgin material production with the recycling and regeneration of waste home appliances, that is, the so—called "pollution reduction and carbon mitigation benefits", this study uses the direct comparison method to calculate the environmental benefits of major recycled materials (copper, aluminum, iron, and plastics). Specifically, for each evaluated environmental impact category (such as Global Warming Potential (GWP), Acidification Potential (AP), etc.), the environmental benefits are calculated according to the following Eq. (33.3):

$$E_B = E_V - E_R \tag{33.3}$$

In this Equation: EB represents the environmental benefits. E_V represents the characterized assessment result of the environmental impact category when producing 1 kg of virgin material (based on the "cradle–to–gate" boundary). This data is sourced from the background database used in this study.E_R represents the characterized assessment result of the same environmental impact category when producing 1 kg of recycled material through the recycling process defined in this study (based on the "gate–to–gate" boundary from waste input to recycled material output). This data is calculated through the LCA model of this study.

If the calculated "environmental benefit" value is positive, it indicates that, compared with the production of virgin materials, the recycling process has a net

Table 33.2 Characterization indicators of environmental impacts of typical home appliance waste treatment

Environmental impact indicators		Characterization value		
Impact category	Unit	Refrigerator	Internal air condition	Drum washing machine
ADPe	kg Sb eq	1.03E−06	5.75E−08	1.71E−06
ADPf	MJ	1.78E+02	1.38E+01	2.78E+02
AP	kg SO2 eq	3.40E−02	2.64E−03	5.60E−02
EP	kg Phosphate eq	4.41E−03	3.59E−04	7.22E−03
GWP	kg CO2 eq	1.68E+01	9.13E−01	2.60E+01
HTP	kg 1,4-DCB eq	1.28E+00	1.13E−01	2.23E+00

reduction effect in this environmental impact category, that is, it generates environmental benefits. Conversely, if the value is negative, it means that in this specific impact category, the environmental load of the evaluated recycling process is higher than that of virgin material production. These calculation results form the basis for the analysis of various environmental benefits shown in Sect. 3.4.

33.4 Results

33.4.1 General Overview of the Overall Environmental Impact of Waste Treatment

Table 33.2 presents the characterization indicator values for six environmental impact categories throughout the complete waste disposal process of one functional unit (one refrigerator, one indoor air conditioning unit, and one washing machine) of typical household appliances. Analysis of the environmental impact assessment characterization indicators in Table six reveals that the impact magnitude increases in the order of indoor air conditioning unit < refrigerator < drum washing machine. The primary reason for this difference is the substantial variation in mass among the three appliance types, with the drum washing machine weighing 66 kg, resulting in relatively higher environmental costs during its disposal process.

33.4.2 Contribution Analysis

Figure 33.3 illustrates the relative contribution (%) of various processes to six environmental impact categories for three appliance types. For refrigerators, the iron recovery stage contributes most significantly across all six environmental impact

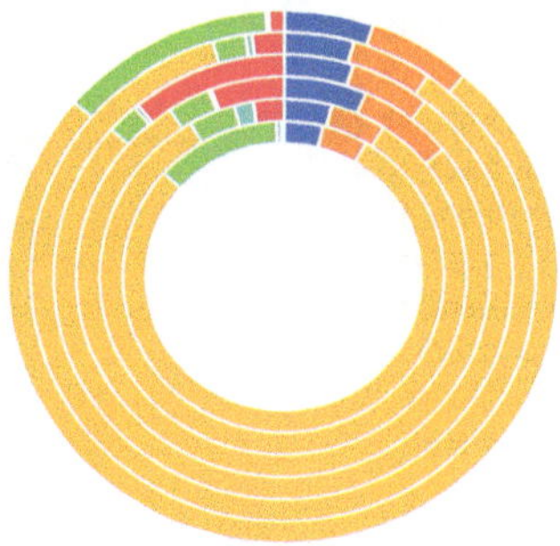

Fig. 33.3 Relative contribution (%) of different stages in typical household appliance waste disposal to six environmental impact categories: **a** refrigerator, **b** indoor air conditioning unit, **c** washing machine. *Note* The environmental impacts from inner to outer rings are ADPe, ADPf, AP, EP, GWP, and HTP

categories, while the aluminum recovery stage shows considerable contribution to HTP. For indoor air conditioning units, disassembly and the recovery of four elements contribute substantially to ADPe, with iron recovery showing the highest contribution and transportation the lowest. Copper recovery contributes most significantly to ADPf. For AP and EP, aluminum and copper recovery contribute minimally, while other stages account for larger proportions, with the transportation stage showing the most significant relative contribution to EP. The iron recovery stage contributes most to GWP, while plastic recovery disproportionately affects the HTP indicator. For drum washing machines, the iron recovery stage demonstrates the most significant environmental impact across all categories; however, compared to refrigerators, the disassembly stage ranks second in contribution to ADPf, AP, EP, and GWP, while transportation and aluminum recovery stages account for larger proportions of ADPe and HTP.

33.4.3 Analysis of Major Environmental Problems

Figure 33.4 presents the normalized indicator values for the refrigerator disposal process. During refrigerator disposal, ADPf and HTP emerge as the most significant environmental concerns relative to Chinese baseline impacts. Across multiple processing stages, particularly the iron recovery (Re.Fe) stage, the normalized values for these two categories substantially exceed those of other environmental impact categories. This is attributed to the substantial consumption of fossil resource-based energy during this stage. Iron recovery involves energy-intensive processes such as high-temperature smelting, necessitating considerable fossil fuel consumption, including coal and natural gas. Additionally, certain components in refrigerators, such as refrigerants and insulation materials, release toxic and hazardous substances

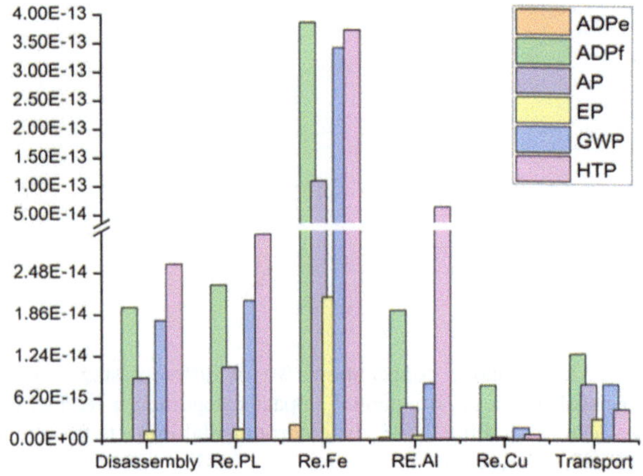

Fig. 33.4 Relative environmental impact values at each stage of refrigerator unit disposal process

during processing, posing potential risks to human health. GWP also shows significant impacts during the disassembly, iron recovery (Re.Fe), and plastic recovery (Re.PL) stages, likely resulting from the relatively high electricity consumption associated with these stages.

Figure 33.5 presents the normalized indicator values for the indoor air conditioning unit disposal process. During the disposal of indoor air conditioning units, HTP and ADPf emerge as the relatively significant environmental concerns compared to Chinese baseline impacts. In the Disassembly and Re.PL stages, the normalized values for these two categories substantially exceed those of other environmental impact categories. Indoor air conditioning units likely contain electronic components with heavy metals and potentially harmful additives in plastic components. During disassembly and plastic recovery processes, these substances may be released, posing potential hazards to human health. The relatively high plastic content in air conditioners also contributes to the elevated environmental impact values during the plastic disassembly process.

Figure 33.6 illustrates the normalized environmental impact indicators for drum washing machines. Normalization analysis reveals that in the disposal process, ADPf and HTP represent the most significant environmental concerns relative to regional/global baseline impacts. During Re.Fe stage, the normalized values for these two categories substantially exceed those of other environmental impact categories; in the Disassembly stage, HTP values are also notably prominent. Additionally, AP shows considerable values during Re.Fe stage, indicating potential emissions of acidifying substances such as sulfur compounds during this phase. GWP also demonstrates significant contributions during the disassembly and Re.Fe stages, suggesting relatively high electricity consumption and possibly direct carbon emissions during the iron recovery phase.

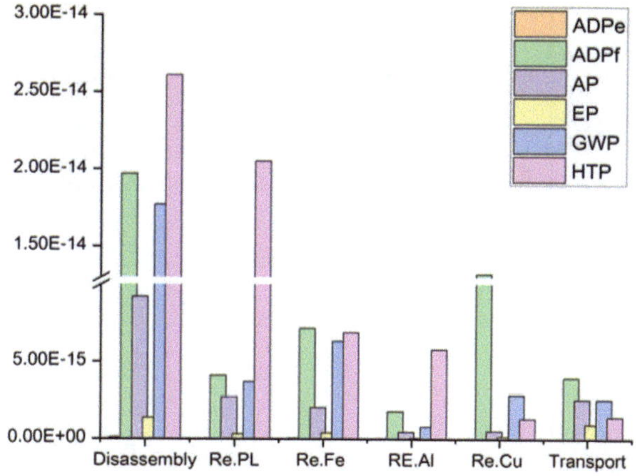

Fig. 33.5 Relative environmental impact values at each stage of indoor air conditioning unit disposal process

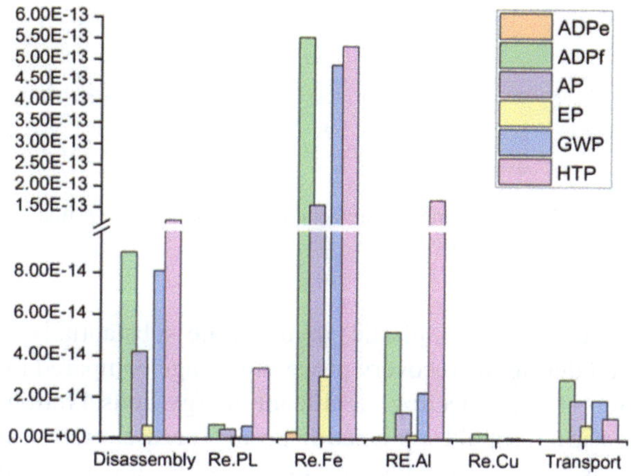

Fig. 33.6 Relative environmental impact values at each stage of drum washing machine disposal process

33.4.4 Analysis of Major Environmental Problems

Figure 33.7 presents the environmental impact comparison between primary and recycled materials. From Fig. 33.7a (ADPf), primary copper exhibits an ADPf value of 17.9 MJ, while recycled copper shows 14.1 MJ, representing an energy saving of 3.8 MJ. Primary aluminum demonstrates an ADPf value of 5.62 MJ, compared to 0.99 MJ for recycled aluminum, yielding an energy saving of 4.63 MJ. Primary iron shows a significantly high ADPf value of 81.2 MJ, while recycled iron registers 23.1 MJ, resulting in an energy saving of 58.1 MJ. Primary plastic exhibits an ADPf value of 9.69 MJ, with recycled plastic at 2.93 MJ, providing an energy saving of 6.76 MJ. With the exception of recycled copper, all other recycled materials

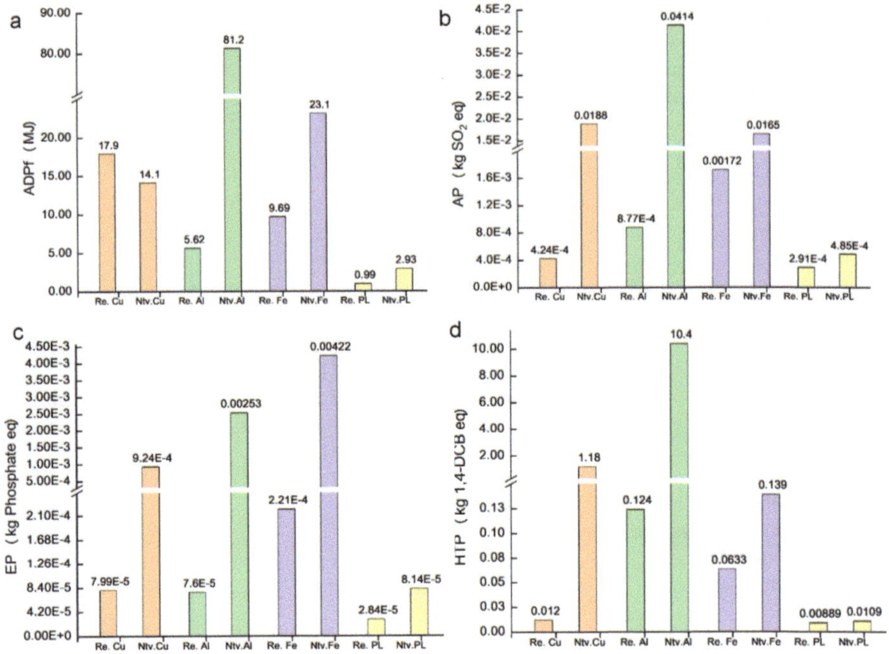

Fig. 33.7 Environmental impact comparison between primary and recycled materials (**a**: ADPf, **b**: AP, **c**: EP, **d**: HTP)

(aluminum, iron, plastic) demonstrate substantially lower fossil energy consumption during the recovery processing stage compared to the primary material production stage, reflecting significant energy conservation benefits. This indicates that waste household appliance disposal processes offer positive implications for resource conservation regarding certain materials, effectively reducing dependency on fossil energy resources.

Examination of Fig. 33.7b (AP), c (EP), and d (HTP) reveals that primary materials generally exhibit higher values than recycled materials on the AP indicator; the EP indicator shows primary iron values significantly exceeding those of recycled iron; and primary materials also demonstrate relatively higher values on the HTP indicator. Recycled materials consistently show lower values across these environmental impact indicators compared to primary materials. Based on these specific comparative values, it is conclusively demonstrated that waste household appliance disposal processes, compared to primary material manufacturing, achieve significant pollution reduction benefits, effectively lowering environmental risks related to acidification, eutrophication, and human toxicity, thus highlighting the dual value of waste disposal systems in both resource conservation and pollution reduction.

Figure 33.8 illustrates the GWP comparison between primary and recycled materials. The GWP data reveals that with the exception of recycled iron, the recycling of the other three materials (copper, aluminum, and plastic) demonstrates

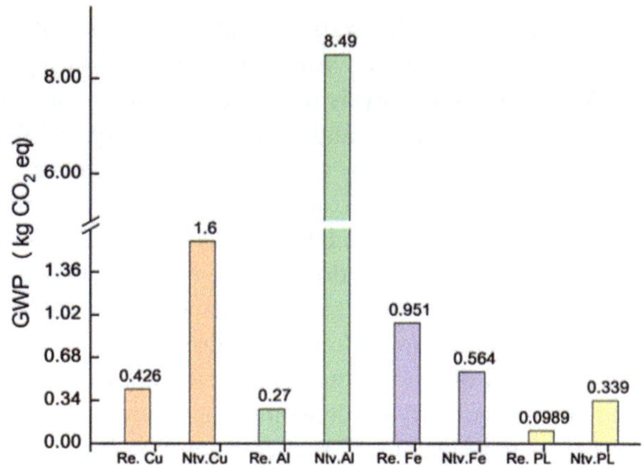

Fig. 33.8 GWP comparison between primary and recycled materials

significant carbon reduction benefits. Notably, recycled iron exhibits a higher GWP value (0.951 kg CO_2eq) compared to primary iron (0.564 kg CO_2eq). This counter-intuitive higher carbon emission from recycled iron may be attributed to several factors: First, recycled iron production involves more complex processes; after pre-treatment and smelting of scrap steel, additional energy-intensive steps such as rolling and heat treatment are required for rebar production. If the electricity for these processes derives from thermal power generation with high-carbon energy structures, coupled with imperfect equipment performance and operational techniques, energy utilization efficiency decreases significantly. Second, recovered scrap steel typically contains numerous impurities, requiring additional energy consumption for purification. Furthermore, if recycled iron production facilities utilize outdated equipment and technologies, lacking mature and efficient technological systems and industrial support, energy utilization efficiency cannot be guaranteed, potentially resulting in increased carbon emissions.

33.5 Discussion and Conclusion

In this study, LCA was presented to conduct a comprehensive and in-depth environmental impact evaluation of typical household appliances at their EoL stage, addressing existing research gaps in this area.

The results demonstrate substantial variations in environmental impacts among different appliance types during their disposal stage. The mass of a product significantly influences the environmental burdens associated with End-of-Life (EoL) processes; among the three appliances evaluated, the drum washing machine exhibited the highest environmental impact, primarily due to its relatively larger mass. This finding underscores the importance of integrating considerations regarding environmental impacts of disposal into product design and material selection at the product design stage. Manufacturers are encouraged to explore lighter yet durable materials to effectively mitigate environmental burdens during disposal processes.

A contribution analysis reveals that different stages of disposal, including transportation, disassembly, and material recovery, exhibit varied contributions across environmental impact categories. Specifically, for refrigerators, iron recovery significantly contributes to most environmental impact indicators; for air conditioners, disassembly processes and the recovery of certain elements have prominent impacts; similarly, iron recovery dominates environmental impacts for washing machines. These identified critical stages (hotspots) can serve as primary targets for implementing focused environmental improvement measures. Efforts should be particularly focused on improving energy efficiency technologies within energy-intensive and environmentally burdensome iron recovery stages, such as optimizing smelting processes to reduce fossil fuel consumption and pollutant emissions.

In the analysis of key environmental issues, ADPf and HTP emerged as prominent concerns across the disposal processes of the three appliances. Elevated ADPf values primarily result from significant fossil fuel consumption during certain disposal processes, particularly in iron recovery stages. Addressing this issue may involve promoting renewable energy technologies within the waste treatment sector, such as solar-powered disassembly equipment or electric vehicles for waste transportation. The relatively high HTP values are mainly attributed to hazardous substance releases during disposal processes; thus, enhancing technologies for the capture and treatment of these substances, such as advanced separation and purification techniques, is crucial.

Environmental benefit assessments indicate that processing discarded appliances can achieve notable energy savings, pollution reduction, and carbon mitigation effects for most materials. However, recycled iron exhibits a higher GWP than primary iron, a situation that needs to be highlighted. Urgent improvements in recycled iron production processes are necessary, suggesting that companies invest in equipment upgrades, optimize production workflows, and enhance energy efficiency. Specifically, adopting advanced electric arc furnace smelting technologies and strengthening waste heat recovery systems can effectively reduce energy consumption and carbon emissions.

In conclusion, this study can help provide robust scientific evidence and comprehensive data to support the optimization of large household appliance disposal processes in China. Policymakers are advised to formulate targeted environmental management strategies based on these findings, such as incentives for disposal enterprises to adopt environmentally friendly technologies and enhance resource recovery efficiency. Simultaneously, appliance manufacturers should actively integrate EoL environmental considerations into product design, emphasizing material recyclability and environmental compatibility, collectively advancing sustainable development and the circular economy within the appliance industry. Future research may further expand the scope to include additional appliance types and diverse regions, thereby generating more comprehensive and representative conclusions.

Acknowledgements This work was supported by the National Key Research and Development Plan (2023YFB3305804), the National Natural Science Foundation of China (52300228), and the R&D Program of Beijing Municipal Education Commission (SM202310005008).

References

Chen J, Wang Z, Wu Y et al (2019) Environmental benefits of secondary copper from primary copper based on life cycle assessment in China. Resour Conserv Recycl 146:35–44
DeMarco M, Fortier MOP (2022) Functional unit choice in space conditioning life cycle assessment: review and recommendations. Energy Build 255:111626
Foelster A-S, Andrew S, Kroeger L et al (2016) Electronics recycling as an energy efficiency measure—a Life Cycle Assessment (LCA) study on refrigerator recycling in Brazil. J Clean Prod 129:30–42
Hou J (2015) Research on the ecological design evaluation of household refrigerators. Qual and Certification 2015(06):53–55
Jiang S, Tian X, Wang L et al (2016) Carbon footprint analysis of the whole life cycle of household air-conditioners. Daily Use Electr Appl 2016(9):46–52+56
Scur G, Barbosa ME (2017) Green supply chain management practices: multiple case studies in the Brazilian home appliance industry. J Clean Prod 141:1293–1302
Tsinghua Media. https://www.tsinghua.edu.cn/info/1182/110598.htm
Web Consulting. https://www.toutiao.com/
Yang H, Cui L, Zhu Q (2016) Life-cycle impact assessment of refrigerators considering the proportion of safe disposal. Syst Eng 34(05):136–143
Zhang Y (2017) Life-cycle assessment of electrolytic aluminum and recycled aluminum production in China. Shandong University, Jinan
Zhang J, Yu S (2010) Environmental impact assessment of the refrigerator production process based on the ERP system (Part 1). Home Appl Technol 2010(02):53–55
Zhou L (2018) Research on the life-cycle assessment of household air-conditioners. Environ Prot Circular Econ 38(08):12–14+18
Zuo T, Nie Z (2003) Foundations of environmental materials, 1st edn. Science Press, Beijing

Part VI
Eco-Environmental Systems Management and Environmental Toxicology

Chapter 34
Microalgae Biofertilizers: Mechanisms of Soil–Plant Interaction and Applications in Sustainable Agriculture

Fuzhou Zhang, Aili Yang, WenJie Li, and Jiaona Guo

Abstract The tension between a growing global demand for food, rising environmental impacts of agricultural intensification, and the uncertainty of future changes in our global climate highlight a crucial need for sustainable alternatives to maintain and increase agricultural productivity. Microalgae are renewable resources with a broad range of options for applications in agricultural settings. This review examines how microalgae-based fertilizers improve soil properties and crop resilience. Microalgae enhance soil organic matter and phosphorus availability. They stabilize soil aggregates and increase water retention. Microalgae modify rhizosphere microbial diversity and boost nutrient cycling. Most studies are limited to short-term laboratory conditions, ignoring dynamic field stresses that inhibit microalgal functions. Key knowledge gaps exist in understanding microalgae-crop molecular interactions and biofilm metabolism in root environments. Future research should prioritize developing cost-effective cultivation techniques, advancing in-situ field monitoring, and screening regionally adaptive strains. Integrating microalgae into soil-microbe synergy frameworks is essential to unlock their potential as a cornerstone of sustainable agriculture, providing theoretical support for applications in diverse soil and climatic regions.

Keywords Microalgae biofertilizers · Soil physicochemical properties · Soil microorganisms · Plant stress tolerance

F. Zhang · A. Yang (✉) · W. Li · J. Guo
Key Laboratory of Environmental Biotechnology (XMUT), Fujian Province University, Xiamen University of Technology, Xiamen, China
e-mail: yangaililoy@xmut.edu.cn

Y. Li et al. (eds.), *16th International Conference on Environmental Science and Development*, Environmental Science and Engineering,
https://doi.org/10.1007/978-3-032-15080-6_34

34.1 Introduction

Future global food security depends critically on agricultural sustainability. While intensification has enhanced productivity, it has concurrently generated significant environmental degradation (Muller et al. 2017; Pingali 2012). These pressures are amplified by dual drivers: a projected 60% increase in production demand from population growth (Alexandratos and Bruinsma 2012) and climate change impacts threatening agricultural systems (Rosenzweig et al. 2014). As a consequence, maintaining high agricultural productivities while simultaneously mitigating environmental impacts and promoting environmental regeneration is an urgent challenge (Steensland 2019).

Algae encompass a polyphyletic group of predominantly aquatic, photosynthetic organisms ranging from unicellular microalgae to multicellular macroalgae (e.g., seaweeds). Within this broad category, microalgae are specifically defined as microscopic (typically <200 μm) photosynthetic microorganisms, including prokaryotic cyanobacteria (commonly termed blue-green algae) and eukaryotic taxa such as Chlorophyta (green algae), Bacillariophyta (diatoms), and Euglenophyta (euglenoids). The adjectival form 'microalgal' denotes materials derived from or associated with these microorganisms, such as microalgal biomass or microalgal-based fertilizers.

Microalgae exhibit a pool of traits with unique value for addressing this challenging agricultural scenario. Microalgae constitute a highly diverse group of primarily photosynthetic microorganisms, comprising cyanobacteria (prokaryotes) and eukaryotes such as green algae, euglenoids, and diatoms (Andersen 2013; Barsanti and Gualtieri 2016; Mata et al. 2010a). In agricultural systems, microalgae enhance soil fertility, promote plant growth and protection, and offer sustainable alternatives to reduce reliance on synthetic fertilizers and pesticides (Holajjer et al. 2013; Prasanna et al. 2014, 2015a). Microalgae are beneficial for soil nutrient cycling and can promote plant growth by improving nutrient availability (Karthikeyan et al. 2007; Prasanna et al. 2015b; Roger and Kulasooriya 1980), producing bioactive substances such as phytohormones (Gheda and Ahmed 2015; Mazhar et al. 2013), forming root associations (Bharti et al. 2019; Bidyarani et al. 2015), or by protecting plants against phytopathogens and pests (Prasanna et al. 2013; Roberti et al. 2015). Microalgae also fix carbon dioxide (CO_2) through photosynthesis for carbon capture and some produce exopolysaccharides (EPS) that improve soil structure (Chamizo et al. 2018; Renuka et al. 2018; Wang et al. 2015).

In this context, microalgae can potentially serve as platforms for developing products such as biofertilizers, organic fertilizers, biostimulants, biocontrol agents, and soil conditioners, which are used for soil improvement and crop production and protection. Nevertheless, fragmented knowledge and insufficient understanding of microalgae's effects and mechanisms on soil–plant systems under varied conditions still restrict their broad-scale utilization in agricultural environments. This review summarizes the effects of microalgae fertilizers on soil quality, microbial communities, and crop yield/stress resistance. It highlights current research limitations and outlines future research needs. The goal is to enhance the value of microalgae as key renewable resources and advance innovative bioeconomy solutions required for sustainable agricultural systems.

34.2 Fundamental Characteristics of Microalgae

As photosynthetic microorganisms with ecological and biotechnological relevance, microalgae exhibit adaptable characteristics, rapid growth potential, and efficient carbon fixation. Their cultivation flexibility (photoautotrophic, mixotrophic, heterotrophic modes), short life cycles, and scalability demonstrate industrial applicability. Microalgae synthesize diverse high-value bioactive compounds, such as lipids, unsaturated fatty acids (EPA/DHA), functional proteins (e.g., phycocyanin) and pigments (e.g., astaxanthin) (Table 34.1) (Buono et al. 2014; Eppink et al. 2017).

Microalgae have become an ideal raw material for industrial processing and a sustainable resource for eco-friendly agriculture. Microalgae biotechnology demonstrates extensive environmental remediation applications, effectively treating coal flue gas, livestock wastewater, sewage, and food effluents while achieving resource recovery (Mukherjee et al. 2016; Wang et al. 2017; Huang et al. 2016). Importantly, microalgal biomass generated during environmental remediation can enhance agricultural productivity, simultaneously achieving environmental protection, cost efficiency, and improved crop quality (Jenkins et al. 2017).

Algae play a crucial role in mitigating agricultural non-point source pollution by assimilating nitrogen, phosphorus, and other nutrients from water bodies, simultaneously enhancing their biomass production (Plöhn et al. 2021) and synthesizing bioactive substances (Stirk et al. 2013). These compounds are subsequently utilized in producing biofuels (Mata et al. 2010b), biofertilizers, animal feed (Atalah et al. 2007; Lu and Xu 2015), biopesticides (Lu and Xu 2015), and soil amendments. Upon application to soil, microalgae can thrive and alter microbial community structures, while their biomass also acts as a natural biostimulant after decomposition to enhance plant growth. Their dual functionality in environmental remediation and agricultural productivity underscores their potential as sustainable resources for circular bioeconomy systems.

Table 34.1 The biological value of microalgae

High-value bioactive compounds	Applications
Proteins	Bait feed, fertilizer, industrial enzymes, bioplastics, surfactants
Astaxanthin	Skin care products, body lotions
Fatty acids	Health products, animal feed
Lipids	Large-scale biodiesel

34.3 Effects of Microalgae Fertilizers on Soil

34.3.1 Effects of Microalgae Fertilizers on Soil Physicochemical Properties

Algae significantly facilitate soil formation by enhancing organic matter content, releasing carbonic acid and organic compounds, weathering parent materials, improving soil structure, facilitating aggregate formation, preventing erosion, and increasing moisture retention capacity. These processes collectively influence critical soil quality indicators, including the development of microbial communities, soil organic matter levels, and enzymatic activity (Chamizo et al. 2018). Microalgae fertilizers exert multifaceted impacts on soil quality. Their organic acids and alkaline substances regulate soil pH toward neutral or optimal ranges, facilitating nutrient uptake by roots. Furthermore, microalgae fertilizers stimulate microbial proliferation and activity, accelerating organic matter decomposition and nutrient cycling, thereby enhancing soil biological properties.

Cyanobacteria serve as key soil amendments. Their mechanisms include: synthesizing plant growth regulators, exuding phosphorus-mobilizing organic acids, and enriching rhizosphere microbiota. These mechanisms collectively optimize soil structure, nutrient utilization efficiency, and plant growth, reinforcing soil ecological functions (Agarwal 1979). Certain nitrogen-fixing cyanobacteria synthesize antimicrobial metabolites. These compounds effectively inhibit pathogen colonization and pest invasion, thereby elevating crop health indices. This mechanism not only bolsters plant resilience but also reduces reliance on synthetic pesticides, supporting sustainable agricultural practices (Pawar et al. 2011). Whitton et al. conducted growth screening of 50 cyanobacterial isolates. All strains showed positive growth in organic phosphorus-specific media, confirming their unique phosphate metabolism. Among them, 47 strains utilized phosphate diesters, while 35 utilized phytate, demonstrating broad adaptability to organic phosphorus utilization—crucial for soil phosphorus management (Mandal et al. 1999). Cyanobacteria convert organic phosphorus to plant-available inorganic forms, facilitating efficient phosphorus uptake through both active release during growth and post-mortem decomposition. This conversion process enhances soil fertility, promotes plant growth, and sustains ecological balance by improving phosphorus cycling (Whitton et al. 1991).

34.3.2 Effects of Microalgae Fertilizers on Soil Microorganisms

Soil-dwelling microalgae enhance soil fertility by modulating microbial metabolic activities. These microalgae secrete phytohormones that promote plant growth and development while strengthening crop resistance to diseases and pests. Collectively, these effects increase yields and reduce reliance on synthetic fertilizers, advancing sustainable agricultural objectives (Sharma et al. 2011). Microalgae, as a vital component of the soil microbiome, function as key drivers of farmland ecosystems by enhancing soil organic matter formation, fostering soil aggregation, and accelerating nutrient cycling (Kallenbach and Grandy 2011). In farmland ecosystems, functionally diverse microbial communities can utilize organic substrates inaccessible to plants, releasing plant-available nutrients and thereby enhancing crop yields (Tautges et al. 2016). Calderón et al. (2017) observed increased populations of bacterial groups, actinomycetes, and fungi in soil following the introduction of Nostoc. Similarly, Acea (2003) demonstrated that field inoculation with cyanobacteria accelerates the proliferation of cellulose-decomposing microorganisms, starch-degrading microbes, ammonifying bacteria, and nitrifying bacteria.

Comprehensive studies show that microalgae fertilizers significantly enhance soil aggregate stability, phosphorus availability, and microbial activity by introducing organic matter and secreting plant hormones and organic acids. However, most existing studies are based on stable laboratory conditions (Table 34.2), ignoring the inhibitory effects of dynamic field stresses on microalgal functions. In practical field applications, microalgae-soil interactions are often influenced by a combination of environmental factors (e.g., intensity of dry–wet cycles, fluctuations in ultraviolet radiation) and biological factors (competition from indigenous microorganisms, interference from crop roots). Existing studies lack quantitative analysis of key issues such as the efficiency of microalgal biofilm formation under dynamic field conditions and the temporal interactions between metabolites and soil components, leading to difficulties in accurately translating laboratory findings into practical applications. Future research should prioritize the development of in-situ field observation technologies, with a primary emphasis on unraveling the coupling mechanisms of the "microalgae-soil-microbe-crop" four-dimensional interaction network under dynamic environmental stresses. Concurrently, efforts are needed to screen regionally adaptive microalgal strains and optimize field application patterns, providing theoretical support for the agricultural use of microalgae in diverse soil types and climatic regions.

34.4 Effects of Microalgae Fertilizers on Crop

34.4.1 Effects of Microalgae Fertilizers on Plant Growth Promotion

The applications of microalgae have significantly improved plant absorption and utilization efficiency of nutrients. Table 34.3 summarizes selected examples of microalgae fertilizer applications enhancing crop quality and yield. Microalgal liquid fertilizers contain active composite cell clusters primarily obtained through collection, separation, and purification from local ecosystems, enabling rapid adaptation

Table 34.2 Examples of improving soil fertility after applying microalgae fertilizers

Microalgal strain (s)	Experimental conditions	Impact on soil	Reference
Chlorella	Pot experiment	Improve total nitrogen, total carbon, available phosphorus content	Alobwede et al. (2019)
Chlorella	Pot experiment	Increase nitrogen, organic carbon, potassium, and phosphorus content, lower soil pH, and increase the number and abundance of microorganisms	Agwa et al. (2017)
Arthrospira platensis	Field experiment	Lower soil pH and increase the number of microorganisms	Suchithra et al. (2022)
Chlorella sp.	Pot experiment	Increase soil organic matter, available P, and available K	Zhang et al. (2024)
Scenedesmus sp.	Field experiment	Increase soil organic matter	Castro et al. (2024)
Chlorella pyrenoidosa	Pot experiment	Increased soil organic matter, ammonium nitrogen, available phosphorus, and available potassium content	Li et al. (2024)
Chlamydomons applanata M9V and *Chlorella vulgaris* S3	Pot experiment	Improve soil organic matter and carbon–nitrogen ratio	Sido et al. (2022)

to abiotic and biotic factors in the environment. Plant roots acquire water, inorganic ions, and organic compounds from soil to sustain photosynthesis and respiration, while rhizosphere redox dynamics critically regulate nutrient uptake efficiency and growth performance. After application to soil, the active components of microalgal liquid fertilizers rapidly form biofilms dominated by microalgae and bacteria on the soil surface. These microalgae and bacteria establish stable symbiotic associations with plant roots, periodically providing metabolically essential products for the plants. Through reciprocal metabolic exchange, plants allocate photosynthates (e.g., sucrose) and host-inaccessible organic compounds to symbionts, which in return mobilize essential nutrients via specialized biochemical pathways, forming obligate cross-kingdom resource interdependencies.

Microalgae are photosynthetic organisms capable of synthesizing and releasing diverse growth-promoting bioactive compounds, including proteins, lipids, carotenoids, vitamins, polyamines, gibberellins, and cytokinins. These compounds play critical roles in regulating plant developmental processes, metabolic pathways, and growth, particularly in enhancing seed germination and seedling development. For seed germination, Saadatnia and Riahi (2009) demonstrated accelerated rice seed germination via aqueous cyanobacterial biomass (0.1 g) soaking, with treated seeds achieving 2-day germination and producing 7-cm shoots/3-cm roots after 20 days, while controls required 5 days for germination and yielded 4-cm shoots/0.5-cm roots. Chittapun et al. (2018) compared cyanobacterial inoculation with conventional

Table 34.3 Examples of enhanced plant growth or performance under laboratory or greenhouse conditions after application

Microalgal strain (s)	Crops	Improved growth/performance parameters	Reference
Microcystis	Wheat	Increase crop yields	Chu et al. (2021)
Chlorella sorokiniana	Wheat	Enhance germination rate and promote plant growth	Kholssi et al. (2019)
Acutodesmus dimorphus	Tomato	Enhance seed germination, increase plant height and flower number and improve crop yields	Garcia-Gonzalez and Sommerfeld (2016)
Chlorella	Spinach	Enhance chlorophyll accumulation	Rupawalla et al. (2021)
Dunaliella salina	Pumpkin	Enhance salt tolerance and productivity	El-Adl et al. (2022)
Chlorella	Tomato	Promote growth, improve quality and extend shelf life	Suchithra et al. (2022)
Asterarcysquadricellulare	Potato	Increase crop yield, chlorophyll content, and soluble sugar content	Cordeiro et al. (2022)
Stoechospermum marginatum	Eggplant	Increase plant height, dry weight, leaf area, chlorophyll content, protein content, soluble sugar content, and ascorbic acid content	Ramya et al. (2015)

fertilizers in rice cultivation, showing that cyanobacteria enhanced root length, shoot length, fresh weight, and dry weight of rice seedlings.

34.4.2 Effects of Microalgae Fertilizers on Plant Stress Resistance Enhancement and Bioactive Compound Accumulation

Microalgae fertilizers contain bioactive compounds which promote plant growth, enhance antioxidant enzyme activity, reduce superoxide radical accumulation, and improve stress resistance (Kapoore et al. 2021). Research has demonstrated that applying *Chlorella* ellipsoida and *Spirulina* maxima to seawater-irrigated wheat significantly increases protein content and antioxidant enzyme activity, thereby enhancing salt tolerance (Abd El-Baky et al. 2010). *Phaeodactylum tricornutum* and *D. salina* enhance salt-stressed pepper (*Capsicum annuum*) seed germination by boosting antioxidant enzyme activity, reducing superoxide radical accumulation, and suppressing lipid peroxidation (Guzmán-Murillo et al. 2013). *Dunaliella salina*'s extracellular polysaccharides, proline, and carotenoids enhance wheat seed germination rate and root growth under salt stress, demonstrating its extract's ability to improve salt tolerance and boost crop yields (El-Arroussi et al. 2016). These findings corroborate that microalgae-derived compounds serve as multifunctional tools for crop protection across diverse abiotic stress scenarios.

Overall, microalgae fertilizers significantly promote seed germination rate, seedling root length, and crop yield by secreting bioactive substances such as gibberellins and cytokinins, while enhancing quality indicators including protein content in rice. However, current studies primarily focus on short-term trials using limited algal species like cyanobacteria and *Chlorella*. The molecular mechanisms governing microalga-crop interactions remain underexplored. Key knowledge gaps persist, particularly regarding biofilm metabolic regulation in root-associated dynamic oxygen gradients. Meanwhile, industrial production is restricted to single algal species like *Chlorella* and *Spirulina*. This results in functional homogeneity of microalgae fertilizers. They exhibit limitations in plant growth promotion and stress resistance induction. Future research should overcome development bottlenecks in algal resource exploitation. Establish sustainable cultivation systems for extremophilic algae (e.g., halotolerant and acidophilic strains). Apply gene editing technologies to enhance functional algal gene expression. Develop closed-loop photobioreactors to advance microalgae fertilizers from generic formulations to soil-crop tailored solutions.

34.5 Conclusion and Outlook

Microalgae fertilizers offer significant benefits for sustainable agriculture by enhancing soil structure, promoting nutrient cycling, and improving crop growth and stress resistance. They increase soil organic matter, stabilize aggregates, and regulate pH, while bioactive compounds like phytohormones boost plant development and tolerance to abiotic stresses.

However, current research has limitations. Most studies are conducted under stable laboratory conditions, ignoring the inhibitory effects of dynamic field stresses such as dry–wet cycles and microbial competition. Key knowledge gaps exist in understanding the molecular mechanisms of microalgae-crop interactions and biofilm metabolism in root environments. Industrial production is also restricted, leading to functional homogeneity in existing microalgae fertilizers.

Future research should focus on developing cost-effective cultivation techniques and in-situ monitoring tools to assess microalgae performance under real-field conditions. Screening for regionally adaptive strains and optimizing application methods will be crucial. Integrating microalgae into soil-microbe synergy frameworks is essential to unlock their full potential, providing practical and theoretical support for sustainable agricultural systems across diverse soil and climatic regions.

References

Abd El-Baky HH, El-Baz FK, El Baroty GS (2010) Enhancing antioxidant availability in wheat grains from plants grown under seawater stress in response to microalgae extract treatments. Sci Food Agric 90:299–303

Acea M (2003) Cyanobacterial inoculation of heated soils: effect on microorganisms of C and N cycles and on chemical composition in soil surface. Soil Biol Biochem 35:513–524

Agarwal A (1979) Blue-green algae to fertilise Indian rice paddies. Nature 279:181–181

Agwa OK, Ogugbue CJ, Williams EE (2017) Field evidence of *C. vulgaris* potentials as a biofertilizer for hibiscus esculentus. Agricult Res 12:181–189

Alexandratos N, Bruinsma J (2012) World agriculture towards 2030/2050: the 2012 revision, agricultural development economics division. FAO, Rome

Alobwede E, Leake JR, Pandhal J (2019) Circular economy fertilization: testing micro and macro algal species as soil improvers and nutrient sources for crop production in greenhouse and field conditions. Geoderma 334:113–123

Andersen RA (2013) The microalgal cell. In: Richmond A, Hu Q (eds) Handbook of microalgal culture: applied phycology and biotechnology. Wiley, Hoboken, pp 3–20

Atalah E, Cruz CMH, Izquierdo MS, Rosenlund G, Caballero MJ, Valencia A, Robaina L (2007) Two microalgae *Crypthecodinium cohnii* and *P. tricornutum* as alternative source of essential fatty acids in starter feeds for seabream (*Sparus aurata*). Aquaculture 270:178–185

Barsanti L, Gualtieri P (2016) Algae: anatomy, biochemistry, and biotechnology, 2nd edn. CRC Press, Boca Raton, pp 1–48

Bharti A, Prasanna R, Kumar G, Kumar A, Nain L (2019) Co-cultivation of cyanobacteria for raising nursery of chrysanthemum using a hydroponic system. J Appl Phycol 31:3625–3635

Bidyarani N, Prasanna R, Chawla G, Babu S, Singh R (2015) Deciphering the factors associated with the colonization of rice plants by cyanobacteria. Basic Microbiol 55:407–419

Buono S, Langellotti AL, Martello A, Rinna F, Fogliano V (2014) Functional ingredients from microalgae. Food Funct 5:1669–1685

Calderón K, Spor A, Breuil M-C, Bru D, Bizouard F, Violle C, Barnard RL, Philippot L (2017) Effectiveness of ecological rescue for altered soil microbial communities and functions. ISME J 11:272–283

Castro IMP, Rosa A, Borges A, Cunha F, Passos F (2024) The effects of microalgae use as a biofertilizer on soil and plant before and after its anaerobic (co-)digestion with food waste. Sci Total Environ 934:173301

Chamizo S, Mugnai G, Rossi F, Certini G, De Philippis R (2018) Cyanobacteria inoculation improves soil stability and fertility on different textured soils: gaining insights for applicability in soil restoration. Front Environ Sci 6:49

Chittapun S, Limbipichai S, Amnuaysin N, Boonkerd R, Charoensook M (2018) Effects of using cyanobacteria and fertilizer on growth and yield of rice, Pathum Thani I: a pot experiment. J Appl Phycol 30:79–85

Chu Q, Lyu T, Xue L, Yang L, Feng Y, Sha Z, Yue B, Mortimer RJG, Cooper M, Pan G (2021) Hydrothermal carbonization of microalgae for phosphorus recycling from wastewater to crop-soil systems as slow-release fertilizers. J Clean Product 283:124627

Cordeiro ECN, Mógor ÁF, Amatussi JO, Mógor G, Marques HMC, De Lara GB (2022) Microalga biofertilizer improves potato growth and yield, stimulating amino acid metabolism. J Appl Phycol 34:385–394

El-Adl MF, Deyab MA, Ghazal MA, Elsadany AY (2022) Impact of the microalga *D. salina* (Dunal) Teodoresco culture and its β-carotene extract on the development of salt-stressed squash (*Cucurbita pepo* L. cv. Mabrouka). Physiol Mol Biol Plants 28:749–762

El-Arroussi H, Elbaouchi A, Benhima R, Bendaou N, Smouni A, Wahby I (2016) Halophilic microalgae *D. salina* extracts improve seed germination and seedling growth of *Triticum aestivum* L. under salt stress. Acta Hortic 12:13–26

Eppink MHM, Olivieri G, Reith H (2017) From current algae products to future biorefinery practices: a review. In: Wagemann K, Tippkötter N (eds) Biorefineries. Springer, Cham, pp 99–123
Garcia-Gonzalez J, Sommerfeld M (2016) Biofertilizer and biostimulant properties of the microalga *Acutodesmus dimorphus*. J Appl Phycol 28:1051–1061
Gheda SF, Ahmed DA (2015) Improved soil characteristics and wheat germination as influenced by inoculation of Nostoc kihlmani and Anabaena cylindrica. Rend Fis Acc Lincei 26:121–131
Guzmán-Murillo MA, Ascencio F, Larrinaga-Mayoral JA (2013) Germination and ROS detoxification in bell pepper (*C. annuum* L.) under NaCl stress and treatment with microalgae extracts. Protoplasma 250:33–42
Holajjer P, Kamra A, Gaur HS, Manjunath M (2013) Potential of cyanobacteria for biorational management of plant parasitic nematodes: a review. Crop Prot 53:147–151
Huang G, Chen F, Kuang Y, He H, Qin A (2016) Current techniques of growing algae using flue gas from exhaust gas industry: a review. Appl Biochem Biotechnol 178:1220–1238
Jenkins SN, Paredes-Valdivia CA, Waite IS, Ayre J, Moheimani N, Barton L, Abbott LK (2017) Assessing the suitability of microalgae biomass produced from piggery waste as a fertilizer. Anim Prod Sci 57:2480
Kallenbach C, Grandy AS (2011) Controls over soil microbial biomass responses to carbon amendments in agricultural systems: a meta-analysis. Agricult Ecosyst Environ 144:241–252
Kapoore RV, Wood EE, Llewellyn CA (2021) Algae biostimulants: a critical look at microalgal biostimulants for sustainable agricultural practices. Biotechnol Adv 49:107754
Karthikeyan N, Prasanna R, Nain L (2007) Evaluating the potential of plant growth promoting cyanobacteria as inoculants for wheat. Soil Biol 43:23–30
Kholssi R, Marks EAN, Miñón J, Montero O (2019) Biofertilizing effect of *Chlorella sorokiniana* suspensions on wheat growth. Plant Growth Regul 38:644–649
Li C, Liang Y, Miao Q, Ji X, Duan P, Quan D (2024) The influence of microalgae fertilizer on soil water conservation and soil improvement: yield and quality of potted tomatoes. Agronomy 14:2102
Lu Y, Xu J (2015) Phytohormones in microalgae: a new opportunity for microalgal biotechnology? Trends Plant Sci 20:273–282
Mandal B, Vlek PLG, Mandal LN (1999) Beneficial effects of blue-green algae and Azolla, excluding supplying nitrogen, on wetland rice fields: a review. Biol Fert Soils 28:329–342
Mata TM, Martins AA, Caetano NS (2010a) Microalgae for biodiesel production and other applications: a review. Renew Sust Energ Rev 14:217–232
Mata TM, Martins AA, Caetano NS (2010b) Microalgae for biodiesel production and other applications: a review. Renew Sustain Energy Rev 14:217–232
Mazhar S, Cohen S, Hasnain S (2013) Auxin producing non-heterocystous cyanobacteria and their impact on the growth and endogenous auxin homeostasis of wheat. Basic Microbiol 53:996–1003
Mukherjee C, Chowdhury R, Sutradhar T, Begam M, Ghosh SM, Basak SK, Ray K (2016) Parboiled rice effluent: a wastewater niche for microalgae and cyanobacteria with growth coupled to comprehensive remediation and phosphorus biofertilization. Algal Res 19:225–236
Muller A, Schader C, El-Hage-Scialabba N, Brüggemann J, Isensee A, Erb K-H, Smith P, Klocke P, Leiber F, Stolze M, Niggli U (2017) Strategies for feeding the world more sustainably with organic agriculture. Nat Commun 8:1290
Pawar GR, Matharu JK, Kumar D (2011) The nitrogen fixation capability of cyanobacteria-a natural source as a bio-fertilizer for arachis hypogaea. IJAR 4:37–43
Pingali PL (2012) Green revolution: impacts, limits, and the path ahead. Proc Natl Acad Sci USA 109:12302–12308
Plöhn M, Spain O, Sirin S, Silva S, Escudero-Oñate C, Ferrando-Climent L, Allahverdiyeva Y, Funk C (2021) Wastewater treatment by microalgae. Physiol Plant 173:568–578
Prasanna R, Chaudhary V, Gupta V, Babu S, Kumar A, Singh R, Shivay YS, Nain L (2013) Cyanobacteria mediated plant growth promotion and bioprotection against Fusarium wilt in tomato. Eur J Plant Pathol 136:337–353

Prasanna R, Sood A, Ratha SK, Singh PK (2014) Cyanobacteria as "green" option for sustainable agriculture. Cyanobact Econ Perspect 15:145–166

Prasanna R, Babu S, Bidyarani N, Kumar A, Triveni S (2015a) Prospecting cyanobacteria-fortified composts as plant growth promoting and biocontrol agents in cotton. Exp Agricult 51:42–65

Prasanna R, Hossain F, Babu S, Bidyarani N, Adak A, Verma S, Shivay YS, Nain L (2015b) Prospecting cyanobacterial formulations as plant-growth-promoting agents for maize hybrids. South Afr J Plant Soil. 32:199–207

Ramya SS, Vijayanand N, Rathinavel S (2015) Foliar application of liquid biofertilizer of brown alga *Stoechospermum marginatum* on growth, biochemical and yield of *Solanum melongena*. Int J Recycl Org Waste Agricult 4:167–173

Renuka N, Guldhe A, Prasanna R, Singh P, Bux F (2018) Microalgae as multi- functional options in modern agriculture: current trends, prospects and challenges. Biotechnol Adv 36:1255–1273

Roberti R, Galletti S, Burzi PL, Righini H, Cetrullo S, Perez C (2015) Induction of defence responses in zucchini (*Cucurbita pepo*) by *Anabaena* sp. water extract. Biol Control 82:61–68

Roger PA, Kulasooriya SA (1980) Blue-green algae and rice. The International Rice Research Institute, Manila

Rosenzweig C, Elliott J, Deryng D, Ruane AC, Müller C, Arneth A, Boote KJ, Folberth C, Glotter M, Khabarov N, Neumann K, Piontek F, Pugh TAM, Schmid E, Stehfest E, Yang H, Jones JW (2014) Assessing agricultural risks of climate change in the 21st century in a global gridded crop model intercomparison. Proc Natl Acad Sci USA 111:3268–3273

Rupawalla Z, Robinson N, Schmidt S, Li S, Carruthers S, Buisset E, Roles J, Hankamer B, Wolf J (2021) Algae biofertilisers promote sustainable food production and a circular nutrient economy: an integrated empirical-modelling study. Sci Total Environ 796:148913

Saadatnia H, Riahi H (2009) Cyanobacteria from paddy fields in Iran as a biofertilizer in rice plants. Plant Soil Environ 55:207–212

Sharma NK, Tiwari SP, Tripathi K, Rai AK (2011) Sustainability and cyanobacteria (blue-green algae): facts and challenges. J Appl Phycol 23:1059–1081

Sido MY, Tian Y, Wang X, Wang X (2022) Application of microalgae *Chlamydomonas applanata* M9V and *C. vulgaris* S3 for wheat growth promotion and as urea alternatives. Front Microbiol 13:1035791

Steensland A (2019) Global agricultural productivity report: productivity growth for sustainable diets, and more, virginia tech college of agriculture and life sciences

Stirk WA, Bálint P, Tarkowská D, Novák O, Strnad M, Ördög V, Van Staden J (2013) Hormone profiles in microalgae: gibberellins and brassinosteroids. Plant Physiol Biochem 70:348–353

Suchithra MR, Muniswami DM, Sri MS, Usha R, Rasheeq AA, Preethi BA, Dineshkumar R (2022) Effectiveness of green microalgae as biostimulants and biofertilizer through foliar spray and soil drench method for tomato cultivation. South Afr J Bot 146:740–750

Tautges NE, Sullivan TS, Reardon CL, Burke IC (2016) Soil microbial diversity and activity linked to crop yield and quality in a dryland organic wheat production system. Appl Soil Ecol 108:258–268

Wang R, Peng B, Huang K (2015) The research progress of CO_2 sequestration by algal bio-fertilizer in China. CO2 Util 11:67–70

Wang J-H, Zhang T-Y, Dao G-H, Xu X-Q, Wang X-X, Hu H-Y (2017) Microalgae-based advanced municipal wastewater treatment for reuse in water bodies. Appl Microbiol Biotechnol 101:2659–2675

Whitton BA, Grainger SLJ, Hawley GRW, Simon JW (1991) Cell-bound and extracellular phosphatase activities of cyanobacterial isolates. Microb Ecol 21:85–98

Zhang Z, Xu M, Fan Y, Zhang L, Wang H (2024) Using microalgae to reduce the use of conventional fertilizers in hydroponics and soil-based cultivation. Sci Total Environ 912:169424

Chapter 35
Spatiotemporal Variations of NDVI and the Effects of Water and Thermal Conditions in the Jiziwan Region of the Yellow River Basin

Feng Wang, Guohe Huang, Yongping Li, Yanfeng Li, Yiting Wei, Changping Chen, Zhipeng Xu, and Zhirui Yang

Abstract This study investigates the spatiotemporal variations of the Normalized Difference Vegetation Index (NDVI) in the Jiziwan region of the Yellow River Basin from 2003 to 2023 and explores the impacts of water and thermal conditions on vegetation dynamics. Results show that NDVI values are highest in summer and lowest in winter, with a general increasing trend from north to south across the region. Over the study period, NDVI exhibited an overall upward trend, particularly in the southern and central areas, indicating improved vegetation conditions. Correlation analysis revealed that groundwater storage is a consistent limiting factor for vegetation growth throughout the year, while the effects of precipitation and soil moisture vary seasonally. Thermal factors, including temperature and net solar radiation, also exhibit complex relationships with NDVI, with temperature generally promoting vegetation growth in spring and summer. These findings highlight the importance of considering both water and thermal conditions in vegetation management and ecological restoration efforts in the region.

Keywords Normalized difference vegetation index · Spatiotemporal variations · Water and thermal conditions

F. Wang
State Key Laboratory of Earth Surface Processes and Resource Ecology, Faculty of Geographical Science, Beijing Normal University, Beijing, China

F. Wang · G. Huang (✉) · Y. Li · Y. Li · Y. Wei · Z. Xu · Z. Yang
State Key Laboratory of Regional Environment and Sustainability, China-Canada Center for Energy, Environment and Ecology Research, UR-BNU, School of Environment, Beijing Normal University, Beijing, China
e-mail: huang@iseis.org

G. Huang
Environmental Systems Engineering Program, University of Regina, Regina, Canada

C. Chen
Quanzhou Normal University, Quanzhou, Fujian, China

Y. Li et al. (eds.), *16th International Conference on Environmental Science and Development*, Environmental Science and Engineering,
https://doi.org/10.1007/978-3-032-15080-6_35

35.1 Introduction

Vegetation, as the primary producer of ecosystems, plays a crucial role not only in global climate change, material exchange between ecosystems, energy transfer, and ecological environment changes (Gao et al. 2016; Zhu et al. 2023) but also has a significant impact on regional agricultural production, soil and water conservation, natural disaster protection, and sustainable human development (Shen et al. 2019; Bento et al. 2018; He et al. 2015). The vegetation coverage (NDVI) in the Jiziwan region of the Yellow River Basin has undergone significant changes, influenced by complex topography, variable climate, and fragile ecological conditions (Liu et al. 2025; Yu et al. 2024). These fluctuations in NDVI have profound implications for the ecological environment and socio-economic development, affecting areas such as soil erosion control, water resource management, and agricultural production (Jian et al. 2022). Understanding the changes in NDVI and their driving factors in this region is crucial for ecological conservation, land-use planning, and climate change adaptation (Liu et al. 2022). Particularly in the context of global climate change and increasing human activities, clarifying the mechanisms by which water and thermal conditions influence vegetation growth can help optimize ecological restoration strategies and promote regional sustainable development (Liu et al. 2021).

Previous studies on NDVI changes in the Jiziwan region of the Yellow River have primarily focused on long-term trends, seasonal variations, and responses to climate change and human activities (Liu et al. 2025; Yu et al. 2024). Numerous studies have shown that in arid and semi-arid regions, precipitation is a key factor controlling NDVI changes (Jian et al. 2022; Liu et al. 2022), as water availability directly impacts vegetation growth and ecosystem stability (Liu et al. 2021). Additionally, thermal factors such as temperature and solar radiation significantly affect vegetation phenology, growth cycles and drought resistance (Sun et al. 2019). Research has also found that human activities, including land-use changes, ecological restoration projects (such as the Grain for Green Program), and soil and water conservation measures, have a significant impact on the spatial patterns and trends of NDVI (Jiapaer et al. 2015; Peng et al. 2012). However, a systematic study on the combined effects of water and thermal conditions on NDVI is still lacking, and further exploration is needed to understand their relative contributions and spatiotemporal variation patterns.

Therefore, this study aims to systematically analyze the key drivers of NDVI changes in the Ji-shaped Bend region of the Yellow River Basin, with a particular focus on the effects of water and thermal conditions on vegetation dynamics. Using remote sensing and climate data combined with Geographic Information System (GIS) technology, this study explores the spatiotemporal variations in NDVI and quantifies the relative contributions of water and thermal factors to vegetation coverage changes. The research will specifically examine the influence of precipitation, temperature, and radiation on NDVI under different seasonal and climatic conditions and further reveal the interaction mechanisms between water and heat in vegetation growth.

Specifically, this study will proceed as follows: First, by analyzing NDVI time series data, it will identify trends in vegetation coverage changes and seasonal variations in the Ji-shaped Bend region and preliminarily analyze the spatiotemporal characteristics of groundwater storage changes, identifying key areas of change. Second, based on climate factors (such as precipitation, temperature, and radiation), a correlation model will be established to assess the individual and combined effects of water and thermal factors on NDVI changes. Finally, the study will incorporate land-use analysis to evaluate the regulatory role of human activities in NDVI variations. Through these analyses, this research aims to fill the existing gap in understanding the interactive effects of water and thermal conditions on NDVI and provide a scientific basis for ecological conservation and land management in this region.

35.2 Data Source and Methods

35.2.1 Dataset

This study utilizes NDVI data from the 250-m resolution China regional Normalized Difference Vegetation Index (NDVI) dataset, spanning the years 2000 to 2023. The dataset is made available by the National Tibetan Plateau/Third Pole Environment Data Center (http://data.tpdc.ac.cn). To ensure consistency with other variables, the original NDVI data was interpolated and aggregated to a coarser spatial resolution of $1° \times 1°$. Meanwhile, meteorological and environmental variables—such as average temperature, total precipitation, soil moisture, and monthly net solar radiation—were obtained from the ERA5 reanalysis dataset by the European Centre for Medium-Range Weather Forecasts (ECMWF), accessible via https://www.ecmwf.int/en/forecasts/dataset/ecmwf-reanalysis-v5?check_logged_in=1. These variables are provided at a spatial resolution of $1° \times 1°$, matching the processed NDVI grid.

The GRACE satellite monitors changes in terrestrial water storage, including the total changes in surface water (lakes and rivers), soil moisture, groundwater, and snow water (Frappart and Ramillien 2018). In this study, the CSR's latest released CSR GRACE/GRACE-FO RL06.2 Mascon Grids data product, based on the Mascon method, was selected to obtain the monthly-scale changes in total land water storage, with a time span from January 2003 to December 2023. The spatial resolution of this data is $0.25° \times 0.25°$. Missing monthly data were processed using linear interpolation.

The GLDAS hydrological model, developed by NASA's Goddard Space Flight Center and NOAA's National Environmental Prediction Center (NCEP) through data assimilation techniques, combines satellite remote sensing data and ground observations to drive land surface models. The GLDAS model provides global data on variables such as air temperature, snow water equivalent, surface runoff, and soil moisture for depths of 0 to 200 cm. The data from this model have become an essential source for global change and water cycle research. To facilitate comparison with the GRACE Mascon model, the GLDAS_NOAH025_M_2.1 dataset was chosen, which has a spatial resolution of 0.25° × 0.25°, a time resolution of monthly scale, and a time span from January 2003 to December 2023, matching the GRACE dataset.

35.2.2 Data Processing

The terrestrial water storage changes extracted from the GRACE data include the total changes in surface water (lakes and rivers), soil water, groundwater, and snow water. Therefore, based on the principle of water balance, the change in groundwater storage can be derived from the following equation (Frappart and Ramillien 2018):

$$\Delta \text{GWS} = \Delta \text{TWS} - \Delta \text{SWS} - \Delta \text{SMS} - \Delta \text{SNS}$$

where ΔGWS represents the change in groundwater, ΔTWS represents the total land water storage change, ΔSWS represents the surface water change, ΔSMS represents the soil moisture change, ΔSNS represents the snow water change. At a global scale, the results simulated by GLDAS show that the changes in snow water storage and surface water storage are very small. Except for the islands of the Arctic Ocean, snow water storage changes are extremely minor elsewhere in the world [24], and therefore, both the snow water and surface water changes can be considered negligible, making their values effectively zero.

35.2.3 Sample Analysis

This study employed the Theil-Sen trend analysis method to evaluate grid-level trends in NDVI and its associated climatic drivers. The Theil-Sen method is a robust, non-parametric approach for detecting trends in time series data (Sun et al. 2019; Wu et al. 2024). It estimates the trend by computing the median of all possible pairwise slopes, making it resistant to outliers and not reliant on the assumption of normality

$$\text{S} = \text{median}\frac{x_\text{j} - x_\text{i}}{j - i}(1 < i < j < n) \tag{35.1}$$

where, x_i and x_j are the NDVI values for year i and year$j$, n is the length of the time series, S is the interannual variation trend of NDVI.

In this study, the Mann–Kendall test was used to evaluate the statistical significance of vegetation trend changes. The formula for calculating the significance test statistic Z is as follows:

$$\text{Z} = \begin{cases} \frac{\beta-1}{\sqrt{\text{var}(\beta)}}, (S > 0) \\ 0, (S = 0) \\ \frac{\beta+1}{\sqrt{\text{var}(\beta)}}, (S < 0) \end{cases} \tag{35.2}$$

$$\beta = \sum_{i=1}^{n-1} \sum_{j=i+1}^{n} \mathrm{sgn}(x_j - x_i) \tag{35.3}$$

If $Z > 0$, the data series exhibits an increasing trend, while if $Z < 0$, the data series shows a decreasing trend. When $|Z| \geq 1.96$, the trend is statistically significant, and when $|Z| \geq 2.58$, the trend is highly significant. β is the trend test statistic, and x_j represents the value in the jth month.

Pearson correlation analysis was used in this study to quantify the linear relationship between NDVI and climate variables. This method computes a correlation coefficient (r) ranging from -1 to 1, where positive values indicate a direct relationship and negative values indicate an inverse relationship. The closer the value is to $\pm$ 1, the stronger the correlation.

35.3 Result and Discussion

35.3.1 The Spatial Distribution Characteristics of NDVI by Season

Figure 35.1 displays the mean distribution of the NDVI across four seasons in the Jiziwan Region from 2003 to 2023. Overall, the NDVI values are relatively low in spring, with most regions appearing in light yellow to orange, indicating less vegetation cover or the initial stage of vegetation growth. In summer, the NDVI values increase significantly, with most areas showing orange to red, reflecting vigorous vegetation growth and higher coverage. The NDVI values decrease in autumn but remain at a relatively high level, with colors ranging from yellow to orange, suggesting that vegetation is entering a mature or declining phase. In winter, the NDVI values are the lowest, with most areas appearing in light yellow, indicating the least vegetation cover, possibly due to seasonal leaf fall or withering. From these seasonal changes, it is evident that the growth status of vegetation is closely related to seasonal changes, with summer being the period of most vigorous vegetation growth and winter being the period of least vegetation cover.

Geographically, the NDVI values in this region show clear spatial differences across different seasons. In the northern areas, including cities such as Hohhot and Baotou, the NDVI values are relatively low in all seasons, likely due to harsh climate conditions that limit vegetation growth. In contrast, the southern areas, such as Yulin and Yan'an, have higher NDVI values in summer and autumn, indicating better vegetation cover and growth conditions. Taking summer as an example, the NDVI values in the northern region mainly range from 0.0 to 0.2, while the central region, including cities like Ordos and Dongshengshi, has higher NDVI values, mainly in the range of 0.6 to 0.8. The southern region, including Yulin and Yan'an, also has high NDVI values, similarly in the range of 0.6 to 0.8. The eastern region, including

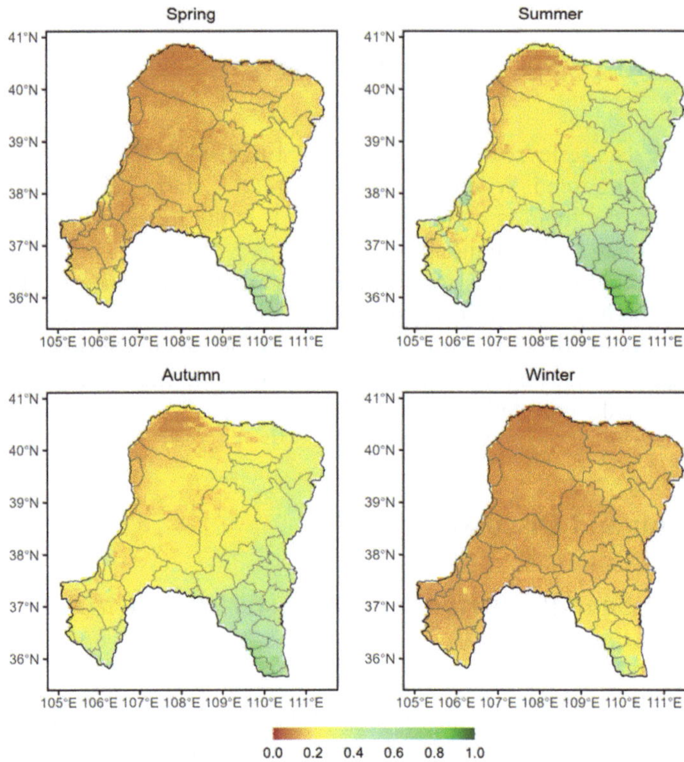

Fig. 35.1 Spatial distribution of seasonal mean NDVI in the Jiziwan region from 2003 to 2023

cities like Shenmu and Fugu, has relatively high NDVI values, mainly in the range of 0.4 to 0.6. Overall, the NDVI values in the Jiziwan region show a trend of gradually increasing from north to south, which may be related to changes in climate conditions, with the northern region being drier and the southern region being more humid, facilitating vegetation growth and cover.

From Fig. 35.2, it can be observed that spring and summer occupy a larger area proportion in the NDVI range of 0–0.2, approaching 40% and 50% respectively. This indicates that in these two seasons, areas with lower vegetation coverage are more extensive. Autumn has the highest area proportion in the NDVI range of 0.2–0.4, about 30%, showing that areas with moderate vegetation coverage are more widespread during this season. Winter has the highest area proportion in the NDVI range of 0.4–0.6, about 15%, which may be related to the slower growth of vegetation or partial dormancy during the winter months. In the NDVI ranges of 0.6–0.8 and 0.8–1, the area proportions are relatively low for all seasons, especially in the range of 0.8–1, which is almost negligible. This suggests that areas with very high vegetation coverage are less common during these seasons. Overall, the chart reflects the changes in vegetation coverage across different seasons, with lower vegetation coverage in spring and summer, and relatively higher coverage in autumn and winter. These

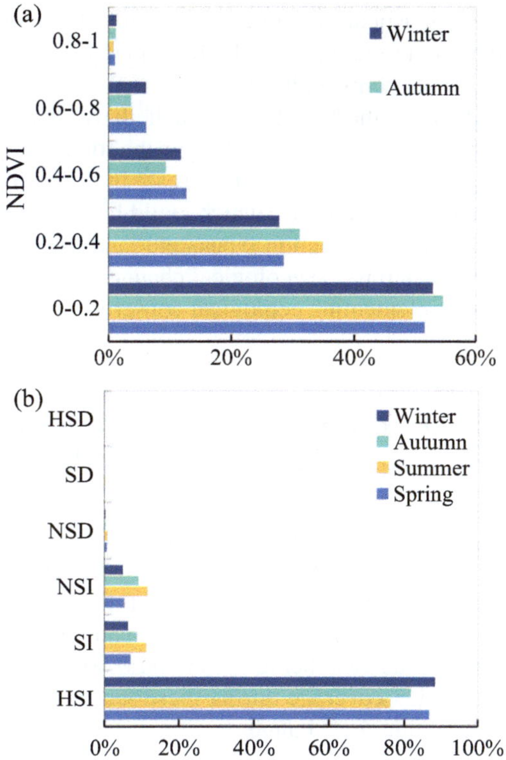

Fig. 35.2 Area proportion of NDVI distribution (**a**) and seasonal change trend (**b**) from 2003 to 2023. (HSD: Highly significant degradation; SD: Significant degradation; NSD: Non-significant degradation; NSI: Non-significant improvement; SI: Significant improvement; HSI: Highly significant improvement.)

seasonal variations may be related to climatic conditions, vegetation growth cycles, and other factors.

35.3.2 Spatial Distribution of Seasonal NDVI Change Trends

We conducted an analysis of the NDVI change trends across different seasons in the Jiziwan region. By employing the Sen's Slope and Mann–Kendall (MK) test methods, we were able to quantitatively and qualitatively describe the multi-year change characteristics of NDVI and identify the regions with the most significant changes. As shown in Fig. 35.3, in spring, the NDVI change rate (S value) in most areas ranged from 0.000 to 0.006, indicating a slight upward trend. Meanwhile, the MK test (Z value) indicated that these changes were statistically significant. During summer, the NDVI change rate in the southern and central regions increased significantly, with some areas reaching an S value of 0.006, and the MK test also showed that the changes in these regions were statistically significant. The patterns in autumn and winter were similar to those in summer, with the NDVI values in the southern and central regions continuing to show an upward trend, and these changes were statistically significant. Spatially, the significant increase in NDVI is observed in the southeastern part of the Jiziwan region, such as Yulin, with the most pronounced increase occurring in summer.

Overall, the NDVI values in the Jiziwan region generally showed an upward trend across all seasons, particularly in the southern and central areas. This suggests that the region has experienced an increase in vegetation cover or improvement in vegetation health over the years, which could be attributed to multiple factors such as climate conditions, vegetation types, and land use practices. These findings are crucial for understanding the ecological changes in the region, formulating land use planning, and implementing ecological conservation measures.

35.3.3 The Spatiotemporal Variations of NDVI and the Correlation with Moisture Conditions

Figure 35.4 illustrates the trends of NDVI, precipitation, soil moisture, groundwater storage, temperature, and net solar radiation in the Jiziwan region, divided by quarter from 2003 to 2023. The data reveals that over the past two decades, the NDVI values in the Ji Zi Wan region have shown an upward trend in spring and summer, with the rates of increase being 0.003 and 0.006 respectively. This indicates an increase in vegetation coverage during these seasons. The NDVI values in autumn and winter exhibit minimal changes.

The potential factors influencing NDVI include two main aspects: moisture and heat. Moisture factors involve precipitation, soil moisture (SM), and groundwater storage. Precipitation shows fluctuating changes in summer, with minimal changes in spring and autumn, and relatively lower levels in winter. Soil moisture is relatively low in summer, likely due to reduced precipitation and increased evaporation. Groundwater storage shows a downward trend across all seasons, with the most significant decline in winter, indicated by a slope of -0.904, which may be related to decreased precipitation and increased demand for agricultural irrigation. Heat factors include temperature and net solar radiation (SSR). Temperature shows an upward trend in summer, while net solar radiation exhibits a slight downward trend (with a rate of change of -5.53×10^{-4}), possibly due to increased cloud cover or air pollution. Net solar radiation changes little in other seasons. These factors collectively affect NDVI changes, with climatic factors such as precipitation and temperature being the main limiting factors affecting vegetation growth.

Figure 35.5 illustrates the spatial distribution of seasonal NDVI constraints by water availability in the Jiziwan region from 2003 to 2023, encompassing three key hydrological factors: precipitation (PRE), soil moisture (SM), and groundwater storage (GW). The analysis reveals seasonal variations and complexities in the limitations imposed by these factors on vegetation growth (Zhang et al. 2023). In spring, NDVI is minimally constrained by water availability. NDVI shows a slight positive correlation with precipitation (average correlation: 0.07), suggesting that increased

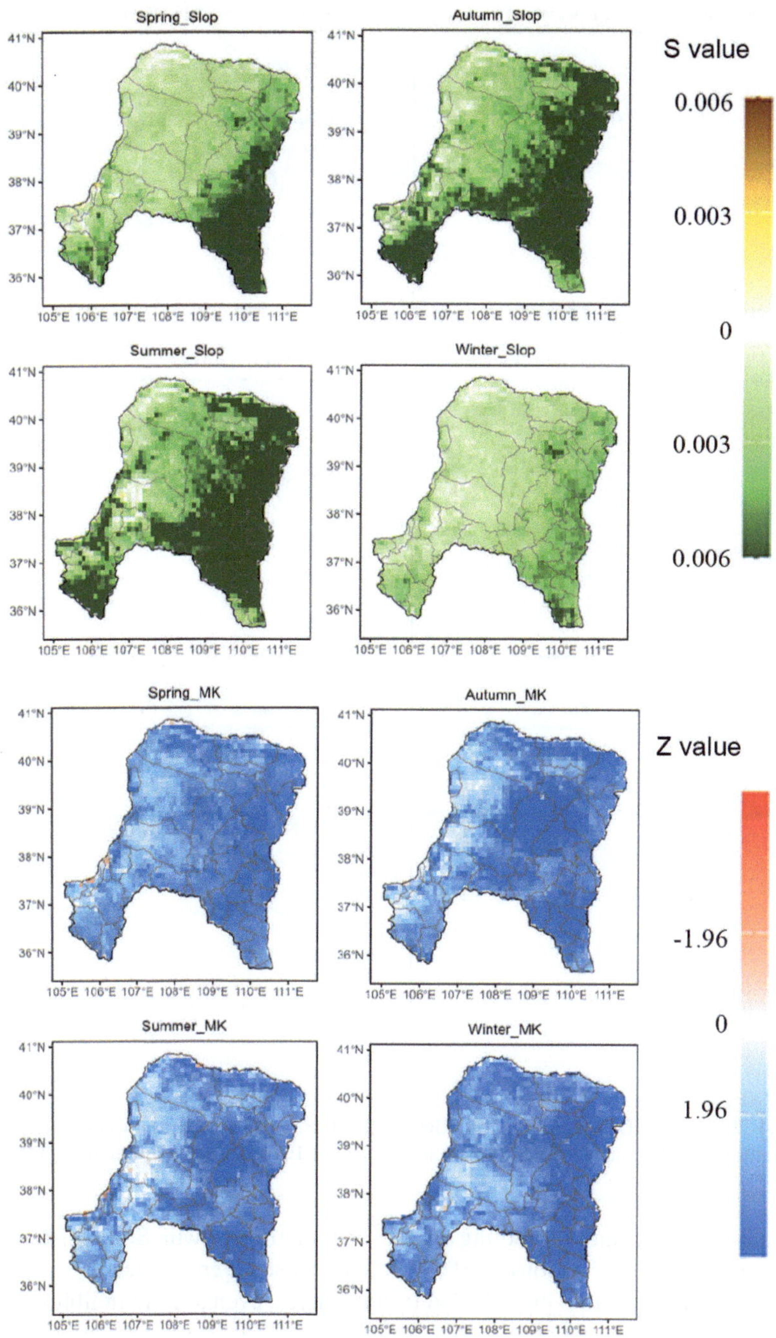

Fig. 35.3 Spatial distribution of seasonal NDVI change trends from 2003 to 2023

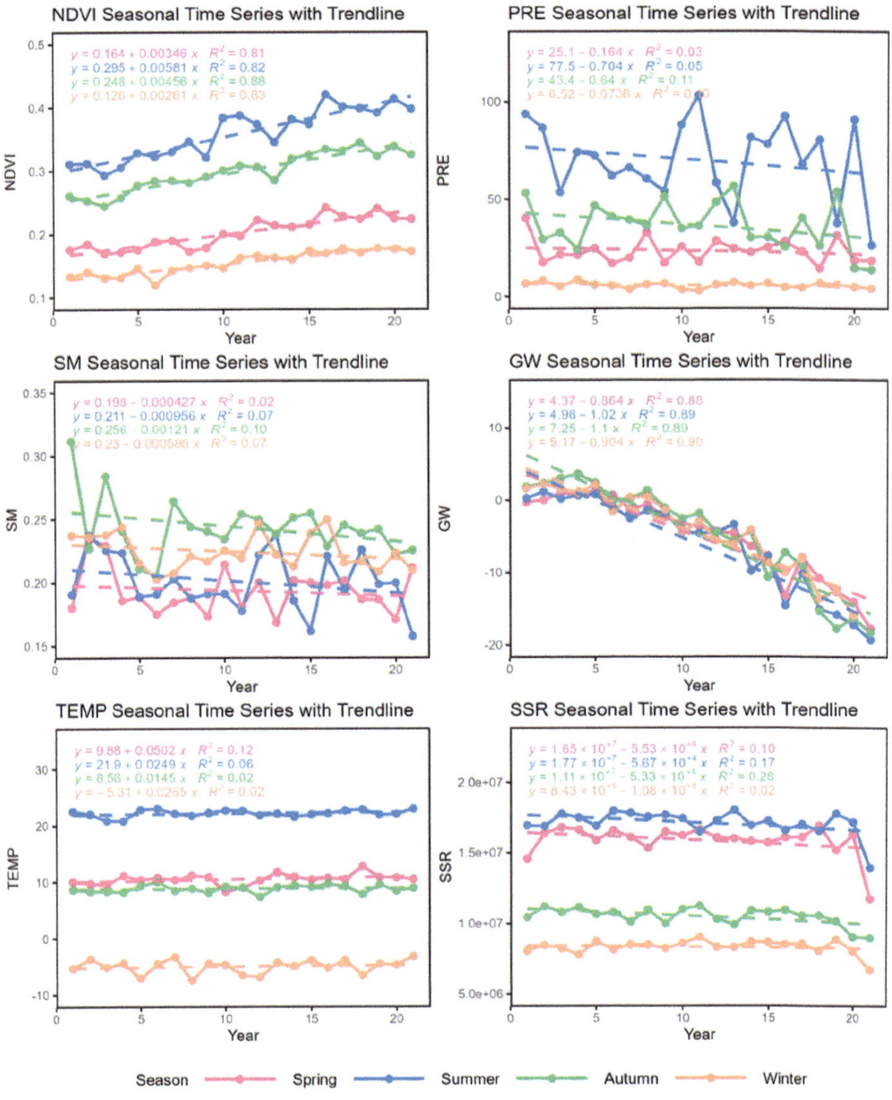

Fig. 35.4 Trends of NDVI, Precipitation (PRE), Soil Moisture (SM), Groundwater Storage (GW), Temperature (TEMP), and Net Solar Radiation (SSR) in the Jiziwan Region by Quarter (2003–2023)

precipitation has a minor promoting effect on vegetation growth. Soil moisture (SM) exhibits an average correlation of −0.04, which is close to zero, indicating an insignificant impact on vegetation growth, and in some areas, even a slight inhibitory effect. However, groundwater storage (GW) shows a strong negative correlation (−0.69), implying that a decline in groundwater resources may impose significant limitations on vegetation growth. In summer, the constraints imposed by water availability become more pronounced, with distinct influences from different hydrological factors. The correlation between NDVI and precipitation increases significantly (0.21), highlighting precipitation as a crucial regulating factor for vegetation growth, where increased precipitation effectively alleviates water limitations. However, soil moisture exhibits a strong negative correlation (–0.22), which may be attributed to intensified evaporation under high temperatures, reducing the effective availability of soil moisture for vegetation growth. Groundwater storage remains negatively correlated (−0.69), indicating that reduced groundwater availability continues to pose substantial constraints on vegetation. During autumn, vegetation dependence on water declines further. The correlation between NDVI and precipitation turns negative (−0.19), indicating that precipitation's contribution to vegetation growth diminishes and may even be unfavorable in certain areas. Soil moisture also exhibits a negative correlation (−0.20), likely due to vegetation entering its senescence phase and reduced transpiration. Groundwater storage maintains a strong negative correlation (−0.66), underscoring its continued role as a limiting factor for vegetation in autumn. In winter, vegetation growth is generally constrained by water availability, but the influence of different factors varies. The correlation between NDVI and precipitation remains negative (−0.18). Soil moisture shows a slight positive correlation (0.03), suggesting a minimal direct limiting effect on vegetation. Groundwater storage continues to exhibit a strong negative correlation (−0.64), reinforcing its role as a significant constraint throughout the year.

Overall, the impact of water availability on NDVI varies across seasons. GW consistently imposes strong constraints on vegetation growth throughout the year, with its negative correlation indicating that declining groundwater levels may be a key long-term factor affecting vegetation. In contrast, PRE and SM have seasonally variable effects, exerting weaker constraints in spring and summer, while playing a more pronounced limiting role in autumn and winter. These findings are crucial for understanding vegetation responses to hydrological factors and provide valuable insights for water resource management and ecosystem conservation.

35.3.4 The Spatiotemporal Variations of NDVI and the Correlation with Thermal Conditions

Figure 35.6 illustrates the spatial correlation between the NDVI and thermal factors (e.g., temperature and net solar radiation) across different seasons. By analyzing the average correlation values, we can better understand how thermal factors influence vegetation growth. In spring, NDVI exhibits a slight negative correlation with net solar radiation (SSR) (average correlation: −0.19), suggesting that increased solar radiation may slightly suppress vegetation growth. This effect could be attributed to

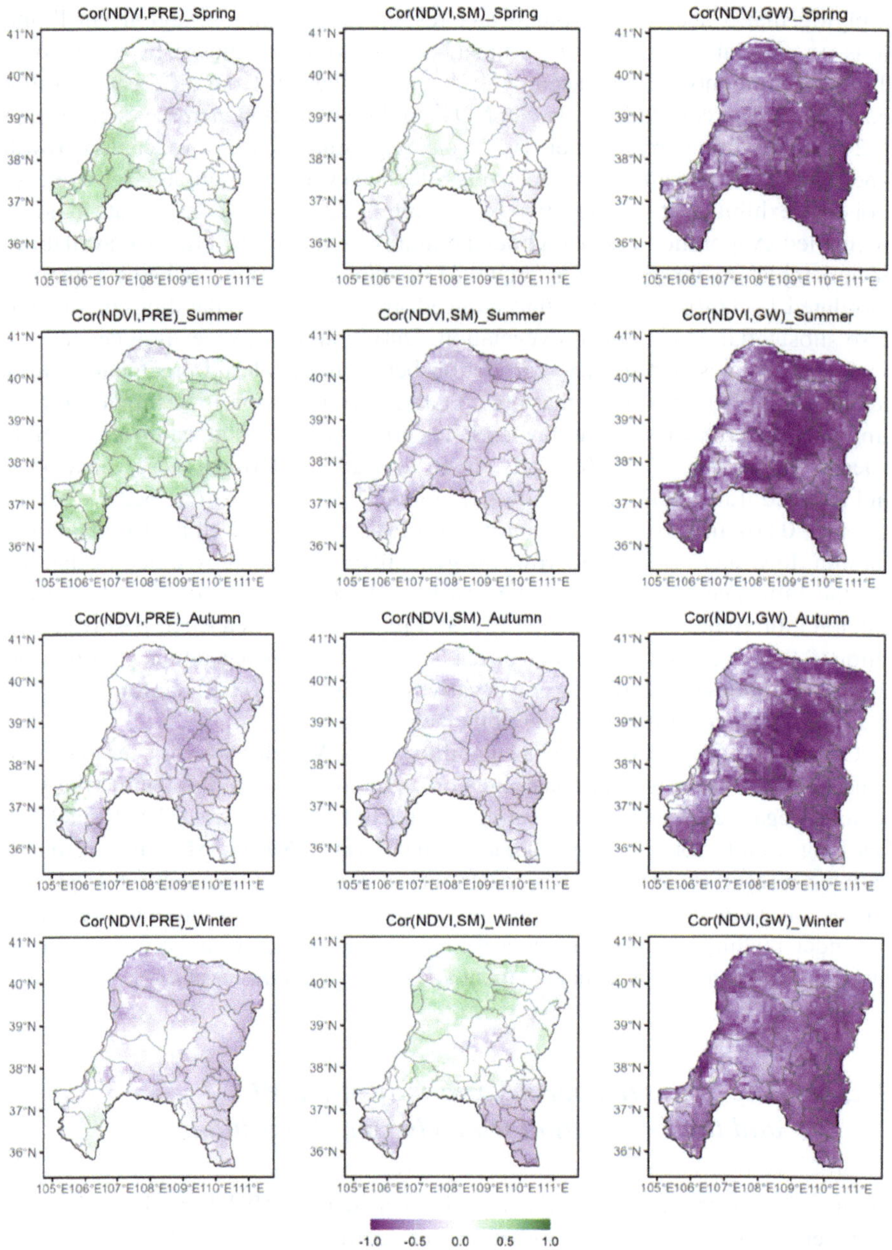

Fig. 35.5 Spatial Distribution of the Correlation Between Seasonal NDVI and Moisture Factors

the relatively low spring temperatures, where excessive solar radiation may accelerate soil moisture evaporation, thereby negatively impacting vegetation growth. Conversely, temperature shows a slight positive correlation with NDVI (0.16), indicating that rising temperatures may promote vegetation growth to some extent. In summer, the correlation between NDVI and net solar radiation becomes more pronounced, with a moderate negative correlation (−0.36). This suggests that excessive solar radiation significantly inhibits vegetation growth, likely due to increased evapotranspiration under high temperatures, which exacerbates water loss. Meanwhile, temperature exhibits a moderate positive correlation with NDVI (0.25), implying that rising temperatures contribute to vegetation growth, potentially by accelerating metabolic processes and extending the growing season. During autumn, NDVI maintains a moderate negative correlation with net solar radiation (−0.28), indicating that excessive radiation continues to have a suppressive effect on vegetation growth. As temperatures gradually decrease in autumn, increased solar radiation may lead to higher evaporation rates, reducing soil moisture availability. Temperature, on the other hand, has a slight positive correlation with NDVI (0.15), suggesting that warmer conditions may still provide some support for vegetation growth. In contrast to other seasons, winter shows a slight positive correlation between NDVI and net solar radiation (0.13). This suggests that, under colder conditions, increased solar radiation may provide the necessary energy to support limited vegetation activity. However, temperature exhibits a near-zero negative correlation with NDVI (−0.01), indicating that temperature fluctuations have little direct impact on vegetation growth in winter and, in some cases, higher temperatures may even slightly suppress it.

Overall, SSR exhibits a negative correlation with NDVI throughout most of the year, with the strongest inhibitory effect occurring in summer. This suggests that excessive solar radiation generally limits vegetation growth, likely due to increased water loss from evaporation. In contrast, the relationship between temperature and NDVI varies by season, indicating that the influence of thermal factors on vegetation growth is not uniform. These findings provide valuable insights for understanding and predicting vegetation responses to climate variables and can support better ecosystem management strategies.

35.4 Conclusions

The study systematically analyzed the spatiotemporal variations of NDVI in the Jiziwan region of the Yellow River Basin from 2003 to 2023 and explored the impacts of water and thermal conditions on vegetation dynamics. Results indicated that NDVI exhibited distinct seasonal patterns, with the highest values in summer and the lowest in winter. Spatially, NDVI values generally increased from north to south, corresponding to climatic gradients. Over the study period, NDVI showed an overall upward trend, particularly in the southern and central regions, suggesting

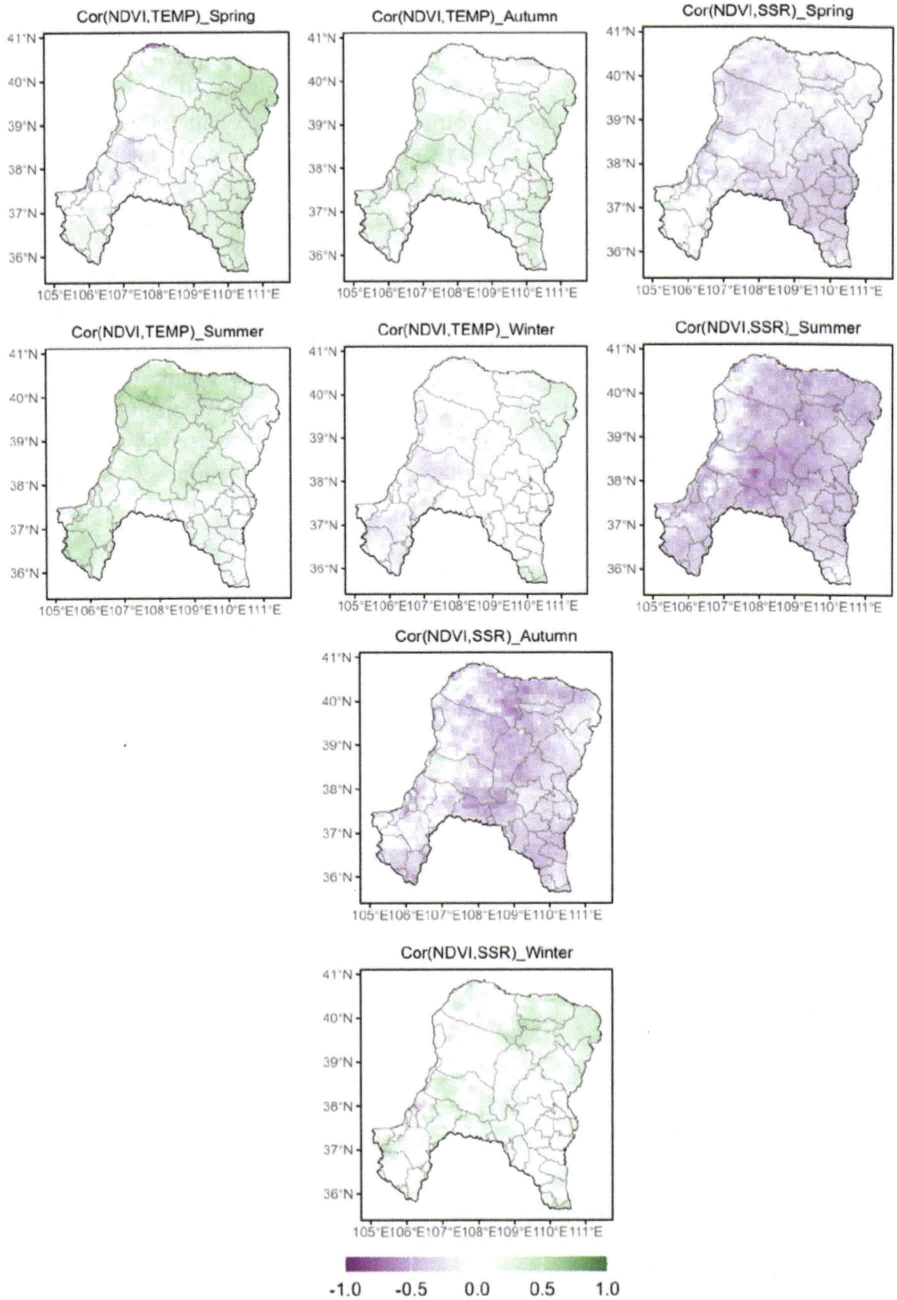

Fig. 35.6 Spatial Distribution of the Correlation Between Seasonal NDVI and Thermal Factors

improved vegetation conditions. The analysis of correlations revealed that groundwater storage was a consistent limiting factor for vegetation growth throughout the year, while the effects of precipitation and soil moisture varied seasonally. Thermal factors, such as temperature and net solar radiation, also exhibited complex relationships with NDVI, with temperature generally promoting vegetation growth in spring and summer, while excessive solar radiation inhibited growth in summer and autumn. These findings highlight the importance of considering both water and thermal conditions in vegetation management and ecological restoration efforts in the region. Future research should focus on integrating more detailed land-use and socio-economic data to further elucidate the drivers of NDVI changes and develop more effective strategies for sustainable land management and ecological conservation. Future research should integrate more detailed land-use and socio-economic data to further elucidate the drivers of NDVI changes and develop effective strategies for sustainable land management and ecological conservation.

Acknowledgements This research was supported by the National Key R&D Program of China (2023YFC3206503) and the Innovative Research Group of the National Natural Science Foundation of China (52221003).

References

Bento VA, Gouveia CM, DaCamara CC, Trigo IF (2018) A climatological assessment of drought impact on vegetation health index. Agric for Meteorol 259:286–295

Frappart F, Ramillien G (2018) Monitoring groundwater storage changes using the gravity recovery and climate experiment (GRACE) satellite mission: a review. Remote Sens 10(6):829

Gao Q, Guo Y, Xu H, Ganjurjav H, Li Y, Wan Y, Qin X, Ma X, Liu S (2016) Climate change and its impacts on vegetation distribution and net primary productivity of the alpine ecosystem in the Qinghai-Tibetan Plateau. Sci Total Environ 554:34–41

He B, Chen AF, Wang HL, Wang QF (2015) Dynamic response of satellite-derived vegetation growth to climate change in the three north shelter forest region in China. Remote Sens 7(8):9998–10016

Jian S, Zhang Q, Wang H (2022) Spatial–temporal trends in and attribution analysis of vegetation change in the Yellow River Basin China. Remote Sens 14(18):4607

Jiapaer G, Liang SL, Yi QX, Liu JP (2015) Vegetation dynamics and responses to recent climate change in Xinjiang using leaf area index as an indicator. Ecol Ind 58:64–76

Lamchin M, Park T, Lee J-Y, Lee W-K (2015) Monitoring of vegetation dynamics in the Mongolia using MODIS NDVIs and their relationship to rainfall by natural zone. J Ind Soc Remote Sens 43:325–337

Liu Y, Tian J, Liu R, Ding L (2021) Influences of climate change and human activities on NDVI changes in China. Remote Sensing 13(21):4326

Liu C, Zhang X, Wang T, Chen G, Zhu K, Wang Q, Wang J (2022) Detection of vegetation coverage changes in the Yellow River Basin from 2003 to 2020. Ecol Ind 138:108818

Liu W, Tang R, Zhang G, Xue J, Xue B, Wang Y (2025) Spatiotemporal changes in evapotranspiration and its influencing factors in the Jiziwan Region of the Yellow River from 1982 to 2018. Remote Sens 17(2):252

Peng J, Dong W, Yuan W, Zhang Y (2012) Responses of grassland and forest to temperature and precipitation changes in Northeast China. Adv Atmos Sci 29(5):1063–1077

Shen X, Xue Z, Jiang M, Lu X (2019) Spatiotemporal change of vegetation coverage and its relationship with climate change in freshwater marshes of Northeast China. Wetlands 39:429–439

Sun W, Wang Y, Fu Y, Xue B, Wang G, Yu G, Zuo D, Xu Z (2019) Spatial heterogeneity of changes in vegetation growth and their driving forces based on satellite observations of the Yarlung Zangbo River Basin in the Tibetan Plateau. J Hydrol 574:324–332

Wu YH, Huang CZ, Huang WW, Chen XJ (2024) Projected changes in temperature and precipitation over Canada in the 21st century. J Environ Inf 44(2):140–154

Yu ZB, Yin S, Bai JH, Wang C, Chen GZ, Wang W, Wang YQ, Cui BS, Liu XH, Li XW (2024) Suaeda salsa in relation to hydrological connectivity: from the view of plant trait networks. J Environ Inf 43(1):41–49

Zhang X, Yan DH, Qin TL, Li CH, Wang H (2023) Differences in China greening characteristics and its contribution to global greening. J Environ Inf 42(2):143–157

Zhu L, Sun S, Li Y, Liu X, Hu K (2023) Effects of climate change and anthropogenic activity on the vegetation greening in the Liaohe River Basin of northeastern China. Ecol Ind 148:110105

Chapter 36
Study on the Coordinated Development of Tourism and Ecological Environment in Gansu Province

Gengxia Yang

Abstract The coordinated development of tourism and the ecological environment is of great significance for maintaining ecological balance. Exploring the coordination relationship between tourism and the ecological environment is conducive to promoting the sustainable development of eco-tourism and achieving the harmonious coexistence between tourism and the environment. Taking Gansu Province as the research object, this paper establishes the comprehensive evaluation index of tourism and ecological environment, uses the entropy method to evaluate the indicators of tourism and ecological environment in Gansu province from 2020 to 2023, and analyzes the coupling degree and coupling coordination degree of tourism and ecological environment in Gansu province using the coupling coordination degree model. It is concluded that tourism and the ecological environment in Gansu province are at the primary coupling level, and the development level of tourism lags behind the development level of the ecological environment. In the process of tourism development and utilization, it is very easy to damage the ecological environment. It is necessary to take into account the ecological environment while developing the tourism industry with high quality, improve the ecological environment management ability of the tourism industry in various aspects, fully grasp the innovative concept of ecological tourism, and develop ecological tourism projects, to better realize the coordinated development of tourism and ecological environment.

Keywords Gansu Province · Tourism. · Ecological environment. · Coordinated development.

G. Yang (✉)
College of Finance and Business, Lanzhou Resources and Environment Voc-Tech University, Lanzhou, China

Y. Li et al. (eds.), *16th International Conference on Environmental Science and Development*, Environmental Science and Engineering,
https://doi.org/10.1007/978-3-032-15080-6_36

36.1 Introduction

There is a close relationship between tourism and the ecological environment. On the one hand, the ecological environment is the basis of tourism development, providing necessary natural and cultural resources for tourism and promoting the development of tourism. On the other hand, the prosperity of tourism provides funds and support for ecological environmental protection, promotes the development of the local economy, and improves the living standards of residents. The impact of tourism activities on the ecological environment is gradually revealed. The coupling and coordinated development of tourism development and the ecological environment is crucial to regional sustainable development. How to coordinate the relationship between the ecological environment and tourism development and achieve high-quality tourism development is the focus of local governments at present. However, with the rapid development of tourism in China, the spatial dislocation of tourist flow, tourist attractions, star-rated hotels, and ecological environment quality caused by differences in tourism resources endowment, location conditions, and infrastructure in different regions has attracted much attention and has become an important issue that needs to be solved urgently in the field of regional tourism research. Therefore, in the development of tourism, we must pay attention to the protection of the ecological environment, adhere to sustainable development, and ensure the harmonious coexistence of tourism development and the ecological environment.

At present, regarding the relationship between tourism and the ecological environment, foreign scholars mainly focus on the measurement of tourism's ecological footprint (Gossling et al. 2002), the sustainable development of tourism and the ecological environment (Tsaur et al. 2006), and the impact mechanism of tourism activities on the ecological environment (Wall and Wright 1977), etc. The research scope covers global large-scale (Gossling 2002) or small-scale tourist attractions (Popovic et al. 2018). Domestic scholars have done a lot of research on them from different spatial scales, research perspectives, and research methods. From the perspective of research scale, it mainly involves the macro level such as the whole country (Fang et al. 2013), the province (Cong et al. 2019), and the prefecture-level city (Wang and Wang 2021), while there are few researchers on the micro level such as the county and scenic spots. From the perspective of research, most domestic scholars have studied the interaction stress and coupling coordination between tourism and the ecological environment from the perspective of system theory, and some scholars have analyzed the ecological environment threshold and tourism carrying capacity. From the perspective of research methods, it mainly includes models and methods such as the coupling coordination degree model, fuzzy mathematical method (Ma 2016), and vulnerability model (He et al. 2022). Based on this, this study constructs an evaluation index system for the adaptability of tourism and the ecological environment in Gansu Province and uses the TOPSIS method of entropy weight to study the coordination between tourism and the ecological environment in Gansu Province from 2019 to 2023, to provide references for the coordinated development of tourism and ecological environment.

36.2 Study Area and Data Source

36.2.1 Overview of the Study Area

Gansu Province is located in northwest China with a total area of 42.58 square kilometers. It borders Shaanxi to the east, Sichuan to the south, Qinghai and Xinjiang to the west, Ningxia and Inner Mongolia to the north, and Mongolia to the north. The terrain of Gansu is complex and diverse, with mountains, plateaus, plains, valleys, deserts, and Gobi. The terrain inclines from southwest to northeast and can be divided into Longnan Mountain, Longzhong Loess Plateau, Hexi Corridor, and the north zone of Hexi Corridor. Climate types include subtropical monsoon climate, temperate continental arid climate, and plateau mountain climate. Gansu has a long history, is an important birthplace of Chinese civilization, an important passage on the Silk Road, known as "the birthplace of Chinese civilization, a museum of natural wonders, a grand view garden of ethnic customs, and a destination of quality tourism", with rich and colorful cultural charm, rich and diverse tourism resources and unique geographical landscape. In 2023, Gansu's cultural tourism economy recovered strongly, receiving 388 million domestic tourists, an increase of 187.8%. The comprehensive tourism revenue reached 274.58 billion yuan, up 312.9% year on year.

36.2.2 Data Sources

On the basis of considering the scientific, accuracy and availability of research data, the coordinated development of tourism and ecological environment in Gansu Province was explored. The research period is determined to be 2020–2023, and the research data come from the Statistical Yearbook of Gansu Province (2020–2023), the statistical data of relevant departments of the Department of Ecology and Environment, the Department of Culture and Tourism of Gansu Province, and the work report of the Gansu Provincial government.

36.3 Index System and Research Design

36.3.1 Index System Construction

Based on the actual situation of Gansu Province and the existing relevant studies, the author, adhering to the principles of objectivity, scientific, systematicity, and guidance, used the literature review method and analysis method to screen and determine the coupling secondary indicators of tourism and ecological environment development, and formed a coupling evaluation index system with two levels of indicators. After index screening, the evaluation index system of Gansu's tourism industry is constructed, which includes nine indexes from four aspects: tourism resources, tourism scale, tourism benefits, and tourism income, and is selected based on supply and demand. The ecological environment evaluation index system of Gansu Province, which includes ten indicators from three aspects of ecological resources, ecological governance, and ecological pollution, is selected with environmental governance as the guide. At the level of indicator properties, "+" is a positive indicator, and "−" is a negative indicator. The larger the value of the former, the better; the smaller the value of the latter, the better (Chen et al. 2022; Fu et al. 2019; Lu and Shi 2018).

36.3.2 Research Design

Research methods. Coupling is used to describe a continuous cooperation phenomenon that occurs when two or more systems interact with each other, and coupling degree is used to express the degree of mutual influence and interaction between various elements of the target system. Further, the coupling model reflects the strength of the interaction between the target systems, and the coupling coordination describes the process of the orderly and harmonious development of the target system based on the coupling state, to reflect the coordination degree of the factors to be studied based on coupling dependence.

Standardized data processing. The selected data usually have differences in order of magnitude and dimension, and the extreme value method is used to standardize the data. To make the data operation processing meaningful, the whole translation of the dimensionless data is carried out to eliminate zero and negative values. In order not to destroy the inherent law of the original data, retain the attributes of the original data to the maximum extent, and σ approach the X_i value as closely as possible, $\sigma = 0.0001$ is adopted in this paper. The processing formula is as follows:

$$Z_i = \begin{cases} \frac{X_i - X_{\min}}{X_{\max} - X_{\min}} + 0.0001 & \textit{Positive indicator} \\ \frac{X_{\max} - X_i}{X_{\max} - X_{\min}} + 0.0001 & \textit{Negative indicator} \end{cases} \tag{36.1}$$

In formula (36.1), X_i is the selected original data, and Z_i is the index value after standardized processing. X_{max} and X_{min} are the maximum and minimum values of each indicator sample. σ is overall translation value.

Indicator Weight Selection. The entropy method is a multi-index evaluation method based on the concept of information entropy, which determines the weight of each index in the comprehensive evaluation by comparing their entropy values, and then realizes the comprehensive evaluation of the evaluation object (Chen et al. 2022). To avoid the influence of subjective factors on the weight of the index system, this study combined the information entropy of the index, the entropy method was used for objective assignment to determine the weight of each index. The specific processing process is as follows:

1. Calculate the proportion of evaluation index value y_{ij} in year i under the j index system

$$y_{ij} = x_{ij} / \sum_{i=1}^{m} x_{ij} \left(0 \le y_{ij} \le 1\right) \tag{36.2}$$

2. Determine the information entropy value e_j of item j

$$e_j = -k \sum_{j=1}^{m} Y_{ij} \times \ln\left(y_{ij}\right) \ \left(k = 1/\ln(m), 0 \le e_j \le 1\right) \tag{36.3}$$

3. Eliminate the interference factors in the selected index and calculate the entropy redundancy value d_j

$$d_j = 1 - e_j \tag{36.4}$$

4. Calculate the weight of item j index W_j

$$W_j = d_j / \sum_{i=1}^{n} d_j \tag{36.5}$$

5. Evaluate the individual indicators

$$S_{ij} = W_j \times X_{ij} \tag{36.6}$$

6. Calculate the overall development level in year i

$$S_i = \sum_{j=1}^{n} S_{ij} \tag{36.7}$$

According to formulas (36.1)–(36.6), the data related to the research indicators were standardized by Excel software the weight value of each indicator was calculated, and the results were obtained as shown in Table 36.1.

Coupling coordination degree model

1. Efficacy function

Table 36.1 The standardization results of the original data of tourism and ecological environment development in Gansu Province and the weights of each index

System layer	Primary index	Secondary index	The year 2020	The year 2021	The year 2022	The year 2023	Index weight	Index property
Tourism system	Tourism resource	Number of A-level tourist attractions	0.2562	0.6974	0.6523	1.0001	0.1387	+
		Number of travel agencies	0.1112	1.0001	0.0001	0.6778	0.1485	+
		Number of star-rated hotels	0.2332	0.7856	0.6543	1.0002	0.0893	+
	Tourism size	Domestic tourist arrivals	0.0145	0.2578	0.1512	0.3245	0.3583	+
		Inbound tourist arrivals	0.0001	0.0023	0.0021	0.0367	0.1392	+
	Tourism benefit	Number of tourism employees	0.0002	0.2876	0.5483	1.0001	0.3432	+
		Share of tourism revenue in GDP	0.0001	0.5723	0.4829	1.0001	0.3573	+
	Tourism income	Foreign exchange earnings from inbound tourism	0.0213	0.0463	0.0372	0.0218	0.3527	+
		Gross domestic tourism income	0.2436	0.2879	0.2153	0.3761	0.2383	+
Ecological environment system	Ecological resources	Per capita wetland area	0.0245	0.2657	0.3545	0.3452	0.2631	+
		Forest coverage rate	0.0001	0.8421	0.9688	1.0001	0.1689	+

(continued)

Table 36.1 (continued)

System layer	Primary index	Secondary index	The year 2020	The year 2021	The year 2022	The year 2023	Index weight	Index property
		Per capita water resources	0.2467	0.0234	0.0547	0.0562	0.2173	+
	Ecological management	Air quality up to standard rate	1.0001	0.0001	0.5317	0.4627	0.0697	+
		Environmental quality index	0.2456	0.5602	0.4879	0.4365	0.0645	+
		Harmless household garbage disposal rate	0.0123	0.0018	0.0456	0.0671	0.0672	+
		Completed investment in waste water and gas project treatment	0.4235	0.3345	0.0432	0.5267	0.0872	+
	Ecological pollution	Discharge of wastewater	0.0234	0.0321	0.0376	0.4521	0.0256	–
		Sulfur and nitrogen oxide emissions	0.0345	0.0456	0.0562	0.0765	0.0175	–
		Dust emission	0.0532	0.0235	0.0453	0.0674	0.0467	–

+ positive indicator and – negative indicator

Set U_i ($i = 1, 2, \ldots m$) is the coupling coordination system parameter of tourism and ecological environment development, and U_{ij} represents the j th index of i th order parameter, whose value is X_{ij} ($j = 1, 2, \ldots n$). α_{ij} and β_{ij} are the upper and lower limits of the parameters of the coupled system of tourism and ecological environment development when it is stable and coordinated. The efficacy function of the two systems when it is stable and coordinated is calculated by eq. (36.7):

$$U_{ij} = \begin{cases} (X_{ij} - \beta_{ij})/(\alpha_{ij} - \beta_{ij}) & U_{ij} \text{ has positive effect} \\ (\alpha_{ij} - X_{ij})/(\alpha_{ij} - \beta_{ij}) & U_{ij} \text{ has negative effect} \end{cases} \tag{36.8}$$

In the formula, U_{ij} reflects the efficacy contribution of variable X_{ij} in the coupling system of tourism and ecological environment development. The closer U_{ij} is to 0, the smaller the efficacy contribution is; the closer U_{ij} is to 1, the greater the efficacy contribution is. Because the tourism system and the ecological environment system are two subsystems that interact with each other and have their characteristics, the author uses the linear weighting method to calculate the contribution value of each parameter, as shown in eq. (36.8).

$$U_i = \sum_{j=1}^{m} (\lambda_{ij\times} U_{ij}) \tag{36.9}$$

In the formula, U_i is the comprehensive evaluation value of the coupled system in year i and λ_{ij} is the weight of each evaluation index.$\sum_{j=1}^{m} \lambda_{ij} = 1$.

2. Coupling model

Referring to the capacity coupling coefficient model in physics, the coupling degree calculation formula of tourism and ecological environment subsystems is as follows:

$$C = \left[(U_1 \times U_2)/(U_1 + U_2)^2\right]^{1/2} \tag{36.10}$$

In the formula, C is the coupling degree and C is between 0 and 1. When $C = 0$, the coupling degree between tourism and ecological environment development is the smallest, indicating that there is no coupling relationship between the two. When $C = 1$, the coupling degree between the two is the largest, indicating a benign coupling development trend between the two. To more accurately classify the coupling coordination degree between tourism and the ecological environment in Gansu province, further refer to the classification setting method of Fu et al. (2019), and set the classification level standard as shown in Table 36.2.

3. Coupling coordination degree model

The coupling degree plays an irreplaceable role in evaluating the coupling degree of the two systems. However, the development characteristics, development efficiency, and development speed of tourism and ecological environment have their particularities, so only the coupling degree is used to judge the synergy between tourism and

Table 36.2 Classification and stage characteristics of the coupling degree between provincial tourism and the ecological environment in Gansu Province

C	Classification stage	Feature
0	Disorderly coupling stage	The two systems are in a state of disordered development, and they are not related to each other and do not affect each other
0–0.3	Low coupling stage	The development of ecological environment is good, the damage degree is low, and it can basically carry the development of tourism
0.3–0.5	Antagonistic stage	The degree of damage to the ecological environment has become stronger, and the capacity to carry tourism development has become weaker
0.5–0.8	Running-in stage	The development of tourism gradually restores the ecological environment, and the two systems enter a benign coupling period
0.8–1	High coupling stage	The two systems are in a state of orderly development and are closely related to each other

ecological environment development, and the probability of errors is high (Lu and Shi 2018). When the two systems are at a low level, the coupling degree between them is relatively high, which indicates that it is difficult to truly reflect the coordination degree and development stage of the two systems only by using the coupling degree as an evaluation index. Therefore, combined with the actual situation of Gansu Province, a coupling model is constructed to evaluate the coordination degree of interaction between the two systems, and the coupling coordination degree is shown in eq. (36.10).

$$\begin{cases} F = a \times U_1 + b \times U_2 \\ D = \sqrt{C \times F} \end{cases} \tag{36.11}$$

In the formula, D represents the coupling coordination degree between tourism and ecological environment. The larger the value of D, the better the coupling coordination degree. According to the value of D, the coupling coordination degree is divided into six levels. C represents the coupling degree, F represents the development index of the two systems of tourism and ecological environment development, a and b are constants, and their values are 0.5 (Li et al. 2023). The specific division is shown in Table 36.3.

Table 36.3 Criteria for evaluating coupling coordination degree

Coordination degree interval	Coordination level	Coordination degree interval	Coordination level
$0 \leq D \leq 0.1$	Extreme disorder	$0.5 < D \leq 0.6$	Forced coordination
$0.1 < D \leq 0.2$	Severe disorder	$0.6 < D \leq 0.7$	Primary coordination
$0.2 < D \leq 0.3$	Moderate disorder	$0.7 < D \leq 0.8$	Intermediate coordination
$0.3 < D \leq 0.4$	Mild disorder	$0.8 < D \leq 0.9$	Good coordination
$0.4 < D \leq 0.5$	Borderline disorder	$0.9 < D \leq 1.0$	Quality coordination

36.4 Empirical Results and Analysis

36.4.1 Analysis of the Overall Development Level of Tourism in Gansu Province

According to formula (36.7), the overall development level of tourism and the ecological environment in Gansu Province from 2020 to 2023 is calculated, and the calculation results are shown in Table 36.4. It can be concluded that the tourism development level of Gansu Province shows a trend of decline and then rise (Fig. 36.1). The period from 2020 to 2022 will be a sharp decline, and the tourism development index will drop from 0.6461 to 0.4833. The reason is that the new coronavirus outbreak at the end of 2019 brought the development of tourism to a standstill, and the tourism development index in 2022 had the lowest value at this stage. With the effective control of the epidemic, the government's policy support for the tourism industry, and the acceleration of the resumption of work of tourism enterprises promoted the gradual recovery of the tourism economy in Gansu, and the level of tourism development increased. In 2023, the tourism industry of Gansu Province will develop rapidly, and the tourism development index will rise from 0.4833 to 0.8132. The reason is that the Gansu Provincial Party Committee and the provincial government issued the Implementation Opinions on Promoting the Deep Integration of Culture and Tourism to Achieve High-quality Development in 2023. Focusing on the deployment of 18 tasks in four aspects "promoting the integration of the use of cultural tourism resources in the whole region", "vigorously promoting the integration and innovation of cultural tourism products", "promoting the high-quality development of the cultural tourism industry through the whole industrial chain", and "comprehensively promoting the cooperation and symbiosis of the cultural tourism market".

Table 36.4 Analysis of coupling coordination between tourism development and ecological protection in Gansu Province

Years	Tourism development level	Ecological environment level	Coupling degree *C*	Coupling coordination degree *D*	Coordination level
2020	0.6461	0.6912	0.5791	0.3562	Moderate disorder
2021	0.5624	0.7418	0.6874	0.4177	Borderline disorder
2022	0.4833	07689	0.6350	0.5236	Forced coordination
2023	0.8132	0.9235	0.8392	0.6375	Primary coordination

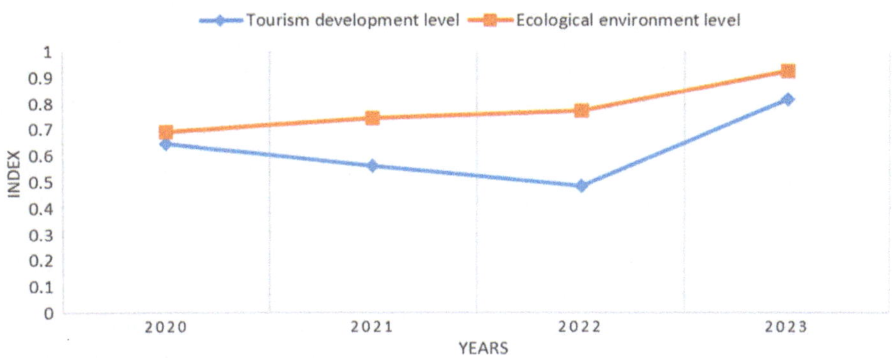

Fig. 36.1 Overall development level evaluation of tourism development and ecological environment in Gansu Province

36.4.2 Analysis of the Overall Development Level of the Ecological Environment in Gansu Province

As shown in Fig. 36.1, in terms of ecological environment, the development level of the ecological environment in Gansu Province from 2020 to 2023 shows a trend of increasing year by year and is in a stage of steady development, with the ecological environment index rising from 0.6912 to 0.9325 in 2023, among which the development is faster in 2023. The reason is as follows: in recent years, Gansu province has achieved remarkable results by implementing a series of ecological and environmental protection measures. Gansu Province has carried out two major campaigns: the protracted war of national land greening and the position war of preventing and controlling desertification. Through the strategy of scientific greening, expanding green, promoting green, and protecting green, has completed a large number of national land greening tasks. In 2023, Gansu Province completed the greening of

13.430,600 mu of land and comprehensively restored 3.981,300 mu of decertified land.

36.4.3 Analysis of Coupling Coordination Types Between Tourism and Ecological Environment in Gansu Province

According to formulas (36.8)–(36.11), the coupling degree and coupling coordination degree of Gansu Province in 2020–2023 are calculated respectively, and the specific results are shown in Table 36.4. It can be seen that the comprehensive evaluation index of the ecological environment in Gansu Province from 2020 to 2023 is higher than the comprehensive evaluation index of tourism development, indicating that the development of the tourism industry in Gansu province lags behind the development of the ecological environment. The coupling degree and coupling coordination degree between the tourism industry and the ecological environment in Gansu Province from 2020 to 2023 are shown in Fig. 36.2. It can be seen that the coupling degree of the two systems rises from 0.5791 to 0.8392, the coupling coordination degree rises from 0.3562 to 0.6375, and the coordination level rises from mild disorder in 2020 to primary coordination in 2023. The coupling degree and coupling coordination degree are gradually improved. But generally speaking, in recent years, Gansu tourism and ecological environment coupling coordination degree is in an obvious growth process. On the one hand, in recent years, the state has repeatedly emphasized the development of an ecological service economy, requiring tourism development not to destroy the ecological environment at a cost. Under this situation, local governments began to take into account tourism development and the ecological environment and promote their coordinated development. On the other hand, along with our country's tourism development begins to seek high-quality development, all places focus on ecological industry development while paying attention to economic interests in tourism development, to realize sustainable development.

36.5 Research Conclusions and Suggestions

36.5.1 Conclusion

Based on the interaction between rural tourism and ecological development, this paper analyzes the mutual promotion and coordinated development mechanism of the two systems, constructs the coupling evaluation index system and coupling evaluation model of the two systems, and replaces the relevant data from the development yearbook of Gansu Province from 2020 to 2023 into the evaluation model. The main conclusions obtained through calculation and analysis are as follows:

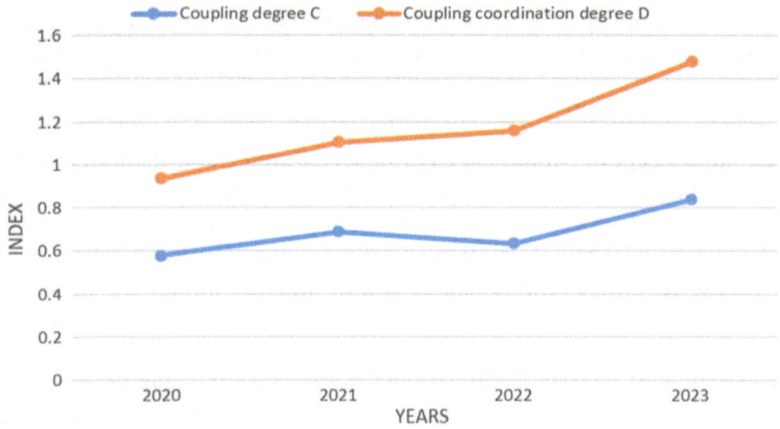

Fig. 36.2 Coupling degree and coupling coordination degree between tourism and ecological environment in Gansu Province

1. The coupling characteristics between tourism and the ecological environment are obvious, and the two interact and influence each other.
2. In the two major ecological and environmental systems of Gansu Province, the development of tourism lags, and the development speed of the ecological environment is stronger than that of tourism. In terms of development trends, the coupling degree between tourism and ecological and environmental development in Gansu Province has increased, and the coordination level has evolved from moderate imbalance in 2020 to primary coordination in 2023.
3. From the principle of system theory, tourism and the ecological environment support and promote each other. In the two systems, tourism becomes the exogenous driving force for ecological environment protection, and the ecological environment is also the source guarantee for the sustainable development of tourism.

36.5.2 Suggestions

1. Grasp the innovative concept of ecological tourism

In the new era, high-quality economic development requires that ecotourism become an important way of sustainable utilization of tourism resources. Therefore, for the coördinated development of tourism and the ecological environment in Gansu Province, it is necessary to fully combine ecological environment and tourism resources, and thoroughly implement the concept of green sustainable coordinated development. For example, for the vast rural areas with relatively rich tourism resources, we can vigorously support the construction of rural tourism and effectively establish a new trend of ecological and sustainable tourism civilization. While giving full play to the advantages of tourism, it can also take into account the ecological environment, and ultimately achieve a win-win situation of economy and ecology.

2. Create eco-tourism projects

Based on the concept of ecological environmental protection, the tourism industry needs to constantly innovate the types of tourism products and services, such as the establishment of ecological cultural tourism projects. Based on the survival of the ecological barrier, the integration of tourism and the ecological environment will be promoted from multiple angles and in multiple directions. Moreover, all regions can make full use of the resource-based complementary advantages, establish high-standard and high-level ecological cultural tourism demonstration zones, form the linkage effect of tourism resources, and truly realize the equal development of tourism and the ecological environment. In addition, for areas with rich tourism resources, based on ensuring the survival of ecological barriers, it is also necessary to develop and cultivate ecological tourism economy pilots while tapping tourism market demand, to continue to enrich related tourism projects.

3. Improving the capacity of tourism to improve the ecological environment

While vigorously developing tourism, the government needs to actively exert the macro-control efficiency of tourism, effectively formulate relevant ecotourism strategies, effectually plan the harmonious development of tourism and the ecological environment, and continue to improve the ecological compensation for tourism development. Based on fully complying with relevant ecological tourism development plans, enterprises can develop and implement more abundant ecological tourism projects to more strictly comply with ecological protection and restoration measures. With the joint efforts of the government and enterprises, we will create a systematic positioning plan for the sustainable development of tourism and the ecological environment, and realize the sustainable development of ecological tourism while focusing on effective management and control.

References

Chen F, Shi HM, Chen Q (2022) The spatiotemporal evolution and influencing factors of the coupling coordination degree of the green urbanization system in the Yangtze River Delta. J Central South Univ Forest Technol (Social Sci) 16(2):26–35

Cong XL, Huang Y, Liu JS (2019) Spatial and temporal evolution of coupled coordination degree of ecotourism and tourism environment of Jilin province. Scientia Geogr Sonica 39(3):496–505

Fang YL, Huang ZF, Duan ZX et al (2013) Coupling and coordinating about Chinese tourism developing and eco environment. Econ Geogr 33(12):195–201

Fu I, Xiong Kang N, Gao Y (2019) Quantitative study on coupling relationship between tourism industry and ecological environment in karst area: a case study of Guizhou province. Ecol Econ 35(1):125–130

Gossling S (2002) Global environmental consequences of tourism. Glob Environ Chang 12(4):283–302

Gossling S, Hansson CB, Horstmeier O et al (2002) Ecological footprint analysis as a tool to assess tourism sustainability. Ecol Econ 43(2):199–211

He XR, Peng KJ, Xu CX (2022) Spatiotemporal evolution and trend prediction of the vulnerability of tourism-economy ecosystem in the Yangtze River economic belt. Acta Ecol Sinica 42(2):487–499

Li MF, Jiang L, Yang XL (2023) Coupling and coordination between tourism development and ecological environment of marine Island tourism destination. J Central South Univ Forest Technol 10(01):187–194

Lu XJ, Shi PF (2018) Research on the coupling of beautiful countryside construction and rural tourism development—a case study of Wuyuan in Jiangxi Province. Fujian Forum Human Social Sci Ed 2:78–82

Ma J (2016) Biodiversity protection strategy under the background of ecological environment threshold and the tourism carrying capacity: according to the core of the world natural heritage Wulingyuan scenic area as an example. Econ Geogr 36(4):195–202

Popovic D, Doijak D, Kuzmanovic D et al (2018) Geopedological evaluation of protected area for recreation and tourism planning evidence from the Bosnia and Herzegovina national park. J Geogr Inst Jovan Civic 68(1):119–131

Tsaur SH, Lin YC, Lin JH (2006) Evaluating ecotourism sustainability from the integrated perspective of resource, community, and tourism. Tour Manag 27(4):640–653

Wall G, Wright C (1977) The environmental impact of outdoor recreation. University of Waterloo, Ontario

Wang ZF, Wang ZY (2021) The impact of environmental regulation on ecological efficiency of the tourism industry and its spatial differentiation: taking the urban agglomeration in the middle reaches of the Yangtze River as an example. Acta Ecol Sinica 41(10):3917–3928

Chapter 37
Field Measurement of Near-Ground Wind Characteristics of Super Typhoon Doksuri

Junyan Zhou, Fangjian Chen, Longzhen Hua, Jing Zhang, Changzhao Qian, and Changping Chen

Abstract To study the pulsation characteristics of the near-ground wind field during the typhoon landfall process, Windcube Doppler laser wind radar is used to measure the landfall process of typhoon "Doksuri" in Xiamen. Twelve layers of observation points are set up between 40 and 290 m in height to obtain the corresponding observation data. Based on the observation results, the wind characteristics such as mean wind speed and direction, wind profile, gust factor, turbulence, and other parameters of the near-ground wind field of "Doksuri" in the coastal area are analyzed. The results show that the wind direction shifted by about 180° before and after the landfall of "Doksuri", and the wind speed time profile showed an "M" shape. The relative ratios of vertical, lateral, and longitudinal turbulence during the observation period were 1:0.88:0.34. The turbulence intensity in the lateral wind direction was consistent with the norms, while the vertical turbulence intensity was significantly lower than the norms. The turbulence intensity decreases with the increase of height, which is large before the typhoon landfall and decreases significantly after the landfall and is significantly affected by geomorphology; the consistency of the gust factor in each direction is good at different heights, and the effect of height difference is not obvious.

Keywords Typhoon "Doksuri" · Field measurements · Near-ground wind characteristics

J. Zhou · F. Chen (✉) · L. Hua · J. Zhang · C. Qian
Xiamen University of Technology, Xiamen, Fujian, China
e-mail: fj_chen@xmut.edu.cn

J. Zhou · F. Chen · L. Hua · J. Zhang · C. Qian · C. Chen
Fujian Provincial Key Laboratory of Wind Disaster and Wind Engineering, Xiamen, Fujian, China

C. Chen
Quanzhou Normal University, Quanzhou, Fujian, China

Y. Li et al. (eds.), *16th International Conference on Environmental Science and Development*, Environmental Science and Engineering,
https://doi.org/10.1007/978-3-032-15080-6_37

37.1 Introduction

The southeastern coastal region of China, as a vital area for economic development, is frequently hit by typhoons due to its unique geographical location. In recent years, intensified global climate change has led to significant changes in the intensity and frequency of typhoons, further exacerbating disaster risks in coastal areas. Statistics show that the region experiences an average of seven typhoon landfalls annually, causing economic losses of approximately $5.6 billion (Elliott et al. 2015). Therefore, high-precision field measurements of near-ground typhoon wind fields are of great scientific and practical significance.

Internationally, the United States and Japan have made significant progress in typhoon observation technology, wind field modeling, and data analysis. In China, recent studies have also achieved notable advancements. Wang et al. (2013a, b); (Wang et al. 2017) monitored wind speed and direction along the Pudong coastline during typhoons "Muifa", "Meari", and "Haikui", investigating turbulence intensity, gust factors, and fluctuating wind power spectra at various heights. Li et al. (2016); Wang et al. (2024) analyzed turbulence integral scales and wind speed power spectral density during typhoon "Soulik", finding consistency with the von Karman spectrum. He et al. (2018) observed that turbulence intensity and gust factors of typhoon "Haima" peaked at landfall, decreasing with height and wind speed, with values below those in Chinese, U.S., and Japanese building codes. He et al. (2020) noted minimal correlation between roughness length, power-law exponents, and wind speed under stable wind direction and terrain during typhoon "Mangkhut". Xia et al. (2021) found turbulence intensity decreased with wind speed, while gust factors followed a Type II GEV distribution during Typhoon "Maria". Zhang et al. (2024) conducted a detailed analysis of near-ground wind field characteristics at various heights and time intervals during typhoon "Soudelor" using wind speed and direction data obtained from an array-based anemometer observation system.

Despite progress, current field measurements remain inadequate due to typhoon complexity and environmental diversity. Expanding observational databases is crucial. This study analyzes "Doksuri" using field data, revealing wind pattern variations with speed and time, offering valuable insights into near-ground wind characteristics during landfall.

37.2 Overview of the Field Measurement Program and Typhoon Process

37.2.1 Typhoon Overview

Typhoon No. 5 in 2023, "Doksuri", formed over the east ocean of The Philippines on July 21. It moved northwestward and intensified into a super typhoon by noon on July 27, located about 415 km South of the Taiwan Strait, near Xiamen, Fujian, with a wind strength of 15. At 9:55 a.m. on July 28, "Doksuri" made landfall on the Coast of Jinjiang, Fujian, as a strong typhoon, with a maximum wind speed of 50 m/s and a minimum Central pressure of 945 hPa. After landfall, it continued northward, and its residual circulation affected North and Northeast China.

37.2.2 Instruments and Observation Points

The observation station for "Doksuri" was located on the West Coast of Dadeng Island. It was 1.1 km away from the Southern coastline of Dadeng Island, and approximately 32 km away from the center of "Doksuri" final landing point in Jinjiang, covered in the influence range of the 12-level wind circle. A Windcube Doppler lidar was installed on a rooftop about 12 m above ground, with observation heights from 40 to 290 m and 12 vertical levels at 40 m, 55 m, 70 m, 80 m, 90 m, 120 m, 150 m, 180 m, 210 m, and 250 m. The system had a wind speed accuracy of 0.1 m/s, direction accuracy of 2°, and a measurable range of 0–60 m/s. The basic geographical information of the observation point is shown in Fig. 37.1.

Typhoon "Doksuri" made landfall at approximately 9:55 am on July 28, 2023, with the most intense wind period concentrated on the same day. Therefore, 24-hour observation data from July 28 were selected for analysis of Doksuri's wind field characteristics, with samples divided into 10-min intervals for detailed examinations.

Fig. 37.1 Geographic location of the observation point

37.3 Measured Data Analysis

37.3.1 Mean Wind Speed and Direction

For Windcube V2 Doppler laser wind radar, three-dimensional wind speed u_x *(t)* 、u_y *(t)* and u_z *(t)* can be measured simultaneously. The 10-min mean wind speed and lateral wind angle (0° for North) at each height of typhoon "Doksuri" observed are shown in Fig. 37.2.

Figure 37.2a reveals that the mean wind speed curve exhibits a distinct typhoon landfall "M" pattern. The first peak zone records a maximum 10-minute wind speed of 22.67 m/s at 40 m, while the second peak reaches 25.11 m/s at the same height. When the 40 m 10-minute mean wind speed exceeds 17.2 m/s, the lateral wind angle is 14.7°, contrasting with 227.9° before the wind speed drops below this threshold. During the typhoon, the lateral wind angle undergoes a continuous variation exceeding 180°. The mean wind speed generally increases with observation height, and the temporal trends of mean wind direction at different measured heights align closely. These observations indicate that the collected data primarily encompass wind condition samples from the eyewall vicinity, eyewall, and periphery of Typhoon "Doksuri".

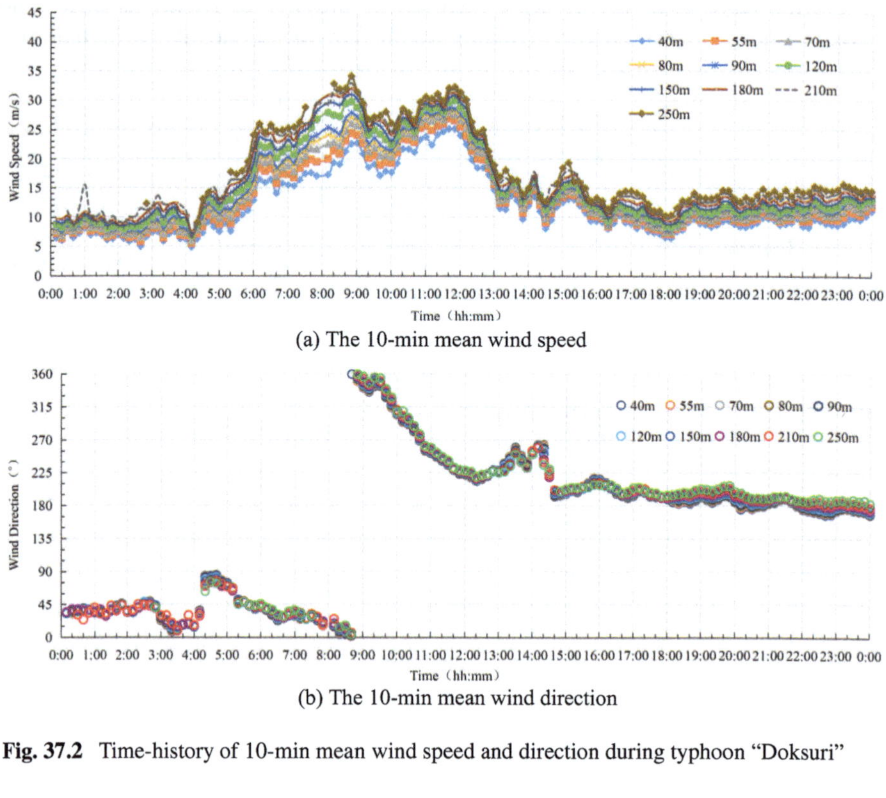

Fig. 37.2 Time-history of 10-min mean wind speed and direction during typhoon "Doksuri"

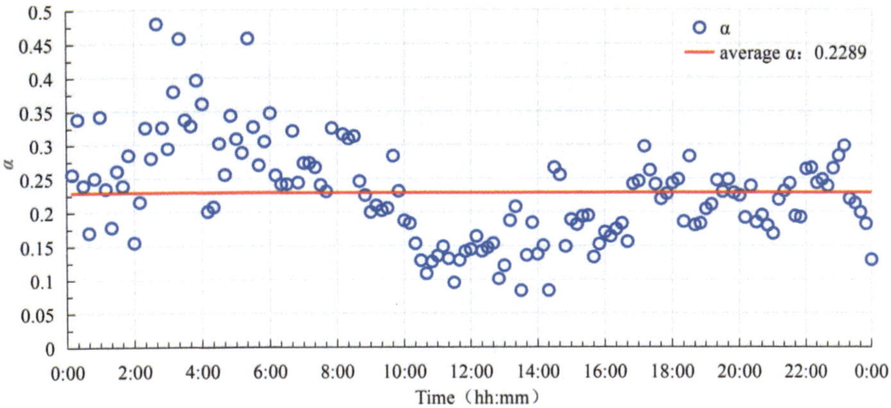

Fig. 37.3 Time-history of ground roughness index

37.3.2 *Wind Profile*

Referring to the exponential form proposed by Davenport, the time-history indices of the mean wind profiles of each subsample are calculated based on the 10-min mean wind speeds at different heights as shown in Fig. 37.3.

The ground roughness index exhibits greater dispersion and larger magnitudes prior to typhoon landfall. Integrating the mean wind direction profile presented in Fig. 37.3, the wind during this period was primarily northeasterly and east-northeasterly, with the incoming flow at the observation site directed toward the mainland of Dadeng Island, characterized by buildings, trees, and hilly terrain. Between 9:00 a.m. and 4:00 p.m., the dominant wind direction underwent a gradual transition from northerly to westerly and then to southerly, accompanied by a corresponding decrease in the surface roughness index. From 11:00 a.m. to 2:00 p.m., southwesterly winds prevailed, with the incoming flow originating from the more open sea area southwest of Dadeng Island. Post-4:00 p.m., the dominant wind shifted to southerly, coinciding with a resurgence in the ground roughness index. The incoming flow at the observation site was obstructed by Kinmen Island, significantly influencing airflow dynamics.

37.3.3 *Wind Attack Angle*

Wind attack angles significantly impact flexible structures, with long-term effects accelerating or exacerbating fatigue damage. Under strong winds, abrupt changes in wind attack angles may induce structural damage or failure. Calculated from lidar-recorded 10-minute mean horizontal and vertical wind speeds at different heights, their temporal variations are shown in Fig. 37.4.

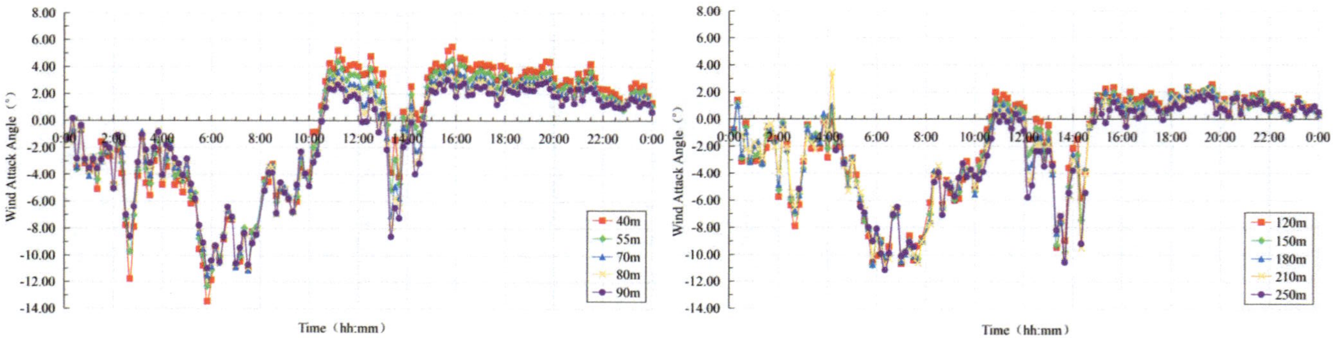

(a) 40m-90m

(b) 120m-250m

Fig. 37.4 Time-history curves of wind attack angles during typhoon "Doksuri"

Figure 37.4 shows that during the 24 hours of July 28, the wind attack angle at each observation height fluctuated greatly, and the rule of change was basically the same. The change amplitude of the wind attack angle decreases with the increase of height, and the change interval of the wind attack angle at 40 m is [−13.45, 5.47], while the change interval of the wind attack angle at 210 m is [−10.74, 3.43]. For the 90 m measuring point at the design height of bridge deck of a sea-cross bridge nearby, the wind attack angle change interval is [−10.97, 3.03]. The negative wind attack angle of "Doksuri" is significantly larger than that of other typhoons that have been published with a range of ± 3° or ± 5°. This is because during the passage of the typhoon, two serious downdrafts occurred, and the harsh observation conditions seriously affected the recognition of suspended particle motion by the lidar, resulting in significant deviations in the observation data and presenting abnormally large negative values in the results. However, this still reflects that Doksuri's wind field may cause negative wind attack angles beyond the normal range of specifications.

37.3.4 Turbulence Intensity

The turbulence intensity I is defined as the ratio of the standard deviation of fluctuating wind speed to the mean wind speed within a 10-min period. Its expression is:

$$I_i = \frac{\sigma_i}{U}, \quad i = u, v, w \tag{37.1}$$

where σ_i denotes σ_u, σ_v, and σ_w, the root mean square of the longitudinal, lateral, and vertical pulsating wind speeds, respectively; and U is the mean wind speed. According to the analysis of the time-course data of wind speed at different observation heights, the time-history results of the turbulence intensity (I_u, I_v, I_w) in the longitudinal, lateral and vertical directions at different heights are shown in Fig. 37.5.

As depicted in Fig. 37.5, turbulence intensity exhibited a notable transition around 4:00 a.m., characterized by an abrupt reduction from 0.4 to 0.1 in the low-altitude region. During this interval, the 10-minute mean wind speed at 40 m was merely 4.87 m/s, suggesting low wind speed scenarios. Overall, turbulence intensity demonstrated a decreasing trend with altitude and was elevated prior to typhoon landfall. Following landfall at 10:00 a.m., it experienced a marked decline. Between 8:00 and 9:00 a.m., turbulence intensity exhibited pronounced fluctuations, with two peaks corresponding to maximum wind speeds at each observation height.

Pre-landfall (8:10 a.m.) turbulence intensity at the eyewall was substantially elevated compared to post-landfall (12:50 p.m.) values. This enhancement was attributed to northeast winds interacting with the undulating terrain of Dadeng Island, which amplified turbulence. Post-landfall, the wind direction shifted to the southwest, transitioning over open sea, resulting in a decrease in turbulence intensity. Turbulence intensity generally decreased with both typhoon progression and increasing height.

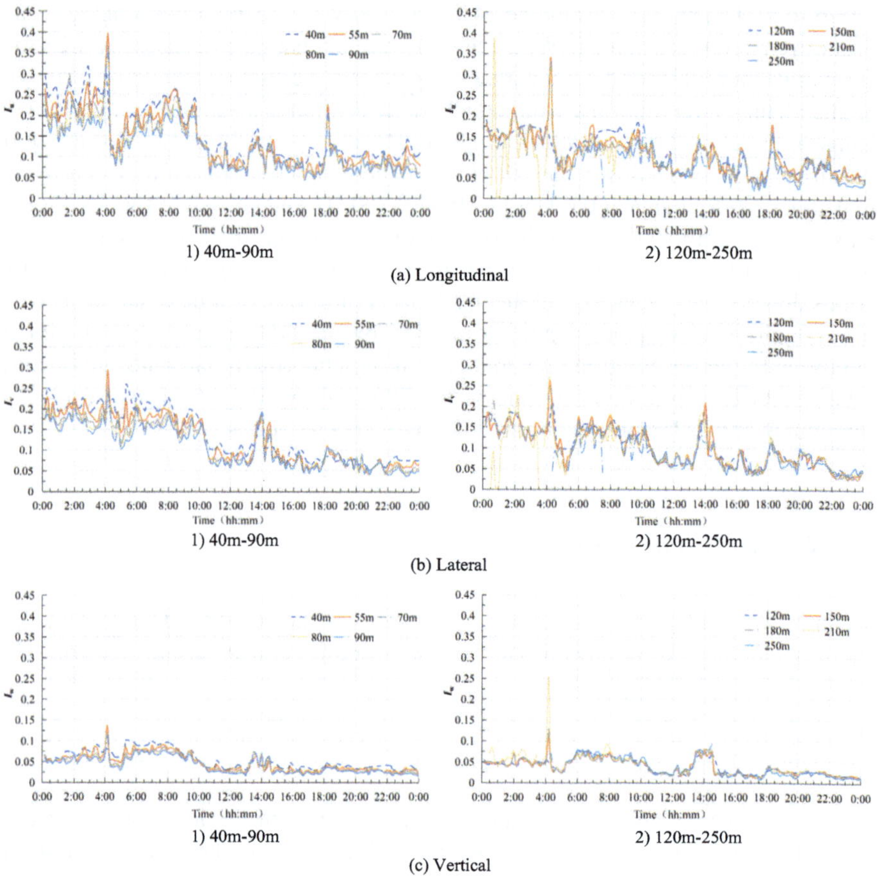

Fig. 37.5 Time-history curves of turbulence intensity at different heights

A significant drop was observed from the eyewall to the rear eyewall. However, a minor turbulence intensity peak occurred around 2:00 p.m.

The temporal variations of longitudinal, lateral, and vertical turbulence intensities at the 40 m observation height are illustrated in Fig. 37.6.

Figure 37.6 illustrates that turbulence intensity across different directions exhibits a high degree of consistency. During the observation period, the mean values of longitudinal, lateral, and vertical turbulence intensities are 0.158, 0.139, and 0.053, respectively, exhibiting a relative ratio of 1:0.88:0.34. Standardized values for highway bridges are generally specified as 1:0.88:0.55. Notably, the lateral turbulence intensity aligns with the specification, whereas the vertical turbulence intensity is significantly lower. At a height of 120 m, the mean tri-directional turbulence intensities are 0.112, 0.102, and 0.039, with a relative ratio of 1:0.92:0.35, indicating that turbulence intensity exhibits minimal variation with height, accompanied by a modest increase in lateral turbulence intensity.

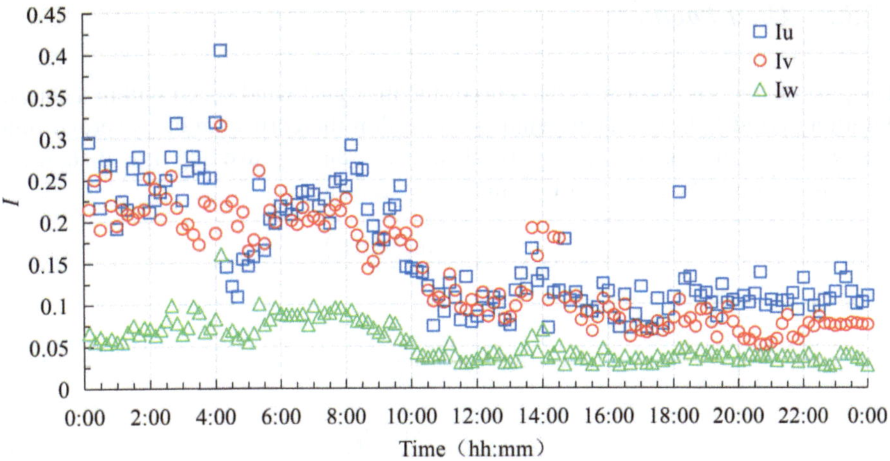

Fig. 37.6 Time-history curve of turbulence intensity components at 40 m

The relationship between tri-directional turbulence intensity and 10-minute mean wind speed is further explored in Fig. 37.7. Longitudinal turbulence intensity reaches its maximum at a height of 40 m (0.291 at 8:10 a.m.), occurring slightly earlier than lateral turbulence intensity (0.227 at 8:00 a.m.), with a ratio of 1:0.78. Lateral turbulence intensity remains relatively low within the typhoon impact zone. Turbulence peaks emerge at the initial stage of typhoon influence when wind speeds exceed 17.2 m/s. The trend line reveals that turbulence intensity decreases with increasing wind speed, with longitudinal intensity exhibiting a more significant reduction compared to lateral intensity.

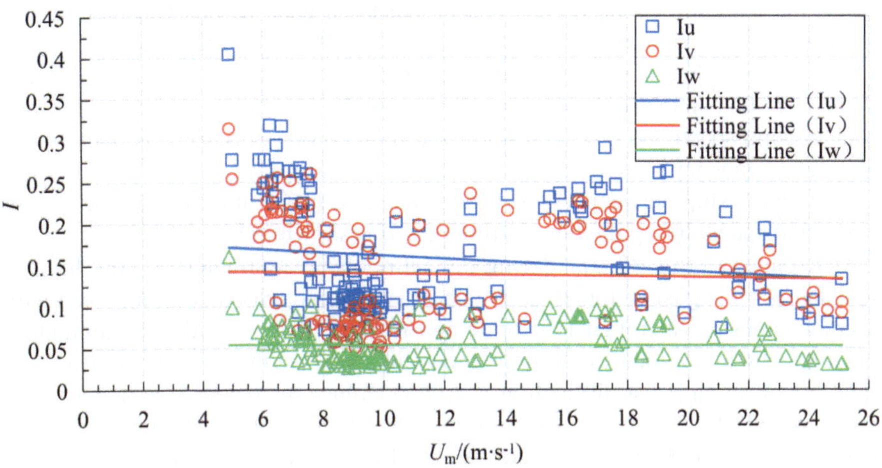

Fig. 37.7 Variation of turbulence intensity components with 10-min mean wind speed at 40 m

37.3.5 Gust Factor

The gust factor G is defined as the ratio of the max gust wind speed within a certain duration period t_g to the mean wind speed within the corresponding average time interval. Here, the gust duration t_g is taken as 3 s, and the average time distance is taken as 10-min. The calculation formulas are as follows:

$$G_{\mathrm{u}}(t_{\mathrm{g}}) = 1 + \frac{\max(\overline{u}(t_{\mathrm{g}}))}{U} \tag{37.2}$$

$$G_{\mathrm{v}}(t_{\mathrm{g}}) = \frac{\max(\overline{v}(t_{\mathrm{g}}))}{U} \tag{37.3}$$

$$G_{\mathrm{w}}(t_{\mathrm{g}}) = \frac{\max(\overline{w}(t_{\mathrm{g}}))}{U} \tag{37.4}$$

The time-history results of the gust factors (G_u、G_v、G_w) at different heights in the longitudinal, lateral and vertical directions are shown in Fig. 37.8.

During the typhoon, gust factors exhibited remarkable consistency across vertical levels but attenuated significantly post-landfall, particularly in the lateral and vertical time histories. The longitudinal gust factor displayed a "W"-shaped temporal profile, reaching peaks prior to landfall (8:10 a.m.), in proximity to the maximum 10-min mean wind speed (12:00 p.m.), and during the typhoon's dissipation phase (3:00 p.m.). Conversely, the lateral gust factor manifested an "M"-shaped pattern, peaking at the time of landfall (10:00 a.m.) and in the late stage (2:00 p.m.), with troughs coinciding with the first (9:00 a.m.) and second wind speed maxima (12:00 p.m.).

Taking the observational data at 40 m altitude as an illustrative case, Fig. 37.9 depicts the temporal evolution of three-dimensional gust factors. Correlating with the time-history variations of mean wind speed and direction presented in Fig. 37.2 of Sect. 3.1, both longitudinal and lateral gust factors underwent pronounced fluctuations preceding and during the typhoon's impact. Following the typhoon's passage, as mean wind speed and direction stabilized (circa 4:00 p.m.), the gust factors also exhibited a tendency toward stabilization.

The relationship between the tri-directional gust factor at 40 m and the 10-min mean wind speed during the observation period is shown in Fig. 37.10a.

As depicted in Fig. 37.10a, during the entire observation interval, both longitudinal and vertical gust factors exhibit a decreasing trend with increasing mean wind speed, whereas the lateral gust factor demonstrates an inverse tendency, augmenting with higher wind velocities. Within low-wind-speed regimes (< 10 m/s) prior to typhoon impact and post-dissipation, a preponderance of smaller gust factor values was recorded. To facilitate a focused analysis of the typhoon's intensive impact phase (6:10 a.m–1:00 p.m., characterized by 10-min mean wind speeds > 14 m/s), Fig. 37.10b presents data for this specific timeframe. During this period, all three directional gust factors displayed a declining relationship with increasing mean wind

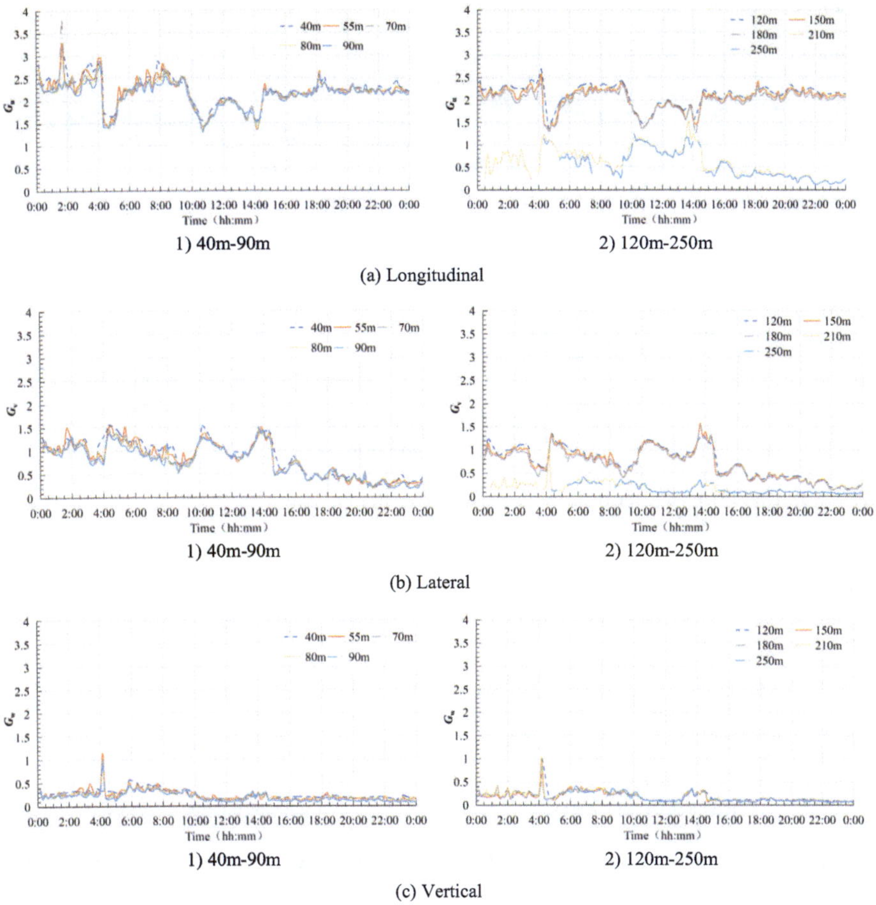

Fig. 37.8 Time-history of gust factors at different heights

speed. Comparative analysis with published typhoon observations in China reveals divergent patterns: typhoons "Muifa" (Wang et al. 2013a), "Meari" (Wang et al. 2013b), "Haima" (He et al, 2020), and "Chan-hom" (Huang and Gu 2019) exhibit gust factors that decrease with rising mean wind speed, whereas typhoon "Hagupit" (Li et al. 2018) demonstrates increasing gust factors over marine surfaces at higher wind velocities. The results highlight distinct wind field characteristics across typhoons, leading to varying correlations between gust factors and mean wind speeds.

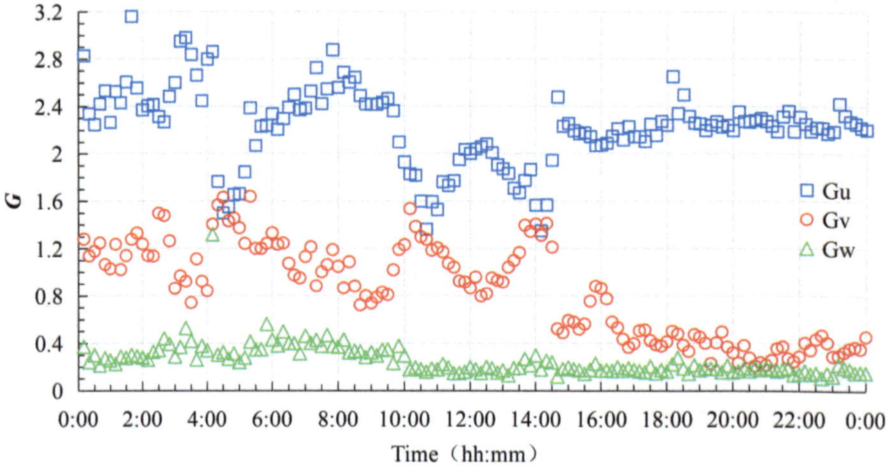

Fig. 37.9 Time-history of gust factor components at 40 m height

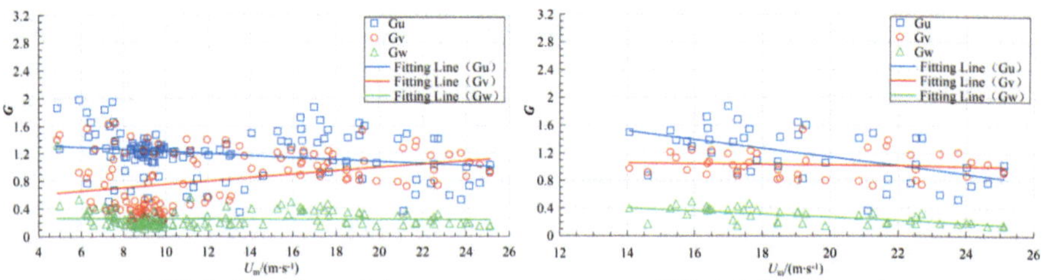

(a) Whole observation period

(b) Period affected by typhoon

Fig. 37.10 Variation of gust factor components with 10-min mean wind speed at 40 m height

37.3.6 Relationship Between Gust Factor and Turbulence Intensity

Considering the relationship between the gust factor and the turbulence intensity, the gust factor and the corresponding turbulence intensity in each direction at different heights (40 m, 90 m, and 180 m) are compared and analyzed in Fig. 37.11.

As shown in Fig. 37.11, the linear relationship between gust factors and turbulence intensity at different heights is similar especially in lateral and vertical directions. In our code, the gust factor is utilized to describe the change of velocity and pressure of the downwind enclosure, defined as:

$$\beta_{zg} = 1 + 2gI_u \tag{37.5}$$

The computational form of the longitudinal gust factor of typhoon is:

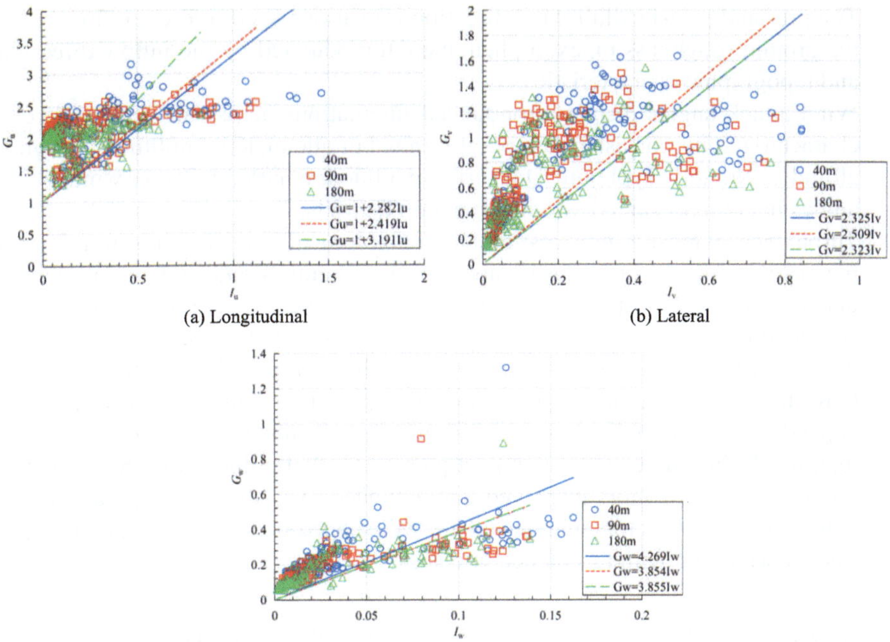

(a) Longitudinal (b) Lateral

(c) Vertical

Fig. 37.11 Variation of gust factors with turbulence intensity

$$G_u = 1 + aI_u \tag{37.6}$$

The coefficients a and g in the two equations have the same physical significance, and both represent the peak factor of pulsating wind pressure change. The fitting coefficient of the longitudinal gust factor of "Doksuri" is about 2.47, which indicates that its longitudinal peak factor is about 2.47.

37.4 Conclusion

Based on the 24-hour observation dataset during the landfall of Typhoon "Doksuri", the following key findings are deduced:

1. The occurrence time of the maximum 10-minute mean wind speed demonstrates a distinct altitude-dependent variation, with earlier manifestations at elevations $\geq$ 150 m compared to $\leq$ 120 m. However, both the temporal distribution of this peak wind speed and the corresponding lateral wind direction angles exhibit marked stochasticity.

2. The pre- and post-landfall wind direction transitions exert a direct influence on the ground roughness index, a phenomenon modulated by the inflow direction and topographic characteristics.
3. Wind attack angles display pronounced fluctuations across all vertical levels, characterized by consistent temporal trends but attenuated amplitude at higher altitudes. Notably, Typhoon "Doksuri"" exhibited remarkably larger wind attack angles in comparison to historical typhoon datasets.
4. Turbulence intensity demonstrates a monotonically decreasing trend with increasing altitude, reaching its apex prior to landfall and undergoing subsequent attenuation post-landfall, primarily due to local topographic effects. All components exhibit consistent temporal variations, with crosswind turbulence being uniform and vertical turbulence remaining below code-specified values.
5. Gust factors exhibit height-invariant characteristics with negligible altitude dependence. Longitudinal and lateral gust factors underwent significant fluctuations during the typhoon's impact phase, stabilizing thereafter. Specifically, longitudinal gust factors displayed a pronounced "W"-shaped temporal profile, while lateral gust factors manifested an "M"-shaped pattern, both maintaining a linear correlation with turbulence intensity.

Acknowledgement The authors would like to acknowledge with great gratitude for the supports of the National Natural Science Foundation of China (Grant No. 52178510), and the Natural Science Foundation of Fujian Province (Grant No. 2022J011253, 2022J05283).

References

Elliott JR, Strobl E, Sun P (2015) The local impact of typhoons on economic activity in China: a view from outer space. J Urban Econ 88(7):50–66

Wang X, Huang P, Gu M (2013a) Field measurement on wind characteristics near ground during typhoon "Muifa". Chin Civil Eng J 46(02):54–61

Wang X, Huang P, Gu M et al (2013b) Field measurement of turbulence characteristics near ground during typhoon 'Meari'. Chin Civil Eng J 28–36

Wang X, Huang C, Huang P et al (2017) Field measurements for characteristics of near ground fluctuating wind during typhoon "Haikui" blowing. J Vib Shock 36(11):199-205+241

Li B, Yang QS, Yang JB (2016) Wind characteristics near ground in South-Eastern Coast area of China based on field measurement. Geomat Nat Haz Risk 7:57–69

Wang X, Yao Z, Guo Y et al (2024) Analysis of the near-ground wind field characteristics during typhoon "Soulik". Appl Sci 14(10)

He HM, Lei X, Nie M et al (2018) Field measurement research of near ground wind field characteristics at landing center during typhoon 'Haima'. J Build Struct 39(10):29–36

He J, He Y, Li Q et al (2020) Observational study of wind characteristics, wind speed and turbulence profiles during super typhoon "Mangkhut". J Wind Eng Ind Aerodyn 206

Xia DD, Dai LM, Lin L et al (2021) A field measurement-based wind characteristics analysis of a typhoon in near-ground boundary layer. Atmosphere 12(7):873

Zhang JG, Wen ZJ, Lei Y (2024) Coastal near ground measured wind field characteristics in typhoon "Soudelor" moving process. J Vib Shock 43(11):272–278+296

Huang Z, Gu M (2019) Estimation of peak factors and gust factors of nonstationary wind speed. J Wind Eng Ind Aerodyn

Li L, Xiao Y, Zhou H et al (2018) Turbulent wind characteristics in typhoon Hagupit based on field measurements. Int J Distrib Sens Netw 14(10)

Chapter 38
Mechanistic Analysis of PVA Microplastics Mitigating Amitriptyline Toxicity to *Chlorella vulgaris*

Wenjie Yuan, Yelin Zhou, Bingjun Shu, and Yicheng Wu

Abstract This study investigated the eco-toxicological impacts of co-exposure to polyvinyl alcohol (PVA) microplastics and amitriptyline on *C. vulgaris*. Amitriptyline exposure at 12 mg/L inhibited algal growth by 72.7% after 8 days, chlorophyll b content decreased by 97.3%, and the Fv/Fm value reduced to 0.22 ± 0.07. PVA mitigated Amitriptyline toxicity to *C. vulgaris,* In *C. vulgaris* exposed to Amitriptyline and PVA co-exposure reduced growth inhibition (36.7–26.8%), elevated chlorophyll a by 17.8%, and enhanced Fv/Fm by 17.2%. The results also showed PVA alleviated oxidative stress of Amitriptyline to *C. vulgaris,* Compared to Amitriptyline single exposure, PVA significantly reduced the malondialdehyde (MDA) content in *C. vulgaris* cells, the activities of superoxide dismutase and catalase increased by 7.8% and 30% respectively. The research results have deepened the understanding of the impact of microplastics on the ecological toxicity of antidepressant drugs, offering insights for microalgae-based remediation of co-contaminated waters.

Keywords Amitriptyline · *Chlorella vulgaris* · Toxicity effect · Microplastics

38.1 Introduction

In modern society's fast-paced lifestyle, the consumption of amitriptyline—a widely prescribed antidepressant for managing depression and chronic pain—has risen substantially in recent years (Pereira et al. 2023). Following human ingestion, 30–50% of unmetabolized amitriptyline is excreted through urine and sweat. Conventional wastewater treatment plants exhibit limited removal efficiency for this compound, leading to its persistent detection in surface waters at concentrations ranging from ng/L to μg/L (Comanescu and Racovita 2024). Amitriptyline exerts multiple toxic effects on aquatic biota, including neurotoxicity, oxidative stress, and

W. Yuan · Y. Zhou · B. Shu · Y. Wu (✉)
College of Environmental Science and Engineering, Xiamen University of Technology, Xiamen, China
e-mail: ycwu@xmut.edu.cn

Y. Li et al. (eds.), *16th International Conference on Environmental Science and Development*, Environmental Science and Engineering,
https://doi.org/10.1007/978-3-032-15080-6_38

impaired reproductive functions in fish and algae (Demin et al. 2017). Owing to its high hydrophobicity, amitriptyline demonstrates significant bioaccumulation potential, thereby posing long-term risks to freshwater ecosystem integrity and human health via trophic transfer (Blanco-Rayón et al. 2021).

Plastics will inevitably fragment into microplastics in the natural environment, thereby affecting the physical and chemical properties of water bodies and the migration and transformation processes of organic pollutants in water (Laskar and Kumar 2019). Compared with traditional microplastics, due to their high surface polarity and rich functional groups, microplastics have a stronger adsorption capacity for organic pollutants, and significantly influence the migration and transformation pathways as well as the bioavailability of pollutants through the carrier effect (Caruso 2019). Given the widespread presence of microplastics in the environment, previous studies have mostly focused on the impacts of MPs on the growth of microalgae. The ecotoxicological effects of the combined exposure of microplastics and the antidepressant drug Amitriptyline, as well as its impacts on the efficiency of algae in remediating water bodies polluted by Amitriptyline, have not been reported yet.

As a primary producer in freshwater ecosystems, *Chlorella vulgaris* is highly sensitive to pollutants and is often used as a model organism for assessing the ecological risks of pollutants in freshwater environments (Rajamanickam and Selvasembian 2025). In this study, freshwater *C. vulgaris* was selected as the research object. By analyzing the dynamic variation patterns of growth curves, the contents of photosynthetic pigments, and the activities of antioxidant enzymes, the ecotoxicological effects of the combined exposure to PVA and Amitriptyline were analyzed, and the regulatory mechanism of microplastics on the toxicity of Amitriptyline was revealed.

38.2 Materials and Methods

38.2.1 Chemicals and Reagent

Amitriptyline hydrochloride (purity > 99%) was purchased from Shanghai Aladdin Biochemical Technology Co., Ltd.; acetonitrile (HPLC grade) and triethylamine (HPLC grade) were purchased from Sinopharm Group Co., Ltd.; methanol (HPLC grade) was purchased from Fisher Chemical, USA. Polyhydroxyalkanoates (PHA) (particle size: approximately 100 μm) were purchased from Jinheng Plastics Co., Ltd.

38.2.2 Determination of Photosynthetic Pigment Content

The contents of chlorophyll a, chlorophyll b and carotenoids were determined by the methanol extraction method (Wood 1985).

38.2.3 *C. vulgaris* Cultivation

C. vulgaris was purchased from Hainan Yuanquan Biotechnology Co., Ltd. The cultivation temperature was maintained at 26 ± 1 °C, the light intensity was set at 4000 lux, and the photoperiod was set as a light/dark ratio of 18 h/6 h.

38.2.4 *Determination of the Combined Toxicity of Microplastics and Amitriptyline*

The absorbance was measured at a wavelength of 680 nm. The concentration of soluble proteins was determined by the Coomassie Brilliant Blue method (Snyder and Desborough 1978). The content of malondialdehyde (MDA), and the activities of superoxide dismutase (SOD) and catalase (CAT) were measured using the kits provided by Nanjing Jiancheng Bioengineering Institute. The chlorophyll fluorescence parameters were determined by a phytoplankton fluorescence meter (PHYTO-PAM, Germany) (Reis et al. 2021).

38.2.5 *Amitriptyline Concentration Determination*

The concentration of Amitriptyline in the solution was determined by high-performance liquid chromatography (LC20A, Shimadzu, Japan). The mobile phase consisted of acetonitrile and water with a volume ratio of 34:66, the flow rate was set at 1 mL/min, and the detection wavelength was 240 nm (Miljković et al. 1997).

38.3 Experimental Results and Analysis

38.3.1 *Effects of Microplastics and Amitriptyline Co-exposure on* C. vulgaris *Growth*

Amitriptyline exposure inhibited the growth of *C. vulgaris*, with the inhibition intensifying as the Amitriptyline concentration increased (Fig. 1). The absorbance of *C. vulgaris* solution at 680 nm increased significantly from 0.82 ± 0.11 to 1.52 ± 0.14 over 8 days, indicating normal growth. Among PVA stress groups, only the 5 mg/L PVA group showed a growth curve overlapping with the control, suggesting no significant growth impact at this concentration. The addition of PVA alleviated the inhibitory effect of low-concentration Amitriptyline on algal growth. After 8 days of exposure, 4 mg/L and 12 mg/L Amitriptyline inhibited algal growth by 36.7% and

Fig. 1 Effects of the combined exposure of PVA and Amitriptyline on the growth of *C. vulgaris*

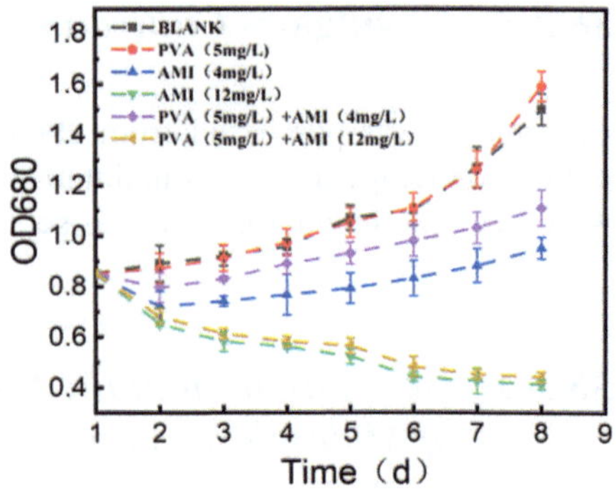

72.7%, respectively. When 5 mg/L PVA was added, the inhibition rates decreased to 26.8% and 70.7%, respectively.

38.3.2 Effects of Microplastics-Amitriptyline Co-exposure on Photosynthetic Pigment Contents in C. vulgaris

PVA exposure alone had little impact on the photosynthetic pigment content of algal cells. In contrast, Amitriptyline exposure significantly inhibited the synthesis of photosynthetic pigments in *C. vulgaris* (Fig. 2). Notably, in algal cells exposed to 12 mg/L Amitriptyline, the content of Chl-b was the lowest, 97.3% lower than that of control cells at the same time (Fig. 2b).

The addition of microplastics mitigated the inhibitory effect of Amitriptyline on the synthesis of photosynthetic pigments in algal cells. After 10 days of exposure to 12 mg/L Amitriptyline, the Chl-a content in algal cells increased from 4.5 to 5.3 mg/L (Fig. 2a), and the carotenoid content increased by 32.5% (Fig. 2c).

38.3.3 Effects of Microplastics and Amitriptyline on C. vulgaris *Photosynthetic Pigment Contents*

Six days after Amitriptyline exposure, the Fv/Fm values of algal cells decreased to varying degrees. Under 4 mg/L Amitriptyline stress, the Fv/Fm value was 0.52 ± 0.12, which increased to 0.62 ± 0.11 after the addition of PVA, 17.2% higher than that of the control group at the same time. Under 4 mg/L Amitriptyline stress, the ΦPSII value was 0.41 ± 0.04; when the exposure concentration increased to 12 mg/L, the ΦPSII value decreased to 0.22 ± 0.07, but increased to 0.35 ± 0.03 after the addition of PVA (Fig. 3).

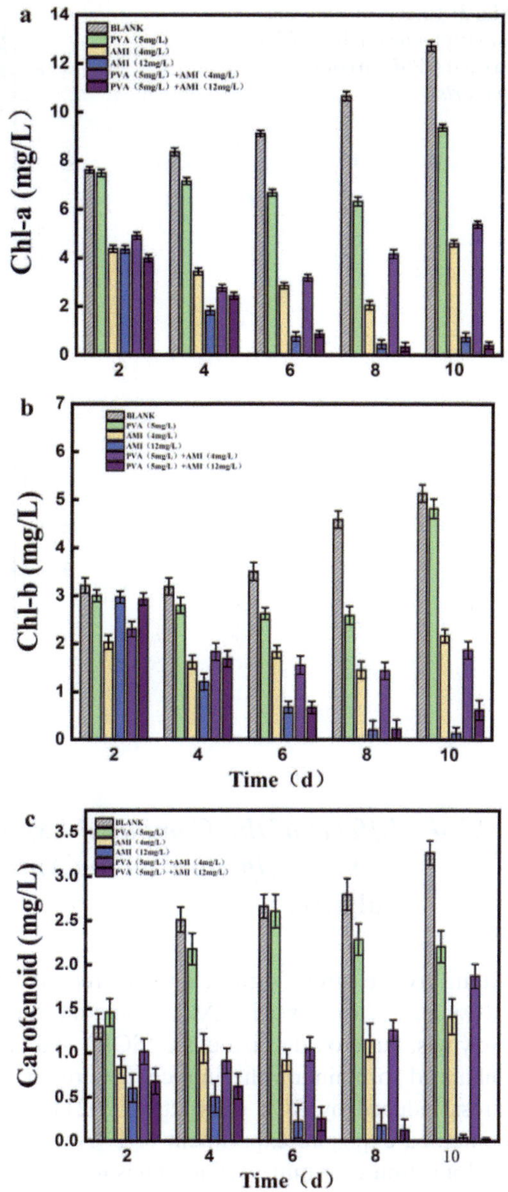

Fig. 2 Effects of PVA and Amitriptyline on the chlorophyll a (**a**), chlorophyll b (**b**) and carotenoids (**c**) of *C. vulgaris*

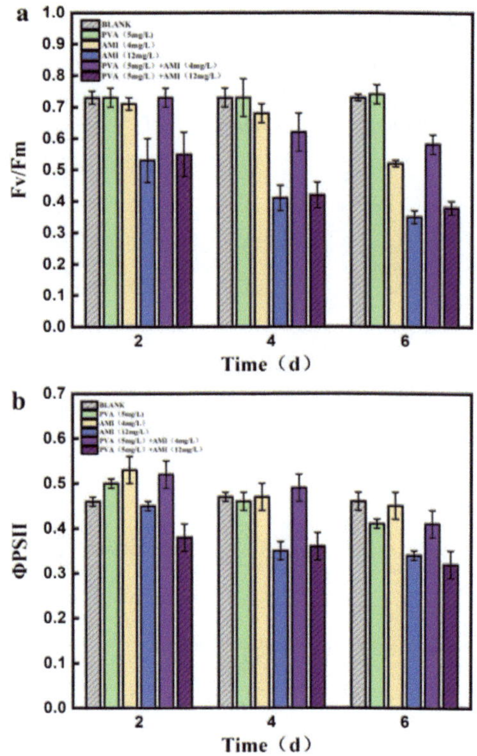

Fig. 3 Effects of PVA and Amitriptyline on the Fv/Fm (**a**) and ΦPSII (**b**) of *C. vulgaris*

38.3.4 Effects of the Combined Exposure of Microplastics and Amitriptyline on the Antioxidant System of **C. vulgaris**

Amitriptyline stress significantly increased the level of reactive oxygen species (ROS) in *C. vulgaris* cells. As a coping mechanism, the activities of two antioxidant enzymes, superoxide dismutase (SOD) and catalase (CAT), in the cells gradually enhanced to maintain the intracellular redox homeostasis and resist the oxidative stress induced by ROS (Gao et al. 2021). As shown in Fig. 4, the results of the combined exposure experiment of PVA and Amitriptyline indicated that with the prolongation of Amitriptyline exposure time and the increase of its concentration, the content of malondialdehyde (MDA) and the activities of antioxidant enzymes in *C. vulgaris* cells showed a synchronous upward trend (Fig. 4b). After 6 days of exposure, in the *C. vulgaris* cells of the 4 mg/L Amitriptyline treatment group, the activity of SOD increased 7.69 times compared with the control group (Fig. 4c), and the activity of CAT increased by 91.21% (Fig. 4d). When the Amitriptyline concentration increased to 12 mg/L, the activities of SOD and CAT increased 10.51 times and

3.25 times respectively compared with the control group. Meanwhile, the contents of total protein (TP) and MDA increased by 28.11% and 562.7% respectively.

Notably, upon the introduction of PVA into the 4 mg/L Amitriptyline treatment group, the total protein content within *C. vulgaris* cells decreased by 7.8% compared to the group exposed to Amitriptyline alone. Additionally, the activities of SOD and CAT also exhibited varying degrees of reduction, indicating that PVA has a certain buffering effect on the activation of antioxidant enzymes induced by Amitriptyline. The addition of PVA microplastics could, to a certain extent, decrease the activities of antioxidant enzymes in the cells. This suggests that microplastics are capable of reducing the oxidative damage caused by Amitriptyline to the cell membrane system (Li et al. 2021).

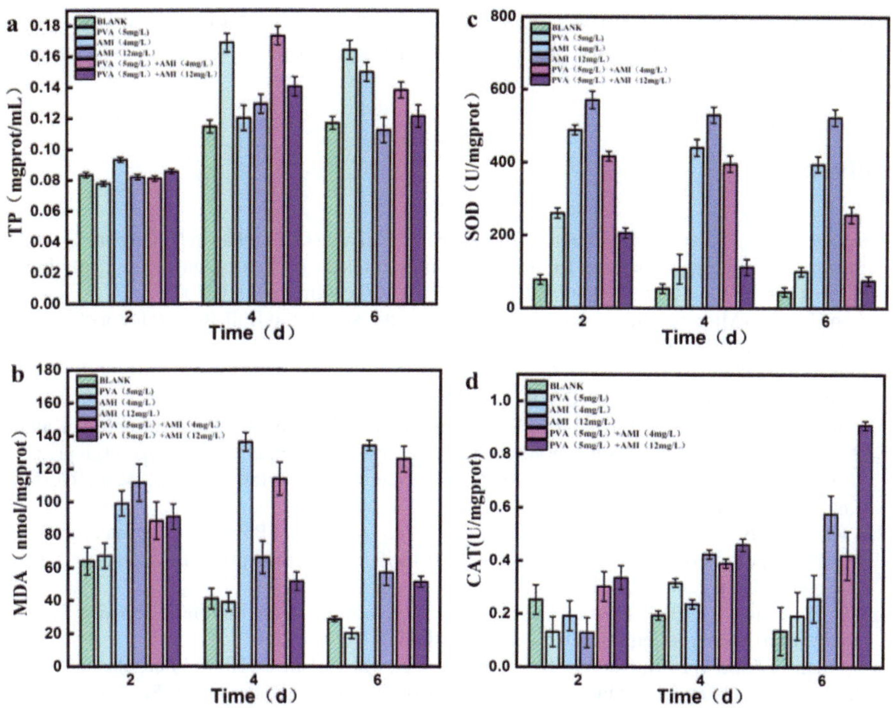

Fig. 4 Effects of PVA and Amitriptyline on the contents of total protein(TP) (**a**) and malondialdehyde (MDA) (**b**), as well as the activities of superoxide dismutase (SOD) (**c**) and catalase (CAT) (**d**) in *C. vulgaris*

38.4 Conclusions

Polyvinyl alcohol (PVA) can alleviate the toxicity of Amitriptyline to *C. vulgaris*, and its mechanism of action is manifested in multiple aspects. In terms of growth, PVA can mitigate the inhibitory effect of low-concentration Amitriptyline on the growth of *C. vulgaris*. Regarding the photosynthetic system, PVA can relieve the inhibition of Amitriptyline on the synthesis of photosynthetic pigments, increase the values of photosynthetic parameters, and protect the photosynthetic system. At the level of oxidative stress, PVA can reduce the oxidative damage caused by Amitriptyline, decrease the intracellular oxidative stress level, and protect the cell membrane system. PVA has been shown to reduce the toxic effects of Amitriptyline on *C. vulgaris* to some extent. This suggests that PVA may interact with pollutants, thereby mitigating their negative impact on aquatic organisms. This finding could provide a novel approach or strategy for environmental remediation.

References

Blanco-Rayón E, Ziarrusta H, Mijangos L, Olivares M, Zuloaga O, Etxebarria N, Izagirre U (2021) Integrated biological response to environmentally-relevant concentration of amitriptyline in Sparus aurata. Ecol Indic 130. https://doi.org/10.1016/j.ecolind.2021.108028

Caruso G (2019) Microplastics as vectors of contaminants. Mar Pollut Bull. https://doi.org/10.1016/j.marpolbul.2019.07.052

Comanescu C, Racovita RC (2024) An overview of degradation strategies for amitriptyline. Int J Mol Sci 25(7). https://doi.org/10.3390/ijms25073822

Demin KA, Kolesnikova TO, Khatsko SL, Meshalkina DA, Efimova EV, Morzherin YY, Kalueff AV (2017) Acute effects of amitriptyline on adult zebrafish: potential relevance to antidepressant drug screening and modeling human toxidromes. Neurotoxicol Teratol. https://doi.org/10.1016/j.ntt.2017.04.002

Gao J-J, Wang B, Peng R-H, Li Z-J, Xu J, Tian Y-S, Yao Q-H (2021) Phytoremediation of multiple persistent pollutants co-contaminated soil by HhSSB transformed plant. Environ Res. https://doi.org/10.1016/j.envres.2021.110959

Laskar N, Kumar U (2019) Plastics and microplastics: a threat to environment. Environ Technol Innovation. https://doi.org/10.1016/j.eti.2019.100352

Li C, Yuan S, Zhou Y, Li X, Duan L, Huang L, Zhou X, Ma Y, Pang S (2021) Microplastics reduce the bioaccumulation and oxidative stress damage of triazole fungicides in fish. Sci Total Environ. https://doi.org/10.1016/j.scitotenv.2021.151475

Miljković B, Pokrajac M, Timotijević I, Varagić V (1997) Solid-phase extraction and high performance liquid chromatographic determination of amitriptyline and nortriptyline in human plasma. J Liq Chromatogr Relat Technol. https://doi.org/10.1080/10826079708010959

Pereira CDS, Cruz JN, Ferreira MKM, Baia-da-Silva DC, Fontes-Junior EA, Lima RR (2023) Global research trends and hotspots analysis of the scientific production of amitriptyline: a bibliometric approach. Pharmaceuticals (Basel) 16(7). https://doi.org/10.3390/ph16071047

Rajamanickam R, Selvasembian R (2025) Insights into the potential of *chlorella* species in the treatment of hazardous pollutants from industrial effluent. World J Microbiol Biotechnol. https://doi.org/10.1007/s11274-025-04351-5

Reis LLD, Alho LOG, Abreu CB, Melao M (2021) Using multiple endpoints to assess the toxicity of cadmium and cobalt for chlorophycean *Raphidocelis subcapitata*. Ecotoxicol Environ Saf 208:111628. https://doi.org/10.1016/j.ecoenv.2020.111628
Snyder JC, Desborough SL (1978) Rapid estimation of potato tuber total protein content with Coomassie brilliant blue G-250. Theor Appl Genet. https://doi.org/10.1007/bf00264747
Wood LW (1985) Chloroform–methanol extraction of chlorophyll a. Can J Fish Aquat Sci. https://doi.org/10.1139/f85-005

Chapter 39 Transcriptomic Analysis Reveals Immunotoxic Effects of Oman Crude Oil Water-Accommodated Fractions on Sea Cucumber (*Apostichopus Japonicus*) Coelomocytes

Xishan Li, Xin Li, Feng Guo, Fu-en Chen, Yongqi Liang, Haonan Wang, and Deqi Xiong

Abstract Oil spills represent a significant threat to marine ecosystems, yet their immunotoxic effects on economically important invertebrates like sea cucumbers remain poorly understood. This study investigated the impacts of Oman crude oil water-accommodated fractions (WAFs) on the immune function of sea cucumber (*Apostichopus japonicus*) coelomocytes using high-throughput RNA sequencing. Sea cucumbers were exposed to WAFs exposure solution at the nominal concentrations of 25 and 75% (*v*/*v*) for 7 d. Transcriptome analysis revealed 4911 and 5405 differentially expressed genes (DEGs) in the 25 and 75% (*v*/*v*) WAF groups, respectively, compared to the control. Gene Ontology (GO) enrichment analysis indicated that these DEGs were prominently involved in biological adhesion, immune system processes, and antioxidant activity. KEGG pathway analysis revealed significant enrichment in immune-related pathways, including NOD-like receptor signaling, Toll-like receptor signaling, and cytosolic DNA-sensing pathways. Notably, the number of immune-related pathways decreased with increasing WAF concentration (6 pathways at 25% (*v*/*v*); 4 pathways at 75% (*v*/*v*)), suggesting a low-concentration stimulation and high-concentration inhibition pattern of immunotoxicity. These findings could provide novel insights into the molecular mechanisms underlying oil

X. Li · X. Li · F. Guo · F. Chen · Y. Liang · H. Wang (✉) · D. Xiong (✉)
College of Environmental Science and Engineering, Dalian Maritime University, Dalian, China
e-mail: wanghnn97@outlook.com

D. Xiong
e-mail: xiongdq@dlmu.edu.cn

Joint Research and Development Center for Sustainable Marine Fuels, Dalian Maritime University, Dalian, China

X. Li
National Marine Environmental Monitoring Center, Dalian, China

Y. Li et al. (eds.), *16th International Conference on Environmental Science and Development*, Environmental Science and Engineering,
https://doi.org/10.1007/978-3-032-15080-6_39

pollution-induced immunotoxicity in marine invertebrates and establish a foundation for developing biomarkers for ecological risk assessment of oil contamination in marine environments.

Keywords Oil spill · Sea cucumber · Immunotoxicity · Transcriptomic analysis

39.1 Introduction

Oil spills and chronic petroleum pollution pose a serious threat to marine ecosystems, releasing complex mixtures of toxic hydrocarbons such as polycyclic aromatic hydrocarbons (PAHs) (Li et al. 2020, 2025, 2019). Exposure to crude oil could cause deleterious effects on various marine animals. For example, in vitro assays with mussel hemocytes showed that exposure to water-accommodated fractions (WAFs) of crude oil (with or without dispersant) caused oxidative stress and immune toxicity (Katsumiti et al. 2019). In copepods, WAFs exposure triggered reactive oxygen species (ROS) generation and developmental toxicity (Han et al. 2014). Fish studies during the *Deepwater Horizon* spill revealed oil-induced immunosuppression that Bayha et al. (2017) found that oil exposure impaired immune function in southern flounder and increased susceptibility to *Vibrio* infection (Bayha et al. 2017). Moreover, reviews of oil toxicology in vertebrates confirmed that crude oil could modulate immune responses and cause hematological and immunological injuries (Bayha et al. 2017). Despite extensive research in vertebrates, the immunotoxicological impacts of oil pollution on marine invertebrate immune systems are still less understood.

Sea cucumbers (*Apostichopus japonicus*) are benthic echinoderms with high ecological and commercial importance (Li et al. 2020, 2021, 2022). They serve as detritivores in nutrient recycling and are widely cultured in coastal East Asia. Sea cucumbers are particularly vulnerable to pollution due to their sessile benthic life stage and continuous processing of sediment and water for food (Li et al. 2022). During oil spills, they cannot escape contaminated areas, making them prone to exposure and bioaccumulation of toxicants. Coelomocytes, the free-circulating immune cells in the coelomic fluid of sea cucumbers, play key roles in non-specific immunity (phagocytosis, pathogen recognition, encapsulation). Previous studies have shown that environmental stressors can alter sea cucumber immunity. For instance, elevated water temperature alters the expression of immune-related genes and suppresses some immune functions in sea cucumbers (Ju et al. 2024). Combined heat stress and oil exposure were reported to exacerbate oxidative stress and apoptosis in sea cucumber tissues (Li et al. 2021). At the cellular level, sea cucumber coelomocytes possess conserved innate immune receptors such as Toll-like receptors (TLRs) and NOD-like receptors (NLRs). Sun et al. (2013) cloned two TLR genes from A. japonicus and demonstrated their up-regulation in response to bacterial and viral immune challenges (Sun et al. 2013). Similarly, NLR family members exist in echinoderms, among which an NLRC4-like receptor and an NLRP10 have been identified in sea cucumber coelomocytes, and these receptors bind pathogen-associated molecular patterns and mediate inflammatory signaling and apoptosis (Chen et al. 2020; Li et al. 2018). The presence of these innate immune pathways suggests sea cucumbers have the molecular machinery to respond to pathogen invasion and stress, but it also raises the question of how pollutants like crude oil impact these systems.

To date, little is known about the transcriptomic changes in sea cucumber immune cells following oil exposure. Our recent work investigated the effects of physically and chemically dispersed Oman crude oil on sea cucumber antioxidant and immune parameters, finding disruptions in antioxidant enzyme activities and immune enzymes (e.g., reduced lysozyme activity) after acute exposure (Li et al. 2022). However, genome-wide expression changes underpinning these physiological effects remain unexplored. Therefore, in this study, we exposed sea cucumbers to Oman crude oil WAFs with different nominal concentrations for 7 d and performed RNA sequencing of coelomocytes to identify differentially expressed genes and affected pathways. By performing GO and KEGG enrichment analyses on DEGs, we aimed to figure out key biological processes and signaling pathways perturbed by oil exposure, focusing on immune-related functions (e.g. NLR signaling, cytokine pathways) that could indicate immunotoxic effects. Our results could provide a comprehensive view of how sea cucumber immunity would be affected at the molecular level by oil pollution, offering insight into the potential health impacts on marine invertebrates in oil spill-polluted environments.

39.2 Materials and Methods

39.2.1 Experimental Design and WAF Preparation

Adult sea cucumbers (average weight ~ 150 g) were obtained from a local aquaculture farm and acclimated in aerated seawater (temperature 16 °C, salinity ~ 30 ppt) for 7 d prior to the exposure tests. Oman crude oil WAFs solutions were prepared using a standard protocol as established by Singer et al. (2000), Singer et al. (2000). Briefly, WAFs stock solutions were prepared with crude oil and pre-filtered natural seawater at an oil loading rate of 25 g/L in 10 L glass aspirator bottles, with being gently stirred (low-energy method) for 23.5 h at 16 °C and then allowed to settle for 0.5 h. The aqueous phase containing dissolved and micro-dispersed oil (WAFs stock solution) was carefully drawn off, avoiding any surface oil layer. Then, the WAFs stock solution was diluted into 25 and 75% (*v*/*v*) WAFs exposure solutions. Sea cucumbers were divided into three groups ($n = 10$ individuals per group): control, 25% (*v*/*v*) WAF, and 75% (*v*/*v*) WAF. Each sea cucumber was held in a 10 L tank with the respective treatment (renewed daily) and maintained at 16 °C for 7 d. Throughout the exposure period, sea cucumbers were monitored for behavior and survival (no mortality occurred). After 7 d, coelomic fluid was withdrawn from each sea cucumber via a sterile syringe. Coelomocytes were collected and immediately pelleting cells by centrifugation (3000 g, 10 min, 4 °C). Cell pellets were flash-frozen in liquid nitrogen and stored at –80 °C until RNA extraction.

39.2.2 RNA Extraction and Quality Assessment

Total RNA was extracted from coelomocytes using the TRIzol reagent method according to the manufacturer's instructions. Briefly, thawed cell pellets were homogenized in 2 mL TRIzol lysis buffer, vortexed, and incubated at room temperature for 5 min to ensure complete cell lysis. The homogenate was centrifuged at 12,000 × g for 5 min at 4 °C, and the supernatant was transferred to a new RNase-free tube. Chloroform/isoamyl alcohol (24: 1, *v/v*) mixture (400 μL) was added, vortexed, and centrifuged at 12,000 × g for 8 min at 4 °C. The aqueous phase was transferred to a fresh tube, mixed with 800 μL isopropanol, and incubated at − 20 °C for 2 h to precipitate RNA. The mixture was centrifuged at 17,500 × g for 25 min at 4 °C, and the supernatant was discarded. The RNA pellet was washed with 1200 μL pre-chilled 75% ethanol, centrifuged at 17,500 × g for 3 min at 4 °C, and air-dried for 10 min. The dried RNA pellet was dissolved in 100 μL DEPC-treated water. RNA concentration, 28S/18S ratio, and RNA Integrity Number (RIN) were assessed using an Agilent Bioanalyzer (Agilent, USA). The average RNA concentration was 302.73 ± 139.66 ng/μL, with an average 28S/18S ratio of 2.51 ± 0.38 and an average RIN value of 8.86 ± 0.51, confirming high-quality RNA suitable for library construction and sequencing.

39.2.3 cDNA Library Construction and RNA-Seq Data Analysis

The cDNA libraries were constructed according to the following procedure: (1) mRNA was enriched using oligo(dT) magnetic beads to capture poly(A) tails; (2) fragmentation buffer was used to break the mRNA into short fragments, followed by reverse transcription using random N6 primers to synthesize cDNA; (3) the resulting double-stranded DNA fragments were end-repaired, 5'-phosphorylated, and 3'-adenylated, followed by ligation to bubble-shaped adapters with protruding 3'-T overhangs; (4) the ligation products were amplified using specific primers; and (5) the PCR products were denatured to form single-stranded DNA and circularized using bridge primers to create a single-stranded circular DNA library. High-throughput RNA-Seq was performed by BGI Genomics Co., Ltd. (Shenzhen, China) using the DNBSEQ platform with three concentration groups and three biological replicates per group. Raw reads obtained from DNBSEQ sequencing were filtered using Perl scripts to remove adapter contamination (reads with more than 5 adapter bases), low-quality reads (reads with more than 15% bases having Phred quality scores < 19), and reads with more than 5% unknown N bases. The filtered data, termed clean reads, were mapped to the sea cucumber reference genome using Bowtie2 (v2.2.5) for genome indexing and TopHat (v2.0.12) for read mapping. New transcripts were predicted using Cufflinks, selecting those with lengths > 200 bp, > 2 exons, and coverage > 3. Single-nucleotide polymorphisms (SNPs), insertions/deletions (InDels), and alternative splicing events were detected. New transcripts with protein-coding potential were incorporated into the reference sequence to create a complete reference for gene expression analysis. Fragments Per Kilobase Million (FPKM) values were calculated to estimate gene expression levels. The DESeq2 (v1.4.5) package was used to identify differentially expressed genes (DEGs) between samples with biological replicates. Cluster analysis and functional enrichment analysis were performed on the identified DEGs.

39.2.4 Gene Ontology (GO) and Kyoto Encyclopedia of Genes and Genomes (KEGG) Enrichment Analysis

The identified DEGs were mapped to the Kyoto Encyclopedia of Genes and Genomes (KEGG) (http://www.genome.jp/kegg/) and Gene Ontology (GO) (http://www.geneontology.org/) databases for classification and pathway enrichment analysis. GO enrichment classification diagrams were generated to visualize the distribution of DEGs across GO terms. KEGG pathway enrichment analysis was conducted using the phyper function in R software, calculating *P*-values and applying false discovery rate (FDR) correction. Pathways with adjusted *P*-values (P_{adj}) ≤ 0.05 were considered significantly enriched. Bubble charts were created to display the top 20 enriched KEGG pathways based on P_{adj} values.

39.3 Results

39.3.1 Genes Differential Gene Expression in Coelomocytes Under Oil Exposure

High-throughput RNA sequencing using the DNBSEQ platform generated approximately 6.45 Gb of data per sample across nine samples (three biological replicates for control and two treatment groups). After filtering, clean reads were mapped to the sea cucumber reference genome with an average mapping rate of 81.38 ± 0.63% to the genome and 41.89 ± 0.91% to the gene set. A total of 29,897 expressed genes were detected, including 27,002 known genes and 2895 predicted new genes. Additionally, 29,188 new transcripts were identified, comprising 26,230 new splice variants of known protein-coding genes and 2958 transcripts of new protein-coding genes. Differential expression analysis using DESeq2 revealed distinct patterns of

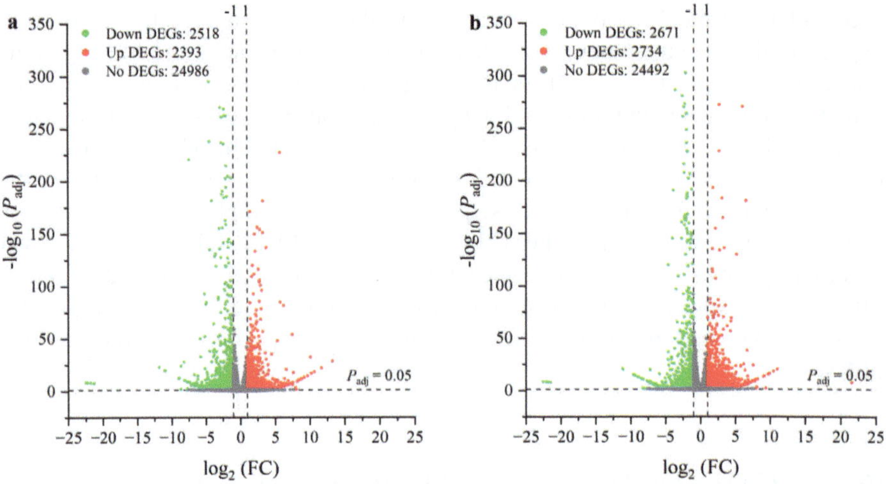

Fig. 39.1 Volcano plots for differentially expressed genes (DEGs) in sea cucumber coelomocytes following 7-d exposure to Oman crude oil water-accommodated fractions (WAFs) with different nominal concentrations (A: 25% (*v*/*v*) WAFs; B: 75% (*v*/*v*) WAFs). Each point represents a gene, with red indicating upregulated DEGs, green indicating downregulated DEGs, and gray representing non-significant genes. Significance thresholds were set at an adjusted *P*-value (P_{adj}) ≤ 0.05 and an $|\log_2$ (fold change (FC))$| \geq 1$

gene expression changes in response to different concentrations of Oman crude oil WAFs (Fig. 39.1). Exposure to 25% (*v*/*v*) WAFs resulted in 4911 DEGs compared to the control, with 2393 (48.7%) up-regulated and 2518 (51.3%) down-regulated genes (Fig. 39.1a). In the 75% (*v*/*v*) WAFs group, 5405 DEGs were identified, including 2734 (50.6%) up-regulated and 2671 (49.4%) down-regulated genes (Fig. 39.1b).

Venn diagram analysis was performed to identify common and unique DEGs between the two WAF exposure groups (Fig. 39.2). Among down-regulated DEGs, 1129 genes were common to both 25% and 75% (*v*/*v*) WAFs groups, while 1389 and 1282 genes were uniquely down-regulated in the 25% and 75% (*v*/*v*) WAFs groups, respectively (Fig. 39.2a). For up-regulated DEGs, 1,243 genes were shared between both exposure groups, with 1150 and 1584 genes uniquely up-regulated in the 25% and 75% (*v*/*v*) WAFs groups, respectively (Fig. 39.2b). The differential regulation pattern indicated that while some cellular responses to oil exposure would be conserved across concentrations, there were also concentration-specific transcriptional changes that might reflect different toxicity mechanisms or compensatory responses at varying exposure levels.

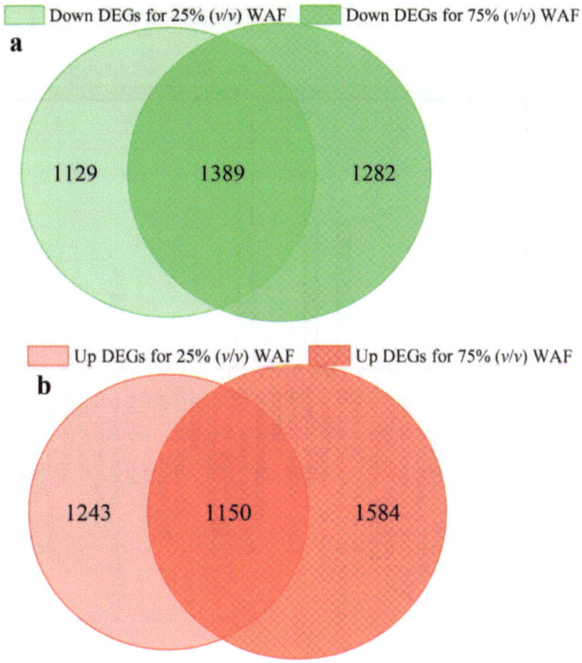

Fig. 39.2 Venn diagrams for the overlap of downregulated **a** and upregulated **b** differentially expressed genes (DEGs) in sea cucumber coelomocytes following 7-d exposure to Oman crude oil water-accommodated fractions (WAFs) with different nominal concentrations (25% and 75% (*v*/*v*) WAFs)

39.3.2 GO Enrichment Analysis

GO enrichment analysis was conducted to elucidate the biological functions of the DEGs induced by Oman crude oil WAFs exposure (Fig. 39.3). In the 25% (*v*/*v*) WAF exposure group, 4911 DEGs were classified into 35 GO terms: 15 (42.9%) in biological process, 2 (5.7%) in cellular component, and 18 (51.4%) in molecular function categories (Fig. 39.3a). The enriched biological process terms included cellular process, metabolic process, biological regulation, response to stimulus, and signaling. The cellular component category included cellular anatomical entity and protein-containing complex. Molecular function terms encompassed binding, catalytic activity, molecular transducer activity, transporter activity, and molecular regulator. In the 75% (*v*/*v*) WAF exposure group, 5405 DEGs were classified into 35 GO terms: 15 (42.9%) in biological process, 2 (5.7%) in cellular component, and 18 (51.4%) in molecular function categories (Fig. 39.3b). Differently, the enriched biological process terms included cellular process, multicellular organismal process, biological regulation, signaling, and locomotion. The cellular component category included protein-containing complex and cellular anatomical entity. Molecular function terms encompassed ATP-dependent activity, catalytic activity, general transcription initiation factor activity, antioxidant activity, and protein folding chaperone.

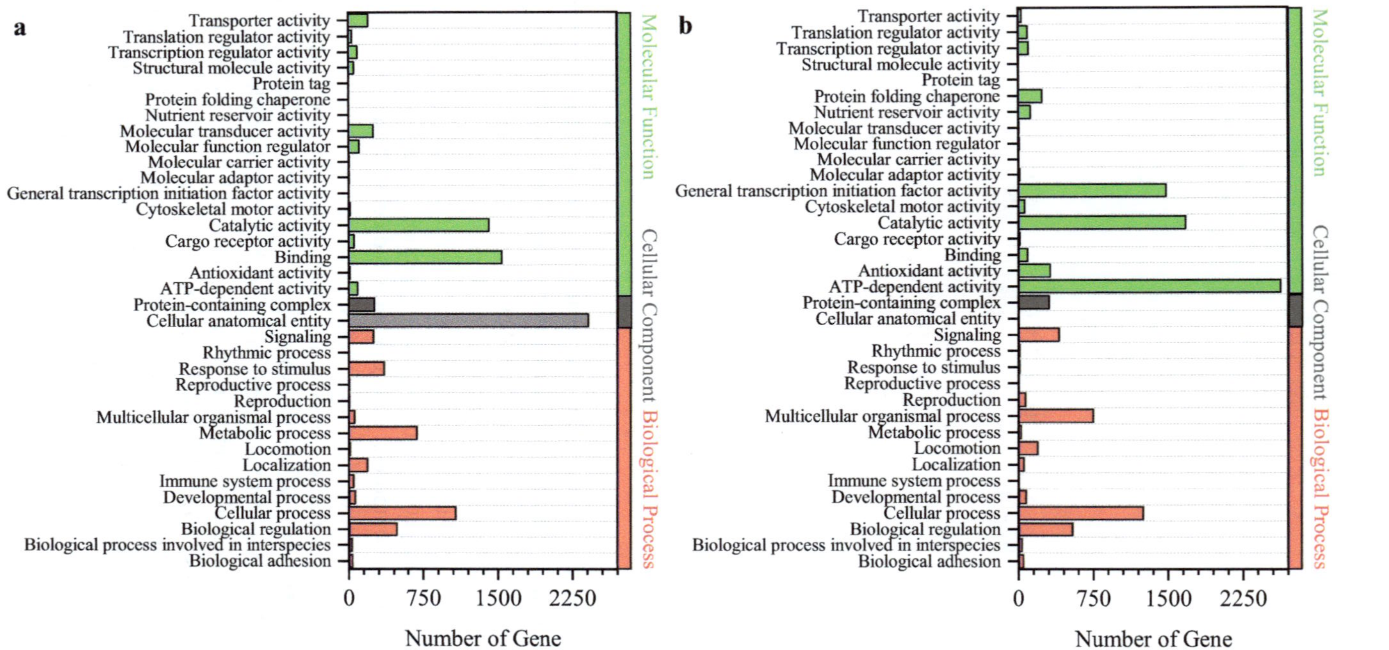

Fig. 39.3 Gene Ontology (GO) enrichment analysis of differentially expressed genes (DEGs) in sea cucumber coelomocytes following 7-d exposure to Oman crude oil water-accommodated fractions (WAFs) with different nominal concentrations (A: 25% (*v*/*v*) WAFs; B: 75% (*v*/*v*) WAFs). DEGs were categorized into Biological Process (BP), Cellular Component (CC), and Molecular Function (MF). Bar charts display the number of DEGs associated with each GO term

39.3.3 KEGG Pathway Enrichment Analysis

KEGG pathway enrichment analysis provided insights into the specific biological pathways affected by Oman crude oil WAFs exposure (Fig. 39.4). In the 25% (*v*/*v*) WAF exposure group, the top 20 significantly enriched KEGG pathways included salivary secretion, NOD-like receptor signaling pathway, and MAPK signaling pathway, which showed the highest significance (Fig. 39.4a). The apoptosis-multiple species pathway exhibited the highest enrichment ratio at 31%. Notably, six immune-related pathways were significantly enriched: NOD-like receptor signaling pathway, cytosolic DNA-sensing pathway, Toll and Imd signaling pathway, complement and coagulation cascades, Toll-like receptor signaling pathway, and platelet activation. In the 75% (*v*/*v*) WAF exposure group, the top 20 significantly enriched KEGG pathways included ribosome, NOD-like receptor signaling pathway, salivary secretion, HIF-1 signaling pathway, and Ras signaling pathway (Fig. 39.4b). The ribosome pathway showed the highest significance and enrichment ratio (41%). Four immune-related pathways were significantly enriched: NOD-like receptor signaling pathway, Toll and Imd signaling pathway, cytosolic DNA-sensing pathway, and Toll-like receptor signaling pathway. Overlap analysis of the top 20 KEGG pathways between different WAF concentration groups identified 10 commonly enriched pathways, with the NOD-like receptor signaling pathway particularly notable among immune-related pathways. This pathway showed the highest number and proportion of DEGs among immune-related pathways across all WAF exposure groups, highlighting its potential importance in the immune response to oil pollution.

39.4 Discussion

This study provided comprehensive insights into the transcriptomic responses of sea cucumber coelomocytes to Oman crude oil WAFs exposure, with a particular focus on immune-related genes and pathways. Our findings revealed complex dose-dependent alterations in gene expression patterns that might underlie the immunotoxic effects of oil pollution on these economically important marine invertebrates. The identification of thousands of differentially expressed genes (4911 and 5405 DEGs in the 25% and 75% (*v*/*v*) WAFs groups, respectively) revealed the profound impact of oil exposure on the transcriptome of sea cucumber coelomocytes. The increasing number of DEGs with higher WAFs concentrations suggested a dose-dependent transcriptional response, consistent with previous studies on other marine organisms exposed to petroleum hydrocarbons (Xu et al. 2017; Jenny et al. 2016). However, our Venn diagram analysis revealed that only approximately half of the DEGs were shared between the two exposure groups, indicating that different concentrations of oil pollutants may trigger distinct molecular mechanisms in addition to common stress responses.

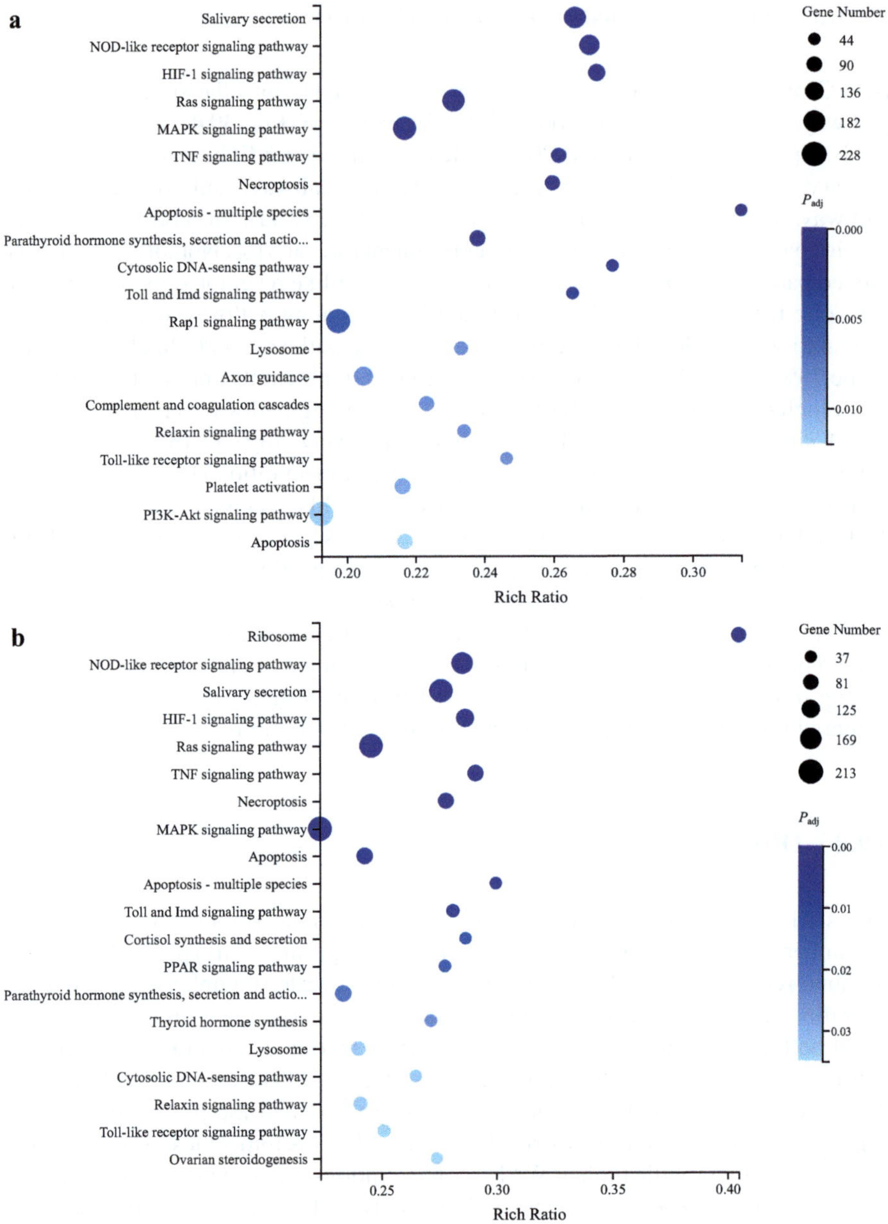

Fig. 39.4 Kyoto encyclopedia of genes and genomes (KEGG) enrichment analysis (TOP 20 enriched pathways) of differentially expressed genes (DEGs) in sea cucumber coelomocytes following 7-d exposure to Oman crude oil water-accommodated fractions (WAFs) with different nominal concentrations (A: 25% (*v*/*v*) WAFs; B: 75% (*v*/*v*) WAFs). Bubble plots illustrate the enrichment (rich) ratio (*x*-axis), number of DEGs (bubble size), and statistical significance (color gradient, adjusted *P*-value (P_{adj})) for each pathway

The GO enrichment analysis provided valuable information about the biological functions affected by oil exposure. The enrichment of immune system process terms in both exposure groups confirms the immunotoxic potential of Oman crude oil WAFs. This finding is particularly significant considering that sea cucumbers rely entirely on innate immunity for defense against pathogens and environmental stressors. Furthermore, the enrichment of antioxidant activity terms suggests the activation of protective mechanisms against oxidative stress, a common consequence of petroleum hydrocarbon exposure in aquatic organisms (Regoli and Giuliani 2014). Among the enriched immune pathways, the NOD-like receptor (NLR) signaling pathway was prominently affected in both exposure groups. NLRs are intracellular pattern recognition receptors that detect pathogen-associated molecular patterns (PAMPs) and damage-associated molecular patterns (DAMPs), triggering inflammatory responses and programmed cell death (Franchi et al. 2009). In vertebrates, NLRs activate inflammasomes, multiprotein complexes that process pro-inflammatory cytokines such as IL-1β and IL-18 (Schroder and Tschopp 2010). Although the exact function of NLRs in sea cucumbers remains to be fully elucidated, our findings suggest that oil exposure may disrupt this critical innate immune pathway, potentially compromising the organism's ability to respond to microbial challenges. The enrichment of the Toll-like receptor (TLR) signaling pathway in both exposure groups is also noteworthy. TLRs are evolutionarily conserved transmembrane receptors that recognize PAMPs and activate signaling cascades leading to the production of antimicrobial peptides, cytokines, and other immune effectors (Kawai and Akira 2010). Our results suggest that oil exposure may interfere with TLR signaling, potentially disrupting the recognition of pathogens and subsequent immune responses. The cytosolic DNA-sensing pathway, which was also enriched in both exposure groups, plays a crucial role in detecting cytosolic DNA, typically derived from invading pathogens or cellular damage, and initiating immune responses (Paludan and Bowie 2013). The enrichment of this pathway suggests that oil exposure may induce cellular damage, leading to the release of DNA into the cytosol, or alternatively, may directly interfere with the DNA-sensing machinery. Either mechanism could potentially disrupt normal immune function in sea cucumber coelomocytes. In addition to immune-related pathways, our analysis revealed the enrichment of apoptosis pathways in both exposure groups, with the apoptosis-multiple species pathway showing the highest enrichment ratio (31%) in the 25% (*v*/*v*) WAFs group. This finding is consistent with previous studies demonstrating that petroleum hydrocarbons can induce apoptosis in various aquatic organisms (Li et al. 2021; Desplat et al. 2023; Moodley et al. 2018). Apoptosis plays a crucial role in immune function by eliminating damaged cells and regulating immune cell homeostasis. Excessive or dysregulated apoptosis, however, can lead to immunosuppression (Elmore 2007). Therefore, the enrichment of apoptosis pathways may represent another mechanism by which oil exposure impairs immune function in sea cucumber coelomocytes. The enrichment of the MAPK signaling pathway, particularly in the 25% (*v*/*v*) WAFs group, is also significant from an immunotoxicological perspective. MAPK pathways transduce signals from various extracellular stimuli to regulate diverse cellular processes, including immune responses (Arthur and Ley 2013). In invertebrates,

MAPK signaling contributes to innate immune responses against pathogens and environmental stressors (Matozzo and Marin 2010). The alteration of this pathway by oil exposure may disrupt normal signal transduction in coelomocytes, potentially impairing their immune functions.

Interestingly, our study also revealed the enrichment of the HIF-1 signaling pathway, particularly in the 25 and 75% (*v*/*v*) WAFs groups. Hypoxia-inducible factor 1 (HIF-1) is a transcription factor that regulates cellular responses to hypoxia and has been increasingly recognized for its role in immunity (Palazon et al. 2014). The enrichment of this pathway suggests that oil exposure may induce hypoxic stress in coelomocytes or directly interfere with oxygen-sensing mechanisms. This finding is consistent with previous studies demonstrating that petroleum hydrocarbons can reduce oxygen availability in aquatic environments and cause hypoxic stress in exposed organisms (Whitehead 2013). The enrichment of the ribosome pathway, which showed the highest significance and enrichment ratio (41%) in the 75% (*v*/*v*) WAFs group, indicates a potential impact on protein synthesis. Ribosomes are essential cellular components responsible for protein translation, and their dysfunction can affect the production of immune effectors such as antimicrobial peptides and complement proteins. The disruption of ribosomal function by high concentrations of oil pollutants may represent another mechanism contributing to immunosuppression in sea cucumber coelomocytes.

Our findings reveal complex transcriptomic changes in sea cucumber coelomocytes in response to Oman crude oil WAFs exposure, with a particular impact on immune-related genes and pathways. The observed pattern of low-concentration stimulation and high-concentration inhibition of immune pathways suggests a hormetic dose–response relationship, which has important implications for ecological risk assessment of oil pollution in marine environments. Future studies should investigate the functional consequences of these transcriptomic changes by examining protein expression, immune cell functionality, and disease susceptibility in oil-exposed sea cucumbers.

Funding This research was funded by the National Natural Science Foundation of China, grant number 42206158; the Liaoning Provincial Natural Science Foundation of China, grant number 2024-BS-016; the National Key Research and Development Program of China, grant number 2023YFC3108303; and the Fundamental Research Funds for the Central Universities, grant number 3132023156.

References

Arthur JSC, Ley SC (2013) Mitogen-activated protein kinases in innate immunity. Nat Rev Immunol 13(9):679–692

Bayha KM, Ortell N, Ryan CN, Griffitt KJ, Krasnec M, Sena J, Ramaraj T, Takeshita R, Mayer GD, Schilkey F et al (2017) Crude oil impairs immune function and increases susceptibility to pathogenic bacteria in southern flounder. PLoS ONE 12(5):e0176559

Chen K, Lv Z, Shao Y, Guo M, Li C (2020) Cloning and functional analysis the first NLRC4-like gene from the sea cucumber *Apostichopus japonicus*. Developm Comparat Immunol 104103541

Desplat Y, Warner JF, Blake EJ, Vijayan N, Cuvelier M, Blackwelder P, Lopez JV (2023) Morphological and transcriptional effects of crude oil and dispersant exposure on the marine sponge *Cinachyrella alloclada*. Sci Total Environ 878162832

Elmore S (2007) Apoptosis: a review of programmed cell death. Toxicol Pathol 35(4):495–516

Franchi L, Warner N, Viani K, Nuñez G (2009) Function of NOD-like receptors in microbial recognition and host defense. Immunol Rev 227(1):106–128

Han J, Won E-J, Hwang D-S, Shin K-H, Lee YS, Leung KM-Y, Lee S-J, Lee J-S (2014) Crude oil exposure results in oxidative stress-mediated dysfunctional development and reproduction in the copepod *Tigriopus japonicus* and modulates expression of cytochrome P450 (*CYP*) genes. Aquatic Toxicol 152308–317

Jenny MJ, Walton WC, Payton SL, Powers JM, Findlay RH, O'Shields B, Diggins K, Pinkerton M, Porter D, Crane DM et al (2016) Transcriptomic evaluation of the American oyster, *Crassostrea virginica*, deployed during the *Deepwater Horizon* oil spill: evidence of an active hydrocarbon response pathway. Marine Environ Res 120166–181

Ju Z, Li X, Yang W, Xiong D (2024) Immune responses of sea cucumber *Apostichopus japonicus* to combined environmental stress from high temperature and oil pollution. Marine Freshwater Res 75(6):MF23161

Katsumiti A, Nicolussi G, Bilbao D, Prieto A, Etxebarria N, Cajaraville MP (2019) *In vitro* toxicity testing in hemocytes of the marine mussel *Mytilus galloprovincialis* (L.) to uncover mechanisms of action of the water accommodated fraction (WAF) of a naphthenic North Sea crude oil without and with dispersant. Sci Total Environ 6701084–1094

Kawai T, Akira S (2010) The role of pattern-recognition receptors in innate immunity: update on toll-like receptors. Nat Immunol 11(5):373–384

Li X, Liao G, Ju Z, Wang C, Li N, Xiong D, Zhang Y (2020) Antioxidant response and oxidative stress in the respiratory tree of sea cucumber (*Apostichopus japonicus*) following exposure to crude oil and chemical dispersant. J Marine Sci Eng 8(8):547

Li X, Wang C, Li N, Gao Y, Ju Z, Liao G, Xiong D (2021) Combined effects of elevated temperature and crude oil pollution on oxidative stress and apoptosis in sea cucumber (*Apostichopus japonicus*, Selenka). Int J Environ Res Public Health 18(2):801

Li X, Zou Y, Xuan H, Yang W, Liao G, Wang C, Xiong D (2022) Impact of physically and chemically dispersed crude oil on the antioxidant defense capacities and non-specific immune responses in sea cucumber (*Apostichopus japonicus*). J Marine Sci Eng 10(10):1544

Li X, Dai Y, Li X, Guo H, Dai J, Wang H, Xiong D, Liao G (2025) Oxidative stress responses and recovery of marine medaka (*Oryzias melastigma*) in early-life stages after acute exposure to crude oil. J Marine Sci Eng 13(5):965

Li LX, Liu XH, Wang H, Wang L, Han B, Chang YQ, Ding J (2018) Molecular characterization and expression of NLRP10 in the antibacterial host defense of the sea cucumber (*Apostichopus japonicus*). Gene 675110–118

Li X, Xiong D, Ding G, Fan Y, Ma X, Wang C, Xiong Y, Jiang X (2019) Exposure to water-accommodated fractions of two different crude oils alters morphology, cardiac function and swim bladder development in early-life stages of zebrafish. Chemosphere 235423–433

Matozzo V, Marin MG (2010) The role of haemocytes from the crab *Carcinus aestuarii* (Crustacea, Decapoda) in immune responses: a first survey. Fish Shellfish Immunol 28(4):534–541

Moodley L, Arnberg M, Dunaevskaya E, Seear P, Bærheim L, Nilsen M, Ramanand SS, Øysæd KB, Westerlund S, Jonsson G et al (2018) Effects of low crude oil chronic exposure on the northern krill (*Meganyctiphanes norvegica*). J Experim Marine Biol Ecol 500120–131

Palazon A, Goldrath Ananda W, Nizet V, Johnson RS (2014) HIF transcription factors, inflammation, and immunity. Immunity 41(4):518–528

Paludan SR, Bowie AG (2013) Immune sensing of DNA. Immunity 38(5):870–880

Regoli F, Giuliani ME (2014) Oxidative pathways of chemical toxicity and oxidative stress biomarkers in marine organisms. Marine Environ Res 93106–117

Schroder K, Tschopp J (2010) The inflammasomes. Cell 140(6):821–832

Singer MM, Aurand D, Bragin GE, Clark JR, Coelho GM, Sowby ML, Tjeerdema RS (2000) Standardization of the preparation and quantitation of water-accommodated fractions of petroleum for toxicity testing. Mar Pollut Bull 40(11):1007–1016

Sun H, Zhou Z, Dong Y, Yang A, Jiang B, Gao S, Chen Z, Guan X, Wang B, Wang X (2013) Identification and expression analysis of two Toll-like receptor genes from sea cucumber (*Apostichopus japonicus*). Fish Shellfish Immunol 34(1):147–158

Whitehead A (2013) Interactions between oil-spill pollutants and natural stressors can compound ecotoxicological effects. Integr Comp Biol 53(4):635–647

Xu EG, Khursigara AJ, Magnuson J, Hazard ES, Hardiman G, Esbaugh AJ, Roberts AP, Schlenk D (2017) Larval red drum (*Sciaenops ocellatus*) sublethal exposure to weathered *Deepwater Horizon* crude oil: developmental and transcriptomic consequences. Environ Sci Technol 51(17):10162–10172

Part VII
Sustainable Urban Design and Healthy Living Environments

Chapter 40
Semantic Urban Streets: Exploring Perception and Streetscape Features of Urban Streets from Panoramic Imagery Using Deep Learning

Salma Shemy, Hany Ayad, and Dina SaadAllah

Abstract The spatial qualities of public spaces shape the livability and experience of urban environments. A well-designed urban environment encourages people to interact, leading to a sense of comfort, interest and staying behaviors. To understand why certain urban spaces are positively or negatively perceived, a set of physical and social factors needs to be evaluated through qualitative and quantitative methods. However, since many urban quality measurement tools do not combine both subjective and objective aspects, it is often difficult to accurately evaluate the quality of streets, and the level of regeneration needed. Recent advancements in various data sources and Machine Learning (ML) have offered new opportunities to investigate the relationship between urban design features and human perception of the built environment. This paper aims to systematically quantify the effect of streetscape design on the visual preference of people in the urban streets of Alexandria. The methodology integrates spatial mapping, panoramic imagery collection, and AI-driven urban scene segmentation for a precise analysis of the streetscape elements. It further investigates the influence of the detected urban elements on the perceived indices of comfort, safety, pleasurability, and accessibility. Results revealed significant positive associations between safety and sidewalk, and between comfort and vegetation, underscoring the value of integrating green and pedestrian infrastructure. Conversely, the negative correlations of car and road presence with multiple perceptual indices emphasize the need to reduce car-centric planning in favor of more human-centered design. This paper contributes to the studies of quantitative methods for linking physical urban features with human perception. It supports the

S. Shemy (✉) · H. Ayad · D. SaadAllah
Department of Architecture, Faculty of Engineering, Alexandria University, Alexandria, Egypt
e-mail: salma.shemy@alexu.edu.eg

H. Ayad
e-mail: hany.m.ayad@alexu.edu.eg

D. SaadAllah
e-mail: dina.saadallah@alexu.edu.eg

Y. Li et al. (eds.), *16th International Conference on Environmental Science and Development*, Environmental Science and Engineering,
https://doi.org/10.1007/978-3-032-15080-6_40

principles of urban sustainability by advancing strategies that prioritize walkability, green infrastructure, and user comfort, which are key components of healthier cities.

Keywords Semantic segmentation · Streetscape perception · Panoramic images

40.1 Introduction

In every city, public spaces are the vital elements that define how people live and experience their daily lives (Mehta 2007; Southworth 2014). In recent decades, there has been a growing recognition towards designing human-centered public spaces (Southworth 2014; Gehl 2013). This notion prioritizes the human experience in space, which is shaped by how people move, perceive and respond to their environments (Lynch 1964; Huang et al. 2023). Thus, a better understanding of how the physical environment influences the visual perception and spatial interaction is fundamental for a better quality of life and designing urban systems that are environmentally sustainable (Zhang et al. 2018). The studies of urban perception evaluation present some practical challenges. Conventional methods deal with costly data-collection processes such as questionnaires, face-to-face interviews, and field studies (Zhang et al. 2018). Recent advancements in digital technology have introduced innovative methods for collecting high-resolution, large-spatial scale urban scene imagery, to support urban perception studies. Publicly accessible street-view images (SVI) such as Google Street View (GSV) and Mapillary have been utilized to assess various aspects of urban settings, enabling data-driven strategies that support sustainable urban design (Aikoh 2023). Accordingly, image-based approaches are used to evaluate urban perception (e.g., safety, comfort, greenery), ranging from real-world street view images to simulated environments. This allows for the assessment of different urban design scenarios while minimizing external factors. Participants are usually presented with two images or scenes and asked to choose which one of them is superior (i.e. wins) for a given perception of interest (Wang et al. 2024). Some researchers have further recognized the importance of semantic information of visual elements (e.g. sidewalk, cars, trees) in these images used for perception evaluation (Qiu et al. 2023). Individuals interpret the scenes they encounter by combining them with their past experiences, resulting in distinct emotional reactions. Visual components can serve as representations of these previous experiences, influencing how people view urban settings. For example, when exploring a new city, individuals may lack direct memories of the area but can connect with recognizable visual features (such as traffic lights), which can affect their sense of safety. Some research has utilized semantic segmentation methods to analyze urban perception (Luo et al. 2022), examining how specific physical elements impact human reactions. In the global south, urban studies have rarely applied image-based perception with deep learning due to limited access to SVI. The lack of open datasets hinders both data collection and model training for such analyses.

The research aims to apply and contextualize a systematic and replicable method for evaluating the experiential quality of urban streets by examining how physical design influences user perception. It proposes an alternative method to manually capture panoramic imagery used in analysis to replace the lack of automated street-level imagery. This approach is used to overcome the problem of availability of open street view imagery, ensuring that such areas are not excluded from data-driven urban studies. Furthermore, it applies a refinement procedure using the Segment Anything Model (SAM) to enhance the segmentation outputs generated by the DeepLabV3+ model trained on Cityscapes classes. This is particularly beneficial for urban classification in contexts where training data specific to local streetscapes is unavailable. It finally determines how different view scenes influence the streetscape evaluation, by examining the impact of segmentation results from both perspective and panoramic images on correlation with visual perception. This methodology supports urban sustainability goals such as walkability, equitable access, greenery, and social well-being. It reflects a holistic view of cities as interconnected systems- where spatial quality, environmental performance and human experience are co-dependent. As such, the framework advances the field of urban ecosystem design, offering practical tools for creating cities that are livable and adaptive. The research questions are as follows:

- To what extent do certain streetscape features (from panoramic and perspective) correlate with visual perception indices?
- What is the spatial variation of perception indices and streetscape features among different urban streets in Alexandria?

40.2 Literature Review

40.2.1 Quantitative Research on Urban Streets

Quantitative analysis of street spaces is crucial for understanding their characteristics and improving the design and management of space. In recent years, research on urban streets evaluation has expanded significantly to incorporate a wider range of methodologies. Traditionally, assessment of street spatial quality relied on manual measurements and subjective evaluations, posing challenges for achieving precise analysis. Both Gehl (2013) and Mehta (2014) adopted similar frameworks to objectively categorize qualities of public spaces. They systematically evaluated the public space in terms of how good it achieved certain uses and social experiences. However, with ongoing advancements, an increasing number of scholars have integrated quantitative approaches into spatial assessment studies. Hillier and Hanson (1984) introduced Space Syntax, marking the beginning of quantitative research on two-dimensional street networks. Rooted in computer technology and graph theory,

Space Syntax enables the analysis of street networks by quantifying and visualizing accessibility, connectivity, and spatial topology, critical parameters for walkable and ecologically responsive urban areas. Data acquisition methods for urban analysis introduced the Point of Interest (POI) data to the scope of quantitative street evaluations. Researchers use Geographic Information System (GIS)-based methods, which rely on road networks and POI data, to quantitively assess streets (Sun et al. 2023). These methods may face some challenges in data processing, limiting their applicability for detailed urban studies.

To address this issue, deep learning techniques, particularly convolutional neural networks (CNNs), offer new possibilities for enhancing the precision of urban street analysis (Zhang et al. 2018; Liu et al. 2017). Deep learning offers applications in diverse domains, including facial recognition, traffic safety, and geographic research. A dominant application of deep learning within the realm of computer vision is image segmentation, which divides an image or video frame into multiple segments based on the distinctions between objects (Liu et al. 2017; Richard 2010). Lately, semantic segmentation has been largely utilized in the urban research field. Semantic segmentation emphasizes labeling each pixel in an image into a semantic category such as cars, buildings, or trees. Relevant work includes studying the built environment through segmentation of satellite and street view imageries. A mass number of worldwide street images has been made accessible by commercial services such as GSV and crowdsourcing platforms such as Mapillary and KartaView (Hou et al. 2024). Accordingly, numerous open street view imagery datasets have been created to support the development of algorithms for analyzing urban environments such as Cityscapes (Marius Cordts 2016), Mapillary Vistas (Gerhard Neuhold 2017), and KITTI-360 (Yiyi Liao and Geiger 2022). Cityscapes and Mapillary Vistas are key datasets for element-level sensing of street objects, with Cityscape providing 19 classes of pixels annotations, and Mapillary Vistas offering 65 classes of street-related dataset. Although 'Mapillary Vistas' dataset covers a broader geographic range, the diverse range of crowdsourced scenes means it contains many non-urban environments. On the other hand, Cityscape focuses on urban streets with precise urban elements labels, though it primarily captures daytime and clear weather conditions. Numerous ready-to-use models, including Pyramid Scene Parsing Network (PSPNet), SegFormer, and DeepLab, are available for image segmentation tasks on the available dataset. A recent widespread segmentation model is Meta's Segment Anything Model (SAM) with two available datasets SAM 1 and SAM 2. The models have been trained on a large dataset of scenes, which includes relevant urban environments and buildings' facades (Kirillov et al. 2023). However, since SAM provides generic masks of objects, it needs to be integrated with a semantic model to be able to label the required classes for research purposes. By enabling the extraction and classification of features such as vegetation, sidewalks, and traffic elements, these models support the creation of human centered streets. Their application in quantitative streetscape analysis contributes to the broader field of urban ecosystem design, offering tools for analyzing sustainability metrics such as thermal comfort and stormwater management.

40.2.2 Perception Evaluation and Subjective Street Quality Measurement

Assessing how people perceive their environment is crucial for urban design research, individual well-being, and overall quality of life (Liu et al. 2017). The factors influencing human perception are complex, often influenced by personal experiences and cultural contexts, making it challenging for researchers to fully understand perception. While traditional approaches like surveys, interviews, and cognitive maps, which include open-ended questions and ranking scales assessments, provide reliable insights, they are often time-consuming, expensive and yield limited data (Zhang et al. 2018). Additionally, response biases can further complicate the analysis, affecting the accuracy of findings. Recent search on public space assessment uses structured indices like the Public Space Index (PSI) (Mehta 2014) and the Public Space Quality Index (PSQI) for evaluating key qualitative aspects such as comfort, safety, and inclusivity. By integrating data from multiple sources, including field observations, participatory approaches, and geospatial analysis, they offer adaptable frameworks for diverse urban contexts. However, data collection remains an issue as it requires extensive field surveys and participatory approaches, which demand extensive financial and human resources. In many cases, data availability may be restricted, as key datasets such as pedestrian movement patterns or socio-demographic statistics may not be publicly accessible.

As cities face increasing pressure to become more resilient and environmentally adaptive, a deeper understanding of human perception becomes integral to advancing urban sustainability and ecosystem design. People experience urban spaces through their senses, primarily through sight, and images of a city serve as a valuable source of information. The development of computer vision and crowdsourcing efforts allowed the achievement of quantitative human perception on a large-scale. In 2013, the Place Pulse 1.0 crowdsourcing project was initiated by MIT Media lab (Salesses 2009). This project allowed volunteers to participate in an urban perception rating experiment across six perceptual attributes: safe, lively, boring, wealthy, depressing, and beautiful (Zhang et al. 2018). Such methods help cities become more sustainable by combining perception analysis with scalable digital tools. They achieve this by giving people useful ways to evaluate public spaces, use evidence-based design, and improve ecological outcomes.

40.3 Data and Methodology

40.3.1 Research Framework

This study contains four main phases: panoramic image collection, visual perception measurement, image segmentation, and correlation analysis (Fig. 40.1). For the image collection phase, an Insta360 nano camera was used to photograph panoramic images

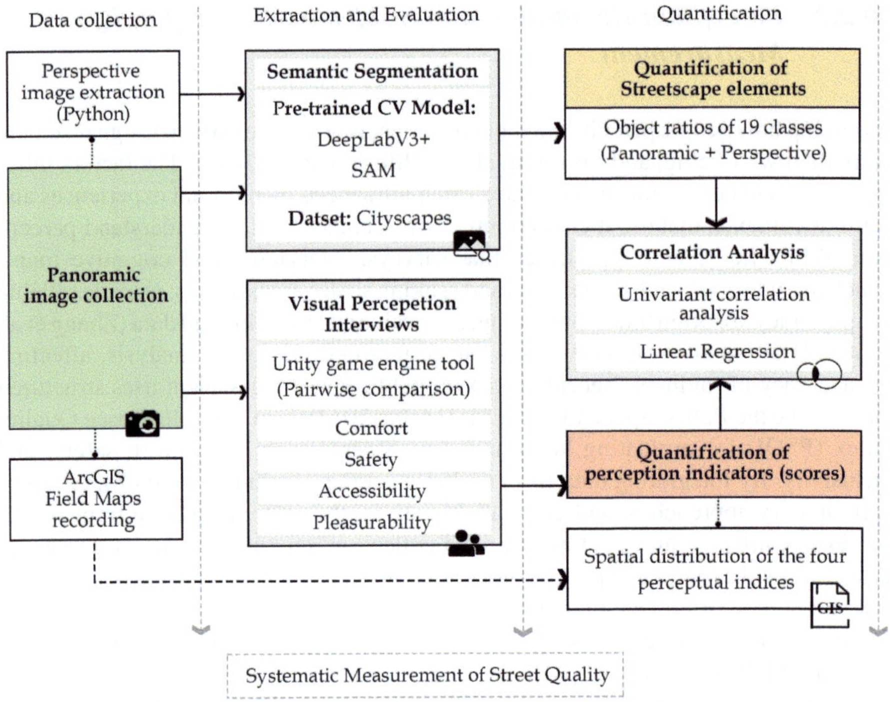

Fig. 40.1 Study framework

across different urban streets in Alexandria. The coordinates of each captured image were recorded using the ArcGIS Field maps application. For the visual perception measurement phase, a tool was developed using the Unity game engine to collect the perceptual ratings of four visual qualities across the captured images. For the image segmentation phase, an image semantic segmentation pipeline was implemented, utilizing Meta's Segment Anything Model and a DeepLabV3 + model pre-trained on the Cityscapes dataset. The pipeline produces the ratios between physical elements (e.g. sidewalk, building, car, etc.) for each image. Finally, univariant correlation analysis was performed between the visual quality scores and the different physical elements. Furthermore, the visual quality scores were visualized using the coordinates stored in ArcGIS to present the spatial variation in values.

40.3.2 Study Area and Data Collection

Alexandria's downtown was chosen as the site for investigation. The downtown district is comprised of a network of urban streets with different forms of goods trading and streetscape. The covered area is 4 km^2 with 20 streets and 110 streetscape

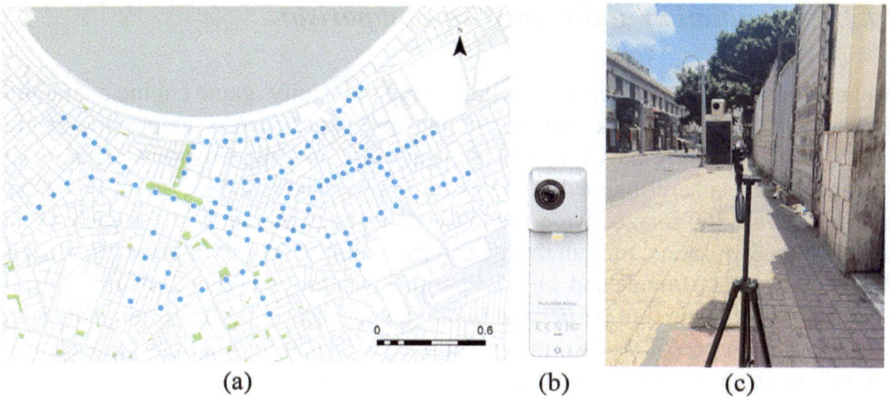

Fig. 40.2 **a** Map showing captured points' locations, **b** Insta360 nano camera, **c** Camera and tripod fixation in streets to capture dataset

points investigated (Fig. 40.2a). Within the study area, there is a vibrant mix of historical and modern architectural character enclosing the streets. Thus, it represents a good environment for future urban design and revitalization projects of commercial areas. The streets selected were similar in the urban street layout and road width ranging from 15 to 20 m. The coverage discontinuity of some paths occurred due to several reasons including ongoing street construction, the presence of religious buildings, or the blockage by informal public transportation. This consistent mapping seeks to systematically quantify the characteristics and streetscape of different urban streets.

40.3.3 Photographs Collection

Photographs were taken using the Insta360 nano camera (Fig. 40.2b) with a resolution of 3040 × 1520. For continuous coverage, the panoramic images were captured every 30–40 m to achieve a systematic representation of each street layout in the study area (Qiu et al. 2023). A total of 110 images were collected. The collection of images took place on clear days between May 7, and June 3, 2024, from 9 a.m. to 11 a.m. This way the images were unobstructed by heavy traffic and backlighting from high sun exposure. The average temperature was 25 °. For each sample point, the image was taken in the middle of the street to maintain the view angle as much as possible. This ensured the consistency of the visual experience and minimized the perspective distortion caused by nearby buildings. The height of the images taken was set to 1.6 m using a tripod to simulate the human viewpoint (Fig. 40.2c). To ensure the self-photographed images don't pose any ethical or data privacy concerns, all sensitive information collected during the field survey, such as faces and license plates, was obscured.

40.3.4 Evaluation and Pair-Wise Comparison

A panoramic street view scenery was built using the 'Unity' game engine containing the stored 110 images. The purpose of the 360 experience is to provide spatial coverage of the entire street environment at each investigated point. Thus, each 360-image was sphere-mapped in the game engine. The interface on unity was set up whereas all 360-images were randomly four-times paired, one for each of the indices: comfort, safety, pleasurability, and accessibility. Thirty raters from Alexandria University (19 female and 11 male), with a background related to urban design education, were invited to participate in the images' scoring. The participants were asked to choose one of the two presented images or identify them as equivalent in perception, in answer to one the following four questions: "which urban scene looks more X" where the X can mean "comfortable", "safe", "pleasurable", and "accessible". The Unity game was projected on a screen where each scene was labeled with the street's name. The raters were able to navigate the scene using the mouse cursors, the arrow keys were used to switch between the two images of each pair, space bar to choose a preferable scene, and 'N' for equal scenes. A total of 23,980 pair-wise comparisons were made across the four indices (5995 for each index). The results of the pair-wise comparison of images were converted into a score for each image using the method followed by Salesses (Salesses 2009). The score of each image was scaled using the ''favorable" and "unfavorable" ratios across all images with which it was compared. The favorable (F) and unfavorable (U) ratios of each image are defined using Eq. (40.1).

$$F_i = \frac{f_i}{f_i + u_i + n_i} \quad , \quad U_i = \frac{u_i}{f_i + u_i + n_i} \tag{40.1}$$

where f_i, u_i, n_i are the number of times an image was selected, not selected, or equalized to another image for each question respectively. From this, the W-score is defined for each image i using Eq. (40.2).

$$W_i = \frac{10}{3}\left(F_i + \frac{1}{f_i}\sum_{n=1}^{f_i} F_n - \frac{1}{u_i}\sum_{m=1}^{u_i} U_n + 1\right) \tag{40.2}$$

where W_i is the corrected ratio from 0 to 10 of each score, f_i is equal to the number of images i was selected over, and u_i is equal to the number of images that were selected over image i. Figure 40.3 shows the four perceptual scores converted from the pair-wise comparison results.

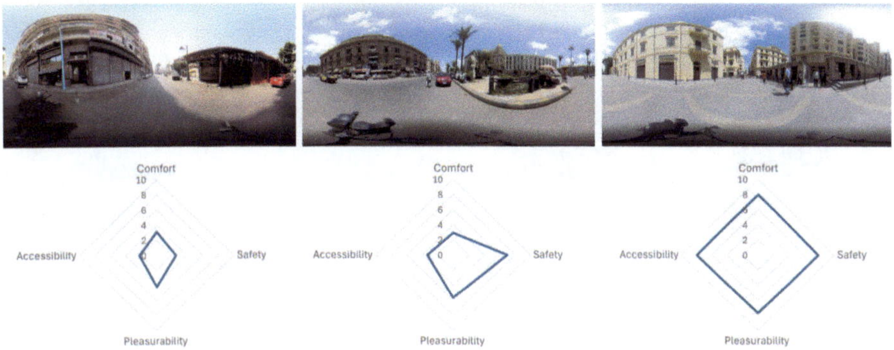

Fig. 40.3 Four scene examples and their converted perceptual indices scores: comfort, safety, pleasurability, accessibility represented using radar chart

40.3.5 Collection of Physical Features

To identify the physical elements in each streetscape point, a pre-trained DeepLabV3 + model obtained from GitHub was used. The model was trained on the Cityscape dataset and is able to perform pixel level semantic labeling for 19 streetscape classes: road, sidewalk, person, rider, car, truck, bus, train, motorcycle, bicycle, building, wall, fence, vegetation, sky, terrain, traffic light, and traffic sign (Fig. 40.4). Additionally, the SAM model by Meta was employed to refine the initial DeepLabV3 + segmentation output. SAM is particularly effective for detecting edges and objects' boundaries as it's a generalist segmentation model (Kirillov et al. 2023). The refinement procedure (Fig. 40.5) consists of iterating through the list of unlabeled masks produced by the SAM prediction and labeling each mask with the Cityscape semantic label, most commonly picked during initial segmentation across the pixels within the mask. The DeepLabV3 + and SAM predictions were executed on a Google Cloud Platform (GCP) runtime with an Nvidia L4 GPU. Once the DeepLabV3 + and SAM predictions were obtained, we were able to execute the refinement procedure locally on a PC (Windows 11, 16 Gb Ram, AMD Ryzen 7).

Minimal manual refinement is done as a final step for some images and consists of manually re-labeling some SAM masks incorrectly labeled by the refinement procedure. This segmentation procedure was applied to both panoramic and perspective images to analyze the difference in results. The perspective images were derived from the panoramic images using the equirectangular-to-perspective conversion equation implemented in Python. The field of view in the equation was set to 120, and each panoramic image was represented by two perspective images: front and back views (Fig. 40.6).

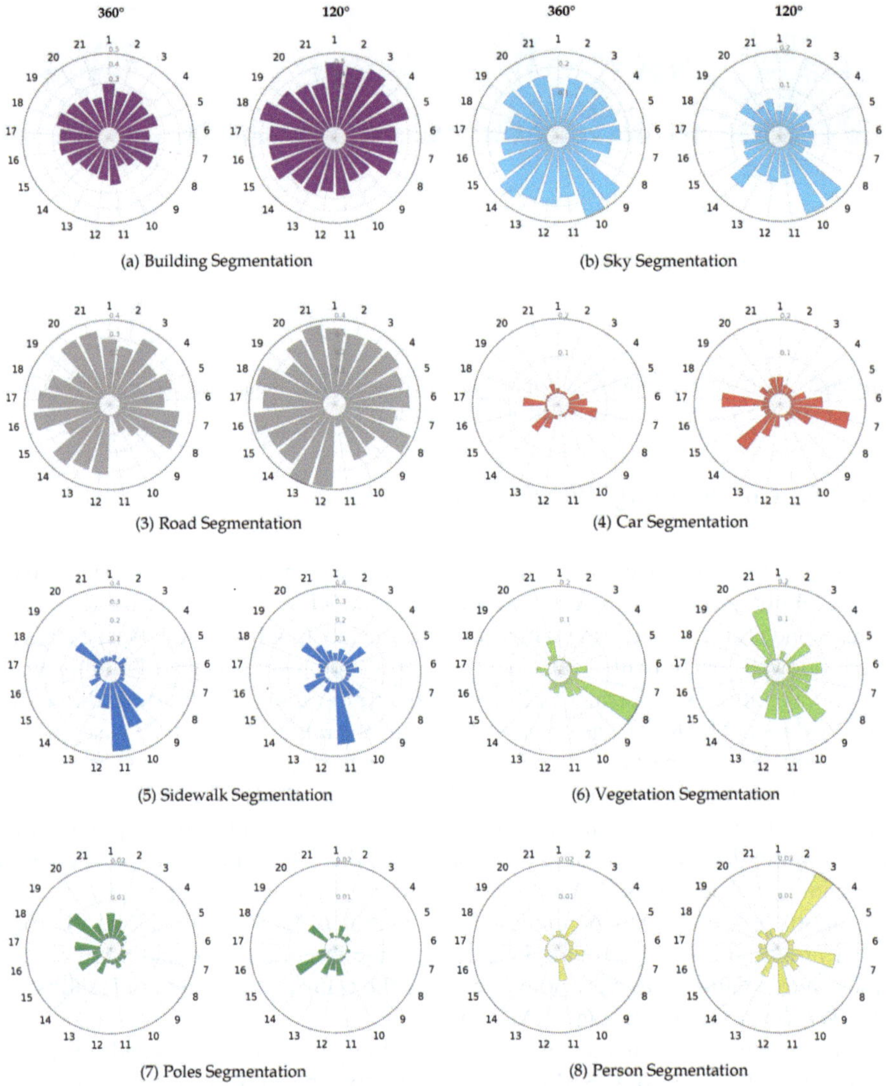

Fig. 40.4 Representation of the segmented 19 Cityscapes classes for panorama images (on left) and perspective images (on right) Rose-diagram of the segmented object ratios for panoramic and perspective datasets

Fig. 40.5 Illustration of the refinement procedure between Deeplabv3 + (bottom left) and SAM (upper left) segmentation results—for four urban streets

Fig. 40.6 Perspective images (front and back views) derived from the panoramic images

40.4 Results and Discussion

40.4.1 Perceptual Rating Results

An average of 600 responses were made by each of the 30 participants for the random pair wise comparison experiment. In this phase ArcGIS Pro 2.3.2 (Esri Inc.) was used to construct the visual spatial distribution and statistical analysis of the perception streetscape scores (Fig. 40.7). Table 40.1 shows the sum of low (<3), medium (3–7), and high (>7) scores of the four visual perception indices among 20 urban streets in Alexandria downtown. According to the perception score bands, the 20 streets can be classified as 8 high-score streetscapes (SUM > 25), 98 medium-score streetscapes (12 < SUM < 25), and 4 low-score streetscapes (SUM < 12), where (SUM) represents the total of the four perception scores. A total of 88% of the streets scenes achieved a combined score higher than 20, indicating that the majority of the participants rated the urban streets in Alexandria's downtown area positively. The average scores of comfort, safety, pleasurability, and accessibility are 4.5, 4.6, 4.8, and 4.8 respectively.

In particular, the highest combined scores for the four indices were 32 and 30 achieved at different sectors of Nabi Danial and Fouad streets respectively. The lowest scores for comfort and safety were 2 and 1.8 respectively, with scene 1 of France

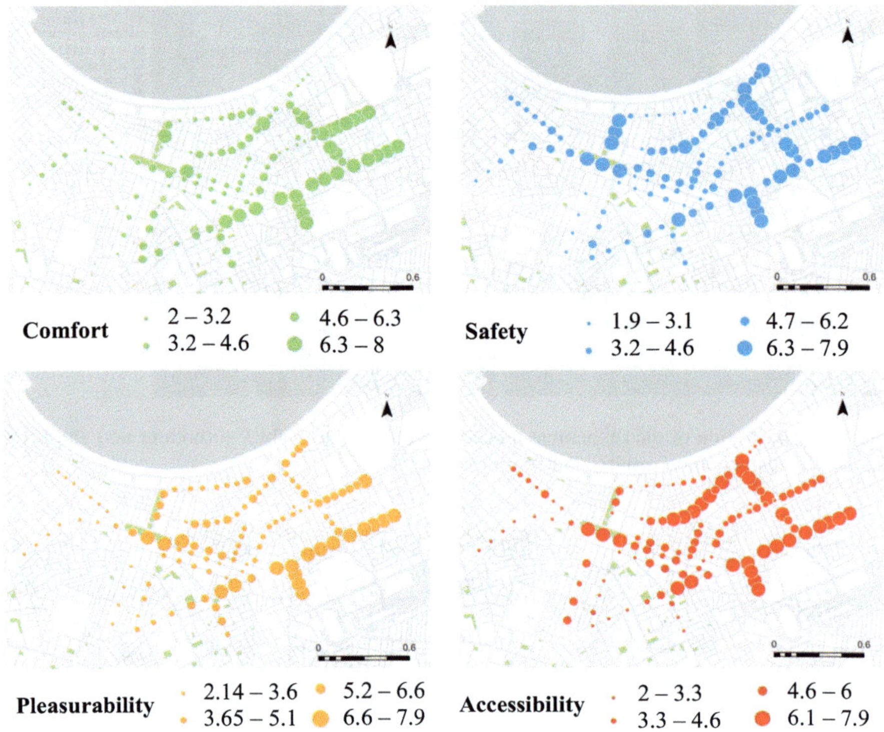

Fig. 40.7 Spatial distribution of the four perception indices scores represented using ArcGIS Pro

Table 40.1 Street scores for each of the four indices (L = low score; M = medium score; H-high score), total indices indicate the streets that are scored low, medium, or high in all four indices

Perceptual index	L	M	H
Comfort	12 streets	84 streets	14 streets
Safety	14 streets	86 streets	10 streets
Pleasurability	17 streets	78 streets	15 streets
Accessibility	16 streets	82 streets	12 streets
Total Indices	4 streets	98 streets	8 streets

street receiving the lowest score for comfort and scene 4 of El-Nasr Street for safety. As for pleasurability and accessibility, the lowest scores were 2.1 and 2 respectively, with scene 3 of El-Nasr Street receiving the lowest score for pleasurability and scene 4 of Ishaak El Nadeem street for accessibility.

40.4.2 Segmentation of Objective Elements Results

The pre-trained DeepLabv3 + model was initially applied to all study images (n = 110), and a set of the 19 Cityscapes segment classes were detected. However, since many segments were misclassified or undefined by clear object boundaries, SAM model was integrated to enhance the segmentation results. The outputs of the two models were overlapped and the object detected by SAM, which had the highest intersection with a specific Cityscapes class, was assigned to that class. Figure 40.8 shows the segmentation results before and after SAM integration, highlighting a noticeable disparity in the ratios between the two images. After calculating the 19 semantic segmentation classes in the resulting panoramic and perspective images from both models, it was discovered that only 8 classes were significantly detected in 98% of the images: sky, building, road, car, sidewalk, vegetation, person, and pole. For the panoramic results, the road represents the highest detected element in the images making 36.2% of the scenes, followed by building at 29% and sky at 21.3%. The segment ratios of the car, sidewalk, and vegetation are 2%, 8.8%, and 2% respectively. On the other hand, the classes: traffic sign, wall, fence, train, motorcycle, truck, and terrain were not detected in 80% of the images, while bus and bicycle were not detected in 90% of the images. As for the perspective images, the ratio for each element is calculated separately for both the front and back parts of the image. Then, the ratios are averaged by adding the front and back ratios and dividing by two. The perspective segmentation results showed that buildings accounted for the largest proportion of detected elements, appearing in 42% of the scenes, followed by roads at 32% and the sky at 9%. The segment ratios of the car, sidewalk, and vegetation are 4%, 7%, and 5% respectively. Rider and traffic light were the least detected classes.

Fig. 40.8 Deeplabv3 + segmentation results before SAM (on left), and after SAM (on right) integration, whereas (A) and (B) shows panorama and perspective results respectively

40.4.2.1 Validation of Segmentation Results

To estimate the impact of the refinement procedure, 30 images were randomly selected and manually annotated using CVAT.ai. The evaluation was conducted using the PASCAL VOC intersection-over-union (IoU) metric, calculated as IoU = TP/(TP + FP + FN), where TP, FP, and FN are the numbers of true positive, false positive, and false negative pixels, respectively, determined over a test set.. The raw DeepLabV3 + segmentation produced an IoU of 61% and the SAM refinement procedure with minimal manual changes produced an IoU of 94%.

40.4.3 Descriptive Statistics

To quantitatively explore the relationship between the perceptual indices and the 19 streetscape elements, a correlation coefficient analysis was conducted using the 110 street data points (n = 110). Subsequently, the significant elements from the univariant analysis were incorporated into a linear regression model. The significance of the regression results obtained was judged at the 5% level.

Panorama results. For the panorama images, the univariant correlation (Table 40.2) revealed that the score of comfort significantly increases with 'vegetation', 'sidewalk', 'person', and 'pole' ratios, achieving positive correlation of 0.328, 0.282, 0.196, and 0.208 respectively. This indicates that tree canopies and dedicated spaces for walking contribute to a more comfortable street environment. In contrast, the 'road' and 'truck' negatively correlate with the comfort index. As for the safety index, 'sidewalk' exhibits a strong positive coefficient of 0.521. Additionally, safety is positively associated with the presence of 'people' and 'pole', showing correlation coefficients of 0.211 and 0.196, respectively. In contrast, the presence of 'car' and 'road' negatively impacts safety scores, indicating they detract from the perception of safety. Similarly, accessibility shows positive correlation with 'sidewalk' (0.37), 'person' (0.222), and 'pole' (0.218). Like safety, accessibility is negatively correlated with the presence of cars and roads. Regarding the pleasurability index, the only notable positive correlation observed was with 'sidewalk' (0.221) and 'vegetation' (0.191). On the other hand, features like truck, car, and building demonstrated negative correlations with pleasurability, suggesting they reduce the overall interest of space (Mehta 2014). Figure 40.9 illustrates the results of univariant correlation analysis between the perception scores and the Cityscapes physical components. In the linear regression analysis, significant elements from the univariant correlation were used for each index. The model identified the key contributing factors for each index as follow: comfort was dominantly influenced by sidewalks ($\beta = 0.864$, $p = 0.04$), vegetation ($\beta = 0.361$, $p = 0$), and poles ($\beta = 0.221$, $p = 0.01$); safety was primarily associated with 'cars' ($\beta = -0.261$, $p = 0.007$); pleasurability was mostly linked to 'buildings' ($\beta = -0.226$, $p = 0.016$), cars ($\beta = -0.19$, $p = 0.043$); and accessibility was also predominantly affected by cars ($\beta = -0.299$, $p = 0.003$).

Table 40.2 Univariant correlation analysis between panorama ratios and visual perception scores where r is correlation coefficient, and standardized beta coefficient (*p < 0.05, **p < 0.005, ***p < 0.001)

	Comfort		Safety		Pleasurability		Accessibility	
	r	P value	r	P value	r	P value	r	P value
Sky	– 0.122	0.21	0.142	0.14	0.125	0.192	– 0.087	0.365
Building	– 0.105	0.279	– 0.082	0.396	– **0.255****	0.007	0.165	0.085
Road	– **0.271****	0.005	– **0.518*****	0	– 0.178	0.063	– **0.333*****	0
Car	– 0.144	0.139	– **0.308****	0.001	– **0.210***	0.028	– **0.391*****	0
Sidewalk	**0.282****	0.003	**0.521****	0	**0.221***	0.021	**0.370*****	0
Vegetation	**0.328****	0.001	0.034	0.722	0.165	0.086	– 0.061	0.525
Person	**0.196***	0.043	**0.212***	0.026	0.109	0.259	**0.222***	0.02
Pole	**0.208***	0.032	**0.196***	0.04	0.093	0.333	**0.218***	0.022
Train	– 0.147	0.13	– 0.114	0.236	– 0.078	0.417	0.048	0.62
Traffic sign	0.077	0.428	– 0.089	0.354	0.034	0.721	– 0.031	0.744
Motorcycle	– 0.117	0.23	– 0.125	0.193	– 0.135	0.158	– 0.045	0.639
Bus	– 0.006	0.955	– 0.064	0.506	– 0.109	0.256	– 0.078	0.419
Wall	0.034	0.73	0.046	0.631	– 0.017	0.858	0.025	0.792
Truck	– **0.199***	0.04	– 0.09	0.351	– **0.225***	0.018	– 0.116	0.226
Bicycle	– 0.007	0.947	– 0.017	0.86	– 0.104	0.278	– 0.095	0.323
Fence	– 0.053	0.588	– 0.015	0.874	– 0.023	0.811	– 0.143	0.137
Terrain	0.17	0.08	– 0.06	0.535	– 0.03	0.756	– 0.029	0.762

Perspective results. As for the perspective images results, the univariant correlation analysis (Table 40.3) revealed significant positive relationships between the comfort index and presence of 'vegetation' and 'sidewalk', with correlation coefficients of 0.478 and 0.461 respectively. Conversely, the 'road' and 'building' showed negative correlations with the comfort index. For the safety index, 'sidewalk', 'sky', and 'pole' displayed a strong positive correlation while roads and cars negatively impacted safety scores. Accessibility index only showed positive correlation with the sidewalk, and negative correlations with the road and car. For the pleasurability index, positive correlations were observed with 'vegetation' and 'sidewalk', while 'motorcycle', 'car', 'road', and 'building' negatively correlated. In the linear regression analysis, significant elements from the correlation were used to model each index. The analysis highlighted that comfort was primarily influenced by 'sidewalk' (0.596) and 'vegetation' (0.551). Safety was most strongly associated with 'sky' ($\beta = 0.222$, $p = 0.004$) and 'sidewalk' ($\beta = 0.527$, $p = 0.007$), pleasurability was

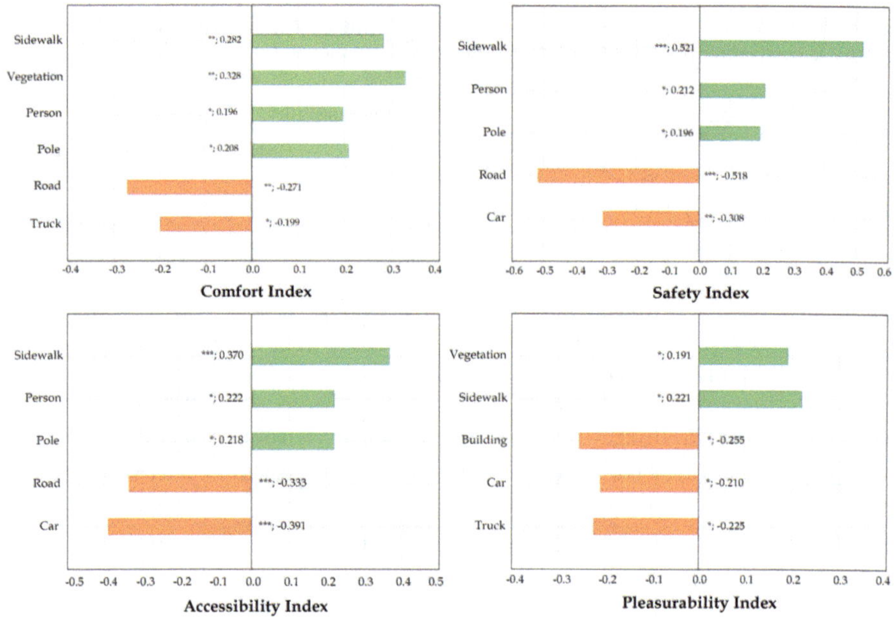

Fig. 40.9 Results of univariant correlation analysis between the perception scores and physical components standardized beta coefficient (*p < 0.05, **p < 0.005, ***p < 0.001)

affected by 'building' ($\beta = -0.295$, $p = 0.04$) and 'sidewalk' ($\beta = 0.654$, $p = 0$), and accessibility was mainly influenced by the presence of 'sidewalk' ($\beta = 0.687$, $p = 0$).

40.5 Discussion

The study offers valuable attempts to explore new perspectives in understanding urban environments through the use of low-cost data collection methods. The application of the panoramic-random pair wise comparison tool enabled a quantitative analysis of four urban perceptual indices. These indices are fundamental dimensions of sustainable urban design as they reflect human-centered qualities of urban spaces. For the pleasurability index, it was observed that participants preferred streetscapes with historic buildings and landmarks over streetscapes with greenery and limited architectural richness. As for comfort, scenes with more tree canopy, wider sidewalks, and less sky-sun exposure were preferred. For safety, active street points with higher pedestrian activity and more sidewalks were selected. For accessibility, scenes with well-maintained sidewalks and minimal pedestrians' obstacles were preferred over narrow obstructed street scenes. The average scores of 88% of the street scenes exceed 20 and the average score of each of the four indices is 4.5, suggesting that

Table 40.3 Univariant correlation analysis between perspective ratios and visual perception scores where standardized beta coefficient (*p < 0.05, **p < 0.005, ***p < 0.001)

	Comfort		Safety		Pleasurability		Accessibility	
	r	P value	r	P value	r	P value	r	P value
Sky	– 0.089	0.356	**0.269****	0.005	0.141	0.142	– 0.021	0.831
Building	– **0.358*****	0	– 0.155	0.107	– **0.331*****	0	0.062	0.522
Road	– **0.364*****	0	– **0.578****	0	– **0.314****	0.001	– **0.456*****	0
Car	– 0.067	0.485	– **0.279****	0.003	– 0.178	0.063	– **0.370*****	0
Sidewalk	**0.461*****	0	**0.644****	0	**0.446*****	0	**0.615*****	0
Vegetation	**0.478*****	0	0.088	0.358	**0.296****	0.002	– 0.008	0.93
Person	0.114	0.238	0.082	0.392	0.001	0.991	0.091	0.344
Pole	0.137	0.152	**0.232***	0.015	0.162	0.09	0.174	0.068
Train	– 0.107	0.265	0.036	0.71	– 0.008	0.935	– 0.043	0.654
Traffic sign	– 0.008	0.932	0.104	0.278	0.034	0.725	0.15	0.118
Motorcycle	– 0.163	0.089	– 0.172	0.072	– **0.194***	0.042	– 0.087	0.364
Bus	– 0.181	0.058	0.033	0.734	– 0.026	0.788	– 0.158	0.1
Wall	– 0.125	0.193	0.058	0.55	– 0.004	0.969	0.066	0.495
Truck	– 0.045	0.639	0.166	0.083	0.052	0.588	– 0.01	0.916
Bicycle	– 0.042	0.664	– 0.064	0.505	– 0.018	0.853	– 0.021	0.826
Fence	– 0.085	0.379	– 0.02	0.84	– 0.027	0.778	– 0.076	0.431
Terrain	– 0.089	0.356	**0.269****	0.005	0.141	0.142	– 0.021	0.831

most of the scenes were positively rated. Generally, the urban scenes with prioritized vehicular movement over pedestrian needs received lower scores. This highlights how visual-perceptual data can inform city planners to promote healthier and less car-centric urban environments, guiding an effective urban ecosystem design. It is important to note that the time required for corresponding with raters and conducting the random-pair-wise comparison was huge, given the large number of pairs. Therefore, more efficient methods could be used for vast dataset such as online survey platforms and crowd-sourcing techniques. The pre-trained DeepLabv3 + model was initially applied to all study images (n = 110) to detect the 19 Cityscapes elements. However, many areas were misclassified or undefined by clear object boundaries specifically in the sky, building, and road classes. The results, both visually and statistically, lacked coherence with the scenes, suggesting considerable inaccuracies will be revealed in the classes' ratios. Thus, the SAM model was integrated to enhance the segmentation results and the outputs of the two models were overlapped. The object detected by SAM, which had the highest intersection with a specific Cityscapes class, was assigned to that class. This pipeline was applied to both panoramic and

Table 40.4 Heatmap visualizing the correlation results of panorama images (A) and perspective images (B)

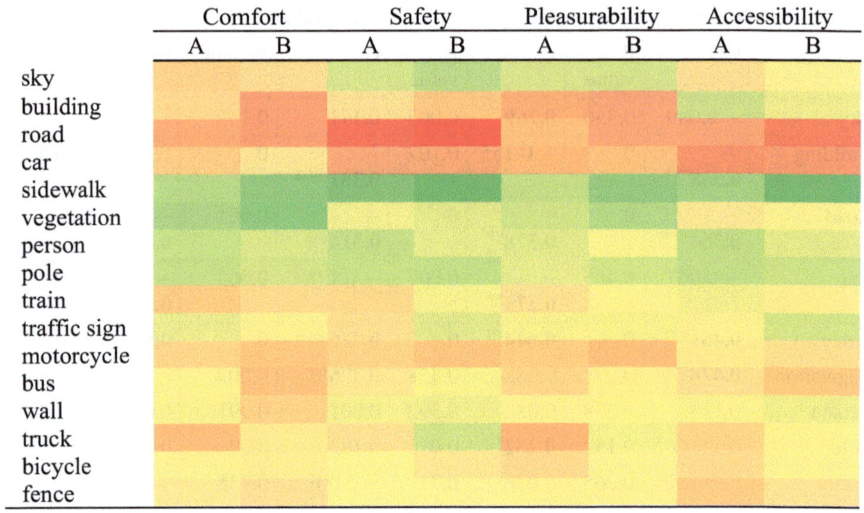

perspective images. For the perspective images, the final class-ratio is calculated by averaging the object ratios of both the front and back views of each image.

The most significant difference between the panorama and perspective segmentation results is visible in the sky and building ratios detected (Table 40.4). This variation offers insights into how visual data perspectives affect environmental interpretation. The 'sky' made 21.3 and 9% of the scenes for the panoramic and perspective images respectively, while the 'building' made 29 and 42% of the panoramic and perspective images respectively. This is attributed to the wider angle of 360-panorama that captures more sky ratio and less building ratio, compared to the perspective images configured with a 120-degree field of view. These distinctions underscore the value of using multiple imaging techniques to capture a fuller range of urban features affecting urban sustainability. Accordingly, the integration of perceptual assessment and deep learning image analysis offers a reliable tool for evaluating and advancing sustainable design. By identifying which streetscape elements are most valued by pedestrians and automating their large-scale classification.

40.6 Conclusion

In this study, a workflow was developed to quantify the built environment-perception relationship based on street-view panoramas and derived perspectives in Alexandria. Our approach proposed a new way to manually collect panoramic images as an alternative to GSV. This method addressed the issue of uncertain image timestamps in GSV by ensuring that all images were captured at a consistent time. The immersive experience offered by the developed panoramic tool to rate the images enhanced the visual experience, enabling a more accurate evaluation of the street scenes. According to the perception score bands, the 110 street points can be classified as 4 low-score scenes, 98 medium-score scenes, and 8 high-score scenes. The Cityscapes dataset was used to obtain the 19 streetscape classes for analysis. However, since the model wasn't trained on Alexandria's streets, it showed great inaccuracies in the initial results. This led us to develop the integrated DeepLabv3 + and SAM workflow to improve the accuracy and reliability of the segmentation results. Therefore, it would be more efficient to develop or train context-based datasets in future similar research. It was revealed that 'sidewalk', 'vegetation', 'people', and 'pole' had the highest positive influence on the selected visual perceptions (comfort, safety, pleasurability, and accessibility). The correlation results also indicated that the 'road', 'car', and 'truck' had the highest negative impact, while 'sky' and 'building' only had significant relationship with safety and pleasurability. These results underscore the value of pedestrian-oriented design and green infrastructure, which are core principles of urban sustainability and eco-system-sensitive planning. Also, the use of the panorama and perspective images allowed us to validate these correlation results and understand the selected urban scenes from both a focused view and larger 360-context. Although the research scope is limited, this study can provide researchers with new methods and techniques to measure the objective and subjective urban design qualities. The results and proposed framework can further be implemented by capturing a larger sample of scenes to generate a more diverse dataset.

References

Aikoh T (2023) Comparing conventional manual measurement of the green view index with modern automatic methods using google street view and semantic segmentation. Urban Forestry Urban Greening 80:127845

Gehl J (2013) Cities for people. Island Press, Washington

Gerhard Neuhold TO, Bulo SR, Kontschieder P (2017) The mapillary vistas dataset for semantic understanding of street scenes

Hillier B, Hanson J (1984) The social logic of space. The social logic of space. Cambridge University Press, Cambridge, pp 52–81

Hou Y et al (2024) Global streetscapes—A comprehensive dataset of 10 million street-level images across 688 cities for urban science and analytics. ISPRS J Photogramm Remote Sens 215:216–238

Huang J et al (2023) A collaborative perception method of human-urban environment based on machine learning and its application to the case area. Eng Appl Artif Intell 119:105746

Kirillov A, Mintun E, Ravi N (2023) Segment anything

Liu L et al (2017) A machine learning-based method for the large-scale evaluation of the qualities of the urban environment. Comput Environ Urban Syst 65:113–125

Luo J et al (2022) Semantic riverscapes: perception and evaluation of linear landscapes from oblique imagery using computer vision. Landsc Urban Plan 228:104569

Lynch K (1964) In: The image of the city. MIT Press

Marius Cordts MO, Ramos S, Rehfeld T (2016) The cityscapes dataset for semantic urban scene understanding

Mehta V (2007) Lively streets. J Plan Educ Res 27(2):165–187
Mehta V (2014) Evaluating public space. J Urban des 19(1):53–88
Qiu W et al (2023) Subjective and objective measures of streetscape perceptions: relationships with property value in Shanghai. Cities 132:104037
Richard S (2010) Computer vision: algorithms and applications
Salesses MP (2009) Place pulse: measuring the collaborative image of the city
Southworth M (2014) Public life, public space, and the changing art of city design. J Urban des 19(1):37–40
Sun H et al (2023) A spatial analysis of urban streets under deep learning based on street view imagery: quantifying perceptual and elemental perceptual relationships. Sustainability 15(20):14798
Wang Y et al (2024) Immersive virtual reality and computer vision for heritage: visual evaluation and perception of the industrial heritage sites along the Yunnan–Vietnam railway (Yunnan section). In: Heritage science, vol 12(1)
Yiyi Liao JX, Geiger A (2022) KITTI-360: a novel dataset and benchmarks for urban scene understanding in 2D and 3D
Zhang F et al (2018) Measuring human perceptions of a large-scale urban region using machine learning. Landsc Urban Plan 180:148–160

Chapter 41
Biophilic Design for Sustainability and Mental Health in Conflict Zones: A Framework and Case Study from Omdurman

Basma Sami Omer Ismail, Hamada Musatafa, Lixing Chen, and Yingzi Zhang

Abstract Conflict zones present severe challenges to mental health and sustainable development. This study proposes a theoretical framework utilizing biophilic design (integrating nature into built environments) as a low-cost strategy to address these interconnected challenges. Despite increasing evidence of nature's restorative effects in stable environments, there is a significant gap in how biophilic principles apply in high-stress, resource-limited conflict zones. The framework, grounded in the biophilia hypothesis, attention restoration theory (ART), and stress reduction theory (SRT), outlines physical, virtual, and socio-cultural interventions adaptable to conflict settings. The methodology employed a mixed-methods approach, combining a needs assessment survey ($n = 53$) in Omdurman, Sudan, followed by an eight-week real indoor plant intervention case study with office workers ($n = 8$). Data collection utilized standardized psychological measures (PSS, STAI) and physiological measurements (blood pressure), with pre-post comparative analysis to evaluate intervention effectiveness. Results from the needs assessment confirmed high levels of conflict-related stress, trauma (72% experienced displacement trauma), and limited nature access, yet a strong perceived need for nature connection. The intervention case study revealed significant improvements in employee well-being across multiple dimensions. Overall, biophilic interventions demonstrated substantial positive effects on participants' mental health and workplace functioning, with significant reductions in physiological stress markers and psychological distress measures. This study concludes that biophilic design offers a practical, sustainable approach for enhancing well-being in conflict-affected populations. We recommend implementing context-specific biophilic interventions as complementary strategies in humanitarian response and post-conflict reconstruction efforts, with particular

B. S. O. Ismail · H. Musatafa · L. Chen · Y. Zhang (✉)
School of Architecture, Southwest Jiaotong University, Chengdu, Sichuan, China
e-mail: zhangyingzi@swjtu.edu.cn

B. S. O. Ismail
School of Architecture, Omdurman Islamic University, Omdurman, Khartoum, Sudan

Y. Li et al. (eds.), *16th International Conference on Environmental Science and Development*, Environmental Science and Engineering,
https://doi.org/10.1007/978-3-032-15080-6_41

attention to cultural relevance, security considerations, and community engagement to maximize effectiveness and sustainability.

Keywords Biophilic design · Conflict zones · Mental health · Sustainability · Theoretical framework · Omdurman

41.1 Introduction

Armed conflict zones, such as Omdurman, Sudan, present significant challenges to human well-being, marked by chronic stress, trauma, and severely limited access to mental health resources. WHO (2023) states that conflict areas have mental disorder rates twice as high as non-conflict zones, with depression and PTSD rates at 15–20% in active conflict areas. At the same time, conflict also damages the environment and hinders sustainable development. Biophilic design, which thoughtfully integrates natural elements and processes into built environments, emerges as a promising, low-cost strategy to address these combined crises. Grounded in established theories like the biophilia hypothesis (innate human affinity for nature) (Wilson 1984), attention restoration theory (ART, nature restores cognitive function) (Kaplan and Kaplan 1989), and stress reduction theory (SRT, nature alleviates stress) (Ulrich 1991), biophilic design offers pathways to psychological restoration and resilience. While its benefits are documented in stable settings (Nieuwenhuis et al. 2014), its application in resource-scarce, high-stress conflict environments requires a tailored approach.

This paper fills a gap by proposing a theoretical framework for applying biophilic design specifically within conflict zones to enhance mental health while promoting sustainability. To validate its relevance, we present findings from a need's assessment survey conducted in Omdurman. Subsequently, we detail a case study implementing the framework through a real indoor plant intervention in an Omdurman office, evaluating its impact on employee well-being. Finally, we discuss the broader implications for sustainable development, particularly concerning Sustainable Development Goals (SDGs) 3 (good health and well-being) and 11 (sustainable cities and communities). See Fig. 41.1.

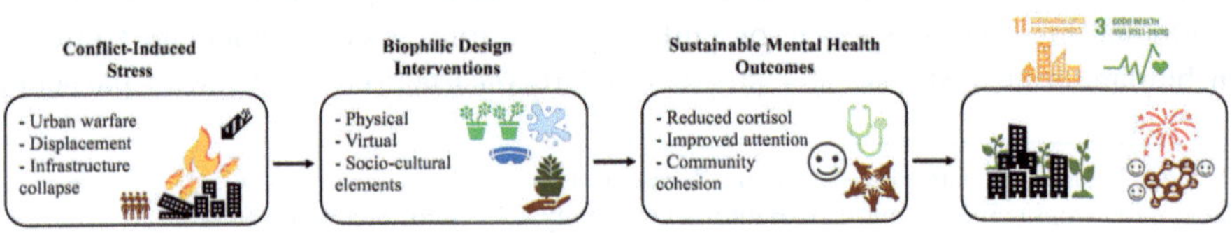

Fig. 41.1 Conceptual diagram linking conflict-induced, biophilic design, and sustainable outcome

41.2 Theoretical Framework and Methodology

41.2.1 Theories Foundation

The framework proposed in this study builds upon and extends several established theoretical foundations in environmental psychology and design. Previous frameworks by Kellert and Calabrese (2015) established biophilic design principles for general built environments, while Browning (2014) developed the 14 patterns of Biophilic Design framework categorizing nature interactions into direct experience, indirect experience, and spatial conditions. Similarly, Gillis and Gatersleben (2015) created a framework connecting biophilic elements to specific psychological benefits. However, these existing frameworks primarily address stable contexts with abundant resources, leaving a critical gap in understanding how biophilic design principles can be adapted for conflict zones with unique constraints. The extension of these frameworks to conflict settings is crucial for several reasons.

(1) Conflict zones present exceptional psychological stressors that require specialized interventions beyond conventional approaches.
(2) Resource limitations and security concerns in conflict areas necessitate adaptations to standard biophilic design practices.
(3) The cultural and contextual specificity of conflict zones demands a framework that integrates local knowledge and preferences.

Our framework addresses these gaps by synthesizing the Biophilia Hypothesis (Wilson 1984), ART (Kaplan and Kaplan 1989), and SRT (Ulrich 1991) with specific adaptations for conflict environments. This integrated theoretical framework identifies three interconnected components: physical, virtual, and socio-cultural elements. As in Fig. 41.2 each component addresses the unique constraints and opportunities present in conflict zones while advancing both mental health and sustainability goals. The framework's innovation lies in its context-specific approach that considers security limitations, resource scarcity, and cultural relevance while maintaining the core psychological mechanisms through which nature exposure benefits human well-being.

(1) Physical elements are direct integration of nature features like low-maintenance, local vegetation, water feature, and natural materials integrated into shelters, community spaces, and temporary structures, considering security, resource limits, and maintenance needs.
(2) Virtual elements are created using technology (such as nature images, soundscapes, virtual reality) to offer healing natural experiences when physical access is restricted by security or resource poverty.
(3) Socio-cultural components by stressing social cohesion, cultural relevance, and community involvement. This includes combining regional environmental knowledge, employing culturally important natural symbols, and designing communal green areas that promote group identity and joint restoration. The

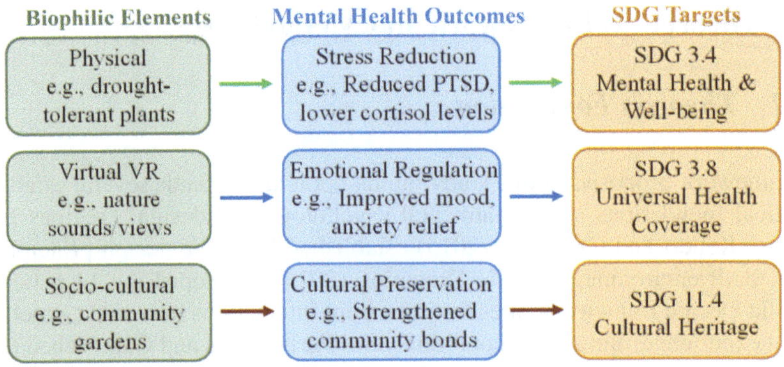

Fig. 41.2 Model mapping biophilic elements to mental health outcomes and SDGs

framework components aid SDG 3.4 (mental health), 3.8 (coverage), 11.7/11.b (green spaces), and 11.4 (heritage) via stress reduction, inclusive access, virtual nature, and cultural preservation, supporting marginalized communities.

41.2.2 Methodology

Data Collection Methods. This study employed a sequential mixed-methods approach. First, a needs assessment survey was conducted in Omdurman, Sudan, to establish the relevance of the theoretical framework in a real conflict setting. Subsequently, a case study implementing a biophilic intervention was carried out to evaluate the framework's effectiveness.

Target Respondents and Sampling Methods. For the needs assessment, 53 participants were recruited from Omdurman using purposive sampling to ensure representation across different demographics affected by conflict. The sample included 96% male participants with an average age of 31 $\pm$ 5 years and an average residency of 18 $\pm$ 2.5 months in the conflict zone. Participants were selected based on their exposure to conflict conditions and willingness to participate in the study. For the intervention case study, eight office workers from a local organization in Omdurman were recruited using convenience sampling. Selection criteria included full-time employment at the office, no pre-existing severe mental health conditions, and commitment to participate for the full eight-week intervention period.

Questionnaires and Measurement Instruments. The needs assessment utilized a structured questionnaire with both closed and open-ended questions covering: (1) Demographic information (age, gender, duration of residence in conflict zone) (2) Current living situation and environmental conditions. (3) Perceived stress and trauma exposure using adapted items from the Perceived Stress Scale (PSS) (Cohen et al. 1983). (4) Access to nature and green spaces. (5) Preferences for nature-based interventions. (6) Cultural significance of local plant species. For the intervention case study, the following standardized instruments were used: (1) Physiological measurements: Systolic and diastolic blood pressure (SBP, DBP) measured using Omron blood pressure monitor. (2) Psychological assessments: 4-item Perceived Stress Scale (PSS) (Cohen et al. 1983) and 4-item State Trait Anxiety Inventory (STAI) (Spielberger et al. 1971).

Intervention Design and Implementation. The intervention consisted of introducing indoor plants into the office environment for eight weeks. Plant selection was based on local availability, cultural significance, and low maintenance requirements. Plants were positioned to maximize visibility from workstations while ensuring they did not interfere with daily activities. Participants received basic care instructions but were not required to maintain the plants themselves.

Data Analysis Methods. For the needs assessment, descriptive statistics were calculated for quantitative data, while thematic analysis was applied to qualitative responses to identify patterns in participants' experiences and preferences. For the intervention case study, pre-post comparisons were conducted using paired t-tests to assess changes in physiological and psychological measures. Statistical significance was set at $p < 0.05$, with effect sizes calculated using Cohen's d to determine the magnitude of observed changes. All statistical analyses were performed using SPSS version 26.0.

41.3 Results and Discussion

41.3.1 Needs Assessment in Omdurman

To ground the theoretical framework and assess the need for biophilic interventions, a survey of 53 participants (96% male, aged 31 ± 5, average residency of 18 ± 2.5 months in conflict zone) highlighted the acute mental health and environmental challenges in Omdurman. The results are presented in a series of figures that illustrate different aspects of participants' experiences and needs.

Figure 41.3 illustrates the current living situation of participants. The data reveals that a significant proportion of respondents 28% are living in temporary shelters, while 22% reside in damaged housing. Only 28% reported living in intact housing, underscoring the severe impact of conflict on basic shelter security. This housing instability creates additional stressors that compound psychological distress and highlights the need for interventions that can be implemented in various living conditions, including temporary and damaged structures (Fig. 41.4). The results show that 58% of respondents are unemployed, 34% are employed, and only 8% student. This high unemployment rate reflects the economic disruption caused by conflict and contributes to financial stress, which can exacerbate mental health challenges.

Figure 41.5 depicts where participants spend most of their time. The majority 36% reported spending most of their time in their home, 23% in their shelters, while

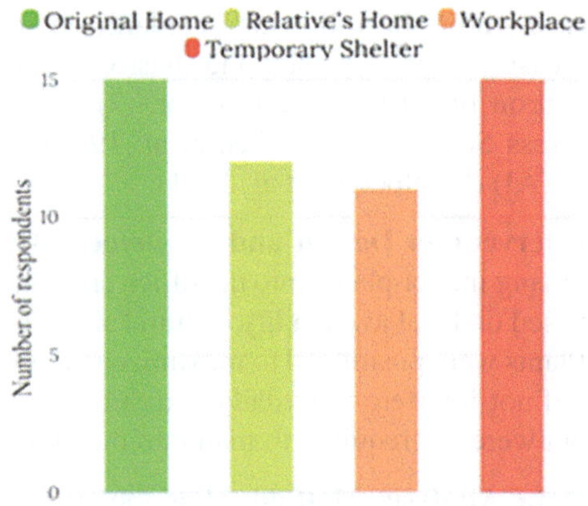

Fig. 41.3 Current living situation

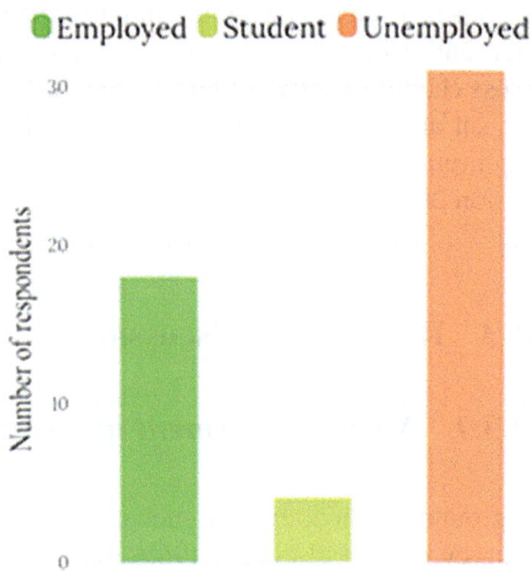

Fig. 41.4 Employment status

28% primarily occupy workspaces, and only 13% spend significant time in relative's home.

This starkly illustrated the conflict's impact by showing high stress and anxiety, 72% experienced trauma on displacement, with 87% describing daily life as "chaotic" or "unsafe" (Fig. 41.6), limited access to nature in participants' environments where only 13% had regular access to green spaces (Fig. 41.7).

In Fig. 41.8, the data shows strong correlations between perceived stress levels and limited nature access, and suggesting potential for social cohesion benefits alongside individual well-being improvements. Participants highlighted culturally plants, like gum arabic trees, acacia, date palms, flowers, and traditional herbs whereas their top preferences for interventions include small indoor plants, nature images and views, natural sounds, shared gardens, VR for nature in shelters, workplaces, homes, or courtyards reflecting security concerns. Data reveals differing community garden

Fig. 41.5 Spending most of the time

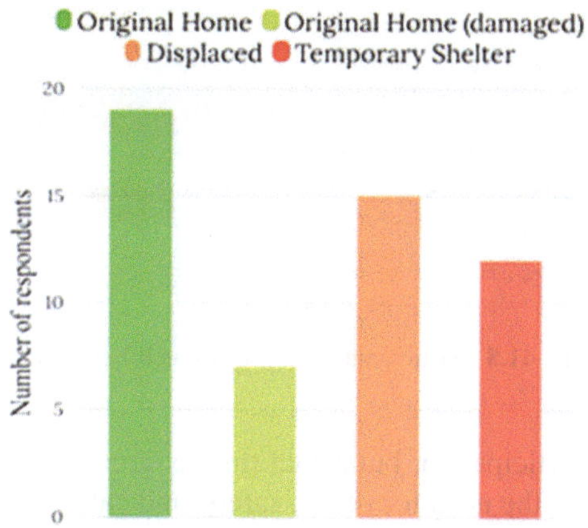

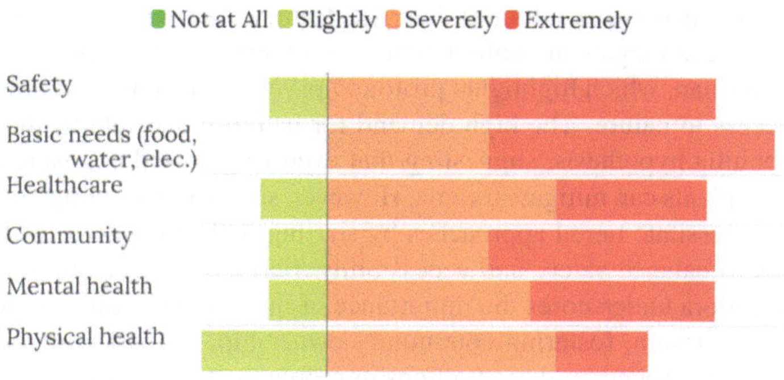

Fig. 41.6 The affection of conflict on daily life

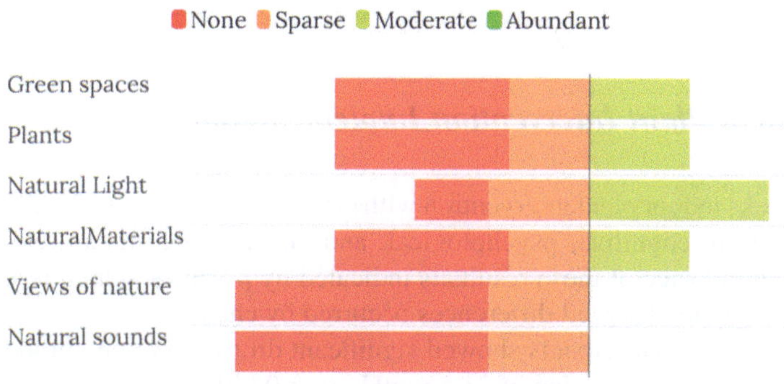

Fig. 41.7 The availability of nature in surrounding environment

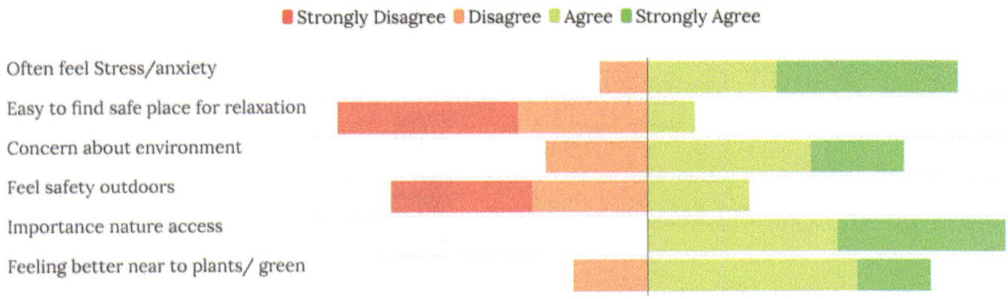

Fig. 41.8 Participant responses to well-being and environmental factors

participation, but social time is crucial for cohesion and personal restoration. Call for the need for peace, safety, and calm, citing issues like "hard to feel calm," "need to escape," "everything is chaos," and "safety is first".

These comprehensive results validate the framework's emphasis on context-specific biophilic strategies in conflict zones, as demonstrated by the needs assessment in Omdurman, which highlights profound psychological distress and severely restricted access to nature. The high demand for nature-based interventions aligns with the biophilia hypothesis, suggesting that even fragmented exposure to nature such as indoor plants can mitigate trauma. However, security constraints in areas like Omdurman necessitate tiered approaches by starting with small-scale and low-risk interventions to balance safety and accessibility. Furthermore, the cultural significance of local flora underscores the importance of integrating traditional ecological knowledge into design, fostering community ownership, resilience, and culturally resonant solutions. Together, these findings reinforce the framework's nature access and urgent need for adaptive, context-driven strategies to reconnect communities with nature.

41.3.2 Real Plant Intervention Implementation

An eight weeks indoor plant intervention with eight office workers yielded significant improvements in cognitive, psychological, and physiological measurements. The statistical significance of these results is indicated by p-values, which represent the probability that the observed differences occurred by chance.

Physiologically, participants showed significant drops in systolic blood pressure (SBP), with an average decline of 21.5 mmHg ($p < 0.001$) and diastolic blood pressure (DBP), falling by 20.5 mmHg ($p < 0.001$), as illustrated in Fig. 41.9. The $p < 0.001$ value here indicates an extremely low probability (less than 0.1%) that these substantial blood pressure reductions occurred by chance, providing strong evidence for the intervention's effectiveness. These results imply that direct relief of cardiac stress could come from exposure to surroundings incorporating nature, hence reducing risks linked with hypertension; a major issue in stressful urban environments. Psychologically, participants reported substantial reductions in perceived

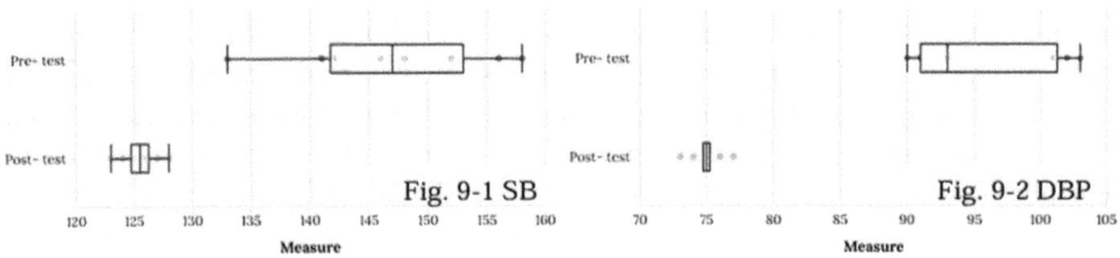

Fig. 41.9 Changes in physiological measurements (blood pressure)

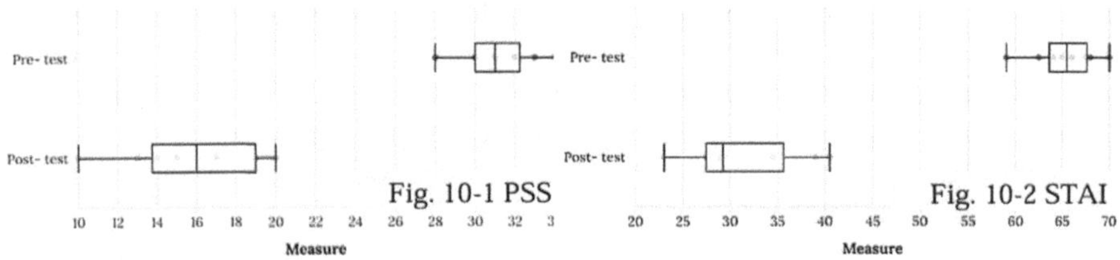

Fig. 41.10 Changes in psychological measurements (PSS, STAI)

stress levels, as measured by the 4-item perceived stress scale (PSS) (Cohen et al. 1983), with scores decreasing by 15.3 points ($p < 0.001$). Likewise, state anxiety scores based on the 4-item state-trait anxiety inventory (STAI) (Spielberger et al. 1971) dropped by 34 points ($p < 0.001$), as shown in Fig. 41.10. These findings highlight how biophilic interventions can reduce acute stress reactions and promote emotional resilience even in settings often defined by persistent stressors.

Cognitively, the interventions enhanced workplace well-being and productivity. Job satisfaction scores improved by 1.9 items ($p < 0.001$) on a standardized scale, while self-reported concentration at work increased by 2.1 items ($p < 0.001$). These results fit attention restoration theory (ART), which holds that environments rich in nature replenish cognitive resources exhausted by extended mental tiredness.

The observed reductions in physiological stress markers (SBP, DBP) and psychological distress (PSS, STAI) are substantial and align with ART (Kaplan and Kaplan 1989) and SRT (Ulrich 1991) and recent research in other settings. For example, a study in dental clinics also found that indoor plants contributed to lower STAI scores and physiological relaxation (Elsadek et al. 2023), while a study highlighted the potential of even small desk plants to reduce office workers' stress (Toyoda et al. 2020). Further supporting these findings, a scoping review confirmed generally beneficial associations between indoor plants and mental health, including reduced stress and depressive symptoms (Han et al. 2022). Meta-analyses also revealed significant benefits of indoor plants on diastolic blood pressure and cognitive function (Han 2024). Although our study was conducted in a high-stress conflict environment, the findings regarding stress reduction and cognitive improvement echo the benefits observed in more stable workplace or healthcare settings. This consistency across contexts suggests that biophilic strategies represent a scalable, non-pharmacological public health tool that can enhance holistic well-being and productivity even under challenging circumstances. Importantly, biophilic interventions typically have a lower environmental footprint and resource requirement compared to traditional mental health infrastructure, making them a particularly sustainable option in resource-constrained conflict settings.

The intervention results demonstrate that even modest biophilic elements can produce measurable improvements in well-being across multiple dimensions. The consistency of positive outcomes across physiological, psychological, and cognitive measures, coupled with the strong statistical significance ($p < 0.001$) of all findings, provides compelling evidence for the framework's effectiveness in real-world conflict settings. These results suggest that biophilic design principles can be successfully adapted to address the unique challenges of conflict zones while maintaining their core psychological benefits.

Crucially, this approach strongly supports sustainable development objectives: SDG 3 (good health and well-being); by providing accessible, non-clinical mental health support, biophilic interventions directly contribute to target 3.4 (promote mental health and well-being) which offers a low-resource method to mitigate stress and build resilience in vulnerable populations, and SDG 11 (sustainable cities and communities); by integrating biophilic design into reconstruction and temporary settlements supports target 11.7 (provide universal access to safe, inclusive and accessible, green and public spaces). Using local, resilient plant species promotes biodiversity (target 11.4) and can contribute to climate adaptation (passive cooling, target 11.b). The socio-cultural component fosters community engagement and strengthens social fabrics, contributing to inclusive and resilient communities.

However, limitations such as the case study´s small sample size (due to the ongoing war disruptions and limited companies work). Nevertheless, between the need assessment, the intervention outcomes and findings from recent comparative studies (Elsadek et al. 2023; Han 2024; Han et al. 2022; Toyoda et al. 2020), provides compelling evidence for the framework´s potential.

41.4 Conclusion

This study has advanced our understanding of how biophilic design principles can be effectively adapted and applied in conflict zones to support mental health and sustainability. Through the development of a comprehensive theoretical framework (Fig. 41.11), needs assessment validation, and empirical testing via a case study intervention, we have demonstrated the viability and impact of this approach in one of the world's most challenging contexts. The integrated findings from our research reveal that biophilic design offers a practical, accessible, and culturally adaptable strategy for addressing the dual challenges of psychological distress and environmental degradation in conflict-affected areas. The framework's three-component

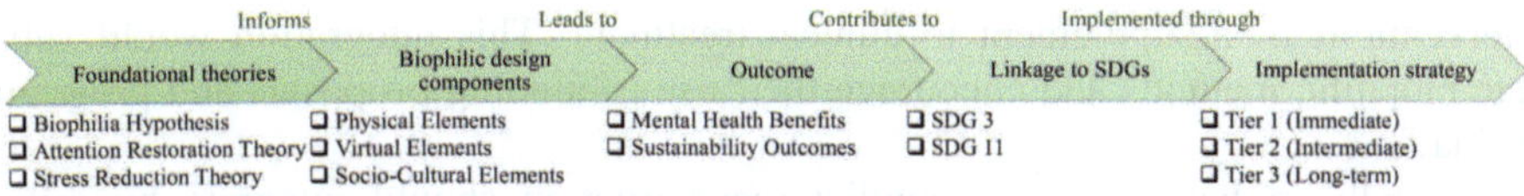

Fig. 41.11 Biophilic design framework for conflict zones

structure: physical, virtual, and socio-cultural, provides flexibility in implementation while maintaining the core psychological mechanisms through which nature exposure benefits human well-being.

Our research contributes to the broader field by bridging previously disconnected domains of environmental psychology, conflict studies, and sustainable development. The successful application of biophilic principles in Omdurman demonstrates that even in resource-constrained, high-security-risk environments, thoughtfully designed nature-based interventions can yield significant benefits for individual and community resilience. The implications of this work extend beyond immediate mental health improvements to include potential long-term benefits for social cohesion, place attachment, and environmental stewardship critical factors in post-conflict recovery and sustainable development. By integrating local ecological knowledge and cultural preferences, biophilic interventions can simultaneously honor traditional relationships with nature while introducing innovative approaches to well-being.

Future research should explore the scalability of these interventions across different conflict contexts, investigate longer-term outcomes, and develop implementation guidelines for humanitarian organizations. Additionally, examining the potential for biophilic design to contribute to conflict resolution through shared environmental stewardship represents a promising direction for further inquiry. In conclusion, this study provides compelling evidence that biophilic design offers a valuable tool for humanitarian aid, post-conflict reconstruction, and advancing sustainable development in the world's most challenging contexts. By recognizing the profound connection between human well-being and the natural environment, even in settings where this connection has been severely disrupted, we can develop more holistic, effective, and sustainable approaches to supporting conflict-affected populations.

41.5 Sustainable Development Recommendations

41.5.1 Policy Recommendations

The framework significantly impacts policy-making in conflict and post-conflict areas. Firstly, international humanitarian organizations should include biophilic design principles in mental health support guidelines and recognize nature-based interventions as a supplement to clinical treatment. This integration would validate biophilic methods and encourage their use in funding proposals and program evaluations.

Secondly, policies for post-conflict reconstruction should integrate biophilic design in public infrastructure, especially in schools, hospitals, and community centers. These criteria may include minimum standards for indoor plants, natural light, views, and green spaces. Integrating biophilic principles into reconstruction helps with psychological and physical redevelopment.

Third, formalizing cross-sector collaborations among conservation, mental health, and urban planning is essential for coordinating biophilic initiatives. These collaborations may use diverse funding, technical expertise, and implementation skills to create programs addressing mental health and environmental sustainability goals together.

41.5.2 Design Guide

Applying theory to practice requires context-specific design guidelines. In conflict zones, directives should focus on cost-effective, low-maintenance, culturally relevant, adaptable interventions. The framework suggests starting with minimal interventions and expanding as resources and stability permit.

Tier 1 interventions focus on affordable, easily accessible elements like desktop plants, nature photography, ambient nature sounds, and sensory gardens even in unstable settings. Tier 2 interventions, depending on security and resources, involve adding significant biophilic elements like green walls in community centers, therapeutic gardens in healthcare, and nature play areas in schools. Tier 3 interventions show full biophilic integration in post-conflict rebuilding, using biophilic urban design, green infrastructure networks, and merging indoor and outdoor public spaces (Fig. 41.12).

41.5.3 Community Engagement

Effective framework implementation requires strong community involvement in design and maintenance. In conflict zones with complex community dynamics, participatory methods ensure biophilic interventions are culturally relevant, inclusive, and sustainable. The framework proposes creating community design committees with diverse stakeholders like mental health professionals, environmental experts, cultural leaders, and marginalized group representatives. These committees can identify locally significant natural elements, prioritize sites for intervention, and create maintenance strategies to promote community ownership. Training local community members in biophilic maintenance can create sustainable care systems and offer livelihood opportunities.

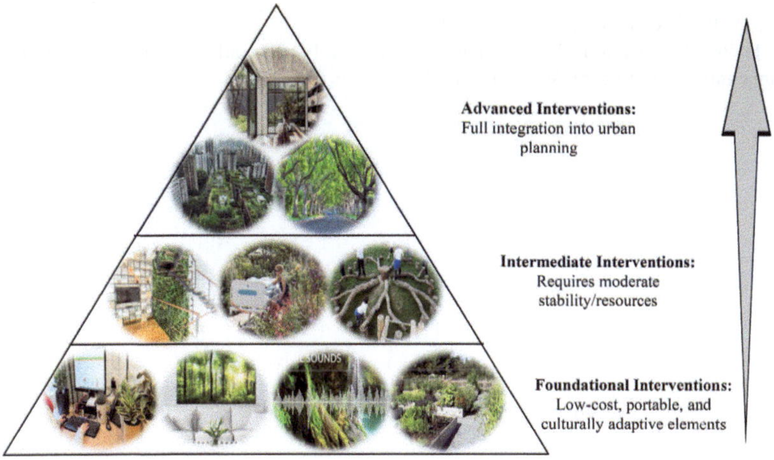

Fig. 41.12 Tiered intervention framework

Acknowledgements This research was funded by National Key Research and Development Program (No.2022YFC3802702) and Southwest Jiaotong University International Graduate Education and Teaching Reform Project (No. GYJG[2023]Y16).

References

Browning W (2014) 14 patterns of biophilic design. Terrapin Bright Green, LCC

Cohen S, Kamarck T, Mermelstein R (1983) A global measure of perceived stress. J Health Soc Behav 385–396

Elsadek M, Xie J, Jie G (2023) The impact of indoor plants on patient recovery: physiological and psychological effects in dental clinics. HortScience 58(11):1376–1382

Gillis K, Gatersleben B (2015) A review of psychological literature on the health and wellbeing benefits of biophilic design. Buildings 5(3):948–963

Han K-T (2024) Effects of indoor plants on well-being. In: Frontiers in psychology, vol 15. Frontiers Media SA, p 1483441

Han K-T, Ruan L-W, Liao L-S (2022) Effects of indoor plants on human functions: a systematic review with meta-analyses. Int J Environ Res Public Health 19(12):7454

Kaplan R, Kaplan S (1989) The experience of nature: a psychological perspective. Cambridge University Press

Kellert S, Calabrese E (2015) The practice of biophilic design. London: Terrapin Bright LLC 3(21):2021–2029

Nieuwenhuis M, Knight C, Postmes T, Haslam SA (2014) The relative benefits of green versus lean office space: three field experiments. J Exp Psychol Appl 20(3):199

Spielberger CD, Gonzalez-Reigosa F, Martinez-Urrutia A, Natalicio LFS, Natalicio DS (1971) The state-trait anxiety inventory. Revista Interamericana de Psicologia/Interam J Psychol 5(3&4)

Toyoda M, Yokota Y, Barnes M, Kaneko M (2020) Potential of a small indoor plant on the desk for reducing office workers' stress. HortTechnology 30(1):55–63

Ulrich RS (1991) Effects of interior design on wellness: theory and recent scientific research. J Health Care Interior Des 3:97–109

Wilson EO (1984) Biophilia. Har, Cambridge, MA
World Health Organization (2023) Mental health of refugees and migrants: risk and protective factors and access to care. World Health Organization

Chapter 42
Investigating the Public Mental Health Implications of Green Infrastructure Through a Neuro-urbanism Approach: An Overview and Bibliometric Analysis

Dalia Shokry Ganzory, Dina Sameh Taha, Jaidaa Mekky, and Manar Talaleldin Mohammed

Abstract Green infrastructure plays a crucial role in enhancing public mental health by fostering restorative urban environments. This study examines the literature on neuroscientific methods and mobile technologies related to green infrastructure within a neuro-urbanism framework, with a particular focus on mental health implications. A comprehensive search of 686 peer-reviewed articles from the Web of Science, PubMed, and SCOPUS identified 51 empirical studies published between 2000 and 2025. Analysis using VOS Viewer uncovered key themes, influential authors, countries, and significant research trends through co-citation analysis, term co-occurrence, and bibliographic coupling. The findings confirm that urban green spaces are essential for mental well-being, supported by neuroscientific evidence, and underscore the need to integrate green infrastructure into urban planning to create healthier, more inclusive, and resilient cities. This review also highlights emerging research directions and methodological advancements, advocating for interdisciplinary collaboration to maximize the mental health benefits of green infrastructure in urban planning and policy.

Keywords Neuroscience · Human experience · Green infrastructure · Neuro-urbanism · Bibliometric analysis · VOS viewer

D. S. Ganzory (✉) · D. S. . Taha · M. T. Mohammed
Department of Architectural Engineering, Faculty of Engineering, Alexandria University, Alexandria, Egypt
e-mail: Dalia.alganzory@alexu.edu.eg

J. Mekky
Department of Neurology, Faculty of Medicine, Alexandria University, Alexandria, Egypt

Y. Li et al. (eds.), *16th International Conference on Environmental Science and Development*, Environmental Science and Engineering,
https://doi.org/10.1007/978-3-032-15080-6_42

42.1 Introduction

Numerous studies have examined the relationship between human well-being and urban natural green spaces, particularly concerning their impacts on child development, physical and mental health, social relationships, and climate regulation (Haluza et al. 2014; Chawla 2015; Sarkar et al. 2014; Debener et al. 2012a; Wolch et al. 2014). However, there is a notable lack of focused research on urban green spaces, especially given the rapid pace of urbanization—currently, over half of the global population resides in urban areas, a figure projected to increase to two-thirds by 2050 (United Nations 2018).

This review aims to address these gaps by emphasizing the significance of experimental, quasi-experimental, and longitudinal studies, which are more effective in tracking changes over time. This emphasis is particularly pertinent for policymakers and urban planners seeking to incorporate mental health considerations into sustainable urban development (Corburn 2004; Branas and MacDonald 2014).

42.2 Materials and Methods

This systematic review aimed to identify, evaluate, and synthesize evidence on the health impacts of urban green infrastructure interventions.

42.2.1 Creating Database

A systematic review evaluated the health impacts of urban green infrastructure interventions. In January 2025, literature searches were performed across multiple databases, including Web of Science (WOS), PubMed, and Scopus. The review was limited to English-language, peer-reviewed articles involving human subjects published between January 2000 and February 2025. Each database search, as shown in Table 42.1, used a tailored strategy with relevant keywords and Boolean operators to capture interdisciplinary studies. The final dataset was compiled after removing duplicates and irrelevant papers, as illustrated in Fig. 42.1, the PRISMA flow diagram.

42.2.2 Literature Review Data Extraction and Coding

Review data collected from each study included: citation information (author(s), date published, sample information (number of participants), method details (mobile technology details/information, neuroscientific method details/information), and mental health outcome.

Table 42.1 Queries used for searching the data

Databases	Keywords
Web of science (WOS)	TS = ("mental health" OR "psychological well-being" OR "mental well-being" OR "stress" OR "emotional well-being" OR "quality of life" OR "subjective well-being") **AND** TS = ("urban environment" OR "urban areas" OR "city" OR "urban planning" OR "built environment" OR "urban spaces" OR "environmental factors") **AND** TS = ("neuroscience" OR "neuroimaging" OR "brain activity" OR "EEG" OR "fMRI" OR "neurofeedback" OR "neuropsychology")
PubMed	("mental health" OR "psychological well-being" OR "mental well-being" OR "stress" OR "emotional well-being" OR "quality of life" OR "subjective well-being") **AND** ("urban environment" OR "urban areas" OR "city" OR "urban planning" OR "built environment" OR "urban spaces" OR "environmental factors") **AND** ("neuroscience" OR "neuroimaging" OR "brain activity" OR "EEG" OR "fMRI" OR "neurofeedback" OR "neuropsychology") **AND** ("mobile technology" OR "mobile health" OR "smartphone" OR "wearable technology" OR "digital health" OR "mHealth" OR "health apps" OR "technology interventions")
Scopus	TITLE-ABS-KEY("mental health" OR "psychological well-being" OR "mental well-being" OR "stress" OR "emotional well-being" OR "quality of life" OR "subjective well-being") **AND** TITLE-ABS-KEY("urban environment" OR "urban areas" OR "city" OR "urban planning" OR "built environment" OR "urban spaces" OR "environmental factors") **AND** TITLE-ABS-KEY("neuroscience" OR "neuroimaging" OR "brain activity" OR "EEG" OR "fMRI" OR "neurofeedback" OR "neuropsychology") **AND** TITLE-ABS-KEY("mobile technology" OR "mobile health" OR "smartphone" OR "wearable technology" OR "digital health" OR "mHealth" OR "health apps" OR "technology interventions")

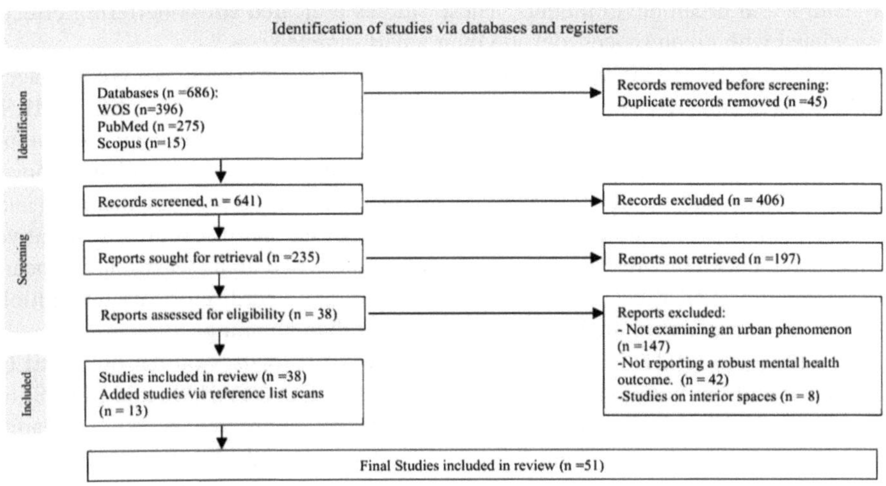

Fig. 42.1 The PRISMA flow diagram

42.2.3 Analysis Using VOS Viewers

Various visualization tools have been used for science mapping and bibliometric analysis to reveal relationships among authors, journals, institutions, countries, and key concepts. This study employed VOS Viewer for creating bibliometric networks. The analyses included co-occurrence, bibliographic coupling by country, and co-citation to identify leading contributors and influential publications.

42.3 Results

Fifty-one studies (published between 2000 and 2025) were included in the review.

About half of the studies (n = 26) were published in the last five years, suggesting an emerging trend of interest in the topic.

42.3.1 Data Extraction

Data extraction was performed in February 2025 on reviewing complete studies, including the abstract, conclusion, and future directions. The data were extracted and categorized into three time periods: 2000–2015, 2016–2020, and 2021–2025.

Between 2000 and 2015, as illustrated in Table 42.2, studies utilizing mobile neuro-technologies such as NIRS and EEG assessed psychophysiological responses to natural and urban environments. These studies indicated stress-buffering effects associated with green spaces, despite their small sample sizes.

Between 2016 and 2020, as shown in Table 42.3, research on urban green spaces and mental health increasingly utilized neurophysiological tools such as EEG, NIRS, fMRI, and POMS to measure real-time responses. These studies consistently demonstrated that natural environments within urban settings reduce stress, enhance mood and cognition, and promote relaxation. Despite employing various methods and modest sample sizes, the research indicated that green spaces produce restorative effects, evidenced by biological markers such as increased alpha waves, lower heart rates, and improved affect. This trend reflects a growing interdisciplinary integration of neuroscience, environmental psychology, and urban planning.

The third era (2021–2025), as listed in Table 42.4, focuses on the use of EEG combined with multimodal approaches (e.g., fNIRS, EDA, PANAS, and POMS) to assess emotional and cognitive reactions in everyday environments. Most research targets urban green areas, contrasting their effects with those of built or indoor environments. Consistent results indicate that exposure to greenery promotes relaxation, reduces stress levels, improves mood, and enhances cognitive functionality. Small green spaces and thoughtfully designed environments significantly

Table 42.2 Summary of first period (2000–2015), N = 4

Sr.	Author (year)	Sample size	Methods	Study settings	Findings
1	Debener et al. (2012b)	16	- EEG	- Participants walked slowly outdoors on a university campus	- Good quality, single-trial EEG data suitable for mobile brain-computer interfaces can be obtained with affordable hardware
2	Roe et al. (2013)	20	- EEG	- The study was conducted in a laboratory setting	- Landscape scenes yield higher engagement and positive preference ratings
3	Aspinall et al. (2015)	12	- EEG	- The study was conducted in urban areas of Edinburgh	- Transitioning from urban to green space reduces arousal and frustration - The study confirms systematic differences in EEG signals across urban areas
4	Joung et al. (2015)	8	- NIRS - POMS - SD	- The study was conducted in a forest/urban area was selected in Daejeon, Korea	- Forest environments have physiological and psychological relaxing effects compared to urban areas

EEG Electroencephalography, *NIRS* Near-infrared spectroscopy, *POMS* Profile of mood states, *SD* Semantic differential

contribute to better mental health. Overall, this period marks a growing integration of environmental psychology and neuroscience.

42.3.2 Evolution of Measurement Methods

The diversity of research methodologies in urban green space and mental health studies has expanded significantly, as shown in Fig. 42.2, with methods increasing from 7 (2000–2015) to 49 (2021–2025). Neurophysiological tools dominate, particularly EEG (40 studies), which has seen significant growth after 2015. Psychometric assessments have diversified from 2 to 13 types. Physiological measurements emerged mainly after 2015, while technological approaches like virtual reality are recent additions. This evolution reflects a shift from basic measures to a multi-method approach, indicating greater interdisciplinary integration among neuroscience, psychology, and urban planning.

Table 42.3 Summary of second period (2016–2020), N = 21

Sr.	Author (year)	Sample size	Methods	Study settings	Findings
1	Chen et al. (2016a)	4	- EEG - EMG - A GPS	- Conducted outdoors at Tongji University, Shanghai	- Identifies urban stressors and stimulators - Data quality varies outdoors, impacting interpretation
2	Chen et al. (2016b)	32	- EEG - POMS	- Participants were exposed to either a nature or urban environment for 20 min	- Nature experiences may induce stronger brain connectivity with enhanced small-world properties compared to stressful urban experiences
3	Tang et al. (2017)	31	- FMRI - PRS	- The research compared urban and natural landscapes' effects on participants in controlled lab	- Natural landscapes restore attention better than urban ones - Water and mountain views offer top restorative benefits
4	Al-Barrak et al. (2017)	40	- EEG	- Participants experienced a café, supermarket, and garden sequentially	- Cafes and gardens show high meditation levels - Alpha = relaxation; beta = mental activity
5	Tilley et al. (2017)	43	- EEG - Self-reported measures	- Participants walked a route combining busy urban streets and green areas	- Walking routes benefit from green, familiar paths - Urban walks affect older adults' moods
6	Neale et al. (2017)	95	- EEG	- Participants walked in urban busy, urban quiet, and urban green settings	- Urban green spaces boost engagement - Busy urban areas raise excitement

(continued)

Table 42.3 (continued)

Sr.	Author (year)	Sample size	Methods	Study settings	Findings
7	Chiang et al. (2016)	80	- EEG - POMS	- The study involved three locations: forest interior, edge, and exterior	- Dense vegetation improves mood and attention - Forests aid stress recovery
8	Demastus and Demastus (2018)	19	- EEG	- The study involved built and natural environments for endurance runners: urban and rural settings	- Urban versus natural settings show clear mental state differences - Nature enhances cognition and lowers anxiety
9	Hassan et al. (2018)	60	- EEG - STAI - SDM	- The study was conducted in a bamboo forest near Chengdu	- Bamboo forests lower blood pressure - Forests increase alpha brainwave activity
10	Olszewska-Guizzo et al. (2018)	29	- EEG	- Data acquisition occurred in a quiet visualization lab on university premises	- Green views enhance positive brainwaves - Upper-floor views with greenery raise right frontal alpha power
11	Elsadek et al. (2019)	25	- EEG - HRV - SD - POMS	- Visual stimulations occurred during daytime hours	- Green façades boost relaxation (↑alpha waves) - Urban green spaces enhance daily emotional well-being
12	Tost et al. (2019)	33	- FMRI	- The study involved young adults living in an urban environment	- More vegetation nearby increases positive emotions (affective valence)
13	Reeves (2019)	36	- EEG - PPG	- The site features biodiverse wetland habitats and an urban environment	- Urban settings raise heart rate in high-stress individuals - Wetlands boost positive mood

(continued)

Table 42.3 (continued)

Sr.	Author (year)	Sample size	Methods	Study settings	Findings
14	Kim et al. (2019)	60	- EEG	- Various landscapes included built urban and nature-dominant settings	- Built nightscapes without adults perceived as least fearful - Nature nightscapes with adults viewed as more fearful
15	Grassini et al. (2019)	32	- EEG	- The study utilized both controlled laboratory experiments and outdoor field studies	- Natural environments require less attentional effort - Nature scenes are more relaxing than urban ones
16	Liu et al. (2019)	50	- EEG - STAI - SDM	- Silent office room at Sichuan Agricultural University	- Landscapes increase relaxation and comfort - Viewing landscapes lowers blood pressure
17	Song et al. (2020)	29	- NIRS - Oxy-Hb (PFC)	- The study was conducted in five forest and urban areas of Japan	- Forest views reduce mental stress (lower oxy-Hb) - Urban green exposure improves mental health
18	Olszewska-Guizzo et al. (2020)	22	- EEG - FAA	- Participants were exposed to six landscape scenes: a park, neighborhood green space, and busy urban down	- Parks show positive mood (frontal alpha asymmetry) - Restorative environments boost alpha-theta activity
19	Chen et al. (2020)	32	- Pretest questionnaire - EEG - NCPCT	- The study was conducted in May in outdoor urban space	- Exposure to restorative environments enhances alpha-theta synchronization and oscillations

(continued)

Table 42.3 (continued)

Sr.	Author (year)	Sample size	Methods	Study settings	Findings
20	Wang et al. (2020)	180	- EEG - Skin Conductance and pulse	- The experiment was conducted in an indoor chamber	- Neat undergrowth enhances relaxation effects
21	Lin et al. (2020)	40	- EEG	- The indoor environment was a classroom - The outdoor space was surrounded by a bamboo forest	- Walking in urban green spaces lowers stress more than sitting - Sitting improves attention restoration more than walking

Note EEG Electroencephalography, *EMG* Electromyography, *fMRI* Functional magnetic resonance imaging, *PRS* Perceived Restorativeness Scale, *NCPCT* Natural Color Preference Choice Task, *FAA* Frontal alpha asymmetry, *NIRS* Near-infrared spectroscopy, *Oxy-Hb (PFC)* Oxyhemoglobin concentrations in the prefrontal cortex, *STAI* State-Trait Anxiety Inventory, *SDM* Spatial decision-making, *POMS* Profile of mood states, *PPG* Photoplethysmography

42.3.3 Thematic Clusters: Term Co-occurrence Analysis

Term co-occurrence analysis was conducted using VOS Viewer, as illustrated in Fig. 42.3. A threshold of 233 words and a minimum of two occurrences were applied, resulting in the identification of ten primary keywords, which are detailed in Table 42.5.

Figure 42.3 shows five thematic clusters: the red cluster (13 terms) focuses on EEG, studies, and demographics; the yellow cluster (5 terms) addresses stress and emotion in mental health; the green cluster (12 terms) explores greenspace and brain activity; the purple cluster (5 terms) highlights students and cognitive benefits; and the blue cluster (6 terms) connects urban populations, landscapes, and happiness. This thematic distribution highlights the expanding interdisciplinary interest in the mental health impacts of green spaces over the past five years (2020–2025). The clusters reflect a multifaceted approach, integrating neuroscience, psychology, and urban studies to better understand how natural environments influence emotional, cognitive, and social well-being.

42.3.4 Thematic Co-occurrence over Time

Over the past decade, research has shifted from navigation and mobility to exploring the psychological, emotional, and physiological effects of green spaces. Figure 42.4

Table 42.4 Summary of third period (2021–2025), N = 26

Sr.	Author (year)	Sample size	Methods	Study settings	Findings
1	Zeng et al. (2021)	360	- EEG	- The experiment was conducted at Sichuan Agricultural University, China, in a classroom with closed curtains	- Moderate vegetation lowers stress expression - Low artificial noise enhances relaxation
2	Zhang et al. (2021)	26	- EEG - EDA - HR	- The study was conducted in a laboratory at Peking University	- Urban street scenes evoke positive emotions - Diverse, less fragmented views boost mood
3	Herman et al. (2021)	17	- EEG	- The study focused on Informal Green Spaces in Warsaw, Poland	- No major emotional difference between informal and formal green spaces
4	Olszewska-Guizzo et al. (2021)	25	- fNIRS	- Participants viewed videos projected on a large screen during the study	- Urban greenery reduces stress and anxiety - More screen time links to less outdoor activity
5	Baena-Extremera et al. (2021)	49	- fMRI	- Participants were divided into outdoor and indoor exercise groups	- Outdoor workouts lower stress more than indoor ones - Green visuals activate motor brain regions
6	Chang et al. (2021)	44	- Rating task - fMRI	- Participants viewed images in a quiet, closed experimental room	- Green spaces may regulate stress via brain pathways
7	Mavros et al. (2022)	42	- EEG - EDA - SAM	- Urban indoor/outdoor settings - Walking or treadmill	- Walking enhances cognitive function - Green areas feel calmer than urban ones
8	Reece et al. (2022)	35	- EEG - UWIST MACL	- Participants were exposed to green, blue, and historic environment at indoor lab	- Greenery increases visual attention (lower alpha waves) - Historic areas improve mood and reduce stress

(continued)

Table 42.4 (continued)

Sr.	Author (year)	Sample size	Methods	Study settings	Findings
9	Olszewska-Guizzo et al. (2022)	79	- CLM - BDI-II - SAM	- The study included three sessions: one in the lab, two outdoors	- Natural settings boost positive emotions more than labs - Quality landscapes improve mood and brain activity
10	Olszewska-guizzo et al. (2022)	68	- EEG - fNIRS - POMS	- The study was conducted in Singapore's urban environments	- Garden exposure benefits both depressed and healthy moods
11	Bailey and Kang (2022)	50	- EEG - State Mindfulness Scale	- The study took place in a large grassy area on campus	- Nature and exercise enhance cognition and mental health
12	Asim et al. (2023)	22	- EEG	- Participants were recruited from campus cafeterias and hostel common rooms	- Natural elements promote restoration and well-being - High building density lowers normalized alpha
13	Olszewska-Guizzo et al. (2023)	74	- CLM - EEG	- The study settings included six naturalistic urban green space environments	- Urban green space design affects mental health - CLM effectively assesses urban green space quality
14	Yuan et al. (2023)	10	- PANAS - fNIRS	- Various urban environments were included: open space, neighborhood, street, green, and blue space	- Blue and green spaces elicit positive emotions - Neurophysiological links to environmental contexts identified
15	Wu et al. (2023)	80	- EEG - POMS - sphygmic Manometer	- The study was conducted in Chengdu, Sichuan Province, China, ginkgo garden and urban landscape forest	- Urban green spaces support emotional well-being - Walking increases relaxation and engagement over sitting

(continued)

Table 42.4 (continued)

Sr.	Author (year)	Sample size	Methods	Study settings	Findings
16	Li et al. (2023)	39	- PANAS scale - EEG	- Four natural environments were selected: Grassplot, Square, Forest, and Lakeside	- EEG data indicated greater relaxation in forests compared to squares
17	Olszewska-Guizzo et al. (2020)	10	- EEG	- Participants navigated through outdoor urban spaces during the experiment	- Outdoor settings complicate brain activity analysis versus indoors
18	Ren et al. (2024)	30	- EEG - VR	- Experimental images included current and transformation scenes of rural streetscapes	- Transformation scenes boosted relaxation and subjective scores - Pz electrode responded strongly to green landscapes
19	McDonnell and Strayer (2024)	92	- EEG	- Participants walked in a nature arboretum or urban medical campus	- A 40-min nature walk improves neural executive control - Nature settings are more restorative than urban ones
20	Hassan and Zhang (2024)	54	- EEG - STAI - SDM	- The room was designed to minimize external disturbances	- Touching real grass lowers blood pressure versus artificial turf - EEG shows greater relaxation with real grass contact
21	Yang et al. (2024)	52	- EEG	- Three forest sites were selected: Caoyutang, Daji, and Meiqi	- Forests support social health, deep thinking, and relaxation better than urban green spaces
22	Li and Du (2024)	36	- EEG - HRV	- Participants completed the experiment in a controlled physical environment	- Move-pass gardens have the highest saccade counts, indicating restorative benefits
23	Naghibi (2024)	25	- EEG - IPA	- Case studies included marginal areas and vacant lots	- Transforming vacant lands into green spaces fosters positive emotions

(continued)

Table 42.4 (continued)

Sr.	Author (year)	Sample size	Methods	Study settings	Findings
24	Naghibi et al. (2024)	25	- POMS-SF - EEG	- The study was conducted in a specific urban setting	- Small urban green spaces (SUGS) are preferred for comfort and relaxation
25	Sadeghpoor et al. (2024)	16	- GHQ - DASS-21 questionnaire - EEG	- Three types of streets were evaluated: green space, motorist-oriented, and pedestrian	- Motorist-oriented streets cause highest stress - Green spaces significantly reduce stress
26	Hou et al. (2024)	240	- POMS - TMD - EEG	- Four types of urban green spaces viewed in closed experimental room	- Preference for green spaces drops with lower visibility

Note EEG Electroencephalography, *EMG* Electromyography, *fMRI* Functional magnetic resonance imaging, *PRS* Perceived Restorativeness Scale, *NCPCT* Natural Color Preference Choice Task, *FAA* Frontal alpha asymmetry, *NIRS* Near-infrared spectroscopy, *Oxy-Hb (PFC)* Oxyhemoglobin concentration in the prefrontal cortex, *STAI* State-Trait Anxiety Inventory, *SDM* Spatial decision-making, *POMS* Profile of mood states, *PPG* Photoplethysmography, *EDA* Electrodermal activity, *CLM* Continuous linear model, *PANAS* Positive and negative affect schedule, *IPA* International Physical Activity Questionnaire, *GHQ* General Health Questionnaire, *DASS-21* Depression Anxiety Stress Scales—21 items, *TMD* Total mood disturbance

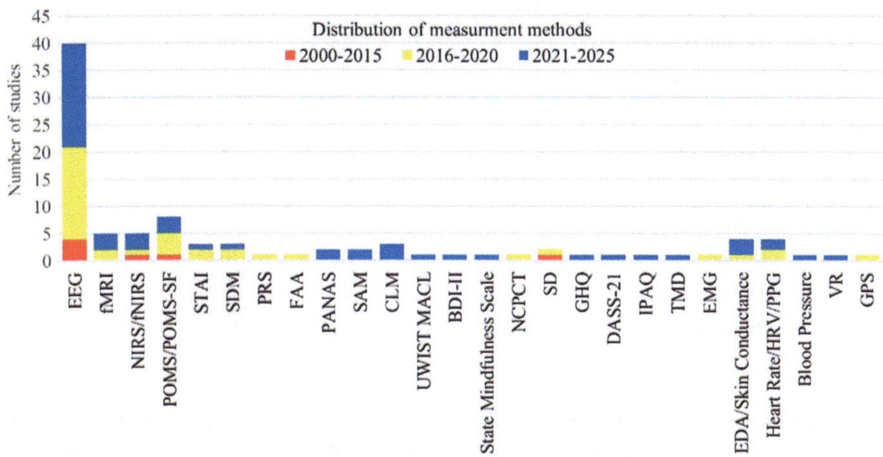

Fig. 42.2 The distribution of measurement methods for the whole study period 2000 to 2025

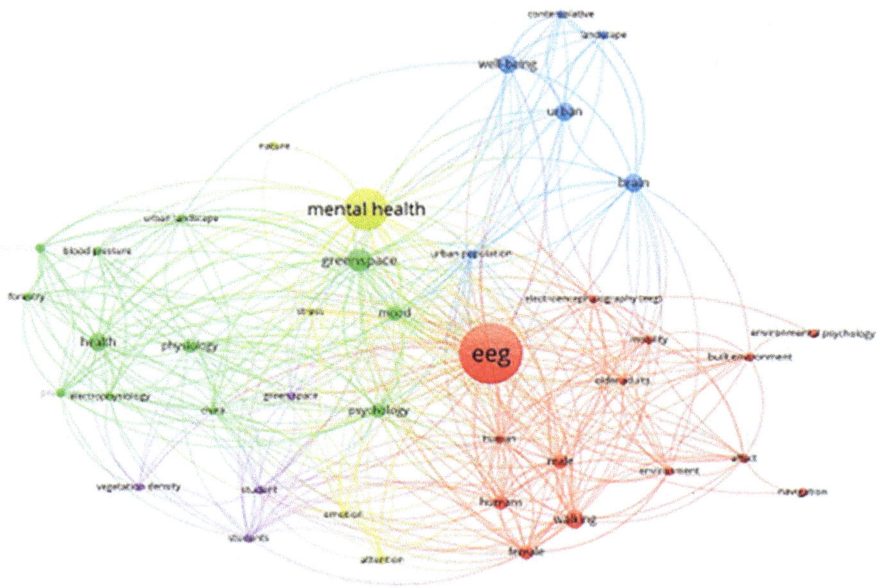

Fig. 42.3 The output of the term co-occurrence analysis for the whole study period 2000 to 2025

Table 42.5 Term co-occurrence analysis (top 10 most frequently occurring terms 2000–2025)

Sr #	Keyword	Occurrences	%	Total link strength
1	EEG	13	5.5	59
2	Mental health	9	3.8	73
3	Green space	7	3	63
4	Walking	5	2.1	51
5	Psychology	4	1.7	34
6	Wellbeing	3	1.2	46
7	Female	3	1.2	36
8	Humans	2	1.2	36
9	Male	2	1.2	36
10	Physiology	2	1.2	34

illustrates keyword trends from 2015 to 2025, highlighting topics such as "well-being," "contemplative landscape," and "environmental psychology," which indicate a growing interest in the mental and physical health benefits of nature.

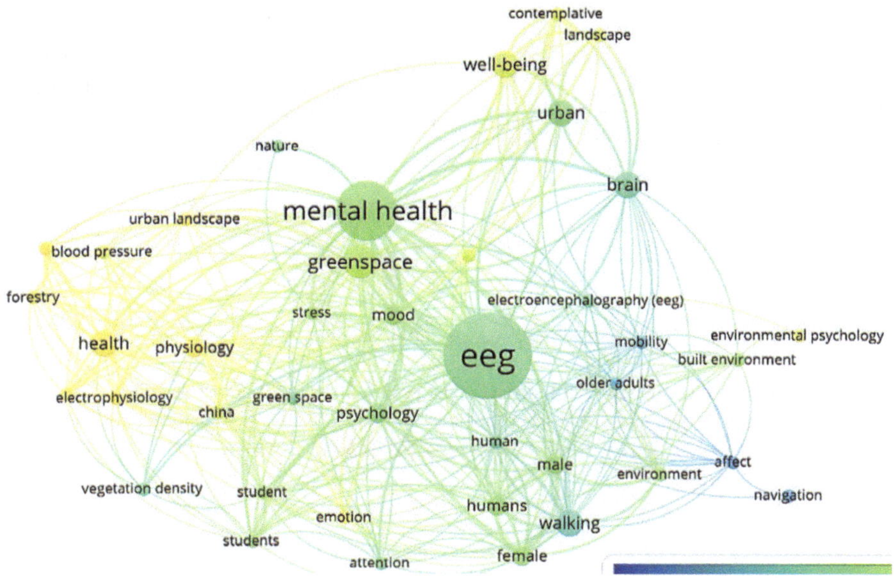

Fig. 42.4 The output of the term co-occurrence analysis for the period 2010 to 2025 last decade (the color gradient representing the latest topics in the field)

42.3.5 Major Contributing Countries

Bibliographic coupling analysis, illustrated in Fig. 42.5, includes inputs from 19 countries, resulting in a total of 51 articles. China is the leading contributor, with 17 studies focused on urban sustainability and environmental psychology. Significant contributions also arise from Singapore, the United Kingdom, and the United States, with Singapore particularly emphasizing neuro-urbanism and green infrastructure. This analysis highlights the limited contributions from countries in the Global South, as shown in Fig. 42.6.

42.3.6 Influential Publications

Table 42.6 presents the most cited papers in the field. Aspinall et al. (2015) from Scotland lead with 584 citations, indicating a significant impact on urban green spaces and mental health. Debener et al. (2012b) from Germany follow with 555 citations, highlighting their influence on neuroarchitecture (Tost et al. 2019), also from Germany, hold third place with 188 citations, emphasizing their contribution to urban spaces and mental health.

Other significant papers include Hassan et al. (2018) with 118 citations, Grassini et al. (2019) with 121 citations, and Tilley et al. (2017) with 109 citations. These

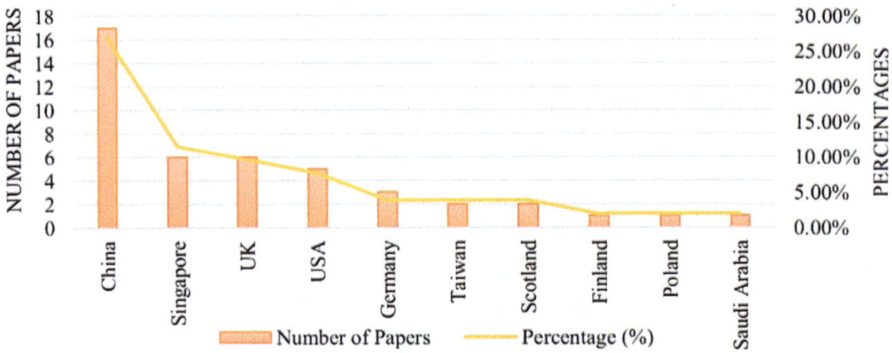

Fig. 42.5 The top 10 contributing countries for the period 2010 to 2025 last decade

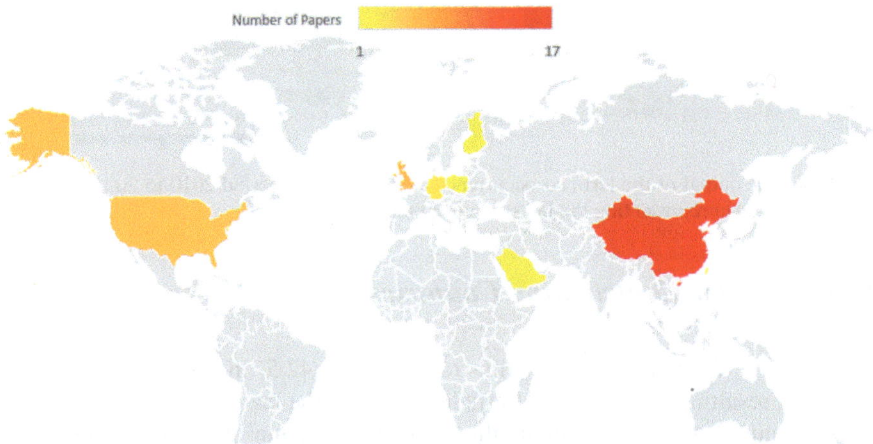

Fig. 42.6 The top 10 contributing countries (the gradient represents the density of documents)

Table 42.6 Analysis of the top 10 cited papers (2015–2025)

Rank #	Cited papers	Year	Citations
1	Aspinall et al. (2015)	2015	584
2	Debener et al. (2012b)	2012	555
3	Tost et al. (2019)	2019	188
4	Chiang et al. (2016)	2017	182
5	Roe et al. (2013)	2013	131
6	Grassini et al. (2019)	2019	121
7	Hassan et al. (2018)	2018	118
8	Elsadek et al. (2019)	2019	117
9	Tilley et al. (2017)	2017	109
10	Tang et al. (2017)	2017	107

studies explore urban green spaces, cognitive functioning, and neurophysiological responses, with contributions from China, the UK, Germany, and Singapore, focusing on the impact of green infrastructure on mental health.

42.4 Discussion

This section summarizes key findings from the review and highlights their implications for neuro-urbanism and urban planning. It integrates insights from neuroscience, psychology, and urban studies.

42.4.1 Implications

This research advances neuro-urbanism by demonstrating how urban green spaces positively influence mental health using EEG, fMRI, and virtual reality. It supports the cognitive architecture model by providing evidence of nature's impact on brain activity, emotional regulation, and stress reduction. A shift from traditional perceptual studies (2000–2015) to advanced approaches (2016–2025) is noted, with significant contributions from China, Singapore, the United Kingdom, and the United States, while developing nations remain underrepresented. These insights highlight the growing relevance of neuroscience in urban studies and the need for interdisciplinary collaboration.

The findings have practical implications for policymakers, architects, and urban planners in designing healthier cities. Green spaces can reduce stress, enhance mood, and boost cognitive function, emphasizing the importance of integrating parks and natural elements into densely populated urban areas. Additionally, emerging technologies like wearable sensors and mobile EEG enable real-time assessments of human responses to urban environments, supporting evidence-based urban design. Collaborative efforts among neuroscience, psychology, urban planning, and architecture are crucial for developing inclusive, sustainable cities that promote mental well-being.

42.4.2 Limitations

This study has several limitations, including geographical imbalance, as most research comes from China, the UK, the USA, and Singapore, limiting generalizability due to socio-economic and cultural disparities.

There is a lack of research from South America and Africa, highlighting the need for greater inclusivity. Additionally, most studies focus on short-term reactions to green spaces, with insufficient exploration of long-term psychological impacts; thus, longitudinal studies are necessary. Advances in EEG and fMRI methods may also lead to inconsistent findings.

42.4.3 Conclusions

This study confirms that urban greens are vital for mental well-being, supported by neuroscientific evidence. It highlights the need to integrate green infrastructure into urban planning to enhance cognitive function, reduce stress, and regulate emotions. Neuro-urbanism advocates for evidence-based urban planning through collaboration among neuroscience, psychology, architecture, and urban planning. Future research should focus on long-term mental health outcomes, diverse geographic areas, and cultural influences. Additionally, designing inclusive cities for vulnerable groups, such as children, older adults, and individuals with mental health issues, is crucial. Integrating neuroscience can help ensure that cities promote well-being, sustainability, and resilience.

Ethical Approval This study utilized published case studies and observational data involving no intervention or risk to subjects. Conducted within ethical guidelines for literature-based research. Therefore, additional ethical approval by an Institutional Review Board (IRB) or ethics committee was not required.

Conflict of Interest Statement Not applicable.

References

Al-Barrak L, Kanjo E, Younis EMG (2017) NeuroPlace: categorizing urban places according to mental states. PLoS One 12(9). https://doi.org/10.1371/journal.pone.0183890

Asim F, Chani PS, Shree V, Rai S (2023) Restoring the mind: a neuropsychological investigation of university campus built environment aspects for student well-being. Build Environ 244. https://doi.org/10.1016/j.buildenv.2023.110810

Aspinall P, Mavros P, Coyne R, Roe J (2015) The urban brain: analysing outdoor physical activity with mobile EEG. Br J Sports Med 49(4):272–276. https://doi.org/10.1136/bjsports-2012-091877

Baena-Extremera A, García JF, Martínez AC, Martín-Pérez C (2021) Sports in Natural Environment, Sports in Urban Environment: An fMRI Study about Stress and Attention/Awareness. J Sports Sci Med 20(4):789–798. https://doi.org/10.52082/jssm.2021.789

Bailey AW, Kang HK (2022) Walking and sitting outdoors: which is better for cognitive performance and mental states? Int J Environ Res Public Health 19(24). https://doi.org/10.3390/ijerph192416638

Branas CC, MacDonald JM (2014) A simple strategy to transform health, all over the place. J Public Health Manag Pract 20(2):157–159. https://doi.org/10.1097/PHH.0000000000000051

Chang DHFF, Jiang B, Wong NHLL, Wong JJ, Webster C, Lee TMCC (2021) The human posterior cingulate and the stress-response benefits of viewing green urban landscapes. Neuroimage 226. https://doi.org/10.1016/j.neuroimage.2020.117555

Chawla L (2015) Benefits of nature contact for children. J Plan Lit 30(4):433–452. https://doi.org/10.1177/0885412215595441

Chen Z, He Y, Yu Y (2016b) Enhanced functional connectivity properties of human brains during in-situ nature experience. PeerJ 7:2016. https://doi.org/10.7717/peerj.2210

Chen Z, Schulz S, He X, Chen Y (2016) A pilot experiment on affective multiple biosensory mapping for possible application to visual resource analysis and smart urban landscape design. In: Proceedings of REAL CORP 2016. International conference on urban planning, regional development and information society (REAL CORP-16), planning times, 22–24 June, vol 2, pp 1–10 [Online]. Available at: https://repository.corp.at/120/

Chen Z, He Y, Yu Y (2020) Attention restoration during environmental exposure via alpha-theta oscillations and synchronization. J Environ Psychol 68:101406. https://doi.org/10.1016/j.jenvp.2020.101406

Chiang YC, Li D, Jane HA (2016) Wild or tended nature? The effects of landscape location and vegetation density on physiological and psychological responses. Landsc Urban Plan 167:72–83. https://doi.org/10.1016/j.landurbplan.2017.06.001

Corburn J (2004) Confronting the challenges in reconnecting urban planning and public health. Am J Public Health 94(4):541–546. https://doi.org/10.2105/AJPH.94.4.541

Debener S, Minow F, Emkes R, Gandras K, de Vos M (2012b) How about taking a low-cost, small, and wireless EEG for a walk? Psychophysiology 49(11):1617–1621. https://doi.org/10.1111/j.1469-8986.2012.01471.x

Debener S, Minow F, Emkes R, Gandras K, Vos MDE (2012) How about taking a low-cost, small, and wireless EEG for a walk?, pp 1–5. https://doi.org/10.1111/j.1469-8986.2012.01471.x

Demastus C, Demastus C (2018) Comparison of psychological states of runners within urban and rural environments

Elsadek M, Liu BY, Lian ZF (2019) Green facades: their contribution to stress recovery and well-being in high-density cities. Urban For Urban Green 46. https://doi.org/10.1016/j.ufug.2019.126446

Grassini S, Revonsuo A, Castellotti S, Petrizzo I, Benedetti V, Koivisto M (2019) Processing of natural scenery is associated with lower attentional and cognitive load compared with urban ones. J Environ Psychol 62:1–11. https://doi.org/10.1016/j.jenvp.2019.01.007

Haluza D, Schönbauer R, Cervinka R (2014) Green perspectives for public health: a narrative review on the physiological effects of experiencing outdoor nature. Int J Environ Res Public Health 11(5):5445–5461. https://doi.org/10.3390/ijerph110505445

Hassan A, Zhang DS (2024) Nature's therapeutic power: a study on the psychophysiological effects of touching ornamental grass in Chinese women. J Health Popul Nutr 43(1). https://doi.org/10.1186/s41043-024-00514-6

Hassan A et al (2018) Effects of walking in bamboo forest and city environments on brainwave activity in young adults. Evidence-Based Complement Alternative Med 2018. https://doi.org/10.1155/2018/9653857

Herman K, Ciechanowski L, Przegalińska A (2021) Emotional well-being in urban wilderness: assessing states of calmness and alertness in informal green spaces (igss) with muse—portable eeg headband. Sustainability (Switzerland) 13(4):1–20. https://doi.org/10.3390/su13042212

Hou J, Wang Y, Zhang X, Qiu L, Gao T (2024) The effect of visibility on green space recovery, perception and preference. Trees Forests People 16. https://doi.org/10.1016/j.tfp.2024.100538

Joung D et al (2015) The prefrontal cortex activity and psychological effects of viewing forest landscapes in autumn season. Int J Environ Res Public Health 12(7):7235–7243. https://doi.org/10.3390/ijerph120707235

Kim M, Cheon SH, Kang Y (2019) Use of electroencephalography (EEG) for the analysis of emotional perception and fear to nightscapes. Sustainability (Switzerland) 11(1). https://doi.org/10.3390/su11010233

Li Y, Du H (2024) Research on the restorative benefits of sky gardens in high-rise buildings based on wearable biosensors and subjective evaluations. Build Environ 260:111691. https://doi.org/10.1016/j.buildenv.2024.111691

Li YB, Zhang JG, Jiang BJ, Li HY, Zhao B (2023) Do all types of restorative environments in the urban park provide the same level of benefits for young adults? A field experiment in Nanjing, China. Forests 14(7). https://doi.org/10.3390/f14071400

Lin W et al (2020) Sitting or walking? Analyzing the neural emotional indicators of urban green space behavior with mobile EEG. J Urban Health 97(2):191–203. https://doi.org/10.1007/s11524-019-00407-8

Liu Y, Jiang M, Hassan A, Chen C (2019) Effects of different landscape visual stimuli on psychophysiological responses in Chinese students

Mavros P, Wälti MJ, Nazemi M, Ong CH, Hölscher C (2022) A mobile EEG study on the psychophysiological effects of walking and crowding in indoor and outdoor urban environments. Sci Rep 12(1). https://doi.org/10.1038/s41598-022-20649-y

McDonnell AS, Strayer DL (2024) Immersion in nature enhances neural indices of executive attention. Sci Rep 14(1). https://doi.org/10.1038/s41598-024-52205-1

Naghibi M (2024) Rethinking small vacant lands in urban resilience: decoding cognitive and emotional responses to cityscapes. CITIES 151. https://doi.org/10.1016/j.cities.2024.105167

Naghibi M, Farrokhi A, Faizi M (2024) Small urban green spaces: insights into perception, preference, and psychological well-being in a densely populated areas of Tehran, Iran. Environ Health Insights 18. https://doi.org/10.1177/11786302241248314

Neale C et al (2017) The aging urban brain: analyzing outdoor physical activity using the Emotiv Affectiv suite in older people. J Urban Health 94(6):869–880. https://doi.org/10.1007/s11524-017-0191-9

Olszewska-Guizzo A, Escoffier N, Chan J, Yok TP (2018) Window view and the brain: effects of floor level and green cover on the alpha and beta rhythms in a passive exposure EEG experiment. Int J Environ Res Public Health 15(11). https://doi.org/10.3390/ijerph15112358

Olszewska-Guizzo A, Sia A, Fogel A, Ho R (2020) Can exposure to certain urban green spaces trigger frontal alpha asymmetry in the brain?—preliminary findings from a passive task EEG study. Int J Environ Res Public Health 17(2):1–10. https://doi.org/10.3390/ijerph17020394

Olszewska-Guizzo A et al (2021) Hemodynamic response to three types of urban spaces before and after lockdown during the COVID-19 pandemic. Int J Environ Res Public Health 18(11). https://doi.org/10.3390/ijerph18116118

Olszewska-guizzo A, Fogel A, Escoffier N, Sia A (2022) Therapeutic garden with contemplative features induces desirable changes in mood and brain activity in depressed adults. Front Psychiatry 13:1–14. https://doi.org/10.3389/fpsyt.2022.757056

Olszewska-Guizzo A, Sia A, Fogel A, Ho R (2022) Features of urban green spaces associated with positive emotions, mindfulness and relaxation. Sci Rep 12(1). https://doi.org/10.1038/s41598-022-24637-0

Olszewska-Guizzo A, Sia A, Escoffier N (2023) Revised Contemplative Landscape Model (CLM): a reliable and valid evaluation tool for mental health-promoting urban green spaces. Urban For Urban Green 86. https://doi.org/10.1016/j.ufug.2023.128016

Reece R, Bornioli A, Bray I, Alford C (2022) Exposure to green and historic urban environments and mental well-being: results from EEG and psychometric outcome measures. Int J Environ Res Public Health 19(20). https://doi.org/10.3390/ijerph192013052

Reeves JP et al (2019) The application of wearable technology to quantify health and wellbeing co-benefits from urban wetlands. Front Psychol 10. https://doi.org/10.3389/fpsyg.2019.01840

Ren H, Wang Y, Zhang J, Zheng Z, Wang Q (2024) Evaluation of the impact of VR rural streetscape enhancement on relaxation–arousal responses based on EEG. Appl Sci (Switzerland) 14(7). https://doi.org/10.3390/app14072996

Roe JJ, Aspinall PA, Mavros P, Coyne R (2013) Engaging the brain: the impact of natural versus urban scenes using novel EEG methods in an experimental setting. Environ Sci (Ruse) 1(2):93–104. https://doi.org/10.12988/es.2013.3109

Sadeghpoor F, Ranjbar E, Esmaeilinasab M, Valiloo MHS, Nieuwenhuijsen MJ (2024) Streets and stress: a pilot study on how quality and design of streets impacts on urban stress. HERD 17(1):224–248. https://doi.org/10.1177/19375867231200584

Sarkar C, Webster C, Gallacher J (2014) Urban built environment configuration and psychological distress in later life: cross-sectional results from the Caerphilly Prospective Study (CaPS). Healthy Cities. https://doi.org/10.4337/9781781955727.00012
Song CR, Ikei H, Kagawa T, Miyazaki Y (2020) Effect of viewing real forest landscapes on brain activity. Sustainability 12(16). https://doi.org/10.3390/su12166601
Tang IC et al (2017) Using functional Magnetic Resonance Imaging (fMRI) to analyze brain region activity when viewing landscapes. Landsc Urban Plan 162:137–144. https://doi.org/10.1016/j.landurbplan.2017.02.007
Tilley S, Neale C, Patuano A, Cinderby S (2017) Older people's experiences of mobility and mood in an urban environment: a mixed methods approach using electroencephalography (EEG) and interviews. Int J Environ Res Public Health 14(2). https://doi.org/10.3390/ijerph14020151
Tost H et al (2019) Neural correlates of individual differences in affective benefit of real-life urban green space exposure. Nat Neurosci 22(9):1389–1393. https://doi.org/10.1038/s41593-019-0451-y
United Nations (2018) World urbanization prospects, vol 12 [Online]. Available at: https://population.un.org/wup/Publications/Files/WUP2018-Report.pdf
Wang Y et al (2020) Physiological and psychological effects of watching videos of different durations showing urban bamboo forests with varied structures. Int J Environ Res Public Health 17(10). https://doi.org/10.3390/ijerph17103434
Wolch JR, Byrne J, Newell JP (2014) Urban green space, public health, and environmental justice: the challenge of making cities 'just green enough.' Landsc Urban Plan 125:234–244. https://doi.org/10.1016/j.landurbplan.2014.01.017
Wu L, Luo S, Li D, Chen Q, Li J, Wen J (2023) Effects of deciduous forests on adolescent emotional health in urban areas: an example from the autumn Ginkgo Forest in Chengdu. Forests 14(6). https://doi.org/10.3390/f14061099
Yang B et al (2024) Quantifying the threshold effects and factors impacting physiological health benefits of forest exposure. Forests 15(3). https://doi.org/10.3390/f15030555
Yuan Y, Wang LT, Wu WJ, Zhong SM, Wang M (2023) Locally contextualized psycho-physiological wellbeing effects of environmental exposures: an experimental-based evidence. Urban For Urban Green 88. https://doi.org/10.1016/j.ufug.2023.128070
Zeng C et al (2021) Electroencephalography (EEG)-based neural emotional response to the vegetation density and integrated sound environment in a green space. Forests 12(10). https://doi.org/10.3390/f12101380
Zhang Z, Zhuo K, Wei W, Li F, Yin J (2021) Emotional responses to the visual patterns of urban streets: evidence from physiological and subjective indicators

Chapter 43
Enhancing Educational Environments: Maximizing School Playground Potential Through Biophilic Design

E. Abdel-Salam, T. Farghaly, H. Abdel-Salam, and A. Hasan

Abstract Increasing disconnection between humans and nature, driven by technological advancements, has negatively impacted health and well-being. The biophilia hypothesis, though relatively new, offers promising solutions to bridge this gap through design. This study investigates the integration of biophilic design in school playgrounds and its influence on children's physical and mental well-being. Using a comparative analysis of three school playgrounds in Alexandria, Egypt—each demonstrating varying levels of biophilic attributes—the research examines how natural elements enhance student experiences and environmental awareness. Biophilic design checklist and student questionnaires were used to evaluate health, environmental awareness, and nature connectedness. Findings highlight the positive impact of biophilic design on children's well-being, reinforcing the need for structured guidelines in educational spaces. This research contributes to the development of a design framework and recommendations for optimizing playgrounds, ensuring a healthier and more enriching learning environment.

Keywords Biophilic design · Physical health · Mental well-being · School playground · Children's development

43.1 Introduction

Children's development is profoundly shaped by their environment, with play serving as a fundamental tool for learning, social interaction, and cognitive growth (Pyle 2002). The following activities, including reading, coloring, creating art projects, listening to music, playing games, and dancing, can all be incorporated into play therapy (Paley 2009). While traditional playgrounds support these activities, research

E. Abdel-Salam (✉) · T. Farghaly · H. Abdel-Salam · A. Hasan
Department of Architectural Engineering, Faculty of Engineering, Alexandria University, Alexandria, Egypt
e-mail: esraa.abdelsalam65@alexu.edu.eg

Y. Li et al. (eds.), *16th International Conference on Environmental Science and Development*, Environmental Science and Engineering,
https://doi.org/10.1007/978-3-032-15080-6_43

highlights that integrating biophilic design—natural elements that foster interaction with nature—can enhance well-being, mental health, and sensory engagement.

Biophilic environments, which incorporate natural light, organic materials, and greenery, have been shown to reduce stress (Tahoun 2019), improve cognitive function, and support mental health (Beyer et al. 2014). Several studies confirm that access to natural elements enhances both physical health and environmental awareness (Zhong et al. 2021), benefiting children, workers, and residents alike. The interconnection between humans and nature is widely acknowledged as a fundamental necessity, given its profound influence on the individual's mental and physical health (Ulrich 1993).

The concept of biophilia, first introduced by Erich Fromm in 1973 and later developed by E. O. Wilson, describes humans' innate affinity for nature (Marcus and Sachs 2013). As urbanization increasingly distances individuals from natural surroundings, designing educational spaces that reinforce this connection is essential for holistic development.

43.1.1 Developing Children's Play Environment

Playground development has evolved significantly, with Dr. Joe Frost highlighting the importance of unstructured play in shaping children's growth (Frost 1992). Early playgrounds, such as the eighteenth century Sand Garden, were inspired by Germany and focused primarily on physical activity. Frost credits Friedrich Froebel with pioneering the integration of play in learning, identifying three evolutionary phases: generated devices, novelty, and modernity (Frost 1992).

Before the mid-twentieth century, playgrounds designed for holistic child development were rare. However, after World War II, new approaches like junk play and adventure play emerged, allowing children to design and engage with their play environments (Frost and Wortham 1988). By the 1950s to 1970s, playgrounds incorporated sculptural elements to spark creativity (Frost 1992).

Since the 1970s, standardized playgrounds have been introduced to address safety concerns, with modular wooden equipment gaining popularity. From the 1990s onward, researchers have focused on integrated playscapes—designs that blend learning, health, and play—to support children's physical, mental, and cognitive development in the twenty-first century (Frost and Wortham 1988).

Playgrounds have developed through the years, yet *biophilic design remains underutilized, presenting an opportunity to enhance sensory engagement and connection to nature.*

43.1.2 Developmental Needs

The environment designed by the architect or landscape architect for children's play should stimulate the child's attention by offering a variety of activity settings and exciting the child in multiple ways. In order to effectively facilitate the growth and development of the child, it is imperative that the environment provides a wide range of stimulating experiences (Eriksen 1985).

Motor and sensory stimulation are vital for a child's development. Sensory experiences like touch, colors, and sounds, along with motor activities that enhance coordination and muscle play, are crucial. Research shows that children with more opportunities for these activities are happier, healthier, and more active (Eriksen 1985; Hourcade 2008).

The need for sensory and motor stimulation is tied to perceptual stimulation, which involves organizing inputs and learning patterns. Exercises that enhance representational skills provide these experiences (Hourcade 2008).

Additionally, emotional stimulation, social interaction, and intellectual stimulation should be taken into consideration in activities that children could do in the playgrounds (Eriksen 1985). *However, this study focuses on physical stimulation that can be achieved while designing the playground* (see Fig. 43.1).

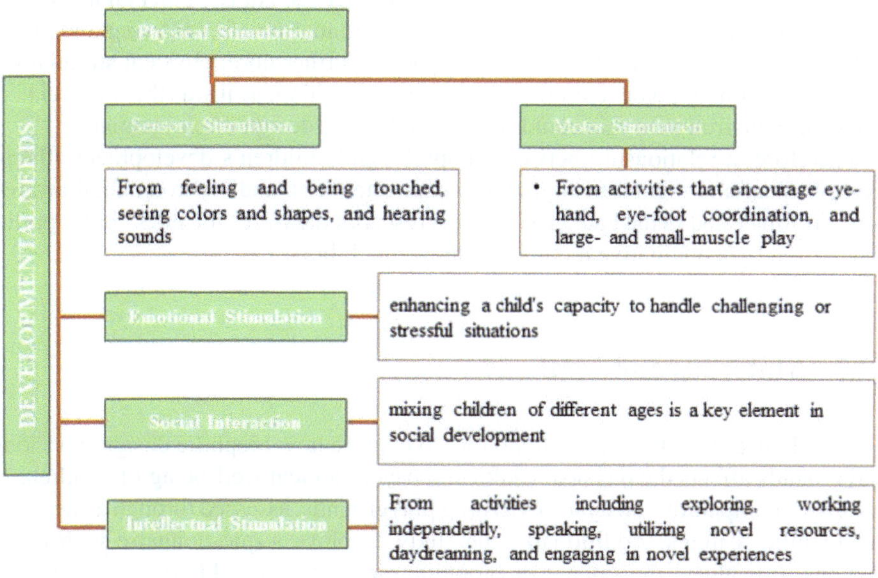

Fig. 43.1 Developmental needs (*source* The researcher 2023)

43.1.3 Biophilic Approach

The biophilia theory suggests regular contact with nature is essential for biological function. Kellert (2008) emphasized the critical link between humans and the environment. Numerous empirical studies have demonstrated a noteworthy correlation between the mental, physical, and psychological well-being of children and their engagement with the natural environment (Louv 2008; Kellert and Wilson 1993).

Biophilic design promotes human and environmental development. Contact with nature benefits human psychology, and bringing nature indoors helps balance modern life with basic needs (Kahn and Kellert 2002).

Particularly in the field of healthcare, positive benefits can be observed. There is a lot of space for development in its generally stressful environment, and multiple studies support the therapeutic power of nature (Ryan et al. 2014).

According to Kellert's six characteristics and features of biophilic design and Terrapin Bright Green's framework, there are two essential categories, which can be considered as a base reference for using a biophilic design (Terrapin Bright Green 2014; Kellert 2008). The framework is created by filtering and extracting suitable playground elements. The items within the Kellert category are organized based on the fourteen biophilic patterns delineated by Browning and Ryan in Terrapin Bright Green's biophilic design framework (see Table 43.1).

The following table illustrates the relationship between developmental needs and how the biophilic attributes could achieve this stimulation after filtering and eliminating some points that refer to interior spaces. Furthermore, physical stimulation was chosen from child developmental needs for matching, as it was the most related to the built environment that could be reflected in the design of playgrounds.

Therefore, a relationship between biophilia and children's development affects their physical and mental health, nature connectedness, and environmental awareness. As illustrated in the background. This research measures the effect of biophilia on children by a qualitative method, as shown in Table 43.2.

43.2 Materials and Methods

This study uses a deductive methodology to explore how biophilic design in school playgrounds affects the physical health and psychological well-being of children. It involves a comparative analysis of existing playgrounds, assessed through a checklist of biophilic features. Additionally, the study employs a questionnaire with three sections to evaluate the impact of biophilic elements on children at the selected sites. The qualitative methodologies, along with the instruments used, experimental design, and underlying rationale, are detailed as shown in Fig. 43.2.

The study measures biophilic elements in three school playgrounds using a checklist of biophilic features. It includes a questionnaire survey to gather data on well-being, health, nature connectedness, and environmental awareness.

Table 43.1 Three key frameworks of biophilic design

Kellert's dimensions	Context	Kellert's elements	Patterns	Kellert's attributes
Organic design/ natural approach	Nature in the space	Environmental features	Visual connection with nature	- Rich color - Existence of water - Existence of fire - Existence of plants - Existence of animals - Preferred views and vistas - Geology and landscape
			Non-visual connection with nature	- Sensory variability (sight, smell, hearing, touch, vision)
		Natural patterns and processes	Dynamic and organic pattern	- Age, change, patina of time - Dynamic balance and tension - Mimicry of organic function and process
			Space and connection	- Integration of parts to wholes - Transitional spaces
		Light and space	Light	- Filtered and diffused light - Light and shadow - Reflected light - Light as shapes and form
			Space	- Spaciousness - Spatial variability - Inside-outside spaces
	Natural analogues	Simulation of natural features	Natural shapes and forms	- Botanical motifs - Animal motifs - Tree and columnar supports - Arches, vaults, domes - Shapes that resist straight lines and right angles

(continued)

Table 43.1 (continued)

Kellert's dimensions	Context	Kellert's elements	Patterns	Kellert's attributes
			Materials connection with nature	- Natural materials - Simulation of natural -texture - Landscape features that define building form
			Complexity and order	- Complementary contrasts - Hierarchically organized ratios and scales
Place-based/ vernacular design	Nature of the space	Place-based relationship	Preservation and placemaking	- Integration of cultural and ecology - Spirit of place
		Human-nature relationship	Prospect and refuge	- Prospect and refuge - Exploration and discovery

Source The researcher (2023), after Terrapin Bright Green (2014) and Kellert (2008)

- **Health and Wellbeing**

The KIDSCREEN-10 index, the shortest form of the KIDSCREEN surveys, assesses general health-related quality of life (HRQoL) with ten items. It can also be computed from the KIDSCREEN-52 or KIDSCREEN-27. This index provides an overall assessment of HRQoL, covering affective symptoms of depression, cognitive attention problems, vitality, energy, and psychosocial aspects of mental health (Ravens-Sieberer et al. 2005).

- **Environmental Awareness**

The New Ecological Paradigm (NEP) stands as the predominant instrument for evaluating environmental concerns and worldviews. It serves as a base or reference point, identifying key areas frequently highlighted in focused studies (Dunlap et al. 2000).

The NEP scale is widely used in many countries, including the United States. It consists of fifteen statements and is a valuable tool for measuring environmental perspectives. (Hawcroft and Milfont 2010; Lalonde and Jackson 2002; Lundmark 2007).

Furthermore, as data and studies increase, more opportunities will be available to evaluate and refine the NEP scale's validity and reliability. The NEP scale stands out as the most optimal choice for evaluating community environmental awareness due to its extensive adoption and application.

Table 43.2 The relationship between biophilia's attributes and developmental needs of children

Physical stimulation	Attributes of biophilic design	
Sensory From feeling and being touched, seeing colors and shapes, and hearing sounds	• Rich color • Existence of water • Existence of fire • Existence of plants • Existence of animals • Preferred views and vistas • Geology and landscape • Diverse habitats and ecosystems • Sensory variability (sight, smell, hearing, touch, vision) • Age, change, patina of time • Dynamic balance and tension • Mimicry of organic function and process • Integration of parts to wholes • Transitional spaces • Botanical motifs	• Animal motifs • Tree and columnar supports • Arches, vaults, domes • Shapes that resist straight lines and right angles • Natural materials • Simulation of natural texture • Landscape features that define building form • Complementary contrasts • Hierarchically organized ratios and scales • Integration of cultural and ecology • Spirit of place • Filtered and diffused light • Light and shadow • Reflected light • Light as shapes and form
Motor From activities that encourage eye-hand, eye-foot coordination, and large- and small-muscle play	• Geology and landscape • Dynamic balance and tension • Mimicry of organic function and process • Transitional spaces • Shapes that resist straight lines and right angles • Filtered and diffused light • Light and shadow	• Reflected light • Light as shapes and form • Spaciousness • Spatial Variability • Inside-outside spaces • Prospect and refuge • Exploration and discovery
Sensory/motor	• Geology and landscape • Dynamic balance and tension • Mimicry of organic function and process • Transitional spaces • Shapes that resist straight lines and right angles	• Filtered and diffused light • Light and shadow • Reflected light • Light as shapes and form

Source The researcher (2023)

However, the New Ecological Paradigm for Children is modeled after the adult versions of NEP and NEP-Revised. It includes ten questions that assess three subscales: "rights of nature," "eco-crisis," and "human exceptionalism," contributing to one's "environmental awareness view." In each school, an instructor assists and monitors the children, encouraging thoughtful responses. Researchers ensured that children understood their information would remain confidential and that they could withdraw from the survey at any time without consequences.

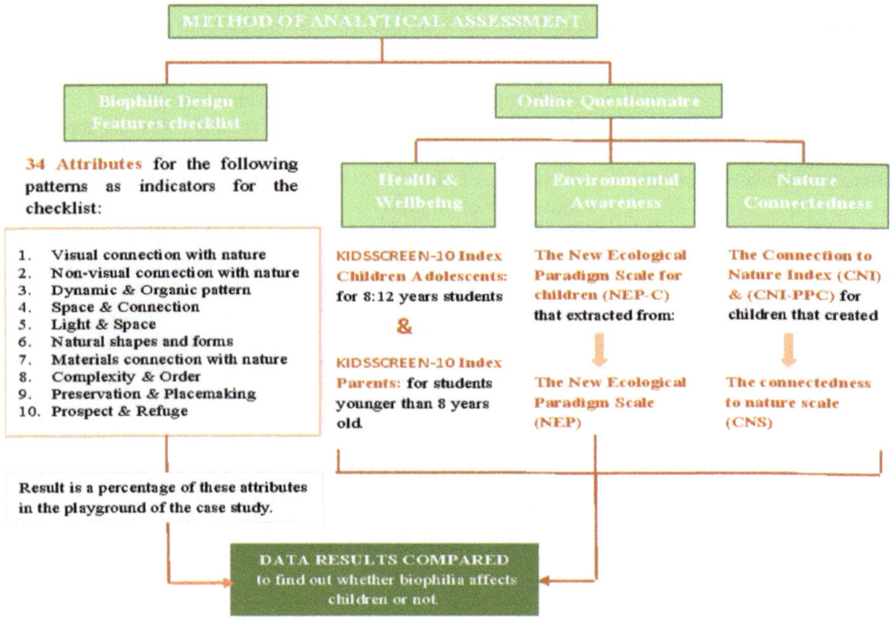

Fig. 43.2 Method of analytical assessment (*source* The researcher 2024)

- **Nature Connectedness**

The Connection to Nature Index (CNI) is a specialized tool for assessing young school children's connection to nature. It measures their feelings towards nature, perceptions of human-nature relationships, and concern for plants and animals. Four themes are included: enjoyment of nature, empathy for creatures, a sense of oneness, and a sense of responsibility (Cheng and Monroe 2012).

Numerous tools have been developed to gauge the extent of an individual's connection to the natural world. The Connection to Nature Index (CNI), created by Cheng and Monroe (2012), is one such questionnaire designed to evaluate the nature connection in children aged eight to ten.

The 24-item CNI (Cheng and Monroe 2012), The Connection to Nature Index (CNI), is one such questionnaire designed to evaluate the nature connection in children aged eight to ten, originally designed for children aged eight to ten, was adapted for urban preschoolers. Unlike Cheng and Monroe's (2012) study, which focused on children outside major metropolitan areas, this study involved younger, urbanized children, many of whom did not attend school. Therefore, a thorough modification of the CNI was necessary.

In the context of early childhood research, parental-proxy measures are commonly employed. Therefore, parents were regarded as potential informants regarding their children's affinity for nature, as indicated by the Connectedness to Nature Index for Parents of Preschool Children (CNI-PPC) (Chan and Wang 2013).

FLS	ESA	ALS
The Egyptian Ministry of Education granted Future Language School its recognition in 2001. To keep kids entertained during playtime and throughout their sports instruction, the school also features a multipurpose playground, courts (basketball, football/soccer, volleyball and sand area), and a traffic circle.	The European Language School was established in 2013. The buildings take approximately 1100 m2 and the school is 4500 m2. The remaining space is split up into kid-friendly zones, green spaces, several athletic courts, secure play areas, a swimming pool, a field for football and a basketball court.	The Alexandria Language School was established in 1997. The amenities of the school are divided into kid-friendly areas, green areas, multiple sports courts, safe play places, a swimming pool, a basketball court, and a football field. Alexandria Language School (ALS) offers about 3000 students a well-rounded education at both of its locations.

El Nabawy El Mohandes Mandara Kebly, Alexandria, Egypt	Soyouf Qebli (Include Izbat Derbanah), Montaza 2, Alexandria, Egypt	El Nabawy El Mohandes Mandara Kebly, Alexandria, Egypt

Fig. 43.3 Selected studied schools in Alexandria, Egypt (The Researcher 2024)

The literature review, combined with a site investigation of school playgrounds, reveals that this national school playground serves as an exemplary environmentally sustainable model for the study, as shown in Fig. 43.3.

43.3 Results

The data obtained from the questionnaire was analyzed to evaluate the hypothesis that a higher score on the survey about environmental awareness and connection to nature, in conjunction with biophilic design, would positively impact on students' environmental awareness and their sense of connection to nature. This increased awareness and connectedness are hypothesized to correlate strongly with improved health and well-being outcomes, as measured by the study scale. Additionally, the data was used to address the three primary research questions of the study:

- First and foremost, can biophilic design, by promoting awareness of the environment and a sense of connection to nature, promoting the health and well-being of children through deliberate environmental design?

- Second essential question: Do various schools have varying health and wellness levels? If so, how does the influence of biophilic environmental design contribute to the difference?
- Third key question: Is there a substantial relationship between community health and well-being and the degree of environmental awareness and natural connectedness?

43.3.1 Biophilic Design Attributes Checklist

The implementation of biophilic design patterns in these three institutions, as outlined in Table 43.1, is detailed in Table 43.4. The provided data demonstrates the incorporation of biophilic design patterns within each institution. Eleven biophilic design patterns were selected from the elements and features listed in Table 43.1, based on criteria such as measurement feasibility, avoidance of redundancy, and relevance to playground design as outdoor spaces. In this regard, some patterns were excluded because they specifically referred to indoor environments, while others were omitted due to being less suitable imitations of others' designs. Analysis of the selected examples confirmed that the eleven patterns applied to the design of school playgrounds, as each of the three schools demonstrated the use of at least one of these patterns.

As previously mentioned, biophilic design encourages a tangible relationship with nature and its inherent processes. This section examines the features that inform, direct, or define the biophilic aspects that are currently present at three schools, as well as the recent and insightful findings that can be obtained from the survey to aid with playground design, which will enhance the well-being and health of children through biophilic design. The data presented in Table 43.3 illustrate the degree of incorporation of biophilic designbiophilic design elements within three educational institutions under study.

The implementation of biophilic design elements is evident across the three schools. Future Language School demonstrated the highest level of biophilic design elements (82.35%) in this survey. Of the three schools, the most exemplary biophilic design attributes are shown in Table 43.4.

43.3.2 Questionnaire (Health and Well-Being, Environmental Awareness, and Nature Connectedness)

The analysis of the online questionnaire involved a comprehensive review of the collected data to address the research objectives. The survey was developed to encompass a variety of responses about Health and Well-being (Table 43.6), Environmental Awareness (Table 43.7), and Nature Connectedness (Table 43.8). It was circulated among students ranging from 4 to 12 years of age (see Table 43.5). Data collection

Table 43.3 Biophilic design attributes checklist

Biophilic patterns				Attributes	Description	FLS	ESA	ALS
Organic design/ natural approach	Nature in the space	Environmental features	Visual connection with nature	Rich color	Earth tones and other natural colors, as well as vivid floral colors, rainbows, sunsets, sparkling water, and blue sky, are examples of colorful natural phenomena	✓	✓	✓
				Existence of water	Existence of artificial or natural bodies of water	✓	✓	✓
				Existence of fire	Existence of artificial fire, such as candles or a fireplace	X	X	X
				Existence of plants	Existence of natural plants	✓	✓	✓
				Existence of animals	Existence of Animals, including aviaries and aquariums Existence of certain animals that are free to wander (like birds) Existence of *Animals kept as pets*	✓	X	X
				Preferred views and vistas	An overview of an avenue or opening from a distance	✓	X	X

(continued)

Table 43.3 (continued)

Biophilic patterns				Attributes	Description	FLS	ESA	ALS
				Geology and landscape	Existence of topographic variation and a thoughtfully planned landscape	✓	✓	✓
			Non-visual connection with nature	Sensory variability (sight, smell, hearing, touch, vision)	A varied and sensual setting for all senses	✓	✓	✓
		Natural pattern and processes	Dynamic and organic pattern	Age, change, patina of time	Natural materials change with time due to aging factors called patina	✓	✓	✓
				Dynamic balance and tension	Combining disparate forces frequently results in a creative tension that gives static shapes an organic appearance	✓	✓	X
				Mimicry of organic function and process	Designs that take inspiration from naturally occurring adaptations, especially those found in different animals. Also referred to as biomimicry	X	X	X

(continued)

Table 43.3 (continued)

Biophilic patterns				Attributes	Description	FLS	ESA	ALS
			Space and connection	Integration of parts to wholes	Discrete elements combine to create an overall whole, particularly when that whole has an emerging characteristic that goes beyond the sum of its parts	✓	✓	✓
				Transitional spaces	Areas that are both within and outside of the constructed environment; include fenestration, doors, gateways, thresholds, and bridges	✓	✓	✓
		Light and space	Light	Filtered and diffused light	Adjusted daylight to cut down on glare	✓	✓	✓
				Light and shadow	The contrast between the dark and bright areas	✓	✓	✓
				Reflected light	Surfaces with light colors, ceilings, and reflecting objects like water reflect light to the source	✓	✓	✓

(continued)

Table 43.3 (continued)

Biophilic patterns				Attributes	Description	FLS	ESA	ALS
				Light as shapes and form	Natural light is manipulated to produce sculptural, dynamic, and capturing shapes	✓	X	X
			Space	Spaciousness	Areas that give off an air of openness, particularly when compared to smaller sections	✓	✓	✓
				Spatial variability	Several areas with alternative proportions and sizes	✓	✓	✓
				Inside-outside spaces	Indoor areas have the appearance of being connected to the outside (such as foyers, colonnades, porches, atriums, and inside gardens)	✓	✓	✓
	Natural analogues	Simulation of natural features	Natural shapes and forms	Botanical motifs	The forms, patterns, and shapes of plants or other vegetation resources	✓	X	X

(continued)

Table 43.3 (continued)

Biophilic patterns				Attributes	Description	FLS	ESA	ALS
				Animal motifs	Mimicry of wildlife. Rather than whole animals, it might just be certain parts (like heads or claws)	X	X	X
				Tree and columnar supports	The general appearance and sculpting of tree-like forms, particularly those with columnar supports	X	X	X
				Arches, vaults, domes	Existence of arches, vaults, or domes for aesthetic or practical reasons	✓	X	X
				Shapes that resist straight lines and right angles	Forms and shapes that are flexible and able to adapt to the stresses and pressures present in the natural world. shapes resistant to angles, straight lines, and mechanical edges	✓	X	X

(continued)

Table 43.3 (continued)

Biophilic patterns				Attributes	Description	FLS	ESA	ALS
			Materials connection with nature	Natural materials	Existence of naturally occurring materials that alter dynamically over time as they weather, age, and acquire a patina	✓	✓	✓
				Simulation of natural texture	Limitation of a natural surface or substance's feel, look, or consistency	✓	✓	✓
				Landscape features that define building form	Manmade environment that is in harmony with its biological surroundings	✓	✓	✓
			Complexity and order	Complementary contrasts	Rendering of seemingly incongruous elements such as open and closed, high and low, and bright and dark	✓	✓	✓

(continued)

Table 43.3 (continued)

Biophilic patterns				Attributes	Description	FLS	ESA	ALS
				Hierarchically organized ratios and scales	Forms that are related to one another hierarchically and occasionally arithmetically or geometrically (such as the Fibonacci ratio and golden proportion, for example)	X	X	X
Place-based/ vernacular design	Nature of the space	Place-based relationship	Preservation and placemaking	Integration of cultural and ecology	Areas that include a region's nature, history, and geography become essential parts of both the person's and the community's identity	X	X	X
				Spirit of place	Constructed settings that take on a life of their own and provide a steady source of inspiration for accountability and stewardship	✓	✓	✓

(continued)

Table 43.3 (continued)

Biophilic patterns				Attributes	Description	FLS	ESA	ALS
		Human-nature relationship	Prospect and refuge	Prospect and refuge	The complementary relationship between providing a safe and protected environment and having the ability to distinguish far-off things, environments, and horizons People can frequently observe (prospect) in environments that satisfy these demands without being seen (refuge)	✓	✓	✓
				Exploration and discovery	Environments that can attract people's desire to explore and learn more since they are rich in information and exhibit restrained depiction	✓	✓	✓
Percentage of biophilic attributes (%)						82.3	64.7	61.76

Source The researcher (2023)

Table 43.4 The most exemplary biophilic design patterns (The Researcher 2024)

Biophilic Design Pattern	Rich color	Existence of water	Existence of plants	Geology & Landscape	Sensory Variability	Age, change, patina of time	Space & connection
FLS							
ESA							
ALS							

Biophilic Design Pattern	Light	Space (Spaciousness, Spatial Variability, Inside-outside spaces)	Materials connection with nature	Complementary contrasts	Spirit of place	Prospect & Refuge
FLS						
ESA						
ALS						

was conducted through an online platform, ensuring efficient and widespread reach. The responses were compiled into a dataset, which was subsequently cleaned and organized to ensure accuracy and completeness.

Table 43.5 Distribution of the studied cases according to the *general information of students*

General information	FLS (n = 50)		ESA (n = 70)		ALS (n = 67)	
	No	%	No	%	No	%
Student's gender						
Female	27	54.0	29	41.4	32	47.8
Male	23	46.0	41	58.6	35	52.2
Student's age						
4 years old	4	8.0	12	17.1	7	10.4
5 years old	6	12.0	19	27.1	14	20.9
6 years old	6	12.0	9	12.9	5	7.5
7 years old	4	8.0	10	14.3	24	35.8
8 years old	3	6.0	6	8.6	2	3.0
9 years old	3	6.0	13	18.6	2	3.0
10 years old	9	18.0	1	1.4	3	4.5
11 years old	3	6.0	0	0.0	2	3.0
12 years old	12	24.0	0	0.0	8	11.9

Table 43.6 Comparison between the three studied language schools according to the score for *Section 1: health and well-being*

Health and wellbeing (average score) (1–5)	FLS	ESA	ALS	p
Children (<8 years old)	*(n = 20)*	*(n = 50)*	*(n = 50)*	
Min–max	3.0–5.0	2.00–5.00	2.50–4.70	0.003^{*}
Mean ± SD	4.03 ± 0.44	3.79 ± 0.60	3.54 ± 0.52	
Median (IQR)	4.0 (3.75–4.30)	3.80 (3.50–4.10)	3.60 (3.10–3.90)	
Sig. bet. school	$p_1 = 0.003^{*}$, $p_2 = 0.244$, $p_3 = 0.059$			
Children (8–12 years old)	*(n = 30)*	*(n = 20)*	*(n = 17)*	
Min–max	3.10–4.40	2.80–5.0	3.10–4.60	0.423
Mean ± SD	3.86 ± 0.32	3.95 ± 0.59	3.76 ± 0.40	
Median (IQR)	3.90 (3.70–4.10)	3.95 (3.50–4.50)	3.80 (3.50–4.0)	
Overall	*(n = 50)*	*(n = 70)*	*(n = 67)*	
Min–max	3.0–5.0	2.0–5.0	2.50–4.70	0.001^{*}
Mean ± SD	3.93 ± 0.38	3.84 ± 0.60	3.60 ± 0.50	
Median (IQR)	3.95 (3.70–4.10)	3.80 (3.50–4.30)	3.60 (3.25–3.90)	
Sig. bet. school	$p_1 = 0.002^{*}$, $p_2 = 0.605$, $p_3 = 0.018^{*}$			

Table 43.7 Comparison between the three studied language schools according to the score for *Section 2: Environmental awareness*

Environmental awareness (average score) (1–5)	FLS	ESA	ALS	p
Children (<8 years old)	*(n = 20)*	*(n = 50)*	*(n = 50)*	
Min–max	3.80–4.60	3.50–4.80	3.40–4.50	0.113
Mean ± SD	4.18 ± 0.32	4.06 ± 0.29	4.01 ± 0.30	
Median (IQR)	4.10 (3.90–4.60)	4.0 (3.90–4.20)	4.05 (3.80–4.20)	
Children (8–12 years old)	*(n = 30)*	*(n = 20)*	*(n = 17)*	
Min–max	3.60–4.60	3.00–4.50	2.40–4.40	0.001*
Mean ± SD	4.08 ± 0.23	3.90 ± 0.42	3.63 ± 0.49	
Median (IQR)	4.05 (4.0–4.20)	3.90 (3.65–4.20)	3.70 (3.50–3.90)	
Sig. bet. school	$p_1 < 0.001^*$, $p_2 = 0.199$, $p_3 = 0.082$			
Overall	*(n = 50)*	*(n = 70)*	*(n = 67)*	
Min–max	3.60–4.60	3.0–4.80	2.40–4.50	0.006*
Mean ± SD	4.12 ± 0.27	4.01 ± 0.34	3.91 ± 0.39	
Median (IQR)	4.10 (3.90–4.20)	4.0 (3.80–4.20)	3.90 (3.70–4.20)	
Sig. bet. school	$p_1 = 0.004^*$, $p_2 = 0.223$, $p_3 = 0.198$			

43.4 Discussion

Findings confirm the positive impact of biophilic design on children's development. Future Language School (FLS) ranked highest in health, environmental awareness, and nature connectedness, significantly surpassing ESA and ALS ($p < 0.05$). The Biophilic Design Checklist further supports this, demonstrating FLS's success in integrating natural elements into play environments with 82.35% (see Table 43.9).

The notable difference in scores suggests that schools with stronger biophilic integration provide enhanced sensory engagement, emotional stability, and environmental awareness for children. FLS's higher incorporation of diverse biophilic attributes aligns with studies emphasizing the role of nature in health and well-being, nature connectedness, and awareness.

Table 43.8 Comparison between the three studied language schools according to score for *overall Section 3: Nature connectedness*

Nature connectedness (average score) (1–5)	FLS	ESA	ALS	p
Children (<8 years old)	*(n = 20)*	*(n = 50)*	*(n = 50)*	
Min–max	3.40–5.0	3.05–4.85	2.90–4.95	0.185
Mean ± SD	4.24 ± 0.50	4.03 ± 0.40	4.03 ± 0.47	
Median (IQR)	4.25 (3.80–4.68)	4.05 (3.75–4.30)	4.05 (3.70–4.25)	
Children (8–12 years old)	*(n = 30)*	*(n = 20)*	*(n = 17)*	
Min–max	3.88–4.94	3.50–5.0	1.94–4.63	0.002^{*}
Mean ± SD	4.45 ± 0.24	4.27 ± 0.47	3.96 ± 0.65	
Median (IQR)	4.50 (4.38–4.63)	4.19 (3.88–4.63)	3.94 (3.88–4.44)	
Sig. bet. school	$p_1 = 0.002^{*}$, $p_2 = 0.345$, $p_3 = 0.092$			
Overall	*(n = 50)*	*(n = 70)*	*(n = 67)*	
Min–max	3.40–5.0	3.05–5.0	1.94–4.95	$< 0.001^{*}$
Mean ± SD	4.37 ± 0.38	4.10 ± 0.43	4.01 ± 0.52	
Median (IQR)	4.44 (4.15–4.63)	4.06 (3.75–4.40)	4.0 (3.73–4.35)	
Sig. bet. school	$p_1 < 0.001^{*}$, $p_2 = 0.005^{*}$, $p_3 = 0.505$			

IQR Inter quartile range; *SD* Standard deviation
p: p value for comparing between the three studied language schools
p_1: p value for comparing between FLS and ALS
p_2: p value for comparing between FLS and ESA
p_3: p value for comparing between ALS and ESA
* Statistically significant at $p \leq 0.05$

Table 43.9 Comparative analysis results

	FLS	ESA	ALS
Biophilic design features checklist	**82.35%**	64.70%	61.76%
Questionnaires' participants	N = 50	N = 70	N = 67
Health and well-being	**3.95**	3.80	3.60
Environmental awareness	**4.10**	4.0	3.90
Nature connectedness	**4.44**	4.06	4.0

Bold highlights that FLS had the highest scores

43.5 Research Limitations

Identifying well-designed playgrounds with distinct biophilic elements was challenging, restricting the selection of case studies with clearly defined biophilic attributes. Additionally, the small participant sample, due to limited student accessibility, may affect the accuracy of the findings.

43.6 Future Directions

Future research should expand sample sizes for greater accuracy, explore diverse biophilic playgrounds to assess long-term impact, conduct longitudinal health and cognitive growth studies, and compare designs globally to understand cultural variations. These directions will deepen insights into biophilic design's role in creating sustainable and health-enhancing educational environments.

43.7 Conclusions

This research project seeks to investigate the effects of the biophilic approach on individuals, with a particular emphasis on children and their developmental requirements. The study is centered on enhancing their physical and mental health, fostering a deeper connection with nature, and promoting an increased awareness of environmental issues. Thus, the survey developed for this study is deemed an appropriate instrument for capturing and analyzing the impacts of biophilic design on the health and well-being of larger communities from a statistical perspective.

The incorporation of biophilic design has been linked to notable benefits, including heightened environmental consciousness and greater attune to nature, as evidenced by elevated scores regarding the survey about environmental awareness and connection to nature. These outcomes have shown a strong correlation with enhanced health and well-being among the individuals involved in the study.

The influence of outstanding biophilic design is responsible for this outcome, if not entirely. Research has shown that nurturing and fulfilling our innate human desire to connect with nature and other living organisms leads to enhanced physical well-being, reduced stress levels, improved cognitive capabilities, and a positive influence on emotions, moods, and preferences. These factors contribute to the overall improvement of community health and welfare.

Consequently, the study emphasizes the importance of considering the biophilic approach in the initial design or upgrading of school playgrounds to enhance their effectiveness for children. The ten patterns in this study could be used as a checklist in the playground to aid in the design, as demonstrated in this study, along with their applicability and examples of application.

References

Beyer KM et al (2014) Exposure to neighborhood green space and mental health: evidence from the survey of the health of Wisconsin. Int J Environ Res Public Health 11(3):3453–3472

Chan CM, Wang WC (2013) Quality of life in overweight and obese young Chinese children: a mixed-method study. Health Qual Life Outcomes 11(1):33

Cheng JC-H, Monroe MC (2012) Connection to nature: children's affective attitude toward nature. Environ Behav 44(1):31–49
Dunlap RE et al (2000) New trends in measuring environmental attitudes: measuring endorsement of the new ecological paradigm: a revised NEP scale. J Soc Issues 56(3):425–442
Eriksen A (1985) Playground design: outdoor environments for learning and development. Olympic Marketing Corporation
Frost JL (1992) Play and playscapes: instructor's guide. Delmar Publishers
Frost JL, Wortham SC (1988) The evolution of American playgrounds. Young Child 43(5):19–28
Hawcroft LJ, Milfont TL (2010) The use (and abuse) of the new environmental paradigm scale over the last 30 years: a meta-analysis. J Environ Psychol 30(2):143–158
Hourcade JP (2008) Interaction design and children. Found Trends Hum–Comput Interact 1(4):277–392
Kahn Jr PH, Kellert SR (2002) Children and nature: psychological, sociocultural, and evolutionary investigations. MIT Press
Kellert SR (2008) Dimensions, elements, and attributes of biophilic design. Biophilic design: the theory, science, and practice of bringing buildings to life, pp 3–19
Kellert SR, Wilson EO (1993) The biophilia hypothesis. Island Press
Lalonde R, Jackson EL (2002) The new environmental paradigm scale: has it outlived its usefulness? J Environ Educ 33(4):28–36
Louv R (2008) Last child in the woods: Saving our children from nature-deficit disorder. Algonquin Books. Updated and Expanded edition
Lundmark C (2007) The new ecological paradigm revisited: anchoring the NEP scale in environmental ethics. Environ Educ Res 13(3):329–347
Marcus CC, Sachs NA (2013) Therapeutic landscapes: an evidence-based approach to designing healing gardens and restorative outdoor spaces. Wiley
Paley VG (2009) A child's work: the importance of fantasy play. University of Chicago Press
Pyle RM (2002) Eden in a vacant lot: special places, species, and kids in the neighborhood of life. Children and nature: psychological, sociocultural, and evolutionary investigations, pp 305–327
Ravens-Sieberer U et al (2005) KIDSCREEN-52 quality-of-life measure for children and adolescents. Expert Rev Pharmacoecon Outcomes Res 5(3):353–364
Ryan CO et al (2014) Biophilic design patterns: emerging nature-based parameters for health and well-being in the built environment. ArchNet-IJAR: Int J Archit Res 8(2):62
Tahoun ZNA (2019) Awareness assessment of biophilic design principles application. In: IOP conference series: earth and environmental science. IOP Publishing
Terrapin Bright Green LLC (2014) 14 patterns of biophilic design: improving health and well-being in the built environment
Ulrich RS (1993) Biophilia, biophobia, and natural landscapes. Biophilia Hypothesis 7:73–137
Zhong W, Schröder T, Bekkering J (2021) Biophilic design in architecture and its contributions to health, well-being, and sustainability: a critical review. Front Archit Res

Chapter 44
Optimizing LID in Small Urban Areas: A Sustainable and Cost-Effective Method with Spatial Uniformity Evaluation

Yafeng Wan, An Xu, Ming Chen, Lei Mao, and Maolong Peng

Abstract To address the challenge of optimizing the spatial layout of Low Impact Development (LID) facilities in small-scale areas, this study established a Storm Water Management Model (SWMM) for a pilot sponge city community in Nanchang City as the primary study area, with an additional area selected for validation. Sensitivity analysis and calibration of model parameters were performed using standardized methods. Furthermore, a method for quantifying the distribution uniformity of LID facilities was proposed. The Non-dominated Sorting Genetic Algorithm II (NSGA-II) was subsequently employed to optimize the LID facility layout. The results indicate a positive correlation between runoff reduction and LID construction costs across different rainfall return periods. Notably, for a 10-year return period rainfall event, a greater runoff reduction was achieved for the same cost compared to higher return periods. Further analysis revealed that dispersed LID layouts exhibited high distribution uniformity but incurred higher costs, while clustered layouts provided advantages in balancing runoff management and cost. Centralized layouts have lower costs but limited runoff management effectiveness. Additionally, carbon emissions from permeable pavement constitute the primary source of total emissions, and total life-cycle carbon emissions are positively correlated with runoff reduction. Considering factors such as cost-effectiveness, LID layout type, and ecological benefits, the optimal implementation strategy for the primary study community was determined to comprise 1.17% green roofs, 3.79% sunken green spaces, and 3.76% permeable pavements. These findings offer technical support for LID implementation in similar areas.

Keywords Sponge city · Low impact development (LID) · SWMM model · NSGA-II algorithm · Uniformity · Carbon emissions

Y. Wan · A. Xu · M. Chen · L. Mao · M. Peng (✉)
School of Civil Engineering and Architecture, Nanchang Jiaotong Institute, Nanchang, China
e-mail: 344568382@qq.com

Y. Li et al. (eds.), *16th International Conference on Environmental Science and Development*, Environmental Science and Engineering,
https://doi.org/10.1007/978-3-032-15080-6_44

44.1 Introduction

Amid the rapid pace of urbanization, the widespread deployment of gray infrastructure has led to a significant increase in impervious surfaces within cities. These surfaces cause rainwater to pool quickly, putting immense strain on urban drainage systems, especially during short bursts of heavy rainfall, which can result in urban flooding. In response to this issue, the "sponge city" concept has emerged.

Previous studies have investigated LID layout strategies, considering site characteristics such as topography and permeability (Martin-Mikle et al. 2015; Chang et al. 2009), facility synergy (Liquete et al. 2015), and the performance of different spatial patterns under varying conditions (Jin et al. 2022). Related work has included the optimization of green space patterns (Yang and Lee 2021) and the use of models like SWMM for evaluation (Dou et al. 2024).

Despite these advancements, research concentrating on optimizing layouts in small-scale areas—particularly those that incorporate a quantitative measure of layout uniformity—remains insufficient. To address this gap, this study proposes a quantitative LID uniformity metric and employs the Storm Water Management Model (SWMM) and the Non-dominated Sorting Genetic Algorithm II (NSGA-II) for multi-objective optimization. This approach aims to minimize costs while maximizing runoff reduction across various LID facility areas, including green roofs, sunken green spaces, permeable pavements in a small-scale pilot community. The analysis explores trade-offs related to layout patterns—specifically dispersed, clustered, and centralized configurations—while considering cost, effectiveness under various rainfall conditions, and life-cycle carbon emissions to identify optimal implementation strategies.

44.2 Materials and Methods

44.2.1 Study Area

The primary study area is a residential community in Nanchang, while a secondary commercial area was utilized for validation. Instances of the SWMM model (Fig. 44.1) were developed using manual delineation of sub-catchment, based on the acquired topographic and land use data.

44.2.2 Model Parameter Sensitivity Analysis

This study employed the Morris method to assess the sensitivity of key parameters within the SWMM model. For the analysis, a design rainfall event with a 10-year return period was selected for the study area. The average imperviousness of the

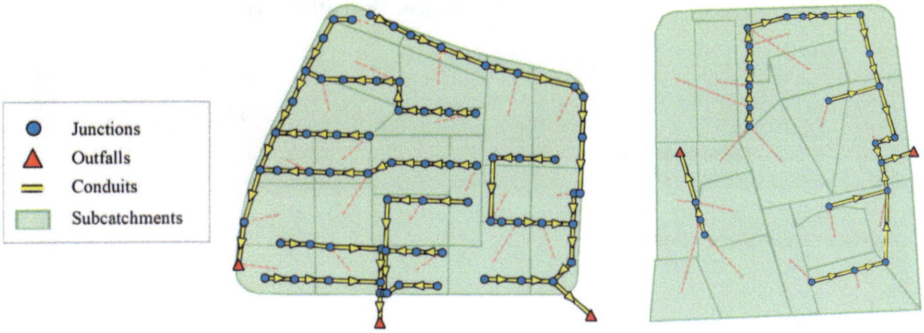

(a) Primary study area (b) Secondary study area

Fig. 44.1 SWMM model

primary study area was calculated using the area weighting method, yielding an average runoff coefficient of 0.59. The total rainfall volume in the study area was 2570.99 m^3, and the SWMM-simulated total runoff was 1765 m^3, resulting in an initial error of 14.06% relative to a reference runoff volume estimated using the average runoff coefficient.

Parameter sensitivity for the SWMM model was assessed using the Morris method under the specified 10-year design rainfall event. The results, including the calculated mean of the elementary effects (μ^*) and the standard deviation of the elementary effects (σ) for each parameter, are visualized in the heatmap presented in Fig. 44.2. From the analysis, key parameters relevant to SWMM (Rossman 2010) were identified, with pervious surface depression storage (D^P), characteristic width (W), pervious surface Manning's coefficient (N^P), and slope (S) as the most influential parameters affecting runoff simulation results in this study area. Figure 44.2 clearly illustrates the relative sensitivity of all examined parameters.

44.2.3 Model Calibration

Key SWMM parameters were calibrated using a Genetic Algorithm (GA) (Ma et al. 2020) (population size 100, 100 iterations) to minimize the percentage error between simulated and observed total runoff (Eq. 44.1). The calibrated parameter values are presented in Table 44.1.

$$f(x) = \frac{|R - R_{\text{observed}}|}{R_{\text{observed}}} \tag{44.1}$$

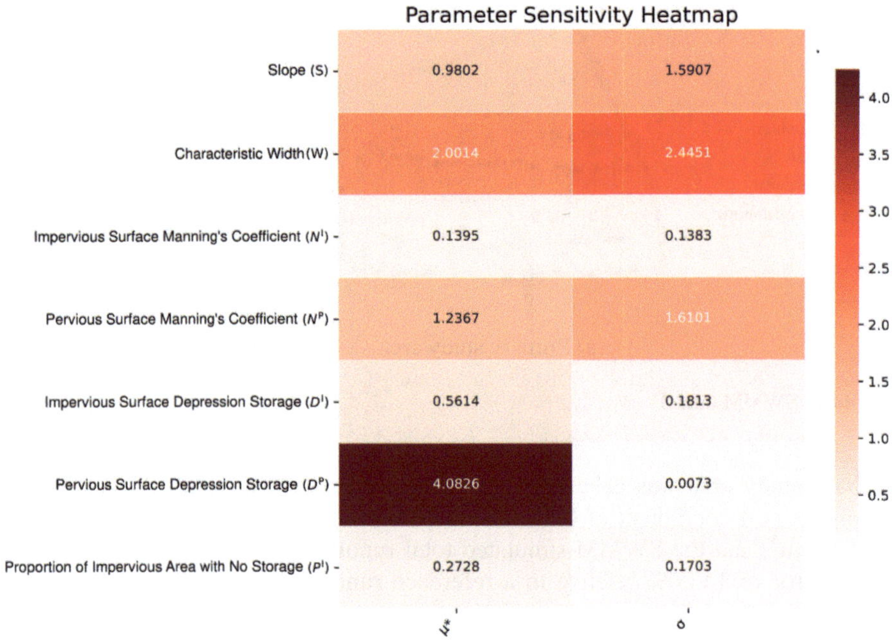

Fig. 44.2 Parameter sensitivity heatmap

Table 44.1 Calibrated parameter values

Parameters	Value
Slope (S)	0.2%
Characteristic width (W)	22
Pervious surface manning's coefficient (N^P)	0.19
Pervious surface depression storage (D^P)	10

44.2.4 Design Rainfall

Design Rainfall. Design rainfall events were generated using the local IDF formula for Xinjian District, Nanchang City:

$$q = \frac{1598 \times (1 + 0.69 \times \lg P)}{(t + 1.4)^{0.64}} \quad (44.2)$$

where q is rainfall intensity (L/(s·ha)), P is return period (years), and t is duration (min). Considering urban vulnerability to extreme events (Yu et al. 2018), representative return periods of 10, 30, and 50 years were selected. A 120-min Chicago hyetograph distribution with a peak coefficient of 0.4 was applied. The resulting hyetographs are shown in Fig. 44.3.

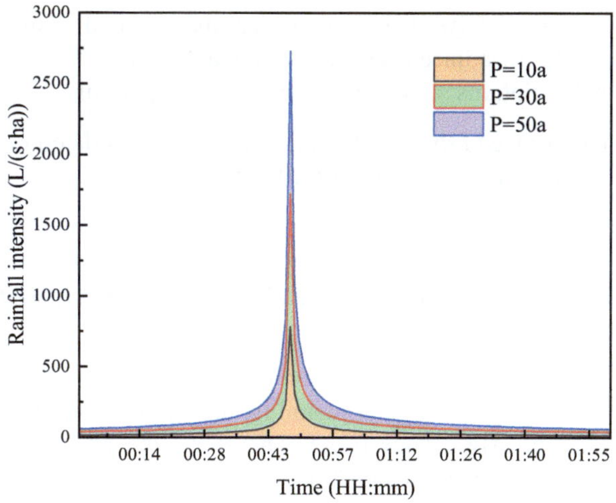

Fig. 44.3 Rainfall intensity for different return periods

44.2.5 LID Scenario Setup

LID facilities were designed to be placed within the primary sub-catchment areas. Appropriate LID types were selected based on the characteristics of the sub-catchment and its underlying surface type. A maximum of three different LID types were permitted per sub-catchment.

Based on examples from the "Case Study of Urban Public Facility Cost Indicators" (Research Institute of Standards and Norms, Ministry of Housing and Urban–Rural Development, Shanghai Municipal Engineering Design Research Institute (Group) Co., Ltd 2021), the unit construction costs for the LID facilities are provided in Table 44.2.

LID Spatial Layout Designs. Based on previous studies and observations of existing LID installations, the spatial arrangement of LID facilities can be broadly categorized into three main types (Jin et al. 2022):

Dispersed Layout. LID facilities are distributed as small-scale units across various sub-catchments or parcels, resulting in an overall uniform distribution.

Table 44.2 Estimated unit construction costs for LID facilities

LID device	Material cost	Labor cost	Machinery cost	Composite fee	Total (RMB/m^2)
Green roof	33.86	113.67	24.84	49.27	221.64
Sunken green space	13.17	48.35	21.47	27.79	110.78
Permeable pavement	45.6	210	16.63	70.9	343.13

Clustered Layout. LID facilities are grouped within a central location serving a cluster of adjacent areas or sub-catchments. This layout is relatively concentrated but maintains some degree of spatial dispersion.

Centralized Layout. LID facilities are consolidated in specific, designated locations, creating a distinct centralized pattern.

44.2.6 Application of the NSGA-II Algorithm

Multi-objective optimization problems, such as those involving trade-offs between different goals in LID planning, are often ill-suited for traditional single-objective Genetic Algorithms (GA). Therefore, the NSGA-II was employed in this study. NSGA-II (Deb et al. 2002) is an evolutionary algorithm based on GA, capable of simultaneously optimizing multiple conflicting objectives (Liu et al. 2022). The NSGA-II parameters were set as follows: population size = 100, number of generations (iterations) = 500, mutation probability = 0.1, and crossover probability = 0.9.

Selection of Optimization Objectives. The optimization objectives defined in this study were the minimization of LID construction cost and the maximization of the total runoff reduction rate. The specific spatial layout pattern (dispersed, clustered, centralized) resulting from the optimization was regarded as a characteristic of the solutions and was analyzed post-optimization, rather than serving as an explicit objective function, thereby ensuring a comprehensive evaluation.

LID Construction Cost. This represents the total economic investment required for the design, installation, and maintenance of the implemented LID facilities. The cost function is expressed as:

$$\text{Min}f(x) = \sum_{i=1}^{n}(A \times 221.64(\text{RMB}) + B \times 110.78(\text{RMB}) + C \times 343.13(\text{RMB})) \tag{44.3}$$

where A is the total area of the green roof arrangement, B is the total area of the sunken green space arrangement, C is the total area of the Permeable pavement, and n is the number of sub-catchments.

Total Runoff Reduction Rate. This quantifies the overall effectiveness of the LID scenario in reducing runoff volume compared to the baseline (pre-LID) condition. The function is expressed as:

$$\text{Max}R_{\text{total}} = \frac{\sum_{i=1}^{n}\left(R_{i,\text{original}} - R_{i,\text{present}}\right)}{R_{i,\text{original}}} \tag{44.4}$$

where R_{total} is the total runoff reduction rate, i represents the number of each sub-catchment, n is the number of sub-catchments, and $R_{i,original}$ and $R_{i,present}$ represent the runoff volume of sub-catchments before and after the setting of LID, respectively.

LID Layout Uniformity. This metric assesses how evenly the LID facilities are distributed across the study area. A combined approach using the Weighted Area Method and the Entropy Method was developed, followed by integration using TOPSIS.

Weighted Area Method. Calculate the distance of each sub-catchment from the other sub-catchments and weight the area of the LID facility by the distance. The specific calculations are as follows:

(1) Calculate the distance matrix.

$$d_{ij} = \sqrt{(x_i - x_j)^2 + (y_i - y_j)^2} \quad (44.5)$$

where d_{ij} denotes the distance between the centroids of sub-catchment i and sub-catchment j, and (x_i, y_i) and (x_j, y_j) are their respective centroid coordinates.

(2) Calculate the weighted influence for each sub-catchment i:

$$A_i = \sum\left(\frac{A_j}{d_{ij}^2}\right) \quad (44.6)$$

where A_j is the total LID area in sub-catchment j, and d_{ij} is the distance between the centroids of sub-catchment i and j. This metric represents the aggregated influence of LIDs in other sub-catchments on sub-catchment i.

Entropy method. Calculate the Shannon entropy of the LID distribution:

$$E = -\sum(p_i \ln(p_i)) \quad (44.7)$$

where p_i is the ratio of the total LID area within sub-catchment i to the area of sub-catchment i itself.

Composite score. Using the TOPSIS method, the weighted area and entropy value are first standardized and weights are set to finally calculate the comprehensive score. The specific steps are as follows:

(1) Standardized decision matrix.
(2) Calculate ideal and anti-ideal solutions.
(3) Calculate the distance to the ideal and anti-ideal solutions.
(4) Calculating similarity

$$C = \frac{D_-}{D_+ + D_-} \quad (44.8)$$

where C is the similarity and D_+ and D_- are the distances to the ideal and anti-ideal solutions, respectively.

The TOPSIS method was employed to calculate a composite uniformity score (C_i, ranging from 0 to 1) for each sub-catchment, based on its standardized weighted influence values and entropy value. A higher C_i score indicates a more desirable uniformity according to the TOPSIS evaluation.

The final composite score for the entire layout is then calculated by summing the individual C_i scores of all sub-catchments. There are 22 sub-catchments in the primary study area, while the secondary study area contains 18. Nevertheless, a constraint exists on the placement of the LID arrangements, limited to approximately 50%, so the range of scoring for the primary study area is 0–11, and the range of scoring for the secondary study area is 0–6. To better judge the significance of the scoring, the study will systematically divide the scoring intervals based on the numerical range of the composite scoring, as shown in Table 44.3.

Life-Cycle Carbon Emissions Calculation. Life-cycle carbon emissions, encompassing material production, transportation, and construction phases, were calculated using Eq. 44.9. Emission factors (EF_k, $kgCO_2e/m^2$) for each LID type (Table 44.4) were sourced from "Low Impact Development Rainwater Facilities for Urban Roads and Open Spaces" (China Academy of Building Standard Design: Low Impact Development Rainwater Facilities for Urban Roads and Open Spaces 2016).

$$C_{\text{total}} = \sum_{k=1}^{n} (A_k \times EF_k) \tag{44.9}$$

where A_k is the area (m^2) of the k-th LID type.

Table 44.3 Scoring intervals for layout uniformity classification

Primary study area scoring range	Secondary study area score range	Layout type
7–11	4–6	Dispersed
4–7	2–4	Clustered
0–4	0–2	Centralized

Table 44.4 Carbon emission factors for LID facilities by life cycle stage

Course of events	Sunken green space (kg/m^2)	Permeable pavement (kg/m^2)	Green roof (kg/m^2)
Material production	10.61	17.68	77.87
Material transportation	2.28	17.68	3.35
Construction	0.21	0.19	4.37
Total	13.1	35.55	85.59

Table 44.5 Maximum allowable LID coverage constraints by land use type

Study area	Land use type	Total area (m^2)	Area percentage (%)	Max LID coverage ratio (%)	Max LID area (m^2)	Max LID area percentage (%)
Primary study area	Roofing	6120.5	22.39	50	3060.28	11.13
	Roads	11,039	40.39	60	6623.95	24.1
	Reforest	10,324	37.77	50	5162.45	18.78
Secondary study area	Roofing	3648.3	20.46	50	1824.16	18.24
	Roads	7187.2	40.31	60	4312.35	43.12
	Reforest	6990.3	39.21	50	3495.17	34.95

Constraints. Maximum allowable LID coverage constraints based on specific land use types were applied during optimization, as detailed in Table 44.5. All LID percentages in this study are relative to the total study area.

44.3 Results and Analysis

44.3.1 Relationship Between Runoff Reductions and LID Construction Costs Under Different Return Periods

As illustrated in Fig. 44.4, there is a positive correlation between runoff reduction and LID construction costs across the different rainfall return periods evaluated. An increase in investment in LID facilities effectively enhances the capacity for stormwater runoff control. Furthermore, the effectiveness of runoff reduction varies with the rainfall return period. LID facilities demonstrate greater effectiveness under lower intensity rainfall conditions (i.e., shorter return periods), whereas achieving comparable reduction levels during higher intensity rainfall events (longer return periods) necessitates greater financial investment. Within the primary study area, for a given cost level, the runoff reduction achieved for the 10-year return period (P = 10a) generally exceeds that for the 30-year (P = 30a) and 50-year (P = 50a) return periods. For example, with an investment of 400,000 RMB, the runoff reduction for P = 10a reaches 26%, compared to approximately 22% for P = 30a and even less for P = 50a.

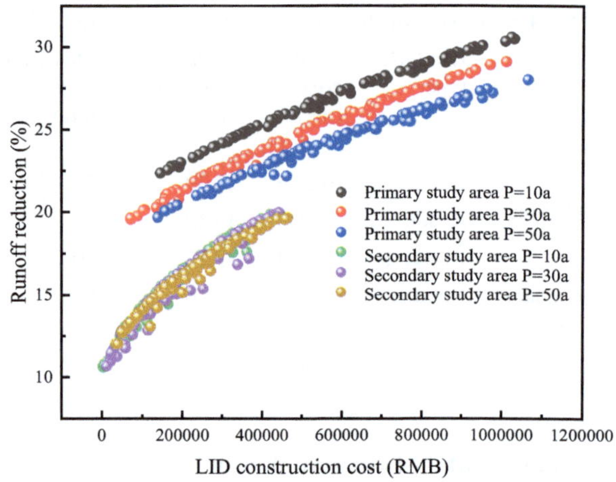

Fig. 44.4 Runoff reduction-cost correlation curve

44.3.2 Analysis of LID Scenarios on the P = 10a Pareto Front

Figure 44.5 illustrates the Pareto front curve, representing the trade-off between construction cost and runoff reduction for the 10-year return period (P = 10a) in the primary study area. Each point on this curve represents a non-dominated solution. The color intensity of the data points corresponds to their calculated Uniformity Score. Following an in-depth analysis of representative solutions from this Pareto front, including those with the highest and lowest construction costs, key data are presented in Table 44.6 for selected non-dominated solutions.

Among these solutions, the maximum cost scenario amounts to 1.0304 million RMB, achieving a runoff reduction of 30.5% with a uniformity score of 9.66. Conversely, the minimum cost scenario is only 145,200 RMB, yielding a 22% runoff reduction and a uniformity score of 4.49.

LID Facility Composition and Impact in Different Cost Scenarios

LID Facility Composition in Extreme Cost Scenarios. The composition of LID facilities varies significantly between the highest-cost and lowest-cost solutions on the P = 10a Pareto front (Fig. 44.6, Table 44.6). The highest-cost scenario achieves maximum runoff reduction at 30.5% and high uniformity (9.66), primarily relying on permeable pavement, which covers 11.55% of the area. This is supplemented by sunken green space and green roofs, accounting for 2.23% and 0.1%, respectively. Conversely, the lowest-cost scenario features limited permeable pavement, covering only 1.23% of the area, with minimal green roofs (0.1%), resulting in a lower runoff reduction of 22% and reduced uniformity (4.49). While this comparison underscores the significant impact of permeable pavement, the overall optimization results suggest the general order of contribution to runoff reduction is as follows: permeable pavement > sunken green space > green roof.

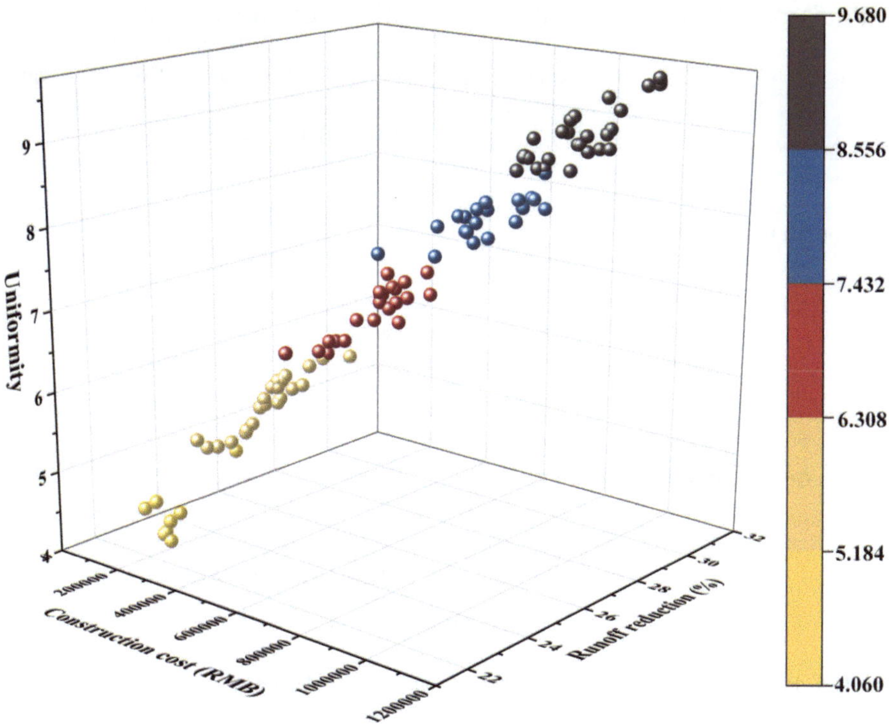

Fig. 44.5 Cost-runoff reduction pareto frontier

Table 44.6 Characteristics of representative non-dominated solutions (P = 10a Scenario)

Scenario	Cost (10 k)	Runoff reduction (%)	Uniformity	Green roof (%)	Sunken green (%)	Permeable pavement (%)	Total LID (%)
Max cost	103.04	30.5	9.66	0.1	2.23	11.55	13.88
Min cost	14.52	22	4.49	0.1	2.18	1.23	3.51

44.3.3 Analysis of Relationships Between LID Layout Pattern, Cost, and Runoff Management Performance

As presented in the NSGA-II optimization results (Fig. 44.7), clear differences in runoff management performance are observed among the different LID layout patterns:

Dispersed Layout. For dispersed layouts, where LID facilities are evenly distributed, the solution points cluster in the region characterized by high uniformity scores (indicated by yellowish colors in the plot). This uniform distribution facilitates relatively even rainwater infiltration and retention across the area, leading to effective runoff

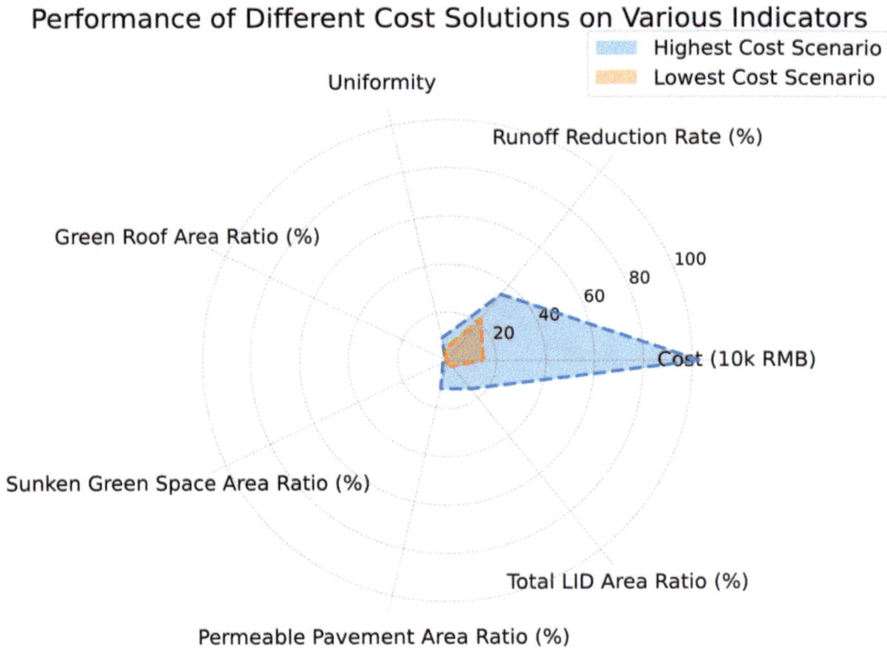

Fig. 44.6 Performance of different cost solutions on various indicators

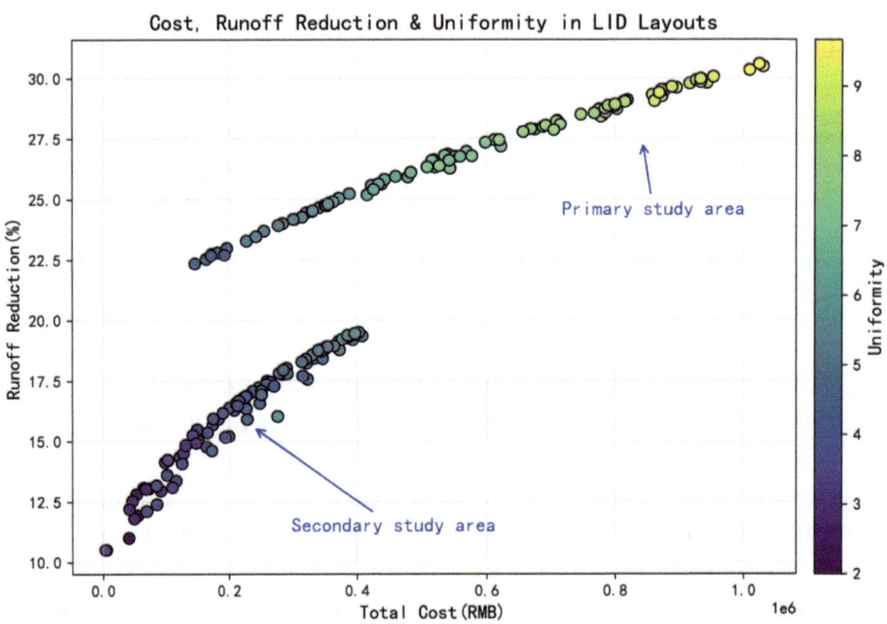

Fig. 44.7 Relationship between total cost, runoff reduction, and uniformity of LID layout methods

reduction. However, implementing facilities across numerous locations results in higher overall construction costs, placing these solutions predominantly in the higher cost range of the Pareto front.

Clustered Layout. In clustered layouts, LID facilities are grouped in central locations while maintaining some spatial dispersion. This arrangement demonstrates stable runoff management performance by balancing centralized treatment with sufficient dispersion to prevent localized overloading. Typically, the costs associated with clustered layouts are lower than those for dispersed layouts, positioning these solutions within the medium cost range while achieving a favorable balance between runoff management effectiveness and cost efficiency.

Centralized Layout. Centralized layouts involve the consolidation of LID facilities in specific locations. These solutions generally incur the lowest costs, with corresponding data points concentrated in the lower cost region of the plot. However, they exhibit lower uniformity scores. During high-intensity rainfall, the concentrated facilities may become overloaded, leading to diminished runoff reduction effectiveness. As a result, solution points for these layouts cluster in the region of low uniformity scores (indicated by purplish colors).

44.3.4 Analysis of Total Life-Cycle Carbon Emissions

The composition of life-cycle carbon emissions was analyzed for the 100 non-dominated solutions obtained for the P = 10a scenario in the primary study area.

Figure 44.8 provides a detailed breakdown of these emissions. Among the scenarios optimized to balance runoff reduction with cost minimization, carbon emissions from permeable pavement dominate, accounting for 87.59% of the total life-cycle emissions (in $kgCO_2e$). In contrast, green roofs contribute only 1.05%, highlighting their relatively low life-cycle carbon emissions. Sunken green spaces account for 11.36%, representing a secondary source of emissions among the evaluated LID types.

Analysis of the Relationship between Total Life-Cycle Carbon Emissions and Runoff Reduction. Furthermore, the relationship between total life-cycle carbon emissions and runoff reduction for the 100 non-dominated solutions was examined, as depicted in Fig. 44.9.

The results of linear regression analysis, summarized in Table 44.7, reveal a highly positive coefficient between total life-cycle carbon emissions and runoff reduction, with a coefficient determination (R-squared) of 0.98938. This indicates a strong linear correlation, where each 1% increase in runoff reduction corresponds to an approximate rise of 12.64 tonnes in total life-cycle carbon emissions.

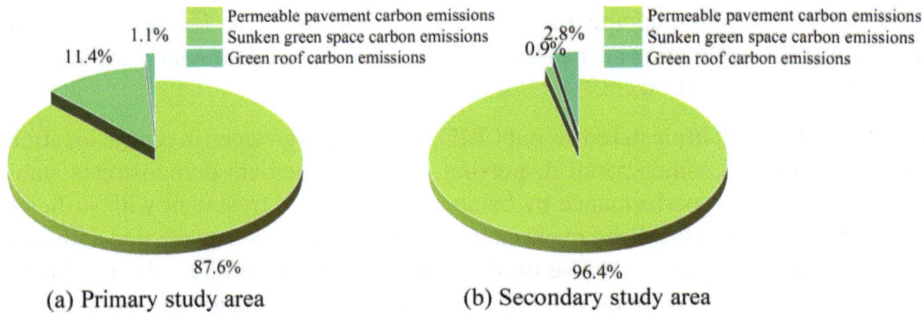

Fig. 44.8 Proportion of each type of carbon emission in total carbon emission

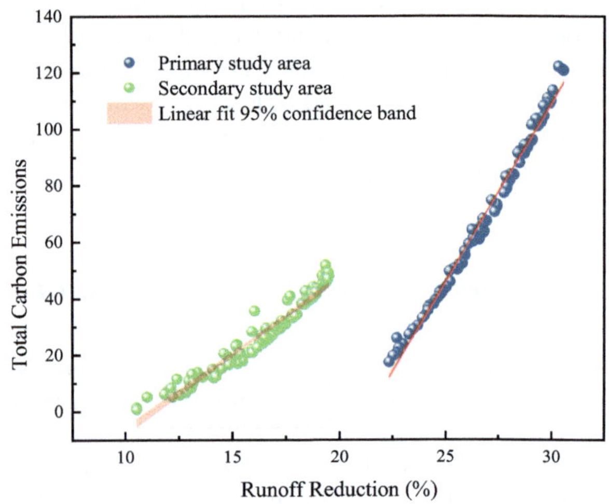

Fig. 44.9 Relationship between total carbon emissions throughout the process and runoff reduction

Table 44.7 Linear fitting data of total carbon emissions throughout the process and runoff reduction

Equation	Intercept	Slope	RSS	R-squared
$y = a + b * x$	-270.64 ± 3.54	12.6416 ± 0.1310	886.46428	0.98938

44.3.5 Optimal Scenario Analysis

For the 10-year return period in the primary study area, five representative LID implementation scenarios were selected from the non-dominated solution set generated by NSGA-II, spanning a range of costs. Total life-cycle carbon emissions were calculated for these scenarios using the method described earlier, with results presented alongside other metrics in Table 44.8.

As shown in Fig. 44.10, the selected scenarios exhibit runoff reductions ranging from 22% to 30.5%, and total life-cycle carbon emissions ranging from 17.32 tonnes to 150.74 tonnes. Across these five scenarios, a clear trend emerges: both runoff

Table 44.8 Data of five sets of LID configuration schemes

scenario name	Green roof (%)	Sunken green space (%)	Permeable pavement (%)	Uniformity	Cost (10 k)
Scenario 1	0.1	2.23	11.55	9.66	103.04
Scenario 2	3.51	4.42	4.62	7.29	73.43
Scenario 3	1.17	3.79	3.76	6.64	50.04
Scenario 4	0.1	2.19	3.03	5.68	32.04
Scenario 5	0.1	2.18	1.23	4.49	14.52

reduction and total life-cycle carbon emissions generally decrease with decreasing construction cost. However, significant variations in cost investment and overall performance are observed among the different LID scenarios.

High-Cost Scenarios (Scenario 1 & 2). These options, costing 1.0304 million RMB (Scenario 1) and 0.7343 million RMB (Scenario 2), represent significant investments. Scenario 1 provides optimal runoff reduction (30.5%) and uniformity (9.66) as a dispersed layout, but its minimal green infrastructure (0.1% green roofs, 2.23% sunken green spaces) limits ecological benefits, and it has high carbon emissions (121.11 tonnes). Scenario 2 offers 23% runoff reduction and high uniformity (7.29) with more green infrastructure (3.51% green roofs, 4.42% sunken green spaces), enhancing ecological potential. However, Scenario 2 is less cost-effective than the balanced Scenario 3, offering similar runoff control at a higher cost and greater carbon emissions.

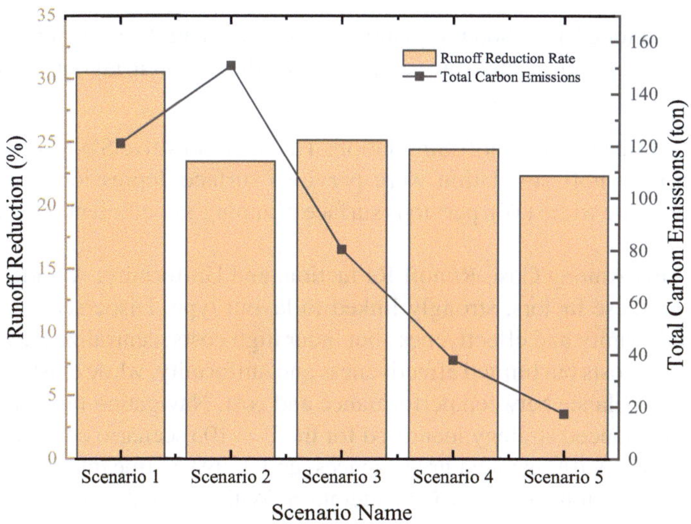

Fig. 44.10 Comparison of runoff reduction rate and total life cycle carbon emissions for the five selected scenarios

Low-Cost Scenarios (Scenarios 4 & 5). The low-cost options (Scenarios 4 and 5) are economically advantageous (costs of 320,400 RMB and 145,200 RMB, respectively) but provide less than 25% runoff reduction. Moreover, the proportions of green roofs and sunken green spaces are low, leading to limited ecological benefits. Their uniformity scores are also lower (5.68 and 4.49, respectively), reflecting their clustered layout patterns. These types of layouts may be susceptible to localized overloading during high-intensity rainfall events and may not fully meet broader goals for urban resilience and synergistic ecological development.

Balanced Scenario (Scenario 3). The balanced option (Scenario 3) achieves a moderate runoff reduction of 25%, with a uniformity score of 6.64, approaching the performance of a dispersed layout. With a construction cost of 500,400 RMB, this option represents a 51.43% reduction in cost compared to Scenario 1, while also reducing carbon emissions by 33.55%. From an ecological perspective, the combination of 1.17% green roofs and 3.79% sunken green spaces offers synergistic benefits, such as adsorbing particulate matter (e.g., PM10), regulating local microclimate, and enhancing the community's aesthetic value.

In summary, Scenario 1 emerges as the optimal choice if stormwater control is prioritized, offering maximum runoff reduction and uniformity. However, when considering a broader range of factors including cost-effectiveness, LID layout implications, ecological co-benefits, and overall sustainability, Scenario 3 emerges as a more suitable and balanced choice for implementation within the community.

44.4 Conclusion

This study developed a method for quantifying the spatial uniformity of LID facilities and conducted a multifaceted investigation into their spatial layout optimization, yielding the following key conclusions:

(1) Key Parameters for Runoff Simulation: The most sensitive SWMM parameters impacting runoff simulation were pervious surface depression storage (D^P), characteristic width (W), pervious surface Manning's coefficient (N^P), and slope (S).
(2) Trade-offs Among Cost, Runoff Reduction, and Uniformity: A trade-off exists between these factors, strongly linked to layout type. Dispersed layouts offer high uniformity and effectiveness but incur high costs; centralized layouts have the lowest costs but limited effectiveness and uniformity; while clustered layouts provide a balance between performance and cost. Navigating this trade-off, the optimal balanced strategy identified for the P = 10a scenario comprised 1.17% green roofs, 3.79% sunken green spaces, and 3.76% permeable pavements.
(3) Carbon Emissions as a Key Consideration: With the goal of minimizing costs and maximizing runoff reductions, permeable pavement is the primary source of life-cycle carbon emissions. Total emissions show a significant positive correlation with runoff reductions, i.e., for every 1% increase in runoff reduction,

total carbon emissions increase by approximately 12.6 tonnes of carbon dioxide equivalent (CO_2e).

These insights provide technical support for optimizing LID layouts in small-scale areas, though future research should encompass more diverse urban contexts.

Acknowledgements This work was supported by the Nanchang Municipal Housing and Urban-Rural Development Science and Technology Program (2024ZJXM022), the Jiangxi Training Program of Innovation and Entrepreneurship for Undergraduates (S202413431011), and the Nanchang Jiaotong Institute Student Research Project Fund (NJX2024-06).

References

Chang C, Lo SL, Huang S (2009) Optimal strategies for best management practice placement in a synthetic watershed. Environ Monit Assess 153:359–364

China Academy of Building Standard Design: Low Impact Development Rainwater Facilities for Urban Roads and Open Spaces (2016) China Planning Press, Beijing

Deb K, Pratap A, Agarwal S et al (2002) A fast and elitist multiobjective genetic algorithm: NSGA-II. IEEE Trans Evol Comput 6(2):182–197

Dou M, Yan JJ, Wang C et al (2024) Assessment of the impact of LID facility layout on urban runoff and pollution loads. J. Zhengzhou Univ (Eng Sci) 45(4):87–94

Jin MX, Tian Y, Lancia M et al (2022) Research of the spatial configuration strategy of sponge facilities under extreme precipitation. Water Resour Hydropower Eng 53(S2):100–106

Liquete C, Kleeschulte S, Dige G et al (2015) Mapping green infrastructure based on ecosystem services and ecological networks: a Pan-European case study. Environ Sci Policy 54:268–280

Liu J, Gao XJ, Jiao YB et al (2022) Multi-objective optimization of the urban drainage system based on SWMM and NSGA-III algorithm. China Rural Water Hydropower 9:90–94

Ma YB, Sheng W, Dai WJ et al (2020) Parameter calibration of SWMM model based on GA. China Rural Water Hydropower 7:46–49

Martin-Mikle CJ, de Beurs KM, Julian JP et al (2015) Identifying priority sites for low impact development (LID) in a mixed-use watershed. Landsc Urban Plan 140:29–41

Research Institute of Standards and Norms, Ministry of Housing and Urban-Rural Development, Shanghai Municipal Engineering Design and Research Institute (Group) Co., Ltd. (2021) Case study of cost indicators for urban public facilities. China Planning Press, Beijing

Rossman LA (2010) Storm water management model user's manual, Version 5.0. U.S. Environmental Protection Agency, Cincinnati, OH

Yang B, Lee D (2021) Urban green space arrangement for an optimal landscape planning strategy for runoff reduction. Land 10(9):897

Yu C, He L, Zhang HD et al (2022) Analysis on characteristics and causes of the extreme rainstorm on 16 July 2018 in Beijing City. J Hohai Univ (Nat Sci) 50(4):42–49, 82

Chapter 45
Assessing Spatial Compactness in Urban Neighborhoods: The Case of Alexandria, Egypt

Hadeel Ayman, Hany M. Ayad, and Dina M. Saadallah

Abstract The field of urban planning has recently adopted various innovative paradigms to foster sustainable cities. Among these, compact urban forms have gained prominence due to their emphasis on creating high-density areas that promote environmental, economic, and social sustainability through diversity and connectivity. Despite the numerous research and studies dedicated to urban compactness, there are still doubts about their veracity in promoting sustainability. The debate remains unresolved. This paper attempts to understand the factors that support or hinder the sustainable development of compact neighborhoods. This goal is achieved through an adopted methodology following the sustainable neighborhood planning strategy proposed by the United Nations Human Settlements Programme (UN-Habitat), as well as conducting a survey to understand residents' experiences and perceptions of living in these neighborhoods. Results revealed that both neighborhoods exhibit compact development patterns, with most urban characteristics approaching, but not fully meeting, UN-Habitat-recommended sustainability benchmarks. Nevertheless, they fail to achieve the desired sustainability progress, exhibiting various shortcomings, particularly in environmental and social aspects.

Keywords Compact urban form · Urban sustainability · UN-Habitat

45.1 Introduction

Emphasizing sustainability, urban forms have changed drastically in the twentieth century. One feature of this change is substituting the compact urban model for the spreading metropolitan one. Lately, the sprawling urban form has drawn criticism. Many studies point to scattered and fractured urban forms as causing traffic congestion, land loss, air pollution, and growing infrastructure expenses (Duany et al. 2000;

H. Ayman (✉) · H. M. Ayad · D. M. Saadallah
Architecture Department, Faculty of Engineering, Alexandria University, Alexandria, Egypt
e-mail: hadeel.ayman@alexu.edu.eg

Y. Li et al. (eds.), *16th International Conference on Environmental Science and Development*, Environmental Science and Engineering,
https://doi.org/10.1007/978-3-032-15080-6_45

Carruthers and Ulfarsson 2002; Burchell and Mukherji 2003). In contrast to sprawling models, compact urban models were well-known for emphasizing urban intensification, limiting urban growth, supporting mixed-use development, and encouraging public transportation (Jenks et al. 1996b). Urban form, as stated by UN-Habitat, influences factors such as housing, employment, accessibility, and safety. They assert that higher urban densities foster environmental sustainability, social cohesion, and prosperity, thereby enhancing quality of life (Dodman et al. 2013).

Indeed, Dodman and Mcgranahan suggest that compact, mixed-use cities represent the most resource-efficient settlement pattern (Dodman et al. 2013).

Nevertheless, still up for contention in sustainability debates are compact cities. Critics claim that higher densities can be detrimental. Lack of sustainable and livable environment policies makes many decision-makers in underdeveloped nations see compact areas as inappropriate and uncomfortable. Projects of recent urban development have mostly embraced the sprawled model, indicating this lack of care. In new city development, Egyptian professionals typically favor low-density cities.

Emphasizing both physical features and the quality of urban planning, design, and resident satisfaction, this study aims to create a thorough assessment approach for small neighborhoods. This assessment will help us to pinpoint the factors and reasons preventing these areas from reaching the intended level of compact forms sustainability.

45.2 Urban Compactness: Definition and Metrics

The concept that would later be known as the 'compact city' was first proposed in the 1930s as a response to the concerns about urban sprawl and gained momentum in the 1960s with figures like Jacobs (1989). The term 'compact city' was coined in 1973 by Dantzig and Saaty (1973). Since then, researchers have investigated various definitions and metrics for urban compactness, resulting in different opinions about what constitutes a compact city.

Burton noted that the common interpretation of a compact city in UK policy refers to a self-contained urban settlement (Jenks et al. 1996a). Another definition provided by Jenks, Williams, and Burton is that compact city is characterized by high density and mixed land use with no sprawl (Jenks et al. 1996a).

A universally accepted definition of a compact city remains ununified; this has led to diverse dimensions and indicators for assessing compactness, which are often shaped by specific study objectives. Researchers have adopted various approaches: Turskis and Zavadskas focused on population density, distribution, and transport (Turskis et al. 2006). Burton used urban density, land use diversity, and intensity across 25 English cities (Burton 2002). Similarly, Abdullahi and Pradhan applied Burton's indicators in Kajang City (Abdullahi et al. 2014), while Liaqat and Waheed added transport and accessibility factors (Liaqat et al. 2017). Zhao and Tang analyzed 160 Chinese cities with points of interest, density, and mixed land use (Zhao et al. 2020). Min and Jin incorporated spatial form (Min et al. 2013). Rahman and Islam focused on the neighborhood level and used the road network connectivity index to link local compactness with overall urban compactness (Rahman et al. 2022).

45.3 Urban Compactness Sustainability Debate

The concept of compact urban development, once controversial, is now widely embraced for its sustainability benefits. Advocates, including UN-Habitat, highlight its positive impacts on resource efficiency, economic growth, public health, social cohesion, and cultural vibrancy (UN-Habitat 2014). Studies indicate that compact cities reduce travel, shorten commutes, decrease car reliance, lower energy consumption and preserve green spaces. Moreover, it has been found that this approach has numerous benefits in terms of resource efficiency, economic growth, public health, social unity, and cultural vitality (Henshilwood and Cullinan 2012; Zhang 2011). Other researchers have shown that the concept of urban compactness can promote sustainability by reducing the amount of travel and shortening commute time, decreasing car dependency, lowering per capita rates of energy use, maintaining the diversity of choice among workplaces, service facilities, and social contacts, and limiting the loss of green and natural areas (Burton 2002; Bibri et al. 2020; Arbury 2005). In general, most researchers advocating for urban compactness agree that this approach contributes positively to environmental, social, and economic outcomes.

However, Critics assert that overly dense urban forms cause noise pollution, higher living expenses, transportation congestion, social exclusion, and lower resident satisfaction (Neuman 2005; Bramley and Power 2009). Compact policies can affect residents, Breheny found that people are unhappy with higher-density housing (Breheny 1997). Accordingly, UN-Habitat recommends well-planned compact neighborhoods that follow sustainable guidelines to balance density and sustainability.

45.4 Sustainable Neighborhood Planning Strategy

The UN-Habitat program supports human settlement development efficiency. It aims to create more compact, socially inclusive, integrated, and connected developments. Thus, it has promoted principles and guidelines for sustainable urban planning to foster livable and efficient neighborhoods. Figure 45.1 summarizes UN-Habitat's principles and the recommended ranges for each indicator.

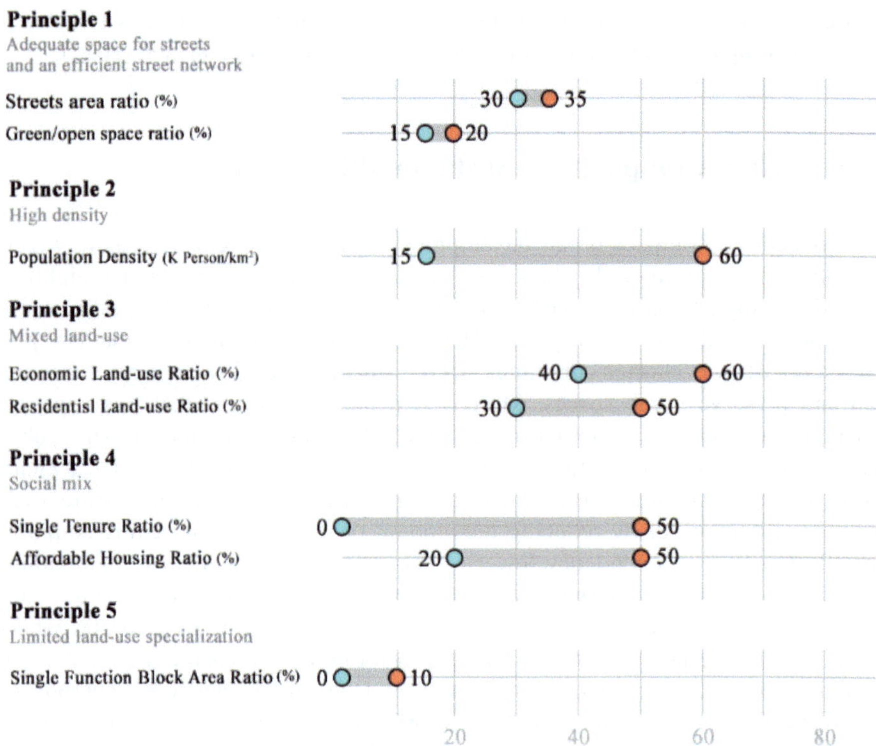

Fig. 45.1 UN-Habitat principles for sustainable compact neighborhood planning, showing recommended value ranges (gray bars) with minimum (blue-green circles) and maximum (orange circles) targets for each indicator. *Source* UN-Habitat (2014)

45.5 Materials and Methods

The compactness assessment methodology comprised three main phases: The first phase measured density, diversity, and connectivity within the studied neighborhoods. The second phase evaluated urban planning quality based on the UN-Habitat's five principles for sustainable, compact, and connected neighborhoods (UN-Habitat 2014). In the third phase, a survey was conducted to test residents' satisfaction with the environmental, social, and economic aspects of the neighborhood.

45.5.1 Phase One: Spatial Compactness Assessment

This section outlines the methodology used to measure the compactness of the selected neighborhoods through a set of selected indicators, the selection of which was guided by a comprehensive literature review.

Urban Form Indicators. To provide a more detailed understanding of the study areas' urban morphology, this research examines urban compactness at the neighborhood scale. Urban Compactness is defined through three core Indicators: density, diversity, and connectivity. The following section provides a comprehensive exposition of the chosen metrics employed for each indicator, elucidating the methodological reasoning for their selection.

Urban Density. In the urban context, density refers to the amount of a particular feature within a defined area, like a block, neighborhood, city, or metropolitan region (Kain et al. 2021). Our study will particularly examine population density, building coverage, and floor area ratio.

(a) *Population Density.*

When conducting urban density analysis, population density is typically the primary concern of researchers. The gross population density formula was utilized to get a precise perception of the population distribution over the total area of the studied neighborhoods.

$$\text{Pop. Density}(\text{person/km}^2) = \frac{P_i}{A_i} \tag{45.1}$$

where P_i = population of neighborhood i; A_i = total area of neighborhood i.

(b) *Building Coverage Ratio.*

Building coverage ratio (BCR) refers to the measurement of the extent of constructed areas within a specific space. This measure determines how crowded the built-up area in a neighborhood appears, as a higher BCR refers to a more crowded built-up area.

$$\text{BCR}(\%) = \frac{B_i}{A_i} \times 100 \tag{45.2}$$

where B_i = total built-up area in neighborhood i; A_i total area of neighborhood i.

(c) *Floor Area Ratio.*

The floor area ratio (FAR) measures a building's floor area relative to the area of its lot. It is expressed as a decimal. FAR effectively assesses the bulk or mass of building volume on a development site and is often used alongside building coverage and lot area to provide a thorough understanding of the building block form. Higher FARs indicate greater building volume.

$$\text{FAR} = \frac{\text{GFA}}{\text{A}_\text{i}} \tag{45.3}$$

where GFA = gross floor area of the built area in the neighborhood i; A_i = total area of neighborhood i.

Diversity. More diversity makes a region more self-sufficient and compact. Land use diversity affects commuting and mobility. Their proximity reduces car dependency. The Simpsons Index measures land use diversity in a region (Borchers and Pieler 2010). Index values range from 0 to 1, with 1 indicating infinite diversity and 0 no diversity. See the Simpsons Index formula below.

(a) Land use diversity

$$\text{Simpson's Index} = 1 - \sum \left(\frac{a}{A}\right)^2 \tag{45.4}$$

where a = The total area of a specific land use category; A = is the total area of all land use categories.

Alongside the computational method, ArcGIS Pro was utilized with weighted overlay analysis to assess land use distribution and proximity. Five categories (residential, commercial, institutional, community facilities, and recreational spaces) were analyzed separately using the Euclidean distance tool. Distances were reclassified from 1 to 5, with 5 indicating high proximity. The combined layers produced a map highlighting areas with diverse land use, with a maximum value of 25 indicating proximity to all five categories.

Design. Urban features like streets, open spaces, and structures shape urban space characteristics. Street network design is crucial for enhancing connectivity and fostering compactness. The ArcGIS network analysis tool was used to measure street connectivity, with the network divided into links, real nodes, and dangle nodes. Key connectivity definitions are outlined in Table 45.1.

(a) Connectivity of road network

The following formulas were used to measure the indices of connectivity (Tresidder and Sharkova 2005):

$$\text{Street Density (street /km}^2) = \frac{X_i}{A_i} \tag{45.5}$$

$$\text{Intersection Density (Nodes/km}^2) = \frac{R_i}{A_i} \tag{45.6}$$

Table 45.1 Connectivity definitions

Phrase	Definition
Link	A roadway or pathway segment between two nodes. A street between two intersections or from a dead end to an intersection
Real node	The endpoint of a link that connects to other links. An intersection
Dangle node	The endpoint of a link that has no other connections. A dead-end or cul-de-sac

Source Tresidder and Sharkova (2005)

$$\text{Connected node ratio}(\%) = \frac{R_i}{K_i} \times 100 \tag{45.7}$$

$$\text{Link-Node Ratio (Connectivity Index)} = \frac{L_i}{K_i} \tag{45.8}$$

where X_i = total street length in the neighborhood i, R_i = the total number of real nodes excluding dangle nodes in the neighborhood i, K_i = total number of nodes including dangle nodes in the neighborhood i, L_i = total number of links in the neighborhood i, A_i = Total area of neighborhood i.

The level of street connectivity is closely linked to the density of intersections and streets. Increased densities of these elements signify a more connected network. When considering the connected node ratio, the maximum value of 100 indicates networks free of dangle nodes, and presumably higher connectivity. Similarly, a higher link-node ratio indicates higher network connectivity. The perfect grid has a ratio of 2.5 (Tresidder and Sharkova 2005).

Data Aggregation, Weighting, and Normalization. The composite compactness index (CCI) for neighborhoods is calculated by summing and aggregating indicator values, with a higher CCI reflecting greater compactness. A multi-criteria decision-making process, specifically the Simple Additive Weighting (SAW) method was used (Afshari et al. 2010). SAW normalizes spatial indicators with varied units, categorizing them as either beneficial or cost criteria based on their impact on compactness. Benefit criteria favor higher compaction values, while cost criteria do the opposite. The normalization formulas used are detailed below (Afshari et al. 2010).

For benefit criteria:

$$n_{ij} = \frac{r_{ij}}{Max(r_{ij})} \tag{45.9}$$

For non-benefit or cost criteria:

$$\mathrm{n_{ij}} = \frac{\mathrm{Min(r_{ij})}}{\mathrm{r_{ij}}} \tag{45.10}$$

where n_{ij} = is the normalised value of the jth indicator from the ith neighborhood; r_{ij} = is the original value for the jth indicator from ith neighborhood.

As a last step, the normalized values were employed in the weighted sum matrix. The CCI was obtained by multiplying the unitless values with the corresponding weights. In the previous studies, various methodologies such as the Analytic Hierarchy Process (AHP), public opinion, and scorecards were utilized to determine the weights of compactness dimensions. However, for the sake of simplicity, this study opted for the equal weight approach, inspired by the research of Burton (2002). The overall CCI was measured by the following equation (Afshari et al. 2010):

$$CCI_i = \sum_{i=1}^{M} wj\ n_{ij} \tag{45.11}$$

where $CCI_i =$ is the composite compactness index for the ith neighborhood; $wj =$ is the weight of the jth indicator; $n_{ij} =$ is the normalized value of jth indicator from ith neighborhood; M = the number of urban compactness assessment indicators.

45.5.2 Phase Two: Strategic Alignment Assessment

The first phase of the analysis, the objective was to measure a set of spatial indicators that reflect the degree of compactness within the two selected neighborhoods. This was followed by the second phase, which involves evaluating these spatial characteristics against the recommended ranges established by the UN-Habitat. It claims that its strategy is a vital framework to create a well-balanced compact neighborhood.

45.5.3 Phase Three: Perception-Based Assessment

To elaborate, asking neighbors about the economic, social, and environmental conditions of their area marks the last phase of the assessment. Including locals in the assessment process is essential since their feedback provides a direct reflection of how well compact design principles translate into real-life experiences. Residents' level of satisfaction depends on their interaction with and perspective of the neighborhood. We created questions to evaluate the three aspects—environmental, social, and financial. On a five-point Likert scale, respondents score each statement strongly agree and strongly disagree. The answers provided will indicate which neighborhood features are appreciated by the local community and which require improvement.

45.5.4 Study Area

The city of Alexandria is in the northern part of Egypt, directly along the Mediterranean Sea. According to the Central Agency for Public Mobilization and Statistics (CAPMAS) The population is more than five and a half million (5,597,171 in August 2024). Our investigation will focus on two primary neighborhoods in the eastern part of Alexandria, Sidi Beshr Bahari, and AL-Mandara Qebly are shown in Fig. 45.2.

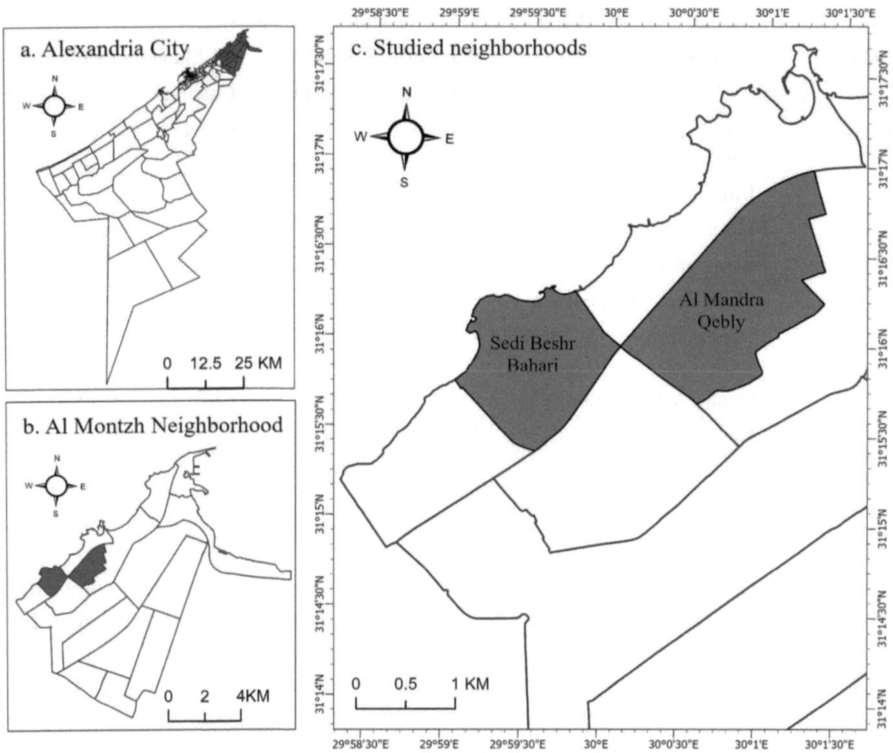

Fig. 45.2 The geographical location and boundaries of the studied neighborhoods. *Source* Author

45.6 Results

45.6.1 Spatial Compactness Assessment

To measure urban compactness, neighborhoods were measured by density, diversity, and connectivity. Except for Al-Mandara, which boasts double Sidi Beshr's population density, both areas have comparable urban densities. Many high-rise buildings reflect high-density development indicated by high building coverage and floor area ratios. (The value summary is in Table 45.2.)

Simpson's Index revealed rather different land use patterns in both regions. This occurred due to the predominance of residential and commercial spaces. With most land uses mixed and close together, the overlaid land use map revealed no land use zoning. Both areas show well-balanced distributions between 17 and 25, suggesting that every component is at least roughly three categories apart. For both sites, Fig. 45.3a–c presents thematic maps of building heights, built-up area, and weighted land use.

Table 45.2 Calculated and normalized values of compactness indicators

Indicators	Metrics	Sidi Beshr Bahari		Al-Mandara Qebly	
		Calculated	Normalized	Calculated	Normalized
Density	Population density (person/km^2)	58,741	0.60	97,245	1.00
	Building coverage ratio (%)	68	1.00	63	0.92
	Floor area ratio	6.11	1.00	5.65	0.92
Mixed-use	Simpson's index	0.60	1.00	0.58	0.97
Connectivity	Street density (street/km^2)	44.08	1.00	43.25	0.98
	Intersection density (nodes/km^2)	410.12	0.52	784.31	1.00
	Connected node ratio (%)	85.24	1.00	85.21	0.99
	Link-node ratio	1.46	1.00	1.39	0.95
CCI Value			0.89		0.97

Source Author

Al-Mandara and Sidi Beshr had practically exact road density, link-node ratio, and connected-node ratio in Fig. 45.3d. Although Al-Mandar had more nodes than Sidi Beshr, both have many roads and nodes, implying a well-connected road system. High values in all measures point to great urban compactness in both neighborhoods. Additionally in the table are the normalized values of these indicators used to derive the CCI. Consequently, both areas are practically as small. Al-Mandara's somewhat higher CCI comes from its higher intersection density than Sidi Bishr.

45.6.2 Analysis of the Current Urban Planning Conditions

Specifically, according to the UN-Habitat urban planning strategy described in Fig. 45.1, the neighborhoods can be described by the following characteristics:

Al-Mandara Qebly and Sidi Beshr Bahari fall short of UN-Habitat minimum of 30% with 26% and 27% street space. Although Al-Mandar has none, both neighborhoods lack green space. Public squares along the seaside promenade in Sidi Beshr are restricted to 6% of the total area. If we are to set up enough social and interactive spaces, this should be raised to 15% or more. Al-Mandara has 97,245 people/km^2, more than what principle 2 allows; Sidi Beshr has 58,741, within the recommended range. In principle 3, residential consumption rises to 57% (Sid Beshr) and 56% (Al-Mandara), the maximum ratio is 50%. To achieve 40%, Al-Mandara's economic area must increase from 38%. In Sidi Beshr, 53% satisfies principle 3 by serving economic needs. Even though the tenure forms of the two areas are distinct, Sidi Beshr has 57% ownership units, which exceeds the recommended 50%. Although neither area has any, the UN recommends 20% affordable houses. According to principle five,

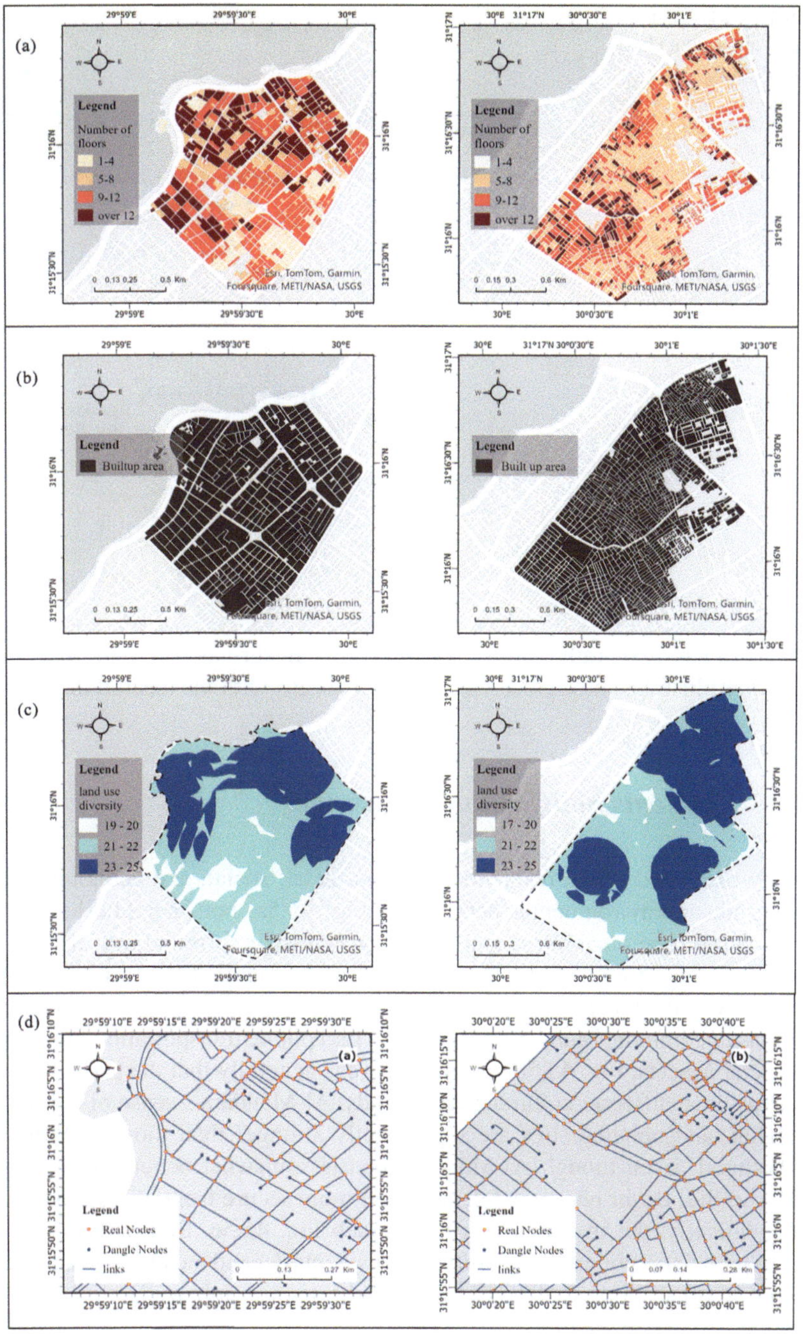

Fig. 45.3 Thematic maps for the studied neighborhoods **a** building heights map, **b** solid and void map, **c** weighted overlay map for the diversity of land uses and **d** road network for studied neighborhoods, Sidi Besh Bahar neighborhood is on the left and Al Mandara Qebly neighborhood is on the right. *Source* Author

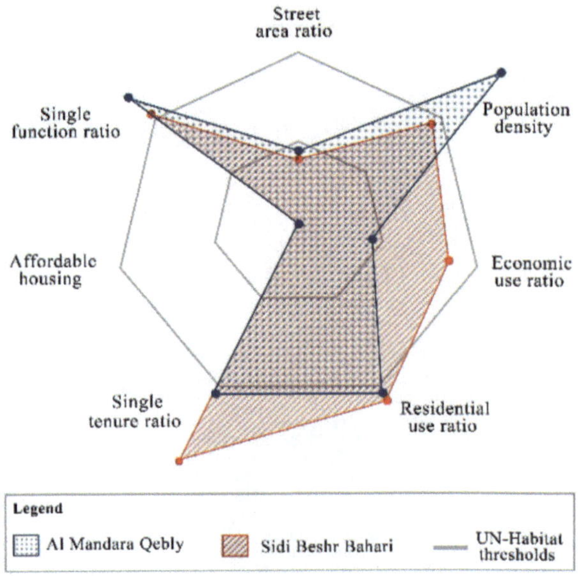

Fig. 45.4 Un-Habitat threshold alignment chart. *Source* Author

Al-Mandara has 12%, rather above the limit; Sidi Beshr has 9% single-use blocks. Figure 45.4 illustrates how closely each region aligns with the desired thresholds.

45.6.3 Residents Satisfaction Assessment

A survey examined urban residents' environmental, social, and economic experiences. From an environmental perspective Fig. 45.5a, both neighborhoods show little variation. The cleanliness and noise level of their surroundings disappointed the people living in both areas. Regarding mobility, many of the residents claimed that the proximity of local facilities motivated walking as their main form of mobility. Cycling is uncommon. Many of the residents find public transportation unfavorable.

In social aspects, Fig. 45.5b, everyone agrees generally that the people living in both areas have different cultural backgrounds. Al-Mandara's sense of community, socializing, and neighbor ties are significantly stronger than those in Sidi Beshr neighborhood. Even though Al-Mandara boasts somewhat better social ties than Sidi Beshr, Sidi Beshr residents felt safer and more secure from crimes and thefts.

Economically, the two areas vary not much Fig. 45.5c. Both areas have good opinions of their neighborhood vitality, and residents depend mostly on nearby stores. Variables in prices options also get some good response. Job availability responses, meantime, are mostly neutral.

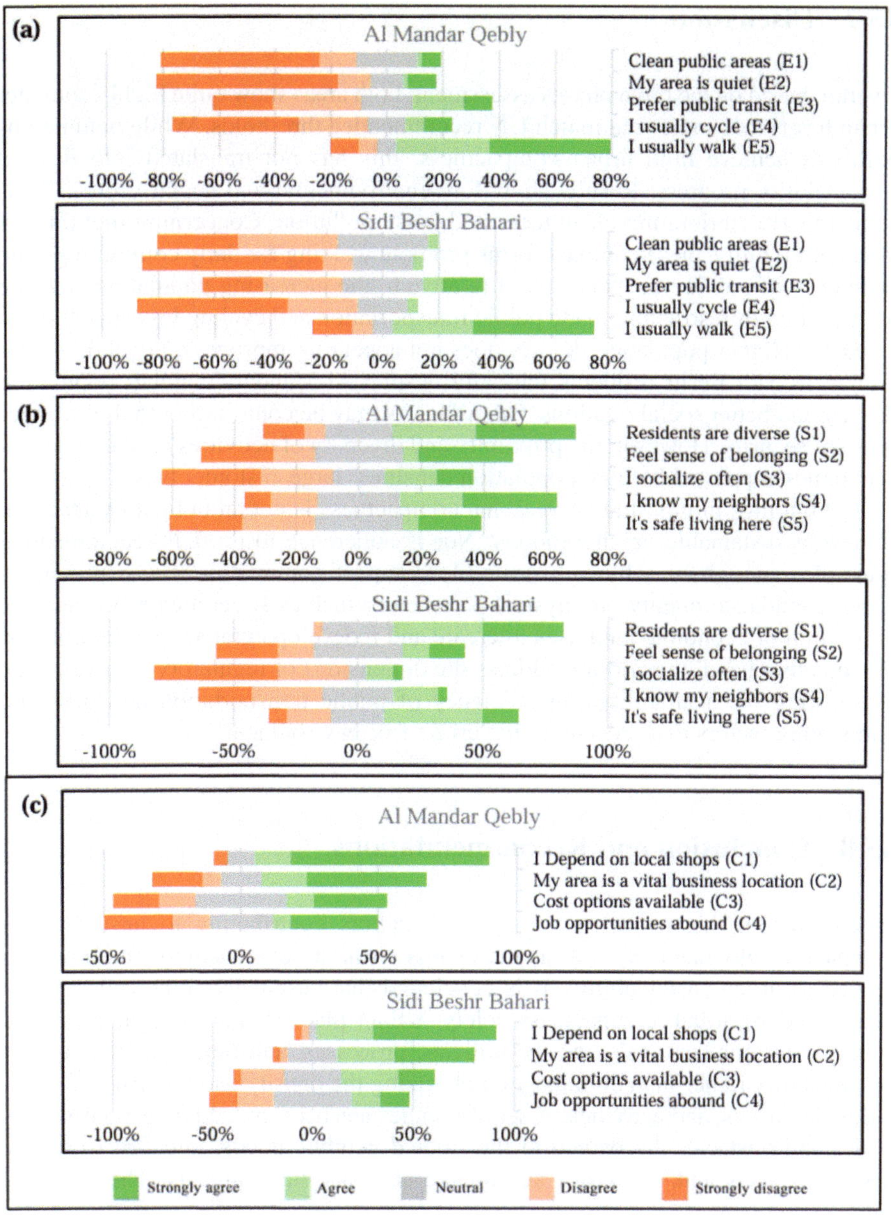

Fig. 45.5 Residents' Survey Responses Residents' views on the environmental, social, and economic conditions in their neighborhood

45.7 Discussion

Overall, based on the three phases assessment, both areas show quite highly compact urban forms that somewhat match UN recommended thresholds. While both neighborhoods achieve high urban compactness, this has not translated into desired sustainability progress. Results show low environmental and social scores. Residents report deteriorating cleanliness and noise pollution. Concerning mobility, in terms of mobility, these compact forms promote walking for daily commuting. The deficiency of public transportation in relative to the increasing population, and the poor cycling infrastructure made public transportation and cycling less appealing to residents. Higher population density does not appear to improve safety and prevent crimes, as Sidi Beshr residents felt safer despite Al-Mandara's higher population density and better social relations. Other factors may be contributing to the stronger sense of security. However, the proximity and diversity of activities revitalized local economies supported by high population densities' large customer base.

The findings in this study reveal that compactness is not an obligatory factor in achieving sustainable neighborhoods. Nor is adherence to the UN-recommended principles enough to achieve a desirable compact community. These principles provide guidance mainly on physical parameters such as street area ratios and the proportion of economic land uses to create and redevelop compact neighborhoods. On the other hand, they do not address the dimensions of regulatory policies, such as ensuring the availability and efficiency of public transportation networks and eliminating factors that prevent a safe, crime-free environment.

45.8 Conclusion and Recommendations

In conclusion, this paper investigates the factors contributing to unsustainable compact developments as well as the claims on urban compactness and how it is related to urban sustainability. It is concluded that sustainable outcomes are not guaranteed by urban compactness solely. Urban planners and legislators should concentrate more on analyzing the policies, processes, and functions that support sustainability rather than on mostly emphasizing the density and intensity of population, buildings, and activities. A set of policies has been created in response to the issues and obstacles discovered in the studied neighborhoods, intended to reduce negative effects and encourage sustainable compact environments. There are the following policies:

- Improving development strategies and urban policies to integrate urban sustainability indicators and improve practices.
- Promoting a balanced mix of diverse housing tenure types.
- Ensuring social diversity by incorporating various ranges of building sizes and costs.

- Developing interconnected, consistent, and safe pavements and preventing surrounding shops or activities from encroaching on the sidewalks.
- Providing connected and safe cycling lanes to encourage cycling and reduce car dependency.
- Creating accessible public parks on available lots and between residential buildings to be accessible to all age groups.
- Implementing density standards in neighborhoods to prevent overcrowding and ensure safety.
- Conducting regular maintenance to ensure the efficiency and cleanliness of public facilities.
- Implementing noise control laws in the neighborhood, including fines for car horns and disruptive activities, while regulating acceptable activities and prohibiting those that disturb, like workshops.
- Developing and expanding public transportation services to accommodate the continuous population growth and prevent rush-hour overcrowding.

References

Abdullahi S, Pradhan B, Jebur MN (2014) GIS-based sustainable city compactness assessment using integration of MCDM, Bayes theorem and RADAR technology. Geocarto Int 30(4):365–387

Afshari A, Mojahed M, Yusuff R (2010) Simple additive weighting approach to personnel selection problem. Int J Innov Manage Technol 1:511–515

Arbury J (2005) From urban sprawl to compact city—an analysis of urban growth management in Auckland

Bibri SE, Krogstie J, Kärrholm M (2020) Compact city planning and development: emerging practices and strategies for achieving the goals of sustainability. Dev Built Environ 4:100021

Borchers A, Pieler T (2010) Exploring the urban form and compactness: a case study of Multan, Pakistan. Genes (Basel) 1(3):413–426

Bramley G, Power S (2009) Urban form and social sustainability: the role of density and housing type. Environ Plann B Plann Des 36(1):30–48

Breheny M (1997) Urban compaction: feasible and acceptable? Cities 14(4):209–217

Burchell R, Mukherji S (2003) Conventional development versus managed growth: the costs of sprawl. Am J Public Health 93:1534–1540

Burton E (2002) Measuring urban compactness in UK towns and cities. Environ Plann B Plann Des 29(2):219–250

Carruthers JI, Ulfarsson GF (2002) Fragmentation and Sprawl: evidence from interregional analysis. Growth Chang 33(3):312–340

Dantzig GB, Saaty TL (1973) Compact city: a plan for a liveable urban environment

Dodman D, Mcgranahan G, Dalal-Clayton B (2013) Integrating the environment in urban planning and management: key principles and approaches for cities in the 21st century

Duany A, Plater-Zyberk E, Speck J (2000) Suburban nation: the rise of sprawl and the decline of the American Dream. Farrar, Straus and Giroux

Henshilwood E, Cullinan M (2012) Urban patterns for a green economy leveraging density. UN-Habitat

Jacobs J (1989) The death and life of great American cities. Knopf Doubleday Publishing Group

Jenks M, Williams K, Burton E (1996a) A sustainable future through the compact city? Urban intensification in the United Kingdom. Environ Des 1:5–20

Jenks M, Burton EJ, Williams K (1996b) The compact city: a sustainable urban form?
Kain J-H et al (2021) Towards a systemic understanding of compact city qualities. J Urban Des 27(1):130–147
Liaqat H et al (2017) Measuring urban sustainability through compact city approach: a case study of Lahore. J Sustain Dev Stud 10
Min H, Jin F, Fath B (2013) Measurement and spatial distribution of urban land use compactness in Chaoyang district of Beijing, China. Chin J Popul Resour Environ 8:3–9
Neuman M (2005) The compact city fallacy. J Plan Educ Res 25(1):11–26
Rahman MH, Islam MH, Neema MN (2022) GIS-based compactness measurement of urban form at neighborhood scale: the case of Dhaka, Bangladesh. J Urban Manage 11(1):6–22
Tresidder M, Sharkova I (2005) Using GIS to measure connectivity: an exploration of issues
Turskis Z, Zavadskas E, Zagorskas J (2006) Sustainable city compactness evaluation on the basis of GIS and Bayes rule. Int J Strateg Prop Manag 10:185–207
UN-Habitat (2014) A new strategy of sustainable neighbourhood planning: five principles, p 8
Zhang XQ (2011) Economic role of cities. In: United Nations human settlements programme. Nairobi UN-Habitat
Zhao F et al (2020) The compactness of spatial structure in Chinese cities: measurement, clustering patterns and influencing factors. Ecosyst Health Sustain 6(1)

Chapter 46
Healthcare Professionals' Views on the Impact of Hospital Design on Emotions and Wellbeing in Patients with Non-communicable Diseases

Nada Yosri Al-Wakil, Hany Mohamed Ayad, and Asmaa Elsayed Hasan

Abstract This study highlights the crucial role of hospital design in patient recovery and wellbeing, particularly for those with non-communicable diseases (NCDs). Through a literature review and interviews with medical professionals in Alexandria, Egypt, it explores the impact of hospital environments on patients' emotions. The findings reveal that thoughtful design can evoke positive emotions, enhance comfort, and promote wellness, ultimately accelerating recovery. By prioritizing patient-centered care, hospitals can create supportive environments that foster physical and emotional healing. The research provides valuable insights for architects and healthcare professionals to create healing environments that prioritize wellbeing and improve patient outcomes. Effective design can reduce stress, promote relaxation, and enhance the overall patient experience. By integrating evidence-based design approaches, hospitals can create spaces that support the physical and emotional needs of patients. This study's findings have significant implications for healthcare design, emphasizing the importance of thoughtful design in improving patient outcomes and enhancing the quality of care. Ultimately, well-designed hospitals can play a vital role in promoting patient wellbeing and recovery.

Keywords Healing environment · Wellbeing · Negative emotions · Physical elements · Non-communicable diseases · Healthcare design

---

N. Y. Al-Wakil (✉) · H. M. Ayad · A. Elsayed Hasan
Architectural Department, Faculty of Engineering, Alexandria University, Al Azaritah WA Ash Shatebi-Qesm Bab Sharqi, Alexandria, Egypt
e-mail: nadaalwakil988@gmail.com

H. M. Ayad
e-mail: Hany.m.ayad@alexu.edu.eg

A. Elsayed Hasan
e-mail: Asmaa.elsayed@alexu.edu.eg

Y. Li et al. (eds.), *16th International Conference on Environmental Science and Development*, Environmental Science and Engineering,
https://doi.org/10.1007/978-3-032-15080-6_46

46.1 Introduction

Chronic non-communicable diseases (NCDs), including cardiovascular disease, cancer, chronic respiratory diseases, and diabetes, are leading causes of death worldwide. These conditions not only affect physical health but also have a profound impact on mental wellbeing (World Health Organization 2023). Prolonged hospital stays and limited mobility can exacerbate stress and anxiety, potentially triggering mental health issues (Faster Capital 2024). Conversely, maintaining a positive mental state can facilitate treatment and prevention, underscoring its importance in NCD care (Faster Capital 2024).Patients with NCDs often experience a range of emotions, including anxiety, depression, stress, anger, and fear (Eman and Hamad 2019). For instance, individuals with heart disease may feel hopeless, helpless, or restless (Eman and Hamad 2019), while cancer patients may encounter shock (National Institutes of Health (NIH) 2023), grief, or despair (Mayo Clinic Health System 2022). Those living with chronic obstructive pulmonary disease (COPD) frequently report feelings of fear and anxiety (American Stroke Association 2024), and managing diabetes can lead to distress, frustration, or guilt (Diabetes UK 2023).

Healthcare environments can evoke fear and uncertainty; however, well-designed spaces can foster positive emotions and improve patient outcomes (Dijkstra 2009). Research in environmental psychology highlights the influence of surroundings on behavior (Gifford 2013), mood (Knez 2001), and cognition (Babin et al. 2003). Healing environments, which integrate physical, psychological, and social aspects, contribute significantly to patient wellbeing and recovery (Bovenberg et al. 2010; van den Berg 2005).

Key elements of healing environments include natural light, ventilation, acoustic design, spatial layout, and color (Aripin 2014). Single-patient rooms, art, and controlled lighting and temperature can also enhance patient experience (Ulrich 1991). By incorporating these elements, healthcare facilities can reduce anxiety, stress, and pain while promoting recovery and wellbeing.

This study explores the correlation between physical environment elements and their impact on alleviating negative emotions in NCD patients, enhancing their wellbeing, and reducing hospital stays. By understanding the relationship between healing environments and patient outcomes, healthcare providers can design spaces that promote recovery, comfort, and wellbeing as shown in Fig. 46.1.

The physical environment is comprised of all tangible physical elements within and surrounding organizations (Mobach 2009). Research has identified various physical elements that positively impact patients' health and wellbeing, including daylight, nature, ventilation, acoustics, spatial layout, and color (Aripin 2014). A comprehensive review of existing studies on the healing-built environment's impact on health outcomes was conducted, focusing on English-language articles published after 2000. The review analyzed 17 literature reviews and identified 67 relevant papers, with 63 meeting the inclusion criteria and 4 excluded due to lack of evidence, as illustrated in Fig. 46.2.

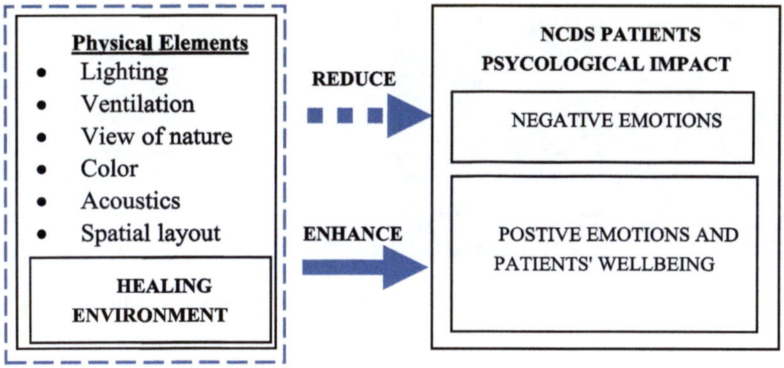

Fig. 46.1 Relation between physical elements and psychological impact on NCDs

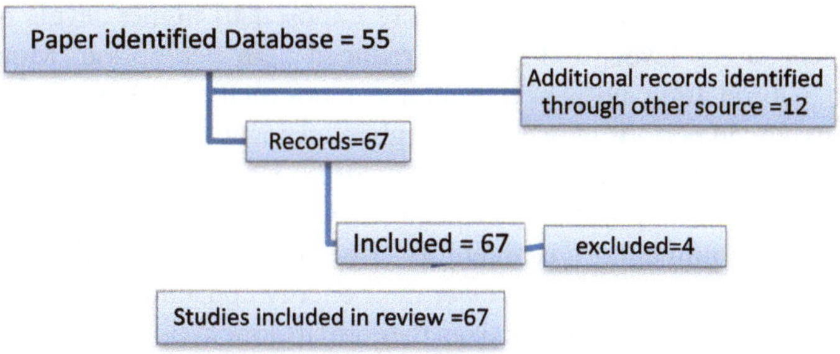

Fig. 46.2 The scope and basic information of literature. *Source* Author

46.2 Physical Elements and Their Therapeutic Impact on NCDs Patients

The following physical elements are explained: Lighting, Natural ventilation, color, Acoustic, view of nature and spatial layout.

46.2.1 Lighting

The recovery of patients who are hospitalized for extended durations can be hindered by various aspects of the hospital environment. Therefore, it is essential to design hospital spaces that prioritize patient comfort and address both mental and physical health needs. Such environments should aim to reduce feelings of depression, stress, tension, and anxiety. Research indicates that incorporating natural light into room designs can accelerate patient recovery and foster a calming atmosphere for both patients and healthcare staff. Long-term patients often perceive sunlight as a powerful

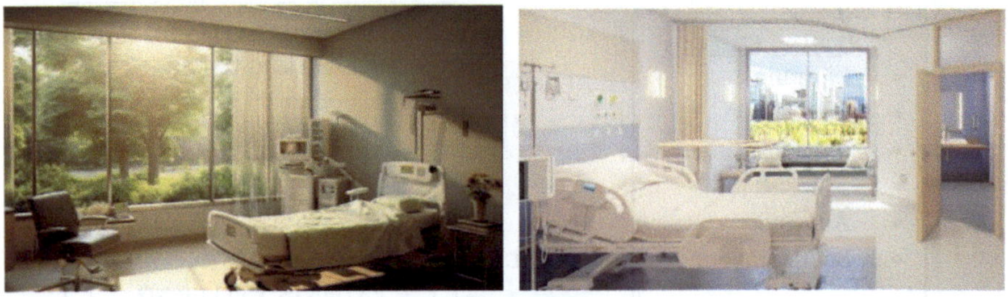

Fig. 46.3 Natural lighting and view of nature in inpatient rooms, adapted from Jadhav (2024)

symbol of hope, healing, and positivity (Khaleghimoghaddam 2023). According to Khaleghimoghaddam (2023), substantial evidence supports the idea that light exposure can relieve depression, ease pain, and enhance sleep quality. Similarly, Joseph (2006) noted that patients who were exposed to higher levels of sunlight reported lower stress levels and slightly reduced pain perception. Moreover, the presence of natural light influences patients' emotional well-being, as remaining indoors for extended periods may lead to heightened negative emotions such as stress, sadness, and anxiety (Morales-Bravo and Navarrete-Hernandez 2022), as illustrated in Fig. 46.3. Beyond its psychological and physiological benefits, natural lighting also supports environmental sustainability. By reducing the reliance on artificial lighting, hospitals can significantly lower their energy consumption, which not only cuts operational costs but also minimizes their environmental footprint, contributing to broader sustainability efforts (Jadhav 2024).

Natural ventilation is advised for hospital wards as an effective infection control strategy, primarily due to its high air exchange rate, which helps prevent the spread of airborne diseases (Qian et al. 2010). In addition to controlling infections, natural ventilation allows patients greater access to fresh air, which contributes to reduced anxiety and stress levels. Research also indicates that patients who stay in rooms with windows offering views of nature often experience significantly shorter postoperative hospital stays compared to those in windowless rooms (Andriani et al. 2021). Furthermore Andriani et al. (2021) highlighted that proper ventilation can influence emotional states, particularly by helping to reduce feelings of anger among patients.

46.2.2 Natural View

Research has demonstrated that spending time in natural environments is associated with a reduced risk of non-communicable diseases (NCDs), lower levels of stress, anxiety, and aggression, as well as increased happiness and overall well-being. Being in nature or even simply viewing natural scenes helps patients feel calmer, happier, and less angry, fearful, or stressed. Exposure to nature also supports both emotional and physical health by decreasing heart rate, blood pressure, muscle tension, and

Fig. 46.4 Head wall patient view and staff view in acute in patients' room (Loneliness, Social Isolation and Nature Connection 2023; Perkins and will 2013)

the release of stress-related hormones (Delagran 2024). It contributes to overall health improvement and fosters a sense of well-being in patients. Additionally, nature exposure has a positive effect on the nervous system by reducing the physiological impact of stress, which aids in stress recovery and decreases pain perception (Attia 2021). According to Loneliness, Social Isolation and Nature Connection (2023), developing a strong connection with nature can help reduce social isolation, alleviate feelings of loneliness, and support overall well-being, as illustrated in Fig. 46.4. Greenery also plays a significant role in sustainable development by enhancing air quality, reducing urban heat, offering sound insulation and improved acoustics, and contributing to infection control measures (Kalender et al. 2023).

46.2.3 Color

According to Khaleghimoghaddam (2023), the color scheme of a recovery room should be designed to evoke feelings of hope and confidence in patients. Different colors have distinct psychological effects; for instance, red is linked to boosting confidence and is therefore used in treating anxiety. In contrast, blue helps calm nervous agitation, while yellow and orange are effective in alleviating depression. Green is considered the most suitable color for managing stress-related conditions. Additionally, neutral tones and colors blended with gray are recognized for their calming properties and their ability to reduce anxiety. Moreover, incorporating reflective surfaces and light color palettes can enhance the effectiveness of natural light. Using light-colored walls and flooring helps to reflect light throughout the space, creating a brighter and more open atmosphere (Jadhav 2024).

46.2.4 Acoustics

Research shows music can help manage patients' mood, reducing anxiety, pain, and stress. Studies also highlight the impact of music on relaxation levels. Additionally, using sound-absorbing materials in rooms can minimize disturbances and create a more peaceful environment (Laursen et al. 2014).

46.2.5 Spatial Layout

Single Patient Rooms. Private rooms revolutionize patient care by bolstering safety through reduced infection risk due to enhanced cleanliness with fewer occupants. This setup also boosts family satisfaction by providing a comfortable environment, allowing extended visits, and mitigating patient isolation. Moreover, single-patient rooms alleviate stress and anxiety, fostering better rest, sleep, and physical well-being (Maben et al. 2015). In contrast, multi-occur conversely, shared rooms have been associated with increased feelings of shame (Chaudhury et al. 2005).

Bathroom. Studies show that shared bathroom facilities heighten infection transmission risks, whereas private bathrooms offer a safer alternative (Gulwadi 2008). Private bathrooms also promote patient privacy, lowering stress and depression levels. Furthermore, they can be designed with safety features to prevent falls, reducing patients' anxiety about falling.

Windows. Windows in older medical buildings served multiple purposes beyond just ventilation and illumination, offering patients a sense of hope and connection to the world outside (Amin et al. 2024).To enable patients to view the outside comfortably, windows were designed with adequate low parapets (van Nijhuis 2017).

Research studies have shown that inappropriate use or lack of physical elements can evoke certain emotions in individuals, particularly in non-communicable diseases (NCDs) patients. These negative emotions significantly impact patients' well-being and ability to manage their life conditions. However, incorporating specific physical elements can improve patients' well-being, healing, and recovery. Lighting will contribute in decreasing patients' anxiety, depression, and stress. Ventilation will contribute in decreasing patients' anxiety and stress. Color will contribute in decreasing patients' anxiety, depression, and stress. Acoustic will contribute in decreasing patients' anxiety and stress. A view of nature can decrease anxiety, depression, stress, anger, fear, and loneliness. Finally, spatial layout can impact patients' anxiety, depression, stress, shame, and loneliness.

NCDs patients experience negative emotions that impact their well-being and ability to manage their condition. Incorporating physical elements can improve their well-being, healing, and recovery. Each element has a specific impact. Lighting affects anxiety, depression, anger, fear, guilt, loneliness, shame, and stress. Ventilation affects anxiety, depression, anger, despair, and stress. Color affects anxiety,

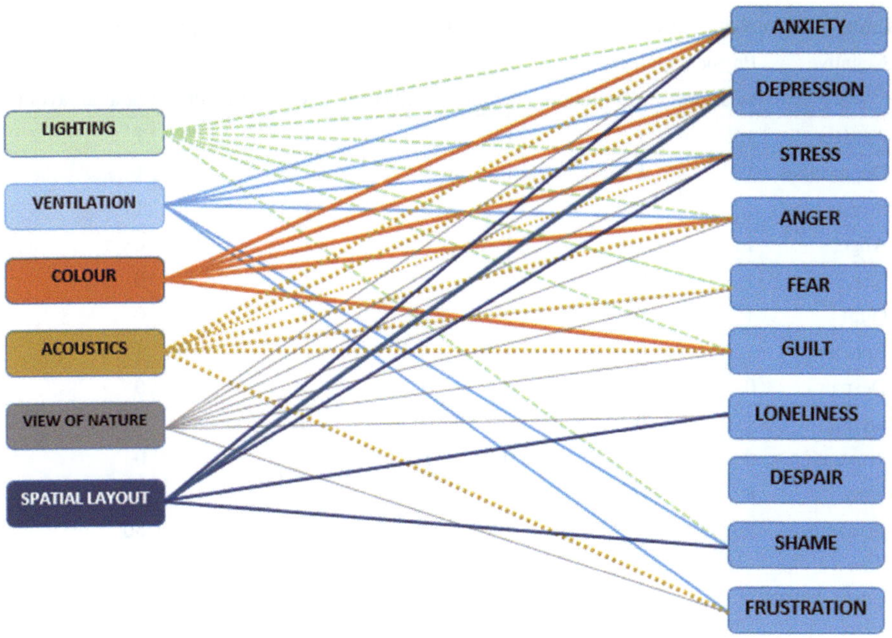

Fig. 46.5 Healing elements contribution in negative emotions. *Source* Author

depression, anger, loneliness, and stress. Acoustics decreases anxiety, depression, stress, anger, fear, guilt, loneliness, and frustration. A view of nature decreases anxiety, depression, stress, anger, fear, loneliness, despair, and frustration. Spatial layout affects anxiety, depression, stress, shame, and loneliness as shown in Fig. 46.5.

The findings indicate that view of nature, acoustics, and lighting have the strongest impact, with an 80% association with reducing negative emotions. Natural ventilation has a 60% impact, while spatial layout and color have a 50% impact, suggesting they may be less influential. These findings are based on a literature review, with some studies using a quantitative coding approach as Some studies used only yes or no coding (1 or 0) respectively through equation (Sum/10 * 100) to measure the impact of physical elements on negative emotions as shown in Table 46.1.

46.3 Materials and Methods

This study employs a mixed-methods approach, combining qualitative literature reviews with quantitative analysis through a phenomenological lens. Experts were interviewed qualitatively to gather insights on healing environments in hospitals. The research also quantitatively assessed the impact of environmental factors on patients'

Table 46.1 Impact of physical elements and negative emotions through frequency coding

Negative emotions	Physical element					
	Lighting	Ventilation	Color	Acoustics	View of nature	Spatial layout
Anxiety	1	1	1	1	1	1
Depression	1	1	1	1	1	1
Stress	1	1	1	1	1	1
Anger	1	1	1	1	1	0
Fear	1	0	0	1	1	0
Guilt	1	0	0	1	0	0
Loneliness	1	0	1	1	1	1
Despair	0	1	0	0	1	0
Shame	1	0	0	0	0	1
Frustration	0	1	0	1	1	0
Sum	8	6	5	8	8	5
Percentage (%)	80	60	50	80	80	50

Source Author

treatment duration, blood pressure control, subjective experience of illness, and the effect of social relationships on their vital signs.

Semi-structured interviews were conducted in English, using a two-part questionnaire. The first part consisted of open-ended questions clustered into 9 blocks with 17 questions, covering topics such as NCDs, patient impact, environmental importance, architectural features, design preferences, patient feedback, collaboration, and long-term benefits. The second part included 10 closed-ended questions. Interviews lasted 30–40 min, except for one that lasted 10 min due to scheduling constraints. Ten medical professionals, including surgeons, ICU specialists, a psychologist, and a cardiology specialist, shared their expertise, providing valuable insights into NCDs and the importance of their roles in healthcare as shown in Table 46.2. Through these interviews, prevalent NCDs among patients were identified.

Table 46.2 List of interviews healthcare professionals

List of interviewed experts	Number
Specialist in general surgery	3
Psychologist	1
ICU specialist	3
Specialist in surgery "vascular surgery"	2
Cardiology specialist	1
Total	10

Source Author

Healthcare professionals reported that heart disease and cancer are the most common NCDs, each accounting for 36% of cases, followed by respiratory diseases at 18%, and stroke at 9%. According to 72% of professionals, cancer tends to emerge in males aged 50 and above, and in females aged 30 and above. However, 18% noted cancer emergence in females aged 25 and above, and 9% stated that cancer can occur at any age. Additionally, 18% of professionals reported cancer emergence in males aged 30 and above. Cancer affects both genders, with varying prevalence across different anatomical sites. Breast and uterine cancers are more common in women, while colon and prostate cancers are more prevalent in men.

NCDs not only cause physical changes but also significantly impact patients' emotional well-being. Healthcare professionals agreed that anxiety, stress, and depression are common emotions experienced by NCDs patients, with some also noting fear, guilt, frustration, and despair. According to the professionals, the prevalence of these emotions is: depression (30%), anxiety (27%), stress (20%), fear (13%), frustration (7%), and despair (3%).

The physical environment of hospitals plays a crucial role in the recovery process of NCDs patients. The most common NCDs reported by healthcare professionals were heart disease (36%), cancer (36%), respiratory disease (18%), and stroke (9%). Cancer tends to emerge in males over 50 and females over 30, although some professionals noted that cancer can occur at any age and affects both genders, with different types being more prevalent in men and women.

According to professionals, respiratory diseases tend to affect men over 40 and women over 50. Heart disease is more common in men over 45, who are more frequently affected than women. Regarding strokes, 91% of professionals said they typically occur in men over 50, while 9% cited cases in men over 25; all agreed that strokes affect women over 60.

Professionals also emphasized the importance of environmental factors in patient comfort and recovery. They highlighted the significance of a well-designed physical environment, citing elements like natural light, ventilation, room layout, color, acoustics, and views of nature, as well as suitable furniture, which can promote relaxation and calmness, ultimately supporting better treatment outcomes and reintegration into daily life.

Interviews with healthcare professionals revealed that certain environmental factors significantly impact patients with non-communicable diseases (NCDs). The top factors were lighting (23.68%), natural ventilation (18.42%), acoustics (13.15%), open views (10.52%), and room occupancy (7.89%), followed by temperature, room size, color, and furniture (5.26%).

Professionals shared case studies illustrating the built environment's positive impact on patient recovery. Examples included patients showing reduced agitation when moved to naturally ventilated rooms with views, a patient opting for a high-risk procedure due to a calming garden view, and a traumatic brain injury patient improving after being relocated to a room with natural light and ventilation. These cases demonstrate the built environment's therapeutic potential in enhancing patient well-being and recovery.

A case study at Al Wakil Hospital in Alexandria, Egypt, explored the link between healing environments and patient well-being using a questionnaire survey. The study examined the relationship between hospital environments (independent variable) and patient well-being (dependent variable), considering factors like gender, age, education, marital status, room occupancy, and room number. Structural equation modeling (SEM) was used to analyze these relationships, while the DASS21 questionnaire measured patients' anxiety, stress, and depression levels during hospitalization.

A survey of 100 patients, with 80 responses, provided insights into the impact of hospital environments on patient well-being. The respondents were mostly female (54.2%), over 55, married (59.7%), and well-educated (68.1%). Chronic respiratory disease (26.4%) and heart disease (23.6%) were the primary reasons for hospitalization. The survey highlighted that anxiety was the biggest concern for patients, followed by depression and stress—with women experiencing more emotional distress than men. Interestingly, the room number itself seemed to affect patients' emotions. On a positive note, a supportive healing environment significantly boosted patient recovery, as reflected in their overall well-being scores.

Patients identified areas for improvement in their hospital environment, with spatial layout being the top priority (28.8%). Other areas included lighting (19.6%), presence of nature (17.6%), ventilation (14.4%), control (7.8%), color (5.9%), and acoustics (0.7%) as illustrated in Table 46.3.

Further analysis confirmed the significant impact of these environmental factors on patient well-being:

- Spatial layout positively affects patient well-being ($\beta = 0.190$).
- Presence of nature enhances patient well-being ($\beta = 0.146$).
- Lighting improves patient well-being ($\beta = 0.203$).
- Color has a positive impact on patient well-being ($\beta = 0.303$).
- Ventilation contributes to patient well-being ($\beta = 0.285$).

Table 46.3 Selection of 3 elements need to be made improvements

Variable (description)	Responses	
	N	Percentage (%)
Layout	44	***28.8***
Light	30	**19.6**
Color	9	5.9
Music	1	0.7
Art	6	3.9
Privacy	8	5.2
Ventilation	22	*14.4*
Presence of nature (outside, inside and in images)	21	13.7
Control (over the light, ventilation, sound comfort)	12	7.8
Total	153	100.0

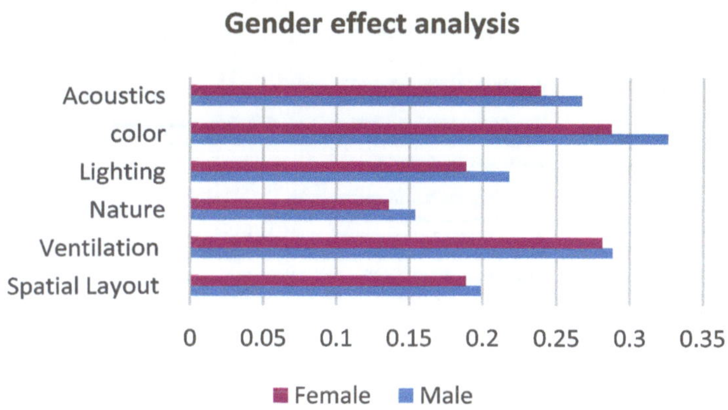

Fig. 46.6 Gender effect analysis on healing elements. *Source* Author

- Acoustics also positively influences patient well-being ($\beta = 0.252$).

These findings suggest that a well-designed hospital environment can significantly support patient recovery and well-being.

Results as shown, shows that the estimated structural model corroborated the 6 hypotheses, as Hospital Healing Environments (Spatial Layout, Nature, Lighting, Color, Ventilation and Acoustics) construct explained 41.2% of Patients' wellbeing Through the following equation.

$$R^2 = 0.412 \tag{46.1}$$

Figure 46.6 illustrates the role of gender in the relationship between spatial layout, nature, lighting, color, ventilation, acoustics, and overall patient wellbeing. The findings indicate that gender serves as a moderating factor specifically in the link between color and patient wellbeing, with male patients reporting the highest wellbeing scores in this context, thereby supporting the presence of a moderating effect.

Figure 46.7 illustrates the influence of age on the relationship between spatial layout, nature, lighting, color, ventilation, acoustics, and overall patient wellbeing. The results show that different age groups emphasize distinct aspects of the physical environment. Participants aged 46–55 reported the highest scores for spatial layout, ventilation, and acoustics, while those aged 36–45 rated nature, lighting, and color the highest—highlighting a variation in environmental preferences across age groups.

Figure 46.8 demonstrates the impact of education level on the relationship between spatial layout, nature, lighting, color, ventilation, acoustics, and patients' overall wellbeing. The findings reveal that individuals with higher levels of education place greater importance on elements such as ventilation, color, lighting, nature, and spatial layout, scoring highest in these areas. Conversely, participants with no formal education scored highest in acoustics, indicating a distinct difference in environmental preferences based on educational background.

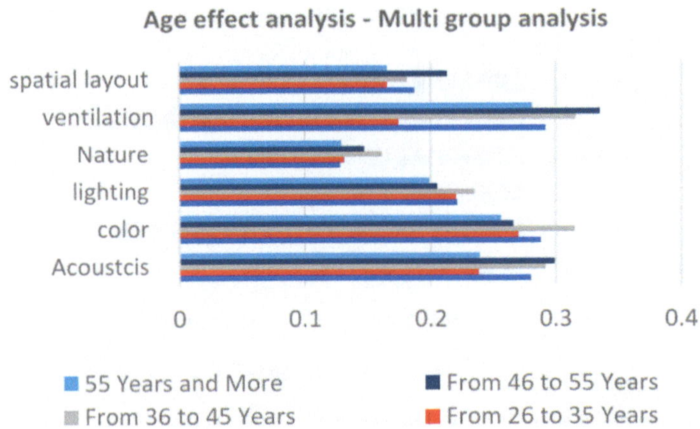

Fig. 46.7 Age effect analysis on healing elements. *Source* Author

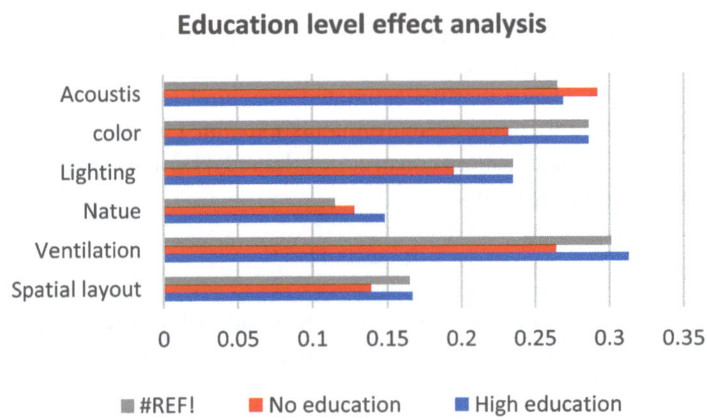

Fig. 46.8 Education level effect analysis on healing elements. *Source* Author

Figure 46.9 illustrates the influence of marital status on the relationship between spatial layout, nature, lighting, color, ventilation, acoustics, and overall patient well-being. The results indicate that marital status affects how individuals prioritize elements of the physical environment. Specifically, legally married individuals placed the highest importance on color; those widowed after marriage prioritized ventilation and acoustics; divorced participants emphasized nature and lighting; while unmarried individuals ranked spatial layout as most important.

Figure 46.10 illustrates the impact of room occupancy on the relationship between spatial layout, nature, lighting, color, ventilation, acoustics, and patients' overall wellbeing. The results show that patients in single rooms reported the highest scores for spatial layout and lighting, while those in double rooms rated ventilation, nature,

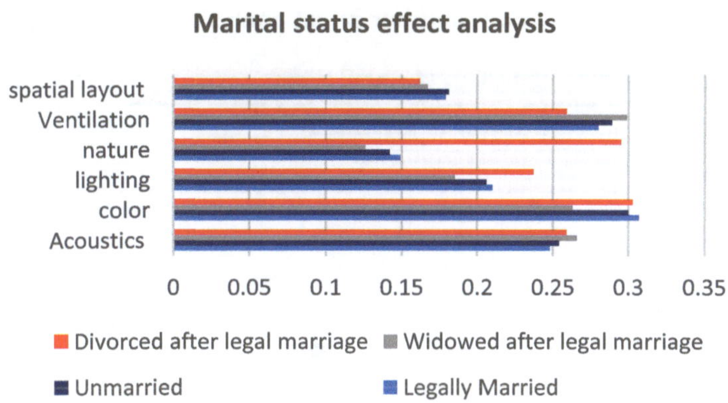

Fig. 46.9 Marital status effect analysis on healing elements. *Source* Author

and acoustics the highest. In contrast, patients in triple rooms gave the highest scores for color, highlighting how room occupancy influences environmental preferences.

Figure 46.11 presents the impact of room number on the relationship between spatial layout, nature, lighting, color, ventilation, acoustics, and overall patient well-being. The findings indicate that different room categories emphasize various environmental factors. Categories 1 and 2 (rooms 202–206 and 201–215) showed the highest scores in ventilation, while Category 3 (rooms 205–208) prioritized nature. Category 5 rated lighting and spatial layout the highest, Category 6 emphasized acoustics, and Category 7 gave the highest importance to color. These results reflect the diverse environmental preferences across different room groups.

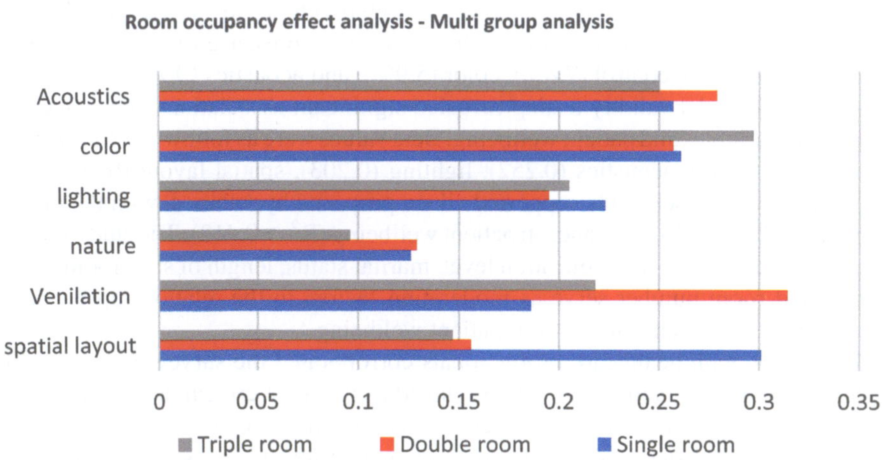

Fig. 46.10 Room occupancy effect analysis on healing elements. *Source* Author

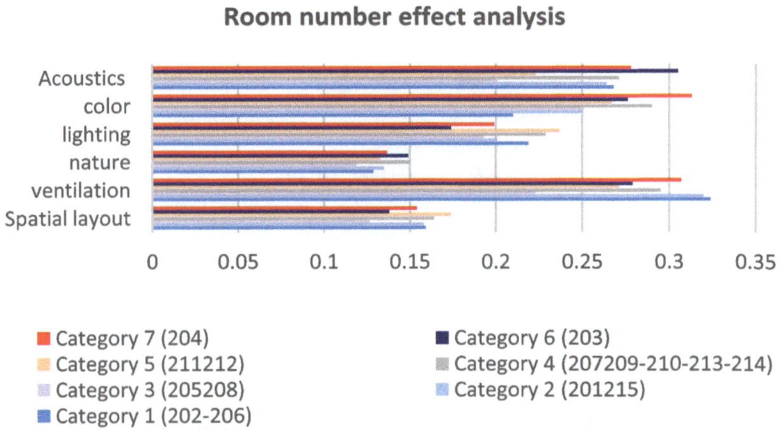

Fig. 46.11 Room number effect analysis on healing elements. *Source* Author

46.4 Conclusion

The central research questions driving this study were: Can healing environments enhance patient wellbeing, and how essential is the creation of such environments for effective hospital treatment? To address these questions, a comprehensive analysis of questionnaire survey data was conducted to examine the causal links between healing environments and patient wellbeing. The analysis revealed that negative emotional outcomes associated with non-communicable diseases (NCDs) such as anxiety, depression, and stress disproportionately affect female patients. Additionally, chronic respiratory disease and heart disease emerged as the most commonly reported conditions, underscoring a potential link between the physical healthcare environment and patient wellbeing. When asked which areas required improvement, the largest proportion of respondents (34%, or 44 individuals) identified spatial layout as a priority. This was followed by lighting (19.6%), presence of nature (17.6%), ventilation (14.4%), control (7.8%), color (5.9%), and acoustics (0.7%).

Structural model validity testing revealed significant associations between environmental variables and patient wellbeing. Beta values were as follows: color (0.303), ventilation (0.285), acoustics (0.252), lighting (0.203), spatial layout (0.190), and nature (0.146). These results supported all six proposed hypotheses, with the model explaining 41.2% of the variance in patient wellbeing ($R^2 = 0.412$). The study further identified that gender, age, education level, marital status, length of stay, room occupancy, and room number serve as moderating factors in the relationship between environmental elements and overall patient wellbeing.

Interviews with healthcare professionals corroborated the survey findings, with lighting, ventilation, and spatial layout ranked as the most critical design elements. These were followed by acoustics, views of nature, color, and temperature, emphasizing a shared understanding of environmental priorities between patients and professionals.

References

American Stroke Association (2024) Emotional and behavioral effects of stroke. https://www.stroke.org/en/about-stroke/effects-of-stroke/emotional-effects. Last accessed 2024

Amin SU, Ullah U, Din M (2024, July) Assessing window design for healing environment in selected hospitals. Ann Hum Soc Sci

Andriani AD, Mulyana D, Dida S, Wahyudin U (2021) The role of healing environment in reducing the stress of patients with non-communicable disease

Aripin S (2014) Healing architecture: a study on the physical aspects, Queensland

Attia DDII (2021) Factors of indoor therapeutic environment and their effects on patients and health care workers

Babin BJ, Hardesty DM, Suter TA (2003) Color and shopping intentions: the intervening effect of price fairness and perceived affect. Bus Res 56:541–551

Bovenberg F, Takkenkamp J, Vennik L, Francken G (2010) Helende omgeving draagt bij aan. Soc Psychiatr 29(94)

Chaudhury H, Mahmood A, Valente M (2005) The use of single patient rooms versus multiple occupancy rooms in acute care environments. Facility Guidelines Institute, Canada

Delagran LMA (2024) How does nature impact our wellbeing? [Online]. Available: https://www.takingcharge.csh.umn.edu/how-does-nature-impact-our-wellbeing

Diabetes UK (2023) Diabetes. https://www.diabetes.org.uk/guide-to-diabetes/emotions. Last accessed 12 Apr 2023

Dijkstra K (2009) Understanding healing environments: effects of physical environmental stimuli on patients' health and wellbeing

Eman M, Hamad A (2019) Heart failure: how to cope with your emotions, 13 Sept 2019. [Online]. Available: https://www.templehealth.org/about/blog/heart-failure-coping-with-emotions

Faster Capital (2024) NCDs and mental health—the link between physical and emotional well-being. https://fastercapital.com/content/NCDs-and-Mental-Health--The-Link-Between-Physical-and-Emotional-Well-being.html. Last accessed 26 June 2024

Gifford R (2013) Environmental psychology matters. Annu Rev Psychol 64:541–579

Gulwadi MPB (2008) The impact of healthcare environmental design on patient falls

Mayo Clinic Health System (2022) Emotional recovery after cancer treatment. https://www.mayoclinichealthsystem.org/hometown-health/speaking-of-health/emotional-recovery-after-cancer-treatment. Last accessed 26 Apr 2022

National Institutes of Health (NIH) (2023) Emotions and cancer. https://www.cancer.gov/about-cancer/coping/feelings. Last accessed 09 Nov 2023

Jadhav DV (2024) The impact of natural light in modern hospital design, 10 June 2024. [Online]. Available: https://www.hospertz.com/the-impact-of-natural-light-in-modern-hospital-design/

Joseph A (2006) Impact of light on outcomes in healthcare settings. [Online]. Available: https://www.healthdesign.org/chd/research/impact-light-outcomes-healthcare-settings

Kalender S, Smajlović S, Dovjak M (2023) Advantages of greenery in hospital environment. Int J Sanitary Eng Res

Khaleghimoghaddam N (2023) Understanding the interplay of light, color, and interior design in healthcare spaces, Konya, Turkey

Knez I (2001) Effects of colour of light on nonvisual psychological processes. J Environ Psychol 21(2):201–208

Laursen J, Danielsen A, Rosenberg J (2014) Effects of environmental design on patient outcome: a systematic review

Loneliness, social isolation and nature connection (2023, May). [Online]. Available: https://silvotherapy.co.uk/articles/loneliness-social-isolation-and-nature-connection

Maben J, Griffiths P, Penfold C, Simon M, Pizzo E, Anderson J, Robert G, Hughes J, Murrells T, Brearley S, Barlow J (2015) Patient experience of the single room ward environment. In: Evaluating a major innovation in hospital design: workforce implications and impact on patient and staff experiences of all single room hospital accommodation, chap 7

Mobach R (2009) Een organisatie van vlees en steen. Van Gorcum, Netherlands

Morales-Bravo J, Navarrete-Hernandez P (2022) Enlightening wellbeing in the home: the impact of natural light design on

Perkins and will (2013) New hospital tower. Rush University Medical Center. ArchDaily

Qian H, Li Y, Seto WH, Ching P, Ching WH, Sun HQ (2010) Natural ventilation for reducing airborne infection in hospitals. [Online]. Available: https://www.ncbi.nlm.nih.gov/pmc/articles/PMC7115780/

Ulrich RS (1991) Effects of interior design on wellness: theory and recent scientific research. J Health Care Inter Design 97–109

van den Berg A (2005) A review of evidence for benefits of nature, daylight, fresh air, and quiet in healthcare settings. In: Health impacts of healing environments. UMCG, Groningen, p 96 (2005)

van Nijhuis J (2017, Mar) Healing environment and patients' well-being, Amersfoort

World Health Organization (2023) Noncommunicable diseases. https://www.who.int/news-room/fact-sheets/detail/noncommunicable-diseases. Last accessed 16 Sept 2023

Chapter 47
Analysis and Test of the Flow Field in the DC Open Wind Tunnel of a Large Climate Laboratory

Zhanpeng Ren, Yajuan Zhang, Bin Li, Jingtao Wu, and Peiqiang Tian

Abstract In order to make the wind tunnel perform wind, rain, and snow experiments in a climate laboratory, this paper designs a movable DC open wind tunnel with a temperature effect for a test section at 4 m from the back end of the tunnel exit. The CFD method is utilized to simulate the 3D flow field of the wind tunnel in the laboratory. The velocity field and pressure field distribution of baseline temperature and low-temperature field at – 30 °C are calculated and analyzed. The differential pressure transmitter and auxiliary equipment are then employed to calibrate the characteristics of the wind tunnel flow field under the corresponding working conditions. The 3D numerical simulation and experiment results indicate that the errors of maximum velocity, velocity stability, evenness, and pressure fluctuation range of specific parts of the simulated wind tunnel flow field are within 10%. The simulation results of the wind tunnel flow field, velocity field, and pressure field are compatible with the experimental data under different temperature conditions. The low-temperature environment deteriorates the air quality of the wind tunnel and increases the pressure fluctuation. The performance indices of simulation and measurement of the wind tunnel meet the design requirements. This study will provide technical support for the quality optimization and performance improvement of the flow field in the DC open wind tunnel.

Keywords Temperature effect · Direct current (DC) open wind tunnel · Numerical simulation · Velocity field · Pressure field

Z. Ren · Y. Zhang (✉) · B. Li · J. Wu · P. Tian
National Key Laboratory of Strength and Structural Integrity, Aircraft Strength Research Institute of China, Xi'an, China
e-mail: Zhang_yj623@126.com

Y. Li et al. (eds.), *16th International Conference on Environmental Science and Development*, Environmental Science and Engineering,
https://doi.org/10.1007/978-3-032-15080-6_47

47.1 Introduction

In order to evaluate the performance of the windshield wiper system of the aircraft, extreme climate environments such as rain, snow, and fog were simulated in the climate laboratory during the takeoff and landing stages of the aircraft. The clarity or visibility of the windshield glass decreases when the aircraft encounters heavy rain, snow, and low-temperature weather. Moreover, the refraction and reflection of light can distort the visual representation of pilots, seriously threatening flight safety. The statistical study of aircraft flight accidents shows that frequent accidents during takeoff and landing may be attributed to heavy rain and snow environment. Aircraft flight will generate airflow in opposite direction of the flight motion, and the direct current (DC) open wind tunnel can produce horizontal strong airflow (Liu et al. 2003; National Defense Science, Technology and Industry Commission 1991). The superimposed environment of strong horizontal airflow and vertical rain and snow is adopted to simulate the service scene of aircraft encountering rain and snow (Gai et al. 2011; Zheng et al. 2012). The new DC open wind tunnel developed by China Aircraft Strength Research Institute can reach a maximum airflow speed of 55 m/s, which can meet the average takeoff and landing speeds of aircraft. During the test, the test aircraft was first towed into the laboratory to be stabilized and fixed according to the requirements, and the wind tunnel and the rain/snow device were installed in the laboratory according to the design requirements. The total control system of the laboratory controlled the test conditions of wind and snow in real-time. The current wind and rain test of the aircraft windshield wiper system can only be performed under natural environmental conditions in the field with a long test cycle and high cost. In order to solve this problem, the real-time simulation of the wind and rain test of the whole aircraft was realized in the laboratory.

There are many reports on flow field analysis and experimental testing in the DC wind tunnel. Ding et al. (2015) designed a DC closed wind tunnel, tested the maximum wind speed in the tunnel and demonstrated the change rule of the flow field through the test results. Hu et al. (2012) evaluated the airflow quality of a wind tunnel's DC low-speed flow field using a hot-film anemometer test device and obtained the velocity and turbulence in the test section. Moonen et al. (2006) and Calautit et al. (2014) analyzed the flow field quality of the wind tunnel through numerical simulations, providing a theoretical basis for determining the improvement method. Zeng et al. (2018) and Wang et al. (2012) analyzed the structure of the DC low-speed wind tunnel by the finite element method and analyzed the turbulence in the test section. Deng (2014) studied the flow field of the DC low-speed wind tunnel experiment platform using the fluent analysis software and obtained flow field characteristics such as turbulence intensity, velocity, and pressure. Burov et al. (2012) and Green (2011) studied the test method of the flow field quality and airflow performance in the DC wind tunnel. Kong et al. (2020) utilized a hot-wire anemometer and other devices to evaluate the wind tunnel performance and determined the functional relationship between the frequency converter's output frequency and wind speed. Lu et al. (2018), Wang et al. (2017), Jiang et al. (2018), Ignatius et al. (2014), Jeffrey and Stephen (2008) and Zhang et al. (2016) studied the design method and flow field simulation of wind tunnels with temperature regulation and control functions, respectively. Tong et al. (2014) and Chen et al. (2013) verified the influence of laboratory environmental conditions on wind speed measurement in the DC wind tunnel. However, there are few reports on the correlation law of the interaction between the flow field characteristics in the DC wind tunnel and the laboratory low-temperature environment.

The McKinley Climate Laboratory in the United States can evaluate the typhoon rain blowing and wind blowing snow. It has a DC open wind tunnel for simulating the ice accumulation on the ground of aircraft. Currently, there is no movable DC open wind tunnel for the rain and snow-blowing tests of the whole platform during the simulated takeoff and landing stages in China. The new DC open wind tunnel (Zhang et al. 2017) constructed by China Aircraft Strength Research Institute is a technological innovation to fill the domestic gap. It is the primary large-scale environmental simulation equipment for domestic climate and environment laboratories. According to the wind tunnel requirements in the closed laboratory room temperature and extremely low-temperature environment, this paper performed the simulation analysis of the wind tunnel airflow velocity field and pressure field characteristics under the required conditions, as well as the wind tunnel airflow characteristics test under the actual use environment. This study can provide technical support for the wind tunnel to perform related experiments in the climate laboratory.

47.2 Technical Specifications of the Wind Tunnel

In 2018, the China Aircraft Strength Research Institute constructed a DC open wind tunnel with a low-temperature environment. The wind tunnel can generate a controlled airflow field in the closed climate environment laboratory and cooperate with the rain and snow fog device of the climate laboratory to simulate the severe weather environment such as heavy rain, snow, and fog during the take-off and landing stages of the aircraft in the laboratory. According to the above usage characteristics, the particular requirements of the environmental wind tunnel are:

(1) Test section location: the test section is located outside the wind tunnel and in the vertical plane 4 m away from the air exit, with an effective section area of 1 m^2 (height × width = 1 m × 1 m). (2) Velocity field of the airflow: the adjustment range of horizontal wind speed in the test section is 5–55 m/s, the wind speed is automatically adjustable, and its unevenness and instability in the test section are $\leq \pm 4\%$ and $\leq \pm 1\%$, respectively. (3) Airflow temperature field: the temperature range of airflow generated by the wind tunnel is 15–30 °C; (4) pressure field: the pressure fluctuation around the wind tunnel should not exceed 60 Pa, and the laboratory pressure field should not be out of control due to strong flow in the wind tunnel, resulting in test risks.

Strong indoor airflow will be generated during the operation of the environmental wind tunnel in a closed laboratory, forming a huge vortex airflow. The irregular flow of air at high velocity will cause many adverse effects on the air intake and ejection of the fan, declining the air quality in the whole test section. Combined with the temperature and pressure requirements of the flow field, numerical analysis, simulation, and testing of flow field characteristics in the wind tunnel were performed.

47.3 Numerical Analysis of Flow Field Characteristics in the Wind Tunnel

The airflow velocity in the wind tunnel is about 55 m/s, and the Mach number is about 0.16, which belongs to low speed without compression. However, considering the temperature change and design index parameters, the continuity equation, momentum equation, and energy equation adopted in the calculation process of this paper are shown in Eqs. (47.1), (47.2), and (47.3), respectively:

$$\frac{\partial \rho}{\partial t} + \frac{\partial}{\partial x_i}(\rho u_i) = 0 \tag{47.1}$$

$$\frac{\partial}{\partial t}(\rho \overline{u_i}) + \frac{\partial}{\partial x_j}(\rho \overline{u_i} \cdot \overline{u_j}) = -\frac{1}{\rho}\frac{\partial \overline{p}}{\partial x_i} + \frac{\partial}{\partial x_j}\left[\mu\left(\frac{\partial \overline{u_i}}{\partial x_j} + \frac{\partial \overline{u_j}}{\partial x_i} - \frac{2}{3}\delta_{ij}\frac{\partial \overline{u_i}}{\partial x_i}\right)\right] + \frac{\partial}{\partial x_j}\left(-\rho\overline{u_i' u_j'}\right) \tag{47.2}$$

$$\frac{\partial}{\partial t}(\rho E) + \nabla(\vec{v}(\rho E + p)) = \nabla\left(k_{eff}\nabla T - \sum_j h_j \vec{J}_j + \overline{\overline{\tau}}_{eff}\vec{v}\right) + S_h \tag{47.3}$$

where ρ is the fluid density; $\overline{u_i}$ is the fluid velocity in the direction i; $\frac{\partial}{\partial t}(\rho \overline{u_i})$ is the time partial derivative of the momentum equation; $\frac{\partial}{\partial x_j}(\rho \overline{u_i} \cdot \overline{u_j})$, $-\frac{1}{\rho}\frac{\partial \overline{p}}{\partial x_i}$, and $\frac{\partial}{\partial x_j}\left[\mu\left(\frac{\partial \overline{u_i}}{\partial x_j} + \frac{\partial \overline{u_j}}{\partial x_i} - \frac{2}{3}\delta_{ij}\frac{\partial \overline{u_i}}{\partial x_i}\right)\right]$ are the convection, pressure, and viscous dissipation terms of the momentum equation, respectively; μ is the dynamic viscosity of the fluid; $\frac{\partial}{\partial x_j}\left(-\rho\overline{u_i' u_j'}\right)$ is the Reynolds stress and is sealed using the standard mode of a vorticity viscosity mode; $\frac{\partial}{\partial t}(\rho E)$ is the time partial derivative of the energy equation; $\nabla(\vec{v}(\rho E + p))$ is the convection term of the energy equation; $\nabla(k_{eff}\nabla T)$ is the energy transferred by heat conduction, and k_{eff} is the heat conduction coefficient; $\nabla\left(\sum_j h_j \vec{J}_j\right)$ is the energy transferred by the heat convection; $\nabla\left(\overline{\overline{\tau}}_{eff}\vec{v}\right)$ is the heat generated by the viscous motion of molecules, and S_h is the energy source term.

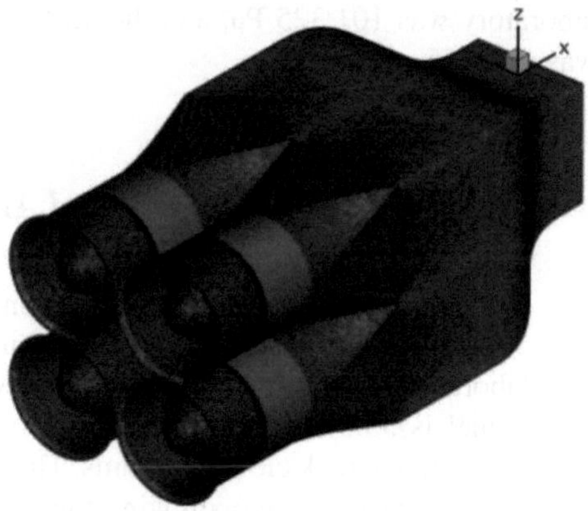

Fig. 47.1 Grid of the DC open environment wind tunnel

47.4 Simulation of Flow Field Characteristics in the Wind Tunnel

This section numerically solves the discrete fluid continuity equation, momentum equation, and energy equation to simulate the fluid movement at different temperatures in a wind tunnel.

47.4.1 Mesh Generation

A hybrid grid was utilized for modeling and meshing using Gambit. Three grid dimensions of the laboratory, environmental wind tunnel, and fan are set. The boundary layer is set for the tunnel wall, and the fan boundary condition is set for the culvert fan in the environmental wind tunnel. Given the direction and pressure increase, the wall boundary condition is set for the adiabatic non-slip wall. Figure 47.1 shows the grid diagram of the DC open environmental wind tunnel.

47.4.2 Parameters Calculation

The solver was set as a pressure-based steady calculation mode, and the SIMPLE algorithm was adopted. The numerical analysis formula of flow field characteristics in the wind tunnel shows the turbulence equation, and the function condition was a standard wall surface. The calculated medium was air, and the relaxation factor was the default value. The model environment is a closed climate laboratory. The most commonly used baseline test temperature of 21 °C and low-temperature extreme temperature of – 30 °C were selected for the temperature field. The pressure in the laboratory was 101,325 Pa, and the airflow velocity in the wind tunnel test section was set at 55 m/s.

47.4.3 Calculation Results and Analysis

(1) Velocity field analysis of the closed environment at 21 °C

As the fan pressure in the wind tunnel increases by 2150 Pa in the closed laboratory environment at 21 °C, the wind speed at the exit of the open wind tunnel is 55.37 m/s. Figure 47.2 shows the airflow velocity distribution and streamlined track cloud diagrams. The airflow velocity gradually decreases after it ejects through the open wind tunnel exit, and the streamline slightly deflects downward. When it approaches the far field laboratory wall, the airflow around it rapidly diffuses. Part of the diffusion airflow enters the open wind tunnel after one week around the plant, and another part enters the open wind tunnel after multiple turns on both sides of the plant. Figure 47.3 shows the velocity distribution in the X direction (4 m after the exit of the open wind tunnel). The velocity in the effective area of the test section corresponding to the open wind tunnel nozzle is 55–56 m/s, and the wind velocity heterogeneity in the test section is below 1.63%.

(2) Pressure field analysis of the closed environment at 21 °C

According to the airflow track analysis in Fig. 47.2, the airflow path around the entrance of the wind tunnel is relatively dense when the wind tunnel runs in a direct-flow environment in the closed laboratory. Therefore, the pressure

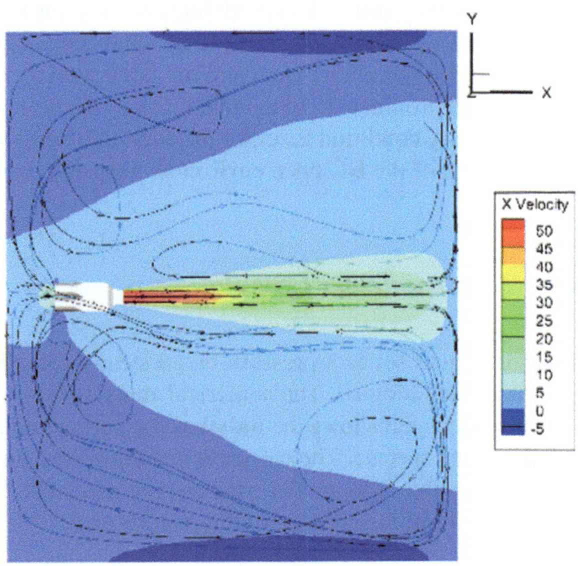

Fig. 47.2 Airflow track and velocity distribution in the open wind tunnel

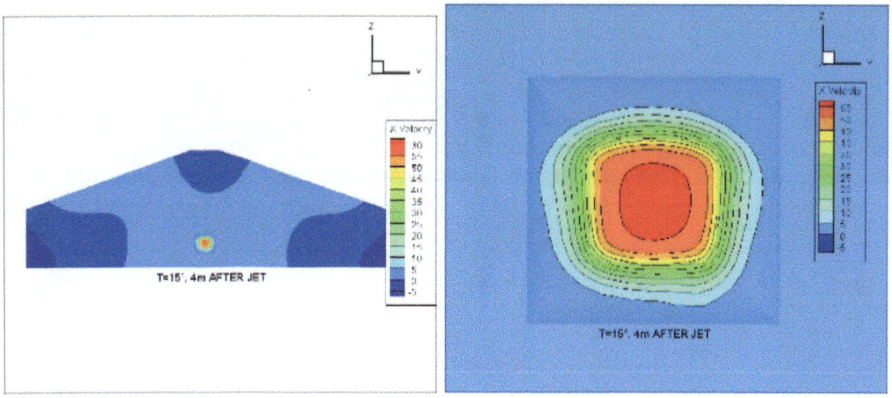

Fig. 47.3 Velocity evenness distribution in the vertical section of the test section in the open wind tunnel

distribution at the section 1 m before the entrance and 4 m before the exit (test section) with significant pressure fluctuation around the wind tunnel was studied. The pressure cloud maps are shown in Figs. 47.4 and 47.5.

Figure 47.5 presents the pressure distribution at the section 4 m after the open wind tunnel exit (test section). The static pressure of the corresponding nozzle position on the section is 3–5 Pa (relative reference value), while the static pressure outside is negative.

(3) Velocity field analysis of the closed environment at – 30 °C

As the fan pressure in the wind tunnel increases by 2600 Pa in the closed laboratory environment at – 30 °C, the wind speed at the exit of the open wind tunnel is 55 m/s. Figure 47.6 shows the airflow velocity distribution and streamline in the X direction. The airflow velocity gradually decreases after it ejects through the exit of the open wind tunnel, and the streamline is slightly deflected to the

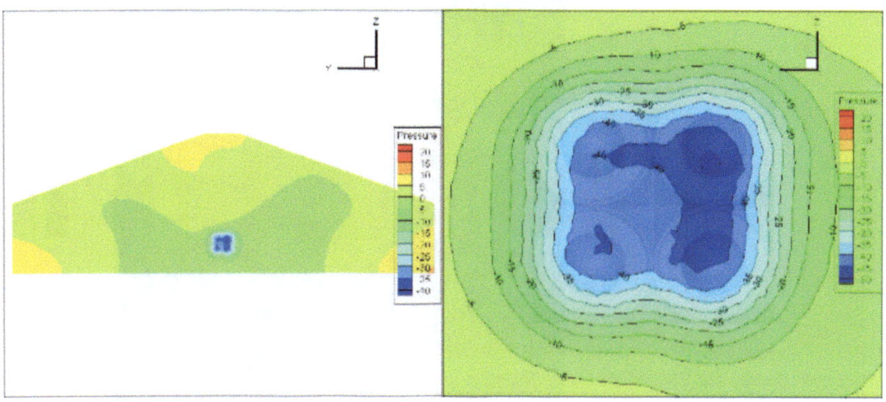

Fig. 47.4 Section pressure distribution at 1 m before the open wind tunnel entrance

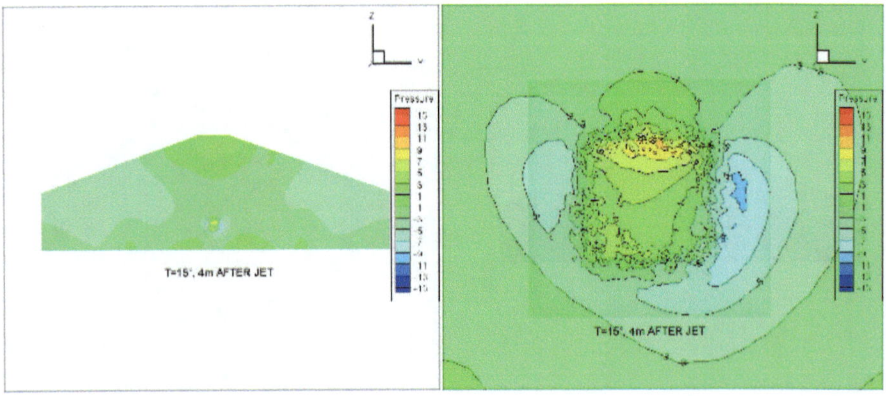

Fig. 47.5 Section pressure distribution in the wind tunnel

lower left in the downstream direction. The airflow direction spreads sharply around when it approaches the far field wall. Part of the diffusion airflow enters the open wind tunnel after one week around the plant, and another part enters the open wind tunnel after multiple turns on both sides of the plant. Figure 47.7 shows the velocity distribution in the X direction (1 m after the exit of the open wind tunnel). The wind speed is 55–56 m/s within 1 m^2 of the corresponding position of the open wind tunnel exit. The unevenness of the wind speed in the test section is 1.92%.

(4) Pressure field analysis of the closed environment at – 30 °C

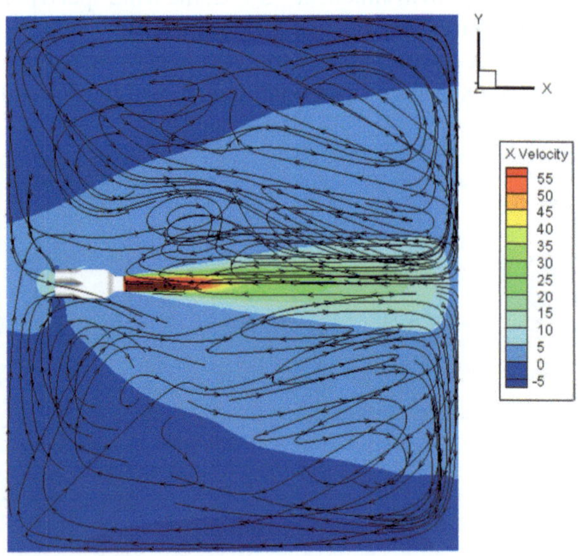

Fig. 47.6 Velocity distribution in the X direction in the horizontal plane of symmetry in the open wind tunnel

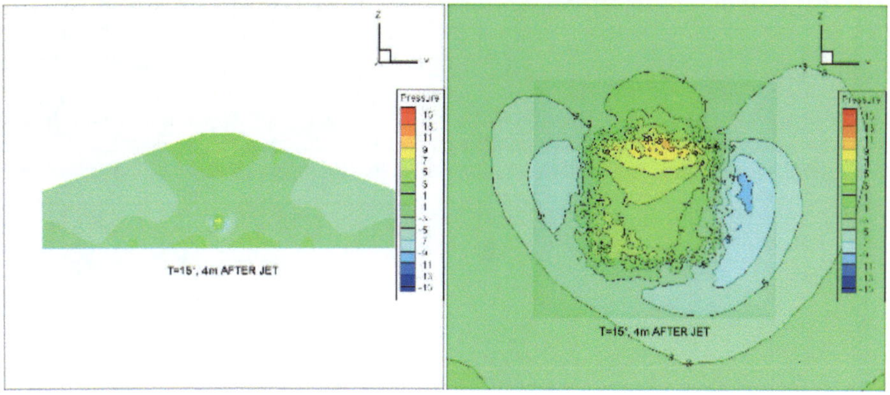

Fig. 47.7 Velocity distribution in the vertical section of the test section in the wind tunnel

The sectional pressure distribution at 1 m before the open wind tunnel entrance is shown in Fig. 47.8. The static pressure in front of the entrance is – 20 to – 60 Pa (relative reference pressure), and the static pressure distribution in each culvert is uneven. When the open wind tunnel runs, the static pressure on both sides rises rapidly. The static pressures of the plant root and top are about 11 Pa and 7 Pa, respectively.

Figure 47.9 shows the sectional pressure distribution 4 m away from the exit of the open wind tunnel. The static pressure of the corresponding exit position is – 20 to – 50 Pa (relative reference value), and the static pressure distribution of each culvert is uneven. The static pressure on both sides of the open wind tunnel rises rapidly. The static pressures of the plant root and top are about 11 Pa and 7 Pa, respectively.

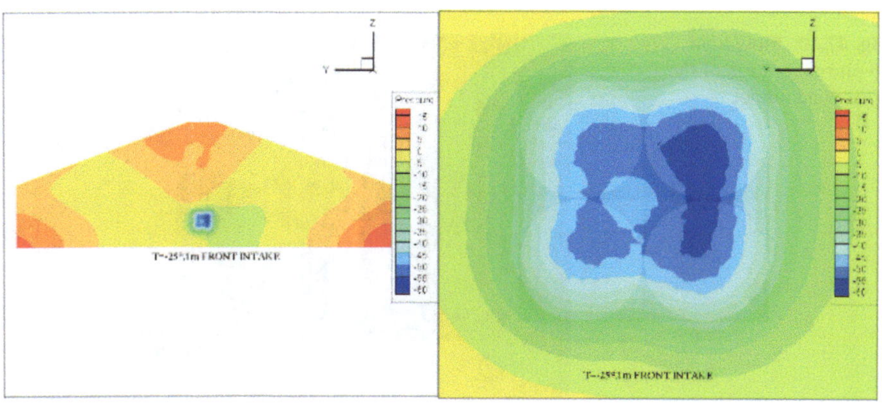

Fig. 47.8 Section pressure distribution at 1 m before the open wind tunnel entrance

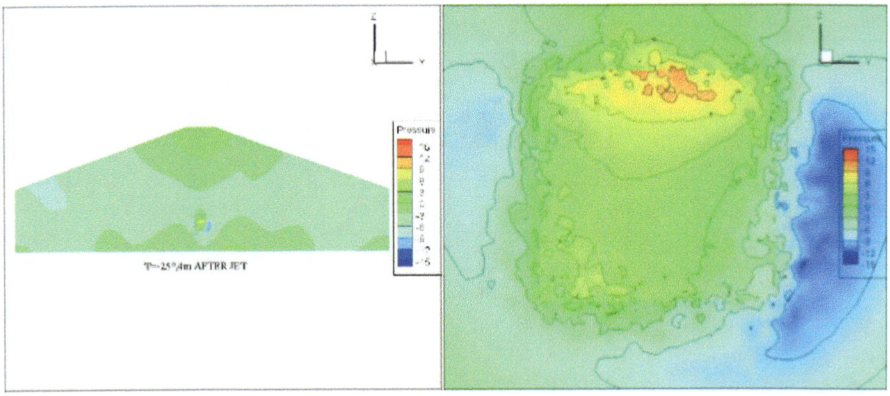

Fig. 47.9 Section pressure distribution in the open wind tunnel

47.5 Calibration of Flow Field Characteristics in the Test Section of the Wind Tunnel

The flow field characteristics in the open wind tunnel were calibrated according to relevant specifications (GJB 1179-91; Wang et al. 2021; GJB 3477-98). A measuring system consisting of a Pitot tube (Li 2017; Zhou 2018), differential pressure transmitter, and inverter was employed to measure the airflow velocity. Figure 47.10 shows the on-site calibration. The performance indices of the velocity and gas pressure fields in the wind tunnel at 21 °C and − 30 °C were compared in the closed laboratory environment, respectively. The results are shown in Tables 47.1 and 47.2.

As shown in Table 47.1, the performance indices of each velocity field in the wind tunnel, such as maximum wind speed, wind speed deviation, velocity unevenness,

Fig. 47.10 Flow field on-site calibration of the open wind tunnel

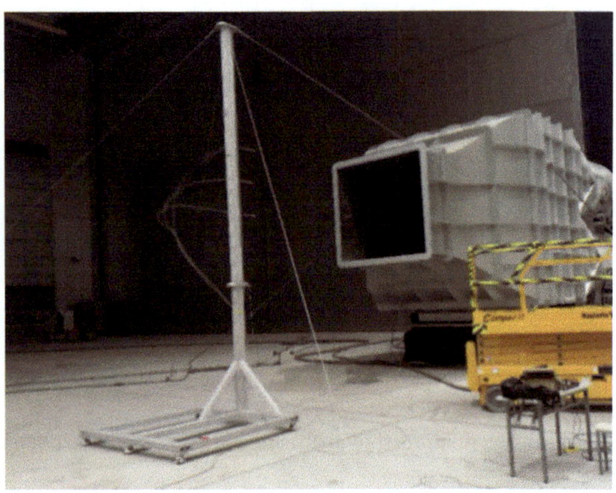

Table 47.1 Calibration results of the velocity field in the effective section of the wind tunnel

Index of the velocity field	Temperature field at 21 °C		Temperature field at – 30 °C	
	Test results	Simulation results	Test results	Simulation results
Mean maximum wind speed	≥ 55.02 m/s	≥ 55.37 m/s	≥ 54.89 m/s	≥ 55 m/s
Wind speed deviation	≤ 0.43%	≤ 1.78%	≤ 0.84%	≤ 1.81%
Velocity unevenness	≤ 0.53%	≤ 1.63%	≤ 0.6%	≤ 1.92%
Velocity instability	≤ 0.4%	–	≤ 0.56%	–

Table 47.2 Test results of the gas pressure field at critical locations in the laboratory around the wind tunnel

Pressure field detection	Temperature field at 21 °C		Temperature field at – 30 °C	
	Test results	Simulation results	Test results	Simulation results
In the laboratory	6–11 Pa	5–10 Pa	8–13 Pa	7–11 Pa
1 m in front of the wind tunnel entrance	– 35 to – 45 Pa	– 30 to – 40 Pa	– 24 to – 58 Pa	– 20 to – 60 Pa
4 m behind the wind tunnel exit	4–9 Pa	3–5 Pa	– 18 to – 49 Pa	– 20 to – 50 Pa

and velocity instability, meet the design requirements and can normally work at room temperature and the required low-temperature environment. Table 47.2 shows that the laboratory pressure fluctuation does not exceed 60 Pa when the open wind tunnel operates at baseline and low temperature, satisfying the design index requirements. By analyzing the response data of the temperature field in Tables 47.1 and 47.2, it is found that the low-temperature environment has varying degrees of influence on both velocity and gas pressure fields, especially on the deviation of maximum wind speed in the wind tunnel and the performance of the air pressure field at the back end of the outlet. A comparison of the simulation results of the wind tunnel flow field with the actual test results indicates that the error of each performance index does not exceed 10%, demonstrating the excellent accuracy of the simulation model.

47.6 Conclusions

Through the analysis of the above calculation results, the following conclusions can be drawn:

(1) The governing equation and discrete scheme for numerical calculation are given according to the conservation law of physics observed by the airflow in the DC wind tunnel.

(2) The CFD method was utilized to simulate the velocity field in the DC open wind tunnel. Accordingly, the airflow track and velocity distribution in the laboratory under different temperature fields in the open wind tunnel, the wind speed unevenness distribution cloud diagram in the vertical section of the wind tunnel, and the pressure field distribution cloud diagram at the critical parts around the wind tunnel were obtained. The simulation results are accurate and meet the technical index requirements.
(3) The technical indicators of the wind tunnel meet the design and use requirements through the actual calibration of the velocity and pressure fields of the critical parts of the open wind tunnel under two temperature fields. Accordingly, the correlation law of the wind tunnel flow field quality change under a multi-field coupling environment is formed.
(4) In order to verify the accuracy and feasibility of the simulation model, the simulation analysis and test verification of the velocity and gas pressure fields in the DC open wind tunnel of the climate laboratory at baseline temperature and extreme low-temperature environment are performed. It demonstrates the theoretical basis and data support for optimal design, model improvement, test accuracy, and quality improvement of the wind tunnel experiment.

References

Burov VV (2012) Measuring and computational system of TsAGIT-128 transonic wind tunnel. Autom Remote Control 72(3):634–641

Calautit JK, Chaudhry HN, Hughes BR et al (2014) A validated design methodology for a closed-loop subsonic wind tunnel. J Wind Eng Ind Aerodyn 125:180–194

Chen Q, Zhang Y, Lu C (2013) Analysis of air temperature and humidity on the impact of mining anemometer with a low-speed wind tunnel verification results. Shanghai Measur Test 3(2):46–48 (in China)

Deng L (2014) Construction of a DC low-speed wind tunnel test platform and research on aerodynamic characteristics of the whole flow field. Wuhan University of Technology

Ding S, Xue X, Lan Y et al (2015) Design and test of NJS-1 plant protection DC closed wind tunnel. J Agric Eng 31(4):76–82

Gai X, Li G, Zhang H (2011) On the design of movable wind tunnel. J Shenyang Aerosp Univ 28(3):27–32

GJB 1179-91. Flow field quality specification for wind tunnel

GJB 3477-98. Flow field quality requirements for shock tunnel

Green J (2011) A short history of the European transonic wind tunnel ETW. Prog Aerosp Sci 47(5):319–368

Hu K, Zhao Z, Qian T (2012) Experiment investigation of the flow field in a simple wind tunnel. Mech Res Appl 2(118):74–75

Ignatius J K, Sathiyavageeswaran S, Chakravarthy SR (2014) Hot-flow simulation of aeroacoustics and suppression by water injection during rocket liftoff. AIAA J 53(1):235–245

Jeffrey SH, Stephen FH (2008) Test capabilities and recent experiences in the NASA Langley 8-foot high temperature tunnel. AIAA 2000-2646

Jiang Y, Zhang L, Zou X et al (2018) Numerical simulation study of water spray cooling at the collecting port of high temperature wind tunnel. J Astronaut 39(6):702–770

Kong W, Du H, Li Q et al (2020) Flow field calibration of small low-speed open DC wind tunnel. J Xihua Univ 39(2):63–69
Li P (2017) Research on the calibration method of hot-wire anemometer under the light wind speed. Hebei University, Baoding
Liu ZC, Liao DX, Dong YX et al (2003) Aerodynamics and structural design of high and low speed wind tunnel. Defense Industry Press, Beijing
Lv C, Zhao Z, Zhang X (2018) Design and performance test of low speed and high temperature wind tunnel. Energy Res Inform 34(1):51–55
Moonen P, Blocken B, Roels S et al (2006) Numerical modeling of the flow conditions in a closed-circuit low-speed wind tunnel. J Wind Eng Ind Aerodyn 94(10):699–723
National Defense Science, Technology and Industry Commission (1991) Quality specification for high-speed wind tunnel and low-speed wind field (GJB1179 91). National Defense Industry Press, Beijing
Tong X, Wang J, Liu C et al (2014) Influence of environment condition on wind speed meterage of direct-current wind tunnel. J PLA Univ Sci Technol (Nat Sci Ed) 15(5):494–498
Wang S, Liu X, Lu L (2012) Simulation and analysis of contraction curve in the contraction section of DC low-speed wind tunnel. Mach Tool Hydraul 40(11):100–104
Wang X, Chen S, Mao H et al (2017) Thermal protection structure design and heat transfer analysis of electric pushrod in low-temperature wind tunnel. Cryog Eng (6):63–69
Wang P, Zhao H, Wang H et al (2021) Simulation study on velocity uniformity in low-speed DC wind tunnel. China Metrol (12):92–98
Zeng Q, Zhou R, Wang H (2018) Finite element analysis of the shell of DC low-speed wind tunnel. Modern Mach (1):11–14
Zhang Z, Chen Z, Nie X, Yao C (2016) Analysis of the thermodynamic characteristics of the diffusion section of the cryogenic wind tunnel based on the fluid-solid thermal coupling. Exp Fluid Mech 30(6):18–23
Zhang Y, Wu J, Meng N (2017) Wind and rain test of aircraft in climatic environment laboratory. Equip Environ Eng 11(14):59–63
Zheng Z, Lei J et al (2012) Test and evaluation of wind flow characteristics of a portable wind tunnel. J Desert Res 6(32):1551–1557
Zhou H (2018) Wind speed sensor based on pitot tube principle. Coal Mine Saf 49(8):99–101

Index

Y. Li et al. (eds.), *16th International Conference on Environmental Science and Development*, Environmental Science and Engineering,
https://doi.org/10.1007/978-3-032-15080-6

E

F

G

H

I

J

L

M

N

MIX
Papier aus verantwortungsvollen Quellen
Paper from responsible sources
FSC® C105338

If you have any concerns about our products,
you can contact us on
ProductSafety@springernature.com

In case Publisher is established outside the EU,
the EU authorized representative is:
Springer Nature Customer Service Center GmbH
Europaplatz 3, 69115 Heidelberg, Germany

Printed by Libri Plureos GmbH
in Hamburg, Germany